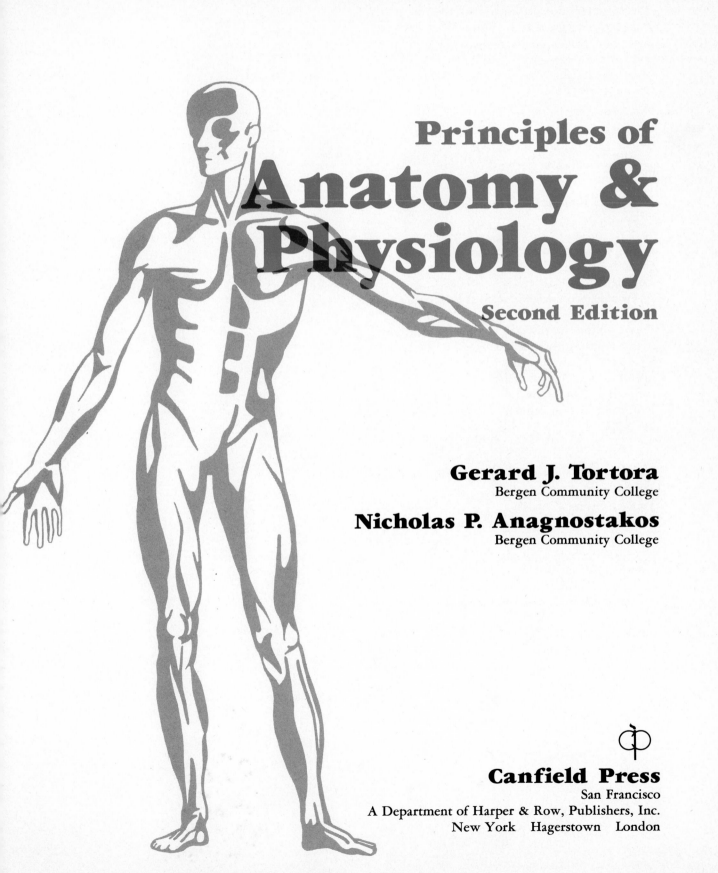

Principles of
Anatomy &
Physiology

Second Edition

Gerard J. Tortora
Bergen Community College

Nicholas P. Anagnostakos
Bergen Community College

Canfield Press
San Francisco
A Department of Harper & Row, Publishers, Inc.
New York Hagerstown London

The Second Edition of *Principles of Anatomy and Physiology* is dedicated to R. Wayne Oler. His keen perception, remarkable foresight, gentle persuasion, and creative genius have contributed immeasurably to the production of a successful textbook and a beautiful friendship.

Cover and Interior Design   Marjorie Spiegelman

**Principles of Anatomy and Physiology:
Second Edition**
Copyright © 1975, 1978
by Gerard J. Tortora and Nicholas P. Anagnostakos

Library of Congress Cataloging in Publication Data

Tortora, Gerard J
    Principles of anatomy & physiology.

    Bibliography: p.
    Includes index.
    1.   Human physiology.   2.   Anatomy, Human.
I.   Anagnostakos, Nicholas Peter, 1924–   joint author.
II.   Title.
QP34.5.T67   1978   612   77–16173
ISBN 0–06–388778–9

78   79   80   81   10   9   8   7   6   5   4   3

Contents in Brief

Contents in Detail

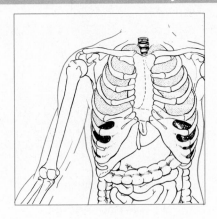
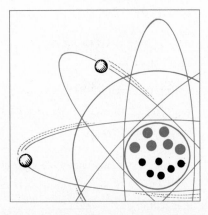

Chapter 3

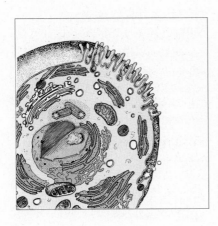

Unit II   Principles of Support and Movement

Chapter 6

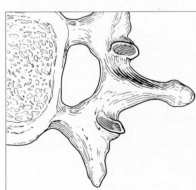

Chapter 7

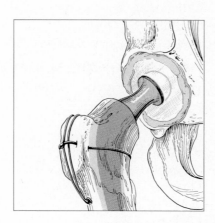
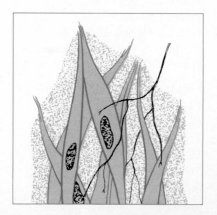

Chapter 11

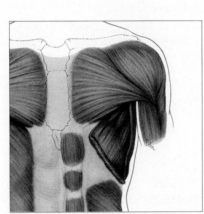

Unit III   Control Systems of the Human Body

Chapter 12

Chapter 13

Chapter 14

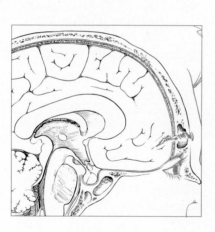

Chapter 15

Chapter 16

Chapter 17

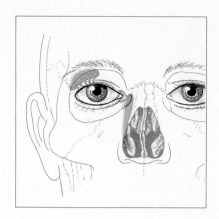

Chapter 18

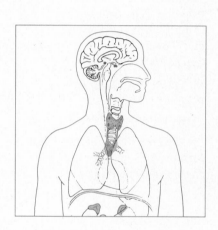

Unit IV   Maintenance of the Human Body

Chapter 19

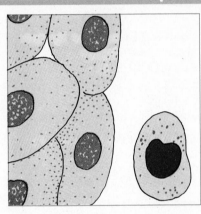

Chapter 20

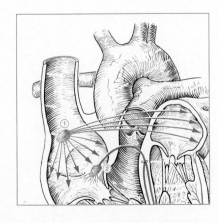

Chapter 21

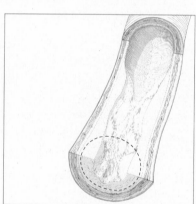

Chapter 22

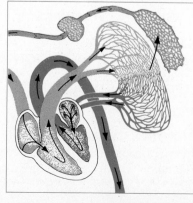

Chapter 23

Chapter 24

Chapter 25

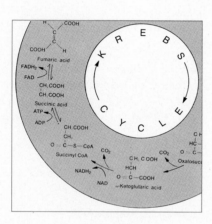

Chapter 26

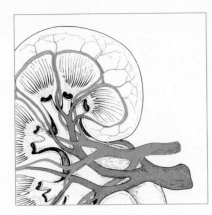

Chapter 27

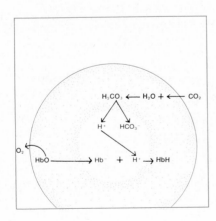

Unit V   Continuity

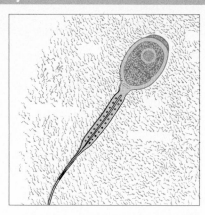

FIGURES

EXHIBITS

Preface

AUDIENCE

Designed for the introductory course in anatomy and physiology, the second edition of *Principles of Anatomy and Physiology* assumes no previous study of the body or of the physicochemical principles underlying body structure and function. The text is geared to students in biological, medical, and health-oriented programs. Among the students specifically served by this volume are those aiming for a career as a nurse, medical assistant, physician's assistant, medical laboratory technologist, radiologic technologist, inhalation therapist, dental hygienist, physical therapist, mortician, or medical record keeper. However, because of the scope of the text, *Principles of Anatomy and Physiology,* Second Edition, is also useful for students in the biological sciences, premedical and predental programs, science technology, liberal arts, and physical education.

OBJECTIVES

The objectives of the second edition of *Principles of Anatomy and Physiology* remain unchanged. As our first objective, we emphasize unifying concepts considered critical to a basic understanding and working knowledge of the human body. To support or explain these concepts, we have selected only data regarded as essential to an introduction to the subject. Our second objective is to present the material at a reading level average students can handle. We do not avoid technical vocabulary or vital, but difficult, concepts. Instead, we develop step-by-step, easy-to-comprehend explanations of each concept while avoiding needlessly difficult nontechnical vocabulary and syntax. These goals eliminate the barriers to a ready comprehension of important concepts and data.

THEMES

Two major themes dominate the second edition of *Principles of Anatomy and Physiology:* homeostasis and pathology. Throughout, the book shows students how dynamic counterbalancing forces maintain normal anat-omy and physiology. Pathology is viewed as a disruption in homeostasis. Accordingly, we present a number of clinical topics and contrast them with specific normal processes.

ORGANIZATION

The book is organized into five principal areas of concentration concerning the human body: organization, support and movement, control systems, maintenance, and continuity.

Unit I, "Organization of the Human Body," provides an understanding of the structural and functional levels of the body from molecules to organ systems. Chapter 1 introduces the concept of homeostasis and defines negative and positive feedback systems. The chapter has been expanded to include the nine abdominopelvic regions and four abdominopelvic quadrants. In this edition the material on the structural plan of the body – anatomical terminology, body cavities, and planes of the body – appears in this introductory chapter. Chapter 2 contains all the physicochemical principles and data required for an understanding of the physiology presented in the text. The first half of the chapter – the atom, chemical reactions, and inorganic compounds – provides information helpful to students who have not taken a chemistry course. The second half – organic compounds – is recommended for all students. The sections on cyclic AMP and prostaglandins are new to this edition.

Chapters 3, 4, and 5 deal with the cellular, tissue, and organ-system levels of organization. In Chapter 3 a generalized animal cell demonstrates the cellular level. A series of illustrations is provided for each cell part. Each series consists of an electron micrograph of the part, an adjacent labeled line drawing of the electron micrograph, and a small illustration of the whole cell where appropriate, with the part under consideration shown in color. The discussion of the cell ends with a description and schematic drawing of protein synthesis. The chapter has been expanded to include cancer and aging. Chapter 4 presents tissue organization through descriptions of the structure, functions, and locations of the principal kinds of epithelium and connective tissue, excluding bone and blood. (The histology of bone, blood, muscle, and nerve tissue is dealt with later, under the relevant organ systems.) The format of the epithelial and connective tissue Exhibits has been improved for this edition. Chapter 4 also discusses inflammation and tissue repair. Chapter 5 utilizes the skin and its accessory organs to acquaint students with the organization of organs and systems. Disorders added to this chapter are impetigo, psoriasis, decubitus, and skin cancer.

Unit II, "Principles of Support and Movement," analyzes the anatomy and physiology of the skeletal system, articulations, and the skeletal muscle system. Chapter 6 considers the histology of osseous tissue, with emphasis on bone structure, formation, growth, and bone disorders, including fractures. In the second edition two chapters cover the skeletal system: Chapter 7 the axial skeleton, and Chapter 8 the appendicular skeleton. Added emphasis is given to the foramina of the skull. Roentgenograms of bones provide students with the opportunity to apply their knowledge to clinical situations. An expanded Chapter 9, on articulations, includes the structural and functional classifications of joints, total hip and total knee replacement techniques, and additional disorders. The physiology of muscle contraction has been updated and expanded in Chapter 10, with the addition of several myograms. Because many students find muscle identification an onerous chore, we provide the following learning aids: The muscle illustrations are duplicates of the drawings used for bone identification, giving students consistent points of reference. Chapter 11 discusses the criteria for naming skeletal muscles. Each muscle Exhibit lists the appropriate prefixes, roots, and suffixes, together with their definitions. In addition, there are numerous surface anatomy photographs of the muscles under consideration.

Unit III, "Control Systems of the Human Body," emphasizes the importance of the nerve impulse in the immediate maintenance of homeostasis, the role of receptors in providing information about the internal and the external environment, and the significance of hormones in maintaining long-range homeostasis. Expanded from four to seven chapters, the unit has been reorganized into more convenient and logical subject areas. In the second edition, three chapters discuss the organs of the nervous system. Chapter 12, on nervous tissue, focuses on membrane potentials, refractory periods, saltatory transmission, excitatory transmission, summation, inhibitory transmission, chemical transmitters, integration, and the organization of neuronal circuits. A separate chapter on the spinal cord and the spinal nerves, Chapter 13 contains additional material on spinal puncture, clinically important reflexes, reflex arcs,

plexuses, dermatomes, peripheral nerve damage, and added disorders such as sciatica and neuritis. Plexuses, like muscles, are treated in Exhibits. The brain and the cranial nerves are also in a separate chapter, Chapter 14. New material expands the discussion of the limbic system, sensory and motor areas of the cerebrum, and cranial nerves. Added central nervous system disorders include dyslexia and Tay-Sachs disease. Two chapters now cover the topic of sensations. Chapter 15—on the sensory, motor, and integrative systems—retains material on sensory pathways, motor pathways, memory, sleep, and wakefulness. Chapter 16, devoted to the autonomic nervous system, contains new sections on biofeedback and meditation. A separate chapter on the special senses, Chapter 17, considers olfaction, gustation, sight, hearing, and equilibrium.

Chapter 18, on the endocrine system, emphasizes the regulation of hormone secretion through feedback systems, with flowcharts for most systems. There is new material on the role of cyclic AMP in endocrine physiology and the control of hormone production by regulating factors. (Discussion of the effects and regulation of the sex hormones is reserved for the chapters on reproduction and development, where these topics fit more logically.) The unit concludes with an explanation of how the general adaptation syndrome operates, how it differs from homeostatic responses, and why it is protective. This explanation provides background for current conceptions about disease and so presents a good review of the substance of the unit.

Unit IV, "Maintenance of the Human Body," illustrates how the body maintains itself on a day-to-day basis through the mechanisms of circulation, respiration, digestion, cellular metabolism, urine production, and buffer systems. This unit, expanded from eight to nine chapters, has been reorganized into an easier-to-follow pattern. For clarity, three chapters discuss the cardiovascular system. Chapter 19 is devoted to the blood: coagulation factors, extrinsic and intrinsic clotting, retraction, fibrinolysis, and prothrombin time. A separate chapter on the heart, Chapter 20, includes new material on the electrocardiogram, cardiac cycle, surface projections, Marey's law, the Bainbridge reflex, and pacemakers. Chapter 21 explains the blood vessels and circulatory routes, with a revised section on the physiology of circulation and an updated discussion of atherosclerosis. A separate chapter treats the lymphatic system in the second edition. Chapter 22 discusses in greater depth the anatomy of the lymph nodes and spleen. It also includes autoimmune diseases and the role of T cells and B cells in immunity. The respiratory system, Chapter 23, now includes coverage of spirometry, elementary gas laws, and the Heimlich maneuver.

Chapter 24 integrates anatomy and physiology by treating digestion regionally. For example, students learn the anatomy of the mouth with its accessory organs and the digestive processes that take place within the mouth before continuing to the next segment of the gastrointestinal tract. Chapter 25 deals with the metabolism of foods. Emphasis is on the relationships of protein and fat metabolism to carbohydrate metabolism. A chart showing integrated metabolism summarizes much of this material. The role of vitamins and minerals in enzyme-coenzyme systems is mentioned. Exhibits list specific vitamins and minerals, their sources, known or suspected functions, and deficiency symptoms. The chapter includes new material on calories, body heat, basal metabolic rate, and temperature regulation.

The restoration and maintenance of fluid balance and blood pH is an important area of knowledge for students considering a career in the health field. Consequently, the last two chapters in the unit pay particular attention to these topics. Chapter 26, on the urinary system, provides new material on cortical and juxtamedullary nephrons, the structure of renal corpuscles, the juxtaglomerular apparatus, and the countercurrent mechanism. The chapter on fluid, electrolyte, and acid-base dynamics, Chapter 27, considers fluid compartments, electrolyte distributions, concentrations, and functions, fluid movements, the regulation of acid–base balance, and acid–base imbalances.

Unit V, "Continuity," emphasizes, in Chapter 28, the relationship of the endocrine system to sexual development, the regulation of menstrual and ovarian cycles, and the maintenance of pregnancy. Chapter 29 concludes the text with the basic concepts of classical genetics to give students background for understanding birth control and the inheritance of genetic disorders.

SPECIAL FEATURES

The book features a number of special learning aids, some of them new to the second edition:

1 Each chapter opens with a comprehensive list of "Student Objectives." Each objective describes a knowledge or skill students should acquire while studying the chapter. To meet these objectives, students should do several things. They must read the chapter carefully and if there are sections not understood after one reading, students should reread them before continuing. In conjunction with the reading, students should pay particular attention to the figures and Exhibits, which are closely coordinated with the narrative.

2 An end-of-chapter "Study Outline" provides a checklist of major topics students should learn. This section consolidates the essential points covered in the chapter so that students can recall and relate the points to one another.

3 End-of-chapter "Review Questions" help students achieve the objectives stated at the beginning of the chapter. After answering the questions, students should reread the objectives to determine whether or not they have met the goals.

4 Health-science students are generally expected to learn a great deal about the anatomy of certain organ systems – specifically, skeletal muscles, blood vessels, and nerves. In these high anatomy areas, the anatomical details do not interrupt the narrative, but appear in tabular form in Exhibits, most of which are closely tied to illustrations. This method organizes the data and deemphasizes the rote learning of concepts presented in the narrative.

5 Numerous clinical applications are grouped in sections entitled "Disorders" at the ends of appropriate chapters. The sections provide a review of normal body processes as well as demonstrating the importance of the study of anatomy and physiology to a career in any of the health fields.

6 The text includes many topics of contemporary interest. These range from aging to biofeedback, from the connection between smoking and cancer to various contraceptive methods.

7 A glossary of selected terms, entitled "Medical Terminology," appears at the ends of appropriate chapters.

8 The line drawings are unusually large, so that details are clearly discerned and labeled. Color is used functionally to differentiate structures and regions. In the second edition we have introduced yellow to denote nerves, blue to distinguish venous blood from arterial blood (shown in red), and a combination of all four colors to delineate the various bones of the skull. New to this edition is the tone-rendering technique for muscles.

9 The photographic facet of the illustration program amplifies the narrative and the line drawings. Numerous photomicrographs, electron micrographs, and scanning electron micrographs enhance the histological discussions. A distinctive feature of the second edition is the inclusion of surface anatomy photographs of the face and muscles. Many roentgenograms have been added to reinforce clinical knowledge.

10 An appendix on "Measuring the Human Body," new to this edition, includes English and metric units of length, mass, and volume.

11 A two-part glossary appears at the end of the book. The first part is a "Glossary of Prefixes, Suffixes, and Combining Forms," together with a pronunciation key. The second part, a comprehensive "Glossary of

Terms," defines the important terms and gives their pronunciations. Where terms are introduced and defined in the text proper, they appear in boldface type.

12 An extensive "Bibliography" suggests readings that correspond to the unit organization of the text.

SUPPLEMENTARY MATERIALS

Ancillary materials available for use with *Principles of Anatomy and Physiology,* Second Edition, include:

1 A complimentary instructor's manual. For each chapter of the text, the manual contains a resume, a list of key instructional concepts, problem-solving essay questions, and suggestions for audiovisual materials, together with a directory of the distributors of these materials. The manual contains a test bank of objective test items—true-false and multiple-choice—for each chapter of the text. The questions are carefully designed to evaluate student understanding of data, concepts, clinical situations and their applications.

2 A complimentary set of 60 transparency masters selected from illustrations in the text.

3 A slide set of 120 35-mm slides of illustrations in the text, together with a script consisting of a brief description and two review questions for each slide.

4 A laboratory manual that contains surface terminology, microscopy, fundamental laboratory techniques, anatomy, and physiology. In the anatomy portion the manual gives directions for the dissection of the cat. By description and illustration, it correlates human anatomy with that of the cat.

5 A laboratory and study manual that guides the student in making microscopic examinations, recording data, and carrying out physiological experiments. This manual is available from Burgess Publishing Company.

Acknowledgments

Canfield Press has made available to us a number of specialists in their respective disciplines to assist in the preparation of the second edition of *Principles of Anatomy and Physiology*. Those to whom we wish to express our deepest gratitude are grouped below according to the nature of their contributions.

For making helpful suggestions on improving the pedagogy: Aaron Appleby, Iona College; Merlyn L. Anderberg, Spokane Falls Community College; David V. Ballard, Utah Technical College; Christine Bowman, Macomb Community College; Gerald R. Dotson, Community College of Denver; Susan K. Gillis, Mercy Central School of Nursing; David J. G. Griffiths, University of Maine; Charles W. Harnsberger, Jr., Macon Junior College; Rex Howard, Amarillo College; Anne B. Knaak, Modesto Junior College; Joseph W. McDaniel, Norwich University; James T. Mullen, Somerset County College; Don Puder, College of Southern Idaho; Ronald M. Reuss, State University College at Buffalo; Eugene Rutheny, Westchester Community College.

For reviewing the outline: Rose A. Bucsi, University of Bridgeport; Robert L. Cole, Erie Community College; Edward A. DeSchuytner, Northern Essex Community College; John L. Frehn, Illinois State University; Frank Gimble, Brookdale Community College; Gary Klein, Cuyahoga Community College; Helen MacAllister, Monmouth College; Sister Nivard Neft, Weber State College; Charles L. Quick, University of Toledo; Norman E. Rich, Golden West College; A. Quinton White, Jacksonville University.

For reviewing the outline and the entire manuscript: Gordon Bradshaw, Phoenix College; Harry S. Reasor, Miami-Dade Community College; Martha Van Bolt, Charles Stewart Mott Community College; Donna M. Van Wynsberghe, University of Wisconsin.

For reviewing the entire manuscript: Ivonna McCabe, Tacoma Community College; Ruth McFarland, Mt. Hood Community College; Jonathan C. Oldham, Metropolitan State College; Charles L. Rutherford, Virginia Polytechnic Institute and State University; John T. Windell, University of Colorado.

For reviewing the line drawings at preliminary and final stages: Ernest Gardner, University of California at Davis; Warren Finke, Diablo Valley College.

For the superb quality of the line drawings: Nelva B. Richardson, Marsha J. Dohrmann, and Helen Gee Jeung, whose talent, inspiration, and knowledge of the subject matter make the illustration program a distinctive feature of the text.

For supplying high-quality photographic materials: Victor B. Eichler, Wichita State University, Steve Harper, the Fisher Scientific Company and S.T.E.M. Laboratories, Inc., for photomicrographs and photographs; John C. Bennett, St. Mary's Hospital in San Francisco, for roentgenograms; Donald Castellaro, Richard Sollazzo, and Deborah J. Massimi, for surface anatomy photographs; Kay Y. James, for researching and obtaining photographic materials. We also wish to acknowledge the many other individuals, publishers, and companies for granting permission to reproduce photographs, photomicrographs, electron micrographs, and scanning electron micrographs.

For the outstanding editorial assistance provided by Canfield Press: Howard E. Boyer, Senior Editor, who personally supervised all phases of the project, providing continuous guidance and encouragement; Pearl C. Vapnek, Production Editor, who coordinated the various facets of the project.

For typing all drafts of the manuscript: Geraldine C. Tortora and Christine Anagnostakos. Geraldine also handled all secretarial duties related to the project.

For thoroughly checking all galley proof and page proof: Joyce Eckelberg, University of Houston; and Harry E. Peery, Tompkins-Cortland Community College, respectively.

As the rather lengthy list above indicates, the participation of numerous individuals of diverse talent and expertise is essential to the production of a textbook of this scope and complexity. For this reason, we would like to invite readers of the second edition to send us their reactions and suggestions to help us in formulating plans for subsequent editions.

Gerard J. Tortora
Nicholas P. Anagnostakos
Biology Department
Bergen Community College
400 Paramus Road
Paramus, NJ 07652

Unit I
Organization of the Human Body

This unit is designed to show you how your body is organized at different levels. After you study the various regions and parts of your body, you will discover the importance of the chemicals that make it up. You will then find out how your cells, tissues, and organs form the systems that keep you alive and healthy.

Chapter 1
An Introduction to the
Human Body

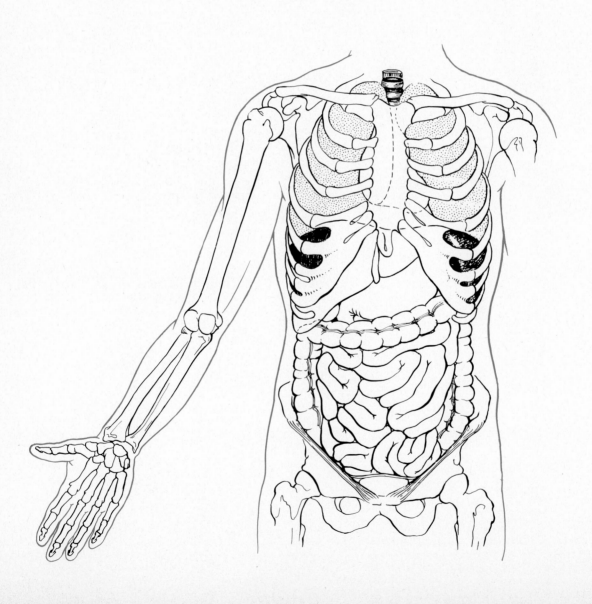

■ Define anatomy, with its subdivisions, and physiology.

■ Determine the relationship between structure and function.

■ Compare the levels of structural organization that make up the human body.

■ Define a cell, a tissue, an organ, a system, and an organism.

■ Define directional terms used in association with the body.

■ List by name and location the principal body cavities and their major organs.

■ Describe the subdivisions of the abdominopelvic cavity into nine regions and four quadrants.

■ Define the anatomical position.

■ Compare common and anatomical terms used to describe the external features of the body.

■ Describe the common anatomical planes of the body.

■ Explain why homeostasis is a state that results in normal body activities and why the inability to achieve homeostasis leads to disorders.

■ Identify the effects of stress on homeostasis.

■ Describe the interrelationships of body systems in maintaining homeostasis.

■ Compare the role of the endocrine and nervous systems in maintaining homeostasis.

■ Define a feedback system and explain its role in homeostasis.

■ Contrast the homeostasis of blood pressure through nervous control and blood sugar level through hormonal control.

You are about to begin a study of the human body in order to learn how your body is organized and how it functions. The study of the human body involves many branches of science. Each contributes to a comprehensive understanding of how your body normally works and what happens when it is injured, diseased, or placed under stress.

ANATOMY AND PHYSIOLOGY DEFINED

Two branches of science that will help you understand your body parts and functions are anatomy and physiology. **Anatomy** (or **morphology**) refers to the study of *structure* and the relationships among structures. Anatomy is a broad science, and the study of structure becomes more meaningful when specific aspects of the science are considered. **Gross anatomy** deals with structures that can be studied without a microscope. Another kind of anatomy, **systemic anatomy,** covers specific systems of the body, such as the system of nerves, spinal cord, and brain or the system of heart, blood vessels, and blood. **Regional anatomy** deals with a specific region of the body, such as the head, neck, chest, or abdomen. **Developmental anatomy** is the study of development from fertilized egg to adult form. **Embryology** is generally restricted to the study of development from the fertilized egg to the end of the eighth week in utero. Other branches of anatomy are **pathological anatomy,** the study of structural changes caused by disease, **histology,** the microscopic study of the structure of tissues, and **cytology,** the study of cells.

Whereas anatomy and its branches deal with structures of the body, **physiology** deals with *functions* of the body parts—that is, how the body parts work. As you will see in later chapters, physiology cannot be completely separated from anatomy. Thus you will learn about the human body by studying its structures and functions together. Each structure of the body is custom-modeled to carry out a particular set of functions. For instance, bones function as rigid supports for the body because they are constructed of hard minerals. Thus the structure of a part often determines the functions it will perform. In turn, body functions often influence the size, shape, and health of the structures. Glands perform the function of manufacturing chemicals, for example, some of which stimulate bones to build up minerals so they become hard and strong. Other chemicals cause the bones to give up minerals so they do not become too thick or too heavy.

LEVELS OF STRUCTURAL ORGANIZATION

The human body consists of several levels of structural organization that are associated with one another in several ways. The lowest level of organization, the **chemical level,** includes all chemical substances essential for maintaining life. All these chemicals are made up of atoms joined together in various ways (Figure 1-1). The chemicals, in turn, are put together to form the next higher level of organization: the **cellular level. Cells** are the basic structural and functional units of the organism. Among the many kinds of cells in your body are muscle cells, nerve cells, and blood cells. Figure 1-1 shows several isolated cells from the lining of the stomach. Each has a different structure, and each performs a different job.

The next higher level of structural organization is the **tissue level. Tissues** are made up of groups of similar cells and their intercellular material that perform a specific function. When the isolated cells shown in Figure 1-1 are joined together, they form a tissue called epithelium, which lines the stomach. Each cell in the tissue has a specific function. Mucous cells produce mucus, a secretion that lubricates food as it passes through the stomach. Parietal cells produce acid in the stomach. Chief cells produce enzymes needed to digest proteins. Other examples of tissues in your body are muscle tissue, connective tissue, and nervous tissue.

In many places in the body, different kinds of tissues are joined together to form an even higher level of organization: the **organ level. Organs** consist of two or more different tissues that perform a specific function. Organs usually have a recognizable shape. Examples of organs are the heart, liver, lungs, brain, and stomach. In Figure 1-1 you see that the stomach is an organ since it consists of two or more kinds of tissues. Three of the tissues that make up the stomach are shown here. The serous layer (also called the serosa) protects the stomach and reduces friction when the stomach moves and rubs against other organs. The muscle tissue layers of the stomach contract to mix food and pass it on to the next digestive organ. The epithelial tissue layer produces mucus, acid, and enzymes.

The next higher level of structural organization in the body is the **system level. A system** consists of an association of organs that have a common function. The digestive system, which functions in the breakdown of food, comprises the mouth, saliva-producing glands called salivary glands, pharynx (throat), esophagus (gullet), stomach, small intestine, large intestine, rectum, liver, gallbladder, and pancreas. All the parts of the body functioning with each other constitute the total **organism**—one living individual.

In the chapters that follow, you will examine the anatomy and physiology of the major body systems. Exhibit 1-1 illustrates these systems, their representative organs, and their general functions. The systems are presented in the Exhibit in the order in which they are discussed in later chapters.

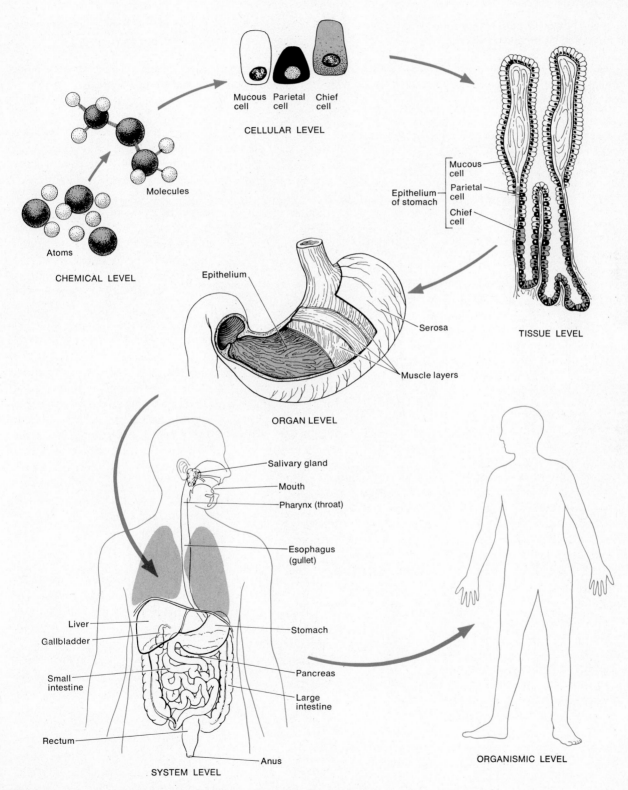

Figure 1-1 Levels of structural organization that compose the body.

Exhibit 1—1 PRINCIPAL SYSTEMS OF HUMAN BODY, REPRESENTATIVE ORGANS, AND FUNCTIONS

1. Integumentary

DEFINITION: The skin and structures derived from it, such as hair, nails, and sweat and oil glands.

FUNCTION: Regulates body temperature, protects the body, eliminates wastes, and receives certain stimuli such as temperature, pressure, and pain.

2. Skeletal

DEFINITION: All the bones of the body, their associated cartilages, and the joints of the body.

FUNCTION: Supports and protects the body, gives leverage, produces blood cells, and stores minerals.

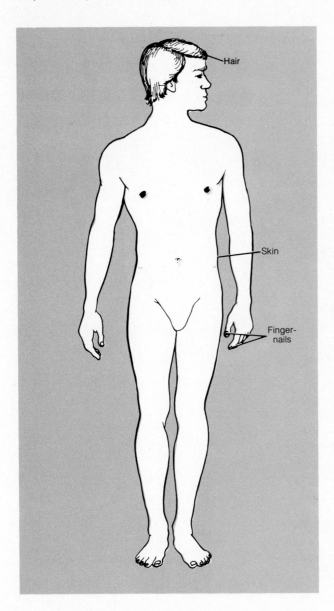

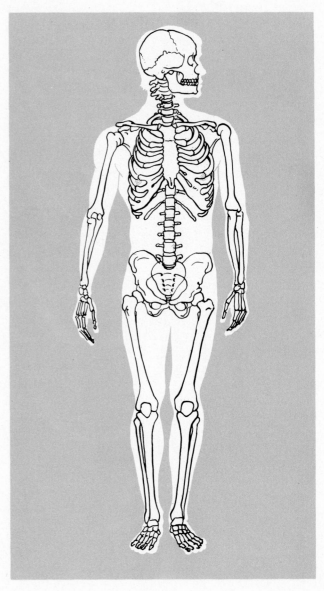

Exhibit 1—1   (cont.)

3. Muscular

DEFINITION: All the muscle tissue of the body, including skeletal, visceral, and cardiac.

FUNCTION: Allows movement, maintains posture, and produces heat.

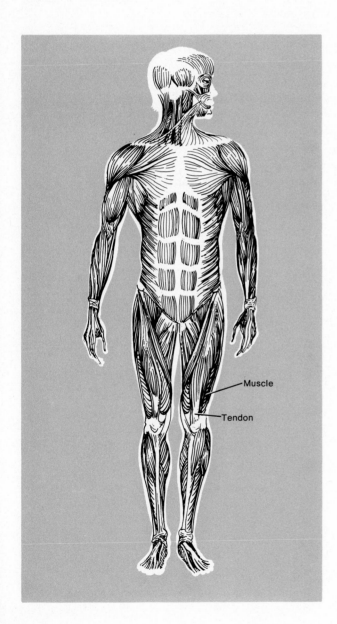

4. Nervous

DEFINITION: Brain, spinal cord, nerves, and sense organs, such as the eye and ear.

FUNCTION: Regulates body activities through nerve impulses.

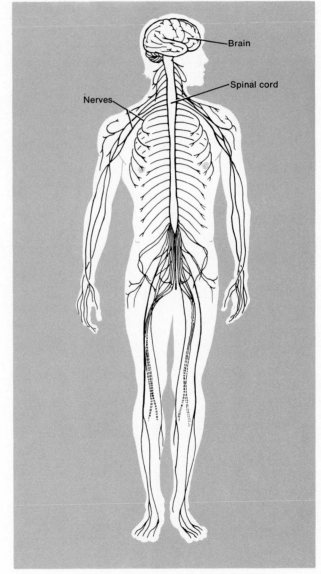

Exhibit 1—1   (cont.)

5. Endocrine

DEFINITION: All glands that produce hormones.

FUNCTION: Regulates body activities through hormones transported by the cardiovascular system.

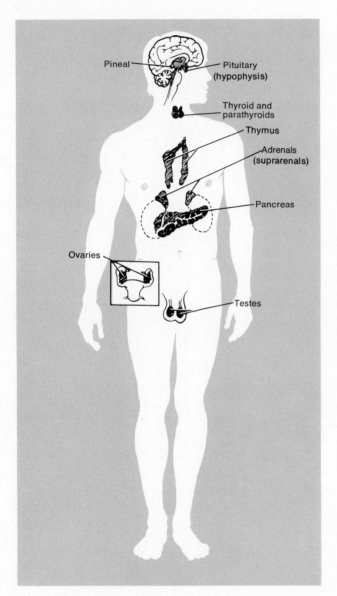

6. Cardiovascular

DEFINITION: Blood, heart, and blood vessels.

FUNCTION: Distributes oxygen and nutrients to cells, carries carbon dioxide and wastes from cells, maintains the acid—base balance of the body, protects against disease, prevents hemorrhage by forming blood clots, and helps regulate body temperature.

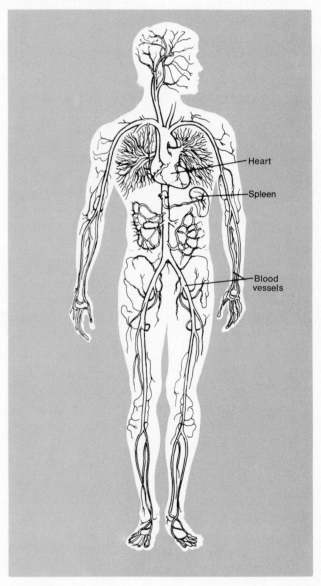

Exhibit 1—1 (cont.)

7. Lymphatic

DEFINITION: Lymph, lymph nodes, lymph vessels and lymph glands, such as the spleen, thymus gland, and tonsils.

FUNCTION: Returns proteins to the cardiovascular system, filters the blood, produces blood cells, and protects against disease.

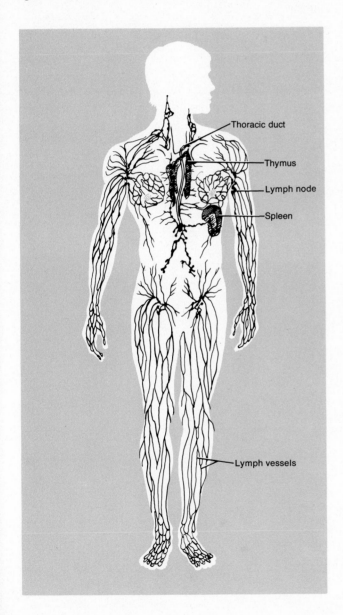

8. Respiratory

DEFINITION: The lungs and a series of passageways leading into and out of them.

FUNCTION: Supplies oxygen, eliminates carbon dioxide, and regulates the acid—base balance of the body.

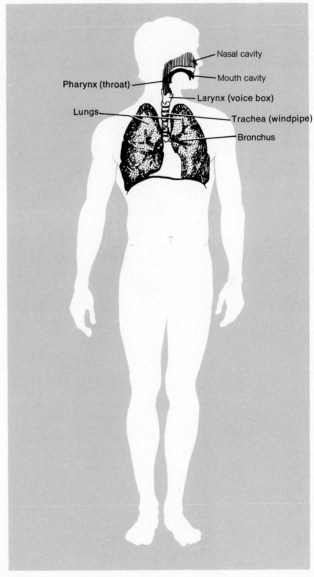

Exhibit 1—1   (cont.)

9. Digestive

DEFINITION: A long canal and associated organs such as the salivary glands, liver, gallbladder, and pancreas.

FUNCTION: Performs the physical and chemical breakdown of food for use by cells and eliminates solid wastes.

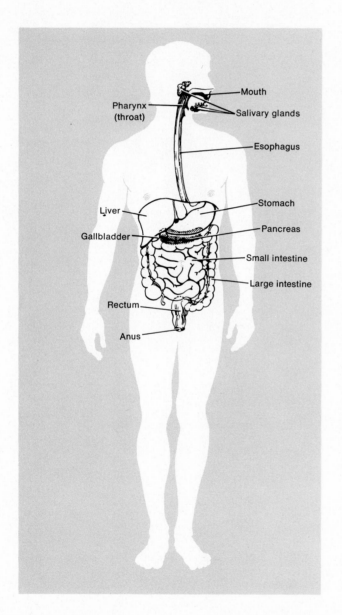

10. Urinary

DEFINITION: Organs that produce, collect, and eliminate urine.

FUNCTION: Regulates the chemical composition of blood, eliminates wastes, regulates fluid and electrolyte balance and volume, and maintains the acid—base balance of the body.

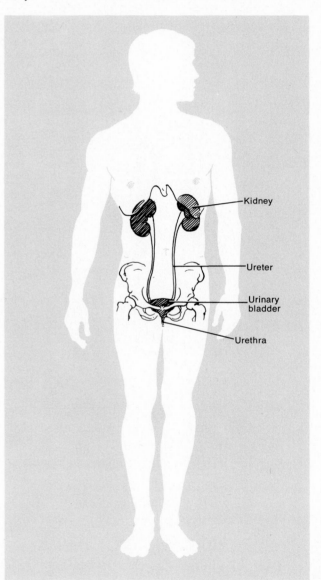

Exhibit 1—1 (cont.)

11. Reproductive

DEFINITION: Organs (testes and ovaries) that produce reproductive cells (sperms and ova) and organs that transport and store reproductive cells.

FUNCTION: Reproduces the organism.

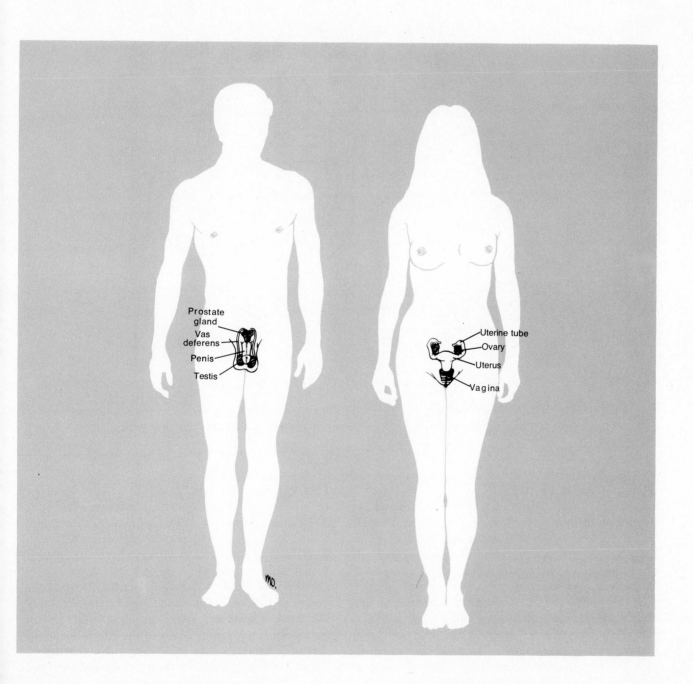

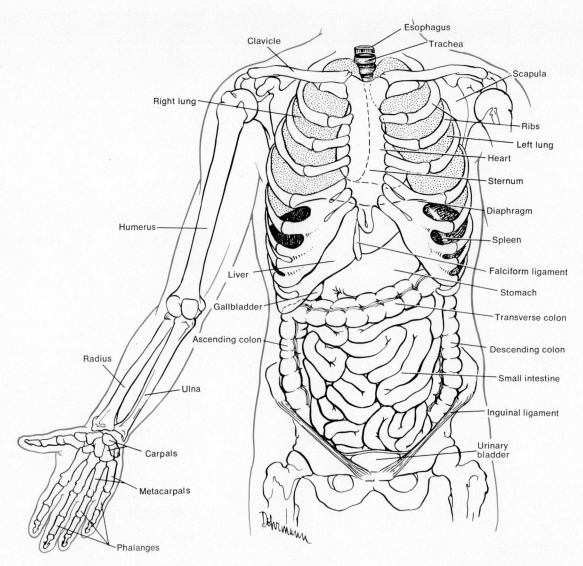

Figure 1-2 Anatomical and directional terms. By studying Exhibit 1-2 with this figure, you should gain an understanding of the meanings of the terms *superior, inferior, anterior, posterior, medial, lateral, proximal,* and *distal.*

STRUCTURAL PLAN

The human organism has certain anatomical characteristics that are easily identifiable and can, therefore, serve as landmarks. For example, humans have a *backbone,* a characteristic that places them in a large group of organisms called vertebrates. Moreover, humans are *bilaterally symmetrical* – the left and right sides of the body are mirror images. Another characteristic of the body's organization is that is resembles a *tube within a tube.* The outer tube is formed by the body wall; the inner tube is the digestive tract. You will need to know such general characteristics as well as the terms used to describe positions and directions in the body.

Directional Terms

In order to explain exactly where a structure of the body is located, anatomists must use certain **directional terms.** If you want to point out the sternum (breastbone) to someone who knows where the clavicle (collarbone) is, you can say that the sternum is inferior (farther away from the head) and medial (toward the middle of the body) to the clavicle. As you can see, using the terms *inferior* and *medial* avoids a great deal of complicated description. Many of the directional terms defined in Exhibit 1-2 may be understood by referring to Figure 1-2. Essential parts of the figure are labeled so you can see the directional relationships among parts.

Exhibit 1–2   DIRECTIONAL TERMS

Term	Definition	Example
Superior (cephalad or cranial)	Toward the head or the upper part of a structure; generally refers to structures in the trunk.	The heart is superior to the liver.
Inferior (caudad)	Away from the head or toward the lower part of a structure; generally refers to structures in the trunk.	The stomach is inferior to the lungs.
Anterior (ventral)	Nearer to or at the front of the body.	The sternum is anterior to the heart.
Posterior (dorsal)	Nearer to or at the back of the body.	The esophagus is posterior to the trachea.
Medial	Nearer the midline of the body or a structure.	The ulna is on the medial side of the forearm.
Lateral	Farther from the midline of the body or a structure.	The ascending colon is lateral to the urinary bladder.
Proximal	Nearer the attachment of an extremity to the trunk.	The humerus is proximal to the radius.
Distal	Farther from the attachment of an extremity to the trunk.	The phalanges are distal to the carpals (wrist bones).
Superficial	Toward or on the surface of the body.	The muscles of the thoracic wall are superficial to the viscera in the thoracic cavity. (See Figure 1-4b.)
Deep (internal)	Away from the surface of the body.	The muscles of the arm are deep to the skin of the arm.
Parietal	Pertaining to the outer wall of a body cavity.	The parietal pleura forms the outer layer of the pleural sacs that surround the lungs. (See Figure 1-4b.)
Visceral	Pertaining to the covering of an organ (viscus).	The visceral pleura forms the inner layer of the pleural sacs and covers the external surface of the lungs. (See Figure 1-4b.)

Body Cavities

Spaces within the body that contain internal organs are called **body cavities.** Specific cavities may be distinguished if the body is viewed after making a *median,* or *midsagittal, section* – that is, after cutting it into right and left halves. Figure 1-3 shows the two principal body cavities. The **dorsal body cavity** is located near the dorsal surface of the body. It is further subdivided into a **cranial cavity,** which is a bony cavity formed by the cranial (skull) bones and contains the brain, and a **vertebral** or **spinal canal,** which is a bony cavity formed by the vertebrae of the backbone and contains the spinal cord and the beginnings of spinal nerves.

The other principal body cavity is the **ventral body cavity.** This cavity, also known as the *coelom,* is located on the ventral aspect of the body. The organs inside the ventral body cavity are called the **viscera.** Its walls are composed of skin, connective tissue, bone, muscles, and serous membrane. Like the dorsal body cavity, the ventral body cavity has two principal subdivisions – an upper portion, called the **thoracic cavity** (or chest cavity), and a lower portion, called the **abdominopelvic cavity.** The anatomical landmark that divides the ventral body cavity into the thoracic and abdominopelvic cavities is the muscular diaphragm. The thoracic cavity, in turn, contains several divisions.

There are two **pleural cavities** (Figure 1-4b), one around each lung, and a **mediastinum,** a mass of tissue between the pleurae of the lungs that extends from the sternum to the vertebral column (Figure 1-4a, b). The

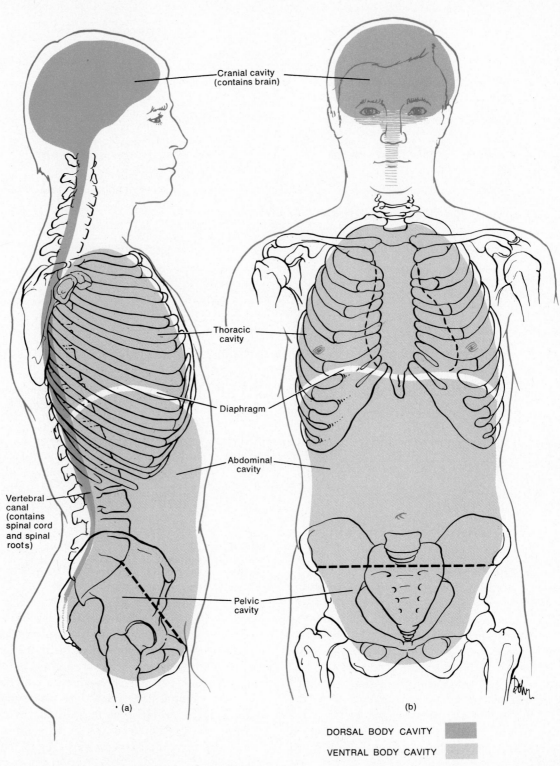

Cranial cavity
(contains brain)

Thoracic
cavity

Diaphragm

Abdominal
cavity

Vertebral
canal
(contains
spinal cord
and spinal
roots)

Pelvic
cavity

(a)

(b)

DORSAL BODY CAVITY

VENTRAL BODY CAVITY

Figure 1-3   Body cavities. (a) Median section through the body to indicate the location of the dorsal and ventral body cavities. (b) Subdivisions of the ventral body cavity.

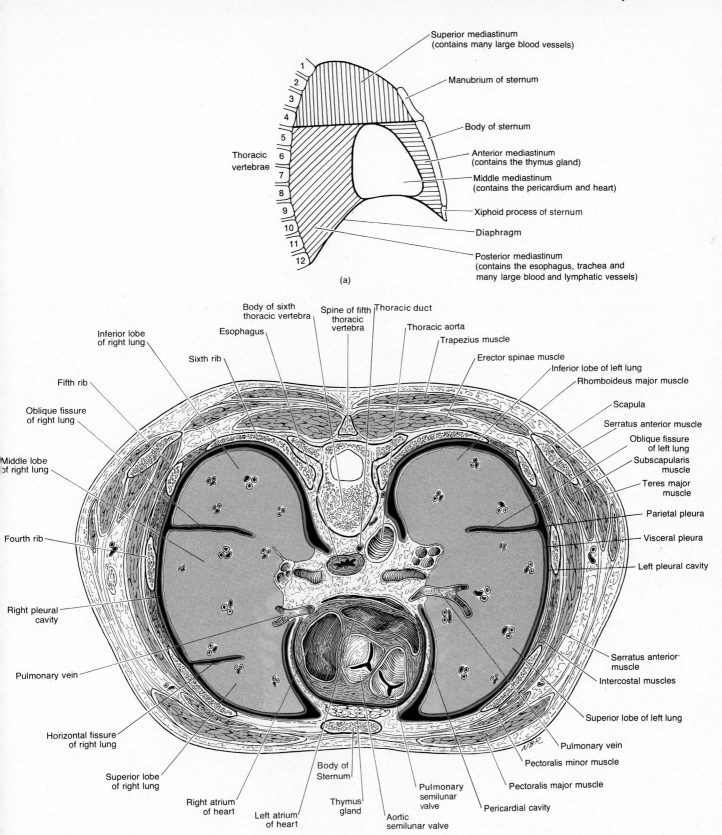

Superior mediastinum
(contains many large blood vessels)

Manubrium of sternum

Body of sternum

Anterior mediastinum
(contains the thymus gland)

Middle mediastinum
(contains the pericardium and heart)

Xiphoid process of sternum

Diaphragm

Posterior mediastinum
(contains the esophagus, trachea and
many large blood and lymphatic vessels)

Thoracic
vertebrae

1 2 3 4 5 6 7 8 9 10 11 12

(a)

Body of sixth
thoracic vertebra

Spine of fifth
thoracic
vertebra

Thoracic duct

Esophagus

Thoracic aorta

Trapezius muscle

Erector spinae muscle

Inferior lobe of left lung

Rhomboideus major muscle

Scapula

Serratus anterior muscle

Oblique fissure
of left lung

Subscapularis
muscle

Teres major
muscle

Parietal pleura

Visceral pleura

Left pleural cavity

Serratus anterior
muscle

Intercostal muscles

Superior lobe of left lung

Pulmonary vein

Pectoralis minor muscle

Pectoralis major muscle

Pericardial cavity

Inferior lobe
of right lung

Sixth rib

Fifth rib

Oblique fissure
of right lung

Middle lobe
of right lung

Fourth rib

Right pleural
cavity

Pulmonary vein

Horizontal fissure
of right lung

Superior lobe
of right lung

Right atrium
of heart

Left atrium
of heart

Body of
Sternum

Thymus
gland

Aortic
semilunar valve

Pulmonary
semilunar
valve

(b)

Figure 1-4 Mediastinum. The mediastinum is the space between the pleurae of the lungs that extends from the sternum to the vertebral column. (a) Subdivisions of the mediastinum seen in right lateral view. (b) Mediastinum seen in a cross section of the thorax. Many of the structures shown and labeled are unfamiliar to you now. They are discussed in detail in later chapters.

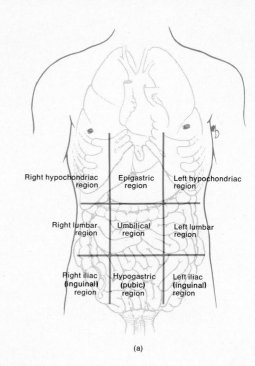

Figure 1-5   Abdominopelvic cavity. (a) The nine regions. The top horizontal line is drawn just below the bottom of the rib cage, and the bottom horizontal line is drawn just below the tops of the hipbones. The two vertical lines are drawn just medial to the nipples. The horizontal and vertical lines divide the area into a larger middle section and smaller left and right sections. (b) Anterior view of the abdomen and pelvis showing the most superficial organs. The greater omentum has been removed.

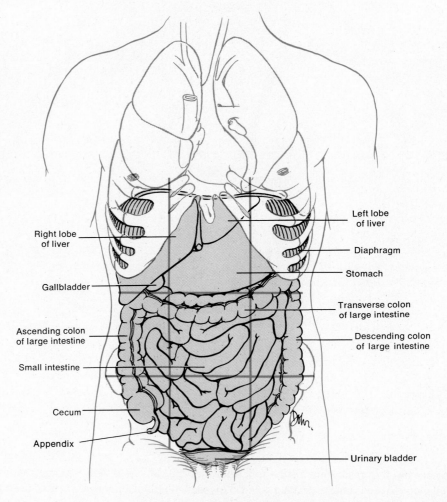

(b)

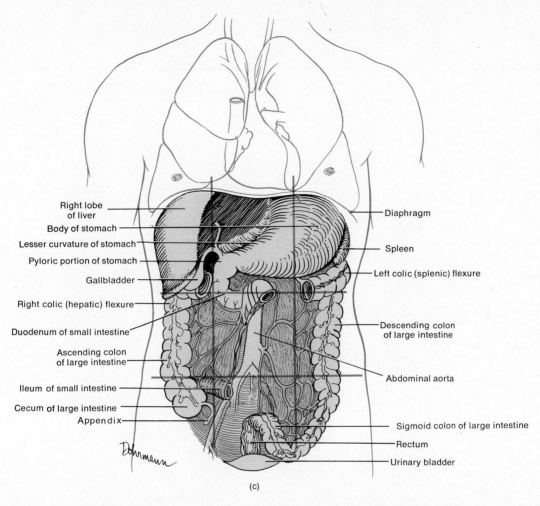

Right lobe of liver
Body of stomach
Lesser curvature of stomach
Pyloric portion of stomach
Gallbladder
Right colic (hepatic) flexure
Duodenum of small intestine
Ascending colon of large intestine
Ileum of small intestine
Cecum of large intestine
Appendix

Diaphragm
Spleen
Left colic (splenic) flexure
Descending colon of large intestine
Abdominal aorta
Sigmoid colon of large intestine
Rectum
Urinary bladder

(c)

Figure 1-5 (cont.)   Abdominopelvic cavity. (c) Anterior view of the abdomen and pelvis in which most of the small intestine and transverse colon have been removed to expose deeper structures.

pericardial cavity is located around the heart (Figure 1-4b). The abdominopelvic cavity, as the name suggests, is divided into two portions, although no wall lies between them (Figure 1-3). The upper portion, the **abdominal cavity,** contains the stomach, spleen, liver, gallbladder, pancreas, small intestine, most of the large intestine, the kidneys, and the ureters. The lower portion, the **pelvic cavity,** contains the urinary bladder, sigmoid colon, rectum, and the internal male or female reproductive organs. One way to demarcate the two cavities is by drawing an imaginary line from the symphysis pubis (anterior joint between hipbones) to the superior border of the sacrum (sacral promontory).

Abdominopelvic Regions

To describe the location of organs easily, the abdominopelvic cavity may be divided into the **nine regions** shown in Figure 1-5. Although some unfamiliar terms

are used in describing the nine regions and their contents, follow the description as best as you can. When the organs are studied in detail in later chapters, they will have more meaning. The *epigastric region* contains the left lobe and medial part of the right lobe of the liver, the pyloric part and lesser curvature of the stomach, the superior and descending portions of the duodenum, the body and upper part of the head of the pancreas, and the two adrenal (suprarenal) glands. The *right hypochondriac region* contains the right lobe of the liver, the gallbladder, and the upper third of the right kidney. The *left hypochondriac region* contains the body and fundus of the stomach, the spleen, the left colic (splenic) flexure, the upper two-thirds of the left kidney, and the tail of the pancreas. The *umbilical region* contains the middle of the transverse colon, the inferior part of the duodenum, the jejunum, the ileum, the hilar regions of the kidneys, and the bifurcations (branching) of the abdominal aorta and inferior vena cava. The *right lumbar*

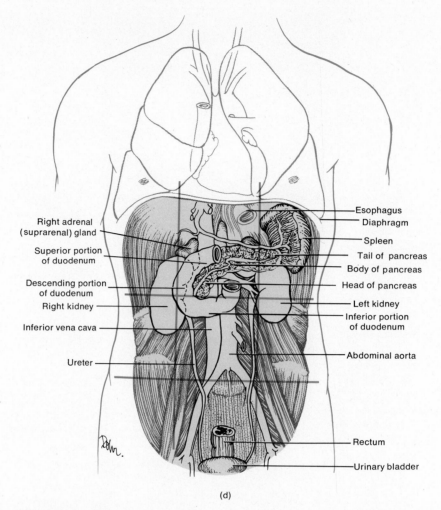

(d)

Figure 1-5 (cont.)   Abdominopelvic cavity. (d) Anterior view of the abdomen and pelvis in which many organs have been removed, exposing the posterior structures.

region contains the superior part of the cecum, the ascending colon, the right colic (hepatic) flexure, the lower lateral portion of the right kidney, and the small intestine. The *left lumbar region* contains the descending colon, the lower third of the left kidney, and the small intestine. The *hypogastric* (or *pubic*) *region* contains the urinary bladder when full, the small intestine, and part of the sigmoid colon. The *right iliac* (or *right inguinal*) *region* contains the lower end of the cecum, the appendix, and the small intestine. The *left iliac* (or *left inguinal*) *region* contains the junction of the descending and sigmoid parts of the colon and the small intestine.

An easier way to divide the abdominopelvic cavity is into four **quadrants** (Figure 1-6). In this method, frequently used by clinicians, a horizontal plane is passed through the umbilicus together with a midsagittal plane. These two planes divide the abdomen into a *right*

superior (upper) quadrant, left superior (upper) quadrant, right inferior (lower) quadrant, and *left inferior (lower) quadrant.* Whereas the nine-region designation is more widely used for anatomical studies, the four-quadrant designation is better suited for locating the site of an abdominopelvic pain, tumor, or other abnormality.

Anatomical Position and Regional Names

When a region of the body is described in an anatomical text or chart, we assume that the body is in the **anatomical position**—erect and facing the observer. The arms are at the sides, and the palms of the hands are turned forward, as in Figure 1-7. The common terms and the anatomical terms, in parentheses, of certain body regions are presented in Figure 1-7 also.

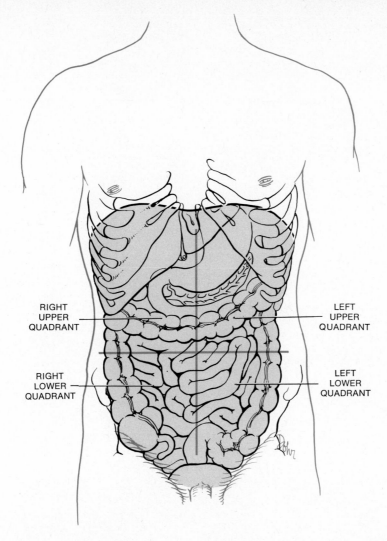

Figure 1-6 Four quadrants of the abdominopelvic cavity. The two lines pass through the umbilicus.

Planes

The structural plan of the human body may also be discussed with respect to **planes** (flat surfaces) that pass through it. Several of these planes are commonly used. Figure 1-8 illustrates the planes described here. A *midsagittal* (or *median*) *plane* through the midline of the body runs vertically and divides the body into equal right and left sides. A *sagittal* (or *parasagittal*) *plane* also runs vertically, but it divides the body into unequal left and right portions. A *frontal* (or *coronal*) *plane* runs vertically and divides the body into anterior and posterior portions. Finally, a *horizontal* (or *transverse*) *plane* runs parallel to the ground and divides the body into superior and inferior portions.

When you examine sections of organs, it is important to know how the section is made so you can understand the anatomical relationship of one part to

another. Figure 1-9 indicates how three different sections—a *cross section,* an *oblique section,* and *longitudinal sections*—are made through a simple tube and through the spinal cord. (See page 22.)

HOMEOSTASIS

The concept of homeostasis is a central theme of human physiology. For this reason, homeostasis is also a central theme of this text. **Homeostasis** is the condition in which the body's internal environment remains relatively constant (*homeo* = same; *stasis* = standing still). For the body's cells to survive, the composition of the surrounding fluids must be controlled precisely at all times. Body fluid outside the body cells is called **extracellular fluid** and is found in two principal places. The fluid filling the microscopic spaces between the cells

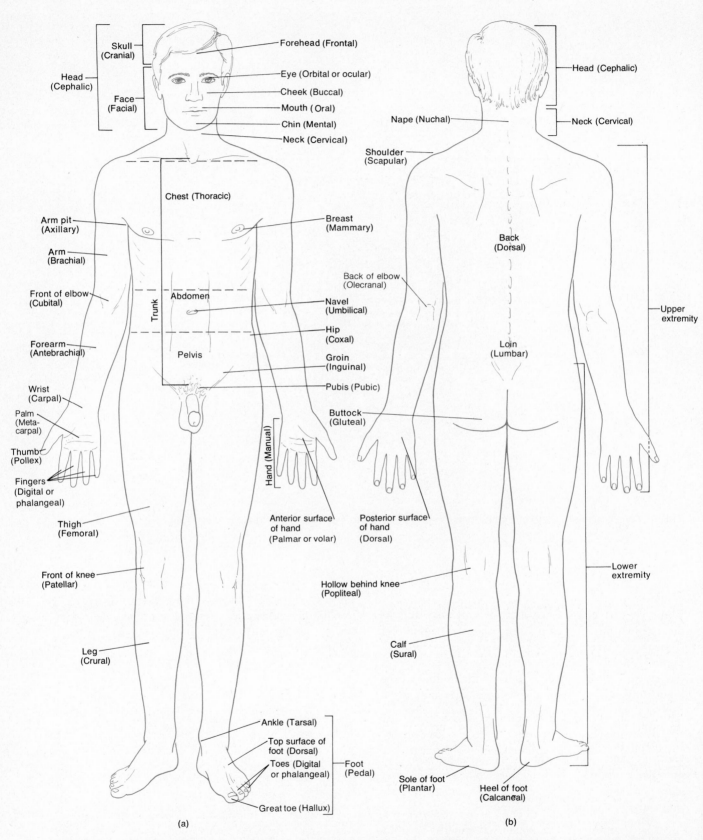

Figure 1-7   Anatomical position. (a) Anterior view. (b) Posterior view. Both common terms and anatomical terms, in parentheses, are indicated for many of the regions of the body.

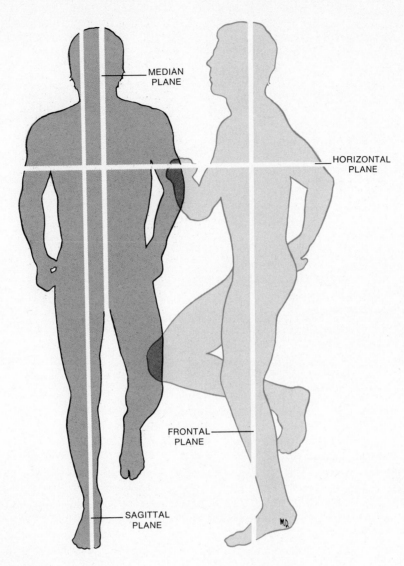

Figure 1-8 Planes of the body.

of tissues is called *intercellular fluid, interstitial fluid* (*inter* = between), or *tissue fluid.* The extracellular fluid in blood vessels is termed *plasma* (Figure 1-10). Fluid within cells is called **intracellular fluid**. Among the substances in extracellular fluid are gases, nutrients, and electrically charged chemical particles called ions – all needed for the maintenance of life. Extracellular fluid circulates through the blood and lymph vessels and from there moves into the spaces between the tissue cells. Thus it is in constant motion throughout the body. Essentially, all body cells are surrounded by the same fluid environment. For this reason, extracellular fluid is often called the body's internal environment.

An organism is said to be in homeostasis when its internal environment (1) contains exactly the optimum concentrations of gases, nutrients, ions, and water, (2) has an optimal temperature, and (3) has an optimal pressure for the health of the cells. When homeostasis is disturbed, ill health results. If the body fluids are not eventually brought back into balance, death may occur.

Stress

Homeostasis in all organisms is continually disturbed by **stress,** which is any stimulus that creates an imbalance in the internal environment. The stress may come from the external environment in the form of heat, cold, loud noises, or lack of oxygen. Or the stress may originate within the body in the form of high blood pressure, pain, tumors, or unpleasant thoughts. Most stresses are mild and routine. Poisoning, overexposure, severe infection, and surgical operations are examples of extreme stress.

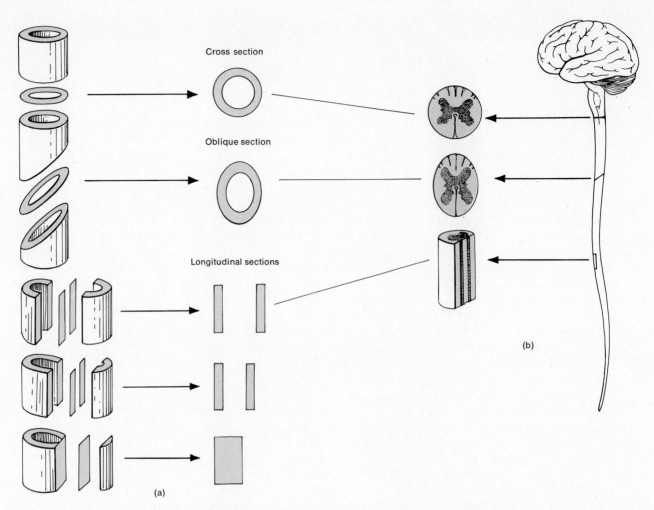

Figure 1-9   How sections are made. (a) Three sections through a tube. (b) The same three sections through the spinal cord.

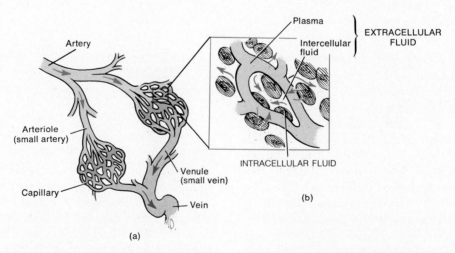

Figure 1-10   Internal environment of the body. (a) Extracellular fluid circulates through arteries and arterioles and then into minute blood vessels called capillaries. From there it moves into the spaces around the body cells. Next the fluid returns to the capillaries, passes through the venules, and is emptied into the veins. (b) Enlarged detail.

Fortunately, the body has many regulating (homeostatic) devices that oppose the forces of stress and bring the internal environment back into balance. High resistance to stress is a striking feature of all organisms. Some people live in deserts where the daytime temperatures easily reach 49°C (120°F). Others work outside all day in subzero weather. Yet everyone's internal body temperature remains near 37°C (98.6°F). Mountain climbers exercise strenuously at high altitudes, where the oxygen content of the air is low. But once they adjust to the new altitude, they do not suffer from oxygen shortage. The extremes in temperature and in oxygen content of the air are external stresses, and the exercise performed is an internal stress, yet the body compensates and remains in homeostasis. Walter B. Cannon (1871–1945), the American physiologist who coined the term homeostasis, noted that the heat produced by the muscles during strenuous exercise would curdle and inactivate the body's proteins if the body did not dissipate heat quickly by sweating. Muscles that are being exercised also produce, in addition to heat, a great deal of lactic acid. If the body did not have a homeostatic mechanism for reducing the amount of the acid, the extracellular fluid would become acidic and destroy the cells.

Every body structure, from the cellular to the system level, contributes in some way to keeping the internal environment within normal limits. One homeostatic function of the cardiovascular system, for example, is to keep the fluids of all parts of the body constantly moving. When we are at rest, fresh blood is circulated throughout the entire body about once every minute. But when we are active and our muscles need nutrients rapidly, the heart quickens its pace and pumps fresh blood to the organs five times a minute. In this way, the cardiovascular system helps compensate for the stress of increased activity.

The respiratory system offers another example of a homeostatic mechanism in the body. The cells need more oxygen and produce more carbon dioxide when they are very active. Therefore, during periods of activity, the respiratory system must work faster to keep the oxygen in the extracellular fluid from falling below normal limits and prevent excessive amounts of carbon dioxide from accumulating.

The digestive system and related organs help maintain the homeostasis involved in providing nutrients and removing wastes. As circulating blood passes through the organs of digestion, the products of digestion are transported to the body fluids so they can be used as nutrients by the cells. The liver, kidneys, endocrine glands, and other organs help to alter or store the products of digestion in various ways. The kidneys also help to remove the wastes produced by cells after the cells have utilized the nutrients.

The homeostatic mechanisms of the body, such as those performed by the cardiovascular, respiratory, and digestive systems, are themselves subject to control by the nervous system and the endocrine system. The nervous system regulates homeostasis by detecting when the body is deviating from its balanced state and by sending messages to the proper organs to counteract the stress. For instance, when muscle cells are active, they take a great deal of oxygen from the blood. They also give off carbon dioxide, which is picked up by the blood. Certain nerve cells detect the chemical changes occurring in the blood and send a message to the brain. The brain then sends a message to the heart to pump blood more quickly to the lungs so the blood can give up its excess carbon dioxide and take on more oxygen. Simultaneously, the brain sends a message to the muscles that control breathing to contract faster. As a result, carbon dioxide can be exhaled and more oxygen can be inhaled.

Homeostasis is also controlled by the endocrine system—a series of glands that secrete chemical regulators, called hormones, into the blood. Whereas nerve impulses coordinate homeostasis rapidly, hormones work slowly. Both means of control are directed toward the same end.

Blood Pressure

Blood pressure is the force exerted by blood against the walls of the blood vessels, especially the arteries. It is determined primarily by three factors: the rate and strength of the heartbeat, the amount of blood, and the resistance offered by the arteries as blood passes through them. The resistance of the arteries results from the chemical properties of the blood and the size of the arteries at the time.

If some stress, either internal or external, causes the heartbeat to speed up, the following sequence occurs (Figure 1-11): As the heart pumps faster, it pushes more blood into the arteries per minute—increasing pressure in the arteries. The higher pressure is detected by pressure-sensitive cells in the walls of certain arteries, which send nerve impulses to the brain. The brain interprets the message and sends impulses to the heart to slow the heart rate.

A **feedback system** is any circular situation in which information is fed back into the system. The nervous control that results in a constant blood pressure is an example. In the case of regulating blood pressure, the body itself may be considered the system. The **input** is the information picked up by the pressure-sensitive cells, and the **output** is the return toward normal blood

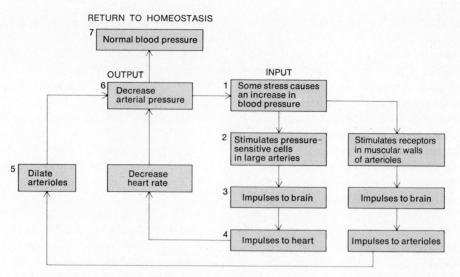

RETURN TO HOMEOSTASIS

7 Normal blood pressure

OUTPUT

6 Decrease arterial pressure

INPUT

1 Some stress causes an increase in blood pressure

5 Dilate arterioles

Decrease heart rate

2 Stimulates pressure-sensitive cells in large arteries

Stimulates receptors in muscular walls of arterioles

3 Impulses to brain

Impulses to brain

4 Impulses to heart

Impulses to arterioles

Figure 1-11 Homeostasis of blood pressure. Note that the output is fed back into the system, and the system continues to lower blood pressure until there is a return to normal conditions.

pressure. Figure 1-11 shows that the system runs in a circle. The pressure-sensitive cells detect when the blood pressure begins to normalize, and this output is fed back into the system, even after the return to homeostasis has begun. In other words, the cells report back to the brain on the changed blood pressure, and, if the pressure is still too high, the brain continues to send out impulses to slow the heartbeat. Since this feedback system reverses the direction of the initial condition from a rising to a falling blood pressure, it is called a **negative feedback system** – the reaction of the body (output) counteracts the stress (input) in order to restore homeostasis. If the brain had signaled the heart to beat even faster and the blood pressure had kept on rising, the system would have been a **positive feedback system.** In a positive feedback system the output *intensifies* the input. Most of the feedback systems of the body are negative feedback systems.

Figure 1-11 shows that a second negative feedback control is involved in maintaining normal blood pressure. Small arteries, called arterioles, have muscular walls that normally "squeeze" the hollow tubes the blood flows through. When the blood pressure increases, sensors in the arterioles send messages to the brain. The brain signals the muscular walls to relax so that the diameter of the small tubes increases. Thus the blood flowing through the arterioles is offered less resistance and blood pressure drops back to normal.

Blood Sugar Level

An example of homeostatic regulation by hormones is the maintenance of blood sugar level. This sugar, called glucose, is found in blood at a certain level and is one of

the body's principal sources of energy. Under normal circumstances, the concentration of sugar in the blood averages about 90 milligrams/100 milliliters of blood.* This level is maintained primarily by two hormones secreted by the pancreas: insulin and glucagon. Suppose you have just eaten some candy. The sugar in the candy is broken down by the organs of digestion and moves from the digestive tract into the blood. The sugar then becomes a stress because it raises the blood sugar level above normal. In response to this stress, the cells of the pancreas are stimulated to secrete insulin (Figure 1-12a). Once insulin enters the blood, it has two principal effects. First, it increases sugar uptake by cells. The blood sugar level is lowered by this action since the sugar moves from the blood into body cells. Second, insulin accelerates the process by which sugar is stored in the liver and muscles. Thus even more sugar is removed from the blood. In essence, insulin decreases blood sugar concentration until it returns to normal.

The other hormone produced by the pancreas, glucagon, has the opposite effect of insulin. Suppose you have not eaten for several hours and your blood sugar level is steadily decreasing. Lack of sugar is now the stress, and, under this condition, other cells of the pancreas are instead stimulated to secrete glucagon (Figure 1-12b). This hormone accelerates the process by which sugar stored in the liver is sent back into the bloodstream. The blood sugar level is thus increased until it returns to normal.

*One milligram (mg) = 0.001 gram (g), 1 g = 0.035 ounce (oz), and 100 milliliters (ml) = 3.38 oz.

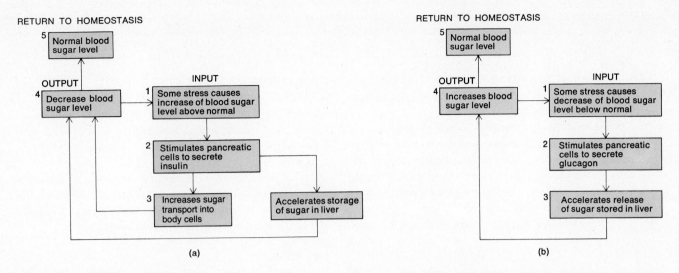

Figure 1-12 Homeostasis of blood sugar level. (a) Mechanism that lowers high blood sugar level. (b) Mechanism that raises low blood sugar level.

MEASURING THE HUMAN BODY

An important concept regarding your body is measurement. Examples of such measurements are organ size, organ weight, and amount of medication to be administered. Measurements involving time, weight, temperature, size, length, and volume are routine to a medical science program. In this book measurements are given in metric units followed by the approximate English equivalent in parentheses. If you are not familiar with English-metric conversions, consult the appendix.

ANATOMY AND PHYSIOLOGY DEFINED

1 Anatomy is the study of structure and how structures are related to each other.

2 Subdivisions of anatomy include gross anatomy (macroscopic), systemic anatomy (systems), regional anatomy (regions), developmental anatomy (development from fertilization to adulthood), embryology (development before eighth week), pathological anatomy (disease), and histology (microscopic study of tissues and cells).

3 Physiology is the study of how structures function.

LEVELS OF STRUCTURAL ORGANIZATION

1 The human body consists of levels of structural organization from the chemical level to the organismic level.

2 The chemical level is represented by all the atoms and molecules in the body. The cellular level consists of cells. The tissue level is represented by tissues. The organ level consists of body organs, and the system level is represented by organs that work together to perform a general function.

3 The human organism is a collection of structurally and functionally integrated systems.

STRUCTURAL PLAN
DIRECTIONAL TERMS

1 Directional terms indicate the relationship of one part of the body to another.

2 Examples of directional terms are superior (toward the head), anterior (near the front), medial (nearer the midline), distal (farther from the attachment of a limb), and external (toward the surface).

BODY CAVITIES

1 Spaces in the body that contain internal organs are called cavities.

2 Dorsal and ventral cavities are the two principal body cavities. The dorsal cavity contains the brain and spinal cord. The organs of the ventral cavity (coelom) are collectively called the viscera.

3 The dorsal cavity is subdivided into the cranial cavity and vertebral canal.

4 The ventral body cavity is subdivided by the diaphragm into an upper thoracic cavity and a lower abdominopelvic cavity.

5 The thoracic cavity contains two pleural cavities and a mediastinum, which includes the pericardial cavity.

ABDOMINOPELVIC REGIONS

1 The abdominopelvic cavity, actually an upper abdominal cavity and a lower pelvic cavity, is divided into nine anatomical regions. It may also be divided into four quadrants.

ANATOMICAL POSITION AND REGIONAL NAMES

1 The position in which the body is studied and described is the anatomical position. The subject stands erect and faces the observer with arms at sides and palms turned forward.

2 Regional names are terms given to specific regions of the body for reference. Examples of regional names include cranial (skull), thoracic (chest), brachial (arm), patellar (knee), cephalic (head), and gluteal (buttock).

PLANES

1 Planes of the body are flat surfaces that divide the body into definite areas. The midsagittal or median plane divides the body into equal right and left sides; the sagittal (parasagittal) plane, into unequal right and left sides, the frontal (coronal) plane, into anterior and posterior portions; the horizontal (transverse) plane, into superior and inferior portions.

2 Sections of organs include cross sections, oblique sections, and longitudinal sections.

HOMEOSTASIS

1 Homeostasis is a condition in which the internal environment of the body (extracellular fluid) remains relatively constant.

2 All body systems attempt to maintain homeostasis.

3 Homeostasis is controlled by the nervous and endocrine systems.

4 Examples include the nervous control of blood pressure and the hormonal control of blood sugar level.

STRESS

1 Stress is any stimulus that creates an imbalance in the internal environment.

2 If a stress acts on the body, homeostatic mechanisms attempt to counteract the effects of the stress and bring the system back to normal.

BLOOD PRESSURE

1 Blood pressure is determined by the rate and force of the heartbeat, the amount of blood, and arterial resistance. A rising blood pressure falls back toward normal because of a negative feedback system of blood pressure regulation.

BLOOD SUGAR LEVEL

1 A normal blood sugar level is maintained by the actions of insulin and glucagon. Insulin causes a rising sugar level to fall back to normal. Glucagon brings a low level up to normal.

MEASURING THE HUMAN BODY

1 Various kinds of measurements are important in understanding the human body.

2 Examples of such measurements include organ size, organ weight, and amount of medication to be administered.

3 Measurements in this book are given in metric units followed by the approximate English equivalents in parentheses.

REVIEW QUESTIONS

1 Define anatomy. How does each subdivision of anatomy help you understand the structure of the human body? Define physiology.

2 Construct a diagram to illustrate the levels of structural organization that characterize the body. Be sure to define each level.

3 Outline the function of each system of the body, and list several organs that compose each system.

4 What does bilateral symmetry mean? Why is the body considered to be a tube within a tube?

5 What is a directional term? Why are these terms important? Can you use each of the directional terms listed in Exhibit 1-2 in a complete sentence?

6 Define a body cavity. List the body cavities discussed, and tell which major organs are located in each. What landmarks separate the various body cavities from each other?

7 Discuss how the abdominopelvic area is subdivided into nine regions. Name and locate each region and list the organs, or parts of organs, in each. Describe how the abdominopelvic cavity is divided into four quadrants and name each quadrant.

8 When is the body in the anatomical position? Why is the anatomical position used?

9 Review Figure 1-7. See if you can locate each region on your own body, and name each by its common and anatomical term.

10 Describe the various planes that may be passed through the body. Explain how each plane divides the body.

11 What is meant by the phrase "a part of the body has been sectioned"? Given an orange, can you make a cross section, oblique section, and longitudinal section with a knife?

12 Define homeostasis. What is extracellular fluid? Why is it called the internal environment of the body?

13 How is stress related to homeostasis? What is a stress? Give several examples.

14 How is homeostasis related to normal and abnormal conditions in the body?

15 Substantiate this statement: "Homeostasis is a cooperative effort of all body parts."

16 What systems of the body control homeostasis? Explain.

17 Discuss briefly how the regulation of blood pressure and blood glucose level are examples of homeostasis.

Chapter 2
The Chemical
Level of
Organization

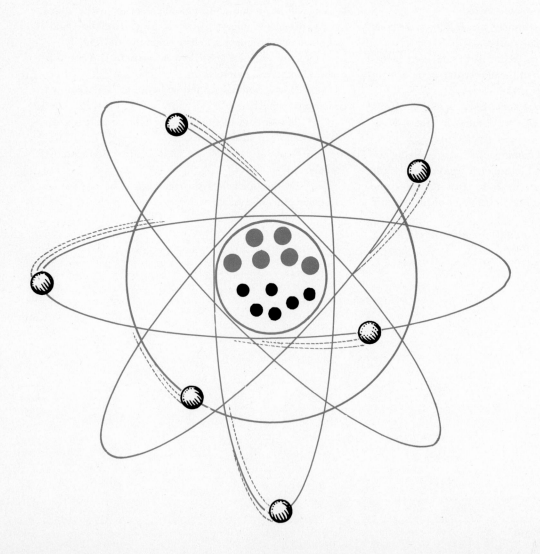

STUDENT OBJECTIVES

■ Identify by name and symbol the principal chemical elements of the human body.

■ Explain, by diagraming, the structure of an atom.

■ Define a chemical reaction as a function of electrons in incomplete outer energy levels.

■ Describe ionic bond formation in a molecule of sodium chloride (NaCl).

■ Discuss covalent bond formation as the sharing of outer energy level electrons and discuss radioactive tracers in medicine.

■ Identify and compare each kind of chemical reaction.

■ Define the type of energy involved when chemical bonds are formed and broken.

■ Define and distinguish between inorganic and organic compounds.

■ Discuss the functions of water as a solvent, suspending medium, chemical reactant, heat absorber, and lubricant.

■ List and compare the properties of acids, bases, and salts.

■ Define pH as the degree of acidity or alkalinity of a solution.

■ Describe the role of a buffer system as a homeostatic mechanism that maintains the pH of a body fluid.

■ Describe and compare the structure and functions of carbohydrates, lipids, and proteins.

■ Differentiate between dehydration synthesis and hydrolysis of organic molecules.

■ Define the homeostatic role of enzymes as catalysts.

■ Contrast the structure of deoxyribonucleic acid (DNA) and ribonucleic acid (RNA).

■ Define the roles of DNA and RNA in heredity and protein synthesis.

■ Identify the function and importance of adenosine triphosphate (ATP), cyclic adenosine-3, 5-monophosphate (cyclic AMP), and prostaglandins (PG).

Many of the common substances you eat and drink—water, sugar, table salt, cooking oil—play vital roles in keeping you alive. In this chapter, you will learn something about how the molecules of these substances function in your body. Fundamental to this study is a knowledge of basic chemistry and chemical processes. To understand the nature of the matter you are made from and the changes this matter goes through in your body, you will need to know which chemical elements are present in the human organism and how they interact.

BASIC CHEMISTRY FOR LIVING SYSTEMS

Chemical Elements

All living and nonliving things consist of **matter,** which is anything that occupies space and possesses mass. Matter may exist in a solid, liquid, or gaseous state. All forms of matter are made up of a limited number of building units called **chemical elements.** At present, scientists can recognize 106 different elements. Elements are designated by letter abbreviations, usually derived from the first or second letter of the Latin or English name for the element. Such letter abbreviations are called *chemical symbols.* Examples of chemical symbols are H (hydrogen), C (carbon), O (oxygen), N (nitrogen), Na (sodium), K (potassium), Fe (iron), and Ca (calcium).

Approximately 24 elements are found in the human organism. Carbon, hydrogen, oxygen, and nitrogen make up about 96 percent of the body's weight. These four elements together with phosphorus and calcium constitute approximately 99 percent of the total body weight. Eighteen other chemical elements, called trace elements, are found in low concentrations and compose the remaining 1 percent.

Atomic Structure

Each element is made up of units of matter called **atoms.** An element is simply a quantity of matter composed of atoms all of the same type. A handful of the element carbon, such as pure coal, contains only carbon atoms. A tank of oxygen contains only oxygen atoms. Measurements indicate that the smallest atoms are less than 1/250,000,000 of an inch in diameter, and the largest atoms are 1/50,000,000 of an inch in diameter. In other words, if 50 million of the largest atoms were placed end to end, they would measure approximately 1 inch in length.

An atom consists of two basic parts: the nucleus and the electrons (Figure 2-1). The centrally located *nucleus* contains positively charged particles called *protons*

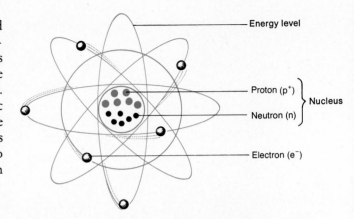

Figure 2-1   Structure of an atom. In this highly simplified version of a carbon atom, note the centrally located nucleus, which contains neutrons and protons. The electrons orbit about the nucleus at varying distances from its center.

(p^+) and uncharged particles called *neutrons* (n). Because each proton has one positive charge, the nucleus itself is positively charged. The second basic part of an atom contains the *electrons* (e^-). These negatively charged particles spin around the nucleus in approximately oval orbits. The number of electrons in an atom always equals the number of protons. Since each electron carries one negative charge, the negatively charged electrons and the positively charged protons balance each other—and the atom is electrically neutral.

What makes the atoms of one element different from those of another? The answer lies in the number of protons. Figure 2-2 shows that the hydrogen atom contains one proton. The helium atom contains two. The carbon atom has six, and so on. Each different kind of atom has a different number of protons in its nucleus. The number of protons in an atom is called the atom's *atomic number.* Therefore we can say that each kind of atom, or element, has a different atomic number.

Atoms and Molecules

When atoms combine with other atoms, or break apart from other atoms, the process is called a **chemical reaction.** Chemical reactions are the foundation of all life processes.

The electrons of the atom actively participate in chemical reactions. The electrons spin around the nucleus in orbits, shown in Figure 2-2 as concentric circles lying at different distances from the nucleus. We call these orbits *energy levels.* Each orbit has a maximum number of electrons it can hold. For instance, the orbit nearest the nucleus never holds more than two electrons, no matter what the element. This orbit can be referred to

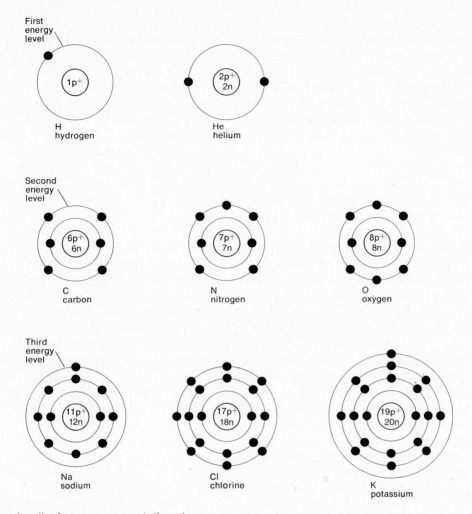

Figure 2-2 Energy levels of some representative atoms.

as the first energy level. The second energy level holds a maximum of eight electrons. The third level also can hold a maximum of eight electrons – and if there is a fourth level, the third level can hold a maximum of 18 electrons.

An atom always attempts to fill its outermost orbit with the maximum number of electrons it can hold. To do this, the atom may give up an electron, take on an electron, or share an electron with another atom – whichever is easiest. Take a look at the chlorine atom. Its outermost energy level, which happens to be the third level, has seven electrons. However, the third level of an atom can hold a maximum of eight electrons. Chlorine can thus be visualized as having a shortage of one electron. In fact, chlorine usually does try to pick up an extra electron. Sodium, by contrast, has only one electron in its outer level. This again happens to be the third energy level. It is much easier for sodium to get rid of the one electron in its third energy level than to fill the

third level by taking on seven more electrons. Atoms of a few elements, like helium, have completely filled outer energy levels and do not need to gain or lose electrons. These are called *inert* elements.

Atoms with incompletely filled outer energy levels, like sodium and chlorine, tend to combine with other atoms in a chemical reaction. During the reaction, the atoms can trade off or share electrons and, thereby, fill their outer energy levels. Atoms that already have filled outer levels generally do not participate in chemical reactions for the simple reason that they do not need to gain or lose electrons. When two or more atoms combine in a chemical reaction, the resulting combination is called a **molecule.** A molecule may contain two atoms of the same kind, as in the hydrogen molecule: H_2. The subscript 2 indicates that there are two hydrogen atoms in the molecule. Molecules may also be formed by the reaction of two or more different kinds of atoms, as in the hydrochloric acid molecule: HCl. Here an atom of hydrogen is attached to an atom of chlorine. A molecule that contains at least two different kinds of atoms is called a **compound.** Hydrochloric acid, which is present in the digestive juices of the stomach, is a compound. A molecule of hydrogen is not.

The atoms in a molecule are held together by forces of attraction called **chemical bonds.** Here we will consider the two basic types of chemical bonds: ionic bonds and covalent bonds.

IONIC BONDS

Atoms are electrically neutral because the number of positively charged protons equals the number of negatively charged electrons. But when an atom gains or loses electrons, this balance is destroyed. If the atom gains electrons, it acquires an overall negative charge. If the atom loses electrons, it acquires an overall positive charge. Such a negatively or positively charged atom or group of atoms is called an **ion.**

Consider the sodium ion (Figure 2-3a). The sodium atom (Na) has 11 protons and 11 electrons, with one electron in its outer energy level. When sodium gives up the single electron in its outer level, it is left with 11 protons and only 10 electrons. The atom now has an overall positive charge of one ($+1$). This positively charged sodium atom is called a sodium ion (written Na^+).

Another example is the formation of the chloride ion (Figure 2-3b). Chlorine has a total of 17 electrons, 7 of them in the outer energy level. Since this energy level can hold 8 electrons, chlorine tends to pick up an electron that has been lost by another atom. By accepting an electron, chlorine acquires a total of 18 electrons. However, it still has only 17 protons in its nucleus. The chloride ion therefore has a negative charge of -1 and is written as Cl^-.

The positively charged sodium ion (Na^+) and the negatively charged chloride ion (Cl^-) attract each other—unlike charges attract one another. The attraction, called an **ionic bond,** holds the two atoms together, and a molecule is formed (Figure 2-3c). The formation of this molecule, sodium chloride (NaCl) or table salt, is one of the most common examples of ionic bonding. Thus an ionic bond is an attraction between atoms in which one atom loses electrons and another atom gains electrons. Generally, atoms whose outer energy level is less than half-filled lose electrons and form positively charged ions called *cations.* Examples of cations are potassium ion (K^+), calcium ion (Ca^{2+}), iron ion (Fe^{2+}), and sodium ion (Na^+). By contrast, atoms whose outer energy level is more than half-filled tend to gain electrons and form negatively charged ions called *anions.* Examples of anions include iodine ion (I^-), chloride ion (Cl^-), and sulfur ion (S^{2-}).

Notice that an ion is always symbolized by writing the chemical abbreviation followed by the number of positive ($+$) or negative ($-$) charges the ion acquires.

Hydrogen is an example of an atom whose outer level is exactly half-filled. The first energy level can hold two electrons, but in hydrogen atoms it contains only one. Hydrogen may lose its electron and become a positive ion (H^+). This is precisely what happens when hydrogen combines with chlorine to form hydrochloric acid (H^+Cl^-). However, hydrogen is equally capable of forming another kind of bond altogether: a covalent bond.

COVALENT BONDS

The second chemical bond to be considered here is the **covalent bond.** This bond is far more common in organisms than is the ionic bond. When a covalent bond is formed, neither of the combining atoms loses or gains an electron. Instead, the two atoms share one, two, or three electron pairs. Look at the hydrogen atom again. One way a hydrogen atom can fill its outer energy level is to combine with another hydrogen atom to form the molecule H_2 (Figure 2-4a). In the H_2 molecule, the two atoms share a pair of electrons. Each hydrogen atom has its own electron plus one electron from the other atom. The shared pair actually circles the nuclei of both atoms. Therefore the outer energy levels of both atoms are filled. When only one pair of electrons is shared between atoms, as in the H_2 molecule, a *single covalent bond* is formed. For convenience, a single covalent bond is expressed as a single line between the atoms (H—H). When two pairs of electrons are shared between two atoms, a *double covalent bond* is formed, which is expressed as two parallel lines ($=$) (Figure 2-4b). A *triple covalent bond,* expressed by three parallel lines ($\equiv$), occurs when three pairs of electrons are shared (Figure 2-4c).

The same principles that apply to covalent bonding between atoms of the same element also apply to atoms of different elements. Methane (CH_4) is an example of covalent bonding between atoms of different elements (Figure 2-4d). The outer energy level of the carbon atom can hold eight electrons but has only four of its own. Each hydrogen atom can hold two electrons but has only one of its own. Consequently, in the methane molecule the carbon atom shares four pairs of electrons. One pair is shared with each hydrogen atom. Each of the four carbon electrons orbits around both the carbon nucleus and a hydrogen nucleus. Each hydrogen electron circles around its own nucleus and the carbon nucleus.

Elements whose outer energy levels are half-filled, such as hydrogen and carbon, form covalent bonds quite easily. In fact, carbon always forms covalent bonds. It never becomes an ion. However, many atoms whose

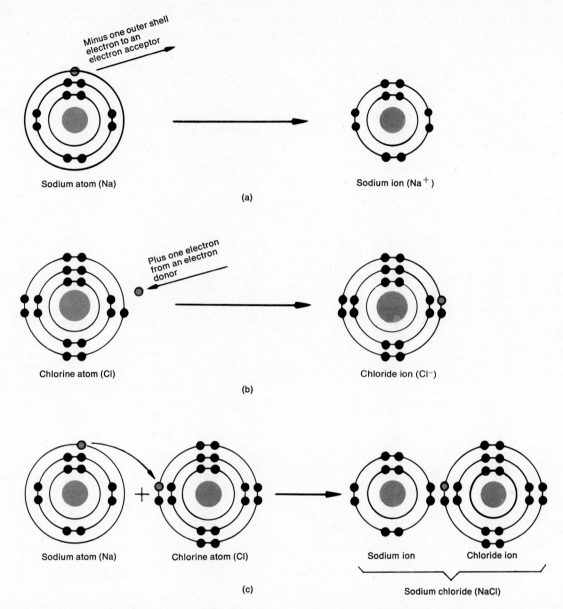

Figure 2-3 Formation of an ionic bond. (a) An atom of sodium attains stability by passing a single electron to an electron acceptor. The loss of this single electron results in the formation of a sodium ion (Na+). (b) An atom of chlorine attains stability by accepting a single electron from an electron donor. The gain of this single electron results in the formation of a chloride ion (Cl−). (c) When the Na+ and the Cl− ions are combined, they are held together by an attraction called an ionic bond, and a molecule of NaCl is formed.

outer energy levels are more than half-filled also form covalent bonds. An example is oxygen. We won't go into the reasons why some atoms tend to form covalent bonds rather than ionic bonds. You should, however, understand the basic principles involved in bond formation. Chemical reactions are nothing more than the making or breaking of bonds between atoms. And these reactions occur continually in all the cells of your body. As you will see again and again, reactions are the

processes by which body structures are built and body functions carried out.

RADIOACTIVE TRACERS

Atoms of an element, although chemically alike, may have different nuclear mass, so the atomic weight assigned to an element is only a mean. Each of the chemically identical atoms of an element with a particular nuclear mass is an *isotope* of that element. Isotopes are

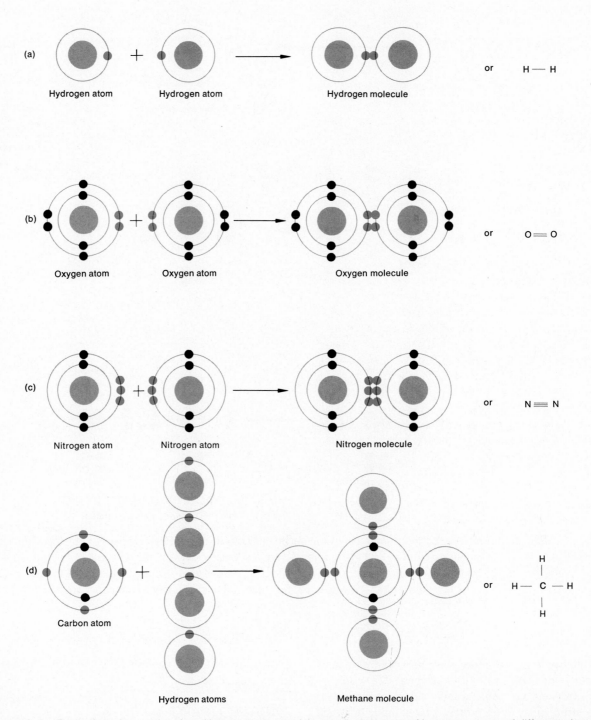

Figure 2-4   Formation of a covalent bond between atoms of the same element and between atoms of different elements. (a) A single covalent bond between two hydrogen atoms. (b) A double covalent bond between two oxygen atoms. (c) A triple covalent bond between two nitrogen atoms. (d) Single covalent bonds between a carbon atom and four hydrogen atoms. The representations on the far right are another way of showing the covalent bonds. Each straight line between atoms is a single covalent bond.

named by a number that indicates their atomic mass, the sum of their neutrons and protons. Certain isotopes called *radioisotopes* are unstable – they "decay" or change their nuclear structure to a more stable configuration. And in decaying they emit radiation that can be detected by instruments. These instruments estimate the amount of radioisotope present in a part of the body, or in a sample of material, and form an image of its distribution.

Radioisotopes of iodine were among the first discovered, and their specificity for the thyroid gland made them the cornerstone for the study of thyroid physiology. Nuclear medicine now uses ^{32}P to treat leukemia and ^{59}Fe for the study of red blood cell production. Moreover, short-lived agents such as technetium-99m pertechnetate (^{99m}Tc) have improved the quality of the images and reduced the patient's radiation dose.

Chemical Reactions

In this section, we look at four basic types of chemical reactions. These reactions are simple yet central to life processes. Once you have learned them, you will be able to understand the chemical reactions discussed later.

SYNTHESIS REACTIONS—ANABOLISM

When two or more atoms, ions, or molecules combine to form new and larger molecules, the process is called a *synthesis reaction.* The word *synthesis* means to put together, and synthesis reactions involve the *forming of new bonds.* Synthesis reactions can be expressed in the following way:

$$A \quad + \quad B \quad \rightarrow \quad AB$$
Atom, ion,   Atom, ion,   Combine to form new
or molecule A   or molecule B   molecule AB

The combining substances, A and B, are called the *reactants;* the substance formed by the combination is the *end product.* The arrow indicates the direction in which the reaction is proceeding. An example of a synthesis reaction is:

$$H^+ \quad + \quad Cl^- \quad \rightarrow \quad HCl$$
Hydrogen   Chloride   Hydrochloric acid
ion   ion   molecule

All the synthesis reactions that occur in your body are collectively called anabolic reactions, or simply *anabolism.* Combining glucose molecules to form glycogen and amino acids to form proteins are two examples of anabolism. The importance of anabolism is considered in detail in Chapter 25.

DECOMPOSITION REACTIONS—CATABOLISM

The reverse of a synthesis reaction is a *decomposition reaction.* The word *decompose* means to break down into smaller parts. In a decomposition reaction, the *bonds are broken.* Large molecules are broken down into smaller molecules, ions, or atoms. A decomposition reaction occurs in this way:

$$AB \quad \rightarrow \quad A \quad + \quad B$$
Molecule AB   Atom, ion, or   Atom, ion,
breaks down   molecule A   or molecule B
into

Under the proper conditions, methane can decompose into carbon and hydrogen:

$$CH_4 \quad \rightarrow \quad C \quad + \quad 4H$$
Methane molecule   One carbon   Four hydrogen
atom   atoms

The subscript 4 on the left-hand side of the reaction equation indicates that four atoms of hydrogen are bonded to one carbon atom in the methane molecule. The number 4 on the right-hand side of the equation shows that four single hydrogen atoms have been set free.

All the decomposition reactions that occur in your body are collectively called catabolic reactions, or simply *catabolism.* The digestion and oxidation of food molecules are examples of catabolism. The importance of catabolism is considered in detail in Chapter 25.

EXCHANGE REACTIONS

All chemical reactions are based on synthesis or decomposition processes. In other words, chemical reactions are simply the making and/or breaking of ionic or covalent bonds. Many reactions, such as *exchange reactions,* are partly synthesis and partly decomposition. An exchange reaction works like this:

$$AB + CD \rightarrow AD + BC$$

Here the bonds between A and B and between C and D are broken in a decomposition process. New bonds are then formed between A and D and between B and C in a synthesis process.

REVERSIBLE REACTIONS

When chemical reactions are reversible, the end product can revert to the original combining molecules. A *reversible reaction* is indicated by two arrows:

$$A + B \underset{\text{breaks down to}}{\overset{\text{combines with}}{\rightleftharpoons}} AB$$

Some reversible reactions occur because neither the reactants nor the end products are very stable. Other reactions reverse themselves only under special conditions:

$$A + B \underset{\text{water}}{\overset{\text{heat}}{\rightleftharpoons}} AB$$

Whatever is written above or below the arrows indicates the special condition under which the reaction occurs. In this case, A and B react to produce AB only when heat is applied, and AB breaks down into A and B only when water is added. Figure 2-5 summarizes the basic chemical reactions that can occur.

METABOLISM

The word **metabolism** represents the sum of all the synthesis and decomposition reactions occurring in the body – the sum of all anabolic and catabolic reactions that go on inside you. When we say that a person has a high metabolism, we mean that the chemical reactions in the body are proceeding at a faster rate than normal. The decomposition reactions are occurring so quickly that foods are broken down completely before the body has a chance to store them. Consequently, the person with a high metabolism can usually eat a great deal without gaining weight. Because of the tremendous amounts of energy produced by rapid decomposition reactions, the person also has a great deal of energy for activities as well as a lot of heat energy. Thus people with high metabolisms appear to have a lot of "nervous" energy and often complain that they are too hot.

Chemical reactions in the bodies of people with low metabolisms proceed more slowly than normal. Food is broken down slowly. Much of it is only partially broken down and then stored. Such people tend to gain weight easily, have little energy, and are often cold. Because their synthesis reactions are also slowed down, their bodies build up new structures very slowly. Wounds, for instance, often take a long time to heal.

ENERGY OF CHEMICAL REACTIONS

Some form of energy is involved whenever bonds between atoms in molecules are formed or broken during the chemical reactions taking place in the body. When a chemical bond is formed, energy is required. When a bond is broken, energy is released. This means that synthesis reactions need energy in order to occur, whereas decomposition reactions give off energy. The building processes of the body – the construction of bones, the growth of hair and nails, the replacement of injured cells – occur basically through synthesis reactions. The breakdown of foods, on the other hand,

occurs through decomposition reactions. When foods are decomposed, they release energy that can be used by the body for its building processes. Released energy can also be used to warm the body by taking the form of *heat energy.* Foods can be partially broken down into compounds that can be stored in the body. Later, when added energy is needed, the body finishes breaking down these reserve compounds. This stored energy is called *potential energy.*

Another form of energy that is important to life processes, *kinetic energy,* is discussed in Chapter 3.

CHEMICAL LIFE PROCESSES

Most of the chemicals in the body exist in the form of compounds. Biologists and chemists divide these compounds into two principal classes: inorganic compounds, which usually lack carbon, and organic compounds, which always contain carbon. **Inorganic compounds** are usually small, ionically bonded molecules that are vital to body functions. They include water, many salts, acids, and bases. **Organic compounds** are held together mostly or entirely by covalent bonds. They tend to be very large molecules and are therefore good building blocks for body structures. Organic compounds present in the body include carbohydrates, lipids, proteins, nucleic acids, and ATP.

Inorganic Substances

WATER

One of the most important, as well as the most abundant, inorganic substances in the human organism is **water.** In fact, with a few exceptions, such as tooth enamel and bone tissue, water is by far the most abundant material in all tissues. About 60 percent of red blood cells, 75 percent of muscle tissue, 92 percent of blood plasma consist of water. Although there is no specific amount of water that must be present in living matter, the average water content is 65 to 75 percent. The following functions of water explain why it is such a vital compound in living systems:

1 Water is an excellent solvent and suspending medium. A *solvent* is a liquid or gas in which some other material (solid, liquid, or gas), called a *solute,* has been dissolved. The combination of solvent plus solute is called a **solution,** and one common example of a solution is salt water. A solute, such as salt in water, never settles out of its solution. The solute can be retrieved only through a chemical reaction or, in some cases, by boiling off the solvent. In a **suspension,** by

GENERAL NATURE   SPECIFIC EXAMPLE

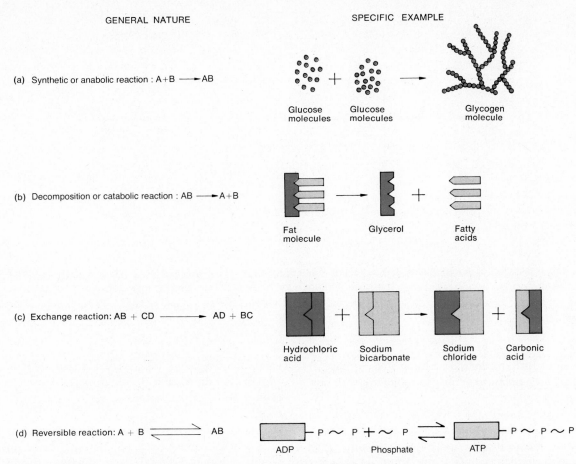

(a) Synthetic or anabolic reaction : A+B ⟶ AB

Glucose molecules   Glucose molecules   Glycogen molecule

(b) Decomposition or catabolic reaction : AB ⟶ A+B

Fat molecule   Glycerol   Fatty acids

(c) Exchange reaction: AB + CD ⟶ AD + BC

Hydrochloric acid   Sodium bicarbonate   Sodium chloride   Carbonic acid

(d) Reversible reaction: A + B ⇌ AB

ADP   Phosphate   ATP

Figure 2-5 Kinds of chemical reactions. (a) *Synthetic or anabolic reaction.* When linked together as shown, molecules of glucose form a molecule of glycogen. Glucose is a sugar that is the primary source of energy. Glycogen is a storage form of that sugar found in the liver and skeletal muscles. (b) *Decomposition or catabolic reaction.* Shown in the example is a molecule of fat breaking down into glycerol and fatty acids. This particular reaction occurs whenever a food containing fat is digested. (c) *Exchange reaction.* In this exchange, atoms of different molecules are exchanging with each other. Shown is a buffer reaction in which the body eliminates strong acids to help maintain homeostasis. (d) *Reversible reaction.* The molecule of ATP (adenosine triphosphate) is an important source of stored energy. When the energy is needed, the ATP breaks down into ADP (adenosine diphosphate) and PO_4 (phosphate group), and energy is released in the reaction. The phosphate group is symbolized as P. The cells of the body reconstruct ATP by using the energy of foods to attach ADP to PO_4.

contrast, the suspended material mixes with the liquid or suspending medium, but it will eventually settle out of the mixture. An example of a suspension is cornstarch and water. If the two materials are shaken together, a milky mixture forms. After the mixture sits for a while, however, the water clears at the top and the cornstarch settles to the bottom.

The solvating property of water is essential to health and survival. For example, water in the blood forms a solution with some of the oxygen you inhale, allowing the oxygen to be carried to your body cells. Water in the blood also dissolves much of the carbon

dioxide that is carried from the cells to the lungs to be exhaled. Furthermore, if the surfaces of the air sacs in your lungs are not moist, oxygen cannot dissolve and, therefore, cannot move into your blood to be distributed throughout your body. Water, moreover, is the solvent that carries nutrients into your body cells and wastes out.

2 Water can participate in chemical reactions. During digestion, for example, water can be added to large nutrient molecules in order to break them down into smaller molecules. This kind of breakdown is necessary if the body is to utilize the energy in nutrients. Water molecules are also used in synthesis reactions—

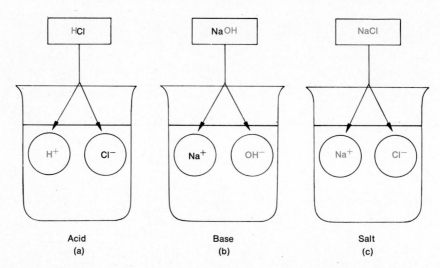

Figure 2-6 Ionization of acids, bases, and salts. (a) When placed in water, hydrochloric acid, HCl, dissociates into H^+ ions. Acids are proton donors. (b) When the base sodium hydroxide, NaOH, is placed in water, it dissociates into OH^- ions. Bases are proton acceptors. (c) When table salt, NaCl, is placed in water, it dissociates into positive and negative ions, neither of which is H^+ or OH^-.

reactions in which smaller molecules are built into larger ones. Such reactions occur in the production of hormones and enzymes.

3 Water absorbs and releases heat very slowly. In comparison to other substances, water requires a large amount of heat to increase its temperature and a great loss of heat to decrease its temperature. Thus water maintains a more constant body temperature than other solvents, despite fluctuations in environmental temperature. Water thus helps to maintain a homeostatic body temperature.

4 Water serves as a lubricant in various regions of the body. It is a major part of mucus and other lubricating fluids. Lubrication is especially necessary in the chest and abdomen, where internal organs touch and slide over each other. It is also needed at joints, where bones, ligaments, and tendons rub against each other. In the digestive tract, water moistens foods to ensure their smooth passage.

ACIDS, BASES, AND SALTS

When molecules of inorganic acids, bases, or salts are dissolved in water in the body cells, they undergo *ionization* or *dissociation*. This is, they break apart into ions. An **acid** may be defined as a substance that dissociates into one or more *hydrogen ions* (H^+) and one or more negative ions or anions. An acid may also be defined as a proton (H^+) donor. A **base**, by contrast, dissociates into one or more *hydroxyl ions* (OH^-) and one or more positive ions or cations. A base may also be viewed as a proton acceptor. Hydroxyl ions, as well as

some other negative ions, have a strong attraction for protons. A **salt,** when dissolved in water, dissociates into cations and anions, neither of which is H^+ or OH^- (Figure 2-6).

Many salts are found in the body. Some occur in cells, whereas others occur in the body fluids, such as lymph, blood, and the extracellular fluid of tissues. The ions of salts are the source of many essential chemical elements. Figure 2-7 shows how salts dissociate into ions that provide these elements. Chemical analyses reveal that sodium and chloride ions are present in higher concentrations than other ions in extracellular body fluids. Inside the cells, phosphate and potassium ions are more abundant than other ions. Chemical elements such as sodium, phosphorus, potassium, or iodine are present in the body only in chemical combination with other elements or as ions. Their presence as free, un-ionized atoms could be instantly fatal. Exhibit 2-1 lists representative elements found in the body.

ACID–BASE BALANCE

The fluids of your body must maintain a fairly constant balance of acids and bases. In solutions such as those found in body cells or in extracellular fluids, acids dissociate into hydrogen ions (H^+) and anions. Bases, on the other hand, dissociate into hydroxyl ions (OH^-) and cations. The more hydrogen ions that exist in a solution, the more acid the solution. Conversely, the more hydroxyl ions that exist in a solution, the more basic, or alkaline, the solution. The term **pH** is used to describe the degree of *acidity* or *alkalinity (basicity)* of a

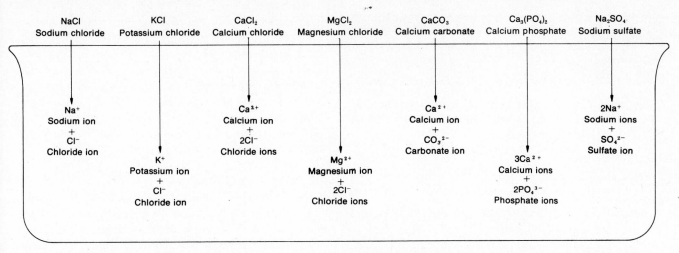

Figure 2-7 Dissociation of representative salts into ions that provide essential chemical elements for the body.

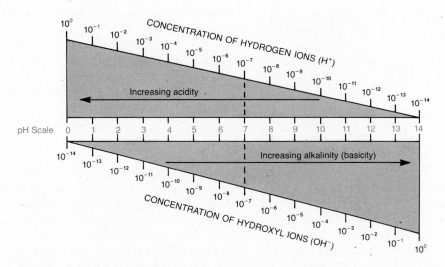

Figure 2-8 pH scale. At pH 7 (neutrality), the concentration of H+ and OH− ions is equal. A pH value below 7 indicates an acid solution; that is, there are more H+ ions than OH− ions. The lower the numerical value of the pH, the more acid the solution is because the H+ ion concentration becomes progressively greater. A pH value above 7 indicates an alkaline (basic) solution—there are more OH− ions than H+ ions. The higher the numerical value of the pH, the more alkaline the solution is because the OH− ion concentration becomes progressively greater. A change in one whole number on the pH scale represents a tenfold change from the previous concentration.

solution. Biochemical reactions—reactions that occur in living systems—are extremely sensitive to even small changes in the acidity or alkalinity of the environments in which they occur. In fact, H+ and OH− ions are involved in practically all biochemical processes, and the functions of cells are modified greatly by any departure from narrow limits of normal H+ and OH− concentrations. For this reason, the acids and bases that are constantly formed in the body must be kept in balance.

A solution's acidity or alkalinity is expressed on a *pH scale* that runs from 0 to 14 (Figure 2-8). The pH scale is based on the number of H+ ions in a solution expressed in chemical units called moles per liter. A pH of 7 means that a solution contains one ten-millionth (0.0000001) of a mole of H+ ions per liter. The number 0.0000001 is written 10^{-7} in exponential form. To convert this value to pH, the negative exponent (-7) is converted into the positive number 7. Thus a solution

Exhibit 2-1   REPRESENTATIVE CHEMICAL ELEMENTS

Chemical Element	Comment
Oxygen (O)	Constituent of water and organic molecules; functions in cellular respiration.
Carbon (C)	Found in every organic molecule.
Hydrogen (H)	Constituent of water, all foods, and most organic molecules.
Nitrogen (N)	Component of all protein molecules and nucleic acid molecules.
Calcium (Ca)	Constituent of bone and teeth; required for blood clotting, hormone synthesis, membrane integrity, and contraction of muscle.
Phosphorus (P)	Component of many proteins, nucleic acids, ATP, and cyclic AMP; required for normal bone and tooth structure; found in nerve tissue.
Chlorine (Cl)	Cl^- is an anion of NaCl, a salt important in water movement between cells.
Sulfur (S)	Component of many proteins, especially the contractile protein of muscle.
Potassium (K)	Required for growth and important in conduction of nerve impulses and muscle contraction.
Sodium (Na)	Na^+ is a cation of NaCl; structural component of bone; essential in blood to maintain water balance; needed for conduction of nerve impulses.
Magnesium (Mg)	Component of many enzymes.
Iodine (I)	Vital to functioning of thyroid gland.
Iron (Fe)	Essential component of hemoglobin and respiratory enzymes.

with 0.0000001 mole of H^+ ions per liter has a pH of 7. A solution with a concentration of 0.0001 (10^{-4}) has a pH of 4, a solution with a concentration of 0.000000001 (10^{-9}) has a pH of 9, and so on.

A solution that is zero on the pH scale has many H^+ ions and few OH^- ions. A solution that rates 14, by contrast, has very many OH^- ions and practically no H^+ ions. The midpoint in the scale is 7, where the concentration of H^+ and OH^- ions is equal. See Exhibit 2-2. Any substance that has a pH of 7, such as pure water,

is neutral. Any solution that has more H^+ ions than OH^- ions is an *acid solution* and has a pH below 7. If a solution has more OH^- ions than H^+ ions, it is a *basic* or *alkaline solution* and has a pH above 7. A change of one whole number on the pH scale represents a tenfold change from the previous concentration. That is, a pH of 2 indicates 10 times fewer H^+ ions than a pH of 1. A pH of 3 indicates 10 times fewer H^+ ions than a pH of 2 and 100 times fewer H^+ ions that a pH of 1.

BUFFER SYSTEMS

Although the pH of body fluids may differ, the normal limits for the various fluids are generally quite specific and narrow. Exhibit 2-3 shows the pH values for certain body fluids compared with common substances. Even though strong acids and bases are continually taken into the body, the pH levels of these body fluids remain relatively constant. The mechanisms that maintain these homeostatic pH values in the body are called **buffer systems.**

The essential function of a buffer system is to react with strong acids or bases in the body and replace them with weak acids or bases that can change the normal pH values only slightly. Strong acids (or bases) ionize easily and contribute many H^+ (or OH^-) ions to a solution. They therefore change the pH drastically. Weak acids (or bases) do not ionize so easily. They contribute fewer H^+ (or OH^-) ions and have little effect on the pH. The chemicals that change strong acids or bases into weak ones are called **buffers** and are found in the body's fluids. Of the several buffer systems, we will examine only the *carbonic acid–bicarbonate buffer system* – the most important one in extracellular fluid.

The carbonic acid–bicarbonate buffer system consists of a *pair* of compounds. One of them is a *weak acid,* and the other is a *weak base*. The weak acid of the buffer pair is *carbonic acid* (H_2CO_3); the weak base is *sodium bicarbonate* ($NaHCO_3$). The carbonic acid is a proton (H^+) donor and the bicarbonate ion of the sodium bicarbonate is a proton acceptor. In solution, the members of this buffer pair dissociate as follows:

Acidic component:
$$H_2CO_3 \rightleftharpoons H^+ + HCO_3^-$$
Carbonic acid   Hydrogen ion   Bicarbonate ion

Basic component:
$$NaHCO_3 \rightleftharpoons Na^+ + HCO_3^-$$
Sodium bicarbonate   Sodium ion   Bicarbonate ion

Each member of the buffer pair has a specific role in helping the body maintain a constant pH. If the body's pH is threatened by the presence of a strong acid, the weak base of the buffer pair goes into operation. If the

Exhibit 2—2 RELATIONSHIP OF pH SCALE TO RELATIVE CONCENTRATIONS OF HYDROGEN (H⁺) AND HYDROXYL (OH⁻) IONS

Concentration of H⁺ ions			pH		Concentration of OH⁻ ions	
1.0	10^0		0		10^{-14}	0.00000000000001
0.1	10^{-1}		1		10^{-13}	0.0000000000001
0.01	10^{-2}		2		10^{-12}	0.000000000001
0.001	10^{-3}		3		10^{-11}	0.00000000001
0.0001	10^{-4}		4		10^{-10}	0.0000000001
0.00001	10^{-5}		5		10^{-9}	0.000000001
0.000001	10^{-6}		6		10^{-8}	0.00000001
0.0000001	10^{-7}	Neutrality	7	Neutrality	10^{-7}	0.0000001
0.00000001	10^{-8}		8		10^{-6}	0.000001
0.000000001	10^{-9}		9		10^{-5}	0.00001
0.0000000001	10^{-10}		10		10^{-4}	0.0001
0.00000000001	10^{-11}		11		10^{-3}	0.001
0.000000000001	10^{-12}		12		10^{-2}	0.01
0.0000000000001	10^{-13}		13		10^{-1}	0.1
0.00000000000001	10^{-14}		14		10^0	1.0

body's pH is threatened by a strong base, the weak acid goes into play.

Consider the following situation. If a strong acid, such as HCl, is added to extracellular fluid, the weak base of the buffer system goes to work and the following reaction occurs:

$$HCl + NaHCO_3 \rightleftharpoons NaCl + H_2CO_3$$

Hydrochloric acid (strong acid); Sodium bicarbonate (weak base of buffer system); Sodium chloride (salt); Carbonic acid (weak acid)

The chloride ion of HCl and the sodium ion of sodium bicarbonate combine to form NaCl, a substance that has no effect on pH. The hydrogen ion of the HCl could greatly lower pH by making the solution more acid, but this H^+ ion combines with the bicarbonate ion (HCO_3^-) of sodium bicarbonate to form carbonic acid, a weak acid that lowers pH only slightly. In other words, because of the action of the weak base of the buffer system, the strong acid (HCl) has been replaced by a weak acid and a salt, and the pH remains relatively constant.

Now suppose a strong base, such as sodium hydroxide (NaOH), is added to the extracellular fluid.

In this instance, the weak acid of the buffer system goes to work and the following reaction takes place:

$$NaOH + H_2CO_3 \rightleftharpoons H_2O + NaHCO_3$$

Sodium hydroxide (strong base); Carbonic acid (weak acid of buffer system); Water; Sodium bicarbonate (weak base)

In this reaction, the OH^- ion of sodium hydroxide could greatly raise the pH of the solution by making it more alkaline. However, the OH^- ion combines with an H^+ ion of carbonic acid and forms water, a substance that has no effect on pH. In addition, the Na^+ ion of sodium hydroxide combines with the bicarbonate ion (HCO_3^-) to form sodium bicarbonate, a weak base that has little effect on pH. Thus, because of the action of the buffer system, the strong base is replaced by water and a weak base, and the pH remains relatively constant.

Whenever a buffering reaction occurs, the concentration of one member of the buffer pair is increased while the concentration of the other decreases. When a strong acid is buffered, for example, the concentration of carbonic acid is increased but the concentration of sodium bicarbonate is decreased. This happens because

Exhibit 2–3 NORMAL pH VALUES OF REPRESENTATIVE SUBSTANCES

Substance	pH Value
Gastric juice (digestive juice of the stomach)	1.2–3.0
Lemon juice	2.2–2.4
Grapefruit	3.0
Cider	2.8–3.3
Pineapple juice	3.5
Tomato juice	4.2
Clam chowder	5.7
Urine	5.5–7.5
Saliva	6.5–7.5
Milk	6.6–6.9
Pure (distilled) water	7.0
Blood	7.35–7.45
Semen (fluid containing sperm)	7.35–7.50
Cerebrospinal fluid (fluid associated with nervous system)	7.4
Pancreatic juice (digestive juice of the pancreas)	7.1–8.2
Eggs	7.6–8.0
Bile (liver secretion that aids in fat digestion)	7.6–8.6
Milk of magnesia	10.0–11.0
Limewater	12.3

carbonic acid is produced and sodium bicarbonate is used up in the acid buffering reaction (see Reaction 1). When a strong base is buffered, the concentration of sodium bicarbonate is increased but the concentration of carbonic acid is decreased. This happens because sodium bicarbonate is produced and carbonic acid is used up on the basic buffering reaction (see Reaction 2). When the buffered substances, HCl and NaOH in this case, are removed from the body via the lungs or kidneys, the carbonic acid and sodium bicarbonate formed as products of the reactions function again as components of the buffer pair.

Organic Substances

Organic substances are chemical compounds that contain carbon and usually hydrogen bond oxygen as well.

Carbon has several properties that make it particularly useful to living organisms. For one thing, it can react with one to several hundred other carbon atoms to form large molecules of many different shapes. This means that the body can build many compounds out of carbon, hydrogen, and oxygen. Each compound can be especially suited for a particular structure or function. The large size of most carbon molecules and the fact that they do not dissolve easily in water make them useful materials for building body structures. Carbon compounds are mostly or entirely held together by covalent bonds, and they tend to decompose easily. This means that organic compounds are also a good source of energy. Ionic compounds are not good energy sources because they form new ionic bonds as soon as the old ones are broken.

CARBOHYDRATES

A large and diverse group of organic compounds found in the body is the **carbohydrates**, also known as sugars and starches. The carbohydrates perform a number of major functions in living systems. A few even form structural units. For instance, one type of sugar is a building block of genes, the molecules that carry hereditary information. Some carbohydrates are converted to proteins and to fats or fatlike substances, which are used to build structures and provide an emergency source of energy. Other carbohydrates function as food reserves. One example is glycogen, which is stored in the liver and skeletal muscles. The principal function of carbohydrates, however, is to provide the most readily available source of energy to sustain life.

Carbon, hydrogen, and oxygen are the elements found in carbohydrates. The ratio of hydrogen to oxygen atoms is always 2 to 1. This ratio can be seen in the formulas for carbohydrates such as ribose ($C_5H_{10}O_5$), glucose ($C_6H_{12}O_6$), and sucrose ($C_{12}H_{22}O_{11}$). Although there are exceptions, the general formula for carbohydrates is $(CH_2O)_n$, where n symbolizes three or more CH_2O units. Carbohydrates can be divided into three major groups: monosaccharides, disaccharides, and polysaccharides.

1 Monosaccharides Simple sugars are called *monosaccharides*. These compounds contain from three to seven carbon atoms and cannot be broken down into simpler sugar molecules. Simple sugars with three carbons in the molecule are called trioses. The number of carbon atoms in the molecule is indicated by the prefix *tri*. There are also tetroses (four-carbon sugars), pentoses (five-carbon sugars), hexoses (six-carbon sugars), and heptoses (seven-carbon sugars). Pentoses and hexoses are exceedingly important to the human organism. The pentose called deoxyribose is a component of genes. The hexose called glucose is the main energy-supplying molecule of the body.

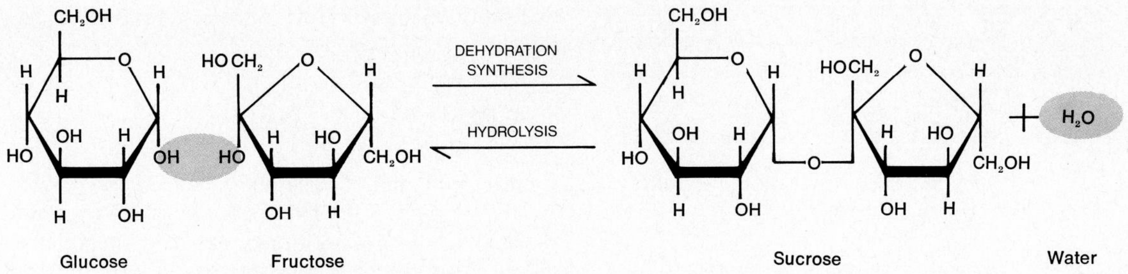

Figure 2-9 Dehydration synthesis and hydrolysis of a molecule of sucrose. In the dehydration synthesis (read from left to right), the two smaller molecules, glucose and fructose, are joined to form a larger molecule of sucrose. Note the loss of a water molecule. In hydrolysis (read from right to left), the larger sucrose molecule is broken down into the two smaller molecules, glucose and fructose. Here, a molecule of water is added to sucrose for the reaction to occur.

2 Disaccharides A second group of carbohydrates, called the *disaccharides,* consists of two monosaccharides joined chemically. In the process of disaccharide formation, two monosaccharides combine to form a disaccharide molecule and a molecule of water is lost. This reaction is known as *dehydration synthesis (dehydration* = loss of water). The following reaction shows disaccharide formation. Molecules of the monosaccharides glucose and fructose combine to form a molecule of the disaccharide sucrose (cane sugar):

$$C_{12}H_{12}O_6 \quad + \quad C_6H_{12}O_6 \quad \rightarrow C_{12}H_{22}O_{11} + H_2O$$

Glucose	Fructose	Sucrose	Water
(monosaccharide)	(monosaccharide)	(disaccharide)	

You may be puzzled to see that glucose and fructose have the same chemical formulas. Actually, they are different monosaccharides since the relative positions of the oxygens and carbons vary in the two different molecules. (See Figure 2-9.) The formula for sucrose is $C_{12}H_{22}O_{11}$ and not $C_{12}H_{24}O_{12}$ since a molecule of H_2O is lost in the process of disaccharide formation. In every dehydration synthesis a molecule of water is lost. Along with this water loss, there is the synthesis of two small molecules, such as glucose and fructose, into one large, more complex molecule, such as sucrose (Figure 2-9). Similarly, the dehydration synthesis of two monosaccharides such as glucose and galactose forms the disaccharide lactose (milk sugar).

Disaccharides can also be broken down into smaller, simpler molecules by adding water. This reverse chemical reaction is called *digestion* or *hydrolysis,* which means to split by using water. A molecule of sucrose, for example, may be digested into its components of glucose and fructose by the addition of water. The mechanism of this reaction also is represented in Figure 2-9.

3 Polysaccharides The third major group of carbohydrates, the *polysaccharides,* consists of eight or more monosaccharides joined together through dehydration synthesis. Polysaccharides have the formula $(C_6H_{10}O_5)_n$. Like disaccharides, polysaccharides can be broken into their constituent sugars through hydrolysis reactions. Unlike monosaccharides or disaccharides, however, they usually lack the characteristic sweetness of sugars like fructose or sucrose and are usually not soluble in water. One of the chief polysaccharides is glycogen.

LIPIDS

A second group of organic compounds that is vital to the human organism is the **lipids.** Like carbohydrates, lipids are composed of carbon, hydrogen, and oxygen, but they do not have a 2:1 ratio of hydrogen to oxygen. Most lipids are insoluble in water, but they readily dissolve in solvents such as alcohol, chloroform, and ether. Since lipids are a large and diverse group of compounds, only one kind, the *fats,* is discussed here. Pertinent information regarding other lipids is provided in Exhibit 2-4.

A molecule of fat consists of two basic components. The first component is called *glycerol;* the second is a group of compounds called *fatty acids* (Figure 2-10). A single molecule of fat is formed when a molecule of glycerol combines with three molecules of fatty acids. This reaction, like the one described for disaccharide formation, is a dehydration synthesis reaction. During hydrolysis, a single molecule of fat is broken down into fatty acids and glycerol.

Fats represent the body's most highly concentrated source of energy. In fact, they provide twice as many Calories (a form of energy) per weight as either carbohydrates or proteins. In general, however, fats are about 10 to 12 percent less efficent as body fuels than are carbohydrates. A great amount of the fat Calorie is wasted and thus not available for the body to use.

PROTEINS

A third principal group of organic compounds is **proteins.** These compounds are much more complex in structure than the carbohydrates or lipids. They are also responsible for much of the structure of body cells and

Exhibit 2—4   RELATIONSHIPS OF REPRESENTATIVE LIPIDS TO HUMAN ORGANISM

Lipids	Relationship	Lipids	Relationship
Fats	Protection, insulation, source of energy.	Bile pigments	Bilirubin, a reddish pigment, and biliverdin, a greenish pigment, are both formed from hemoglobin and are responsible for the brown color of feces.
Phospholipids			
Lecithin	Major lipid component of cell membranes; constituent of plasma.	Cytochromes	Coenzymes involved in the respiration of all cells.
Cephalin and sphingomyelin	Found in high concentrations in nerves and brain tissue.	**Other lipid substances**	
Steroids		Carotenes	Pigment in egg yolk, carrots, and tomatoes; vitamin A is formed from carotenes; retinene, also formed from vitamin A, is a photoreceptor in the retina of the eye.
Cholesterol	Constituent of all animal cells, blood, and nervous tissue; suspected relationship to heart disease and "hardening of the arteries."		
Bile salts	Substances that emulsify or suspend fats before their digestion.	Vitamin E	"Antisterility" vitamin in rats; necessary for the synthesis of connective tissue in wound healing in humans.
Vitamin D	Produced in skin on exposure to ultraviolet radiation; necessary for bone growth and development.	Vitamin K	Vitamin that promotes blood clotting and prevents excessive bleeding.
Estrogens	Sex hormones produced in large quantities by females.	Prostaglandins	Membrane-associated lipids that stimulate smooth muscle to contract, raise and lower blood pressure, and regulate metabolism.
Androgens	Sex hormones produced in large quantities by males.		
Porphyrins (lipid portions of organic molecules)			
Hemoglobin	Oxygen-transporting pigment in red blood cells.		

are related to many physiological activities. For example, proteins in the form of enzymes speed up many essential biochemical reactions. Other proteins assume a necessary role in muscular contraction. Antibodies are proteins that provide the human organism with defenses against invading microbes. And some hormones that regulate body functions are also proteins. A classification of proteins on the basis of function is shown in Exhibit 2-5.

Chemically, proteins always contain carbon, hydrogen, oxygen, and nitrogen. Many proteins also contain sulfur and phosphorus. Just as monosaccharides are the building units of sugars and fatty acids and glycerol are the building units of fats, *amino acids* are the building blocks of proteins. In protein formation, amino acids combine to form more complex molecules, while water molecules are lost. The process is a dehydration synthesis reaction, and the bonds formed between amino acids are called *peptide links* (Figure 2-11).

When two amino acids combine, a *dipeptide* results. Adding another amino acid to a dipeptide produces a *tripeptide*. Further additions of amino acids result in the formation of *polypeptides*, which are large protein molecules. At least 20 different amino acids are found in proteins. A great variety of proteins is possible because each variation in the number or sequence of amino acids can produce a different protein. The situation is similar to using an alphabet of 20 letters to form words. Each letter could be compared to a different amino acid, and each word would be a different protein.

ENZYMES

Proteins that are produced by living cells to catalyze reactions in the body are **enzymes**. Protein in the form of an enzyme acts as a *catalyst* — any chemical substance that affects the speed of a reaction without being permanently altered by the reaction.

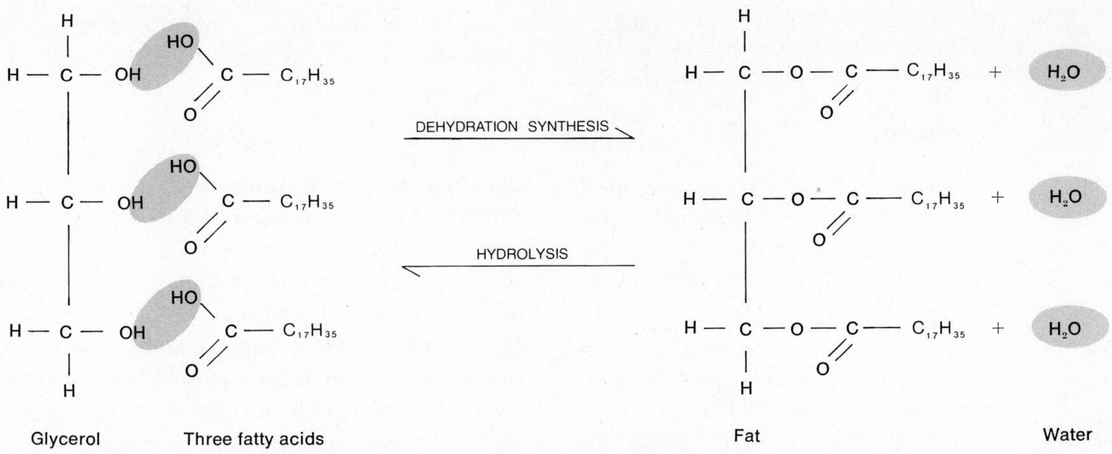

Figure 2-10 Dehydration synthesis and hydrolysis of a fat. In the dehydration synthesis (read from left to right), one molecule of glycerol combines with three fatty acid molecules, and there is a loss of three molecules of water. In hydrolysis (read from right to left), a molecule of fat is broken down into a single molecule of glycerol and three fatty acid molecules upon the addition of three molecules of water. The fatty acid shown is stearic acid, a component of corn oil, coconut oil, beef fat, and pork fat.

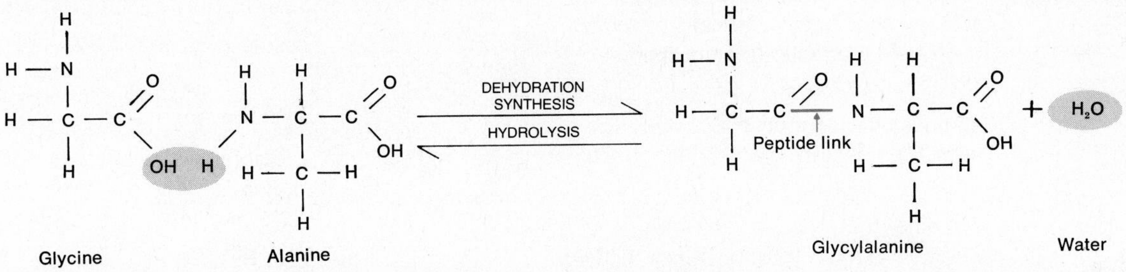

Figure 2-11 Protein formation. When two or more amino acids are chemically united, the resulting bond between them is called a peptide link. In the example shown, glycine and alanine, the two amino acids, are joined to form the dipeptide, glycylalanine. At the point where water is lost, the peptide link is formed.

For a chemical or biochemical reaction to occur, a certain amount of energy is required: the *activation energy.* As stated earlier, energy can be transformed from one state to another. When heat energy is added to molecules, some of the heat is transformed to kinetic energy and the energy of the molecules is increased. One way to speed up molecules to activation-level energy is to heat them. Unfortunately, moderate heat denatures (destroys the activity of) many body proteins. The role of an enzyme, then, is to decrease the amount of energy needed to start the reaction.

Exactly how enzymes lower activation energies is not fully understood. However, it is known that an enzyme attaches itself to one of the reacting molecules, called a *substrate,* and the two form a temporary *enzyme–substrate complex.* Many enzymes exist, but each kind can attach to only one kind of substrate. Apparently, the enzyme molecule must fit perfectly with the substrate molecule like pieces of a jigsaw puzzle (Figure 2-12). If the enzyme and substrate do not fit properly, no reaction occurs.

When an enzyme–substrate complex is formed, the substrate molecule can react with other molecules in a decomposition or synthesis reaction. After the reaction is completed, the products of the reaction move away from the enzyme and the enzyme is free to attach to another substrate molecule. The whole process of attachment, reaction, and detachment takes place very quickly. Most enzymes are capable of interacting with up to 5 million substrate molecules per minute at 0°C and, within limits, about double that for every 10°C increase in temperature above 0°C.

Many enzymes are named by adding the suffix *ase* to the name of their substrates. For example, the enzyme involved in breaking down sucrose is called *sucrase.* Enzymes that hydrolyze (break down) proteins are classified as *proteases.* Other enzymes are named by their action. *Dehydrogenases,* for example, are enzymes that remove hydrogen atoms from a substrate.

Enzymes are clearly essential to your body's overall homeostasis. They quickly catalyze chemical reactions, and they also govern the reactions that occur.

Exhibit 2—5 CLASSIFICATION OF PROTEINS BY FUNCTION

Nature of Protein	Description
Structural	Proteins that form the structural framework of various parts of the body. Examples: keratin in the skin, hair, and fingernails and collagen in connective tissue.
Regulatory	Proteins that function as hormones and regulate various physiological processes. Examples: insulin, which regulates blood sugar, and adrenalin, which regulates the diameter of blood vessels.
Contractile	Proteins that serve as contractile elements in muscle tissue. Examples: myosin and actin.
Immunological	Proteins that serve as antibodies to protect the body against invading microbes. Example: gamma globulin.
Transport	Proteins that transport vital substances throughout the body. Example: hemoglobin, which transports oxygen.
Catalytic	Proteins that act as enzymes and function by controlling biochemical reactions. Examples: salivary amylase, pepsin, and lactase.

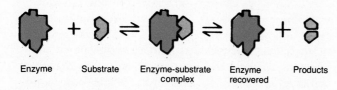

Enzyme Substrate Enzyme–substrate complex Enzyme recovered Products

Figure 2-12 Enzyme action in a decomposition reaction. The enzyme and substrate molecules combine to form an enzyme–substrate complex. During combination, the substrate is changed into products. Once the products are formed, the enzyme is recovered (moves away from the products) and may be used again to catalyze a similar reaction.

NUCLEIC ACIDS

Exceedingly large organic molecules containing carbon, hydrogen, oxygen, nitrogen, and phosphorus are **nucleic acids,** compounds first discovered in the nuclei of cells. Whereas the basic structural units of proteins are amino acids, the basic units of nucleic acids are *nucleotides.* Nucleic acids are divided into two principal kinds: **deoxyribonucleic acid (DNA)** and **ribonucleic acid (RNA).**

A molecule of DNA is a chain composed of repeating nucleotide units. Each nucleotide of DNA consists of three basic parts (Figure 2-13a):

1 It contains one of four possible *nitrogen bases,* which are ring-shaped structures containing atoms of C, H, O, and N. The nitrogen bases found in DNA are named adenine, thymine, cytosine, and guanine.
2 It contains a pentose sugar called *deoxyribose.*
3 It also contains *phosphate groups.*

The nucleotides are named according to the nitrogen base that is present. Thus a nucleotide containing thymine is called a *thymine nucleotide.* One containing adenine is called an *adenine nucleotide,* and so on.

The chemical composition of the DNA molecule was known before 1900, but it was not until 1953 that a model of the organization of the chemicals was constructed. This model was proposed by J. D. Watson and F. H. C. Crick on the basis of data from many investigations. Figure 2-13b shows the following structural characteristics of the DNA molecule:

1 The molecule consists of two strands with crossbars. The strands twist about each other in the form of a *double helix* so that the shape resembles a twisted ladder.
2 The uprights of the DNA ladder consist of alternating phosphate groups and the deoxyribose portions of the nucleotides.
3 The rungs of the ladder contain paired nitrogen bases. As shown, adenine always pairs off with thymine and cytosine always pairs off with guanine.

Cells contain hereditary material called genes, each of which is a segment of a DNA molecule. Our genes determine which traits we inherit, and they control all the activities that take place in our cells throughout a lifetime. When a cell divides, its hereditary information is passed on to the next generation of cells. The passing of information is possible because of DNA's unique structure.

RNA, the second principal kind of nucleic acid, differs from DNA in several respects. RNA is single-stranded; DNA is double-stranded. The sugar in the RNA nucleotide is the pentose ribose. And RNA does not contain the nitrogen base thymine. Instead of thymine, RNA has a nitrogen base called uracil. At least three different kinds of RNA have been identified in cells. Each type has a specific role in reacting with DNA to help regulate protein synthesis reactions.

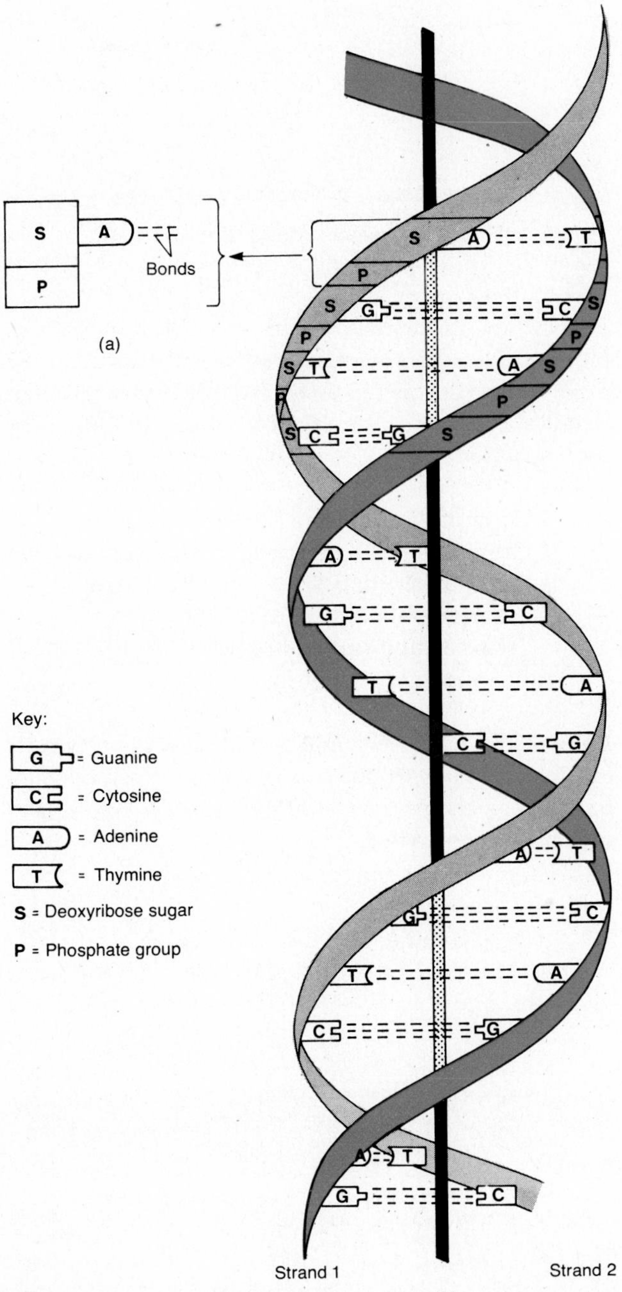

ADENOSINE TRIPHOSPHATE (ATP)

A molecule that is indispensable to the life of the cell is **adenosine triphosphate (ATP).** This substance is found universally in living systems and performs the essential function of storing energy for various cellular activities. Structurally, ATP consists of three phosphate groups and an adenosine unit composed of adenine and the five-carbon sugar ribose (Figure 2-14). ATP is regarded as a high-energy molecule because of the total amount of usable energy it releases when it is broken down by the addition of a water molecule (hydrolysis).

When the terminal phosphate group, symbolized P, is hydrolyzed, the reaction liberates a great deal of energy. This energy is used by the cell to perform its basic activities. Removal of the terminal phosphate group leaves a molecule called **adenosine diphosphate (ADP).** This reaction may be represented as follows:

$$\text{ATP} \rightleftharpoons \text{ADP} + \text{P} + \text{E}$$

Adenosine   Adenosine   Phosphate   Energy
triphosphate diphosphate

The energy supplied by the catabolism of ATP into ADP is constantly being used by the cell. Since the supply of ATP at any given time is limited, a mechanism exists to replenish it—a phosphate group is added to ADP to manufacture more ATP. Logically, energy is required to manufacture ATP. The reaction may be represented as follows:

$$\text{ADP} + \text{P} + \text{E} \rightleftharpoons \text{ATP}$$

Adenosine   Phosphate   Energy   Adenosine
diphosphate   triphosphate

The energy required to attach a phosphate group to ADP is supplied by various decomposition reactions taking place in the cell, particularly by the decomposition of glucose. The body does not use the energy from the decomposition of glucose because glucose cannot be stored in cells. Some glucose can be converted to glycogen and stored in the liver, but the rest is decomposed immediately. ATP, by contrast, can be stored in every cell, where it provides potential energy that is not released until needed.

CYCLIC AMP

A chemical substance closely related to ATP is **cyclic adenosine—3, 5—monophosphate (cyclic AMP).** Essentially it is a molecule of adenosine monophosphate with the phosphate attached to the ribose sugar at two places (Figure 2-15). The attachment forms a ring-

Key:
G = Guanine
C = Cytosine
A = Adenine
T = Thymine
S = Deoxyribose sugar
P = Phosphate group

Figure 2-13 DNA molecule. (a) Adenine nucleotide. (b) Portion of an assembled DNA molecule.

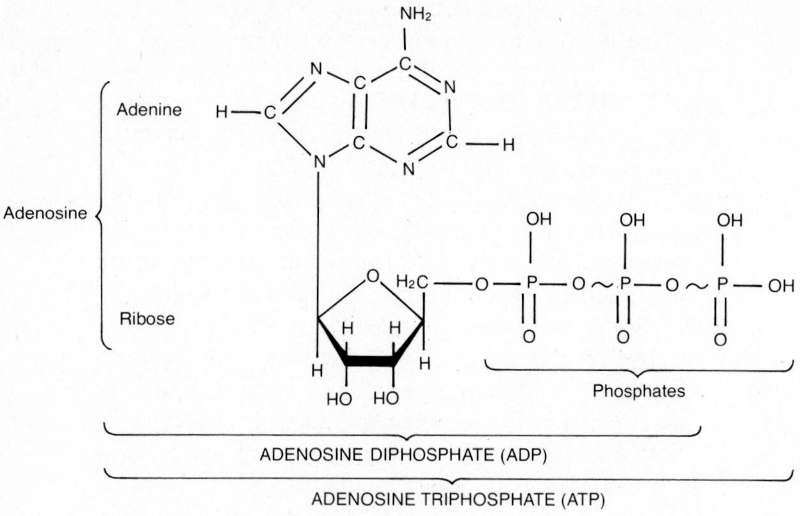

Figure 2-14   Structure of ATP.

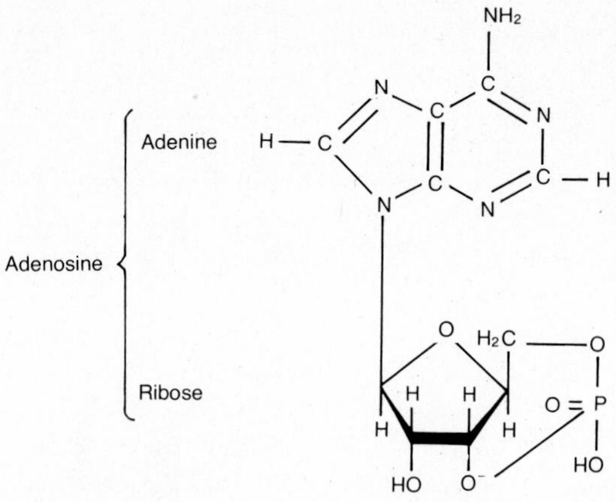

Figure 2-15   Structure of cyclic AMP.

shaped structure – thus the name cyclic AMP. Cyclic AMP is formed from ATP by the action of a special enzyme, called *adenyl cyclase,* located in the cell membrane. Although cyclic AMP was discovered in 1958, only recently has its function in cells become clear. One function is related to the action of hormones, a topic we explore in detail in Chapter 18.

PROSTAGLANDINS

Prostaglandins (PG) are a large group of membrane-associated lipids composed of 20 carbon fatty acids containing a cyclopentane ring. (A cyclopentane contains five carbon atoms joined to form a ring.) Prostaglandins were first discovered in prostate gland secretions, but they are now known to be present in many body tissues. Prostaglandins are produced in cell membranes in minute quantities and are rapidly decomposed by catabolic enzymes. One function of prostaglandins is to control the activity of adenyl cyclase and thus the production of cyclic AMP. This function is discussed in Chapter 18.

Although synthesized in very small quantities, prostaglandins are potent substances and exhibit a wide variety of effects on the body. Basically, prostaglandins mimic hormones. Among the effects produced by prostaglandins are:

1 Lowering and raising blood pressure.
2 Stimulating and inhibiting uterine contractions.
3 Causing abortion.
4 Inducing labor.
5 Transmitting nerve impulses.
6 Regulating metabolism.

Investigations are now under way to determine the role of prostaglandins in preventing peptic ulcers, opening bronchial passageways, clearing nasal passages, and inducing menstruation.

STUDY OUTLINE

BASIC CHEMISTRY FOR LIVING SYSTEMS
CHEMICAL ELEMENTS

1 Matter is anything that occupies space and has mass. It is made of building units called chemical elements.

2 Carbon, hydrogen, oxygen, and nitrogen make up 96 percent of body weight. These elements together with phosphorus and calcium make up 99 percent of total body weight.

ATOMIC STRUCTURE

1 Units of matter of all chemical elements are called atoms.

2 Atoms consist of a nucleus, which contains protons and neutrons, and orbiting electrons moving in energy levels.

3 The total number of protons of an atom is its atomic number. This number is equal to the number of electrons in the atom.

ATOMS AND MOLECULES

1 The electrons are the part of an atom that actively participate in chemical reactions. An atom attempts to fill its outer energy level with electrons through bonding.

2 A molecule is the smallest unit of two or more combined atoms. A molecule containing two or more different kinds of atoms is a compound.

3 In an ionic bond, outer energy level electrons are transferred from one atom to another. The transfer forms ions, whose unlike charges attract each other and form ionic bonds.

4 In a covalent bond, there is a sharing of pairs of outer energy level electrons.

5 Radioactive isotopes are valuable in the study, diagnosis, and treatment of many disorders of the body.

CHEMICAL REACTIONS

1 Synthesis reactions involve the combination of reactants to produce a new molecule. The reactions are anabolic: bonds are formed.

2 In decomposition reactions, a substance breaks down into other substances. The reactions are catabolic: bonds are broken.

3 Exchange reactions involve the replacement of one atom or atoms by another atom or atoms.

4 In reversible reactions, end products can revert to the original combining molecules.

5 When chemical bonds are formed, energy is needed. When bonds are broken, energy is released. This is known as chemical bond energy.

CHEMICAL LIFE PROCESSES

1 Inorganic substances usually lack carbon, contain ionic bonds, resist decomposition, and dissolve readily in water.

2 Organic substances always contain carbon and usually hydrogen. Most organic substances contain covalent bonds and are insoluble in water.

INORGANIC SUBSTANCES

1 Water is the most abundant substance in the body. It is an excellent solvent and suspending medium, participates in chemical reactions, absorbs and releases heat slowly, and lubricates.

2 Acids, bases, and salts ionize into ions in water. An acid ionizes into H^+ ions; a base ionizes into OH^- ions. A salt ionizes into neither H^+ nor OH^- ions. Cations are positively charged ions; anions are negatively charged ions.

3 The pH of different parts of the body must remain fairly constant for the body to remain healthy. On the pH scale, 7 represents neutrality. Values below 7 indicate acid solutions, and values above 7 indicate alkaline solutions.

4 The pH values of different parts of the body are maintained by buffer systems, which usually consist of a weak acid and a weak base. Buffer systems eliminate excess H^+ ions and excess OH^- ions in order to maintain pH homeostasis.

ORGANIC SUBSTANCES

1 Carbohydrates are sugars or starches that provide most of the energy needed for life. They may be monosaccharides, disaccharides, or polysaccharides. Carbohydrates, and other organic molecules, are joined together to form larger molecules with the loss of water by a process called dehydration synthesis. In the reverse process, called hydrolysis or digestion, large molecules are broken down into smaller ones upon the addition of water.

2 Lipids are a diverse group of compounds that includes fats, steroids, pigments, and vitamins. Fats protect, insulate, provide energy, and are stored.

3 Proteins are constructed from amino acids. They give structure to the body, regulate processes, provide protection, help muscles to contract, and transport substances.

4 Enzymes are proteins produced by the body. They catalyze, or speed up, chemical reactions. Enzymes act on substrates by lowering the activation energy needed for reaction.

5 DNA and RNA are nucleic acids consisting of nitrogen bases, sugar, and phosphate groups. DNA is a double helix and is the primary chemical in genes. RNA differs slightly in structure and chemical composition from DNA and is concerned largely with protein synthesis reactions.

6 The principal energy-storing molecule in the body is ATP. When its energy is liberated, it is decomposed to ADP. ATP is manufactured from ADP using the energy supplied by the food that is eaten.

7 Cyclic AMP is closely related to ATP and assumes a function in certain hormonal reactions.

8 Prostaglandins are lipids that control the production of cyclic AMP. They mimic the effects of hormones.

REVIEW QUESTIONS

1 What is the relationship of matter to the body?

2 Define a chemical element. List the chemical symbols for 10 different chemical elements. Which chemical elements make up the bulk of the human organism?

3 What is an atom? Diagram the positions of the nucleus, protons, neutrons, and electrons in an atom of oxygen and an atom of nitrogen. What is an atomic number?

4 What is meant by an energy level?

5 How are chemical bonds formed? Distinguish between an ionic bond and a covalent bond. Give at least one example of each.

6 Can you determine how a molecule of $MgCl_2$ is ionically bonded? Magnesium has two electrons in its outer shell. Construct a diagram to verify your answer.

7 Refer to Figure 2-4b and c. See if you can determine why there is a double covalent bond between atoms in an oxygen molecule (O_2) and a triple covalent bond between atoms in a nitrogen molecule (N_2).

8 Explain how radioactive isotopes work. What specific parts of the body can easily be examined using this technique?

9 What are the four principal kinds of chemical reactions? How are anabolism and catabolism related to synthesis and decomposition reactions, respectively? How is energy related to chemical reactions?

10 Identify what kind of reaction each of the following represents:

(a) $H_2 + Cl_2 \rightarrow 2HCl$

(b) $3NaOH + H_3PO_4 \rightarrow Na_3PO_4 + 3H_2O$

(c) $CaCO_3 + CO_2 + H_2O \rightarrow Ca(HCO_3)_2$

(d) $HNO_3 \rightarrow H^+ + NO_3^-$

(e) $NH_3 + H_2O \rightleftharpoons NH_4^+ + OH^-$

11 How do inorganic compounds differ from organic compounds? List and define the principal inorganic and organic compounds that are important to the human body.

12 What are the essential functions of water in the body? Distinguish between a solution and a suspension.

13 Define an acid, a base, and a salt. How does the body acquire some of these substances? List some functions of the chemical elements furnished as ions of salts.

14 What is pH? Why is it important to maintain a relatively constant pH? What is the pH scale? List the normal pH values of some common fluids, biological solutions, and foods.

15 Refer to Exhibit 2-3, and select the two substances whose pH values are closest to neutrality. Is the pH of milk or the pH of cerebrospinal fluid closer to 7? Is the pH of bile or the pH of urine farther from neutrality? If there are 100 OH^- ions at a pH of 8.5, how many OH^- ions are there at a pH of 9.5?

16 What are the components of a buffer system? What is the function of a buffer pair? Diagram and explain how the carbonic acid–bicarbonate buffer system of extracellular fluid maintains a constant pH even in the presence of a strong acid or strong base. How is this an example of homeostasis?

17 Why are the reactions of buffer pairs more important with strong acids and bases than with weak acids and bases?

18 Define a carbohydrate. Why are carbohydrates essential to the body? How are carbohydrates classified?

19 Compare dehydration synthesis and hydrolysis. Why are they significant?

20 How do lipids differ from carbohydrates? What are some relationships of lipids to the body?

21 Define a protein. What is a peptide linkage? Discuss the classification of proteins on the basis of function.

22 Distinguish between an enzyme and a substrate. List some principal characteristics of enzymes. Relate the concept of activation energy to enzyme action. How do enzymes maintain homeostasis?

23 What is a nucleic acid? How do DNA and RNA differ with regard to chemical composition, structure, and function?

24 What is ATP? What is the essential function of ATP in the human body? How is this function accomplished?

25 How is cyclic AMP related to ATP? What is the function of cyclic AMP?

26 Define a prostaglandin. List some physiological effects of prostaglandins.

Chapter 3
The Cellular
Level of
Organization

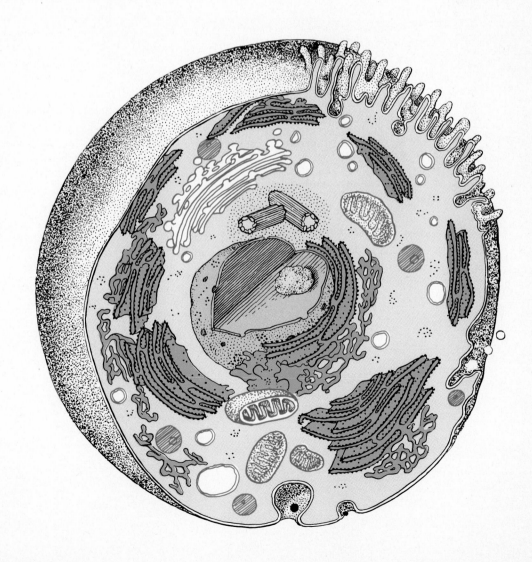

STUDENT OBJECTIVES

■ Define and list a cell's generalized parts.

■ Describe the structure and molecular organization of the plasma membrane.

■ List the factors related to semipermeability of the plasma membrane.

■ Define diffusion, facilitated diffusion, osmosis, filtration, dialysis, active transport, phagocytosis, and pinocytosis.

■ Describe the structure and function of several modified plasma membranes.

■ Describe the chemical composition and list the functions of cytoplasm.

■ Describe two general functions of a cell nucleus.

■ Distinguish between agranular and granular endoplasmic reticulum.

■ Define the function of ribosomes.

■ Describe the role of the Golgi complex in the synthesis, storage, and secretion of glycoproteins.

■ Describe the function of mitochondria as "powerhouses of the cell."

■ Explain why a lysosome in a cell is called a "suicide packet," and describe microtubules.

■ Describe the structure and function of centrioles in cellular reproduction.

■ Differentiate between cilia and flagella.

■ Define cell inclusion and give several examples.

■ Define extracellular material. Give examples.

■ Describe the stages and events involved in cell division.

■ Discuss the significance of cell division.

■ Define a gene.

■ Describe the sequence of events involved in protein synthesis.

■ Describe the relationship of cancer and aging to cells.

■ Define medical terminology associated with cells.

The study of the body at the cellular level of organization is important because many activities essential to life occur in cells and many disease processes originate there. A **cell** may be defined as the basic, living, structural and functional unit of the body and, in fact, of all organisms. It is the smallest structure capable of performing all the activities vital to life. **Cytology** is the branch of science concerned with the study of cells. This chapter concentrates on the structure, functions, and reproduction of cells.

GENERALIZED ANIMAL CELL

A **generalized animal cell** is a composite of many different cells in the body. Examine the generalized cell illustrated in Figure 3-1, and keep in mind that no such single cell actually exists.

For convenience, we can divide the generalized cell into four principal parts:

1 Plasma (or cell) membrane   The outer, limiting membrane separating the cell's internal parts from the extracellular fluid and external environment.

2 Cytoplasm   The substance between the nucleus and cell surface in which organelles are embedded.

3 Organelles   The cellular components that are highly specialized for specific cellular activities.

4 Inclusions   The secretions and storage areas of cells.

Extracellular materials, which are substances external to the cell surface, will also be examined in connection with cells.

PLASMA (CELL) MEMBRANE

The exceedingly thin structure that separates one cell from other cells and from the external environment is called the **plasma membrane,** or **cell membrane.** Electron microscopy studies have shown that the plasma membrane ranges from 65 to 100 angstroms (Å) in thickness, a dimension below light-microscope limits. Scientists have known for a long time that plasma membranes are composed of phospholipid and protein molecules. Recent studies of membrane structure suggest a new concept regarding the arrangement of phospholipid and protein molecules. This new concept of membrane structure is called the *fluid mosaic hypothesis.* These studies seem to indicate that the phospholipid molecules, which account for about half the mass of the membrane, are arranged in two parallel rows. This double row of phospholipid molecules is termed a bilayer (Figure 3-2). The protein molecules are arranged somewhat differently. Some lie at or near the inner and outer surface of the membrane. Others penetrate the membrane partway, completely, singly, or in pairs. Such an arrangement suggests that membranes are not static but that the proteins and phospholipids have a considerable degree of movement. It is quite possible that many key functions of membranes will be fully explained once the reasons for protein and phospholipid movements are understood. The fluid mosaic hypothesis may explain how specific receptor sites on membranes attach to hormones (for example, insulin), to transmitter substances produced by axons, and to antigens on the surfaces of red blood cells. Other features of the membrane structure are areas that appear as breaks along the membrane surface. These breaks occur at intervals and range in size from 7 to 10 Å in diameter. Researchers suspect they may be pores. (See Figure 3-2b.)

The basic functions of the plasma membrane are to enclose the components of the cell and to serve as a boundary through which substances must pass to enter or exit the cell. One important characteristic of the plasma membrane is that it permits certain ions and molecules to enter or exit the cell but restricts the passage of others. For this reason, plasma membranes are described as **semipermeable, differentially permeable,** or **selectively permeable.** In general, plasma membranes are freely permeable to water. In other words, they let water into and out of the cell. However, they act as barriers to the movement of almost all other substances. The ease with which a substance passes through a membrane is called the membrane's *permeability* to that substance. The permeability of a plasma membrane appears to be a function of several factors:

1 *Size of molecules*   Large molecules cannot pass through the plasma membrane. Water and amino acids are small molecules and can enter and exit the cell easily. However, most proteins, which consist of many amino acids linked together, seem to be too large to pass through the membrane. Many scientists believe that the giant-sized molecules do not enter the cell because they are larger than the diameters of the suspected membrane pores.

2 *Solubility in lipids*   Substances that dissolve easily in lipids pass through the membrane more readily than other substances since a major part of the plasma membrane consists of lipid molecules.

3 *Charge on ions*   The charge of an ion attempting to cross the plasma membrane can determine how easily the ion enters or leaves the cell. The protein portion of the membrane is capable of ionization. If an ion has a charge opposite that of the membrane, it is attracted to the membrane and passes through more readily. If the

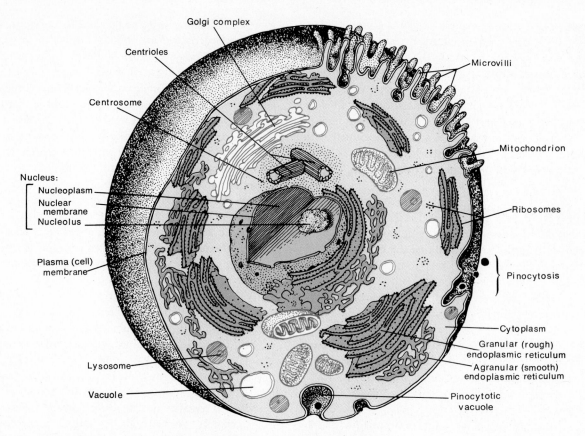

Figure 3-1 Generalized animal cell based on electron microscope studies.

ion attempting to cross the membrane has the same charge as the membrane, it is repelled by the membrane and its passage is restricted. This phenomenon conforms to the rule of physics that opposite charges attract, whereas like charges repel each other.

4 *Presence of carrier molecules* Plasma membranes contain special molecules called carriers that are capable of attracting and transporting substances across the membrane regardless of size, ability to dissolve in lipids, or membrane charge.

Movement of Materials Across Plasma Membranes

The mechanisms whereby substances move across the plasma membrane are essential to the life of the cell. Certain substances, for example, must move into the cell to support life, whereas waste materials or harmful substances must be moved out. The processes involved in these movements may be classed as either passive or active. In *passive processes,* substances move across plasma membranes without assistance from the cell. The substances move, on their own, from an area where their concentration is greater to an area where their concentration is less. The substances could also be forced across

the plasma membrane by pressure from an area where the pressure is greater to an area where it is less. In *active processes,* the cell contributes energy in moving the substance across the membrane.

PASSIVE PROCESSES

■ *Diffusion* A passive process called **diffusion** occurs when there is a *net* or greater movement of molecules or ions from a region of high concentration to a region of low concentration. The movement from high to low concentration continues until the molecules are evenly distributed. At this point, they move in both directions at an equal rate. This point of even distribution is called *equilibrium.* The difference between high and low concentrations is called the *concentration gradient.* Molecules moving from the high-concentration area to the low-concentration area are said to move *down* or *with* the concentration gradient. If a dye pellet is placed in a beaker filled with water, the color of the dye is seen immediately around the pellet. At increasing distances from the pellet, the color becomes lighter (Figure 3-3). Later, however, the water solution will be a uniform color. The dye molecules possess kinetic energy, which causes them to move about at random throughout the entire area. The dye molecules move down the concen-

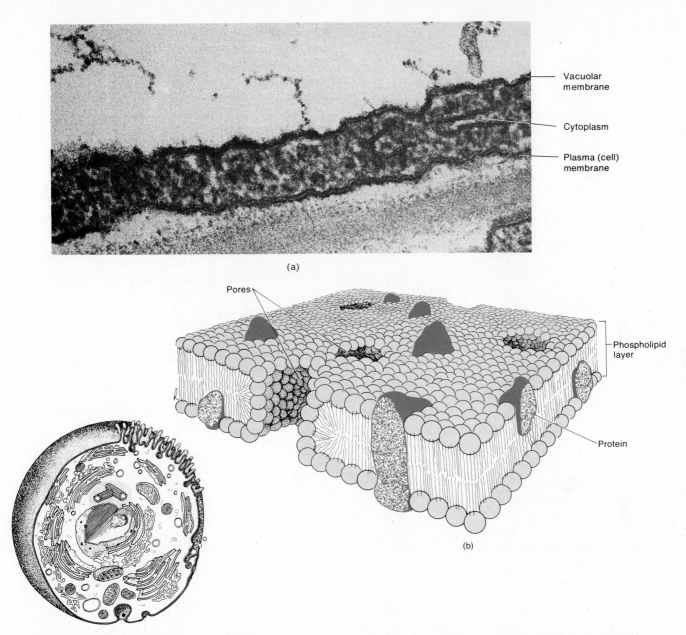

Vacuolar membrane

Cytoplasm

Plasma (cell) membrane

(a)

Pores

Phospholipid layer

Protein

(b)

Figure 3-2 Plasma membrane. (a) Electron micrograph of the plasma membrane at a magnification of 225,000X. (Courtesy of Myron C. Ledbetter, Brookhaven National Laboratory.) (b) Enlargement of the plasma membrane showing the latest concept regarding the relationship of the phospholipid layer and protein molecules.

tration gradient from an area of high concentration to an area of low concentration. The water molecules also move from a high-concentration to a low-concentration area. When dye molecules and water molecules are evenly distributed among themselves, equilibrium is reached and diffusion ceases, even though molecular movements continue.

In the example cited, no membranes were involved. Diffusion may occur, however, through semipermeable membranes in the body. A good example of this kind of diffusion in the human body is the

movement of oxygen from the blood into the cells and the movement of carbon dioxide from the cells back into the blood.

■ *Facilitated Diffusion* Another example of diffusion through a semipermeable membrane occurs by a process called **facilitated diffusion**. Although some chemical substances are insoluble in lipids, they can still pass through the plasma membrane. Among these are different sugars, especially glucose. In the process of facilitated diffusion, glucose combines with a carrier

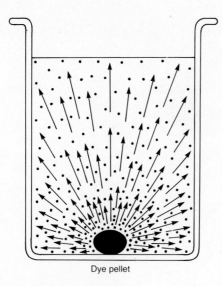

Dye pellet

Figure 3-3   Principle of diffusion.

substance. The combined glucose-carrier is soluble in the lipid layer of the membrane, and the carrier transports the glucose to the inside of the membrane. At this point, the glucose separates from the carrier and enters the cell. The carrier then returns to the outside of the membrane to pick up more glucose and transport it inside. The carrier makes the glucose soluble in the lipid portion of the membrane so it can pass through the membrane. By itself, glucose is insoluble and cannot penetrate the membrane. You can get a fairly good idea as to how a carrier works by inspecting Figure 3-6a. In the process of facilitated diffusion, the cell does not expend energy, and the movement of the substance is from a region of higher to lower concentration.

The rate of facilitated diffusion depends on (1) the difference in concentration of the substance on either side of the membrane, (2) the amount of carrier available to transport the substance, and (3) how quickly the carrier and substance combine. The process is greatly accelerated by insulin, a hormone produced by the pancreas. One of insulin's functions is to lower the blood-glucose level by accelerating the transportation of glucose from the blood into body cells. This transportation, as we have just seen, is by facilitated diffusion.

■ *Osmosis*   Another passive process by which materials move across membranes is **osmosis**. Unlike diffusion, this process specifically refers to the net movement of water molecules through a semipermeable membrane from an area of high water concentration to an area of lower water concentration. Once again, a simple apparatus may be used to demonstrate the process. The apparatus shown in Figure 3-4 consists of a tube constructed from cellophane, a semipermeable mem-

brane. The cellophane tube is filled with a colored, 20 percent sugar (sucrose) solution. The upper portion of the cellophane tube is plugged with a rubber stopper through which a glass tubing is fitted. The cellophane tube is placed into a beaker containing pure water. Initially, the concentrations of water on either side of the semipermeable membrane are different. There is a lower concentration of water inside the cellophane tube than outside. Because of this difference, water moves from the beaker into the cellophane tube. The force with which the water moves is called osmotic pressure. Very simply, **osmotic pressure** is the force under which a solvent moves from a solution of lower solute concentration to a solution of higher solute concentration when the solutions are separated by a semipermeable membrane. There is no movement of sugar from the cellophane tube inside the beaker, however, since the cellophane is impermeable to molecules of sugar — sugar molecules are too large to go through the pores of the membrane. As water moves into the cellophane tube, the sugar solution becomes increasingly diluted and begins to move up the glass tubing. In time, the water that has accumulated in the cellophane tube and the glass tubing exerts a downward pressure that forces water molecules back out of the cellophane tube and into the beaker. When water molecules leave the cellophane tube and enter the tube at the same rate, equilibrium is reached.

■ *Isotonic, Hypotonic, and Hypertonic Solutions*
Osmosis may also be understood by considering the effects of different water concentrations on red blood cells. If the normal shape of a red blood cell is to be maintained, the cell must be placed in an **isotonic solution** (Figure 3-5a). This is a solution in which the concentrations of water molecules and solute molecules are the same on both sides of the semipermeable membrane. The concentrations of water and solute in the extracellular fluid outside the red blood cell must be the same as the concentration of the intracellular fluid. Under ordinary circumstances, a 0.85 percent NaCl solution is isotonic for red blood cells. In this condition, water molecules enter and exit the cell at the same rate, allowing the cell to maintain its normal shape. A different situation results if red blood cells are placed in a solution that has a lower concentration of solutes and, therefore, a higher concentration of water. This is called a **hypotonic solution**. In this condition, water molecules enter the cells faster than they can leave — causing the red blood cells to swell and eventually burst (Figure 3-5b). The rupture of red blood cells in this manner is called *hemolysis* or *laking*. Distilled water is a strongly hypotonic solution. On the other hand, a **hypertonic solution** has a higher concentration of solutes and a lower concentration of water than the red blood cells.

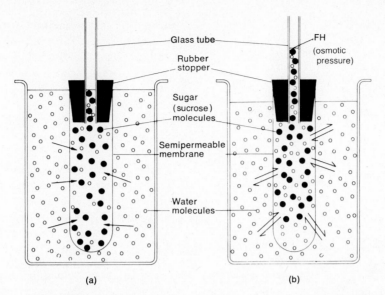

Figure 3-4 Principle of osmosis. (a) Apparatus at the start of the experiment. (b) Apparatus at equilibrium. In (a), the cellophane tube contains a 20 percent sugar solution and is immersed in a beaker of distilled water. The arrows indicate that water molecules can pass freely into the tube, but that sugar (sucrose) molecules are held back by the semipermeable membrane. As water moves into the tube by osmosis, the sugar solution is diluted and the volume of the solution in the cellophane tube increases. This increased volume is shown in (b), with the sugar solution moving up the glass tubing. The final height reached (FH) occurs at equilibrium and represents the osmotic pressure. At this point, the number of water molecules leaving the cellophane tube is equal to the number of water molecules entering the tube.

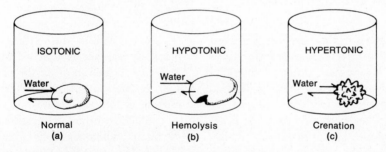

Figure 3-5 Principle of osmosis applied to red blood cells. Shown here are the effects on red blood cells when placed in (a) an isotonic solution, in which they maintain normal shape; (b) a hypotonic solution, in which they undergo hemolysis; and (c) a hypertonic solution, in which they undergo crenation.

One example of a hypertonic solution is a 10 percent NaCl solution. In such a solution, water molecules move out of the cells faster than they can enter. This situation causes the cells to shrink (Figure 3-5c). The shrinkage of red blood cells in this manner is called *crenation*. Red blood cells may be greatly impaired or destroyed if placed in solutions that deviate significantly from the isotonic state.

■ *Filtration* A third passive process involved in moving materials in and out of cells is **filtration**. This process involves the movement of solvents such as water and dissolved substances such as sugar across a semipermeable membrane by mechanical pressure. Such a movement is always from an area of higher pressure to an area of lower pressure and continues as long as a pressure difference exists. Most small to medium-sized molecules can be forced through a cell membrane. An example of filtration occurs in the kidneys, where the blood pressure supplied by the heart forces water and urea through thin cell membranes of tiny blood vessels and into the kidney tubule cells. In this basic process, protein molecules are retained by the body since they are too large to be forced through the cell membranes of the blood vessels. Harmful substances such as urea are small enough to be forced through and eliminated, however.

■ *Dialysis* The final passive process to be considered is **dialysis**. Essentially, dialysis involves the separation of small molecules from large molecules by diffusion of the

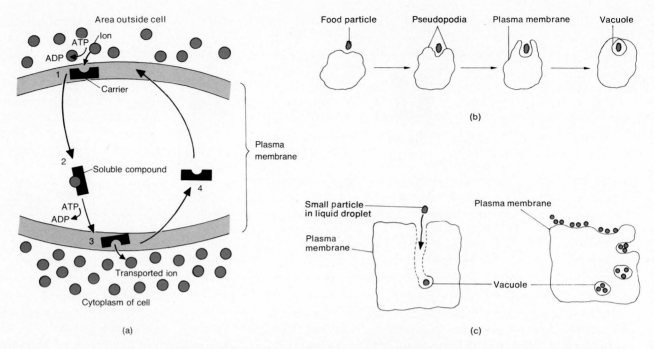

Figure 3-6   Active processes. (a) Mechanism of active transport. (b) Phagocytosis. (c) Two variations of pinocytosis. In the variation on the left, the ingested substance enters a channel formed by the plasma membrane and becomes enclosed in a vacuole at the base of the channel. In the variation on the right, the ingested substance becomes enclosed in a vacuole that forms and detaches at the surface of the cell.

smaller molecules through a semipermeable membrane. Suppose a solution containing molecules of various sizes is placed in a tube that is permeable only to the smaller molecules. The tube is then placed in a beaker of distilled water. Eventually, the smaller molecules will move from the tube into the water in the beaker and the larger molecules will be left behind. This principle of dialysis is employed in artificial kidneys. The patient's blood is passed into a dialysis tube outside the body. The dialysis tube takes the place of the kidneys. As the blood moves through the tube, waste products pass from the blood into a solution surrounding the dialysis tube. At the same time, certain nutrients are passed from the solution into the blood. The blood is then returned to the body.

ACTIVE PROCESSES

When cells actively participate in moving substances across membranes, they must expend energy. Cells can even move substances against a concentration gradient. The active processes considered here are active transport, phagocytosis, and pinocytosis.

■ *Active Transport*   The process by which substances, usually ions, are transported across plasma membranes from an area of lower concentration to an area of higher concentration is called **active transport** (Figure 3-6a). Although the exact mechanism is not known, the following sequence is believed to occur:

1 An ion in the extracellular fluid outside the plasma membrane is attached to an enzymelike carrier molecule located in or on the plasma membrane.

2 The ion-carrier complex is soluble in the lipid portion of the membrane.

3 The compound moves toward the interior of the membrane where it is split by enzymes.

4 The ion is then transported into the intracellular fluid of the cell, and the carrier returns to the outer surface of the membrane to pick up another ion.

The energy for the attachment and release of the carrier molecule is supplied by ATP.

■ *Phagocytosis*   Another active process by which cells take in substances across the plasma membrane is called **phagocytosis,** or "cell eating" (Figure 3-6b). In this process, projections of cytoplasm, called *pseudopodia,* engulf solid particles external to the cell. Once the particle is surrounded, the membrane folds inwardly, forming a membrane sac around the particle. This newly formed sac, called a *digestive vacuole,* breaks off from the outer cell membrane, and the solid material inside the vacuole is digested. Indigestible particles and cell products are removed from the cell by a reverse phagocytosis. This process is important because molecules and particles of material that would normally be restricted from crossing the plasma membrane can be brought into or removed from the cell. The phagocytic white blood cells of the body constitute a vital defense mechanism. Through phagocytosis, the white blood cells destroy bacteria and other foreign substances. (See Figure 4-4.)

■ *Pinocytosis* In **pinocytosis,** or "cell drinking," the engulfed material consists of a liquid rather than a solid (Figure 3-6c). Moreover, no cytoplasmic projections are formed. Instead, the liquid is attracted to the surface of the membrane. The membrane folds inwardly, surrounds the liquid, and detaches from the rest of the intact membrane. Few cells are capable of phagocytosis, but many cells may carry on pinocytosis. Examples include cells in the kidneys and urinary bladder.

When both phagocytosis and pinocytosis involve the inward movement of materials, they are together referred to as *endocytosis.* However, since phagocytosis and pinocytosis can also work in reverse by exporting materials from cells in vacuoles, they are also referred to as *exocytosis.*

Modified Plasma Membranes

Electron microscope studies have revealed that plasma membranes of certain cells contain a number of modifications. That is, they have different structures for very specific purposes. For example, the membranes of some cells lining the small intestine have small, cylindrical projections called **microvilli** (Figure 3-7a, b). These fingerlike projections enormously increase the absorbing area of the cell surface. A single cell may have as many as 3,000 microvilli, and a 1 sq mm (0.0394 sq inch) area of intestine may contain as many as 200 million microvilli.

Another membrane modification is found in the rods and cones of the eye. They serve as photoreceptors, or light-receiving cells. The upper portion of each rod contains two-layered, disc-shaped membranes called **sacs** that contain the pigments involved in vision (Figure 3-7c, d). Another example of a membrane modification is the **stereocilia.** They are found only in cells lining a duct (epididymis) of the male reproductive system. They appear by light microscopy as long, slender, branching processes at the free surfaces of the lining cells (Figure 3-7e, f). Electron micrographs show stereocilia to be microvilli. A final example of a membrane modification is the **myelin sheath** that surrounds portions of certain nerve cells (Figure 3-7g, h). It is thought that the myelin sheath increases the velocity of impulse conduction, protects the portion of the nerve cell it surrounds, and is related to the nutrition of the nerve cell.

CYTOPLASM

The living matter inside the cell's plasma membrane and external to the nucleus is called **cytoplasm** (Figure 3-8a, b). It is the matrix or ground substance in which various cellular components are found. Physically, cytoplasm may be described as a thick, semitransparent, elastic fluid containing suspended particles. Chemically, cytoplasm is 75 to 90 percent water plus solid components. Proteins, carbohydrates, lipids, and inorganic substances compose the bulk of the solid components. The inorganic substances and most carbohydrates are soluble in water and are present as a true solution. The majority of organic compounds, however, are found as colloids — particles that remain suspended in the surrounding ground substance. Since the particles of a colloid bear electrical charges that repel each other, they remain suspended and separated from each other.

Functionally, cytoplasm is the substance in which chemical reactions occur. The cytoplasm receives raw materials from the external environment and converts them into usable energy by decomposition reactions. Cytoplasm is also the site where new substances are synthesized for cellular use. It packages chemicals for transport to other parts of the cell or other cells of the body and facilitates the excretion of waste materials.

ORGANELLES

Despite the myriad chemical activities occurring simultaneously in the cell, there is little interference of one reaction with another. The cell has a system of compartmentalization provided by structures called **organelles.** These structures are specialized portions of the cell that assume various roles in growth, maintenance, repair, and control.

Nucleus

The **nucleus** is generally a spherical or oval organelle (Figure 3-8a, b). In addition to being the largest structure in the cell, the nucleus contains hereditary factors of the cell, called genes, which control cellular structure and direct many cellular activities. Certain cells, such as mature red blood cells, do not have nuclei. These cells carry on only limited chemical activity and are not capable of growth or reproduction.

The nucleus is separated from the cytoplasm by a double membrane called the *nuclear membrane* (Figure 3-8c). Between the two layers of the nuclear membrane is a space called the *perinuclear cisterna.* This arrangement of the nuclear membrane resembles the structure of the plasma membrane. Minute pores in the nuclear membrane allow the nucleus to communicate with a membranous network in the cytoplasm called the endoplasmic reticulum. Substances entering and exiting the nucleus are believed to pass through the tiny pores. Three prominent structures are visible within the nuclear membrane. The first of these is a gel-like fluid that fills

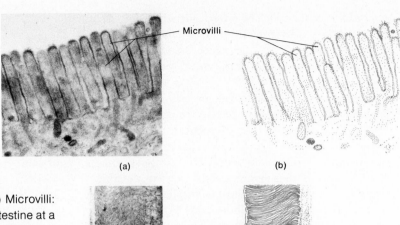

(a)

(b)

Figure 3-7 Modified plasma membranes. (a) Microvilli: electron micrograph of a portion of the small intestine at a magnification of 20,000×. (b) Microvilli: labeled diagram of the electron micrograph. (c) Rod sacs: electron micrograph of a portion of a rod of the eye at a magnification of 2,000×. (d) Rod sacs: labeled diagram of the electron micrograph. (e) Stereocilia: photomicrograph of stereocilia projecting from the lining cells of the ductus epididymis at a magnification of 200×. (f) Stereocilia: labeled diagram of the photomicrograph. (g) Myelin sheath: electron micrograph of a myelin sheath in cross section at a magnification of 20,000×. (h) Myelin sheath: labeled diagram of the electron micrograph. (Electron micrographs courtesy of E. B. Sandburn, University of Montreal. Photomicrograph courtesy of Victor B. Eichler, Wichita State University.)

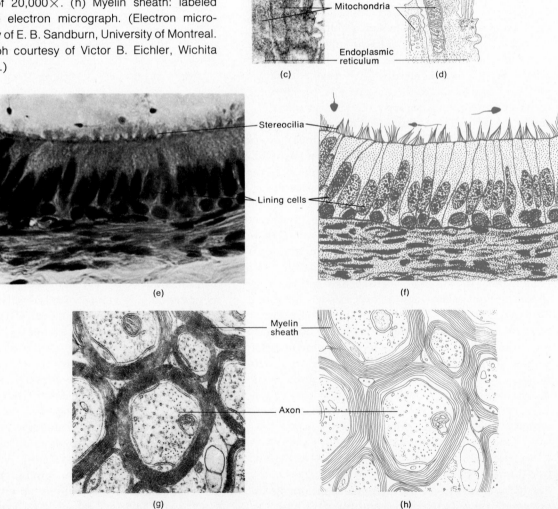

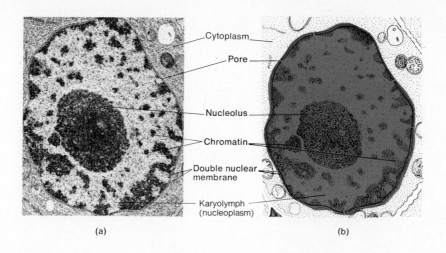

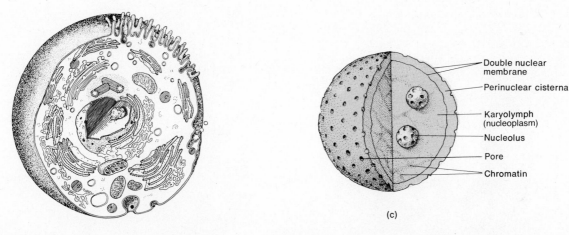

Figure 3-8 Cytoplasm and nucleus. (a) Electron micrograph of cytoplasm and the nucleus at a magnification of 31,600×. (Courtesy of Myron C. Ledbetter, Brookhaven National Laboratory.) (b) Diagram of the electron micrograph. (c) Diagram of a nucleus with two nucleoli.

the nucleus called *karyolymph (nucleoplasm).* Spherical bodies called the *nucleoli* are also present. These structures are composed primarily of RNA and assume a function in protein synthesis. Finally, there is the *genetic material* consisting principally of DNA. When the cell is not reproducing, the genetic material appears as a threadlike mass called *chromatin.* Prior to cellular reproduction the chromatin shortens and thickens into rod-shaped bodies called *chromosomes.*

Endoplasmic Reticulum and Ribosomes

Within the cytoplasm, there is a system consisting of pairs of parallel membranes enclosing narrow cavities of varying shapes. This system is known as the **endoplasmic reticulum,** or **ER** (Figure 3-9). The ER, in other words, is a network of canals running through the entire cytoplasm. These canals are continuous with both the

plasma membrane and the nuclear membrane. It is thought that the ER provides a surface area for chemical reactions, a pathway for transporting molecules within the cell, and a storage area for synthesized molecules. And, together with the Golgi complex, the ER is thought to secrete certain chemicals. Attached to the outer surfaces of the ER are exceedingly small, dense, spherical bodies called **ribosomes.** The ER in these areas is referred to as *granular,* or *rough,* reticulum. Portions of the ER that lack ribosomes are called *agranular,* or *smooth,* reticulum. Ribosomes are thought to serve as the sites of protein synthesis in the cell.

Golgi Complex

Another structure found in the cytoplasm is the **Golgi complex.** This structure usually consists of four to eight flattened channels, stacked upon each other with ex-

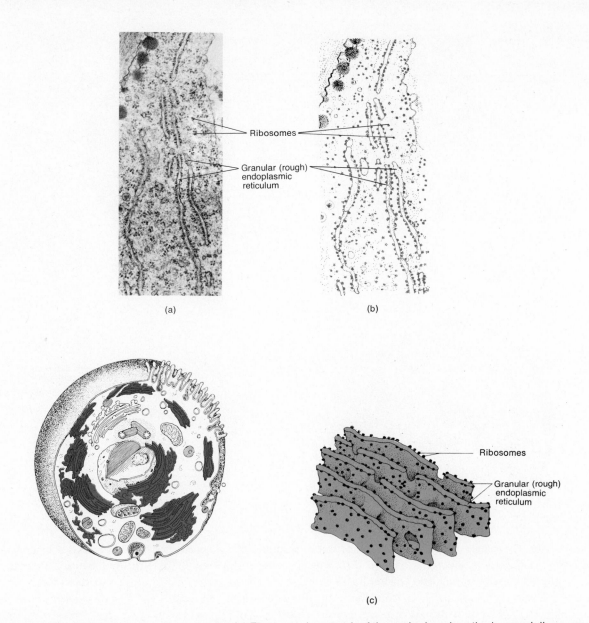

Figure 3-9   Endoplasmic reticulum and ribosomes. (a) Electron micrograph of the endoplasmic reticulum and ribosomes at a magnification of 76,000×. (Courtesy of Myron C. Ledbetter, Brookhaven National Laboratory.) (b) Diagram of the electron micrograph. (c) Diagram of the endoplasmic reticulum and ribosomes. See if you can find the agranular (smooth) endoplasmic reticulum in Figure 3-10.

panded areas at their ends. The stacked elements are called *cisternae,* and the expanded, terminal areas are *vesicles* (Figure 3-10). Generally, the Golgi complex is located near the nucleus and is directly connected, in parts, to the ER. One function of the Golgi complex is the secretion of proteins. *Secretion* is the production and release from the cell of a fluid that usually contains a variety of substances. Proteins synthesized by the ribosomes associated with granular ER are transported into the ER tubules. They migrate along the ER tubules until they reach the Golgi complex. As proteins accumulate in

the cisternae of the Golgi complex, the cisternae expand to form vesicles. After a certain size is reached, the vesicles pinch off from the cisternae. The protein and its associated vesicle is referred to as a *secretory granule.* The secretory granule then moves toward the surface of the cell where the protein is secreted. Cells of the digestive tract that secrete protein enzymes utilize this mechanism. The vacuole prevents "digestion" of the cytoplasm of the cell as it moves toward the cell surface.

Another function of the Golgi complex is associated with lipid secretion. It occurs in essentially the

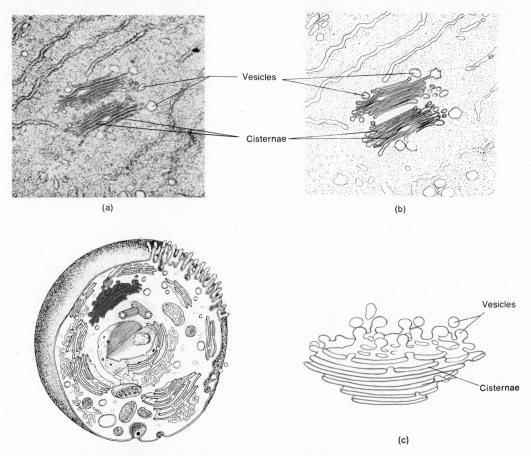

Figure 3-10 Golgi complex. (a) Electron micrograph of two Golgi complexes at a magnification of 78,000×. (Courtesy of Myron C. Ledbetter, Brookhaven National Laboratory.) (b) Diagram of the electron micrograph. (c) Diagram of the Golgi complex.

same way as protein secretion, except the lipids are synthesized by the agranular ER. The lipids pass through the ER into the Golgi complex. As in the mechanism just described, the lipids migrate into the cisternae and vesicles and are discharged at the surface of the cell. In the course of moving through the cytoplasm, the vesicle may release lipids into the cytoplasm before being discharged from the cell. These appear in the cytoplasm as lipid droplets. Among the lipids secreted in this manner are the steroids. (See Exhibit 2-4.)

The Golgi complex also functions in the synthesis of carbohydrates. Recent evidence indicates that the carbohydrates synthesized by the Golgi complex are combined with proteins synthesized by the ribosomes associated with granular ER to form carbohydrate-protein complexes. These complexes of carbohydrate and protein are called *glycoproteins.* As glycoproteins are assembled, they accumulate in the flattened channels of the Golgi complex. The channels expand and form vesicles. After a critical size is reached, the vesicles pinch off from the channel, migrate through the cytoplasm, and pass out of the cell through the plasma membrane. Outside the plasma membrane, the vesicles rupture and release their contents. Essentially, the Golgi complex synthesizes carbohydrates and combines them with proteins. It then packages the resulting glycoprotein and secretes it from the cell. The Golgi complex is well developed and highly active in secretory cells such as those found in the pancreas and the salivary glands.

Mitochondria

Small, spherical, rod-shaped, or filamentous structures called **mitochondria** appear throughout the cytoplasm. When sectioned and viewed under an electron microscope, each reveals an elaborate internal organization (Figure 3-11). A mitochondrion consists of a double membrane similar in structure to the plasma membrane. The outer mitochondrial membrane is smooth, but the inner membrane is arranged in a series of folds called *cristae.* The center of the mitochondrion is called the *matrix.* Because of the nature and arrangement of the

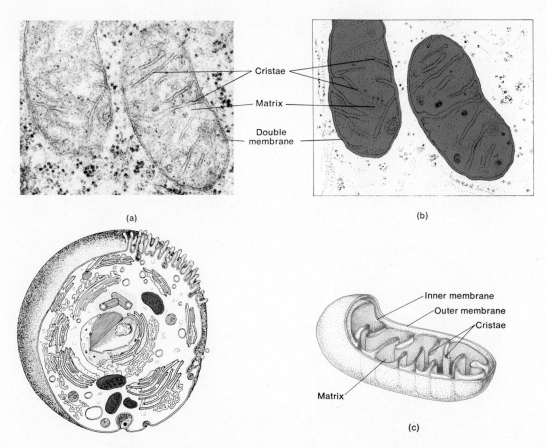

Figure 3-11   Mitochondria. (a) Electron micrograph of two mitochondria at a magnification of 20,000×. (Courtesy of Myron C. Ledbetter, Brookhaven National Laboratory.) (b) Diagram of the electron micrograph. (c) Diagram of a mitochondrion.

cristae, the inner membrane provides an enormous surface area for chemical reactions. Enzymes involved in energy-releasing reactions that form ATP are located on the cristae. Mitochondria are frequently called the "powerhouses of the cell" because of their central role in the production of ATP. Active cells such as muscle cells have a large number of mitochondria because of their high energy expenditure.

Lysosomes

When viewed under the electron microscope, **lysosomes** appear as membrane-enclosed spheres. They are formed from Golgi complexes (Figure 3-12). Unlike mitochondria, lysosomes have only a single membrane and lack detailed structure. Moreover, they contain powerful digestive enzymes capable of breaking down many kinds of molecules. These enzymes are also capable of digesting bacteria that enter the cell. White blood cells, which ingest bacteria by phagocytosis, contain

large numbers of lysosomes. Scientists have wondered why these powerful enzymes do not also destroy their own cells. Perhaps the lysosome membrane in a healthy cell is impermeable to enzymes so they cannot move out into the cytoplasm. However, when a cell is injured, the lysosomes release their enzymes. The enzymes then promote reactions that break the cell down into its chemical constituents. The chemical remains are either reused by the body or excreted. Because of this function, lysosomes have been called "suicide packets."

Microtubules

Small tubules made of protein and called **microtubules** are found in most cells. They range from 200 to 270 Å in diameter and are remarkably uniform in size. They do not branch. Microtubules are thought to be internal "skeletons" that afford cell shape and stiffness to the regions they occupy. They are also believed to function as intracellular channels along which substances move

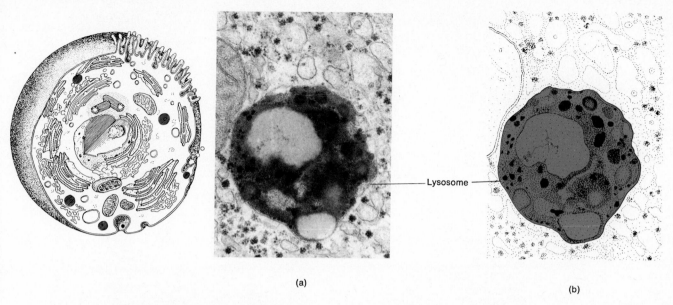

(a)

(b)

Figure 3-12 Lysosome. (a) Electron micrograph of a lysosome at a magnification of 55,000×. (Courtesy of F. Van Hoof, Université Catholique de Louvain.) (b) Diagram of the electron micrograph.

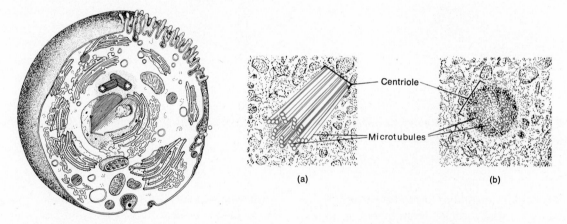

(a)

(b)

Figure 3-13 Centrosome and centrioles. (a) Centriole in longitudinal section. (b) Centriole in cross section.

from place to place. Microtubules make up the internal structure of cilia and flagella and the spindle fibers that appear in dividing cells.

Centrosome and Centrioles

A dense area of cytoplasm, generally spherical and located near the nucleus, is called the **centrosome** or **centrosphere**. Within the centrosome is a pair of cylindrical structures: the **centrioles** (Figure 3-13). Each centriole is composed of a ring of nine evenly spaced bundles surrounding two central tubules. Each bundle, in turn, consists of three microtubules. The two centrioles are situated so that the long axis of one is at right angles to the long axis of the other. Centrioles assume a role

in cell reproduction. Certain cells, such as most mature nerve cells, do not have a centrosome and so do not reproduce. This is why they cannot be replaced if destroyed.

Flagella and Cilia

Some body cells possess projections for moving the entire cell or for moving substances along the surface of the cell. These projections contain cytoplasm and are bounded by the plasma membrane. If the projections are few and long in proportion to the size of the cell, they are called **flagella**. An example of a flagellum is the tail of a sperm cell, used for locomotion. If the projections are numerous and short, resembling many hairs, they are called **cilia**. In humans, ciliated cells of the respiratory

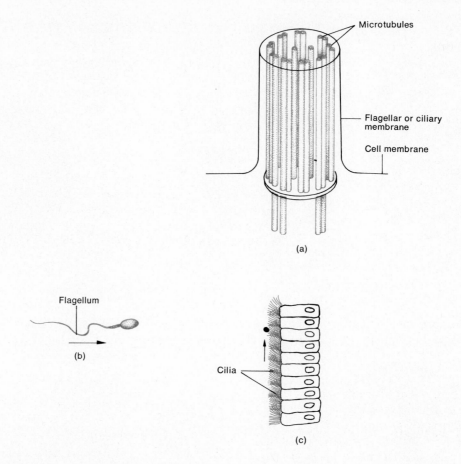

Figure 3-14   Flagella and cilia. (a) Structure of a flagellum or cilium. (b) Flagellum of a sperm cell. (c) Cilia of the respiratory tract moving a particle upward toward the mouth.

tract move lubricating fluids over the surface of the tissue and trap foreign particles. Electron microscopy has revealed no fundamental structural difference between cilia and flagella (Figure 3-14). Both consist of nine pairs of microtubules that form a ring around two solitary microtubules in the center.

CELL INCLUSIONS

The **cell inclusions** are a large and diverse group of chemical substances. These products are principally organic and may appear or disappear at various times in the life of the cell. *Hemoglobin* lies inside red blood cells. It performs the function of attaching to oxygen molecules and carrying the oxygen to other cells. *Melanin* is a pigment stored in the cells of the skin, hair, and eyes. It protects the body by screening out harmful ultraviolet rays from the sun. *Glycogen* is a polysaccharide that is stored in liver and skeletal muscle cells. When the body requires quick energy, liver cells can break down the glycogen

into glucose and release it. *Lipids,* which are stored in fat cells, may be decomposed when the body runs out of carbohydrates for producing energy. A final example of an inclusion is *mucus,* which is produced by cells that line organs. Its function is to provide lubrication.

The major parts of the cell and their functions are summarized in Exhibit 3-1.

EXTRACELLULAR MATERIALS

The substances that lie outside cells are called **extracellular materials.** They include the body fluids, which provide a medium for dissolving, mixing, and transporting substances. They include secreted inclusions like mucus. And they also include special substances that form the matrix in which some cells are embedded.

The matrix materials are produced by certain cells and deposited outside their plasma membranes. The matrix supports the cells, binds them together, and gives strength and elasticity to the tissue. Some matrix materials are *amorphous* – they have no specific shape. These include hyaluronic acid and chondroitin sulfate. *Hyaluronic acid* is a viscous, fluidlike substance that binds cells together, lubricates joints, and maintains the shape of the eyeballs. *Chondroitin sulfate* is a jellylike substance that provides support and adhesiveness in cartilage, bone, heart valves, the cornea of the eye, and the umbilical cord. Other matrix materials are *fibrous,* or threadlike. Fibrous materials provide strength and support for tissues. Among these are **collagen,** or *collagenous fibers.* Collagen is found in all types of connective tissue, especially in bones, cartilage, tendons, and ligaments. **Reticulin,** also called *reticular fibers,* is a matrix material that forms a network around fat cells, nerve fibers, muscle cells, and blood vessels. It forms the framework or stroma for many soft organs of the body such as the spleen. **Elastin,** found in *elastic fibers,* gives elasticity to skin and to tissues forming the walls of blood vessels.

CELL DIVISION

Most of the cell activities mentioned thus far maintain the life of the cell on a day-to-day basis. However, cells become damaged, diseased, or worn out and then die. Thus new cells must be produced for growth.

    Cell division is the process by which cells reproduce themselves. Cell division or, more appropriately, nuclear division, may be of two kinds. In the first kind a single parent cell duplicates itself. This process is known as mitosis (nuclear division) and cytokinesis (cytoplasmic division). The process ensures that each new daughter cell has the same *number* and *kind* of chromosomes as the original parent cell. After the process is complete, the two daughter cells have the same hereditary material and genetic potential as the parent cell. This kind of cell division results in an increase in the number of body cells. Mitosis and cytokinesis are the means by which dead or injured cells are replaced and new cells are added for body growth. The second kind of division is a mechanism by which sperm and egg cells are produced. This process, meiosis, is the mechanism that enables the reproduction of an entirely new organism. The process of meiosis is discussed in detail in Chapter 29.

Exhibit 3—1 CELL PARTS AND THEIR FUNCTIONS

Part	Functions
Plasma membrane	Protects and allows substances to enter or exit the cell through diffusion, facilitated diffusion, osmosis, filtration, dialysis, active transport, phagocytosis, and pinocytosis.
Cytoplasm	Serves as the ground substance in which chemical reactions occur.
Organelles	
Nucleus	Contains genes and controls cellular activities.
Endoplasmic reticulum	Provides a surface area for chemical reactions; provides a pathway for transporting chemicals; serves as a storage area.
Ribosomes	Act as sites of protein synthesis.
Golgi complex	Synthesizes carbohydrates, combines carbohydrates with proteins, packages materials for secretion, and secretes lipids and glycoproteins.
Mitochondria	Produce ATP.
Lysosomes	Digest chemicals and foreign microbes.
Microtubules	Serve as intracellular "skeleton" and passageway and component of cilia, flagella, and spindle fibers.
Centrioles	Help organize spindle fibers during cell division.
Flagella and cilia	Allow movement of cell or movement of particles along surface of cell.
Inclusions	Involved in overall body functions. Include materials retained in cell (hemoglobin), reserve materials (glycogen, fats), and secretions (mucus).

Mitosis

In a 24-hour period, the average adult loses about 500 million cells from different parts of the body. Obviously, these cells must be replaced. Cells that have a short life span – the cells of the outer layer of skin, the cornea of the eye, the digestive tract – are continually being replaced. The succession of events that takes place during mitosis and cytokinesis is plainly visible under a microscope after the cells have been stained in the laboratory.

    When a cell reproduces, it must replicate its chromosomes so its hereditary traits may be passed on to succeeding generations of cells. A **chromosome** is a highly coiled DNA molecule. This is the molecule that contains your hereditary information. Each human chromosome consists of about 20,000 genes.

The process called **mitosis** is the replication of chromosomes and the distribution of the two sets of chromosomes into two separate and equal nuclei. For convenience, biologists divide the process into four stages: prophase, metaphase, anaphase, and telophase. There are arbitrary classifications. Mitosis is actually a continuous process, one stage merging imperceptibly into the next. Interphase is the stage that occurs between consecutive cell divisions.

INTERPHASE

When a cell is carrying on every life process except division, it is said to be in **interphase** (Figure 3-15a). One of the principal events of interphase is the replication of DNA. When DNA replicates, its helical structure partially uncoils (Figure 3-16). Those portions of DNA that remain coiled stain darker than the uncoiled portions. This unequal distribution of stain causes the DNA to appear as a granular mass called **chromatin.** (See Figure 3-15a). During uncoiling, DNA separates at the points where the nitrogen bases are connected. Each exposed nitrogen base then picks up a complementary nitrogen base (with associated sugar and phosphate group) from the cytoplasm of the cell. This uncoiling and complementary base pairing continues until each of the two original DNA strands is matched and joined with two newly formed DNA strands. The original DNA molecule has become two DNA molecules.

During interphase the cell is also synthesizing most of its RNA and proteins. It is producing chemicals so that all cellular components can be doubled during division. A microscopic view of a cell during interphase shows a clearly defined membrane, nucleoli, karyolymph, and chromatin. As interphase progresses, a pair of centrioles appears. The centrioles duplicate and the resulting two pairs of centrioles separate. Once a cell completes its activities during interphase, mitosis begins.

PROPHASE

During **prophase** (Figure 3-15b) the centrioles move apart and project a series of radiating fibers called *astral rays.* The centrioles move to opposite poles of the cell and become connected by another system of fibers called *spindle fibers.* The astral rays and spindle fibers consist of microtubules. Together, the centrioles, astral rays, and spindle fibers are referred to as the *mitotic apparatus.* Simultaneously, the chromatin has been shortening and thickening into chromosomes. The nucleoli have become less distinct, and the nuclear membrane has disappeared. Each prophase "chromosome" is actually composed of a pair of separate structures called *chroma-*

tids. Each chromatid is a complete chromosome consisting of a double-stranded DNA molecule. Each chromatid is attached to its chromatid pair by a small spherical body called the *centromere.* During prophase, the chromatid pairs move toward the equatorial plane region or equator of the cell.

METAPHASE

During **metaphase** (Figure 3-15c), the second stage of mitosis, the chromatid pairs line up on the equatorial plane of the spindle fibers. The centromere of each chromatid pair attaches itself to a spindle fiber.

ANAPHASE

Anaphase (Figure 3-15d), the third stage of mitosis, is characterized by the division of the centromeres and the movement of complete identical sets of chromatids, now called chromosomes, to opposite poles of the cell. During this movement, the centromeres attached to the spindle fibers seem to drag the trailing parts of the chromosomes toward opposite poles.

TELOPHASE

Telophase (Figure 3-15e), the final stage of mitosis, consists of a series of events nearly the reverse of prophase. By this time, two identical sets of chromosomes have reached opposite poles. New nuclear membranes begin to enclose them. The chromosomes start to assume their chromatin form. Nucleoli reappear, and the spindle fibers and astral rays disappear. The centrioles also replicate so that each cell has two centriole pairs. The formation of two nuclei identical to those of cells in interphase terminates telophase. A mitotic cycle has thus been completed (Figure 3-15f).

Cytokinesis

Division of the cytoplasm, called **cytokinesis,** often begins during late anaphase and terminates at the same time as telophase. Cytokinesis begins with the formation of a *cleavage furrow* that runs around the cell's equator. The furrow progresses inward, resembling a constricting ring, and cuts completely through the cell to form two separate portions of cytoplasm (Figure 3-15d to f).

GENE ACTION: PROTEIN SYNTHESIS

Mitosis and cytokinesis are vital processes because they ensure that the daughter cells have identical sets of chromosomes and, therefore, identical genes. **A gene** is a group of nucleotides on a DNA molecule that serves as the master mold for manufacturing a specific protein.

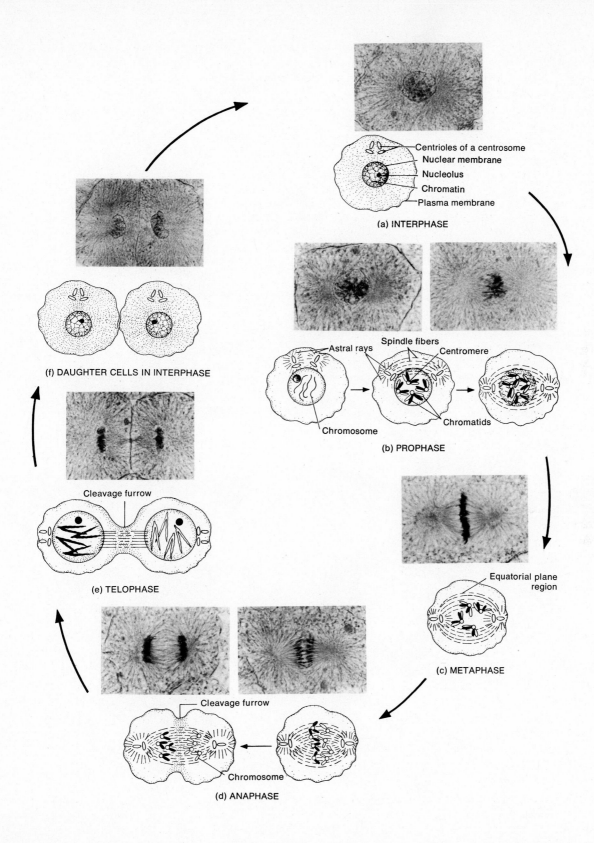

Figure 3-15 Cell division: mitosis and cytokinesis. Photomicrographs and diagrammatic representations of the various stages of division in whitefish eggs. Read the sequence starting at (a), and move clockwise until you complete the cycle. (Courtesy of Carolina Biological Supply Company.)

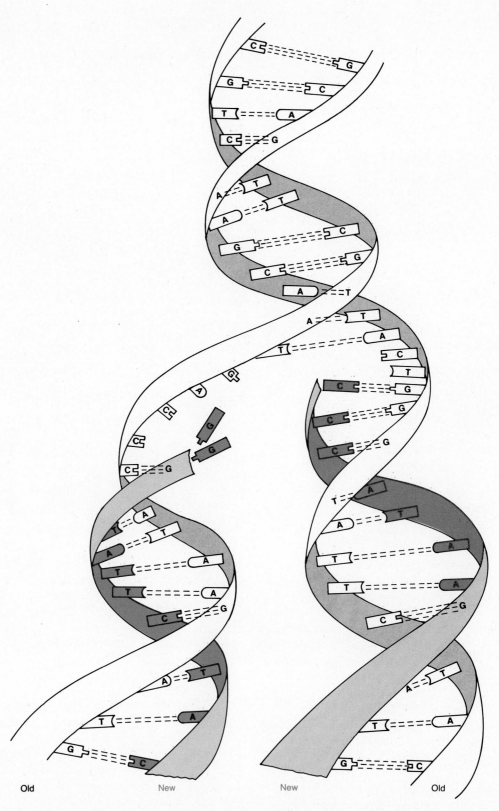

Figure 3-16 Replication of DNA. Each strand of the double helix separates by breaking the bonds between nucleotides. New nucleotides attach at the proper sites, and two new strands of DNA are paired off with the two old strands. After replication, the two DNA molecules, each consisting of a new and an old strand, return to their helical structure.

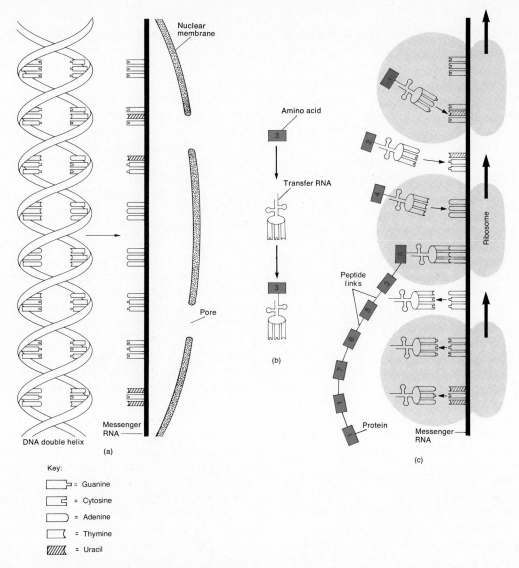

Key:

⬜⊐ = Guanine

⬜⊏ = Cytosine

⬜ = Adenine

⬜⌐ = Thymine

▨ = Uracil

Figure 3-17 Protein synthesis. (a) In the nucleus, the DNA helix separates, and one strand of the helix serves as a mold for the formation of messenger RNA. The messenger RNA leaves the nucleus and moves to a ribosome. (b) In the cytoplasm, transfer RNA molecules attach to amino acids. Each specific kind of transfer RNA attaches to a specific amino acid. (c) Transfer RNAs bring their amino acids to the messenger RNA on the ribosome. As the ribosome moves along the messenger RNA strand, the transfer RNAs temporarily bond to the messenger RNA according to a specific pairing of nitrogen bases. In the process, the amino acids are brought into close contact and peptide links form between them. The chain of bonded amino acids is the synthesized protein.

Each gene averages about 1,000 pairs of nucleotides, which appear in a *specific sequence* on the DNA molecule. No two genes have exactly the same sequence of nucleotides. This fact is the key to heredity. Gene action is thought to occur in the following way.

Each gene directs the manufacture of a particular kind of RNA. The gene segments of DNA manufacture RNA much as the entire DNA molecule replicates itself during mitosis (Figure 3-17a). The major differences between DNA and RNA synthesis are:

1 In RNA synthesis, the sugar ribose is used instead of deoxyribose, and the nitrogen base uracil is used instead of thymine.

2 Each RNA molecule is molded by only a segment of the DNA molecule. The RNA molecule, therefore, does not have nearly the number of nucleotides as the DNA molecule has.

Several types of RNA are manufactured by genes. Ribosomes, the sites of protein synthesis, consist of a type of RNA called **ribosomal RNA (rRNA)**. Two other kinds of RNA are messenger RNA and transfer RNA. After a molecule of **messenger RNA (mRNA)** is produced from DNA, it leaves the nucleus and travels to a ribosome. Meanwhile, a number of **transfer RNA (tRNA)** molecules also produced from DNA leave the nucleus. Each *t*RNA molecule migrates into the cytoplasm and attaches to an amino acid. Each kind of *t*RNA attaches to only one kind of amino acid (Figure 3-17b). The *t*RNA molecules then migrate with their amino acids to the *m*RNA that is attached to a ribosome (Figure 3-17c). The *m*RNA serves as a mold for the manufacture of a protein—as DNA serves as a mold for the production of RNA. Each group of nitrogen bases on the *m*RNA attracts one kind of *t*RNA with its amino acid and no other. The sequence and kind of bases in the *m*RNA group determine which *t*RNA molecules will be attracted. Thus the *m*RNA determines the order in which the amino acids line up. As the *t*RNAs pair off with their *m*RNA groups, a line of amino acids forms alongside the *m*RNA molecule as the ribosome moves along the *m*RNA strand. Peptide linkages (bonds) are then formed between the amino acids, and a protein is made. The role of the ribosome is to move along the *m*RNA strand. As it does, a *t*RNA attached to its specific amino acid fits into place on the *m*RNA strand, leaving behind its amino acid. This process continues until the protein is synthesized.

Remember that the base sequence of the gene determines the sequence of the bases in the *m*RNA. The sequence of the bases in the *m*RNA then determines the order and kind of amino acids that will form the protein. Thus each gene is responsible for making a particular protein.

CELLS AND AGING

As Frédéric Verzar, the Swiss dean of gerontologists, once said: "Old age is not an illness; it is a continuation of life with decreasing capacities for adaptation." Only recently has his view of aging as a progressive failure of the body's homeostatic adaptive responses gained wide acceptance. There has been a strong tendency to confuse aging with many diseases frequently associated with it—especially cancer and atherosclerosis. Each, in fact, probably accelerates the other.

The obvious characteristics of aging are well known: graying and loss of hair, loss of teeth, wrinkling of skin, decreased muscle mass, and increased fat deposits. The physiological signs of aging are gradual deterioration in function and capacity to respond to environmental stress. Thus basic kidney and digestive metabolic rates decrease, as does the ability to maintain a constant internal environment despite changes in temperature, diet, and oxygen supply. These manifestations

of aging are related to a decrease in the actual number of cells in the body (100,000 brain cells are lost each day) and to the disordered functioning of the cells that remain.

The extracellular components of tissues also change with age. *Collagen* fibers, responsible for the strength in tendons, increase in number and change in quality with aging. These changes in arterial walls are as much responsible for the loss of elasticity as those in arteriosclerosis. *Elastin,* another constituent of the intercellular matrix, is responsible for the elasticity of blood vessels and skin. It thickens, fragments, and acquires a greater affinity for calcium with age—changes that may be associated with the development of arteriosclerosis.

Several kinds of cells in the body—heart cells, skeletal muscle cells, neurons— are incapable of replacement. Recent experiments have proved that certain other cell types are limited when it comes to cell division. Cells grown outside the body divided only a certain number of times and then stopped. The number of divisions correlated with the donor's age. The number of divisions also correlated with the normal lifespan of the different species from which the cells were obtained—strong evidence for the hypothesis that cessation of mitosis is a normal, genetically programed event.

Just as the factors that limit the life of an individual cell are unknown, so are those that restrict the growth or life of a tissue or organ. At menopause, the ovary ceases to function. Its cells die long before the rest of the female body. Perhaps similar mechanisms determine longevity.

Some investigators have studied aging from the standpoint of immunology. The ability to develop antibodies is said to diminish with age. Senescence, according to researchers, results in the older person's immunological system having a "shotgun," rather than a specific, response to foreign protein. This shotgun response may include an autoimmune reaction that attacks and gradually destroys the individual's own normal tissue and organs.

The following generalizations on aging can be made:

1 Aging is a general process that produces observable changes in structure and function.

2 Aging produces increased vulnerability to environmental stress and disease.

3 Evidence suggests that the life span is about 110 years but maximum life expectancy is no greater than 85 years.

4 The mechanism behind aging is not known. Improvements in life expectancy may be due to an overall reduction in the number of life-threatening situations or to modification of the aging process.

5 Eventually the aging process may be modified and life span and life expectancy lengthened.

DISORDERS

Cancer is not a single disease but many. The human body contains more than a hundred different types of cells, each of which can malfunction in its own distinctive way to cause cancer. When cells in some area of the body duplicate unusually quickly, the excess of tissue that develops is called a growth, or *tumor.* Tumors may be cancerous and fatal or quite harmless. A cancerous growth is called a *malignant tumor, or malignancy.* A noncancerous growth is called a *benign growth.*

Cells of malignant growths all have one thing in common: They duplicate continuously and often very quickly. This growth continues until the victim dies or until every malignant cell is removed or destroyed. As the cancer grows, it expands and begins to compete with normal tissues for space and nutrients. Eventually, the normal tissues regress and die. The organ functions less and less efficiently until finally it ceases to function altogether. Cancer cells may spread from the original, or *primary,* growth. Cancer of the breast, for instance, has a tendency to spread to the lungs. The spread of cancer to other regions of the body is called *metastasis.* Metastasis occurs when a malignant cell breaks away from the growth, enters the bloodstream, and is carried through the body. Wherever the cell comes to rest, it establishes another tumor: a *secondary* growth. Usually death is caused when a vital organ regresses because of competition with the cancer cells for room and nutrients. Pain develops when the growth impinges on nerves or blocks a passageway so that secretions build up pressure.

Cancer cells multiply without control. Individual cells vary in size and shape, and the orderly orientation of normal cells is replaced by disorganization which may be so extensive that no recognizable structures remain.

At present, cancers are classified by their microscopic appearance and the body site from which they arise. At least 100 different cancers have been identified in this way. If finer details of appearance are taken into consideration, the number can be increased to 200 or more. The name of the cancer is derived from the type of tissue in which it develops. *Sarcoma* is a general term for any cancer arising from connective tissue. *Osteogenic sarcomas* (*osteo* = bone; *genic* = origin), the most frequent type of childhood cancer, destroy normal bone tissue and eventually spread to other areas of the body. *Myelomas* (*myelos* = marrow) are malignant tumors, occurring in middle-aged and older people, that interfere with the blood-cell-producing function of bone marrow and cause anemia. *Chondrosarcoma* is a cancerous growth of the cartilage (*chondro* = cartilage).

Benign tumors are composed of cells that do not metastasize—that is, the growth does not spread to other organs. Removing all or part of a tumor to determine whether it is benign or malignant is called a biopsy. A benign tumor may be removed if it impairs a normal body function or causes disfiguration.

Cancer has been observed in all species of vertebrates. In fact, cellular abnormalities that resemble cancer—such as crown gall of tomatoes—have been observed in plants as well. Cell masses that resemble cancers of higher animals have also been produced and studied in insects.

What triggers a perfectly normal cell to lose control and become abnormal? Scientists are uncertain. First there are environmental agents: substances in the air we breathe, the water we drink, the food we eat. The World Health Organization estimates that these agents—called carcinogens—may be associated with 60 to 90 percent of all human cancer. Examples of carcinogens are the hydrocarbons found in cigarette tar. Ninety percent of all lung cancer patients are smokers. Another environmental factor is radiation. Ultraviolet light from the sun, for example, may cause genetic mutations in exposed skin cells and lead to cancer, especially among light-skinned people.

Viruses are a second cause of cancer, at least in animals. These agents are tiny packages of nucleic acids without life of their own that are capable of infecting cells and converting them to virus-producers. Virologists have linked tumor viruses with cancer in many species of birds and mammals, including primates. Since these experiments have not been performed on humans, there is no absolute proof that viruses cause human cancer. Nevertheless, with over 100 separate viruses identified as carcinogens in many species and tissues of animals, it is also probable that at least some cancers in humans are due to virus.

A reminder: As mentioned in the preface, each chapter in this text that discusses a major system of the body is followed by a glossary of **medical terminology.** Both normal and pathological conditions of the system are included in these glossaries. You should familiarize yourself with the terms since they will play an essential role in your medical vocabulary. Some of these disorders, as well as disorders discussed in the text, are referred to as local or systemic. A **local disease** is one that affects one part or a limited area of the body. A **systemic disease** affects either the entire body or several parts.

MEDICAL TERMINOLOGY

Autoimmune disease (*auto* = self)   A disease due to immunological action of one's own cells or antibodies on components of the body.

Biopsy   The removal and microscopic examination of tissue from the living body for diagnosis.

Carcinogen   A chemical or other environmental agent that produces cancer.

Carcinoma   A malignant tumor or cancer made up of epithelial cells.

Geriatrics   The branch of medicine devoted to the medical problems and care of elderly persons.

Gerontology   The study of old age.

Immunotherapy   Treatment by production of immunity.

Neoplasm   Any abnormal formation or growth, usually a malignant tumor.

Progeny   Offspring or descendants.

Senescence   The process of growing old.

STUDY OUTLINE

GENERALIZED ANIMAL CELL

1 A cell is the basic, living, structural and functional unit of the body.

2 A generalized cell is a composite that represents various cells of the body.

3 Cytology is the science concerned with the study of cells.

4 The principal parts of a cell are the plasma membrane, cytoplasm, organelles, and inclusions. Extracellular materials are manufactured by the cell and deposited outside the plasma membrane.

PLASMA (CELL) MEMBRANE

1 The plasma membrane, or cell membrane, surrounds the cell and separates it from other cells and the external environment.

2 The plasma membrane is composed of proteins and a bilayer of lipids. It is believed that the membrane contains pores.

3 The membrane's semipermeable nature restricts the passage of certain substances. Substances can pass through the membrane depending on their molecular size, lipid solubility, electrical charges, and the presence of carriers.

MOVEMENT OF MATERIALS ACROSS PLASMA MEMBRANES

1 Passive processes involve the kinetic energy of individual molecules.

2 Diffusion is the net movement of molecules or ions from an area of higher concentration to an area of lower concentration until an equilibrium is reached.

3 In facilitated diffusion, certain molecules like glucose combine with a carrier to become soluble in the lipid portion of the membrane.

4 Osmosis is the movement of water through a semipermeable membrane from an area of higher water concentration to an area of lower water concentration.

5 Osmotic pressure is the force under which a solvent moves from a solution of lower solute concentration to a solution of higher solute concentration when the solutions are separated by a semipermeable membrane.

6 Filtration is the movement of water and dissolved substances across a semipermeable membrane by pressure.

7 Dialysis is the separation of small molecules from large molecules by diffusion through a semipermeable membrane.

8 Active processes involve the use of ATP by the cell.

9 Active transport is the movement of ions across a cell membrane from lower to higher concentration. This process relies on the participation of carriers.

10 Phagocytosis is the ingestion of solid particles by pseudopodia. It is an important process used by white blood cells to destroy bacteria that enter the body.

11 Pinocytosis is the ingestion of a liquid by the plasma membrane. In this process, the liquid becomes surrounded by a vacuole.

MODIFIED PLASMA MEMBRANES

1 The membranes of certain cells are structured for specific functions.

2 Microvilli are microscopic fingerlike projections of the plasma membrane that increase the surface area for absorption.

3 Rod and cone cells of the eye contain sacs of light-sensitive pigments.

4 The myelin sheath of nerve cells protects, aids impulse conduction, and provides nutrition.

5 Stereocilia are long, slender, branching cells which line the epididymis.

CYTOPLASM

1 The cytoplasm is the living matter inside the cell that contains organelles and inclusions.

2 It is composed mostly of water plus proteins, carbohydrates, lipids, and inorganic substances. The chemicals in cytoplasm are either in solution or in a colloid (suspended) form.

3 Functionally, cytoplasm is the medium in which chemical reactions occur.

ORGANELLES

Organelles are specialized structures that carry on specific activities.

NUCLEUS

1 Usually the largest organelle, the nucleus controls cellular activities.

2 Cells without nuclei, such as mature red blood cells, do not grow or reproduce.

3 The parts of the nucleus include the nuclear membrane, nucleoplasm, nucleoli, and genetic material (DNA), comprising the chromosomes.

ENDOPLASMIC RETICULUM AND RIBOSOMES

1 The ER is a network of parallel membranes continuous with the plasma membrane and nuclear membrane.

2 It functions in chemical reactions, transportation, and storage.

3 Granular or rough ER has ribosomes attached to it. Agranular or smooth ER does not contain ribosomes. Ribosomes are small spherical bodies that serve as sites of protein synthesis.

GOLGI COMPLEX

1 This structure consists of four to eight flattened channels vertically stacked on each other.

2 In conjunction with the ER, the Golgi complex synthesizes glycoproteins and secretes lipids.

3 It is particularly prominent in secretory cells such as those in the pancreas or salivary glands.

MITOCHONDRIA

1 These structures consist of a smooth outer membrane and a folded inner membrane. The inner folds are called cristae.

2 The mitochondria are called "powerhouses of the cell" because ATP is produced in them.

LYSOSOMES

1 Lysosomes are spherical structures containing digestive enzymes.

2 They are found in large numbers in white blood cells, which carry on phagocytosis.

3 If the cell is injured, lysosomes release enzymes and digest the cell. Thus they are called "suicide packets."

MICROTUBULES

1 Microtubules consist of small protein tubules.

2 They form intracellular "skeletons" and intracellular channels and compose cilia, flagella, and spindle fibers.

CENTROSOME AND CENTRIOLES

1 The dense area of cytoplasm containing the centrioles is called a centrosome.

2 Centrioles are paired cylinders arranged at right angles to one another. They assume an important role in cell reproduction.

FLAGELLA AND CILIA

1 These cell projections have the same basic structure and are used in movement.

2 If projections are few and long, they are called flagella. If they are numerous and hairlike, they are called cilia.

3 The flagellum on a sperm cell moves the entire cell. The cilia on cells of the respiratory tract move foreign matter along the cell surfaces toward the throat for elimination.

CELL INCLUSIONS

1 These chemical substances are produced by cells. They may be stored, may participate in chemical reactions, and may have recognizable shapes.

2 Examples of cell inclusions are glycogen, hemoglobin, mucus, and melanin.

EXTRACELLULAR MATERIALS

1 These are all the substances that lie outside the cell membrane.

2 They provide support and a medium for the diffusion of nutrients and wastes.

3 Some, like hyaluronic acid, are amorphous or have no shape. Others, like collagen, are fibrous or threadlike.

CELL DIVISION

1 Cell division that results in the formation of new cells is called mitosis and cytokinesis. Nuclear division that results in the production of sperm and egg cells is termed meiosis.

2 Mitosis and cytokinesis replace and add body cells. Prior to mitosis and cytokinesis, the DNA molecules, or chromosomes, replicate themselves so the same chromosomal complement can be passed on to future generations of cells.

MITOSIS

1 Mitosis – division of the nucleus – consists of prophase, metaphase, anaphase, and telophase.

2 A cell carrying on every life process except division is said to be in interphase.

CYTOKINESIS

1 Cytokinesis – division of the cytoplasm – begins in late anaphase and terminates in telophase.

GENE ACTION: PROTEIN SYNTHESIS

1 Genes consist of about 1,000 pairs of nucleotides on a DNA molecule.

2 DNA directs the manufacture of proteins through RNA as follows:

(a) DNA unwinds and the unpaired nucleotides serve as a mold for the synthesis of RNA. This RNA is called messenger RNA.

(b) Messenger RNA travels to a ribosome that consists of ribosomal RNA. Another kind of RNA, called transfer RNA, attaches to amino acids and moves to the ribosome.

(c) At the ribosome, transfer RNA attaches the amino acids to the messenger RNA according to the sequence determined by the DNA in the gene.

(d) Proteins thus synthesized may serve as enzymes or structural proteins.

CELLS AND AGING

1 It has been said that "old age is not an illness; it is a continuation of life with decreasing capacities for adaptation."

2 Many theories of aging have been proposed, but none successfully answers all the experimental objections.

REVIEW QUESTIONS

1 Define a cell. What are the four principal portions of a cell? What is meant by a generalized cell?

2 Discuss the structure of the plasma membrane. What determines the permeability of the plasma membrane? How are plasma membranes modified for various functions?

3 What are the major differences between active processes and passive processes in moving substances across plasma membranes?

4 Define and give an example of each of the following: diffusion, facilitated diffusion, osmosis, filtration, active transport, phagocytosis, pinocytosis.

5 Compare the effect on red blood cells of an isotonic, hypertonic, and hypotonic solution. What is osmotic pressure?

6 Describe the structure and function of microvilli, rod sacs, myelin sheaths, and stereocilia as membrane modifications.

7 Discuss the chemical composition and physical nature of cytoplasm. What is its function?

8 What is an organelle? By means of a diagram, indicate the structure and describe the function of the following organelles: nucleus, endoplasmic reticulum, ribosomes, Golgi complex, mitochondria, lysosomes, microtubules, centrosome, centrioles, cilia, flagella.

9 Define a cell inclusion. Provide examples and indicate their functions.

10 What is an extracellular material? Give examples and the functions of each.

11 How does DNA replicate itself?

12 Discuss mitosis and cytokinesis with regard to stages. What are the characteristics of each stage, the relative duration, and the importance?

13 Summarize the steps involving gene action in protein synthesis.

14 List the five generalizations on aging that can be made based on current information.

Chapter 4
The Tissue
Level of
Organization

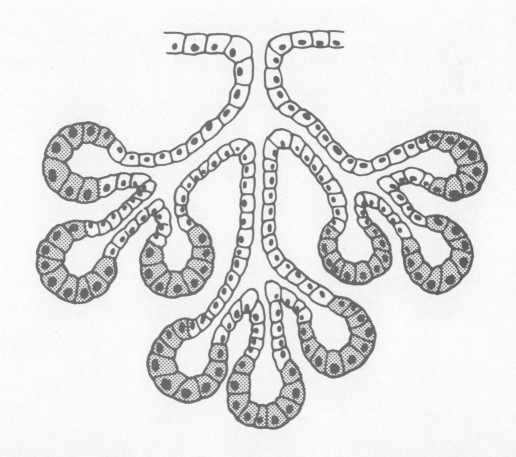

STUDENT OBJECTIVES

■ Define a tissue.

■ Classify the tissues of the body into four major types.

■ Describe the distinguishing characteristics of epithelial tissue.

■ Contrast the structural and functional differences of covering, lining, and glandular epithelium.

■ Compare the shape of cells and the layering arrangements of covering and lining epithelium.

■ List the structure, function, and location of simple, stratified, and pseudostratified epithelium.

■ Define a gland.

■ Distinguish between exocrine and endocrine glands.

■ Classify exocrine glands according to structural complexity and physiology.

■ Describe the distinguishing characteristics of connective tissue.

■ Contrast the structural and functional differences between embryonic and adult connective tissues.

■ Describe the ground substance, fibers, and cells that constitute connective tissue.

■ List the structure, function, and location of loose connective tissue, adipose tissue, and dense, elastic and reticular connective tissue.

■ List the structure, function, and location of the three types of cartilage.

■ Define an epithelial membrane.

■ List the location and function of mucous, serous, synovial, and cutaneous membranes.

■ List and describe the symptoms of tissue inflammation.

■ Outline the stages of the inflammatory response.

■ Describe the conditions necessary for tissue repair.

■ Describe the importance of nutrition, adequate circulation, and age to tissue repair.

Cells are highly organized units, but they do not function in isolation. They work together in a group of similar cells called a tissue.

TYPES OF TISSUES

A **tissue** is a group of similar cells and their intercellular substance functioning together to perform a specialized activity. Certain tissues function in moving body parts. Others move food through body organs. Some tissues protect and support the body. Others produce chemicals such as enzymes and hormones. Depending on their function and structure, the various tissues of the body are classified into four principal types:

1 **Epithelial tissue,** which covers body surfaces, lines body cavities, and forms glands.

2 **Connective tissue,** which protects and supports the body and its organs and binds organs together.

3 **Muscular tissue,** which is responsible for movement.

4 **Nervous tissue,** which initiates and transmits nerve impulses that coordinate body activities.

EPITHELIAL TISSUE

The tissues falling into this main category perform many activities in the body, ranging from protection to secretion. **Epithelial tissue,** or more simply **epithelium,** may be divided into two subtypes: (1) *covering and lining epithelium* and (2) *glandular epithelium.* (See Exhibit 4-1.) Covering and lining epithelium forms the outer covering of external body surfaces and the outer covering of some internal organs. It lines the body cavities and the interiors of the respiratory and digestive tracts, blood vessels, and ducts. It comprises, along with nervous tissue, the parts of the sense organs that are sensitive to stimuli such as light and sound. Glandular epithelium constitutes the secreting portion of glands.

Both types of epithelium consist largely or entirely of closely packed cells with little or no intercellular material between adjacent cells. Such materials are collectively called the matrix. In addition, the epithelial cells are arranged in continuous sheets that may be either single or multilayered. Nerves may run through these sheets, but blood vessels do not. Thus they are *avascular.* The vessels that supply nutrients and remove wastes are located in underlying connective tissue. Epithelium overlies and adheres firmly to the connective tissue, which holds the epithelium in position and prevents it from being torn. The surface of attachment between the epithelium and connective tissue is a thin layer of modified epithelial and connective tissue called the **basement membrane.** Since all epithelium is subjected to a certain degree of wear, tear, and injury, its cells must divide and produce new cells to replace those that are destroyed. These general characteristics are found in both types of epithelial tissue.

Covering and Lining Epithelium

ARRANGEMENT OF LAYERS

Covering and lining epithelium is arranged in several different ways related to location and function. If the epithelium is specialized for absorption or filtration and is in an area that has minimal wear and tear, the cells of the tissue are arranged in a single layer. Such an arrangement is called **simple epithelium.** If the epithelium is not specialized for absorption or filtration and is found in an area with a high degree of wear and tear, then the cells are stacked in several layers. This tissue is referred to as **stratified epithelium.** A third, less common, arrangement of epithelium is called **pseudostratified.** Like simple epithelium, pseudostratified epithelium has only one layer of cells. However, some of the cells do not reach the surface—an arrangement that gives the tissue a multilayered, or stratified, appearance. The pseudostratified cells that do reach the surface either secrete mucus or contain cilia that move mucus and foreign particles for eventual elimination from the body.

CELL SHAPES

In addition to classifying covering epithelium according to the number of its layers, it may also be categorized by cell shape. The cells may be flat, cubelike, or columnar or may resemble a cross between shapes. **Squamous** cells are flattened, scalelike, and attached to each other like a mosaic. **Cuboidal** cells are usually cube-shaped in cross section. They sometimes appear as hexagons. **Columnar** cells are tall and cylindrical, appearing as rectangles set on end. **Transitional** cells often have a combination of shapes and are found where there is a great degree of distention or expansion in the body. Transitional cells on the bottom layer of an epithelial tissue may range in shape from cuboidal to columnar. In the intermediate layer, they may be cuboidal or polyhedral. Transitional cells in the superficial layer may range from cuboidal to squamous, depending on how much they are pulled out of shape during certain body functions.

CLASSIFICATION

Considering layers and cell type in combination, we may classify covering and lining epithelium as follows:

Simple
1 Squamous
2 Cuboidal
3 Columnar
Stratified
1 Squamous
2 Cuboidal
3 Columnar
4 Transitional
Pseudostratified

SIMPLE EPITHELIUM

■ *Simple Squamous Epithelium*   This type of simple epithelium consists of a single layer of flat, scalelike cells. The surface of this epithelium resembles a tiled floor. The nucleus of each cell is centrally located and oval or spherical. Since simple squamous epithelium has only one layer of cells, it is highly adapted to diffusion, osmosis, and filtration. Thus it lines the air sacs of the lungs, where oxygen is exchanged with carbon dioxide. It is present in the part of the kidney that filters the blood. It is also found in delicate structures such as the crystalline lens of the eye and the lining of the eardrum. Simple squamous epithelium is found in body parts that have little wear or tear. A tissue similar to simple squamous epithelium is endothelium. **Endothelium** lines the heart, the blood vessels, and the lymph vessels and forms the walls of capillaries. The term **mesothelium** is applied to another simple squamous epithelium-like tissue that lines the thoracic, abdominal, and pelvic cavities and covers the viscera within them.

■ *Simple Cuboidal Epithelium*   From above, the cells of simple cuboidal epithelium appear as closely fitted polygons. The cuboidal nature of the cells is obvious only when the tissue is sectioned at right angles. Like simple squamous epithelium, these cells possess a central nucleus. Simple cuboidal epithelium covers the surfaces of the ovaries and lines the inner surfaces of the cornea. In the kidneys, where it forms the kidney tubules and contains microvilli, it functions in water reabsorption. It also lines the smaller ducts of some glands and the secreting units of glands, such as the thyroid. This tissue performs the functions of secretion and absorption. *Secretion,* usually a function of epithelium, is the production and release by cells of a fluid that may contain a variety of substances such as mucus, perspiration, or enzymes. *Absorption* is the intake of fluids or other substances by cells of the skin or mucous membranes.

■ *Simple Columnar Epithelium*   The surface view of simple columnar epithelium is similar to that of simple cuboidal tissue. When sectioned at right angles to the surface, however, the cells appear somewhat rectangular. The nuclei are located near the bases of the cells. Simple columnar epithelium is modified in several ways, depending on location and function. Simple columnar epithelium lines the stomach, the small and large intestines, the digestive glands, and the gallbladder. In such sites, the cells protect the underlying tissues. Many of them are also modified to aid in food-related activities. In the small intestine especially, the plasma membranes of the cells are folded into microvilli. (See Figure 3–7a.) The microvilli arrangement increases the surface area of the plasma membrane and thereby allows digested nutrients and fluids to diffuse into the body at a faster rate. Interspersed among the typical columnar cells of the intestine are other modified columnar cells called *goblet cells.* These cells, which secrete mucus, are so named because the mucus accumulates in the upper half of the cell, causing the area to bulge out. The whole cell resembles a goblet or wine glass. The secreted mucus serves as a lubricant between the food and the walls of the digestive tract. A third modification of columnar epithelium is found in cells with hairlike processes called **cilia.** In some portions of the upper respiratory tract, ciliated columnar cells are interspersed with goblet cells. Mucus secreted by the goblet cells forms a film over the respiratory surface. This film traps foreign particles that are inhaled. The cilia, which wave in unison, move the mucus and foreign particles toward the throat, where it can be swallowed or eliminated. Thus air is filtered by this process before entering the lungs. Ciliated columnar epithelium, combined with goblet cells, is also found in the uterus and uterine tubes of the female reproductive system.

STRATIFIED EPITHELIUM

In contrast to simple epithelium, stratified epithelium consists of at least two layers of cells. Thus it is durable and can protect underlying tissues from the external environment and from wear and tear. Some stratified epithelium cells also produce secretions. The name of the specific kind of stratified epithelium depends on the shape of the surface cells.

■ *Stratified Squamous Epithelium*   In the layers of this type of epithelium, the superficial cells are flat whereas the deep cells vary in shape from cuboidal to columnar. The basal, or bottom, cells are continually multiplying by cell division. As new cells grow, they compress the cells on the surface and push them outward. According to this growth pattern, basal cells continually shift upward and outward. As they move farther from the deep layer and their blood supply, they become dehydrated, shrink, and grow harder. At the surface, the cells are rubbed off. New cells continually emerge, are sloughed off, and replaced.

One form of stratified squamous epithelium is called *nonkeratinized stratified squamous epithelium.* This tissue is found on wet surfaces that are subjected to considerable wear and tear and do not perform the function of absorption – the linings of the mouth, the esophagus, and the vagina. Another form of stratified squamous epithelium is called *keratinized stratified squamous epithelium.* The surface cells of this type are modified into a tough layer of material containing keratin. **Keratin** is a protein that is waterproof, resistant to friction, and impervious to bacterial invasion. The outer layer of skin consists of keratinized tissue.

■ *Stratified Cuboidal Epithelium* This rare type of epithelium is found in the ducts of the sweat glands of adults. It sometimes consists of more than two layers of cells. Its function is mainly protective.

■ *Stratified Columnar Epithelium* Like stratified cuboidal epithelium, this type of tissue is also uncommon in the body. Usually the basal layer or layers consist of shortened, irregularly polyhedral cells. Only the superficial cells are columnar in form. This kind of epithelium lines part of the male urethra and some larger excretory ducts such as lactiferous ducts in the mammary glands. It functions in protection and secretion.

■ *Transitional Epithelium* This kind of epithelium is very much like nonkeratinized stratified squamous epithelium. The distinction is that the outer layer of cells in transitional epithelium tend to be large and rounded rather than flat. This feature allows the tissue to be stretched without the outer cells breaking apart from one another. When stretched, they are drawn out into squamouslike cells. Because of this arrangement, transitional epithelium lines hollow structures that are subjected to expansion from within, such as the urinary bladder. Its function is to help prevent a rupture of the organ.

PSEUDOSTRATIFIED EPITHELIUM

The third category of covering and lining epithelium is called pseudostratified epithelium. The nuclei of the cells in this kind of tissue are at varying depths. Even though all the cells are attached to the basement membrane in a single layer, some do not reach the surface. This feature gives the impression of a multilayered tissue, the reason for the designation *pseudo*stratified epithelium. It lines the larger excretory ducts of many glands and parts of the male urethra. Pseudostratified epithelium may be ciliated and may contain goblet cells. In this form, it lines most of the upper respiratory passages and certain ducts of the male reproductive system.

Glandular Epithelium

The function of glandular epithelium is secretion, accomplished by glandular cells that lie in clusters deep to the covering and lining epithelium. A **gland** may consist of one cell or a group of highly specialized epithelial cells that secrete substances into ducts or into the blood. The production of such substances always requires active work by the glandular cells and results in an expenditure of energy.

On the basis of this distinction, all glands of the body are classified as exocrine or endocrine (Exhibit 4-1). **Exocrine glands** secrete their products into ducts or tubes that empty at the surface of the covering and lining epithelium. The product of an exocrine gland may be released at the skin surface or into the lumen of a hollow organ. The secretions of exocrine glands include enzymes, oil, and sweat. Examples of exocrine glands are sweat glands, which eliminate perspiration to cool the skin, and salivary glands, which secrete a digestive enzyme. **Endocrine glands,** by contrast, are ductless and, consequently, must secrete their products directly into the blood. The secretions of endocrine glands are always hormones. The pituitary, thyroid, and adrenal glands are endocrine glands.

STRUCTURAL CLASSIFICATION OF EXOCRINE GLANDS

Exocrine glands are classified into two structural types: unicellular and multicellular. **Unicellular glands** are single-celled. A good example of a unicellular gland is a goblet cell (Exhibit 4-1). Goblet cells are found in the epithelial lining of the digestive, respiratory, urinary, and reproductive systems. They produce mucus to lubricate the free surfaces of these membranes. **Multicellular glands** are many-celled glands and occur in several different forms (Figure 4-1). If the secretory portions of the gland are tubular, they are referred to as *tubular glands.* If flasklike, they are called *acinar glands.* If the gland contains both tubular and flasklike secretory portions, it is called a *tubuloacinar* gland. Further, if the duct of the gland does not branch, it is referred to as a *simple gland;* if the duct does branch, it is called a *compound gland.* By combining the shape of the secretory portion with the degree of branching of the duct, we arrive at the following structural classification for exocrine glands:

Unicellular One-celled gland that secretes mucus. Example: Goblet cell of the digestive system.

Exhibit 4—1   EPITHELIAL TISSUES

Covering and Lining Epithelium

Simple Squamous (130×)

DESCRIPTION: Single layer of flat, scalelike cells; large, centrally located nucleus.

LOCATION: Lines air sacs of lungs, glomerular capsule of kidneys, crystalline lens of eyes, and eardrum. Called endothelium when it lines heart, blood, and lymphatic vessels, and forms capillaries. Called mesothelium when it lines the ventral body cavity and covers viscera as part of a serous membrane.

FUNCTION: Filtration, absorption, and secretion in serous membranes.

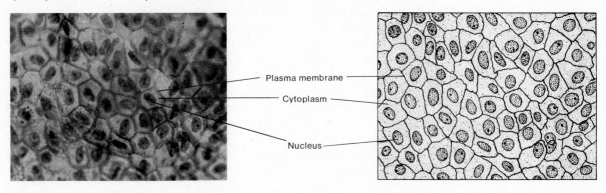

Plasma membrane
Cytoplasm
Nucleus

Simple Cuboidal (130×)

DESCRIPTION: Single layer of cube-shaped cells; centrally located nucleus.

LOCATION: Covers surface of ovary; lines inner surface of cornea and lens of eye, kidney tubules, and smaller ducts of many glands.

FUNCTION: Secretion and absorption.

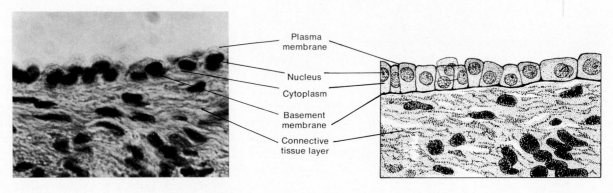

Plasma membrane
Nucleus
Cytoplasm
Basement membrane
Connective tissue layer

Multicellular   Many-celled glands.
  Simple   Single, nonbranched duct.
  1 **Tubular**   The secretory portion is straight and tubular. Example: Crypts of Lieberkuhn of intestines.
  2 **Branched tubular**   The secretory portion is branched and tubular. Examples: Gastric and uterine glands.
  3 **Coiled tubular**   The secretory portion is coiled. Examples: Sudoriferous (sweat) glands.

  4 **Acinar**   The secretory portion is flasklike. Example: Seminal vesicle glands.
  5 **Branched acinar**   The secretory portion is branched and flasklike. Example: Sebaceous (oil) glands.

  Compound   Branched duct.
  1 **Tubular**   The secretory portion is tubular. Examples: Bulbourethral glands, testes, and liver.
  2 **Acinar**   The secretory portion is flasklike. Exam-

Exhibit 4—1 (cont.)

Simple Columnar (Nonciliated) (130×)

DESCRIPTION: Single layer of nonciliated rectangular cells; contains goblet cells; nuclei at bases of cells.

LOCATION: Lines stomach, small and large intestines, digestive glands, and gallbladder.

FUNCTION: Secretion and absorption.

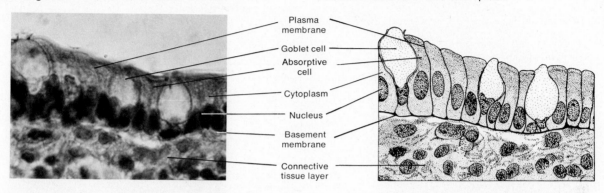

Simple Columnar (Ciliated) (130×)

DESCRIPTION: Single layer of ciliated rectangular cells; contains goblet cells; nuclei at bases of cells.

LOCATION: Lines some portions of upper respiratory tract, uterine (fallopian) tubes, and uterus.

FUNCTION: Moves mucus by ciliary action.

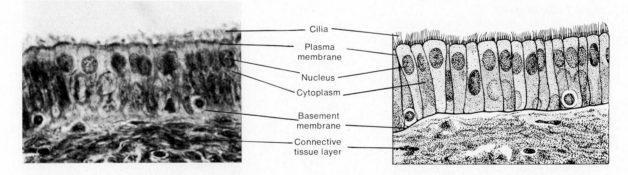

ples: Mammary and salivary glands (sublingual and submandibular).

3 **Tubuloacinar** The secretory portion is both tubular and flasklike. Examples: Salivary glands (parotid) and pancreas.

FUNCTIONAL CLASSIFICATION OF EXOCRINE GLANDS

The functional classification of exocrine glands is based on how the gland releases its secretion. The three recognized categories are holocrine, merocrine, and apocrine glands. **Holocrine glands** accumulate a secretory product in their cytoplasm. The cell then dies and is discharged with its contents as the glandular secretion (Figure 4-2a). The discharged cell is replaced by a new cell. One example of a holocrine gland is a sebaceous gland of the skin. **Merocrine glands** are those that produce secretions that do not contain any portion of the secretory cell (Figure 4-2b). The secretion is simply formed and discharged by the cell. An example of a merocrine gland is the pancreas. Another is the salivary glands. **Apocrine glands** are those whose secretory product accumulates at the apical (outer) margin of the secreting cell. The apical region of the cell and its secretory contents pinch off from the rest of the cell to form the secretion (Figure 4-2c). The remaining part of the cell repairs itself and repeats the process. An example of an apocrine gland is the mammary gland.

Exhibit 4—1   (cont.)

Stratified Squamous (65×)

DESCRIPTION: Several layers of cells; deeper layers are cuboidal to columnar; superficial layers are flat and scale-like; basal cells replace surface cells as they are lost.

LOCATION: Nonkeratinizing variety lines wet surfaces such as mouth, esophagus, part of epiglottis, and vagina. Keratinizing variety forms outer layer of skin.

FUNCTION: Protection.

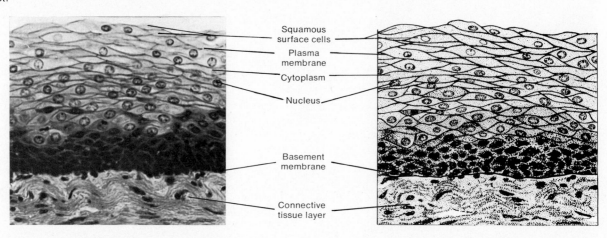

Squamous surface cells
Plasma membrane
Cytoplasm
Nucleus
Basement membrane
Connective tissue layer

Stratified Cuboidal (80×)

DESCRIPTION: Two or more layers of cube-shaped cells.

LOCATION: Ducts of adult sweat glands.

FUNCTION: Protection.

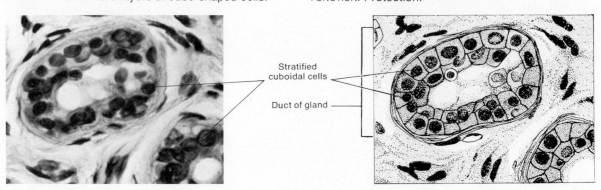

Stratified cuboidal cells
Duct of gland

CONNECTIVE TISSUE

The most abundant tissue in the body is **connective tissue.** This binding and supporting tissue usually has a rich blood supply. Thus it is *vascular.* The cells are widely scattered, rather than closely packed, and there is considerable intercellular material comprising the matrix. In contrast to epithelium, connective tissues do not occur on free surfaces, such as the surfaces of a body cavity or the external surface of the body. The general functions of connective tissues are protection, support, and the binding together of various organs.

The intercellular substance in a connective tissue largely determines the tissue's qualities. These substances are nonliving and may consist of fluid, semifluid, or mucoid (mucuslike) material. In cartilage, the intercellular material is firm. In bone, it is quite rigid. The living parts of connective tissue are the cells, which produce the intercellular substances. The cells may also

Exhibit 4—1 (cont.)

Stratified Columnar (130×)

DESCRIPTION: Several layers of polyhedral cells; only superficial layer is columnar.

LOCATION: Lines part of male urethra and some larger excretory ducts.

FUNCTION: Protection and secretion.

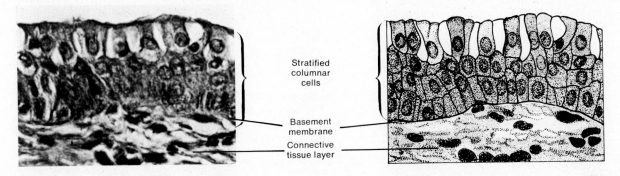

Stratified
columnar
cells

Basement
membrane

Connective
tissue layer

Stratified Transitional (110×)

DESCRIPTION: Resembles stratified squamous nonkeratinizing tissue, except superficial cells are larger and more rounded.

LOCATION: Lines urinary tract.

FUNCTION: Permits distention.

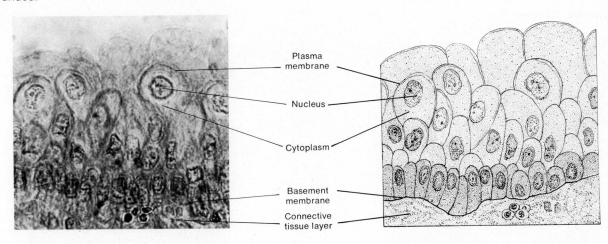

Plasma
membrane

Nucleus

Cytoplasm

Basement
membrane

Connective
tissue layer

store fat, ingest bacteria and cell debris, form anticoagulants, or give rise to antibodies that protect against disease.

Loose and Dense Connective Tissue

Before classifying and studying connective tissue, it will be helpful to distinguish between loose and dense connective tissues. **Loose connective tissue** refers to the arrangement of intercellular substance. That is, the fibers in the intercellular substance are neither abundant nor arranged to prevent stretching. In addition, the intercellular substance is soft or jellylike in consistency. By contrast, **dense connective tissue** is characterized by the close packing of fibers and less intercellular substance. In areas of the body where tensions are exerted in various directions, the fiber bundles are interwoven and without regular orientation. Such a dense connective tissue is

Exhibit 4—1   (cont.)

Pseudostratified (170×)

DESCRIPTION: Not a true stratified tissue; nuclei of cells are present at different levels; some cells do not reach surface, but all sit on the basement membrane.

LOCATION: Lines larger excretory ducts of many large glands and male urethra; ciliated variety with goblet cells lines most of the upper respiratory passageways and some ducts of male reproductive system.

FUNCTION: Secretion and movement of mucus by ciliary action.

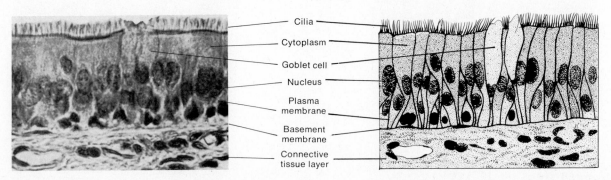

Cilia
Cytoplasm
Goblet cell
Nucleus
Plasma membrane
Basement membrane
Connective tissue layer

Glandular Epithelium

Exocrine Gland (110×)

DESCRIPTION: Secretes products into ducts.

LOCATION: Sweat, oil, and wax glands of the skin; digestive glands such as salivary glands which secrete into mouth cavity.

FUNCTION: Produce perspiration, oil, wax, and digestive enzymes.

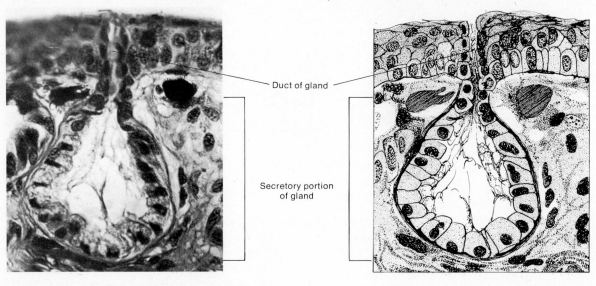

Duct of gland

Secretory portion of gland

Exhibit 4—1   (cont.)

Endocrine Gland (45×)

LOCATION: Pituitary gland at base of brain; thyroid gland near larynx; adrenal (suprarenal) glands above kidneys.

DESCRIPTION: Secretes hormones directly into blood.

FUNCTION: Produce hormones that regulate various body activities.

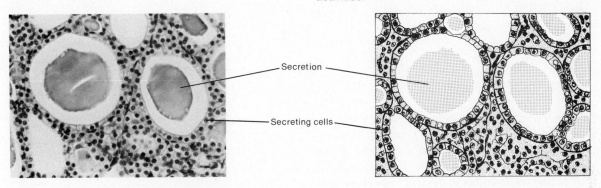

Secretion

Secreting cells

Photomicrographs courtesy of Victor B. Eichler, Wichita State University, except for pseudostratified tissue, which is courtesy of Donald I. Patt, from Comparative Vertebrate Histology, *by Donald I. Patt and Gail R. Patt, Harper & Row, Publishers, Inc., New York, 1969.*

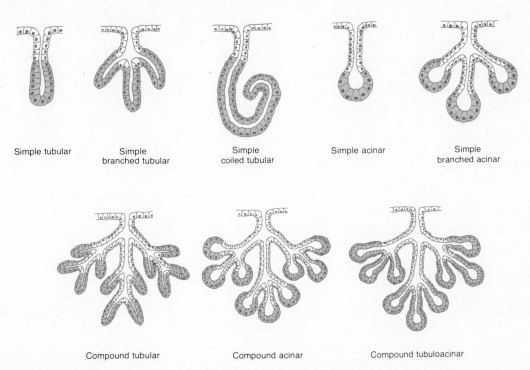

Simple tubular

Simple branched tubular

Simple coiled tubular

Simple acinar

Simple branched acinar

Compound tubular

Compound acinar

Compound tubuloacinar

Figure 4-1   Structural types of multicellular exocrine glands. The secretory portions of the glands are indicated in color. The uncolored areas represent the ducts of the glands.

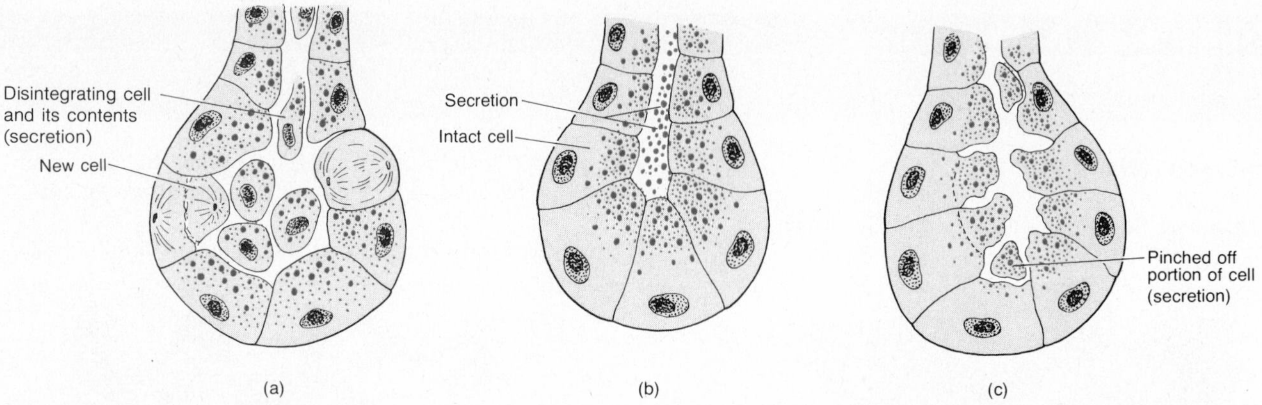

Figure 4-2 Functional classification of multicellular exocrine glands. (a) Holocrine gland. (b) Merocrine gland. (c) Apocrine gland.

referred to as *irregularly arranged.* Dense, irregularly arranged connective tissue occurs in sheets and forms most fasciae, the dermis of the skin, the fibrous capsules of some organs (testes, liver, lymph nodes), the periosteum of bone, and the perichondrium of cartilage. In other areas of the body, dense connective tissue is adapted for tension in one direction and the fibers have an orderly, parallel arrangement. Such a connective tissue is known as *regularly arranged.* Dense, regularly arranged connective tissue comprises tendons, ligaments, and aponeuroses.

Loose connective tissue contains each of the three types of fibers – collagenous, elastic, and reticular. Dense, irregularly arranged connective tissue also contains each of these three types, but collagenous fibers predominate. The dense, regularly arranged connective tissue of many ligaments and aponeuroses is composed of collagenous fibers. Such ligaments are known as collagenous ligaments. However, some ligaments are composed of elastic fibers and are known as yellow elastic ligaments. Examples of these include the ligamenta flava of the vertebrae, the suspensory ligament of the penis, and the true vocal cords.

Classification

Connective tissue may be classified in several ways. We will classify them as follows:

Embryonic connective tissue
Adult connective tissue
   Connective tissue proper
    1 Loose connective (areolar) tissue
    2 Adipose tissue
    3 Dense connective (collagenous) tissue
    4 Elastic connective tissue
    5 Reticular connective tissue

   Cartilage
    1 Hyaline cartilage
    2 Fibrocartilage
    3 Elastic cartilage
   Bone (osseous) tissue
   Vascular (blood) tissue

Embryonic Connective Tissue

Connective tissue that is present primarily in the embryo or fetus is called **embryonic connective tissue** (Exhibit 4-2). Whereas the term *embryo* refers to a developing human from fertilization through the first 2 months of pregnancy, a *fetus* is regarded as a developing human from the third month of pregnancy to birth. One example of embryonic connective tissue found almost exclusively in the embryo is *mesenchyme* – the tissue from which all other connective tissues eventually arise. Mesenchyme may be observed beneath the skin and along the developing bones of the embryo. Some mesenchymal cells are scattered irregularly throughout adult connective tissue, most frequently around blood vessels. Here mesenchymal cells differentiate into fibroblasts that assist in wound healing. Another kind of embryonic connective tissue is *mucous connective tissue,* found only in the fetus. This tissue, also called *Wharton's jelly,* is located in the umbilical cord of the fetus where it supports the wall of the cord.

Adult Connective Tissue

Adult connective tissue is connective tissue that exists in the newborn and that does not change after birth. It is subdivided into several kinds.

CONNECTIVE TISSUE PROPER

Connective tissue that has a more or less fluid intercellular material and a fibroblast as the typical cell is

Exhibit 4–2   CONNECTIVE TISSUES

Embryonal

Mesenchymal (130×)

DESCRIPTION: Consists of highly branched mesenchymal cells embedded in a fluid substance.

LOCATION: Under skin and along developing bones of embryo; some mesenchymal cells are found in adult connective tissue, especially along blood vessels.

FUNCTION: Forms all other kinds of connective tissue.

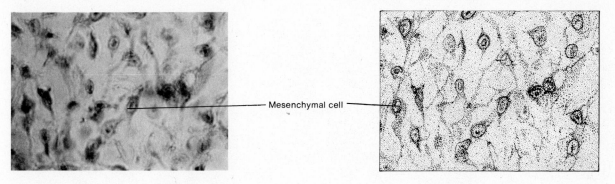

Mesenchymal cell

Mucous (130×)

DESCRIPTION: Consists of flattened or spindle-shaped cells embedded in a mucuslike substance containing fine collagenous fibers.

LOCATION: Umbilical cord of fetus.

FUNCTION: Support.

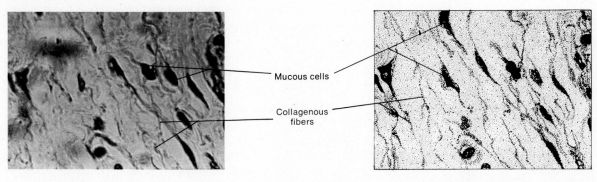

Mucous cells

Collagenous fibers

---

termed **connective tissue proper.** Five examples of such tissues may be distinguished.

■ *Loose Connective (Areolar) Tissue* This type of tissue is one of the most widely distributed connective tissues in the body. Structurally, it consists of fibers and several kinds of cells embedded in a semifluid intercellular substance. This intercellular substance consists of a viscous material called **hyaluronic acid.** The thick consistency of this material may impede the movement of substances through the tissue. However, if an enzyme called hyaluronidase is injected into the tissue, hyaluronic acid changes to a water consistency.

This feature is of clinical importance because the reduced viscosity hastens the absorption and diffusion of injected drugs and fluids through the tissue and thus can lessen tension and pain.

The three types of fibers embedded between the cells of loose connective tissue are collagenous fibers, elastic fibers, and reticular fibers. **Collagenous,** or **white, fibers** are very tough and resistant to a pulling force, yet are somewhat flexible because they are usually wavy. These fibers often occur in bundles. They are composed of many minute fibers called fibrils lying parallel to one another. The bundle arrangement affords a great deal of strength. Chemically, collagenous fibers

Exhibit 4—2   (cont.)

Adult

Loose or Areolar (65×)

DESCRIPTION: Consists of fibers (collagenous and elastic) and several kinds of cells (fibroblasts, macrophages, plasma cells, and mast cells) embedded in a semifluid ground substance.

LOCATION: Subcutaneous layer of skin, mucous membranes, blood vessels, nerves, and body organs.

FUNCTION: Strength, elasticity, support, phagocytosis, produces antibodies, and produces an anticoagulant.

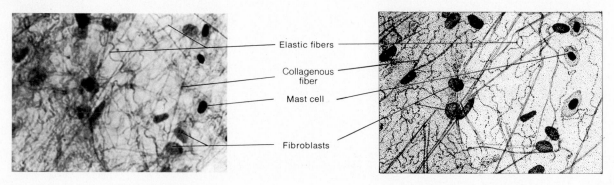

Elastic fibers
Collagenous fiber
Mast cell
Fibroblasts

Adipose (90×)

DESCRIPTION: Contains fibroblasts specialized for fat storage; cells have a "signet-ring shape."

LOCATION: Subcutaneous layer of skin, around heart and kidneys, marrow of long bones, padding around joints.

FUNCTION: Reduces heat loss through skin, serves as a food reserve, supports, and protects.

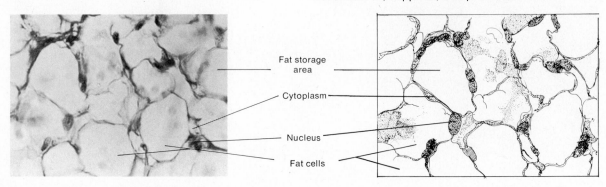

Fat storage area
Cytoplasm
Nucleus
Fat cells

consist of the protein collagen. **Elastic, or yellow, fibers,** by contrast, are smaller than collagenous fibers and freely branch and rejoin one another. Elastic fibers consist of a protein called elastin. These fibers also provide strength and have great elasticity, up to 50 percent of their length. **Reticular fibers** also consist of collagen, but they are very thin fibers that branch extensively. Some authorities believe that reticular fibers are immature collagenous fibers. Like collagenous fibers, reticular fibers provide support and form the *stroma* (framework) of many soft organs.

The cells in loose connective tissue are numerous and varied. Most are **fibroblasts** – large, flat cells with branching processes. If the tissue is injured, the fibroblasts are believed to form collagenous fibers. Some evidence suggests that fibroblasts also form elastic fibers and the viscous ground substance. Mature fibroblasts are referred to as **fibrocytes.** The basic distinction between the two is that fibroblasts (or *"blasts"* of any form of cell)

Exhibit 4—2 (cont.)

Dense or Collagenous (65×)

DESCRIPTION: Collagenous, or white, fibers predominate and are arranged in bundles; fibroblasts are in rows between bundles.

LOCATION: Forms tendons, ligaments, aponeuroses, membranes around various organs, and fasciae.

FUNCTION: Provides strong attachment between various structures.

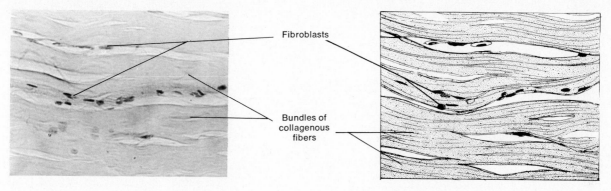

Fibroblasts

Bundles of collagenous fibers

Elastic (65×)

DESCRIPTION: Elastic, or yellow, fibers predominate and branch freely; fibroblasts present in spaces between fibers.

LOCATION: Lung tissue, cartilage of larynx, walls of arteries, trachea, bronchial tubes, true vocal cords, and ligamenta flava of vertebrae.

FUNCTION: Allows stretching of various organs.

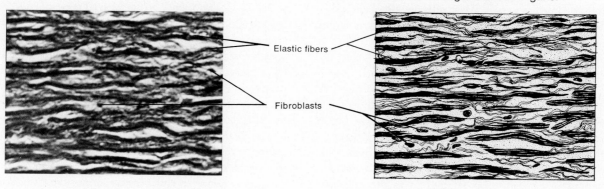

Elastic fibers

Fibroblasts

are involved in the formation of immature tissue or repair of mature tissue, and fibrocytes (or *"cytes"* of any form of cell) are involved in maintaining health of mature tissue. Other cells found in loose connective tissue are called **macrophages, or histiocytes.** They are irregular in form with short branching processes. These cells are derived from undifferentiated mesenchymal cells called hemocytoblasts and are capable of engulfing bacteria and cellular debris by the process of phagocytosis. Thus they provide a vital defense for the body. A third kind of cell

in loose connective tissue is the **plasma cell.** These cells are small and either round or irregular. Plasma cells probably develop from a type of white blood cell called a lymphocyte. They give rise to antibodies and, accordingly, provide a defensive mechanism through immunity. Plasma cells are found in many places of the body, but most are found in connective tissue, especially that of the digestive tract and the mammary glands. Another cell in loose connective tissue is the **mast cell.** It may develop from another type of white blood cell called a

Exhibit 4—2   (cont.)

Reticular (65×)

DESCRIPTION: Consists of a network of interlacing fibers; cells are thin and flat and wrapped around fibers.

LOCATION: Liver, spleen, and lymph nodes.

FUNCTION: Forms stroma, or framework, of organs; binds together smooth muscle tissue cells.

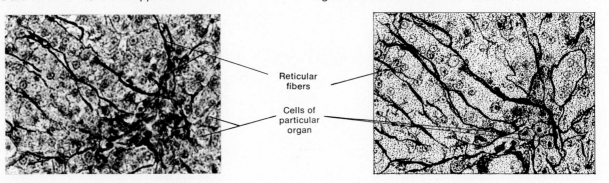

Reticular fibers

Cells of particular organ

---

basophil. The mast cell is somewhat larger than a basophil and is found in abundance along blood vessels. It forms heparin, an anticoagulant that prevents blood from clotting the vessels. Mast cells are also believed to produce histamine, a chemical that dilates, or enlarges, small blood vessels. Other cells in loose connective tissue include **melanocytes,** or pigment cells, fat cells, and white blood cells.

Loose connective tissue is continuous throughout the body. It is present in all mucous membranes and around all blood vessels and nerves. And it occurs around body organs. Combined with adipose tissue, it forms the *subcutaneous layer*—the layer of tissue that attaches the skin to underlying tissues and organs. The subcutaneous layer is also referred to as the *superficial fascia.*

■ *Adipose Tissue*   This kind of tissue is basically a form of loose connective tissue in which the fibroblasts are specialized for fat storage. Adipose cells have the shape of a "signet ring" because the cytoplasm and nucleus are pushed to the edge of the cell by a large droplet of fat. Adipose tissue is found wherever loose connective tissue is located. Specifically, it is in the subcutaneous layer below the skin, around the kidneys, at the base and on the surface of the heart, in the marrow of long bones, as a padding around joints, and behind the eyeball in the orbit. Adipose tissue is a poor conductor of heat and therefore reduces heat loss

through the skin. It is also a major food reserve and generally supports and protects various organs.

■ *Dense Connective (Collagenous) Tissue*   This tissue has a predominance of collagenous fibers, or white fibers, arranged in bundles. The cells found in collagenous connective tissue are fibroblasts, which are placed in rows between the bundles. The tissue is silvery white, tough, yet somewhat pliable. Because of the great strength of this dense, regularly arranged tissue, it is the principal component of *tendons,* which attach muscles to bones; many *ligaments,* which hold bones together at joints; and *aponeuroses,* which are flat bands connecting one muscle with another or with a bone. Dense, irregularly arranged collagenous connective tissue forms *membrane capsules* around the kidneys, heart, testes, liver, and lymph nodes. It also forms the *deep fasciae*—sheets of connective tissue wrapped around muscles to hold them in place.

■ *Elastic Connective Tissue*   Unlike collagenous connective tissue, elastic connective tissue has a predominance of freely branching elastic fibers. These fibers give the tissue a yellowish color. Fibroblasts are present only in the spaces between the fibers. Elastic connective tissue can be stretched and will snap back into shape. It is a component of the cartilages of the larynx, the walls of elastic arteries, the trachea, bronchial tubes to the lungs, and the lungs themselves. Elastic connective tissue

Exhibit 4—2 (cont.)

Hyaline Cartilage (30×)

DESCRIPTION: Also called gristle; appears as a bluish white, glossy mass; contains numerous chondrocytes; is the most abundant type of cartilage.

LOCATION: Ends of long bones, ends of ribs, nose, parts of larynx, trachea, bronchi, bronchial tubes, and embryonic skeleton.

FUNCTION: Provides movement at joints, flexibility, and support.

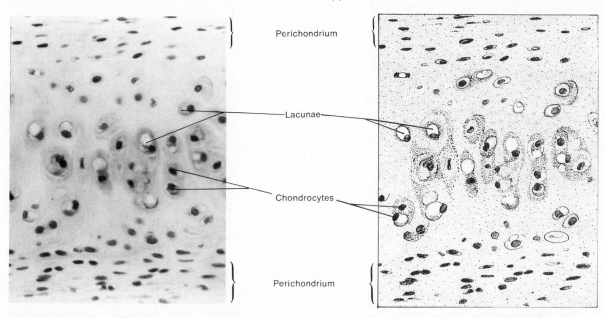

Perichondrium

Lacunae

Chondrocytes

Perichondrium

---

provides stretch and strength, allowing structures to perform their functions efficiently. Yellow elastic ligaments, composed mostly of elastic fibers, comprise the ligamenta flava of the vertebrae and the true vocal cords.

■ *Reticular Connective Tissue*   This kind of connective tissue consists of interlacing reticular fibers. It helps to form the framework, or body, of many organs, including the liver, spleen, and lymph nodes. Reticular connective tissue also helps to bind together the cells of smooth muscle tissue. It is especially adapted to providing strength and support.

CARTILAGE

One type of connective tissue is capable of enduring considerably more stress than the tissues just discussed. **Cartilage** consists of a dense network of collagenous

fibers and elastic fibers firmly embedded in a gel-like substance. The cells of mature cartilage, called *chondrocytes,* occur singly or in groups within spaces called *lacunae* in the intercellular substance. The surface of cartilage is surrounded by irregularly arranged dense connective tissue called the *perichondrium* (*chondro* = cartilage; *peri* = around). Three kinds of cartilage are recognized: hyaline cartilage, fibrocartilage, and elastic cartilage (Exhibit 4-2).

■ *Hyaline Cartilage*   This cartilage, also called gristle, appears as a bluish white, glossy, homogeneous mass. The collagenous fibers, although present, are not visible, and the prominent chondrocytes are found in lacunae. Hyaline cartilage is the most abundant kind of cartilage in the body. It is found at joints over the ends of long bones (where it is called *articular cartilage*) and forms the *costal cartilages* at the ventral ends of the ribs. Hyaline cartilage also helps to form the nose, larynx, trachea,

Exhibit 4—2   (cont.)

Fibrocartilage (65×)

DESCRIPTION: Consists of chondrocytes scattered among bundles of collagenous fibers.

LOCATION: Symphysis pubis and intervertebral discs.

FUNCTION: Support and fusion.

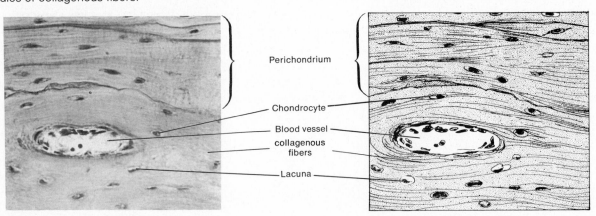

Perichondrium

Chondrocyte

Blood vessel

collagenous fibers

Lacuna

Elastic Cartilage (65×)

DESCRIPTION: Consists of chondrocytes located in a threadlike network of elastic fibers.

LOCATION: Epiglottis, parts of larynx, external ear, and Eustachian (auditory) tubes.

FUNCTION: Gives support and maintains shape.

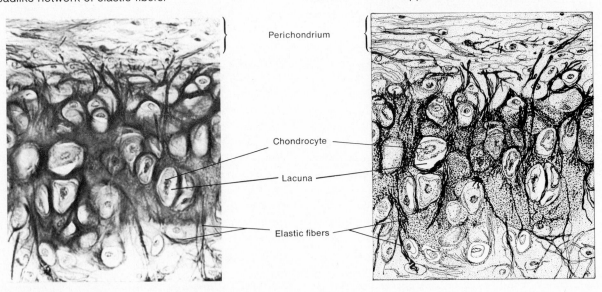

Perichondrium

Chondrocyte

Lacuna

Elastic fibers

Photomicrographs courtesy of Victor B. Eichler, Wichita State University.

bronchi, and bronchial tubes leading to the lungs. Most of the embryonic skeleton consists of hyaline cartilage. It affords flexibility and support.

■ *Fibrocartilage* Chondrocytes scattered through many bundles of visible collagenous fibers are found in this type of cartilage. Fibrocartilage is found at the symphysis pubis, the point where the coxal (hip) bones fuse anteriorly at the midline. It is also found in the discs between the vertebrae. This tissue combines strength and rigidity.

■ *Elastic Cartilage* In this tissue, chondrocytes are located in a threadlike network of elastic fibers. Elastic cartilage provides strength and maintains the shape of certain organs – the larynx, the external part of the ear (the pinna), and the Eustachian (auditory) tubes (the internal connection between the middle ear cavity and the upper throat).

BONE TISSUE
The details of bone, or osseous, tissue, another kind of connective tissue, are discussed in Chapter 6 as part of the skeletal system.

VASCULAR TISSUE
This kind of connective tissue, also known as blood, is treated in Chapter 19 as a component of the circulatory system.

MUSCLE TISSUE AND NERVOUS TISSUE
Epithelial and connective tissue can take a variety of forms to provide a variety of body functions. They are all-purpose tissues. By contrast, the third major type of tissue, **muscle tissue**, consists of highly modified cells that perform one basic function: contraction. The fourth major type, **nervous tissue**, is specialized to conduct electrical impulses.

MEMBRANES
The combination of an epithelial layer and an underlying connective tissue layer constitutes an **epithelial membrane**. The principal epithelial membranes of the body are mucous membranes, serous membranes, and the cutaneous membrane or skin. Another kind of membrane, a synovial membrane, is also significant.

Mucous Membranes
Mucous membranes line the body cavities that open to the exterior. Examples include the membranes lining the

entire digestive, respiratory, excretory, and reproductive tracts. The surface tissue of a mucous membrane may vary in type – it is stratified squamous epithelium in the esophagus and simple columnar epithelium in the intestine. The epithelial layer of a mucous membrane secretes mucus, which prevents the cavities from drying out. It also traps dust in the respiratory passageways and lubricates food as it moves through the digestive tract. In addition, the epithelial layer is responsible for the secretion of digestive enzymes and the absorption of food. The connective tissue layer of a mucous membrane binds the epithelium to the underlying structures. It is referred to as the *lamina propria* and allows some flexibility of the membrane. It holds the blood vessels in place, protects underlying muscles from abrasion or puncture, provides the epithelium covering it with oxygen and nutrients, and removes wastes.

Serous Membranes
Serous membranes line the body cavities that do not open to the exterior, and they cover the organs that lie within those cavities. They consist of thin layers of loose connective tissue covered by a layer of mesothelium. Serous membranes are in the form of double-walled sacs. The part attached to the cavity wall is called the *parietal* portion. The part that covers the organs inside these cavities is the *visceral* portion. The serous membrane lining the thoracic cavity and covering the lungs is called the pleura. The membrane lining the heart cavity and covering the heart is the pericardium (*cardio* = heart). The serous membrane lining the abdominal cavity and covering the abdominal organs and some pelvic organs is called the peritoneum.

The epithelial layer of a serous membrane secretes a lubricating fluid that allows the organs to glide easily against each other or the wall of the cavities. The connective tissue layer of the serous membrane consists of a relatively thin layer of loose connective tissue.

Cutaneous Membrane
The **cutaneous membrane,** or skin, constitutes an organ of the integumentary system and is discussed in the next chapter.

Synovial Membranes
Synovial membranes line the cavities of the joints. Like serous membranes, they line structures that do not open to the exterior. Unlike mucous and serous membranes, synovial membranes do not contain epithelium. They

are composed of loose connective tissue with elastic fibers and varying amounts of fat. Synovial membranes, therefore, are not epithelial membranes. Synovial membranes secrete *synovial fluid,* which lubricates the ends of bones as they move at joints.

TISSUE INFLAMMATION

When cells are damaged, the injury sets off an **inflammatory response**. The injury, which may be viewed as a stress, can have various causes. It could result from mechanical trauma, such as a clean knife incision during surgery. Bacteria that give off toxic (poisonous) chemicals could enter through the nose, pores in skin, or by way of a splinter or nail. Cells can be injured if the blood supply is cut off, causing the cells to "starve."

Symptoms

Inflammation is usually characterized by four fundamental symptoms: *redness, pain, heat,* and *swelling.* A fifth symptom can be the loss of function of the injured area. Whether loss of function occurs depends on the site and extent of the injury. In addition to these effects on the body, the inflammatory response, in apparent contradiction to the symptoms observed, serves a protective and defensive role. It attempts to neutralize and destroy toxic agents at the site of injury and prevent their spread to other organs. Thus the inflammatory response is an attempt to restore tissue homeostasis.

The immediate inflammatory response to tissue injury consists of a complicated sequence of physiological and anatomical adjustments. Various body components are involved in the initial response: blood vessels, intercellular fluid mixed with parts of injured cells (called the *exudate*), the cellular components of blood, and the surrounding epithelial and connective tissues. Other factors that affect the inflammatory response are the individual's age and general state of health. Healing processes of all types exert a great demand on the body's store of nutrients. Thus nutrition plays an essential role in healing.

Stages

The inflammatory response is one of the body's internal systems of defense (Figure 4-3a). The response of a tissue to a rusty nail wound is similar to other inflammatory responses such as that resulting from the bacterial invasion which causes a sore throat. These are the six basic stages of response:

1 The injury stimulates tissue in the damaged area to release a chemical called *histamine,* a derivative of the amino acid histidine. Histamine is believed to dilate

(increase the diameter of) the blood vessels. It also increases the permeability of the plasma membranes of the blood vessel cells. This dilation increases the amount of blood that can enter the area. The increase in permeability allows defensive substances in the blood to pass through the vessel walls and into the injured tissue. Such defensive substances include white blood cells, antibodies, oxygen, and scab-forming chemicals. The increased blood supply also removes toxic products and dead cells, preventing them from complicating the injury. These toxic substances include waste products released by invading microorganisms.

2 The body may also respond by increasing the metabolic rate and quickening the heartbeat so that more blood circulates to the injured area per minute. Within minutes after the injury, the quickened metabolism, circulation, and especially the dilation and increased permeability of capillaries produce heat, redness, and swelling. The heat results from the large amount of warm blood that accumulates in the area and, to a certain extent, from the heat energy produced by the metabolic reactions. The large amounts of blood in the area are also responsible for the redness. The increased permeability of the capillary walls allows quantities of fluid to move out of the blood and into the intercellular spaces of the tissue. Since the fluid moves into the intercellular spaces faster than it can be drained off, it accumulates in the tissue, causing it to swell. The swelling is called *edema.*

3 Pain, whether immediate or delayed, is a cardinal symptom of inflammation. It can result from an injury to nerve fibers or from an irritation caused by the release of toxic chemicals from microorganisms. Pain may also be due to increased pressure from an accumulation of extracellular fluid within the tissues.

4 Shortly following the injury, white blood cells rush to the site from throughout the body. The white blood cell, or leucocyte, is the body's first line of defense against the effects of the injury. The first white blood cells to arrive at the site of injury are neutrophils and monocytes (Figure 4-3b). These are actively phagocytic cells that can engulf foreign particles their own size (Figure 4-4). When neutrophils die, their lysosomes release enzymes that cause the decomposition of injured cells and bacteria in the affected area. Additional neutrophils then ingest the resulting debris. Other white blood cells, called lymphocytes, are thought to be involved either in the formation or release of *antibodies*—proteins that render invading bacteria and their chemicals harmless.

5 Nutrients stored in the body are used to support the defensive cells. They are also used in the increased metabolic reactions of the cells under attack.

6 The blood contains a soluble protein called *fibrinogen.* The increased permeability of capillaries due to histamine causes leakage of fibrinogen to tissues.

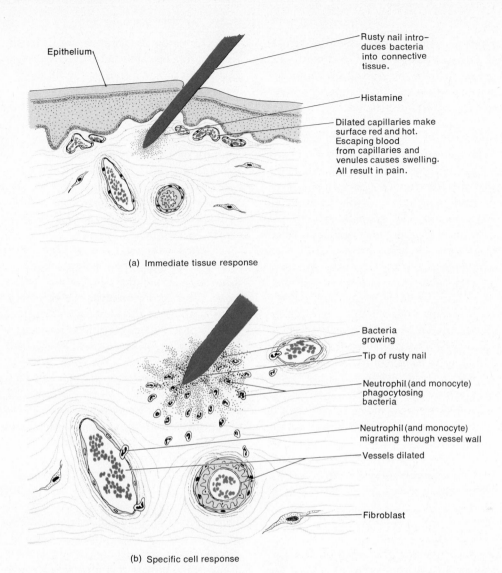

Epithelium

Rusty nail intro-
duces bacteria
into connective
tissue.

Histamine

Dilated capillaries make
surface red and hot.
Escaping blood
from capillaries and
venules causes swelling.
All result in pain.

(a) Immediate tissue response

Bacteria
growing

Tip of rusty nail

Neutrophil (and monocyte)
phagocytosing
bacteria

Neutrophil (and monocyte)
migrating through vessel wall

Vessels dilated

Fibroblast

(b) Specific cell response

Figure 4-3   Tissue and cell response to an injury. (a) Immediate tissue response. (b) Specific cell response.

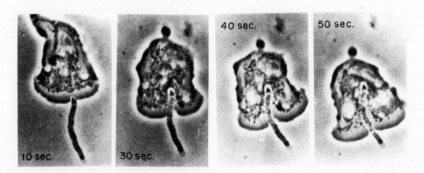

10 sec.   30 sec.   40 sec.   50 sec.

Figure 4-4   Phagocytosis. In this time-lapse photography sequence, a bacterial cell (the rodlike structure) is ingested and destroyed by a human neutrophil. (Courtesy of James G. Hirsch, from *Cell Biology,* by Robert M. Dowben, Harper & Row, Publishers, Inc., New York, 1971.)

Fibrinogen is then converted to an insoluble, thick network called fibrin, which localizes and traps the invading organisms, thereby preventing their spread. This network eventually forms a blood clot that prevents hemorrhage and isolates the infected area.

In all but very mild inflammations, a substance called pus is discharged. **Pus** is a thick fluid that contains living, as well as nonliving, white blood cells plus debris from other dead tissue. If the pus cannot drain out of the body, an *abscess* develops, which is simply an excessive accumulation of pus in a confined space. When inflamed tissue is shed many times, it produces an open sore, called an *ulcer,* on the surface of an organ or tissue. Ulcers may result from a prolonged inflammatory response to a continuously injured tissue. For instance, overproduction of digestive acids in the stomach may cause a steady erosion of the epithelial tissue lining the stomach. People with poor circulation are susceptible to ulcers in the tissues of their legs. The ulcers develop when the tissues are continuously damaged by a shortage of oxygen and nutrients.

White blood cells are the body's first line of defense against injury. The blood supply drains off and blocks the spread of harmful materials. This removal cannot restore the tissues to normalcy, however, especially if the cells have been damaged. Other processes replace the injured areas with new tissues and restore cells that are damaged through normal use.

TISSUE REPAIR

The process by which tissues replace dead or damaged cells is called repair. **Tissue repair** begins during the active phase of inflammation, but it can be completed only after all harmful substances have been neutralized or removed from the site of the injury. New cells originate by cell duplication from the **stroma,** the supporting connective tissue, or from the **parenchymal cells,** which form the organ's functioning part. The epithelial cells that secrete and absorb are the parenchymal cells of the intestine, for example. The restoration of an injured organ or tissue to normal structure and function depends entirely on which type of cell — parenchymal or stromal — is active in the repair. If only parenchymal elements accomplish the repair, a perfect or near-perfect reconstruction of the injured tissue may occur. However, if the fibroblast cells of the stroma are active in the repair, the tissue will be replaced with new connective tissue called scar tissue. This condition is known as **fibrosis.** Since scar tissue is not specialized to perform the functions of the parenchymal tissue, its function is impaired.

The cardinal factor in tissue repair lies in the capacity of parenchymal tissue to regenerate. This capacity, in turn, depends on the ability of the parenchymal cells to replicate quickly.

Repair Process

If injury to a tissue is slight, repair may sometimes be accomplished with the drainage and reabsorption of pus, followed by parenchymal regeneration. When the area of skin loss is great, fluid moves out of the capillaries and the area becomes dry. Fibrin seals the open tissue by hardening into a *scab.*

When tissue and cell damage are extensive and severe, as in large, open wounds, both the connective tissue stroma and the parenchymal cells are active in repair. This repair involves the rapid cell division of many fibroblasts, the manufacture of new collagenous fibers to provide strength, and an increase by cell division of the number of small blood vessels in the area. All these processes create an actively growing, connective tissue called **granulation tissue.** This new granulation tissue forms across a wound or surgical incision to provide a framework (stroma). The framework supports the epithelial cells that migrate into the open area and fill it. The newly formed granulation tissue also secretes a fluid that kills bacteria.

Conditions Affecting Repair

Three factors affect tissue repair: nutrition, blood circulation, and age. Nutrition is vital in the healing process since a great demand is placed on the body's store of nutrients. Protein-rich diets are important since most of the cell structure is made from proteins. Vitamins also play a direct role in wound healing. The following vitamins are thought to be essential to proper wound repair:

1 Vitamin A is essential in the replacement of epithelial tissues, especially in the respiratory tract.

2 The B vitamins — thiamine, nicotinic acid, riboflavin — are needed by many enzyme systems in cells. They are needed especially for the enzymes involved in decomposing carbohydrates to glucose, which is crucial to both heart and nervous tissue. These vitamins may relieve pain in some cases.

3 Vitamin C directly affects the normal production and maintenance of intercellular substances. It is required for the manufacture of cementing elements of connective tissues, especially collagen. Vitamin C also strengthens and promotes the formation of new blood vessels. With vitamin C deficiency, even superficial

wounds fail to heal, and the walls of the blood vessels become fragile and are easily ruptured.

4 Vitamin D is necessary for the proper absorption of calcium from the intestine. Calcium gives bones their hardness and is necessary for the healing of fractures.

5 Vitamin E is believed to promote healing of injured tissues and may prevent scarring.

6 Vitamin K assists in the clotting of blood and thus prevents the injured person from bleeding to death.

In tissue repair, proper blood circulation is indispensable. It is the blood that transports oxygen, nutrients, antibodies, and many defensive cells to the site of injury. The blood also plays an important role in the removal of tissue fluid, blood cells that have been depleted of oxygen, bacteria, foreign bodies, and debris. These elements would otherwise interfere with healing.

Generally, tissues heal faster and leave less obvious scars in the young than in the aged. When you are young, tissues have a better blood supply. The young body is generally in a much better nutritional state, and the cells of younger people have a faster metabolic rate. Thus cells can duplicate their materials and divide more quickly.

STUDY OUTLINE

TYPES OF TISSUES

1 Certain tissues of the body function in moving body parts, while others move food through organs.

2 Some tissues protect and support the body, and still others produce chemicals such as enzymes and hormones.

3 Depending on their function and structure, the various tissues of the body are classified into four principal types: epithelial, connective, muscular, and nervous.

EPITHELIAL TISSUE

1 A tissue is a group of similar cells and their intercellular substance specialized for a particular function.

2 Epithelium covers and lines body surfaces and forms glands.

3 Epithelium has many cells, little intercellular material, and no blood vessels. It is attached to connective tissue by a basement membrane. It can replace itself.

COVERING AND LINING EPITHELIUM

1 Simple epithelium is a single layer of cells adapted for absorption or filtration.

2 Stratified epithelium has several layers of cells and is adapted for protection.

3 Epithelial cell shapes include squamous (flat), cuboidal (cubelike), columnar (rectangular), and transitional (variable).

4 Simple squamous epithelium is adapted for diffusion and filtration and is found in lungs and kidneys. Endothelium lines the heart and blood vessels. Mesothelium lines body cavities and covers internal organs.

5 Simple cuboidal epithelium is adapted for secretion and absorption in kidneys and glands.

6 Simple columnar epithelium lines the digestive tract. Specialized cells containing microvilli perform absorption. Goblet cells perform secretion. In some portions of the respiratory tract, the cells are ciliated to move foreign particles out of the body.

7 Stratified squamous epithelium is protective. It lines the upper digestive tract and forms the outer layer of skin.

8 Stratified cuboidal epithelium is found in adult sweat glands.

9 Stratified columnar epithelium protects and secretes. If is found in the male urethra and excretory ducts.

10 Transitional epithelium lines the urinary bladder and is capable of stretching.

11 Pseudostratified epithelium has only one layer but gives the appearance of many. It lines excretory and respiratory structures where it protects and secretes.

GLANDULAR EPITHELIUM

1 A gland is a single cell or a mass of epithelial cells adapted for secretion.

2 Exocrine glands (sweat, oil, and digestive glands) secrete into ducts.

3 The structural classification includes unicellular and multicellular glands.

4 The functional classification includes holocrine, merocrine, and apocrine glands.

5 Endocrine glands secrete hormones directly into the blood.

CONNECTIVE TISSUE

1 Connective tissue is the most abundant body tissue. It has few cells, an extensive matrix, and a rich blood supply.

2 General types include loose and dense connective tissues.

3 It protects, supports, and binds organs together.

LOOSE AND DENSE CONNECTIVE TISSUE

1 Loose connective (areolar) tissue is widely distributed. It contains three kinds of fibers (collagenous, elastic, and reticular). It also has several kinds of cells (fibroblasts, macrophages, plasma cells, mast cells, and white blood cells). Loose connective tissue forms the subcutaneous layer under the skin. It is present in mucous membranes and around blood vessels, nerves, and body organs. When the fibroblasts in loose connective tissue become infiltrated with fat, the tissue is then known as adipose tissue.

2 Dense connective (collagenous) tissue forms tendons, ligaments, and deep fasciae. All three are usually primarily composed of closely packed collagenous fibers. In tendons and ligaments, collagenous fibers are arranged in parallel bundles. In fasciae, these fibers are interwoven at various angles with each other. A few ligaments are composed of closely packed elastic fibers.

CLASSIFICATION

Connective tissue may be classified in several ways. We classify them as follows:

    Embryonic connective tissue
    Adult connective tissue
      Connective tissue proper
        1 Loose connective (areolar) tissue
        2 Adipose tissue
        3 Dense connective (collagenous) tissue
        4 Elastic connective tissue
        5 Reticular connective tissue
      Cartilage
        1 Hyaline cartilage
        2 Fibrocartilage
        3 Elastic cartilage
      Bone (osseous) tissue
      Vascular (blood) tissue

EMBRYONIC CONNECTIVE TISSUE

1 Mesenchyme forms all other connective tissue.

2 Mucous connective tissue is found only in the umbilical cord of the fetus, where it gives support.

ADULT CONNECTIVE TISSUE

1 Adult connective tissue is connective tissue that exists in the newborn and that does not change after birth. It is subdivided into several kinds: connective tissue proper, cartilage, bone tissue, and vascular tissue.

2 Connective tissue proper has a more or less fluid intercellular material, and a typical cell is the fibroblast. Five examples of such tissues may be distinguished.

3 Loose connective (areolar) is one of the most widely distributed connective tissues in the body.

4 Adipose tissue is a form of loose connective tissue in which the fibroblasts are specialized for fat storage.

5 Dense connective (collagenous) tissue has a predominance of collagenous or white fibers arranged in bundles.

6 Elastic connective tissue has a predominance of freely branching elastic fibers that give it a yellow color.

7 Reticular connective tissue consists of interlacing reticular fibers.

8 Cartilage has a gel-like matrix of collagenous and elastic fibers that contains chondrocytes.

9 Hyaline cartilage is found at the ends of bones, in the nose, and in respiratory structures. It is flexible, allows movement, and provides support.

10 Fibrocartilage connects the pelvic bones and the vertebrae. It provides strength.

11 Elastic cartilage maintains the shape of organs such as the external ear.

MUSCLE TISSUE AND NERVOUS TISSUE

1 Muscle tissue performs one major function: contraction.

2 Nervous tissue is specialized to conduct electrical impulses.

MEMBRANES

1 An epithelial membrane is an epithelial layer overlying a connective tissue layer. Examples are mucous, serous, and cutaneous.

2 Mucous membranes line cavities that open to the exterior, such as the digestive tract.

3 Serous membranes (pleura, pericardium, peritoneum) line closed cavities and cover the organs in the cavities. These membranes consist of parietal and visceral portions.

4 The cutaneous membrane is the skin.

5 Synovial membranes line joint cavities and do not contain epithelium.

TISSUE INFLAMMATION

1 Damage to a tissue causes an inflammatory response characterized by redness, pain, heat, and swelling.

2 The inflammatory response is initiated by histamine released by the damaged tissue.

3 Further cell injury is prevented by phagocytic neutrophils and by antibody-producing lymphocytes.

TISSUE REPAIR

1 Tissue repair is the replacement of damaged or destroyed cells by healthy ones.

REPAIR PROCESS

1 If the injury is superficial, tissue repair involves pus removal (if pus is present), scab formation, and parenchymal regeneration.

2 If damage is extensive, granulation tissue is involved.

CONDITIONS AFFECTING REPAIR

1 Nutrition is important to tissue repair. Various vitamins and a protein-rich diet are needed.

2 Adequate circulation of blood is needed.

3 The tissues of young people repair rapidly and efficiently; the process slows down with aging.

REVIEW QUESTIONS

1 Define a tissue. What are the four basic kinds of human tissue?

2 What characteristics are common to all epithelium? Distinguish covering and lining epithelium from glandular epithelium.

3 Discuss the classification of epithelium based on layering and cell type.

4 For each of the following kinds of epithelium, briefly describe the microscopic appearance, location in the body, and functions: simple squamous, simple cuboidal, simple columnar, stratified squamous, stratified cuboidal, stratified columnar, transitional, and pseudostratified.

5 What is a gland? Distinguish between endocrine and exocrine glands. Describe the classification of exocrine glands according to structural complexity and function and give at least one example of each.

6 Enumerate the ways in which connective tissue differs from epithelium. Distinguish between loose and dense connective tissues. How are connective tissues classified?

7 Compare embryonal connective tissue with adult connective tissue.

8 Describe the following connective tissues with regard to microscopic appearance, location in the body, and function: loose (areolar), adipose, dense (collagenous), elastic, reticular, hyaline cartilage, fibrocartilage, and elastic cartilage.

9 Define the following kinds of membranes: mucous, serous, cutaneous, and synovial. Where is each located in the body? What are their functions?

10 Below are some descriptive statements for various tissues. For each statement, name the tissue described:

A stratified epithelium that permits distention.
A single layer of flat cells concerned with filtration and absorption.
Forms all other kinds of connective tissue.
Specialized for fat storage.
An epithelium with waterproofing qualities.
Forms the framework of many organs.
Produces perspiration, wax, oil, and digestive enzymes.
Cartilage that shapes the external ear.
Contains goblet cells and lines the intestine.
Most widely distributed connective tissue.
Forms tendons, ligaments, and aponeuroses.
Specialized for the secretion of hormones.
Provides support in the umbilical cord.
Lines kidney tubules and is specialized for absorption and secretion.
Permits extensibility of lung tissue.

11 Describe the principal physiological responses associated with inflammation. What is the immediate tissue response to injury? What is the response to tissue injury by neutrophils and lymphocytes?

12 What is meant by tissue repair? How does the repair take place? What conditions affect tissue repair?

Chapter 5
The Integumentary System

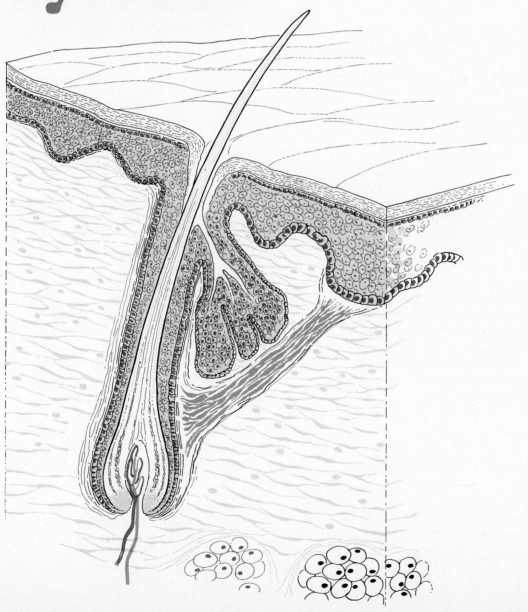

STUDENT OBJECTIVES

■ Define the skin as the organ of the integumentary system.

■ Explain how the skin is structurally divided into epidermis and dermis.

■ List the structural layers of the epidermis and describe their functions.

■ Explain the composition and functions of the dermis.

■ Contrast the structure and functions of derivatives of the epidermis such as hair, glands, and nails.

■ Describe the role of the skin in maintaining normal body temperature.

■ Contrast the causes and treatment for skin disorders such as acne, impetigo, lupus erythematosus, psoriasis, decubitus, and skin cancer.

■ Describe the effects of a burn.

■ Classify burns into first, second, and third degrees.

■ Define the "rule of nines" for estimating the extent of a burn.

■ Define medical terminology associated with the integumentary system.

An aggregation of tissues that performs a specific function is an **organ**. The next higher level of organization is a **system** – a group of organs operating together to perform specialized functions. The skin and its derivatives, such as hair, nails, glands, and several specialized receptors, constitute the **integumentary system** of the body. First, let us consider the skin as an organ.

SKIN

The **skin** or **cutis** is an organ because it consists of tissues structurally joined together to perform specific activities. Not just a simple thin covering that keeps the body together and gives it protection, the skin is quite complex in structure and performs several vital functions. In fact, this organ is essential for survival.

The skin is the largest organ of the body. For the average adult, the skin occupies a surface area of approximately 7,620 sq cm (3,000 sq inches). It varies in thickness from .05 to 3 mm (0.0197–0.1182 inch) and is somewhat thicker on the dorsal and extensor surfaces than on the ventral and flexor surfaces. It covers the body and protects the underlying tissues, not only from bacterial invasion but also from drying out and from harmful light rays. Moreover, the skin helps to control body temperature, prevents excessive loss of inorganic and organic materials, receives stimuli from the environment, stores chemical compounds, excretes water and salts, and synthesizes several important compounds, including vitamin D.

Structure

Structurally, the skin consists of two principal parts (Figure 5-1). The outer, thinner portion, which is composed of epithelium, is called the **epidermis**. The epidermis is cemented to the inner, thicker, connective tissue part called the **dermis**. Beneath the dermis is a *subcutaneous layer* of tissues (*sub* = under). This layer, also called the *superficial fascia*, consists of areolar and adipose tissues. Fibers from the dermis extend down into the superficial fascia and anchor the skin to the subcutaneous layer. The superficial fascia, in turn, is firmly attached to underlying tissues and organs.

Epidermis

The **epidermis** is composed of stratified epithelium in four or five cell layers, depending on its location in the body. (See Figure 5-1.) Where exposure to friction is greatest, such as the palms and soles, the epidermis has five layers. In all other parts it has four layers. The names of the layers from the inside outward are as follows:

1 Stratum basale This single layer of columnar cells is capable of continued cell division. As these cells multiply, they push up toward the surface. Their nuclei degenerate, and the cells die. Eventually, the cells are shed in the top layer of the epidermis.

2 Stratum spinosum This layer of the epidermis contains 8 to 10 rows of polygonal (many-sided) cells that fit closely together. The surfaces of these cells may assume a prickly appearance when prepared for microscope examination (*spinosum* = prickly). The stratum basale and stratum spinosum are sometimes collectively referred to as the *stratum germinativum* to indicate the layers where new cells are germinated.

3 Stratum granulosum This third layer of the epidermis consists of two or three rows of flattened cells that contain darkly staining granules of a substance called keratohyaline. This compound is involved in the first step of keratin formation. Keratin is a waterproofing protein found in the top layer of the epidermis.

4 Stratum lucidum This layer is quite pronounced in the thick skin of the palms and soles. It consists of three to four rows of clear, flat, dead cells that contain droplets of a translucent substance called eleidin. Eleidin is formed from keratohyaline and is eventually transformed to keratin.

5 Stratum corneum This layer consists of 25 to 30 rows of flat, dead cells containing keratin. These cells are continuously shed and replaced. The stratum corneum serves as an effective barrier against light and heat waves, bacteria, and many chemicals. Be sure to examine the photomicrograph of the skin in Figure 5-2.

The color of the skin is due to a pigment called melanin. The amount of melanin varies the skin color from yellow to black. This pigment is found throughout the basale and spinosum layers and in the granulosum of all Caucasian people. In Negroes melanin is found in all epidermal layers. Melanin is synthesized in cells called *melanocytes.* They produce the pigment and pass it on to the epidermal cells. When the skin is exposed to ultraviolet radiation, both the amount and the darkness of melanin increase, tanning and further protecting the body against radiation. Thus melanin serves a vital protective function. In mammals, the melanocyte-stimulating hormone (MSH) produced by the anterior pituitary gland causes increased melanin synthesis. The exact role of MSH in humans is not clear. Another pigment called carotene is found in the corneum and fatty areas of the dermis in Oriental people. Together carotene and melanin account for the yellowish hue of their skin. The pink color of Caucasian skin is due to blood vessels in the dermis. The redness of the vessels is not heavily masked by pigment. The epidermis has no blood vessels, a characteristic of all epithelia.

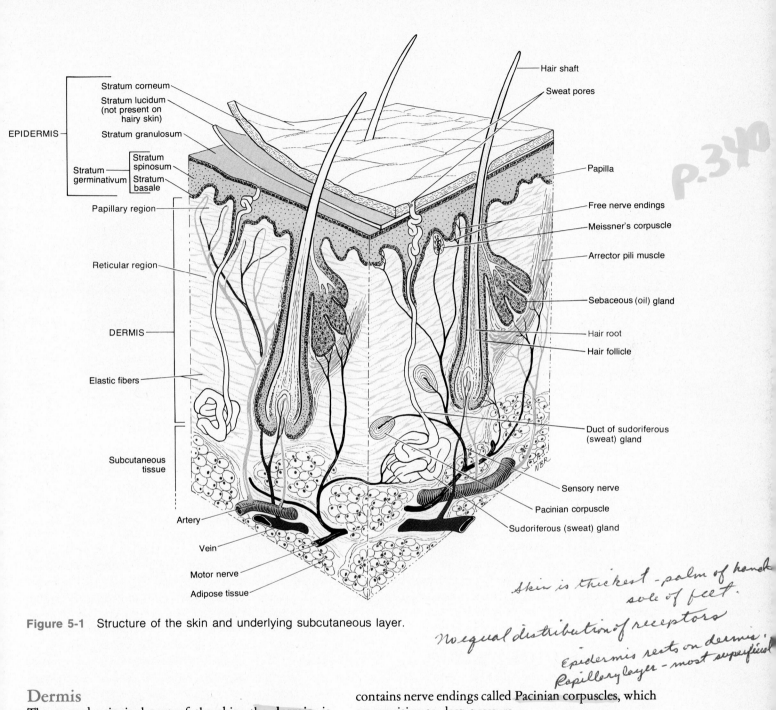

EPIDERMIS
- Stratum corneum
- Stratum lucidum (not present on hairy skin)
- Stratum granulosum
- Stratum germinativum
  - Stratum spinosum
  - Stratum basale

Papillary region

Reticular region

DERMIS

Elastic fibers

Subcutaneous tissue

Artery

Vein

Motor nerve

Adipose tissue

Hair shaft

Sweat pores

Papilla

Free nerve endings

Meissner's corpuscle

Arrector pili muscle

Sebaceous (oil) gland

Hair root

Hair follicle

Duct of sudoriferous (sweat) gland

Sensory nerve

Pacinian corpuscle

Sudoriferous (sweat) gland

Figure 5-1 Structure of the skin and underlying subcutaneous layer.

[handwritten annotations:] P.340
Skin is thickest - palm of hand sole of feet.
No equal distribution of receptors
Epidermis rests on dermis.
Papillary layer - most superficial

Dermis

The second principal part of the skin, the **dermis**, is composed of connective tissue containing collagenous and elastic fibers. (See Figure 5-1.) It is about 0.5 to 2.5 mm (0.0197–0.1085 inch) thick. Numerous blood vessels, nerves, glands, and hair follicles are also embedded in the dermis. The upper region of the dermis, about one-fifth of the total layer, is named the *papillary region* – its surface area is greatly increased by small, fingerlike projections called **papillae.** These structures project into the epidermis and contain loops of capillaries. Some papillae contain Meissner's corpuscles – nerve endings sensitive to light touch. The dermis also

contains nerve endings called Pacinian corpuscles, which are sensitive to deep pressure.

The ridges marking the external surface of the epidermis are caused by the size and arrangement of the papillae in the dermis. Some ridges cross at various angles and can be seen on the back of your hand. Other ridges on your palms and fingertips prevent slipping. The ridge patterns on the fingertips and thumbs are different in each individual.

The remaining portion of the dermis is called the *reticular region.* It is a dense, irregular, collagenous connective tissue. The irregular arrangement of fibers permits flexibility and strength in all directions. This

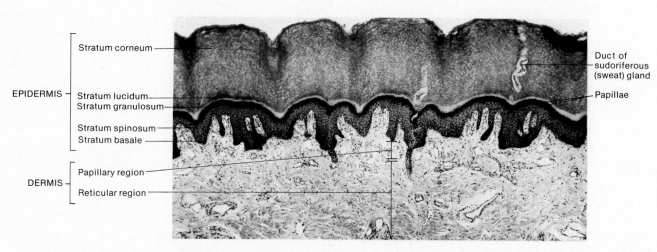

EPIDERMIS
- Stratum corneum
- Stratum lucidum
- Stratum granulosum
- Stratum spinosum
- Stratum basale

DERMIS
- Papillary region
- Reticular region

Duct of sudoriferous (sweat) gland

Papillae

Figure 5-2   Photomicrograph of the skin of the palmar surface of the hand at a magnification of 65×. (Courtesy of Edward J. Reith, from *Atlas of Descriptive Histology,* by Edward J. Reith and Michael H. Ross, Harper & Row, Publishers, Inc., New York, 1970.)

area of the dermis contains many blood vessels and also collagenous and elastic fibers. The spaces between the interlacing fibers are occupied by adipose tissue and sweat glands. The reticular zone is attached to the organs beneath it, such as bone and muscle, by the subcutaneous layer.

EPIDERMAL DERIVATIVES

Organs derived from the skin – hair, glands, nails – perform functions that are necessary and sometimes vital. Hair and nails protect the body. The sweat glands help regulate body temperature.

Hair

Growths of the epidermis variously distributed over the body are **hairs** or **pili.** The primary function of hair is protection. Though the protection is limited, hair guards the scalp from injury and the sun's rays. Eyebrows and eyelashes protect the eyes from foreign particles. Hair in the nostrils and external ear canal protects these structures from insects and dust.

Hairs are composed of a number of parts (Figure 5-3). Each hair consists of a free shaft and a root. The *shaft* is the portion most of which projects above the surface of the skin. The *root* is the portion below the surface that penetrates deep into the dermis. Surrounding the root is the *hair follicle,* which is made up of an internal zone of epithelium (the *internal root sheath*) and an external zone of epithelium (the *external root sheath*). The base of each follicle is enlarged into an onion-shaped structure, the *bulb.* This structure contains an indentation, the *connective tissue papilla,* filled with loose connective tissue. The papilla contains many blood vessels and provides nourishment for the growing hair. The base of the bulb also contains a region of cells called the *matrix,* a germinal layer. The cells of the matrix produce new hairs by cell division when older hairs are shed. This replacement occurs within the same follicle. Hair grows about 1mm (0.0394 inch) every 3 days. Hair loss in an adult is about 70 to 100 hairs a day. But the rate of growth may be altered by illness.

Sebaceous glands and bundles of smooth muscle are also associated with hair. These smooth muscles, called *arrector pili,* extend from the dermis of the skin to

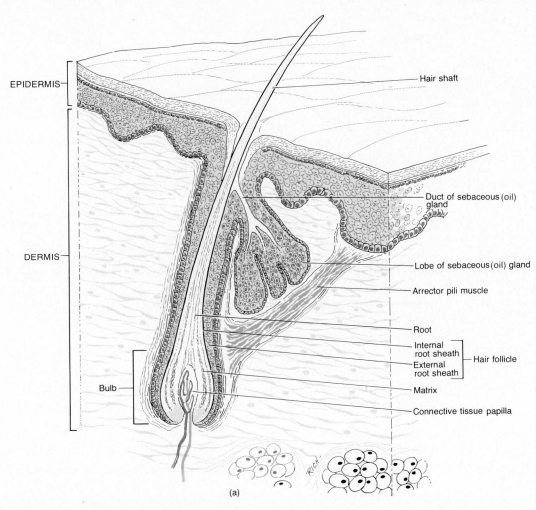

Figure 5-3   Principal parts of a hair and associated structures. (a) Diagram.

the side of the hair follicle. In its normal position hair is arranged at an angle to the surface of the skin. The arrector pili muscles contract under stresses of fright and cold and pull the hairs into a vertical position. This contraction results in "goosebumps" or "gooseflesh" because the skin around the shaft forms slight elevations.

Glands

Two kinds of glands associated with the skin are sebaceous glands and sweat glands.

SEBACEOUS (OIL) GLANDS

Sebaceous (oil) glands, with few exceptions, are connected to hair follicles. (See Figure 5-3). They are simple branched acinar glands connected directly to the follicle by a short duct. Absent in the palms and soles, they vary in size and shape in other regions of the body. For example, they are small in most areas of the trunk and extremities but large in the skin of the breasts, face,

neck, and upper chest. The sebaceous glands secrete an oily substance called _sebum,_ a mixture of fats, cholesterol, proteins, and inorganic salts. These glands keep the hair from drying and becoming brittle and also form a protective film that prevents excessive evaporation of water from the skin. The sebum keeps the skin soft and pliable, too. When sebaceous glands of the face become enlarged because of accumulated sebum, blackheads develop. Since sebum is nutritive to certain bacteria, pimples or boils often result. The color of blackheads is due to oxidized oil, not dirt.

secretes sebum

SUDORIFEROUS (SWEAT) GLANDS   *secretes sweat*

Sudoriferous (sweat) glands (*sudor* = sweat; *ferre* = to bear) are distributed throughout the skin except on the nail beds of the fingers and toes, margins of the lips, eardrums and tips of the penis and clitoris. In contrast to sebaceous glands, sudoriferous glands are most numerous in the skin of the palms and the soles. They are also found in abundance in the armpits and forehead.

of waste, sweat, antibacterial

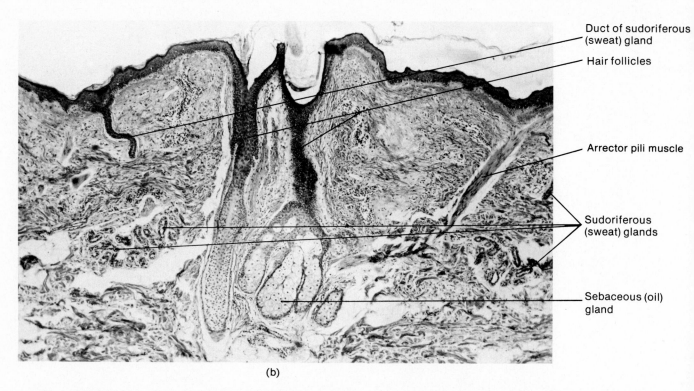

Figure 5-3 (cont.)   Principal parts of a hair and associated structures. (b) Photomicrograph of a section from the skin of the face at a magnification of 45×.

Each gland consists of a coiled end embedded in the subcutaneous tissue and a single tube that projects upward through the dermis and epidermis. This tube, the excretory duct, terminates in a pore at the surface of the epidermis. (See Figure 5-1). The base of each sudoriferous gland is surrounded by a network of small blood vessels. In the axillary region, sudoriferous glands are of the simple branched tubular type. Elsewhere they are simple coiled tubular glands.

Perspiration, or *sweat,* is the substance produced by sudoriferous glands. It is a mixture of water, salts (mostly NaCl), urea, uric acid, amino acids, ammonia, sugar, lactic acid, and ascorbic acid. Its principal function is to help regulate body temperature. It also helps to eliminate wastes.

Nails

Modified horny cells of the epidermis are referred to as **nails.** The cells form a clear, solid covering over the dorsal surfaces of the terminal portions of the fingers and toes. Each nail (see Figure 5-4) consists of a *nail body,* a *free edge,* and the *nail root.* Most of the nail body is pink because of the underlying vascular tissue. The whitish semilunar area at the proximal end of the body is called

the *lunula.* It appears whitish because the vascular tissue underneath does not show through.

The *eponychium* ("cuticle") is a narrow band of epidermis that extends from the margin of the nail wall (lateral border), adhering to it. It occupies the proximal border of the nail and consists of stratum corneum.

The epithelium of the posterior part of the nail bed is known as the *nail matrix.* Its function is to bring about the growth of nails. Essentially, growth occurs by the transformation of superficial cells of the matrix into nail cells. In the process, the outer, harder layer is pushed forward over the stratum germinativum. The average growth in the length of fingernails is about 1 mm (0.0394 inch)/week. The growth rate is somewhat slower in toenails.

HOMEOSTASIS

Humans, like other mammals, are *homeotherms* – warm-blooded organisms. This means that we are able to maintain a remarkably constant internal body temperature of 37°C (98.6°F) even though the environmental temperature may vary over a broad range. Suppose you are participating in an athletic event where the temperature is 37.8°C (100°F). A sequence of events is set into

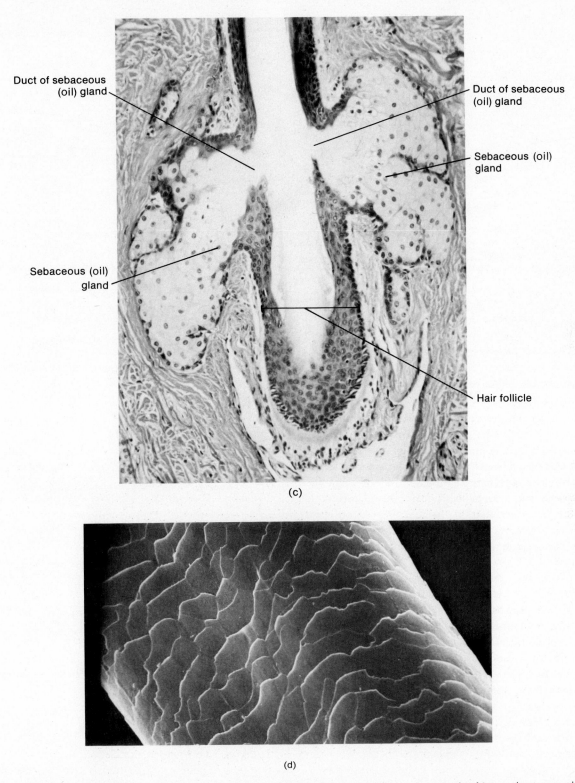

Duct of sebaceous (oil) gland

Duct of sebaceous (oil) gland

Sebaceous (oil) gland

Sebaceous (oil) gland

Hair follicle

(c)

(d)

Figure 5-3 (cont.) Principal parts of a hair and associated structures. (c) Photomicrograph of two sebaceous glands opening into a hair follicle at a magnification of 160×. (d) Scanning electron micrographs of the surface of a hair shaft at a magnification of 1,000×. (Photomicrographs courtesy of Edward J. Reith, from *Atlas of Descriptive Histology,* by Edward J. Reith and Michael H. Ross, Harper & Row, Publishers, Inc., New York, 1970. Electron micrograph courtesy of Fisher Scientific Company and S.T.E.M. Laboratories, Inc., Copyright 1975.)

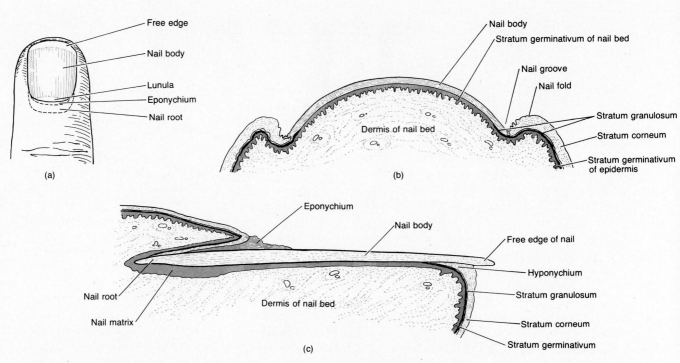

Figure 5-4   Structure of nails. (a) Fingernail viewed from above. (b) Cross section of a fingernail and the nail bed. (c) Longitudinal section of a fingernail.

operation to counteract this above-normal temperature, which may be considered a stress. Sensing devices in the skin called receptors pick up the stimulus – in this case, heat – and activate nerves that send the message to your brain. A temperature-regulating area of the brain then sends nerve impulses to the sudoriferous glands, which produce more perspiration. As the perspiration evaporates from the surface of your skin, it is cooled and your temperature is lowered. This sequence of events is shown in Figure 5-5.

Note that temperature regulation by the skin involves a feedback system because the output (cooling of the skin) is fed back to the skin receptors and becomes part of a new stimulus-response cycle. In other words, after the sudoriferous glands are activated, the skin receptors keep the brain informed about the external temperature. The brain, in turn, continues to send messages to the sudoriferous glands until the temperature returns to 37°C (98.6°F). Like most of the body's feedback systems, temperature regulation is a negative feedback system – the output, cooling, is the opposite of the original condition, overheating.

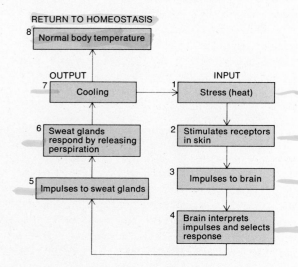

Figure 5-5   Role of the skin in maintaining the homeostasis of body temperature.

DISORDERS

ACNE

Acne is an inflammation of sebaceous glands and usually begins at puberty. In fact, a few comedones (blackheads) on the face may be the first signs of approaching puberty.

Common acne is called *acne vulgaris.* It occurs mostly in individuals between the ages of 14 and 25 and affects more than 80 percent of teenagers. It appears to affect an increasing number of older people, however, and women in their 20s who had little or none as teenagers. This latter variety, *acne cosmetica,* has been blamed on the prolonged use of makeup and other cosmetics.

At puberty the sebaceous glands, under the influence of androgens (male hormones), grow in size and increase production of their complex lipid product, sebum. Although testosterone, a male hormone, appears to be the most potent circulating androgen for sebaceous cell stimulation, adrenal and ovarian androgens can stimulate sebaceous secretions as well. Acne occurs predominantly in sebaceous follicles in which the sebaceous gland dominates and the hair is rudimentary. The sebaceous follicles are rapidly colonized by microorganisms that thrive in the lipid-rich environment of the follicles. When this occurs, the cyst or sac of connective tissue cells could destroy and displace epidermal cells, resulting in permanent scarring. Care must be taken to avoid squeezing, pinching, or scratching the lesions. The four basic types of acne lesions in order of increasing severity are comedones (open blackheads, closed whiteheads), papules, pustules, and cysts.

IMPETIGO

Impetigo is a superficial skin infection caused by staphylococci or streptococci. It is characterized by isolated pustules that become crusted and rupture. Occurring principally around mouth, nose, and hands, the inflammation is located in the papillary layer of the skin involving the capillary network and stratum corneum. The disease is most common in children, and epidemics in nurseries (due to staphylococci) may be serious.

SYSTEMIC LUPUS ERYTHEMATOSUS (SLE, LUPUS)

Systemic lupus erythematosus (SLE, Lupus) is an inflammatory connective tissue disorder occurring mostly in young women during their 20s. Although SLE is variously labeled a collagen-vascular, an immune complex, or a connective tissue disease, its cause is still unknown. It has replaced syphilis as "the great imitator"—it affects virtually every organ system with symptoms that range from mild arthralgias to fulminating glomerulonephritis.

Lupus is not contagious. The body attacks its own tissues, failing to differentiate between what is foreign and what is not. This inflammatory process occurs in collagenous connective tissue throughout the body. Symptoms include low-grade fever, aches, fatigue, and sometimes an eruption across the bridge of the nose and cheeks called a "butterfly rash." Other cutaneous lesions may occur with blistering and ulceration. For some reason the SLE victim produces antibodies which, finding no foreign invaders to attack, turn on DNA, the nucleic acid substance in the cells. The patient may have taken medicine, such as penicillin, sulfa, or tetracycline, or may have been exposed to excessive sunlight, injury, emotional upset, infection, or other stress. These triggering factors, once recognized, are to be avoided by the patient in the future.

The number of SLE cases is somewhere between 200,000 and 500,000 per year in the United States. Although SLE is not thought to be hereditary, there seems to be a strong incidence of other connective tissue disorders—especially rheumatoid arthritis and rheumatic fever—in relatives of SLE victims. The two most serious complications involve inflammation of the kidneys and the central nervous system.

PSORIASIS

Psoriasis is a chronic, occasionally acute, relapsing, papulosquamous skin disease borne by 6 to 8 million people in the United States. In its severe forms, psoriasis is a disabling and disfiguring affliction. It may begin at any age, although it is most severe between ages 10 and 50. The cause is unknown. It is thought, however, to be hereditary in at least one-third of the patients.

Psoriasis is characterized by distinct, reddish, slightly raised plaques or papules covered with scales. Itching is seldom severe, and the lesions heal without scarring. Psoriasis ordinarily involves the scalp, the elbows and knees, the back, and buttocks. Occasionally the disease is generalized. Psoriasis involves a high rate of mitosis, either as the primary defect or as secondary to other phenomena. Any antimitotic drug is effective against this disease.

DECUBITUS

Decubitus (bedsore, pressure sore, trophic ulcer) is caused by a constant deficiency of blood to tissues overlying a bony prominence that has been subjected to prolonged pressure against an object such as a bed, cast, or splint. It causes tissue ulceration and tissue destruction and is seen most frequently in patients who are bedridden for long periods of time.

The most common areas involved are the skin over the sacrum, heels, ankles, buttocks, and other large bony promi-

Hypothalmus gland - temp. regulator

nences. The chief causes are protracted pressure from infrequent turning of the patient, trauma and maceration of the skin, and malnutrition. It begins with damage to the sensitive subcutaneous and deeper tissues; later the skin dies. Small breaks in the epidermis become infected. Maceration of the skin often follows soaking of bed and clothing by perspiration, urine, or feces.

SKIN CANCER

Excessive sun exposure can result in **skin cancer**. Some 600,000 people have diagnosed skin cancer in the United States and 300,000 more will have diagnosed skin cancer this year. About 9,300 of these new cases will be melanoma.

Everyone, regardless of skin pigmentation, is a potential victim of skin cancer if exposure to sunlight is sufficiently intense and continuous. Natural skin pigment can never give complete protection. If you must be in direct sunlight for long periods of time, use a suitable sunscreen. One of the best agents for protection against overexposure to sunlight is para-aminobenzoic acid (PABA). The alcohol preparations of PABA are best because the active ingredient binds to the stratum corneum of the skin.

A number of widely prescribed drugs (tetracyclines, sulfa drugs, and others) and constituents of foods (such as riboflavins) are potential photosensitizers. When activated in the body by light, they may produce substances that can damage the tissues in sensitive individuals. A typical response is the appearance of a rash on parts of the body exposed to the sun.

BURNS

Tissues may be damaged by thermal (heat), electrical, radioactive, or chemical agents. These agents can destroy the proteins in the exposed cells and cause cell injury or death. Such damage results in a **burn**. The tissues that are directly or indirectly in contact with the environment, such as the skin or the linings of the respiratory and digestive tracts, are affected. Generally, however, the systemic effects of a burn are a greater threat to life than the local effects. *Systemic* effects occur throughout the body. *Local* effects occur in one area of the body. The systemic effects of burn include:

1 A large loss of water, plasma, and plasma proteins, which causes shock.
2 Bacterial infection.
3 Reduced circulation of blood.
4 A decrease in urine production.

CLASSIFICATION

Burns are classified into three types: first degree, second degree, and third degree. In *first-degree* burns, the damage is restricted to the epidermal layers of the skin. Symptoms are limited to local effects such as redness, tenderness, pain, and edema—the cardinal signs of inflammation. In *second-degree* burns, the epidermal and portions of the dermal layers of the skin are damaged, but rapid regeneration of epithelium is still possible. Blisters containing elements of blood and lymph form on the skin surface or beneath the epidermis. Blisters beneath or within the epidermis are called *bullae*. In *third-degree* burns, both the epidermis and the dermis are destroyed. The skin surface may be charred or white or both. It is lifeless and insensitive to touch. The regeneration of epithelium originates from the wound edges. Regeneration is slow, and much granulation tissue forms before being covered by epithelium. Even if skin grafting is quickly begun, these wounds frequently contract and produce disfiguring or disabling scars.

"RULE OF NINES"

A fairly accurate method for estimating the extent of a burn is to apply the "*rule of nines*":

1 If the anterior and posterior surfaces of the head and neck are affected, the burn covers 9 percent of the body surface.
2 The anterior and posterior surfaces of each shoulder, arm, forearm, and hand also comprise 9 percent of the body surface.
3 The anterior and posterior surfaces of the trunk, including the buttocks, constitute 36 percent.
4 The anterior and posterior surfaces of each foot, leg, and thigh as far up as the buttocks total 18 percent.
5 The perineum consists of 1 percent. The perineum includes the anal and urogenital regions.

TREATMENT

A severely burned individual should be moved as quickly as possible to a hospital. Treatment may then include:

1 Cleansing the burn wounds thoroughly.
2 Removing all dead tissue so antibacterial agents can directly contact the wound surface and thereby prevent infection.
3 Replacing lost body fluids.
4 Covering wounds with grafts as soon as possible.

Receptors
Cutaneous senses
Meissner's corpuscles ⎫ *fine or*
Merkel's disks ⎭ *light touch*
Pacinian corpuscles - deep pressure
End bulb of Krause - cold
End organs of Ruffinni - heat

Hormones & nerves cause vasoconstriction — less blood, less loss of heat
vasodialation — more blood, more loss of heat

5 The Integumentary System   113

MEDICAL TERMINOLOGY

Albinism (*alb* = white; *ism* = condition) Congenital (existing at birth) absence of pigment from the skin, hair, and parts of the eye.

Anhidrosis (*an* = without; *hidr* = sweating; *osis* = condition) Inability to sweat.

Callus An area of hardened and thickened skin that is usually seen in palms and soles and is due to pressure and friction.

Carbuncle A hard, round, deep, painful inflammation of the subcutaneous tissue that causes necrosis (deadness) and pus formation (abscess).

Carcinogenesis Production of cancer.

Comedo A collection of sebaceous material and dead cells in the hair follicle and excretory duct of the sebaceous gland. Usually found over the face, chest, and back, and more commonly during adolescence. Also called blackhead or whitehead.

Cyst (*cyst* = sac containing fluid) A sac with a distinct connective tissue wall, containing a fluid or other material.

Decubitus ulcer A bedsore. An ulcer formed due to continual pressure over the skin.

Dermatome An instrument for excising areas of skin to be used for grafting.

Epidermophytosis (*epi* = upon; *derm* = skin; *phyto* = pertaining to plants; *osis* = condition) Any fungus infection of the skin producing scaliness with itching. Called athlete's foot when it affects the feet.

Erythema Congestive or exudative redness of the skin caused by engorgement of capillaries in lower layers of the skin. It occurs with any skin injury, infection, or inflammation.

Furuncle A boil; an abscess resulting from infection of a hair follicle.

Hypodermic (*hypo* = under; *derm* = skin) The area beneath the skin.

Impetigo An inflammatory skin disease.

Intradermal (*intra* = within) Within the skin. Also called intracutaneous.

Maceration Softening of a solid by soaking; wasting away.

Melanoma (*melano* = dark-colored; *oma* = tumor) A cancerous tumor consisting of melanocytes that produce skin pigment.

Nevi Round, pigmented, flat, or raised skin areas that may be present at birth or develop later. Varying in color from yellow-brown to black. Also called moles or birthmarks.

Nodule A large cluster of cells raised above the skin but extending deep into the tissues.

Papilloma Tumors, such as warts, in the skin or the lining of internal organs.

Papule A small, round skin elevation varying in size from a pinpoint to that of a split pea. One example is a pimple.

Pathogenesis Origination and development of a disease.

Polyp A tumor on a stem found especially on mucous membranes.

Pustule A small, round elevation of the skin containing pus.

Subcutaneous (*sub* = under; *cutis* = skin) Beneath the skin.

Topical Pertaining to a definite area; local.

Wart A common, contagious, noncancerous epithelial tumor caused by a virus.

STUDY OUTLINE

SKIN

1 The skin and its derivatives of hair, glands, and nails constitute the integumentary system.

2 The skin is the largest body organ. It performs the functions of protection, maintaining body temperature, picking up stimuli, and excretion.

3 The principal parts of the skin are the outer epidermis and inner dermis. The dermis overlies the subcutaneous layer.

4 The epidermal layers, from the inside outward, are the stratum basale, spinosum, granulosum, lucidum, and corneum. The basale and spinosum undergo continuous cell division and produce all other layers.

5 The dermis consists of a papillary region and a reticular region. The papillary region is connective tissue containing blood vessels, nerves, oil glands, hair follicles, and papillae. The reticular region is connective tissue containing fat and sweat glands.

EPIDERMAL DERIVATIVES

1 The epidermal derivatives of the skin are hair, sebaceous glands, sudoriferous glands, and nails.

2 Hairs consist of a shaft above the surface, a root anchored in the dermis, and a hair follicle.

3 Sebaceous glands are connected to hair follicles. They secrete sebum, which moistens hair and waterproofs the skin.

4 Sudoriferous glands produce perspiration, which carries wastes to the surface and assists temperature regulation.

5 Nails are modified epidermal cells. The principal parts of a nail are the body, free edge, root, eponychium, and matrix.

HOMEOSTASIS

1 Normal body temperature is 37°C (98.6°F).

2 If environmental temperature is high, skin receptors sense the stimulus (heat) and generate impulses that are transmitted to the brain. The brain then causes the sweat glands to produce sweat. As the sweat evaporates, the skin is cooled.

3 The skin-cooling response is a negative feedback mechanism.

REVIEW QUESTIONS

1 Define an organ. In what respect is the skin an organ? What is the integumentary system?

2 List the principal functions of the skin.

3 Compare the structures of epidermis and dermis. What is the subcutaneous layer?

4 List and describe the epidermal layers from the inside outward. What is the importance of each layer?

5 How is the dermis adapted to receive stimuli for touch, pressure, or pain?

6 Describe the structure of a hair. How are hairs moistened? What produces "goosebumps" or "gooseflesh"?

7 Contrast the locations and functions of sebaceous glands and sudoriferous glands. What are the name and chemical components of the secretions of each?

8 From what layer of the skin do nails form? Describe the principal parts of a nail.

9 Explain with a labeled diagram how the skin helps maintain normal body temperature.

10 What is a feedback system? A negative feedback system? Relate these definitions to the maintenance of normal body temperature.

Unit II
Principles of
Support
and Movement

This unit considers two primary themes—support and movement. You will study the various ways in which the body is supported and the different movements it can perform. Both support and movement are made possible by the cooperative effort of bones, joints, and muscles.

Chapter 6
Skeletal Tissue

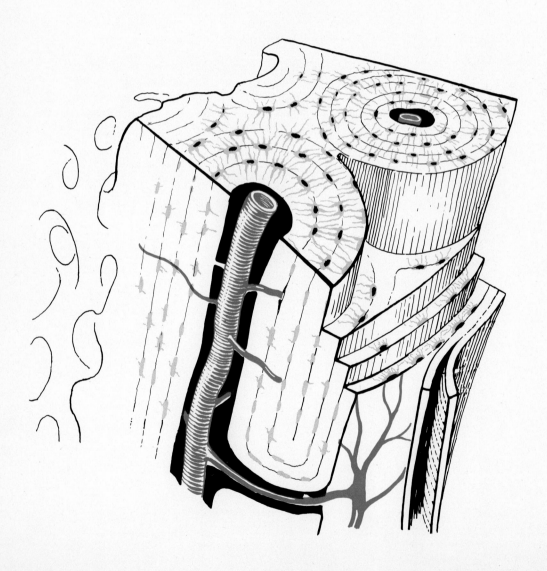

STUDENT OBJECTIVES

■ Describe the components of the skeletal system.

■ Describe the functions of the skeletal system.

■ Describe the gross features of a long bone.

■ Describe the histological features of dense bone tissue.

■ Compare the histological characteristics of spongy and dense bone.

■ Define ossification.

■ Contrast the steps involved in intramembranous and endochondral ossification.

■ Interpret roentgenograms of normal ossification.

■ Describe bone construction and destruction as a homeostatic mechanism.

■ Describe the conditions necessary for normal bone growth.

■ Define rickets and osteomalacia as vitamin deficiency disorders.

■ Contrast the causes and clinical symptoms associated with osteoporosis.

■ Define a fracture and list 13 kinds of fractures.

■ Describe the sequence of events involved in fracture repair.

■ Define medical terminology associated with the skeletal system.

Without the skeletal system we would be unable to make coordinated movements, such as walking or grasping. The slightest jar to the head or chest could damage the brain or heart. It would even be impossible to chew food. The framework of bones that protects our organs and allows us to move is called the **skeletal system**. Besides bone, the skeletal system consists of cartilage in the nose, larynx, outer ear, and at bone attachments. The points where bones attach to each other are called *joints* or *articulations*.

FUNCTIONS

The skeletal system performs several basic functions. First, it *supports* the soft tissues of the body so that the form of the body and an erect posture can be maintained. Second, the system *protects* delicate structures – the brain, the spinal cord, the lungs, the heart, the major blood vessels in the thoracic cavity. Third, the bones serve as *levers* to which the muscles of the body are attached. When the muscles contract, the bones acting as levers produce *movement*. Fourth, the bones serve as *storage areas* for mineral salts – especially calcium and phosphorus. A fifth major feature of the skeletal system is *blood-cell production,* which occurs in the red marrow of the bones. This process is referred to as hematopoiesis.

HISTOLOGY

Structurally, the skeletal system consists of two types of connective tissue: cartilage and bone. In Chapter 4, we discussed the microscopic structure of cartilage. Here our attention will be directed to discussing the microscopic structure of bone tissue.

Like other connective tissues, **bone,** or **osseous tissue,** contains a great deal of intercellular substance surrounding widely separated cells. Unlike other connective tissues, the intercellular substance of bone contains abundant mineral salts, primarily calcium phosphate and calcium carbonate. These salts are responsible for the hardness of bone, which is thus said to be ossified. Embedded in the intercellular substance are collagenous fibers that reinforce the tissue.

The microscopic structure of bone may be analyzed by considering the anatomy of a long bone such as the humerus (arm bone). As shown in Figure 6-1a, a typical long bone consists of the following parts:

1 **Diaphysis,** which is the shaft or long, main portion.

2 **Epiphyses,** which are the extremities or ends of the bone.

3 **Articular cartilage,** a thin layer of hyaline cartilage covering the epiphysis where the bone forms a joint with another bone.

4 **Periosteum,** a dense, white, fibrous membrane covering the remaining surface of the bone. The periosteum (*peri* = around; *osteo* = bone) consists of two layers. The outer, *fibrous layer* is composed of connective tissue containing blood vessels, lymphatic vessels, and nerves that pass into the bone. The inner *osteogenic layer* contains elastic fibers, blood vessels, and **osteoblasts** – cells responsible for forming new bone during growth and repair (Figure 6-1c). The word *blast* means a germ or bud. It denotes an immature cell or tissue that later develops into a specialized form. The periosteum is essential for bone growth, repair, and nutrition. It also serves as a point of attachment for ligaments and tendons. A photomicrograph of the periosteum is shown in Figure 6-2.

5 **Medullary** (or **marrow**) **cavity,** which is the space within the diaphysis that contains the fatty *yellow marrow* in adults.

6 **Endosteum,** a layer of osteoblasts that lines the medullary cavity and contains scattered osteoclasts – cells that may assume a role in the removal of bone.

Bone is not a solid, homogeneous substance. In fact, all bone is porous. The pores contain living cells and blood vessels that supply the cells with nutrients. The pores also make bones lighter. Think of how much more energy you would expend if you had to drag around solid bones. Depending on the degree of porosity, the regions of a bone may be categorized as spongy or compact (Figure 6-1b). **Spongy,** or **cancellous,** bone tissue contains many large spaces filled with marrow. It makes up most of the bone tissue of short, flat, and irregularly shaped bones and most of the epiphyses of long bones. Spongy bone tissue also provides a storage area for marrow. **Compact,** or **dense,** bone tissue, by contrast, contains few spaces. It is deposited in a thin layer over the spongy bone tissue. The layer of compact bone is thicker in the diaphyses than the epiphyses. Compact bone tissue provides protection and support and helps the long bones resist the stress of weight placed on them (Figure 6-3).

We can compare the differences between spongy and compact bone tissue by looking at the highly magnified, transverse sections in Figure 6-1. One main difference is that adult compact bone has a concentric-ring structure, whereas spongy bone does not. Blood vessels and nerves from the periosteum penetrate the compact bone through *Volkmann's canals* (Figure 6-1c). The blood vessels of these canals connect with blood vessels and nerves of the medullary cavity and those of

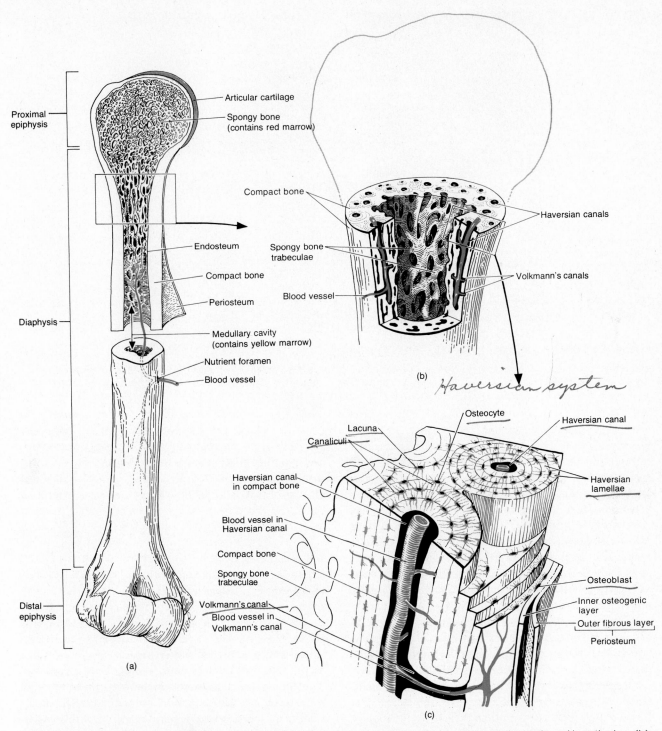

Figure 6-1   Osseous tissue. (a) Macroscopic appearance of a long bone that has been partially sectioned lengthwise. (b) Histological structure of bone. (c) Enlarged aspect of Haversian systems in compact bone.

the *Haversian canals.* The Haversian canals run longitudinally through the bone. Around them are *lamellae*—concentric rings of hard, calcified, intercellular substance. Between the lamellae are small spaces called *lacunae,* where osteocytes are found. **Osteocytes** are mature osteoblasts that have lost their ability to produce new bone tissue. Radiating in all directions from the lacunae are minute canals called *canaliculi,* which con-

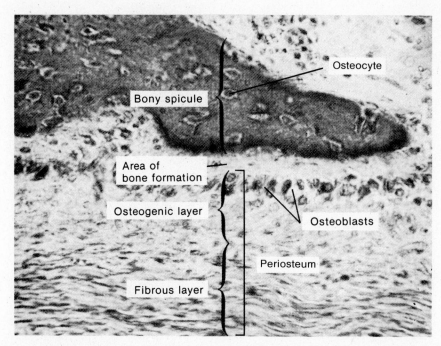

Figure 6-2 Photomicrograph of compact bone showing the osteogenic and fibrous layers of the periosteum at a magnification of 300×. (Courtesy of Donald I. Patt, from *Comparative Vertebrate Histology,* by Donald I. Patt and Gail R. Patt, Harper & Row, Publishers, Inc., New York, 1969.)

nect with other lacunae and, eventually, with the Haversian canal. Thus an intricate network is formed throughout the bone. This branching network provides numerous routes for blood vessels so that nutrients can reach the osteocytes and wastes can be removed. Each Haversian canal, with its surrounding lamellae, lacunae, osteocytes, and canaliculi, is called a **Haversian system** or **osteon.** Haversian systems are characteristic of adult bone. The areas between Haversian systems contain *interstitial lamellae.* These also possess lacunae with osteocytes and canaliculi, but their lamellae are usually not connected to the Haversian systems.

In contrast to compact bone, spongy bone does not contain true Haversian systems. It consists of an irregular latticework of thin plates of bone called *trabeculae.* The spaces between the trabeculae of some bones are filled with *red marrow.* Within the trabeculae lie the small spaces called lacunae, which contain the osteocytes. Blood vessels from the periosteum penetrate the spongy bone, and the osteocytes in the trabeculae are nourished directly from the blood circulating through the marrow cavities. The cells of red marrow are responsible for producing new blood cells.

Most people think of bone as a very hard, white material. Yet the bones of an infant are not hard at all. It is common knowledge that it is dangerous to drop an infant, especially on its head. Its bones are "soft," and the fall may change the shape of its head or damage its brain.

Moreover, most of us know that a child's bones are generally more pliable than those of an adult. The final shape and hardness of adult bones require many years to develop and depend on a complex series of chemical changes. Let us now see how bones are formed and how they grow.

OSSIFICATION

The process by which bone forms in the body is called **ossification** or **osteogenesis.** The "skeleton" of a human embryo is composed of fibrous membranes and hyaline cartilage. Both are shaped like bones and provide the medium for ossification. Ossification begins around the sixth week of embryonic life and continues until adulthood. Two kinds of bone formation are recognized. The first is called intramembranous ossification. This term refers to the formation of bone directly on or within the fibrous membranes (*intra* = within; *membranous* = membrane). The second kind, endochondral (intracartilaginous) ossification, refers to the formation of bone in cartilage (*endo* = within; *chondro* = cartilage). These two kinds of ossification do *not* lead to differences in the structure of mature bones. They simply indicate different methods of bone formation.

The first stage in the development of bone is the migration of embryonic connective tissue cells (mesenchymal cells) into the area where bone formation is

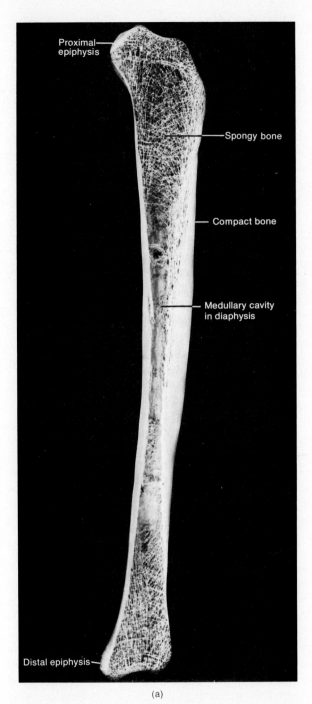

Proximal
epiphysis

Spongy bone

Compact bone

Medullary cavity
in diaphysis

Distal epiphysis

(a)

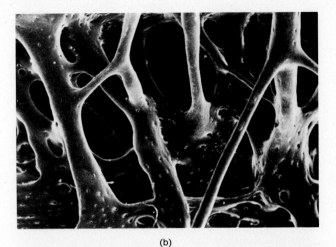

(b)

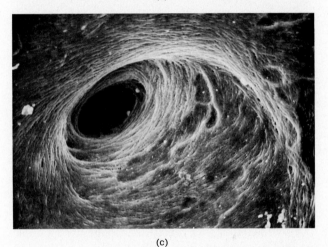

(c)

Figure 6-3 Spongy and compact bone. (a) Photograph of a longitudinal section through the tibia to illustrate the differences between spongy and compact bone. (Courtesy of Cornelius Rosse and D. Kay Clawson, *Introduction to the Musculoskeletal System,* Harper & Row, Publishers, Inc., New York, 1970.) (b) Scanning electron micrograph of spongy bone trabeculae at a magnification of 25×. (c) Scanning electron micrograph of a view inside a Haversian canal of compact bone at a magnification of 250×. (Electron micrographs courtesy of Fisher Scientific Company and S.T.E.M. Laboratories, Inc., Copyright 1975.)

about to begin. Soon these cells increase in number and size. In some skeletal structures they become chondroblasts; in others some become osteoblasts. The chondroblasts will be responsible for cartilage formation. The osteoblasts will form bone tissue by intramembranous or endochondral ossification.

Intramembranous Ossification

Of the two types of bone formation, the simpler and more direct is **intramembranous ossification.** The flat bones of the roof of the skull, mandible (lower jawbone), and probably part of the clavicles are formed in this way. The essentials of this process are as follows.

Osteoblasts formed from mesenchymal cells cluster in the fibrous membrane. The site of such a cluster is called a *center of ossification.* The osteoblasts then secrete intercellular substances. These substances are partly composed of collagenous fibers that form a framework, or matrix, in which calcium salts are quickly deposited. The deposition of calcium salts is called *calcification.* When a cluster of osteoblasts is completely surrounded by the calcified matrix, it is called a *trabecula*. As trabeculae form in nearby ossification centers, they fuse into the open latticework characteristic of spongy bone. With the formation of successive layers of bone, some osteoblasts become entrapped in the minute spaces called lacunae. The entrapped osteoblasts lose their ability to form bone and are called osteocytes. The spaces between the trabeculae fill with red marrow. The original connective tissue that surrounds the growing mass of bone then becomes the periosteum. The ossified area has now become true spongy bone. Eventually, the surface layers of the spongy bone will be reconstructed into compact bone. Much of this newly formed bone will be destroyed and reformed so the bone may reach its final adult size and shape.

Endochondral Ossification

The replacement of cartilage by bone is called **endochondral ossification.** Most bones of the body, including the skull, are formed in this way. Since this type of ossification is best observed in a long bone, we will investigate the tibia, or shin bone (Figure 6-4).

Early in embryonic life, a cartilage model of the future bone is laid down. This model is covered by a membrane called the *perichondrium.* Midway along the shaft of this model a blood vessel penetrates the perichondrium, stimulating the cells in the internal layer of the perichondrium to enlarge and become osteoblasts. The cells begin to form a collar of spongy bone around the middle of the diaphysis of the cartilage model. Once the perichondrium starts to form bone, it is called the periosteum. Simultaneously with the appearance of the bone collar and the penetration of blood vessels, changes occur in the cartilage in the center of the diaphysis. In this area, the *primary ossification center,* cartilage cells hypertrophy (increase in size) – probably because they accumulate glycogen for energy and produce enzymes to catalyze future chemical reactions. When the hypertrophied cells burst, there is a change in extracellular pH to a more alkaline pH causing the intercellular substance to become *calcified* – that is, minerals are deposited within it. Once the cartilage becomes calcified, nutritive materials required by the cartilage cells can no longer diffuse through the intercellular substance and the cartilage cells die. Then the intercellular substance begins to degenerate, leaving large cavities in the cartilage model. The blood vessels grow along the spaces where cartilage cells were previously located and enlarge the cavities further. Gradually, these spaces in the middle of the shaft join with each other, and the marrow cavity is formed.

As these developmental changes are occurring, the osteoblasts of the periosteum deposit successive layers of bone on the outer surface so that the collar thickens, becoming thickest at the diaphysis. The cartilage model continues to grow at its ends, steadily increasing in length. Eventually, blood vessels enter the epiphyses and secondary ossification centers appear in the epiphyses and also lay down spongy bone. In the tibia, one secondary ossification center develops in the proximal epiphysis soon after birth. The other center develops in the distal epiphysis during the child's second year.

After the two secondary ossification centers have formed, bone tissue has completely replaced cartilage, except in two regions. Cartilage continues to cover the articular surfaces of the epiphyses, where it is called articular cartilage. It also remains as a plate between the epiphysis and diaphysis, in which case it is called the **epiphyseal plate.** The epiphyseal plate allows the bone to increase in length until early adulthood. As the child grows, cartilage cells are produced by mitosis on the epiphyseal side of the plate. Cartilage cells are then destroyed and the cartilage is replaced by bone on the diaphyseal side of the plate. In this way, the thickness of the epiphyseal plate remains fairly constant but the bone on the diaphyseal side increases in length. Growth in diameter occurs along with growth in length. In this process, the bone lining the marrow cavity is destroyed so that the cavity increases in diameter. At the same time, osteoblasts from the periosteum add new osseous tissue around the outer surface of the bone. Initially, diaphyseal and epiphyseal ossification produce only spongy bone. Later, by reconstruction, the outer region of spongy bone is reorganized into compact bone.

Around the age of 17, the epiphyseal cartilage cells stop duplicating and the entire cartilage is slowly replaced by bone. Bone growth stops. The remnant of the epiphyseal plate is called the **epiphyseal line.** Ossification of all bones is usually completed by age 25. Figure 6-5 consists of a series of **roentgenograms** (photographs taken with x rays) that show ossification in the epiphyses of two long bones at the knee. Bones undergoing either intramembranous or endochondral ossification are continually remodeling from the time that initial calcification occurs until the final structure appears. Compact bone is formed by the transformation of spongy bone. The diameter of a long bone is increased by the destruction of the bone closest to the marrow cavity and the construction of new bone around the outside of the diaphysis. However, even after bones have reached their adult shapes and sizes, old bone is perpetually destroyed and new osseous tissue is formed in its place.

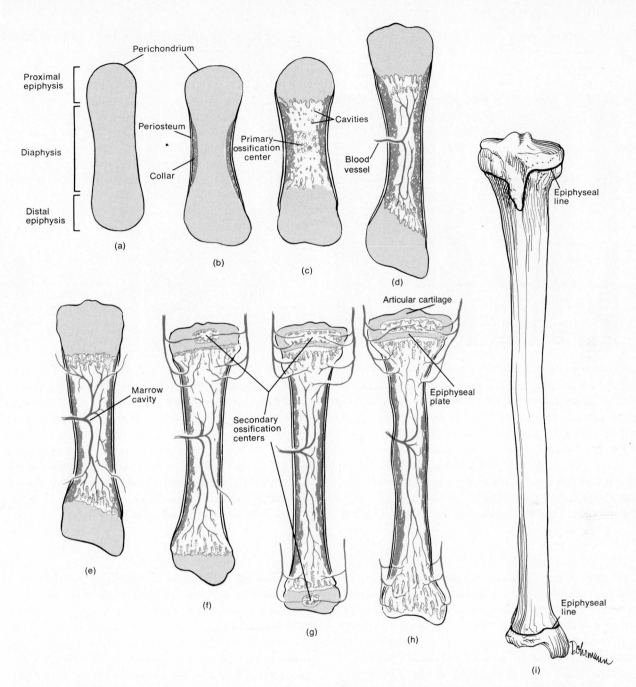

Figure 6-4 Endochondral ossification of the tibia. (a) Cartilage model. (b) Collar formation. (c) Development of primary ossification center. (d) Entrance of blood vessels. (e) Marrow-cavity formation. (f) Thickening and lengthening of the collar. (g) Formation of secondary ossification centers. (h) Remains of cartilage as the articular cartilage and epiphyseal plate (i) Formation of the epiphyseal line.

HOMEOSTASIS

Bone shares with skin the unique feature of replacing itself throughout adult life. This **remodeling** takes place at different rates in various body regions. The distal portion of the femur (thighbone) is replaced about every 4 months. By contrast, bone in certain areas of the shaft will not be completely replaced during the individual's life. Remodeling allows worn or injured bone to be removed and replaced with new tissue. It also allows bone to serve as the body's storage area for calcium. Many other tissues in the body need small amounts of calcium in order to perform their functions. For exam-

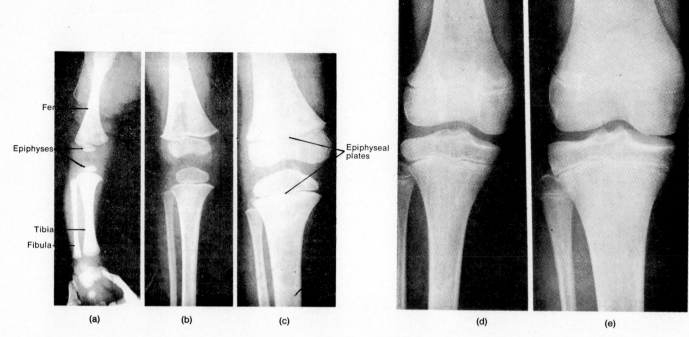

Figure 6-5 Roentgenograms of normal ossification at the knee. (a) One-month-old infant. The epiphyses at the knee are mostly cartilage. Ossification centers have formed for the femur and tibia. The space between the ends of the bones is occupied by epiphyseal cartilage. (b) Two-year-old child. Centers of ossification have grown. The white transverse zones marking the ends of the shafts are areas where mineral salts are deposited temporarily around the degenerating cartilage cells. (c) Five-year-old child. The epiphyses have assumed the shape of adult bones. The epiphyseal plates are clearly visible between the epiphyses and diaphyses of all three bones. (d) Eight-year-old child. The epiphyseal plates are still distinct as ossification continues. (e) Twelve-year-old child. The epiphyses have ossified almost completely. The epiphyseal plates are assuming the character of epiphyseal lines. (Courtesy of Lester W. Paul and John H. Juhl, *The Essentials of Roentgen Interpretation,* 3d ed., Harper & Row, Publishers, Inc., New York, 1972.)

ple, muscle needs calcium in order to contract. The muscle cells take their calcium from the blood. However, the blood itself needs calcium in order to clot. The blood continually trades off calcium with the bones, removing calcium when other tissues are not receiving enough of this element and resupplying the bones with calcium to keep them from losing too much bone mass.

The cells believed to be responsible for the destruction of bone tissue are called **osteoclasts** (*clast* = break). In the healthy adult, a delicate homeostasis is maintained between the action of the osteoclasts and the removal of calcium, on one hand, and the action of the bone-making osteoblasts and the deposition of calcium on the other. Should too much new tissue be formed, the bones become abnormally thick and heavy. If too much calcium is deposited in the bone, the surplus may form thick bumps, or spurs, on the bone that interfere with movement at joints. A loss of too much tissue or calcium weakens the bones and allows them to break easily.

Normal bone growth in the young and bone replacement in the adult depend on several factors. First, sufficient quantities of calcium and phosphorus, components of the primary salt that makes bone hard, must be included in the diet. Second, the individual must obtain sufficient amounts of vitamins A, C, and D. These substances are particularly responsible for the proper utilization of calcium and phosphorus by the body. Third, the body must manufacture the proper amounts of the hormones responsible for bone tissue activity.

Certain hormones are responsible for the general growth of bones. Too much or too little of these hormones during childhood makes the adult abnormally tall or short. Other hormones specialize in regulating the osteoclasts. And still others, especially the sex hormones, aid osteoblastic activity and thus promote the growth of new bone. The sex hormones act as a double-edged sword. They aid in the growth of new bone, but they also bring about the degeneration of all the cartilage cells in the epiphyseal plates. Because of the sex hormones, the typical adolescent experiences a spurt of growth during puberty, when sex hormone levels start to increase. The individual then quickly completes the growth process as the epiphyseal cartilage disappears. Premature puberty can actually prevent one from reaching an average adult height because of the simultaneous premature degeneration of the plates. Still another kind of hormone, produced by the parathyroid glands in the neck, determines whether the blood will deposit calcium and phosphorus in osseous tissue or whether it will remove these elements from the bones.

DISORDERS

Many bone disorders result from deficiencies in vitamins or minerals or from too much or too little of the hormones that regulate bone homeostasis. Infections and tumors are also responsible for certain bone disorders.

VITAMIN DEFICIENCIES

Vitamin D is important to normal bone growth and maintenance. It is essential for the synthesis of a protein that transports the calcium obtained from foods across the lining of the intestine and into the extracellular fluid. When the body lacks this vitamin, it is unable to absorb calcium and phosphorus. A deficiency of vitamin D produces rickets in children and osteomalacia in adults.

RICKETS

In the condition called **rickets,** epiphyseal cartilage cells cease to degenerate and new cartilage continues to be produced. Epiphyseal cartilage thus becomes wider than normal. At the same time, the soft matrix laid down by the osteoblasts in the diaphysis fails to calcify. As a result, the bones stay soft. When the child walks, the weight of the body causes the bones in the legs to bow. Malformations of the head, chest, and pelvis also occur.

OSTEOMALACIA

A deficiency of vitamin D in the adult causes the bones to give up excessive amounts of calcium and phosphorus. This loss, called *demineralization,* is especially heavy in the bones of the pelvis, legs, and spine. Demineralization caused by vitamin D deficiency is called **osteomalacia** (*malacia* = softness). After the bones demineralize, the weight of the body produces a bowing of the leg bones, a shortening of the backbone, and a flattening of the pelvic bones. Osteomalacia mainly affects women who live on poor cereal diets devoid of milk, are seldom exposed to the sun, and have repeated pregnancies that deplete the body of calcium.

OSTEOPOROSIS

Osteoporosis or **osteopenia** is a bone disorder affecting the middle-aged and elderly—white women more than men of black or white ancestry and black women not at all. Between puberty and the middle years, the sex hormones maintain osseous tissue by stimulating the osteoblasts to form new bone. After menopause, however, women produce smaller amounts of sex hormones. During old age, both men and women produce smaller amounts. As a result, the osteoblasts become less active and there is a decrease in bone mass. Osteoporosis affects the entire skeletal system, especially the spine, legs, and feet. As the spine collapses and curves, the breasts sag and the ribs fall on the pelvic rim. This condition leads to gastrointestinal distension, which causes distress and an overall decrease in muscle tone.

Among the factors implicated in bone loss is a decrease in estrogen, calcium deficiency and malabsorption, vitamin D deficiency, loss of muscle mass, inactivity, and high-protein diets. Osteoporosis seems to be an inevitable accompaniment of aging, except in black women.

FRACTURES

In simplest terms, a **fracture** is any sudden break in a bone. Sometimes the fracture is restored to normal position by manipulation without surgery. This procedure of setting a fracture is called *closed reduction.* In other cases, the fracture must be exposed by surgery before the break is rejoined. This procedure is known as *open reduction.* Although fractures of bones of the extremities may be classified in several different ways, the following scheme is useful (see Figure 6-6):

1 **Partial** The break across the bone is incomplete.
2 **Complete** The break occurs across the entire bone. The bone is completely broken into two pieces.
3 **Simple** or **closed** The fractured bone does not break through the skin.
4 **Compound** or **open** The broken ends of the fractured bone protrude through the skin.
5 **Comminuted** The bone is splintered at the site of impact and smaller fragments of bone are found between the two main fragments.
6 **Greenstick** A partial fracture in which one side of the bone is broken and the other side bends.
7 **Spiral** The bone has been twisted apart.

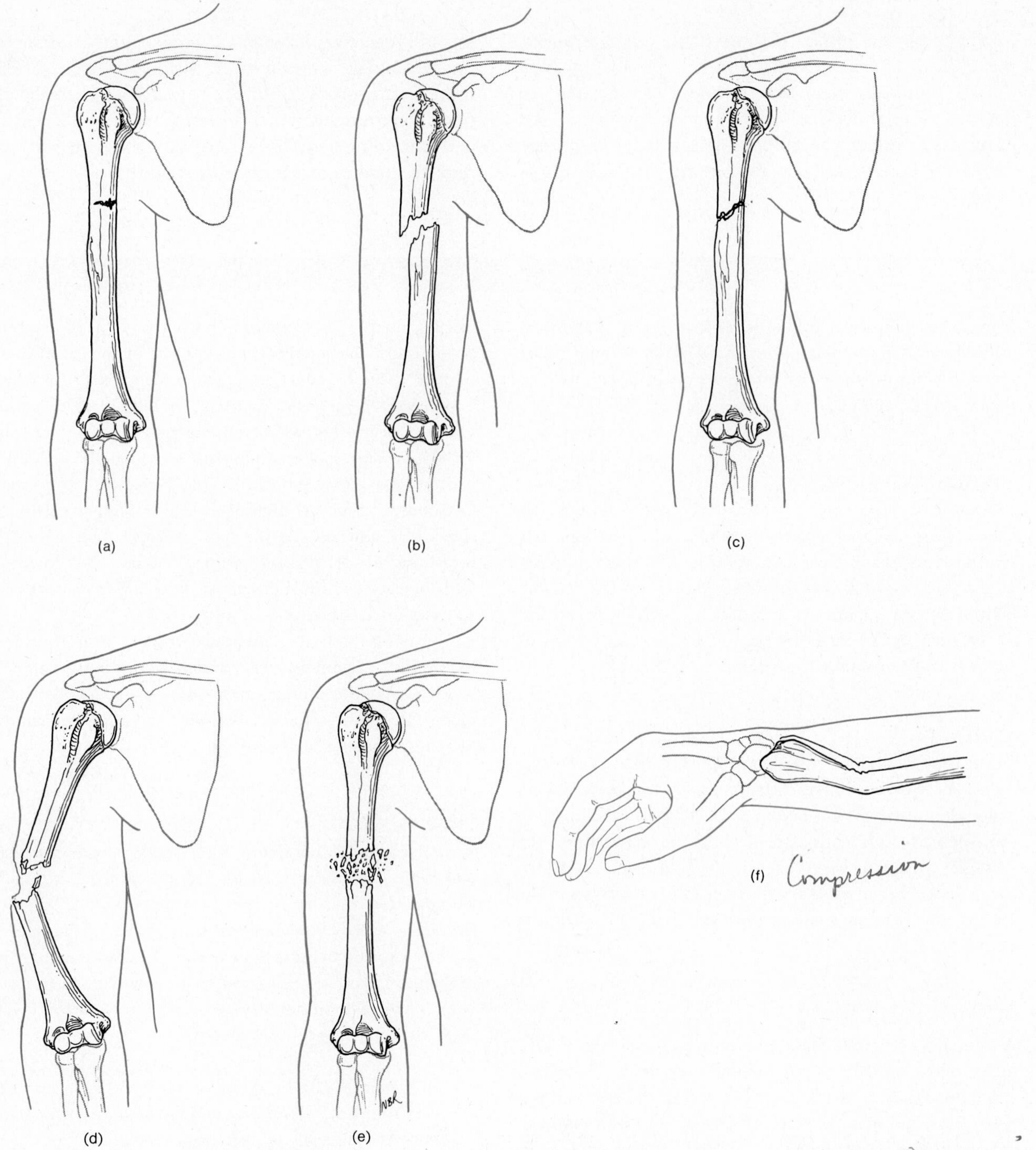

(a)

(b)

(c)

(d)

(e)

(f) Compression

Compression

Figure 6-6 Types of fractures. (a) Partial. (b) Complete. (c) Simple. (d) Compound. (e) Comminuted. (f) Greenstick.

8 Transverse A fracture at right angles to the long axis of the bone.

9 Impacted A fracture in which one fragment is firmly driven into the other.

10 Pott's A fracture of the distal end of the fibula, with serious injury of the distal tibial articulation.

11 Colles' A fracture of the distal end of the radius in which the distal fragment is displaced posteriorly.

12 Displaced A fracture in which the anatomical alignment of the bone fragments is not preserved.

13 Nondisplaced A fracture in which the anatomical alignment of the bone fragments has not been disrupted.

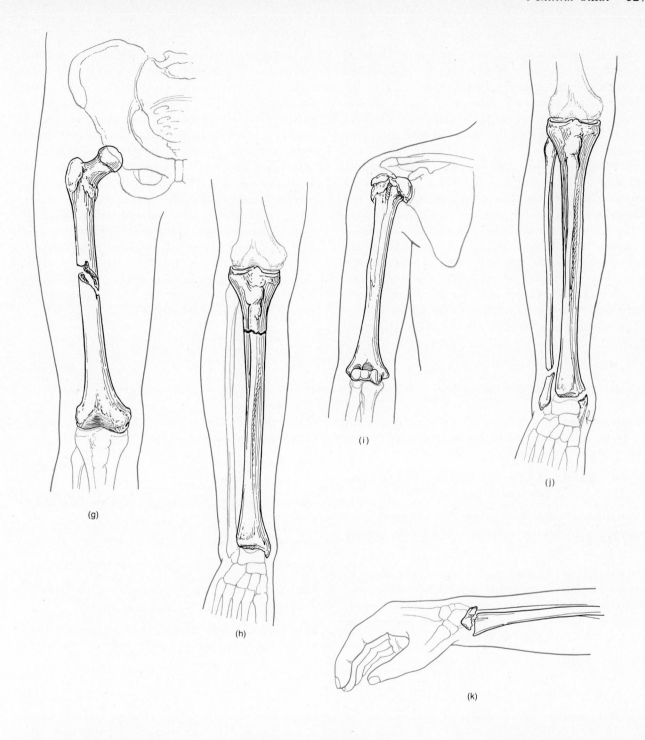

Figure 6-6 (cont.) Types of fractures. (g) Spiral. (h) Transverse. (i) Impacted. (j) Pott's. (k) Colles'

Unlike the skin, which may repair itself within days, or muscle, which may mend in weeks, a bone sometimes requires months to heal. A fractured femur, for example, may take 6 months to heal because sufficient calcium to strengthen and harden new bone is deposited only gradually. Bone cells also grow and reproduce slowly. Moreover, the blood supply to

bone is poor, which explains the difficulty in the healing of an infected bone.

The following steps occur in the repair of a fracture:

1 As a result of the fracture, blood vessels crossing the fracture line are broken. These vessels are found in the

periosteum, Haversian systems, and marrow cavity. As blood pours from the torn ends of the vessels, it coagulates and forms a clot in and about the site of the fracture. This clot, called a **fracture hematoma,** usually occurs 6 to 8 hours after the injury. Since the circulation of blood ceases when the fracture hematoma forms, bone cells and periosteal cells at the fracture line die.

2 A growth of new bone tissue – a **callus** – develops in and around the fractured area. It forms a bridge between separated areas of bone. The callus that forms from the osteogenic cells of the torn periosteum and develops around the outside of the fracture is called an *external callus.* The callus that forms from the osteogenic cells of the endosteum and develops between the two ends of bone fragments and between the two marrow cavities is called the *internal callus.*

Approximately 48 hours after a fracture occurs, the cells that ultimately repair the fracture become actively mitotic. These cells come from the osteogenic layer of the periosteum,

the endosteum of the marrow cavity, and the bone marrow. As a result of their accelerated mitotic activity, the cells of the three regions grow toward the fracture. During the first week following the fracture, the cells of the endosteum and bone marrow form new trabeculae in the marrow cavity near the line of fracture. This is the internal callus. During the next few days, osteogenic cells of the periosteum form a collar around each bone fragment. The collar, or external callus, is replaced by trabeculae. The trabeculae of the calli are joined to living and dead portions of the original bone fragments.

3 The final phase of fracture repair is the **remodeling** of the calli. Dead portions of the original fragments are gradually resorbed. Compact bone replaces spongy bone around the periphery of the fracture. In some cases, the healing is so complete that the fracture line is undetectable, even by x ray. However, a thickened area on the surface of the bone usually remains as evidence of the fracture site.

MEDICAL TERMINOLOGY

Achondroplasia (*a* = without; *chondro* = cartilage; *plasia* = growth)   Imperfect ossification within cartilage of long bones during fetal life; also called fetal rickets.

Brodie's abscess   Infection in the spongy tissue of a long bone, with a small inflammatory area.

Craniotomy (*cranium* = skull; *tome* = a cutting)   Any surgery that requires cutting through the bones surrounding the brain.

Necrosis (*necros* = death; *osis* = condition)   Death of tissues or organs; in the case of bone, necrosis results from deprivation of blood supply; could result from fracture, extensive removal of periosteum in surgery, exposure to radioactive substances, and other causes.

Osteitis   Inflammation or infection of bone.

Osteoarthritis (*arthro* = joint)   A degenerative condition of bone and also the joint.

Osteoblastoma (*oma* = tumor)   A benign tumor of the osteoblasts.

Osteochondroma   A benign tumor of the bone and cartilage.

Osteoma   A benign bone tumor.

Osteomyelitis   Infection that involves bone marrow.

Osteosarcoma (*sarcoma* = connective tissue tumor)   A malignant tumor composed of osseous tissue.

Pott's disease   Inflammation of the backbone, caused by the microorganism that produces tuberculosis.

STUDY OUTLINE

FUNCTIONS

1 The skeletal system consists of all bones attached at joints, cartilage between joints, and cartilage found elsewhere (nose, larynx, and outer ear).

2 The functions of the skeletal system include support, protection, leverage, mineral storage, and blood-cell formation.

HISTOLOGY

1 Parts of a typical long bone are the diaphysis (shaft), epiphyses (ends), articular cartilage, periosteum, medullary (marrow) cavity, and endosteum.

2 Cancellous, or spongy, bone has many marrow-filled pores and does not contain Haversian systems. It consists of trabeculae containing osteocytes and lacunae.

3 Dense, or compact, bone has fewer pores and fewer networks of passageways called Haversian systems. Dense bone lies over spongy bone and composes most of the bone tissue of the diaphyses.

OSSIFICATION

1 Bone forms by a process called osteogenesis, which begins when mesenchymal cells become transformed into osteoblasts.

2 Intramembranous ossification occurs within fibrous membranes of the embryo and the adult.

3 Endochondral ossification occurs within a cartilage model.

4 The primary ossification center of a long bone is in the diaphysis. Cartilage degenerates, leaving cavities that merge to form the marrow cavity. Osteoblasts lay down bone. Next ossification occurs in the epiphyses, where bone replaces cartilage, except for the epiphyseal plate.

5 In both types of ossification, spongy bone is laid down first. Dense bone is later reconstructed from spongy bone.

HOMEOSTASIS

1 The homeostasis of bone growth depends on a balance between bone formation and destruction.

2 Old bone is constantly destroyed by osteoclasts while new bone is constructed by osteoblasts. This process is called remodeling.

3 Normal growth depends on calcium, phosphorus, and vitamins (A, C, and D) and is controlled by hormones that are responsible for bone mineralization and reabsorption.

REVIEW QUESTIONS

1 Define the skeletal system. What are its five principal functions?

2 Diagram the parts of a long bone, and list the functions of each part. What is the difference between compact and spongy bone tissue? Diagram the microscopic structure of bone.

3 What is meant by ossification? When does the process begin and end?

4 Distinguish between the two principal kinds of ossification.

5 Outline the major events involved in intramembranous and endochondral ossification.

6 What is a roentgenogram?

7 List the primary factors involved in bone growth.

8 How does osteoblast activity in balance with osteoclast activity demonstrate the homeostasis of bone?

Chapter 7
The Skeletal System:
The Axial Skeleton

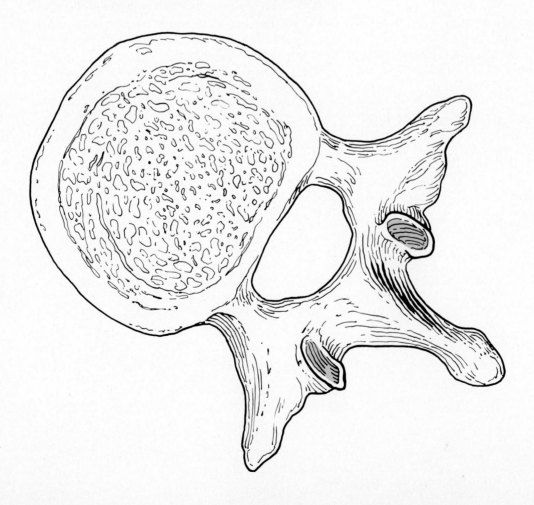

STUDENT OBJECTIVES

■ Define the four principal types of bones in the skeleton.

■ Explain the relationship between bone structure and function.

■ Describe the various markings on the surfaces of bones.

■ Relate the structure of the marking to its function.

■ Describe the components of the axial and appendicular skeleton.

■ Identify the bones of the skull and the major markings associated with each.

■ Identify the sutures and fontanels of the skull.

■ Identify the paranasal sinuses of the skull in projection diagrams and roentgenograms.

■ Identify the bones of the vertebral column and their principal markings.

■ List the defining characteristics and curves of each region of the vertebral column.

■ Identify the bones of the thorax and their principal markings.

The **skeletal system** forms the framework of the body. For this reason, a familiarity with the names, shapes, and positions of individual bones will help you to understand some of the other organ systems. For example, movements such as throwing a ball, typing, and walking require the coordinated use of bones and muscles. To understand how muscles produce different movements, you need to learn the parts of the bones to which the muscles attach. The respiratory system is also highly dependent on bone structure. The bones in the nasal cavity form a series of passageways that help clean, moisten, and warm inhaled air. Furthermore, the bones of the thorax are specially shaped and positioned so the chest can expand during inhalation. Many bones also serve as landmarks to students of anatomy as well as to surgeons. Blood vessels and nerves often run parallel to bones. These structures can be located more easily if the bone is identified first. The superior portions of the lungs are located just inferior to the clavicle. The bottom of the rib cage can be used to locate the diaphragm and liver.

TYPES OF BONES

The bones of the body may be classified into four principal types: long, short, flat, and irregular. **Long bones** have greater length than width and consist of a diaphysis (shaft) and two epiphyses (extremities). They are slightly curved for strength. A curved bone is structurally designed to absorb the stress of the body weight at several different points so the stress is evenly distributed. If such bones were straight, the weight of the body would be unevenly distributed and the bone would easily fracture. Examples of long bones include bones of the thighs, legs, toes, arms, forearms, and fingers. Figure 6-1a shows the parts of a long bone. **Short bones** are somewhat cube-shaped and nearly equal in length and width. Their texture is spongy throughout, except at the surface, where there is a thin layer of compact bone. Examples of short bones are the wrist and ankle bones. **Flat bones** are generally thin and composed of two more or less parallel plates of compact bone enclosing a layer of spongy bone. The term *diploe* is applied to the spongy bone of the cranial bones. Flat bones afford considerable protection and provide extensive areas for muscle attachment. Examples of flat bones include the cranial bones, which protect the brain, the sternum and ribs, which protect organs in the thorax, and the scapulas. **Irregular bones** have complex shapes and cannot be grouped into any of the three categories just described. They also vary in the amount of spongy and compact bone present. Such bones are the vertebrae and certain facial bones.

Besides these four principal types of bones, two other kinds are recognized. **Wormian,** or **sutural,** bones are small clusters of bones between the joints of certain cranial bones. Their number varies greatly from person to person. **Sesamoid bones** are small bones in tendons where considerable pressure develops—for instance, in the wrist. These bones, like the Wormian bones, are also variable in number. Two sesamoid bones, the patellas, or kneecaps, are present in all individuals.

SURFACE MARKINGS

The surfaces of bones reveal various **markings.** The structure of many of these markings indicates their functions. Long bones that bear a great deal of weight have large, rounded ends that can form sturdy joints. Other bones have depressions that receive the rounded ends. Rough areas serve for the attachment of muscles, tendons, and ligaments. Grooves in the surfaces of bones provide for the passage of blood vessels, and openings occur where blood vessels and nerves pass through the bone. Exhibit 7-1 describes the different markings and their functions.

DIVISIONS OF THE SKELETAL SYSTEM

The adult human skeleton consists of approximately 206 bones grouped in two principal divisions: the **axial** and the **appendicular.** The *axis,* or center, of the human body is a straight line that runs vertically along the body's center of gravity. This imaginary line runs through the head and down to the space between the feet. The midsagittal section is drawn through this line. The axial division of the skeleton consists of the bones that lie around the axis: ribs, breastbone, the bones of the skull, and backbone. The appendicular division contains the bones of the free *appendages,* which are the upper and lower extremities, plus the bones called *girdles,* which connect the free appendages to the axial skeleton. The 80 bones of the axial division and the 126 bones of the appendicular division are typically grouped as follows:

Axial skeleton
Skull
1 Cranium 8
2 Face 14
Hyoid (above the larynx) 1
*Auditory ossicles,** 3 in each ear 6
Vertebral column 26

*Although the auditory ossicles are not considered part of the axial or appendicular skeleton, but rather as a separate group of bones, they are placed with the axial skeleton for convenience.

Exhibit 7—1 BONE MARKINGS

Marking	Description	Example
Depressions and openings		
Fissure	A narrow, cleftlike opening between adjacent parts of bones through which blood vessels and nerves pass.	Superior orbital fissure of the sphenoid bone (Figure 7-2).
Foramen (*foramen* = hole)	A rounded opening through which blood vessels, nerves, and ligaments pass.	Infraorbital foramen of the maxilla (Figure 7-2).
Meatus (canal)	A tubelike passageway running within a bone.	External auditory meatus of the temporal bone (Figure 7-2).
Paranasal sinus (*sin* = cavity)	An air-filled cavity within a bone connected to the nasal cavity.	Frontal sinus of the frontal bone (Figure 7-8).
Groove or sulcus (*sulcus* = ditchlike groove)	A furrow or groove that accommodates a soft structure such as a blood vessel, nerve, or tendon.	Intertubecular sulcus of the humerus (Figure 8-4).
Fossa (*fossa* = basinlike depression)	A depression in or on a bone.	Mandibular fossa of the temporal bone (Figure 7-4).
Process	Any prominent, roughened projection.	Mastoid process of the temporal bone (Figure 7-2).
Processes that form joints		
Condyle (*condylus* = knucklelike process)	A large, convex articular prominence.	Medial condyle of the femur (Figure 8-10).
Head	A rounded articular projection supported on the constricted portion (neck) of a bone.	Head of the femur (Figure 8-10).
Facet	A smooth, flat surface.	Articular facet for the tubercle of rib on a vertebra (Figure 7-14).
Processes to which tendons, ligaments, and other connective tissues attach		
Tubercle (*tuber* = knob)	A small, rounded process.	Greater tubercle of the humerus (Figure 8-4).
Tuberosity	A large, rounded, usually roughened process.	Ischial tuberosity of the hipbone (Figure 8-8).
Trochanter	A large, blunt projection found only on the femur.	Greater trochanter of the femur (Figure 8-10).
Crest	A prominent border or ridge on a bone.	Iliac crest of the hipbone (Figure 8-7).
Line	A less prominent ridge.	Linea aspera of the femur (Figure 8-10).
Spinous process (spine)	A sharp, slender process.	Spinous process of a vertebra (Figure 7-12).
Epicondyle (*epi* = above)	A prominence above a condyle.	Medial epicondyle of the femur (Figure 8-10).

Thorax

1	Sternum	1
2	Ribs	<u>24</u>
		80

Appendicular skeleton

Shoulder girdles

1	Clavicle	2
2	Scapula	2

Upper extremities

1	Humerus	2
2	Ulna	2
3	Radius	2
4	Carpals	16
5	Metacarpals	10
6	Phalanges	28

Pelvic girdle

1	Coxal, hip, or pelvic bone	2

Lower extremities

1	Femur	2
2	Fibula	2
3	Tibia	2
4	Patella	2
5	Tarsals	14
6	Metatarsals	10
7	Phalanges	<u>28</u>
		126

Now that you understand how the skeleton is organized into axial and appendicular divisions, refer to Figure 7-1 to see how the two divisions are joined to form the skeleton. The bones of the axial skeleton are shown in gray. Be certain to locate the following regions of the skeleton: skull, cranium, face, hyoid bone, vertebral column, thorax, shoulder girdle, upper extremity, pelvic girdle, and lower extremity.

SKULL

The **skull,** which contains 22 bones, rests on the superior end of the vertebral column and is composed of two sets of bones: cranial bones and facial bones. The **cranial bones** enclose and protect the brain and the organs of sight, hearing, and balance. The eight cranial bones are the frontal bone, parietal bones (two), temporal bones (two), the occipital bone, sphenoid, and ethmoid. There are 14 **facial bones:** the nasal bones (two), maxillae (two), zygomatic bones (two), mandible, lacrimal bones (two), palatine bones (two), inferior nasal conchae (two), and vomer. Be sure you can locate all the cranial and facial bones in the anterior, lateral, and medial views of the skull. See Figure 7-2.

Sutures

A **suture,** meaning seam, is an immovable joint found only between skull bones. Very little connective tissue is found between the bones of the suture. Four prominent skull sutures include the:

1 **Coronal suture** between the frontal bone and the two parietal bones.
2 **Sagittal suture** between the two parietal bones.
3 **Lambdoidal suture** between the parietal bones and the occipital bone.
4 **Squamosal suture** between the parietal bones and the temporal bones.

Refer to Figures 7-2 and 7-3 for the locations of these sutures. Several other sutures are also shown. Their names are descriptive of the bones they connect. For example, the frontonasal suture is between the frontal bone and the nasal bones. These sutures are indicated in Figures 7-2 to 7-5.

Fontanels

The "skeleton" of a newly formed embryo consists of cartilage or fibrous membrane structures shaped like bones. Gradually the cartilage or fibrous membrane is replaced by bone. At birth, membrane-filled spaces called **fontanels,** meaning fountains, are found between cranial bones. See Figure 7-3. These "soft spots" are areas where the bone-making process is not yet complete. They allow the skull to be compressed during birth. Physicians find the fontanels helpful in determining the position of the infant's head prior to delivery. Although an infant may have many fontanels at birth, the form and location of six of them are fairly constant.

The **anterior (frontal) fontanel** is located between the angles of the two parietal bones and the two segments of the frontal bone. This fontanel is roughly diamond-shaped, and it is the largest of the six fontanels. It usually closes in 18 to 24 months.

The **posterior (occipital) fontanel** is situated between the two parietal bones and the occipital bone. This fontanel is considerably smaller than the anterior fontanel. It is diamond-shaped and generally closes about 2 months after birth.

The **anterolateral (sphenoidal) fontanels** are paired. One is located on each side of the skull at the junction of the frontal, parietal, temporal, and sphenoid bones. These fontanels are quite small and irregular in shape. They normally close by the third month after birth.

The **posterolateral (mastoid) fontanels** are also paired. One is situated on each side of the skull at the junction of the parietal, occipital, and temporal bones.

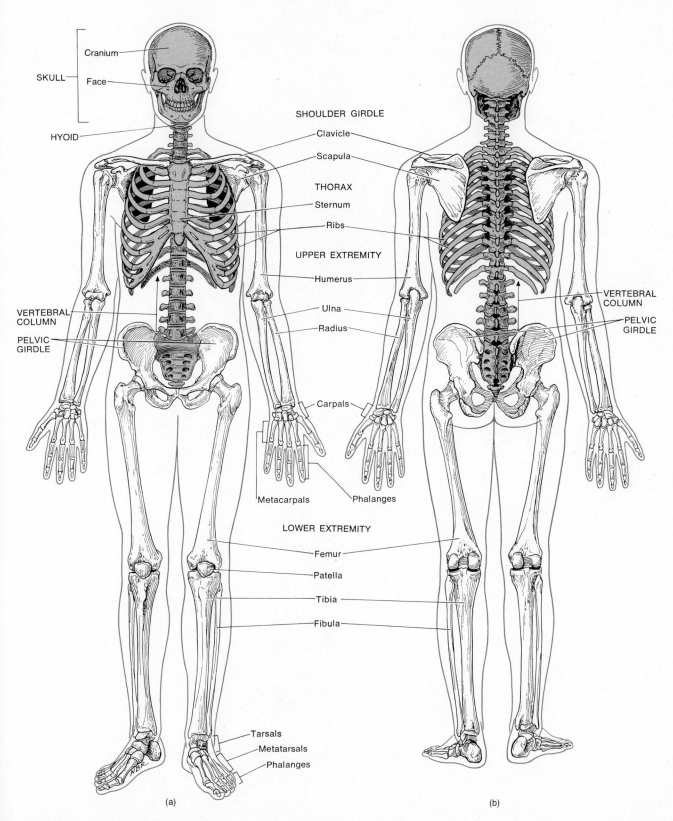

SKULL — Cranium

— Face

HYOID

VERTEBRAL COLUMN

PELVIC GIRDLE

SHOULDER GIRDLE

Clavicle

Scapula

THORAX

Sternum

Ribs

UPPER EXTREMITY

Humerus

Ulna

Radius

Carpals

Metacarpals

Phalanges

LOWER EXTREMITY

Femur

Patella

Tibia

Fibula

Tarsals

Metatarsals

Phalanges

VERTEBRAL COLUMN

PELVIC GIRDLE

(a)

(b)

Figure 7–1   Divisions of the skeletal system. (a) Anterior view. (b) Posterior view.

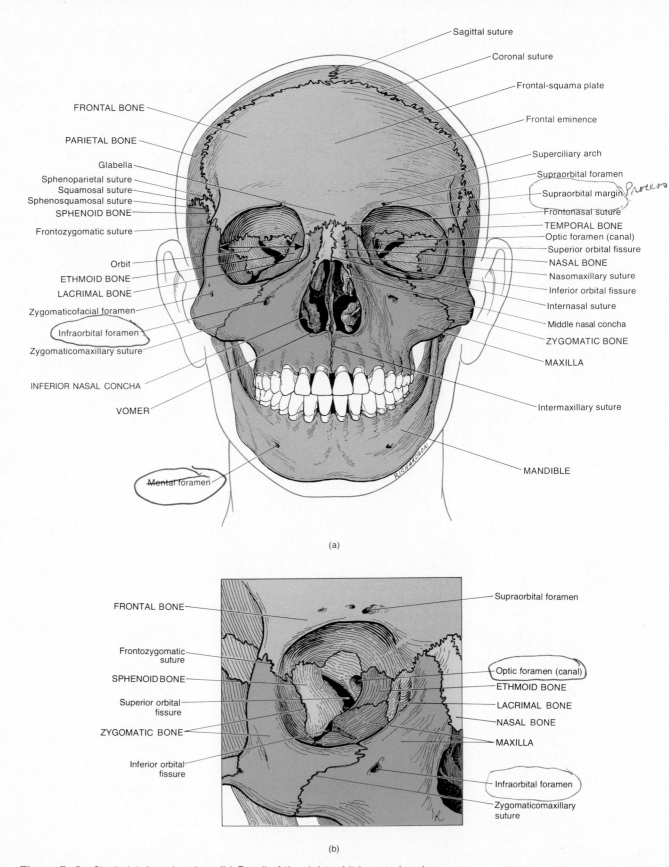

Figure 7–2   Skull. (a) Anterior view. (b) Detail of the right orbit in anterior view.

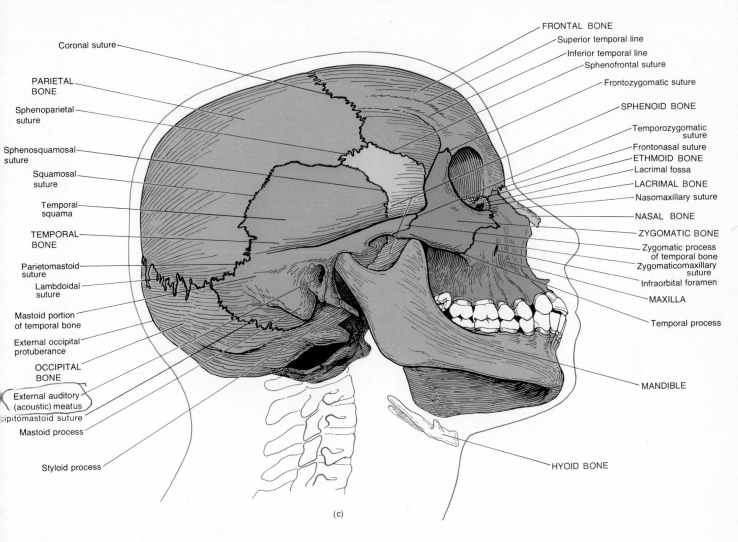

Figure 7–2 (cont.) Skull. (c) Right lateral view.

These fontanels are irregularly shaped. They begin to close 1 or 2 months after birth, but closure is not generally complete until the age of 1 year.

Frontal Bone

The **frontal bone** forms the forehead, the anterior part of the cranium; the superior portion of the *orbits* (eye sockets); and most of the anterior part of the cranial floor. Soon after birth the left and right parts of the frontal bone unite.

If you examine the anterior and lateral views of the skull in Figure 7-2, you will note the **squama,** or vertical plate (*squam* = scale). This scalelike plate gradually slopes down from the coronal suture, then turns abruptly downward. It projects slightly above its lower edge on either side of the midline to form the **frontal eminences.** Inferior to each eminence is a horizontal ridge the **superciliary arch,** caused by the projection of the frontal sinuses posterior to the eyebrow. Between the eminences and the arches just superior to the nose is a flattened area, the **glabella.** A thickening of the frontal bone inferior to the superciliary arches is called the **supraorbital margin.** From this margin the frontal bone extends posteriorly to form the roof of the orbit and part of the floor of the cranial cavity. Within the supraorbital margin, slightly medial to its midpoint, is a hole called the **supraorbital foramen** (*foramen* = passageway). The supraorbital nerve and artery pass through this foramen. The **frontal sinuses** lie deep to the superciliary arches. These mucus-lined cavities act as sound chambers to give the voice resonance.

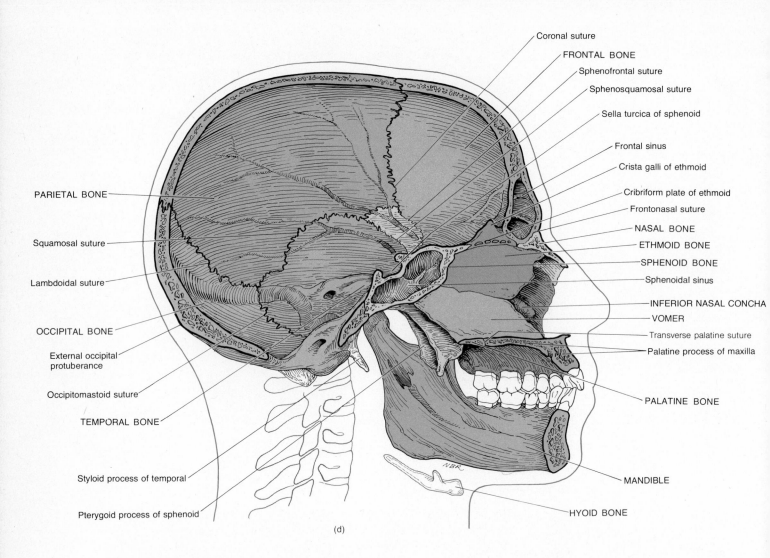

Coronal suture
FRONTAL BONE
Sphenofrontal suture
Sphenosquamosal suture
Sella turcica of sphenoid
Frontal sinus
Crista galli of ethmoid
Cribriform plate of ethmoid
Frontonasal suture
NASAL BONE
ETHMOID BONE
SPHENOID BONE
Sphenoidal sinus
INFERIOR NASAL CONCHA
VOMER
Transverse palatine suture
Palatine process of maxilla
PALATINE BONE
MANDIBLE
HYOID BONE

PARIETAL BONE
Squamosal suture
Lambdoidal suture
OCCIPITAL BONE
External occipital protuberance
Occipitomastoid suture
TEMPORAL BONE
Styloid process of temporal
Pterygoid process of sphenoid

(d)

Figure 7–2 (cont.)   Skull. (d) Median view.

Parietal Bones

The two **parietal bones** (*paries* = wall) form the greater portion of the sides and roof of the cranial cavity.

The external surface contains two slight ridges that may be observed by looking at the lateral view of the skull in Figure 7-2. These are the **superior temporal line** and a less conspicuous **inferior temporal line** below it. The internal surface has many eminences and depressions that accommodate the blood vessels supplying the outer meninx (covering) of the brain called the dura mater.

Temporal Bones

The two **temporal bones** form the inferior sides of the cranium and part of the cranial floor. The term *tempora* pertains to the temples.

In the lateral view of the skull in Figure 7-2, notice the **squama** or **squamous portion**—a thin, large, expanded area that forms the anterior and superior part

of the temple. Projecting from the inferior portion of the squama is the **zygomatic process,** which articulates with the temporal process of the zygomatic bone. The zygomatic process of the temporal bone together with the temporal process of the zygomatic bone constitutes the **zygomatic arch.** At the floor of the cranial cavity, shown in Figure 7-5, is the **petrous portion** of the temporal bone. This portion is triangular and located at the base of the skull between the sphenoid and occipital bones. The petrous portion contains the internal ear, the essential part of the organ of hearing. It also contains the **carotid foramen (canal)** through which the internal carotid artery passes (see Figure 7-4). Posterior to the carotid foramen and anterior to the occipital bone is the **jugular foramen (fossa)** through which the internal jugular vein and the glossopharyngeal nerve (IX), vagus nerve (X), and accessory nerve (XI) pass. Between the squamous and petrous portions is a socket called the **mandibular fossa.** This part of the temporal bone

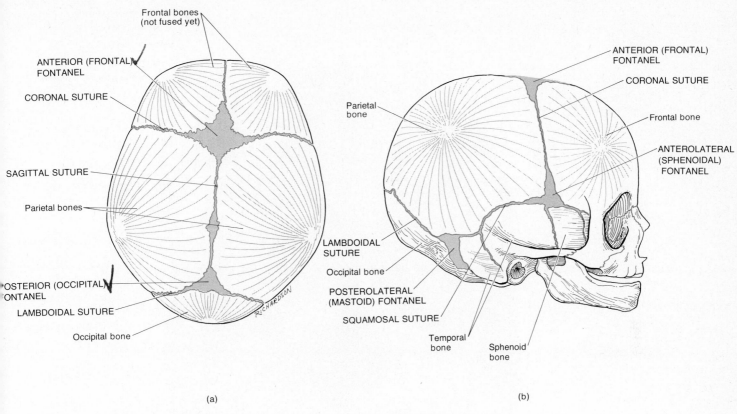

Figure 7–3 Fontanels of the skull at birth. (a) Superior view. (b) Right lateral view.

articulates with the condyle of the lower jaw. It is seen best in Figure 7-4. In the lateral view of the skull in Figure 7-2, you will see the **mastoid portion** of the temporal bone, located posterior and inferior to the external auditory meatus, or ear canal. In the adult, this portion of the bone contains a number of **mastoid air "cells."** These air spaces are separated from the brain only by thin bony partitions. If *mastoiditis,* the inflammation of these bony cells, occurs, the infection may spread to the brain or its outer covering. The mastoid air cells do not drain as do the paranasal sinuses. The **mastoid process** is a rounded projection of the temporal bone posterior to the external auditory meatus. It serves as a point of attachment for several neck muscles. Near the posterior border of the mastoid process is the **mastoid foramen** through which a vein to the transverse sinus and a small branch of the occipital artery to the dura mater pass. The **external auditory meatus** is the canal in the temporal bone that leads to the middle ear. The **internal acoustic meatus** is superior to the jugular foramen. It transmits the facial and acoustic nerves and the internal auditory artery. The **styloid process** projects downward from the undersurface of the temporal bone and serves as a point of attachment for

muscles and ligaments of the tongue and neck. Between the styloid process and the mastoid process is the **stylomastoid foramen,** which transmits the facial nerve (VII) and stylomastoid artery (see Figure 7-4).

Occipital Bone

The **occipital bone** forms the posterior part and a good portion of the base of the cranium (see Figure 7-4).

The **foramen magnum** is a large hole in the inferior part of the bone through which the medulla oblongata and its membranes, the accessory nerve (XI), and the vertebral and spinal arteries pass. The **occipital condyles** are oval processes with convex surfaces, one on either side of the foramen magnum, which articulate with depressions on the first cervical vertebra. At the base of the condyles is the **hypoglossal canal (fossa)** through which the hypoglossal nerve (XII) passes (see Figure 7-5). The **external occipital protuberance** is a prominent projection on the posterior surface of the bone just superior to the foramen magnum. You can feel this structure as a definite bump on the back of your head, just above your neck. The protuberance is also visible in Figure 7-2c.

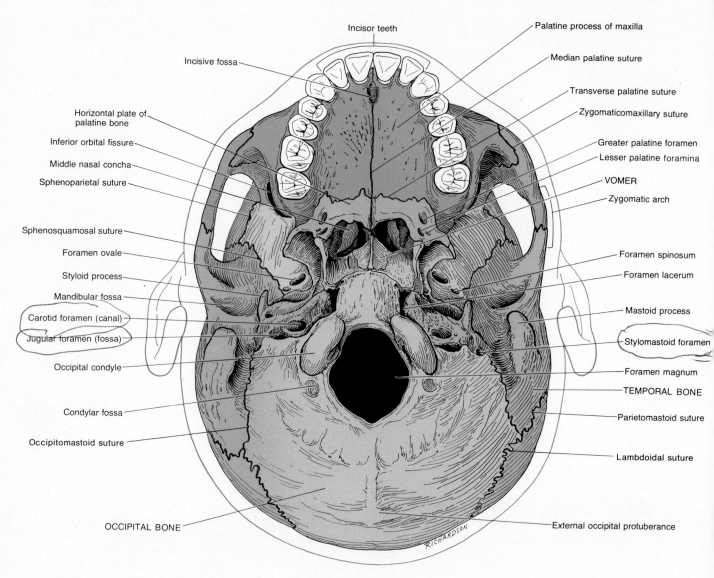

Figure 7—4 Skull in inferior view.

Sphenoid Bone

The **sphenoid bone** is situated at the anterior part of the base of the skull (see Figure 7-5). The combining form *spheno* means wedge. This bone is referred to as the keystone of the cranial floor because it binds the other cranial bones together. If you view the floor of the cranium from above, you will note that the sphenoid articulates with the temporal bones laterally, the ethmoid and frontal bones anteriorly, and the occipital bone posteriorly. It lies posterior and slightly superior to the nasal cavities and forms part of the floor and sidewalls of the eye socket. The shape of the sphenoid is frequently described as a bat with outstretched wings.

The **body** of the sphenoid is the cubelike central portion between the ethmoid and occipital bones. It contains a large air space, the **sphenoidal sinus**, which drains into the nasal cavity. (See Figure 7-8). On the superior surface of the sphenoid body is a depression called the **sella turcica**, meaning Turk's Saddle. This depression houses the pituitary gland. The **greater wings** of the sphenoid are lateral projections from the body and form the anterolateral floor of the cranium. The greater wings also form part of the lateral wall of the skull just anterior to the temporal bone. The **lesser wings** are anterior and superior to the greater wings. They form part of the floor of the cranium and the

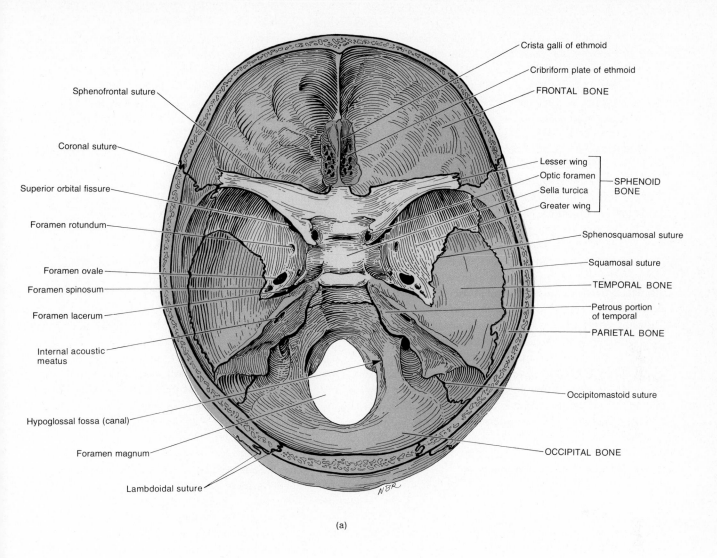

(a)

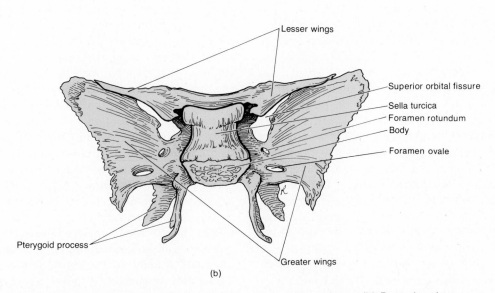

(b)

Figure 7–5 Sphenoid bone. (a) Viewed in the floor of the cranium from above. (b) Posterior view.

posterior part of the orbit, or eye socket. Between the body and lesser wing, you can locate the **optic foramen** through which the optic nerve (II) and ophthalmic artery pass. Lateral to the body between the greater and lesser wings is a somewhat triangular slit called the **superior orbital fissure.** It is an opening for the oculomotor nerve (III), trochlear nerve (IV), ophthalmic branch of the trigeminal nerve (V), and abducens nerve (VI). This fissure may also be seen in the anterior view of the skull in Figure 7-2. On the inferior part of the sphenoid bone you can see the **pterygoid processes.** These structures project inferiorly from the points where the body and greater wings unite. The pterygoid processes form part of the lateral walls of the nasal cavity. At the base of the lateral pterygoid process in the greater wing is the **foramen ovale** through which the mandibular branch of the trigeminal nerve (V) passes. Another foramen, the **foramen spinosum,** at the posterior angle of the sphenoid transmits the middle meningeal vessels. The **foramen lacerum** is bounded anteriorly by the sphenoid bone, and medially by the sphenoid and occipital bones. It transmits the internal carotid artery and the meningeal branch of the ascending pharyngeal artery. A final foramen associated with the sphenoid bone is the **foramen rotundum** through which the maxillary branch of the trigeminal nerve (V) passes. It is located at the junction of the anterior and medial parts of the sphenoid bone.

Ethmoid Bone

The **ethmoid bone** is a light, spongy bone located anterior to the sphenoid and posterior to the nasal bones (see Figure 7-6). The combining form *ethmos* means sieve. This bone forms part of the anterior portion of the cranial floor, the medial wall of the orbits, the superior portions of the nasal septum, or partition, and most of the sidewalls of the nasal roof. The ethmoid is the principal supporting structure of the nasal cavity.

Its **lateral masses** or **labyrinths** compose most of the wall between the nasal cavity and the orbits. They contain several air spaces, or "cells," which together form the **ethmoidal sinuses.** The sinuses are shown in Figure 7-8. The **perpendicular plate** (Figure 7-7) forms the superior portion of the nasal septum. The **cribriform plate,** or **horizontal plate,** lies in the anterior floor of the cranium and forms the roof of the nasal cavity. The cribriform plate contains the **olfactory foramina** through which the olfactory nerves (I) pass. These nerves function in smell (see Figure 7-5). Projecting upward from the horizontal plate is a triangular process called the **crista galli,** which means Cock's Comb. This

structure serves as a point of attachment for the membranes that cover the brain. The labyrinths contain two thin, scroll-shaped bones on either side of the nasal septum. These are called the **superior nasal concha** and the **middle nasal concha.** The conchae allow for the efficient circulation and filtration of inhaled air before it passes into the lungs. See Figure 7-2 also.

Nasal Bones

The paired **nasal bones** are small, oblong bones that meet at the middle and superior part of the face. Their fusion forms the superior part of the bridge of the nose. The inferior portion of the nose, indeed the major portion, consists of cartilage. See Figures 7-7 and 7-2.

Maxillae

The paired maxillary bones unite to form the upper jawbone (see Figure 7-7). The **maxillae** articulate with every bone of the face except the mandible, or lower jawbone. They form part of the floor of the orbits, part of the roof of the mouth, and part of the lateral walls and floor of the nasal cavities. The two portions of the maxillary bones unite, and the fusion is normally completed before birth. If the palatine processes of the maxillary bones do not unite before birth, a condition called **cleft palate** results. Another form of this condition, called **harelip,** involves a split in the upper lip. Harelip is often associated with cleft palate. Depending on the extent and position of the cleft, speech and swallowing may be affected.

Each maxillary bone contains a **maxillary sinus (antrum of Highmore)** that empties into the nasal cavity. See Figure 7-8. The **alveolar process** (*alveolus* = hollow) contains the bony sockets into which the teeth are set. The **palatine process** is a horizontal projection of the maxilla that forms the anterior and larger part of the hard palate, or anterior portion of the roof of the oral cavity. The **infraorbital foramen,** which can be seen in the anterior view of the skull in Figure 7-2, is an opening in the maxilla inferior to the orbit. The infraorbital nerve and artery are transmitted through this opening. Another prominent fossa in the maxilla is the **incisive fossa** just posterior to the incisor teeth. Through it pass branches of the descending palatine vessels and the nasopalatine nerve. A final fossa associated with the maxilla and sphenoid bone is the **inferior orbital fissure.** It is located between the greater wing of the sphenoid and the maxilla (see Figure 7-4). It transmits the maxillary branch of the trigeminal nerve (V) and the infraorbital vessels.

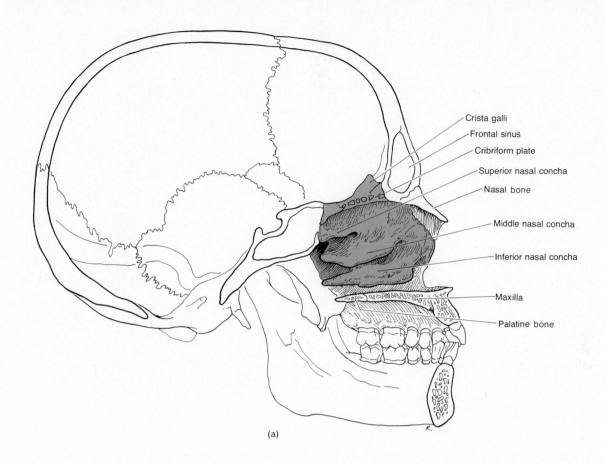

Crista galli
Frontal sinus
Cribriform plate
Superior nasal concha
Nasal bone
Middle nasal concha
Inferior nasal concha
Maxilla
Palatine bone

(a)

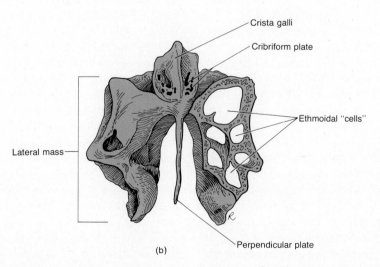

Crista galli
Cribriform plate
Ethmoidal "cells"
Lateral mass
Perpendicular plate

(b)

Figure 7–6 Ethmoid bone. (a) Median section showing the ethmoid bone on the inner aspect of the left part of the skull. (b) Anterior view.

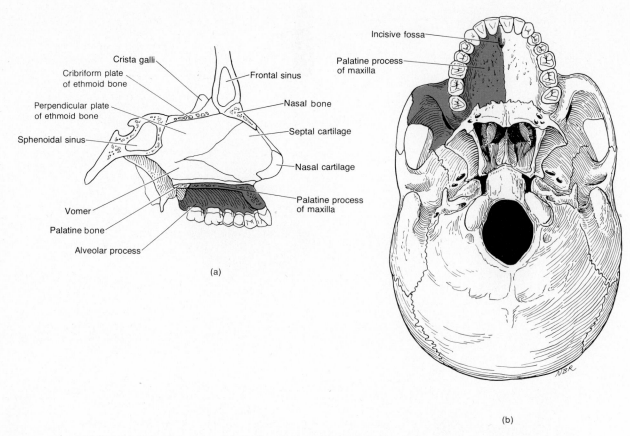

Figure 7–7 Maxillae. (a) Median view of the left maxilla. (b) Inferior view of the skull, showing the maxillae.

Paranasal Sinuses

Cavities, called **paranasal sinuses,** are located in certain bones near the nasal cavity (see Figure 7-8). The paranasal sinuses are lined with mucous membranes that are continuous with the lining of the nasal cavity. Cranial bones containing paranasal sinuses are the frontal bone, the sphenoid, the ethmoid, and the maxillae. The ethmoid sinus consists of a series of small cavities called ethmoid "cells," which range in number from 3 to 18.

Zygomatic Bones

The two **zygomatic bones (malars),** commonly referred to as the cheekbones, form the prominences of the cheeks and part of the outer wall and floor of the orbits (Figure 7-2c).

The **temporal process** of the zygomatic bone projects posteriorly and articulates with the zygomatic process of the temporal bone. These two processes form the **zygomatic arch.** A foramen associated with the zygomatic bone is the **zygomaticofacial foramen** near the center of the bone (see Figure 7-2). It transmits the zygomaticofacial nerve and vessels.

Mandible

The **mandible** or lower jawbone is the largest, strongest facial bone (see Figure 7-9). It is the only movable bone in the skull.

In the lateral view you can see that the mandible consists of a curved, horizontal portion called the **body** and two perpendicular portions called the **rami.** The **angle** of the mandible is the area where each ramus meets the body. Each ramus has a **condylar process** that articulates with the mandibular fossa of the temporal bone. It also has a **coronoid process** to which the temporalis muscle attaches. The depression between the coronoid and condylar processes is called the **mandibular notch.** The **mental foramen** (*mentum* = chin) is approximately below the first molar tooth. The mental nerve and vessels pass through this opening. Dentists inject anesthetics through this foramen. The **alveolar process,** like that of the maxillae, is an arch containing the sockets for the teeth. Another foramen associated with the mandible is the **mandibular foramen** on the medial surface of the ramus. It transmits the inferior alveolar nerve and vessels.

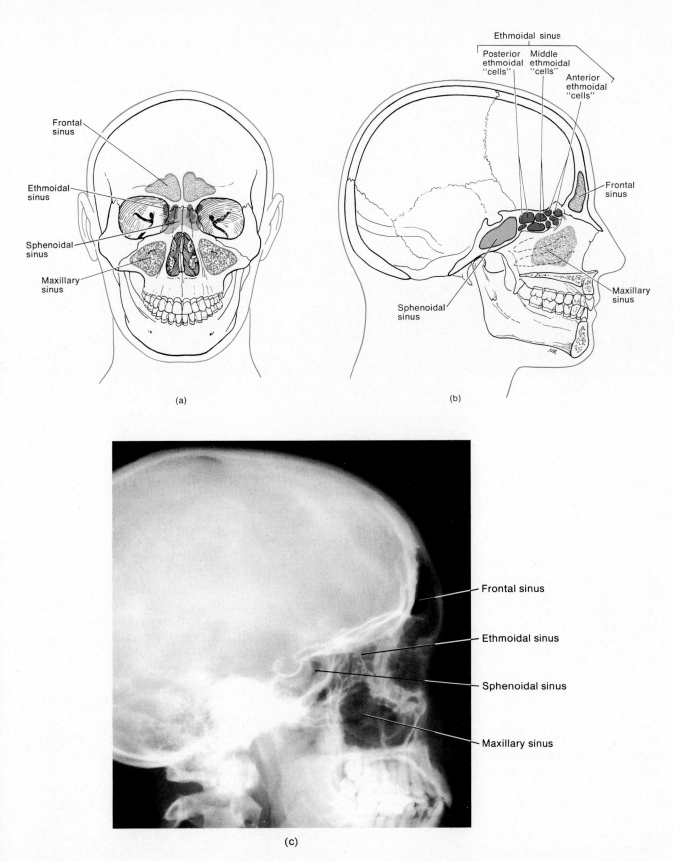

Figure 7–8 Paranasal sinuses. (a) Anterior view. (b) Right lateral view. (c) Lateral view of the skull. (Courtesy of Eastman Kodak Company.)

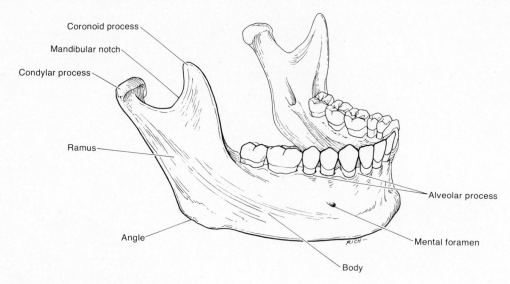

Figure 7–9   Mandible in right lateral view.

Lacrimal Bones

The paired **lacrimal bones** are thin bones roughly resembling a fingernail in size and shape (*lacrima =* tear). These bones are posterior and lateral to the nasal bones in the medial wall of the orbit. They can be seen in the anterior and lateral views of the skull in Figure 7-2. The lacrimal bones form a part of the medial wall of the nasal cavity. They also contain the **lacrimal foramina** through which the tear ducts pass into the nasal cavity (see Figure 7-2). The lacrimal bones are the smallest bones of the face.

Palatine Bones

The two **palatine bones** are L-shaped and form the posterior portion of the hard palate, part of the floor and lateral walls of the nasal cavities, and a small portion of the floor of the orbit. The posterior portion of the hard palate is formed by the **horizontal plates** of the palatine bones. These can be seen in Figures 7-4 and 7-2d. Two foramina associated with the palatine bones are the greater and lesser palatine foramina. The **greater palatine foramen,** at the posterior angle of the hard palate, transmits the greater palatine nerve and descending palatine vessels (see Figure 7-4). The **lesser palatine foramina,** usually two or more on each side, are posterior to the greater palatine foramina. They transmit the lesser palatine nerve. See Figure 7-4.

Inferior Nasal Conchae

Refer to the views of the skull in Figure 7-2a and 7-6a. The two **inferior nasal conchae** (*concha =* shell) are scroll-like bones that project into the nasal cavity inferior to the superior and middle nasal conchae of the ethmoid

bone. They serve the same function as the superior and middle nasal conchae. That is, they allow for the circulation and filtration of air before it passes into the lungs. The inferior nasal conchae are separate bones and not part of the ethmoid.

Vomer

The **vomer,** which means plowshare, is a roughly triangular bone that forms the inferior part of the nasal septum. It is clearly seen in the anterior view of the skull in Figure 7-2. Its inferior border articulates with the cartilage septum that divides the nose into a right and left nostril. Its superior border articulates with the perpendicular plate of the ethmoid bone. Thus the structures that form the nasal septum, or partition, are the perpendicular plate of the ethmoid, the septal cartilage, and the vomer. If the vomer is pushed to one side – that is, deviated – the nasal chambers are of unequal size. See the skull viewed from below (Figure 7-4) for another view of the vomer.

Before continuing your study of the bones of the axial skeleton, refer to Exhibit 7-2 on page 150, which contains a summary of the foramina of the skull.

HYOID BONE

The single **hyoid bone** (*hyoid =* U-shaped) is a unique component of the axial skeleton because it does not articulate with any other bone (see Figure 7-10). Rather, it is suspended from the styloid process of the temporal bone by ligaments. The hyoid is located in the neck between the mandible and larynx. It supports the tongue and provides attachment for some of its muscles.

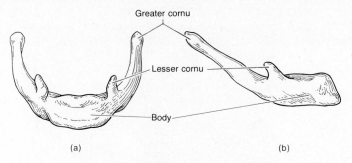

Figure 7-10 Hyoid bone. (a) Anterior view. (b) Right lateral view.

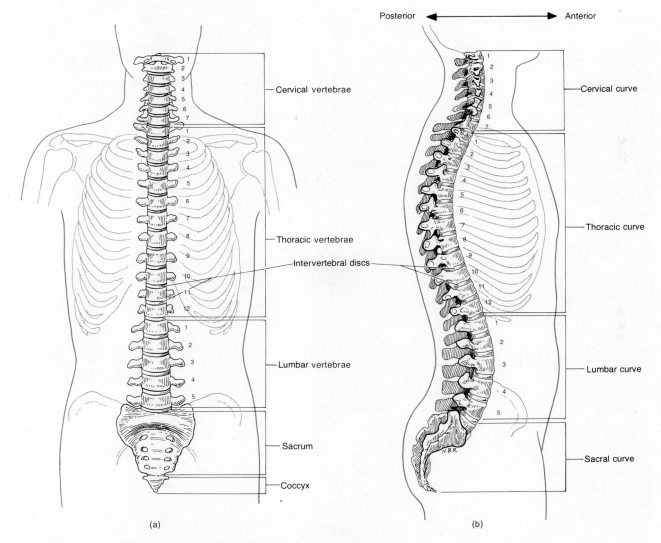

Figure 7-11 Vertebral column. (a) Anterior view. (b) Right lateral view.

Refer to the anterior and lateral views of the skull in Figure 7-2 to see the position of the hyoid bone.

The hyoid consists of a horizontal **body** and paired projections called the **lesser cornu** and the **greater cornu.** Muscles and ligaments attach to these paired projections.

VERTEBRAL COLUMN

The **vertebral column**, or **spine**, together with the sternum and ribs, constitutes the skeleton of the **trunk** of the body (see Figure 7-11). The vertebral column is composed of a series of bones called **vertebrae**. In the average adult, the column measures about 71 cm (28

Seventh cervical vertebra —

First thoracic vertebra —

First cervical vertebra

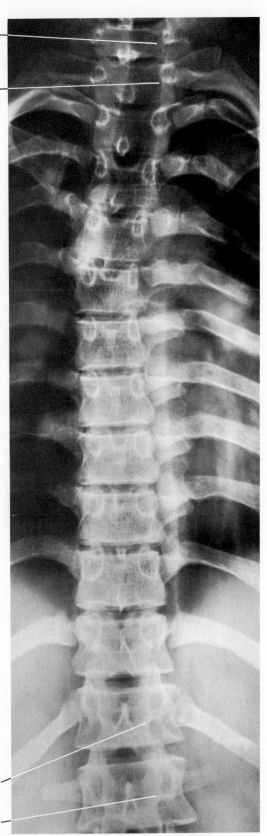

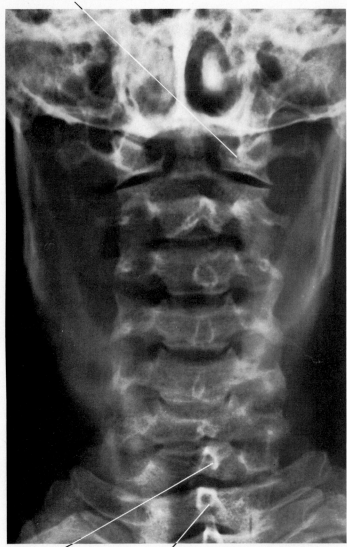

Seventh cervical vertebra   First thoracic vertebra

(c)

Twelfth thoracic vertebra —

First lumbar vertebra —

(d)

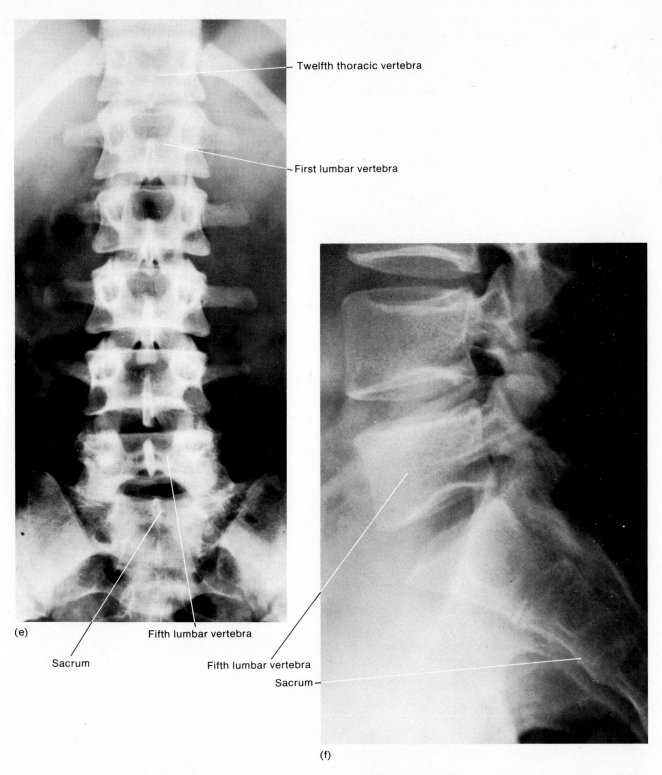

— Twelfth thoracic vertebra

—First lumbar vertebra

(e)

Fifth lumbar vertebra

Sacrum

Fifth lumbar vertebra

Sacrum—

(f)

Figure 7–11 (cont.) Vertebral column. (c) Anteroposterior projection of the cervical vertebrae. (d) Anteroposterior projection of the thoracic vertebrae. (e) Anteroposterior projection of the lumbar vertebrae. (f) Lateral view of the sacrum. (Courtesy of Eastman Kodak Company.) The term *projection* as used here and in subsequent legends of roentgenograms refers to which part of the body the x-ray beam enters and then exits. A *view,* by contrast, is determined by the side of the body closest to the film.

Exhibit 7—2   FORAMINA OF SKULL

Foramen	Location	Structures Passing Through
Carotid (Figure 7–4)	Petrous portion of temporal bone.	Internal carotid artery.
Greater palatine (Figure 7–4)	Posterior angle of hard palate.	Greater palatine nerve and descending palatine vessels.
Hypoglossal (Figure 7–5)	Superior to base of occipital condyles.	Hypoglossal nerve and branch of ascending pharyngeal artery.
Incisive (Figure 7–7)	Posterior to incisor teeth.	Branches of descending palatine vessels and nasopalatine nerve.
Infraorbital (Figure 7–2a)	In maxilla inferior to orbit.	Infraorbital nerve and artery.
Jugular (Figure 7–4)	Posterior to carotid canal between petrous portion of temporal and occipital bones.	Internal jugular vein, glossopharyngeal nerve (IX), vagus nerve (X), and accessory nerve (XI).
Lacerum (Figure 7–5)	Bounded anteriorly by sphenoid, posteriorly by petrous portion of temporal, and medially by the sphenoid and occipital bones.	Internal carotid artery and branch of ascending pharyngeal artery.
Lacrimal (Figure 7–2c)	Lacrimal bone.	Lacrimal (tear) duct.
Lesser palatine (Figure 7–4)	Posterior to greater palatine foramen.	Lesser palatine nerves.
Magnum (Figure 7–4)	Occipital bone.	Medulla oblongata and its membranes, the accessory nerve (XI), and the vertebral and spinal arteries.
Mandibular	Medial surface of ramus of mandible.	Inferior alveolar nerve and vessels.
Mastoid (Figure 7–4)	Posterior border of mastoid process of temporal bone.	Vein to transverse sinus and branch of occipital artery to dura mater.
Mental (Figure 7–9)	Inferior to second premolar tooth in mandible.	Mental nerve and vessels.
Olfactory (Figure 7–5)	Cribriform plate of ethmoid bone.	Olfactory nerve (I).
Optic (Figure 7–5)	Between upper and lower portions of lesser wing of sphenoid bone.	Optic nerve (II) and ophthalmic artery.
Ovale (Figure 7–5)	Greater wing of sphenoid bone.	Mandibular branch of trigeminal nerve (V).
Rotundum (Figure 7–5)	Junction of anterior and medial parts of sphenoid bone.	Maxillary branch of trigeminal nerve (V).
Spinosum (Figure 7–5)	Posterior angle of sphenoid bone.	Middle meningeal vessels.
Stylomastoid (Figure 7–4)	Between styloid and mastoid processes of temporal bone.	Facial nerve (VII) and stylomastoid artery.
Supraorbital (Figure 7–2a)	Supraorbital margin of orbit.	Supraorbital nerve and artery.
Zygomaticofacial (Figure 7–2a)	Zygomatic bone.	Zygomaticofacial nerve and vessels.

Chapter 8
The Skeletal System:
The Appendicular Skeleton

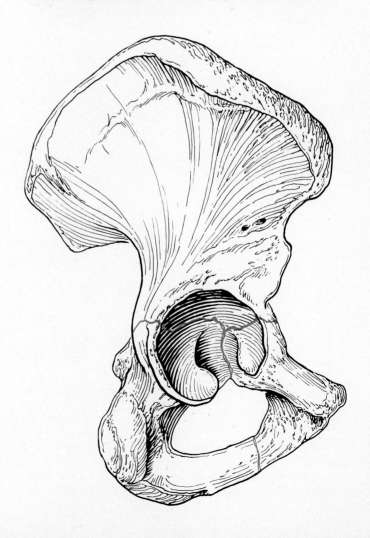

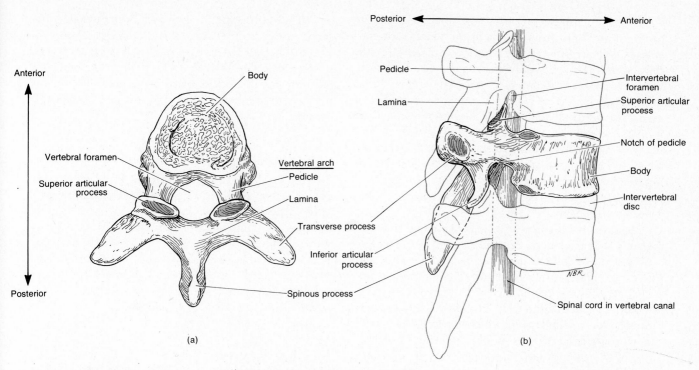

Figure 7–12   A typical vertebra. (a) Superior view. (b) Right lateral view.

inches). In effect, the vertebral column is a strong, flexible rod that moves anteriorly, posteriorly, and laterally. It encloses and protects the spinal cord, supports the head, and serves as a point of attachment for the ribs and the muscles of the back. Between the vertebrae are openings called **intervertebral foramina.** The nerves that connect the spinal cord to various parts of the body pass through these openings.

The adult vertebral column typically contains 26 vertebrae. These are distributed as follows: 7 **cervical vertebrae** in the neck region; 12 **thoracic vertebrae** posterior to the thoracic cavity; 5 **lumbar vertebrae** supporting the lower back; 5 **sacral vertebrae** fused into one bone called the **sacrum;** and usually four **coccygeal vertebrae** fused into one or two bones called the **coccyx.** Prior to the fusion of the sacral and coccygeal vertebrae, the total number of vertebrae is 33. Between the vertebrae are fibrocartilaginous **intervertebral discs.** These discs form strong joints and permit various movements of the column.

When viewed from the side, the vertebral column shows four **curves.** From the anterior view, these are alternately convex, meaning they curve out toward the viewer, and concave, meaning they curve away from the viewer. The curves of the column, like the curves in a long bone, are important because they increase its strength, help maintain balance in the upright position, absorb shocks from walking, and help protect the column from fracture.

In the fetus, the four curves of the vertebrae are not present. There is only a single curve that is anteriorly concave. At approximately the third postnatal month when an infant begins to hold its head erect, the **cervical curve** develops. Later, when the child stands and walks, the **lumbar curve** develops. The cervical and lumbar curves are convex anteriorly. Because they are modifications of the fetal positions, they are called **secondary curves.** The other two curves, the **thoracic curve** and the **sacral curve,** are anteriorly concave. Since they retain the anterior concavity of the fetus, they are referred to as **primary curves.**

Typical Vertebra

All the vertebrae of the column are basically similar in structure (see Figure 7-12). But there are differences in size, shape, and detail. A typical vertebra consists of the following components:

1 The **body** is the thick, disc-shaped anterior portion that is the weight-bearing part of a vertebra. Its superior and inferior surfaces are roughened for the attachment of intervertebral discs. The anterior and lateral surfaces contain nutrient foramina for blood vessels.

2 The **vertebral arch (neural arch)** extends posteriorly from the body of the vertebra. With the body of the vertebra, it surrounds the spinal cord. It is formed

by two short, thick processes, the **pedicles,** which project posteriorly from the body to unite with the laminae. The **laminae** are the flat parts that join to form the posterior portion of the vertebral arch. The space that lies between the vertebral arch and body contains the spinal cord. This space is known as the **vertebral foramen.** The vertebral foramina of all vertebrae together form the **vertebral,** or **spinal, canal.** The pedicles are notched superiorly and inferiorly in such a way that, when they are arranged in the column, there is an opening between vertebrae on each side of the column. This opening, the **intervertebral foramen,** permits the passage of the spinal nerves.

3 Seven **processes** arise from the vertebral arch. At the point where a lamina and pedicle join, a **transverse process** extends laterally on each side. A single **spinous process** or **spine** projects posteriorly and inferiorly from the junction of the laminae. These three processes serve as points of muscular attachment. The remaining four processes form joints with other vertebrae. The two **superior articular processes** of a vertebra articulate with the vertebra immediately superior to it. The two **inferior articular processes** of a vertebra articulate with the vertebra inferior to it.

Cervical Region

When viewed from above, it can be seen that the bodies of **cervical vertebrae** are smaller than those of the thoracic vertebrae (see Figure 7-13). The arches, however, are larger. The spinous processes of the second through sixth cervical vertebrae are often *bifid*—that is, with a cleft. Each cervical transverse process contains an opening, the **transverse foramen.** The vertebral artery and its accompanying vein and nerve fibers pass through it.

The first two cervical vertebrae differ considerably from the others. The first cervical vertebra, the **atlas,** is named for its support of the head. Essentially, the atlas is a ring of bone with **anterior** and **posterior arches** and large **lateral masses.** It lacks a body and a spinous process. The superior surfaces of the lateral masses, called **superior articular surfaces,** are concave and articulate with the occipital condyles of the occipital bone. This articulation permits the movement seen when nodding the head. The inferior surfaces of the lateral masses, the **inferior articular surfaces,** articulate with the second cervical vertebra. The transverse processes and **transverse foramina** of the atlas are quite large.

The second cervical vertebra, the **axis,** does have a **body.** A peglike process called the **dens,** or **odontoid process,** projects up through the ring of the atlas. The dens makes a pivot on which the atlas and head rotate. This arrangement permits side-to-side rotation of the head.

The third through sixth cervical vertebrae correspond to the structural pattern of the typical cervical vertebra shown. The seventh cervical vertebra, however, is somewhat different. It is called the **vertebra prominens** and is marked by a large, nonbifid spinous process that may be seen and felt at the base of the neck. (See Figure 7-11.)

Thoracic Region

Viewing a typical **thoracic vertebra** from above, you can see that it is considerably larger and stronger than a vertebra of the cervical region (see Figure 7-14). In addition, the spinous process on each vertebra is long, pointed, and directed inferiorly. Thoracic vertebrae also have longer and heavier transverse processes than cervical vertebrae. Except for the eleventh and twelfth thoracic vertebrae, the transverse processes have **facets** for articulating with the tubercles of the ribs.

Lumbar Region

The **lumbar vertebrae** are the largest and strongest in the entire column (see Figure 7-15). Their superior articular processes are directed medially instead of superiorly. And their inferior articular processes are directed laterally instead of inferiorly. Their various projections are short and thick, and the spinous process is heavy for the attachment of the large back muscles.

Sacrum and Coccyx

The **sacrum** is a triangular bone formed by the union of five sacral vertebrae (see Figure 7-16). These are indicated in the figure as S1 through S5. It serves as a strong foundation for the pelvic girdle. It is positioned at the posterior portion of the pelvic cavity between the two hipbones. Anterior and posterior views of the bone are shown here. The concave anterior side of the sacrum faces the pelvic cavity. It is smooth and contains four **transverse lines** that mark the joining of the vertebral bodies. At the ends of these lines are four pairs of **pelvic foramina.** The convex, posterior surface of the sacrum is irregular. It contains a **median sacral crest,** a **lateral sacral crest,** and four pairs of **dorsal foramina.** These foramina communicate with the pelvic foramina through which nerves and blood vessels pass. The **sacral canal** is a continuation of the vertebral canal. The laminae of the fifth sacral vertebra, and sometimes the fourth, fail to meet. This leaves an inferior entrance to the vertebral canal called the **sacral hiatus.** The superior border of the sacrum exhibits an anteriorly projecting border, the **sacral promontory.** It is an obstetrical landmark for measurements of the pelvis. An imaginary line running from the superior surface of the symphysis

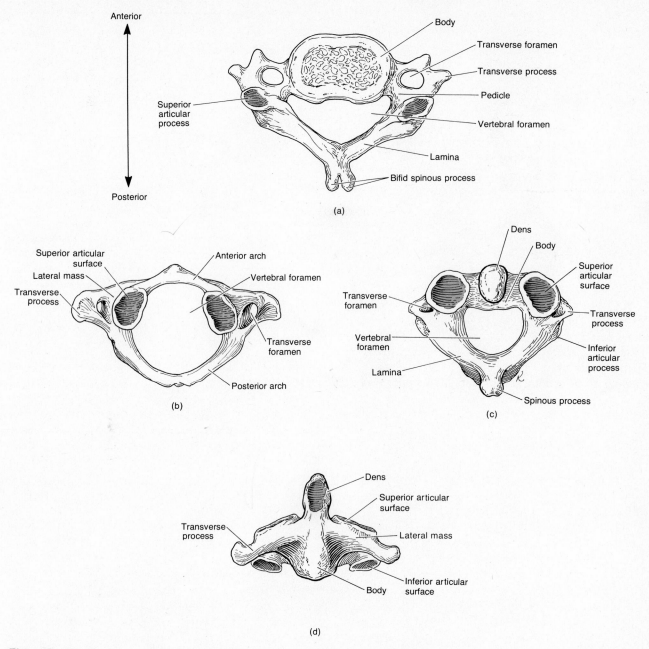

Figure 7–13 Cervical vertebrae. (a) Superior view of a cervical vertebra. (b) Superior view of the atlas. (c) Superior view of the axis. (d) Anterior view of the axis.

pubis to the sacral promontory separates the abdominal and pelvic cavities. Laterally, the sacrum has a large **auricular surface** for articulating with the ilium of the hipbone. Its superior articular process articulates with the fifth lumbar vertebra.

The **coccyx** is also triangular in shape and is formed by the fusion of the coccygeal vertebrae, usually the last four. These are indicated in Figure 7-16 as Co1 through Co4. It articulates superiorly with the sacrum. The coccyx is the most rudimentary part of the column, representing the vestige of a tail.

THORAX

The term **thorax** refers to the chest (see Figure 7-17). Its skeleton is a bony cage formed by the sternum, costal cartilage, ribs, and the bodies of the thoracic vertebrae. It is shown here in the anterior view. The thoracic cage is roughly cone-shaped, the narrow portion being superior and the broad portion inferior. It is flattened from front to back. The thoracic cage encloses and protects the organs in the thoracic cavity. It also provides support for the bones of the shoulder girdle and upper extremities.

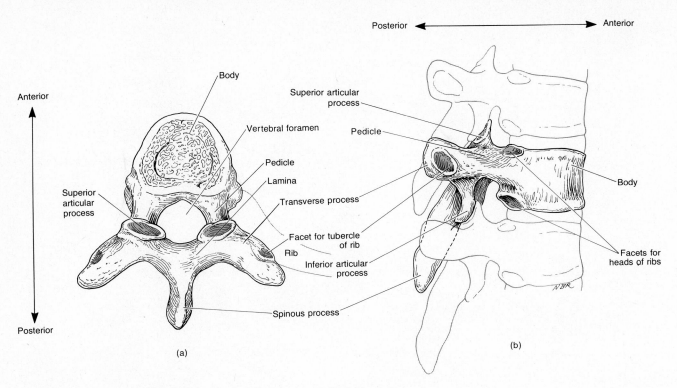

Figure 7–14   Thoracic vertebrae. (a) Superior view. (b) Right lateral view.

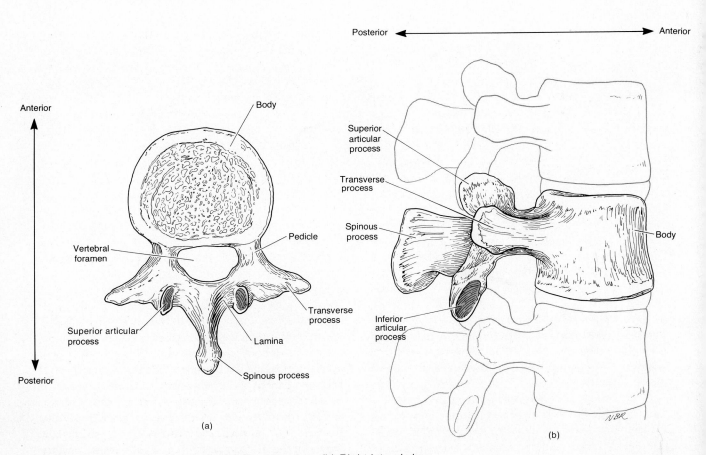

Figure 7–15   Lumbar vertebrae. (a) Superior view. (b) Right lateral view.

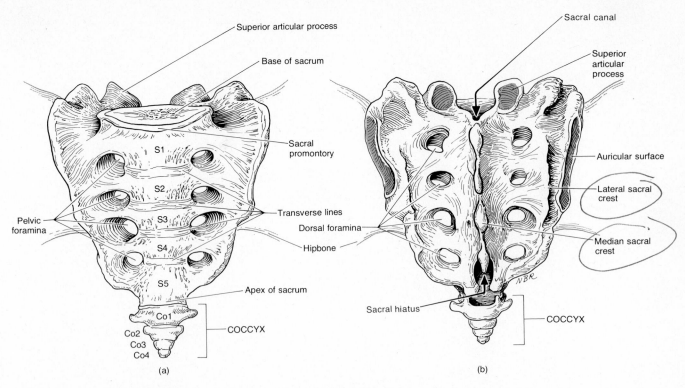

Figure 7–16 Sacrum and coccyx. (a) Anterior view. (b) Posterior view.

Sternum

The **sternum**, or breastbone, is a flat, narrow bone measuring about 15 cm (6 inches) in length. It is located in the median line of the anterior thoracic wall.

The sternum (see Figure 7-17) consists of three basic portions: the **manubrium**, which is a triangular, superior portion; the **body**, which is the middle, largest portion; and the **xiphoid process**, which is the inferior, smallest portion. The manubrium has a depression on its superior surface called the **jugular (suprasternal) notch**. On each side of the jugular notch are **clavicular notches** that articulate with the medial ends of the clavicles. The manubrium also articulates with the first and second ribs. The body of the sternum articulates directly or indirectly with the second through tenth ribs. The xiphoid process has no ribs attached to it but provides attachment for some abdominal muscles.

Ribs

Twelve pairs of **ribs** make up the sides of the thoracic cavity. (See Figure 7-17.) The ribs increase in length from the first through seventh. Then they decrease in length to the twelfth rib. Each rib articulates posteriorly with its corresponding thoracic vertebra. The first through seventh ribs are also attached directly to the sternum by a strip of hyaline cartilage, called **costal cartilage** (*costa* = rib). These ribs are called **true ribs**. The remaining five pairs of ribs are referred to as **false**

ribs because their costal cartilages do not attach directly to the sternum. Instead the cartilages of the eighth, ninth, and tenth ribs attach to each other and then to the cartilage of the seventh rib. The eleventh and twelfth ribs are also designated as **floating ribs** because their anterior ends do not attach even indirectly to the sternum. They attach to the muscles of the body wall instead.

Although there is some variation in rib structure, we will examine the parts of a typical rib when viewed from the right side and from behind (see Figure 7-18). The **head** of a typical rib is a projection at the posterior end of the rib. The **neck** is a constricted portion just lateral to the head. A knoblike structure on the posterior surface where the neck joins the body is called a **tubercle**. The **body**, or **shaft**, is the main part of the rib. The inner surface of the rib has a **costal groove** that protects blood vessels and a nerve, artery, and vein. The posterior portion of the rib is connected to a vertebra by its head and tubercle. The head fits into a facet on the body of a vertebra, and the tubercle articulates with the transverse process of the vertebra. Each of the second through ninth ribs articulates with the bodies of two adjacent vertebrae. The first, tenth, eleventh, and twelfth ribs articulate with only one vertebra each. On the eleventh and twelfth ribs, there is no articulation between the tubercles and the transverse processes of their corresponding vertebrae. Spaces between ribs are called **intercostal spaces**.

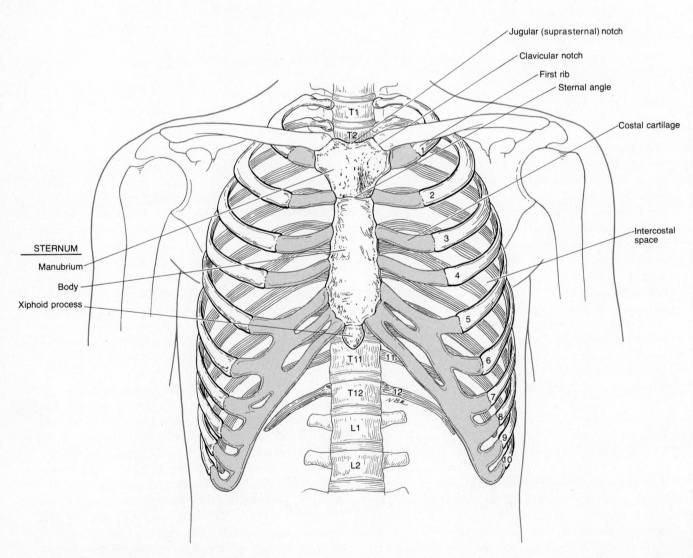

Figure 7–17 Bony thorax in anterior view.

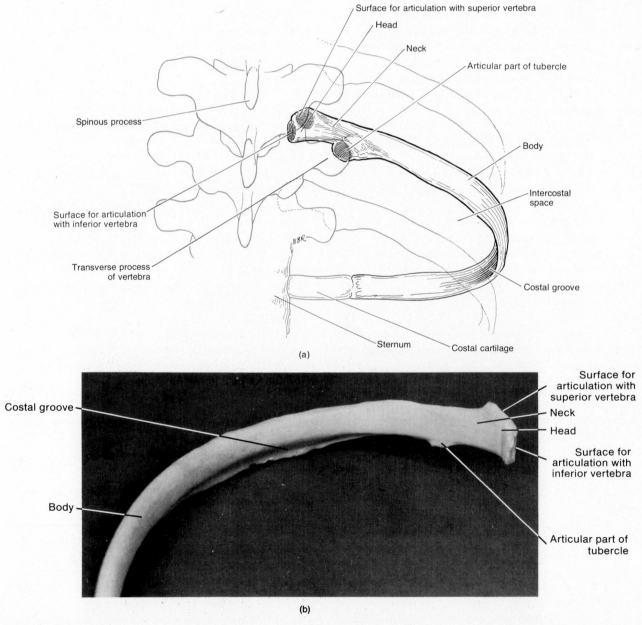

Surface for articulation with superior vertebra

Head

Neck

Articular part of tubercle

Body

Spinous process

Intercostal space

Surface for articulation with inferior vertebra

Transverse process of vertebra

Costal groove

Sternum

Costal cartilage

(a)

Surface for articulation with superior vertebra

Neck

Head

Surface for articulation with inferior vertebra

Costal groove

Body

Articular part of tubercle

(b)

Figure 7–18   A typical rib. (a) A right rib viewed from below and behind. (b) Photograph of the inner aspect of a portion of the fifth right rib. (Courtesy of Vincent P. Destro, Mayo Foundation.)

TYPES OF BONES

1 On the basis of shape, bones are classified as long, short, flat, or irregular.

2 Wormian or sutural bones are found between the sutures of certain cranial bones. Sesamoid bones develop in tendons or ligaments.

SURFACE MARKINGS

1 Markings are definitive areas on the surfaces of bones.

2 Each marking is structured for a specific function—joint formation, muscle attachment, or as a passageway for nerves and blood vessels.

3 Terms that describe markings include fissure, foramen, meatus, fossa, process, condyle, head, facet, tuberosity, crest, and spine.

DIVISIONS OF THE SKELETAL SYSTEM

1 The axial skeleton consists of bones arranged along the longitudinal axis. The parts of the axial skeleton are the skull, hyoid bone, auditory ossicles, vertebral column, sternum, and ribs.

2 The appendicular skeleton consists of the bones of the girdles and the upper and lower extremities. The parts of the appendicular skeleton are the shoulder girdle, the bones of the upper extremities, the pelvic girdle, and the bones of the lower extremities.

SKULL

1 Sutures are immovable joints between bones of the skull. Examples are coronal, sagittal, lambdoidal, and squamosal sutures.

2 Fontanels are membrane-filled spaces between the cranial bones of fetuses and infants. The major fontanels are the anterior, posterior, anterolateral, and posterolateral.

3 The skull consists of the cranium and the face. It is composed of 22 bones.

4 The eight cranial bones include the frontal, parietal (two), temporal (two), occipital, sphenoid, and ethmoid.

5 The 14 facial bones are the nasal (two), maxillae (two), zygomatic (two), mandible, lacrimal (two), palatine (two), inferior nasal conchae (two), and vomer.

6 Paranasal sinuses are cavities in bones of the skull that communicate with the nasal cavity. They are lined by mucous membranes.

7 The cranial bones containing the paranasal sinuses are the frontal, sphenoid, ethmoid, and maxilla.

8 The mastoid sinus is located in the temporal bone.

VERTEBRAL COLUMN

1 The vertebral column, the sternum, and the ribs constitute the skeleton of the trunk.

2 The bones of the adult vertebral column are the cervical vertebrae (7), thoracic vertebrae (12), lumbar vertebrae (5), the sacrum, and the coccyx.

3 The vertebral column contains primary curves (thoracic and sacral) and secondary curves (cervical and lumbar). These curves give strength, support, and balance.

THORAX

1 The thoracic skeleton consists of the sternum, the ribs and costal cartilages, and the thoracic vertebrae.

2 The thorax protects vital organs in the chest area.

1 What are the four principal types of bones? Give an example of each.

2 What are surface markings? Describe and give an example of each.

3 Distinguish between the axial and appendicular skeletons. What subdivisions and bones are contained in each?

4 What bones compose the skull? The cranium? The face?

5 Define a suture. What are the four principal sutures of the skull? Where are they located?

6 What is a fontanel? Describe the location of the six major fontanels.

7 What is a paranasal sinus? Give examples of cranial bones that contain paranasal sinuses.

8 Identify each of the following: sinusitis, cleft palate, harelip, and mastoiditis.

9 What is the hyoid bone? In what respect is it unique? What is its function?

10 What bones form the skeleton of the trunk? Distinguish between the number of nonfused vertebrae found in the adult vertebral column and that of a child.

11 What is a curve in the vertebral column? How are primary and secondary curves differentiated? What is a curvature? Give three examples of curvatures.

12 What bones form the skeleton of the thorax? What are the functions of the thoracic skeleton? What is a sternal puncture?

STUDENT OBJECTIVES

■ Identify the bones of the shoulder girdle and their major markings.

■ Identify the upper extremity, its component bones, and their markings.

■ Identify the components of the pelvic girdle and their principal markings.

■ Identify the lower extremity, its component bones, and their markings.

■ Define the structural features and importance of the arches of the foot.

■ Compare the principal structural differences between male and female skeletons.

This chapter discusses the bones of the girdles and extremities and compares the differences between male and female skeletons.

SHOULDER GIRDLES

The **shoulder,** or **pectoral, girdles** attach the bones of the upper extremities to the axial skeleton (see Figure 8-1). Structurally, each of the two shoulder girdles consists of two bones: a clavicle and a scapula. The shoulder girdles have no articulation with the vertebral column. The clavicle is the anterior component of the shoulder girdle and articulates with the sternum at the sternoclavicular joint. The posterior component, the scapula, which is positioned freely by complex muscle attachments, articulates with the clavicle and humerus. Although the shoulder joints are weak, they allow movement in many directions and are thus freely movable.

Clavicles

The **clavicles,** or collarbones, are long slender bones with a double curvature (see Figure 8-2). The two bones lie horizontally in the superior and anterior part of the thorax superior to the first rib.

The medial end of the clavicle, the **sternal extremity,** is rounded and articulates with the sternum. The broad, flat, lateral end, the **acromial extremity,** articulates with the acromion process of the scapula. This joint is called the **acromioclavicular joint.** Refer to Figure 8-1 for a view of these articulations. The **conoid tubercle** on the inferior surface of the lateral end of the bone serves as a point of attachment for a ligament.

Scapulae

The **scapulae,** or shoulder blades, are large, triangular, flat bones situated in the dorsal part of the thorax between the levels of the second and seventh ribs (see Figure 8-3). Their medial borders are located about 5 cm (2 inches) from the vertebral column.

A sharp ridge, the **spine,** runs diagonally across the posterior surface of the flattened, triangular **body.** The end of the spine projects as a flattened, expanded process called the **acromion.** This process articulates with the clavicle. Inferior to the acromion is a depression called the **glenoid cavity.** This cavity articulates with the head of the humerus to form the shoulder joint. The thin edge of the body near the vertebral column is the **medial** or **vertebral border.** The thick edge closer to the arm is the **lateral** or **axillary border.** The medial and lateral borders joint at the **inferior angle.** The superior edge of the scapular body is called the **superior border.** At the lateral end of the superior border is a projection of the anterior surface called the **coracoid process** to which muscles attach. Above and below the spine are two fossae: the **supraspinatous fossa** and the **infraspinatous fossa,** respectively. Both serve as surfaces of attachment for shoulder muscles. On the anterior surface is a slightly hollowed-out area called the **subscapular fossa,** also a surface of attachment for shoulder muscles.

UPPER EXTREMITIES

The **upper extremities** consist of 60 bones. The skeleton of the right upper extremity is shown in Figure 8-1. It includes a humerus in each arm, an ulna and radius in each forearm, carpals, or wrist bones, metacarpals, which are the palm bones, and phalanges in the fingers of each hand.

Humerus

The **humerus,** or arm bone, is the longest and largest bone of the upper extremity (see Figure 8-4). It articulates proximally with the scapula and distally at the elbow with both ulna and radius.

The proximal end of the humerus consists of a **head** that articulates with the glenoid cavity of the scapula. It also has an **anatomical neck,** which is an oblique groove just distal to the head. The **greater tubercle** is a lateral projection distal to the neck. The **lesser tubercle** is an anterior projection. Between these tubercles runs an **intertubercular sulcus (bicipital groove).** The **surgical neck** is a constricted portion just distal to the tubercles and is named because of its liability to fracture. The **body** or shaft of the humerus is cylindrical at its proximal end. It gradually becomes triangular and is flattened and broad at its distal end. Along the middle portion of the shaft, there is a roughened, V-shaped area called the **deltoid tuberosity.** This area serves as a point of attachment for the deltoid muscle. The following parts are found at the distal end of the humerus. The **capitulum** is a rounded knob that articulates with the head of the radius. The **radial fossa** is a depression that receives the head of the radius when the forearm is flexed. The **trochlea** is a pulleylike surface that articulates with the ulna. The **coronoid fossa** is an anterior depression that receives part of the ulna when the forearm is flexed. The **olecranon fossa** is a posterior depression that receives the olecranon of the ulna when the forearm is extended. The **medial epicondyle** and **lateral epicondyle** are rough projections on either side of the distal end.

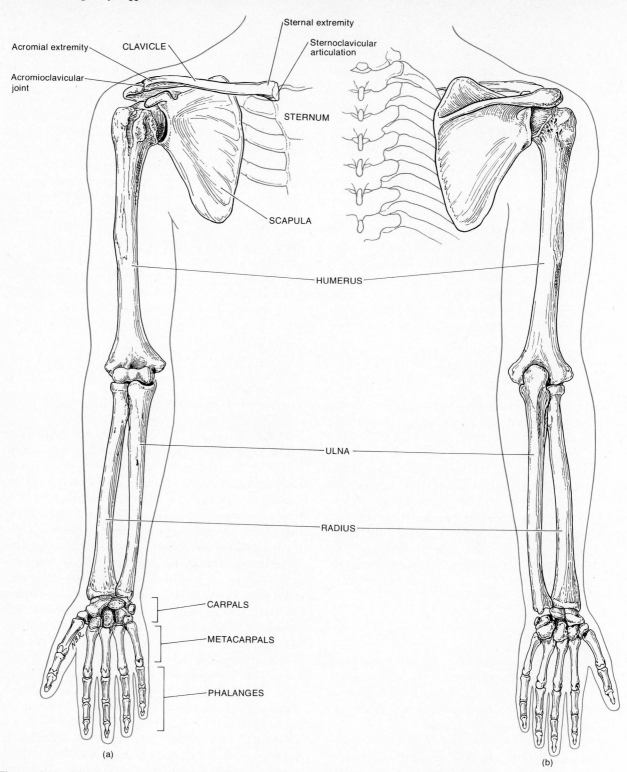

Acromial extremity

CLAVICLE

Acromioclavicular joint

Sternal extremity

Sternoclavicular articulation

STERNUM

SCAPULA

HUMERUS

ULNA

RADIUS

CARPALS

METACARPALS

PHALANGES

(a)

(b)

Figure 8–1   Right shoulder girdle and upper extremity. (a) Anterior view. (b) Posterior view.

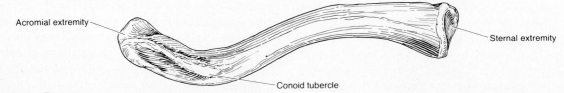

Acromial extremity

Sternal extremity

Conoid tubercle

Figure 8–2   Right clavicle viewed from below.

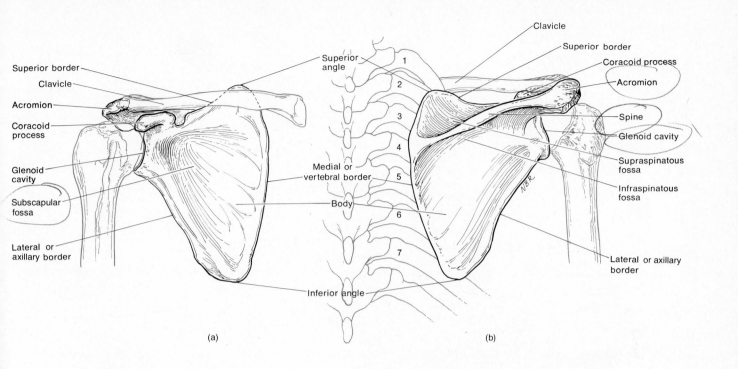

(a)

(b)

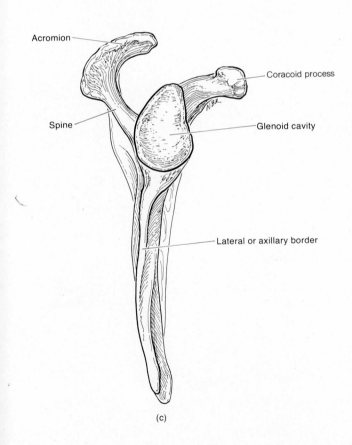

(c)

Figure 8–3 Right scapula. (a) Anterior view. (b) Posterior view. (c) Lateral border view.

Ulna and Radius

The **ulna** is the medial bone of the forearm (see Figure 8-5). In other words, it is located on the small finger side.

The proximal end of the ulna presents an **olecranon (olecranon process),** which forms the prominence of the elbow. The **coronoid process** is an anterior projection that, together with the olecranon, receives the trochlea of the humerus. The **trochlear notch (semilunar notch)** is a curved area between the olecranon and the coronoid processes. The trochlea of the humerus fits into this notch. The **radial notch** is a depression located laterally and inferiorly to the trochlear notch. It receives the head of the radius. The distal end of the ulna consists of a **head** that is separated from the wrist by a fibrocartilage disc. A **styloid process** is on the posterior side of the distal end.

The **radius** is the lateral bone of the forearm. That is, it is situated on the thumb side.

The proximal end of the radius has a disc-shaped **head** that articulates with the capitulum of the humerus and radial notch of the ulna. It also has a raised, roughened area on the medial side called the **radial tuberosity.** This is a point of attachment for the biceps muscle. The shaft of the radius widens distally to form a concave inferior surface that articulates with two bones of the wrist called the lunate and navicular bones. Also at the distal end is a **styloid process** on the lateral side and a medial, concave **ulnar notch** for articulation with the distal end of the ulna. A common fracture of the radius called a *Colles' fracture* occurs along the shaft about 2 ⅓ cm (1 inch) from the distal end of the bone.

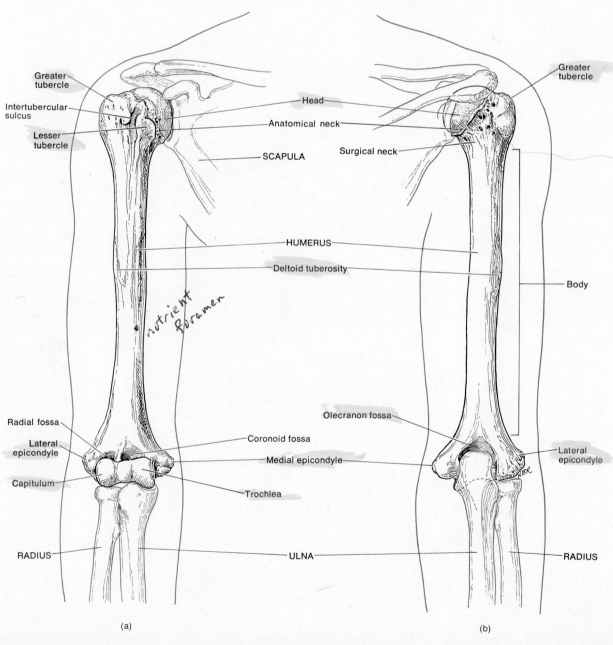

Figure 8-4   Right humerus. (a) Anterior view. (b) Posterior view.

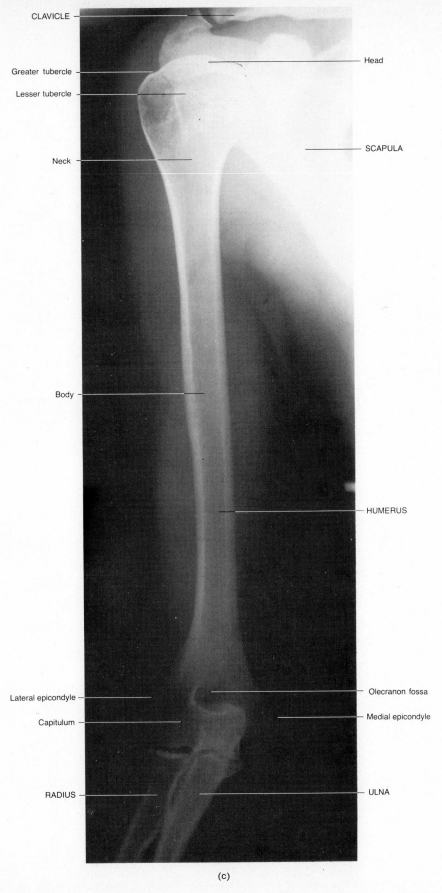

CLAVICLE

Greater tubercle

Lesser tubercle

Neck

Body

Lateral epicondyle

Capitulum

RADIUS

Head

SCAPULA

HUMERUS

Olecranon fossa

Medial epicondyle

ULNA

(c)

Figure 8–4 (cont.)   Right humerus. (c) Anteroposterior projection. (Courtesy of Harvey Peck, Good Samaritan Hospital, Suffern, New York, and Daniel Sorrentino.)

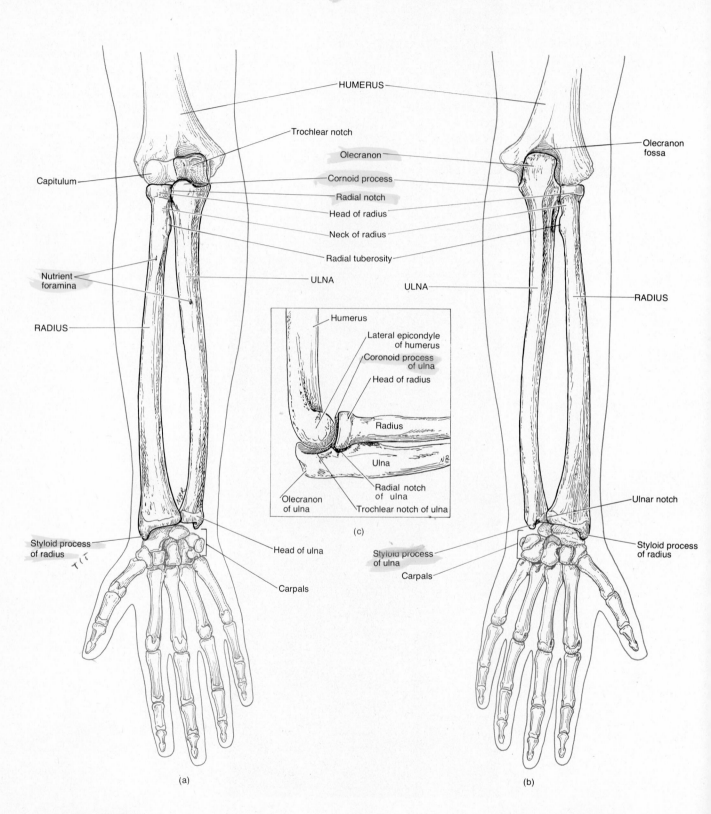

HUMERUS

Trochlear notch

Olecranon

Cornoid process

Radial notch

Head of radius

Neck of radius

Radial tuberosity

Capitulum

Nutrient foramina

RADIUS

ULNA

Styloid process of radius

Head of ulna

Carpals

Olecranon fossa

ULNA

RADIUS

Ulnar notch

Styloid process of radius

Styloid process of ulna

Carpals

Humerus

Lateral epicondyle of humerus

Coronoid process of ulna

Head of radius

Radius

Ulna

Radial notch of ulna

Olecranon of ulna

Trochlear notch of ulna

(c)

(a)

(b)

Figure 8–5   Right ulna and radius. (a) Anterior view. (b) Posterior view. (c) Lateral view of the right elbow.

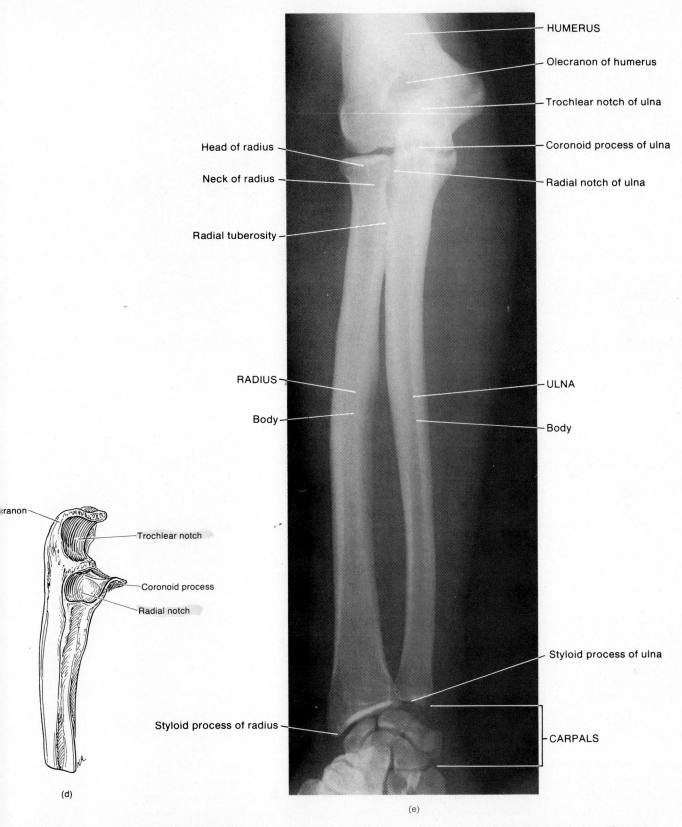

HUMERUS

Olecranon of humerus

Trochlear notch of ulna

Head of radius

Coronoid process of ulna

Neck of radius

Radial notch of ulna

Radial tuberosity

RADIUS

ULNA

Body

Body

ranon

Trochlear notch

Coronoid process

Radial notch

Styloid process of ulna

Styloid process of radius

CARPALS

(d)

(e)

Figure 8–5 (cont.) Right ulna and radius. (d) Details of proximal end of ulna. (e) Anteroposterior projection. (Courtesy of John C. Bennett, St. Mary's Hospital, San Francisco.)

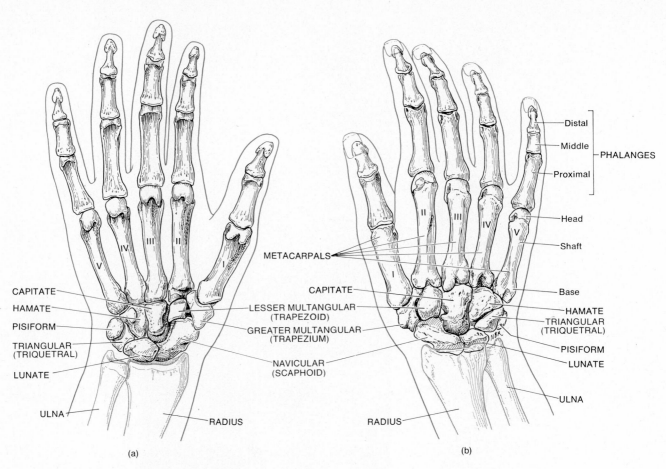

Figure 8—6   Right wrist and hand. (a) Anterior view. (b) Posterior view.

Carpus, Metacarpus, and Phalanges

The **carpus,** or wrist, consists of eight small bones united to each other by ligaments (see Figure 8-6). The bones are arranged in two transverse rows, with four bones in each row. The proximal row, from the lateral to medial position, consists of the following bones: **navicular (scaphoid), lunate, triangular (triquetral),** and **pisiform.** In about 70 percent of the cases involving carpal fractures, only the navicular is involved. The distal row of bones, from lateral to medial position, consists of the following: **greater multangular (trapezium), lesser multangular (trapezoid), capitate,** and **hamate.**

The five bones of the **metacarpus** constitute the palm of the hand. Each metacarpal bone consists of a proximal **base,** a **shaft,** and a distal **head.** The metacarpal bones are numbered I to V, starting with the lateral bone. The bases articulate with the distal row of carpal bones and with one another. The heads articulate with the proximal phalanges of the fingers. The heads of the metacarpals are commonly called the "knuckles" and are readily visible when the fist is clenched.

The **phalanges,** or bones of the fingers, number 14 in each hand. There are two phalanges in the first digit, called the thumb or pollex, and three phalanges in each of the remaining four digits. The first row of phalanges, the **proximal row,** articulates with the metacarpal bones and second row of phalanges. The second row of phalanges, the **middle row,** articulates with the proximal row and the third row. The third row of phalanges, the **distal row,** articulates with the middle row. A single finger bone is referred to as a **phalanx.** The thumb has no middle phalanx.

PELVIC GIRDLE

The **pelvic girdle** consists of the two **coxal bones,** commonly called the pelvic, innominate, or hipbones (see Figure 8-7). The pelvic girdle provides a strong and stable support for the lower extremities on which the weight of the body is carried. The coxal bones are united to each other anteriorly at the symphysis pubis. They unite posteriorly to the sacrum.

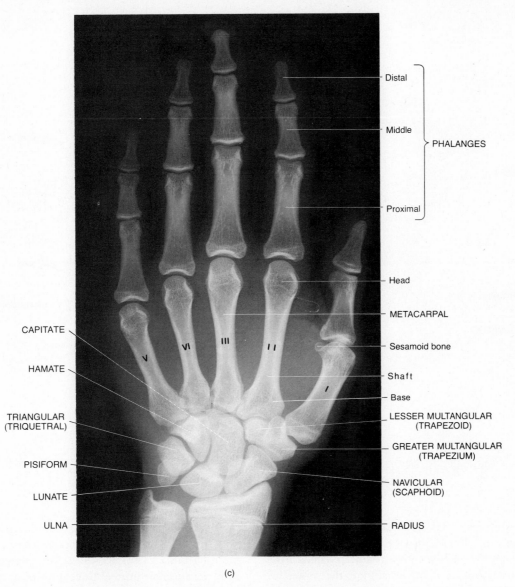

- Distal
- Middle — PHALANGES
- Proximal

- Head
- METACARPAL
- Sesamoid bone
- Shaft
- Base
- LESSER MULTANGULAR (TRAPEZOID)
- GREATER MULTANGULAR (TRAPEZIUM)
- NAVICULAR (SCAPHOID)
- RADIUS

CAPITATE
HAMATE
TRIANGULAR (TRIQUETRAL)
PISIFORM
LUNATE
ULNA

(c)

Figure 8–6 (cont.)   Right wrist and hand. (c) Anteroposterior projection. Note the sesamoid bone. (Courtesy of Harvey Peck, Good Samaritan Hospital, Suffern, New York, and Daniel Sorrentino.)

Together with the sacrum and coccyx, the pelvic girdles form the basinlike structure called the **pelvis.** The pelvis is divided into a greater pelvis and a lesser pelvis. The **greater pelvis (false pelvis)** is the expanded portion situated superior to the narrow bony ring called the **brim of the pelvis.** The greater pelvis consists laterally of the two ilia and posteriorly of the superior portion of the sacrum. There is no bony component in the anterior aspect of the greater pelvis. Rather, the front is formed by the walls of the abdomen. The **lesser** or **true pelvis** is inferior and posterior to the pelvic brim. It is formed by parts of the ilium, pubis, sacrum, and coccyx. The lesser pelvis contains a superior opening called the **pelvic inlet** and an inferior opening called the **pelvic outlet.** *Pelvimetry* is the measurement of the size of the inlet and outlet of the birth canal.

Coxal Bones

The two **coxal bones (os coxae)** of a newborn consist of three components: a superior **ilium,** an inferior and

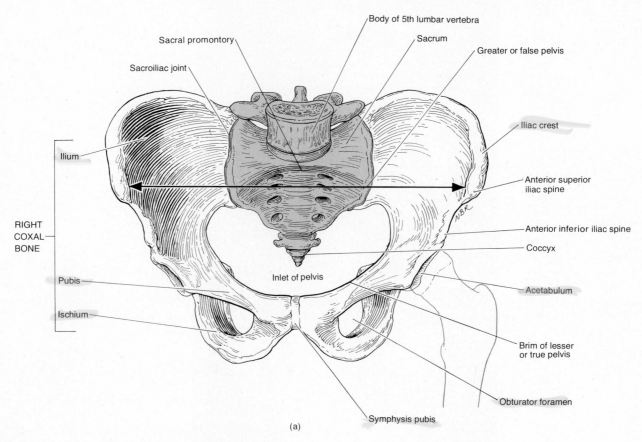

Body of 5th lumbar vertebra

Sacral promontory

Sacrum

Sacroiliac joint

Greater or false pelvis

Ilium

Iliac crest

RIGHT
COXAL
BONE

Anterior superior
iliac spine

Anterior inferior iliac spine

Coccyx

Pubis

Inlet of pelvis

Acetabulum

Ischium

Brim of lesser
or true pelvis

Obturator foramen

Symphysis pubis

(a)

Figure 8–7   Pelvic girdle. (a) Anterior view.

anterior **pubis,** and an inferior and posterior **ischium** (see Figure 8-8). Eventually, the three separate bones fuse into one. The area of fusion is a deep, lateral fossa called the **acetabulum.** This structure is the socket for the head of the femur. Although the adult coxae are both single bones, it is common to discuss the bones as if they still consisted of three portions.

The ilium is the largest of the three subdivisions of the coxal bone. Its superior border, the **iliac crest,** ends anteriorly in the **anterior superior iliac spine.** The **anterior inferior iliac spine** is located inferior to the anterior superior spine. Posteriorly, the iliac crest ends in the **posterior superior iliac spine.** The **posterior inferior iliac spine** is just inferior. The spines serve as points of attachment for muscles of the abdominal wall. Just inferior to the posterior inferior iliac spine is the **greater sciatic notch.** The internal surface of the ilium seen from the medial side is the **iliac fossa.** It is a concavity where the iliacus muscle attaches. Posterior to this fossa are the **iliac tuberosity,** a point of attachment for the sacroiliac ligament, and the **auricular surface,**

which articulates with the sacrum. The other conspicuous markings of the ilium are three arched lines on its gluteal (buttock) surface called the **posterior gluteal line,** the **anterior gluteal line,** and the **inferior gluteal line.** The gluteal muscles attach to the ilium between these lines.

The ischium is the inferior, posterior portion of the coxal bone. It contains a prominent **ischial spine,** a **lesser sciatic notch** below the spine, and an **ischial tuberosity.** The rest of the ischium, the **ramus,** joins with the pubis and together they surround the **obturator foramen.**

The pubis is the anterior and inferior part of the coxal bone. It consists of a **superior ramus,** an **inferior ramus,** and a **body** that contributes to the formation of the symphysis pubis. The **symphysis pubis** is the joint between the two coxal bones. It consists of fibrocartilage (Figure 8-7). The **acetabulum** is the socket formed by the ilium, ischium, and pubis. Two-fifths of the acetabulum is formed by the ilium, two-fifths by the ischium, and one-fifth by the pubis.

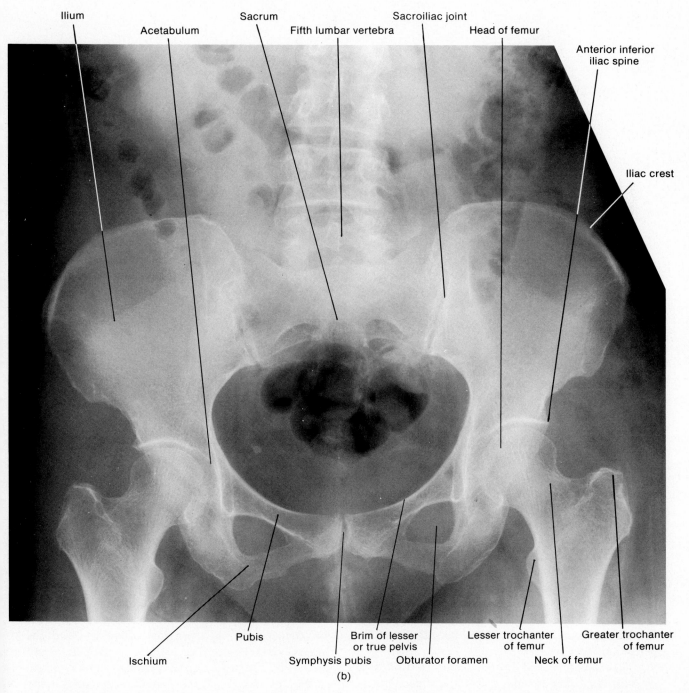

Ilium
Acetabulum
Sacrum
Fifth lumbar vertebra
Sacroiliac joint
Head of femur
Anterior inferior iliac spine
Iliac crest

Pubis
Brim of lesser or true pelvis
Lesser trochanter of femur
Greater trochanter of femur
Ischium
Symphysis pubis
Obturator foramen
Neck of femur

(b)

Figure 8–7 (cont.)   Pelvic girdle. (b) Anteroposterior projection. (Courtesy of John C. Bennett, St. Mary's Hospital, San Francisco.)

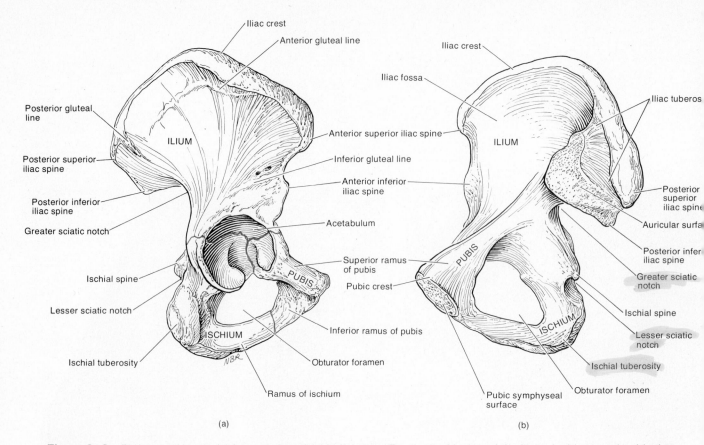

Figure 8-8 Right coxal bone. (a) Lateral view. (b) Medial view. The lines of fusion of the ilium, ischium, and pubis that are shown in color in (a) are not actually visible in an adult bone.

LOWER EXTREMITIES

The **lower extremities** are composed of 60 bones (see Figure 8-9). These include the femur of each thigh, each kneecap, the fibula and tibia in each leg, the ankle bones in each ankle, and the metatarsals and phalanges of each foot.

Femur

The **femur,** or thigh bone, is the longest and heaviest bone in the body (see Figure 8-10). Its proximal end articulates with the coxal bone. Its distal end articulates with the tibia. The shaft of the femur bows medially so that it approaches the femur of the opposite thigh. As a result of this convergence, the knee joints are brought nearer to the body's line of gravity. The degree of convergence is greater in the female because the female pelvis is broader.

The proximal end of the femur consists of a rounded **head** that articulates with the acetabulum of the coxal bone. The **neck** of the femur is a constricted region distal to the head. A fairly common fracture in the elderly occurs at the neck of the femur. Apparently the neck becomes so weak that it fails to support the body. The **greater trochanter** and **lesser trochanter** are projections that serve as points of attachment for some of the thigh and buttock muscles. Between the trochanters on the anterior surface is a narrow **intertrochanteric line.** Between the trochanters on the posterior surface is an **intertrochanteric crest.**

The shaft of the femur contains a rough vertical ridge on its posterior surface called the **linea aspera.** This ridge serves for the attachment of several thigh muscles.

The distal end of the femur is expanded and includes the **medial condyle** and **lateral condyle.** These articulate with the tibia. A depressed area between the condyles on the posterior surface is called the **intercondylar fossa.** The **patellar surface** is located between the condyles on the anterior surface. Lying superior to the condyles are the **medial epicondyle** and **lateral epicondyle.**

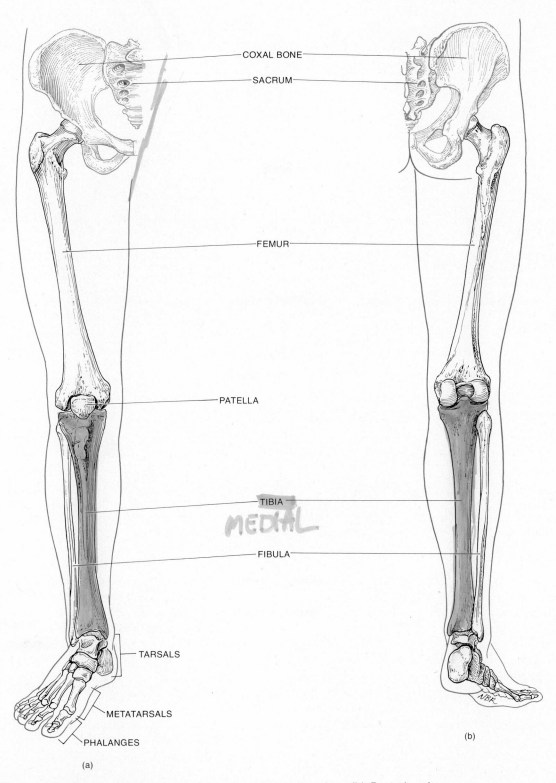

COXAL BONE

SACRUM

FEMUR

PATELLA

TIBIA

FIBULA

TARSALS

METATARSALS

PHALANGES

(a)

(b)

Figure 8–9 Right pelvic girdle and lower extremity. (a) Anterior view. (b) Posterior view.

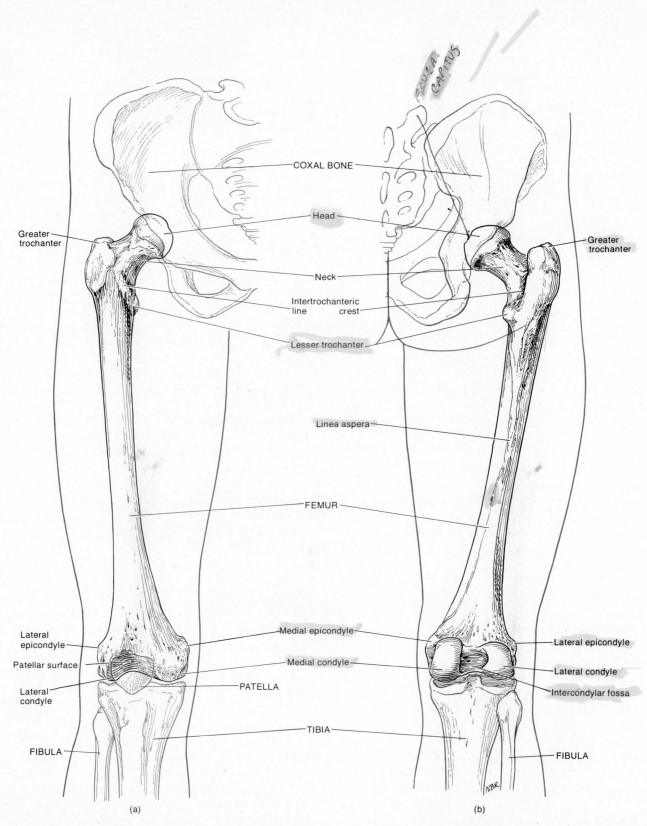

Figure 8–10   Right femur. (a) Anterior view. (b) Posterior view. A roentgenogram of the proximal end of the femur is shown in Figure 8–7b.

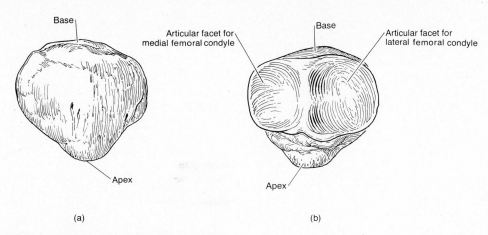

Figure 8–11   Right patella. (a) Anterior view. (b) Posterior view.

Patella

The **patella,** or kneecap, is a small, triangular bone anterior to the knee joint (see Figure 8-11). It develops in the tendon of the quadriceps femoris muscle. A bone that forms in a tendon, such as the patella, is called a sesamoid bone. The broad superior end of the patella is called the **base.** The pointed inferior end is the **apex.** The posterior surface contains two articular surfaces. These are the **articular facets** for the medial and lateral condyles of the femur.

Tibia and Fibula

The **tibia,** or shinbone, is the larger, medial bone of the leg (see Figure 8-12). It bears the major portion of the weight on the leg. The tibia articulates at its proximal end with the femur and at its distal end with the fibula of the leg and talus of the ankle.

The proximal end of the tibia is expanded into a **lateral condyle** and a **medial condyle.** These articulate with the condyles of the femur. The slightly concave condyles are separated by an upward projection called the **intercondylar eminence.** The **tibial tuberosity** on the anterior surface is a point of attachment for the patellar ligament.

The medial surface of the distal end of the tibia forms the **medial malleolus.** This structure articulates with the talus bone of the ankle and forms the prominence that can be felt on the medial surface of your ankle. The **fibular notch** articulates with the fibula.

The **fibula** is parallel, and lateral, to the tibia. It is considerably smaller than the tibia.

The **head** of the fibula, the proximal end, articulates with the lateral condyle of the tibia below the level of the knee joint. The distal end has a projection called the **lateral malleolus,** which articulates with the talus

bone of the ankle. This forms the prominence on the lateral surface of the ankle. The inferior portion of the fibula also articulates with the tibia at the **fibular notch.** A fracture of the lower end of the fibula with injury to the tibial articulation is called a **Pott's fracture.**

Tarsus, Metatarsus, and Phalanges

The **tarsus** is a collective designation for the seven bones of the ankle (see Figure 8-13). The term *tarsos* pertains to a broad, flat surface. The **talus** and **calcaneus** are located on the posterior part of the foot. The anterior part contains the **cuboid, navicular,** and three **cuneiform** bones called the first (medial), second (intermediate), and third (lateral) cuneiform. The talus, the uppermost tarsal bone, is the only bone of the foot that articulates with the fibula and tibia. It is surrounded on one side by the medial malleolus of the tibia and on the other side by the lateral malleolus of the fibula. During walking, the talus initially bears the entire weight of the extremity. About half the weight is then transmitted to the calcaneus. The remainder is transmitted to the other tarsal bones. The calcaneus, or heel bone, is the largest and strongest tarsal bone.

The **metatarsus** consists of five metatarsal bones numbered I to V from the medial to lateral position. The metatarsals articulate proximally with the first, second, and third cuneiform bones and with the cuboid. Distally, they articulate with the proximal row of phalanges. The first metatarsal is thicker than the others because it bears more weight.

The **phalanges** of the foot resemble those of the hand both in number and arrangment. The great (big) toe, or hallux, has two large, heavy phalanges. The other four toes each have three phalanges. These are the proximal, middle, and distal phalanges.

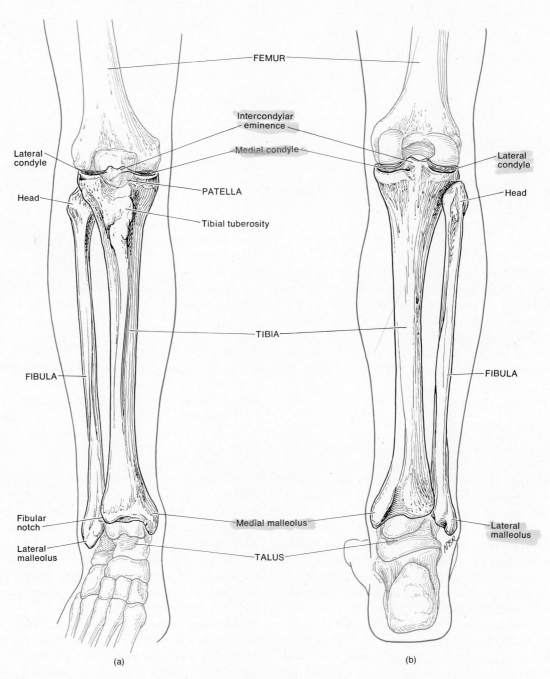

Figure 8–12 Right tibia and fibula. (a) Anterior view. (b) Posterior view.

Arches of the Foot

The bones of the foot are arranged in two **arches** (see Figure 8-14). These arches enable the foot to support the weight of the body and provide leverage while walking.

The arches are not rigid. They yield as weight is applied and spring back when the weight is lifted.

The **longitudinal arch** has two parts. Both consist of tarsal and metatarsal bones arranged to form an arch

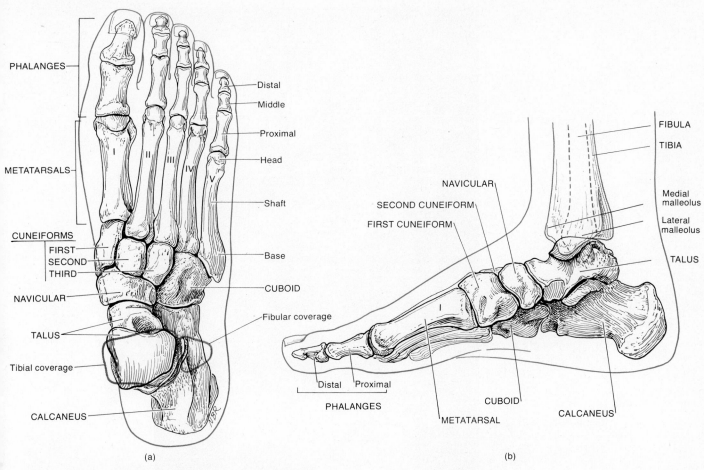

Figure 8–13 Right foot. (a) Superior view. (b) Medial view.

from the anterior to the posterior part of the foot. The **medial,** or inner, part of the longitudinal arch originates at the calcaneus. It rises to the talus and descends anteriorly through the navicular, the three cuneiforms, and the three medial metatarsals. The talus is the keystone of this arch. The **lateral,** or outer, part of the longitudinal arch also begins at the calcaneus. It rises at the cuboid and descends to the two lateral metatarsals. The cuboid is the keystone of the arch.

The **transverse arch** is formed by the calcaneus, navicular, cuboid, and the posterior parts of the five metatarsals.

The bones composing the arches are held in position by ligaments and tendons. If these ligaments and tendons are weakened, the height in the longitudinal arch may decrease or "fall." The result is *flatfoot*. A *bunion* is an abnormal lateral displacement of the big toe from its natural position. This condition produces an inflammatory reaction of the bursae that results in the formation of abnormal tissue.

MALE AND FEMALE SKELETONS

The bones of the male are generally larger and heavier than those of the female. The articular ends are also thicker when compared with the shafts. In addition, certain muscles of the male are larger than those of the female. Consequently, the male skeleton has larger tuberosities, lines, and ridges for the attachment of larger muscles.

One marked difference between male and female skeletons is the structure of the pelvis. The main differences between the male and female pelvis concern adaptations directly relating to childbearing (Figure 8-15). The female pelvis is wider, shallower, and lighter in structure than that of the male. The ilia of the female flare laterally to broaden the hips. The inlet of the true pelvis in the female is nearly oval, whereas that of the male is triangular or heart-shaped. The sacrum of the female is shorter, wider, and less curved than that of the male. The female coccyx is also more movable. The sciatic notches are wider and shallower in the female.

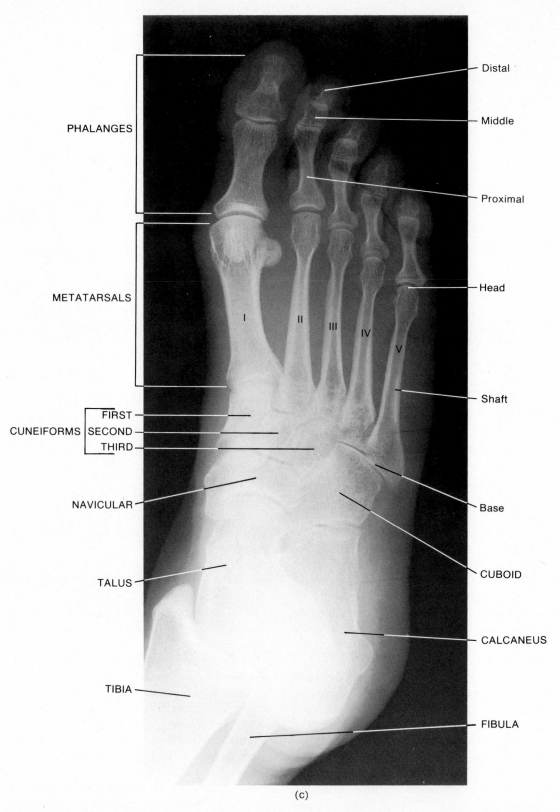

PHALANGES

Distal

Middle

Proximal

METATARSALS

Head

I   II   III   IV   V

CUNEIFORMS FIRST
            SECOND
            THIRD

Shaft

NAVICULAR

Base

TALUS

CUBOID

CALCANEUS

TIBIA

FIBULA

(c)

Figure 8–13 (cont.)   Right foot. (c) Plantar view. (Courtesy of John C. Bennett, St. Mary's Hospital, San Francisco.)

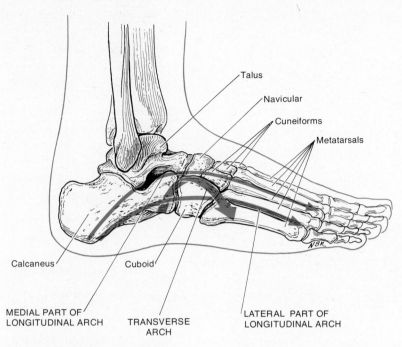

Figure 8–14 Labels: Talus, Navicular, Cuneiforms, Metatarsals, Calcaneus, Cuboid, MEDIAL PART OF LONGITUDINAL ARCH, TRANSVERSE ARCH, LATERAL PART OF LONGITUDINAL ARCH

Figure 8–14 Arches of the right foot in lateral view.

The ischial spines and tuberosities of the female turn outward and are further apart than in the male. The pubic arch thus forms an obtuse angle rather than an acute angle as in the male. The pubic arch is the angle at which the right and left pubic portions of the coxal bones meet.

All characteristics contribute to the wider outlet of the true pelvis in the female. This feature, of course, accommodates the birth of the child. In addition, the ligaments of the sacroiliac joint stretch during pregnancy and childbirth. Additional space is provided for the developing fetus, and delivery is facilitated.

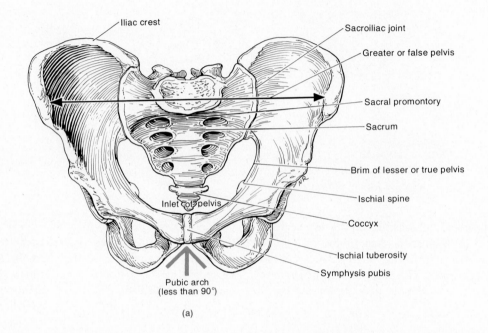

Iliac crest

Sacroiliac joint

Greater or false pelvis

Sacral promontory

Sacrum

Brim of lesser or true pelvis

Ischial spine

Coccyx

Inlet of pelvis

Ischial tuberosity

Symphysis pubis

Pubic arch
(less than 90°)

(a)

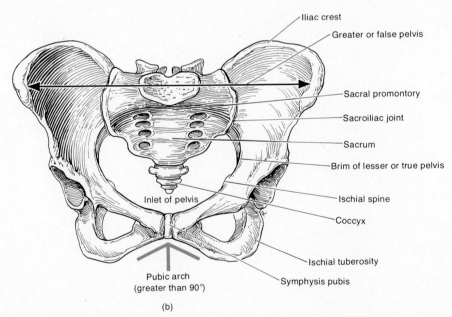

Iliac crest

Greater or false pelvis

Sacral promontory

Sacroiliac joint

Sacrum

Brim of lesser or true pelvis

Ischial spine

Coccyx

Inlet of pelvis

Ischial tuberosity

Symphysis pubis

Pubic arch
(greater than 90°)

(b)

Figure 8–15 Pelvis. (a) Male pelvis in anterior view. (b) Female pelvis in anterior view.

STUDY OUTLINE

SHOULDER GIRDLES

1 Each shoulder girdle or pectoral girdle consists of a clavicle and scapula.

2 Each attaches the upper extremity to the trunk.

UPPER EXTREMITIES

1 The bones of each upper extremity include the humerus, ulna, radius, carpals, metacarpals, and phalanges.

PELVIC GIRDLE

1 The pelvic girdle consists of two coxal bones or hipbones.

2 It attaches the lower extremities to the trunk.

LOWER EXTREMITIES

1 The bones of each lower extremity include the femur, tibia, fibula, tarsus, metatarsus, and phalanges.

2 The arches of the foot are bones arranged for support and leverage.

3 The two parts of the longitudinal arch are the higher medial and lower lateral arches. The other arch is the transverse arch.

MALE AND FEMALE SKELETONS

1 Male bones are generally larger than female bones.

2 The female pelvis is adapted for pregnancy and childbirth.

REVIEW QUESTIONS

1 What is a shoulder girdle? What are the bones of the upper extremity? What is a pelvic girdle? What are the bones of the lower extremity?

2 Define a Colles' fracture and a Pott's fracture.

3 Define an arch of the foot. What is its function? Distinguish between a longitudinal arch and a transverse arch.

4 How do bunions and flatfeet arise?

5 What are the principal structural differences between male and female skeletons?

Chapter 9
Articulations

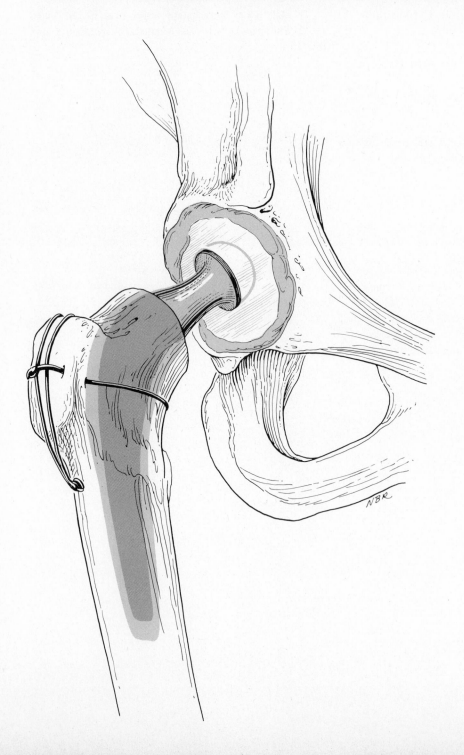

■ Define an articulation and identify the factors that determine the degree of movement at a joint.

■ Contrast the structure, kind of movement, and location of fibrous, cartilaginous, and synovial joints.

■ Describe the detailed structure of a synovial joint.

■ Discuss and compare the movements possible at various synovial joints.

■ Describe the causes and symptoms of common joint disorders, including arthritis, rheumatism, rheumatoid arthritis, osteoarthritis, gouty arthritis, bursitis, and tendinitis.

■ Define dislocation.

■ Define a sprain.

■ Describe the conditions that may cause a slipped disc.

■ Define medical terminology associated with joints.

Bones are much too rigid to bend. Fortunately, the skeletal system consists of many separate bones, which are held together at joints by flexible connective tissue. All movements that change the positions of the bony parts of the body, such as the extremities, occur at joints. You can understand the importance of joints if you imagine how a cast over the knee joint prevents flexing the leg or how a splint on a finger limits the ability to manipulate small objects.

The term **articulation** or **joint** refers to a point of contact between bones or between cartilage and bones. The joint's structure determines its function. Some joints permit no movement. Others permit slight movement. Still others afford unrestricted movement. In general, the more closely the bones fit together, the stronger the joint. At tightly fitted joints, however, movement is restricted. The greater the movement, the looser the fit. Unfortunately, loosely fitted joints are prone to dislocation. Movement at joints is also determined by the flexibility of the connective tissue that binds the bones together and by the position of ligaments, muscles, and tendons.

CLASSIFICATION

Functional

The functional classification of joints takes into account the degree of movement they permit. Functionally, joints are classified as **synarthroses,** which are immovable joints; **amphiarthroses,** which are slightly movable joints; and **diarthroses,** which are freely movable joints.

Structural

The structural classification of joints is based on the presence or absence of a joint cavity (a space between the bones) and the kind of connective tissue that binds the bones together. Structurally, joints are classified as **fibrous,** in which there is no joint cavity and the bones are held together by fibrous connective tissue; **cartilaginous,** in which there is no joint cavity and the bones are held together by cartilage; and **synovial,** in which the joint contains a synovial cavity.

FIBROUS JOINTS

Fibrous joints lack a joint cavity, and the articulating bones are held close together by fibrous connective tissue. They allow little or no movement. The two types of fibrous joints are sutures and syndesmoses. **Sutures** are found between bones of the skull. (See Figure 7-2.) Some sutures consist of interlocking, jagged margins of

bone that fit together like a jigsaw puzzle. In other sutures, the margins of the bones overlap. In either case, the bones are barely separated by a thin layer of fibrous tissue. Such joints are immovable and are classified as synarthroses. Some sutures, present during growth, are replaced by bone in the adult. In this case, they become **synostoses,** in which there is complete fusion of bone across the suture line. One synostosis is the joint between the left and right sides of the frontal bone. Synostoses are synarthrotic. The bone surfaces of a **syndesmosis** are united by dense fibrous tissue. The joint is slightly movable because the bones are more separated from each other than they are in a suture. Thus it is an amphiarthrotic joint. An example of a syndesmosis-type joint is the distal articulation of the tibia with the fibula. (See Figure 8-12.)

CARTILAGINOUS JOINTS

Another joint that has no joint cavity is the **cartilaginous joint.** Here the articulating bones are tightly connected by cartilage. Like fibrous joints, they allow little or no movement (Figure 9-1). A **synchondrosis** is a cartilaginous joint in which the connecting material is hyaline cartilage, as in the epiphyseal plate. Such a joint is found between the epiphysis and diaphysis of a growing bone and is immovable. Thus it is synarthrotic. Since the hyaline cartilage is eventually replaced by bone when growth ceases, the joint is temporary. It is replaced by a synostosis. A **symphysis** is a cartilaginous joint in which the connecting material is a broad, flat disc of fibrocartilage. This joint is found between bodies of vertebrae. A portion of the intervertebral disc is cartilaginous material. The symphysis pubis between the anterior surfaces of the coxal bones is another example of a symphysis-type joint. These joints are slightly movable, or amphiarthrotic.

SYNOVIAL JOINTS

When a joint cavity is present, the articulation is called a **synovial joint** (Figure 9-2). The cavity, called a **synovial** or **joint cavity,** is a space between the articulating bones. Because of this cavity and because no tissue exists between the articulating surfaces of the bones, synovial joints are freely movable. Synovial joints are diarthrotic by function. Synovial joints are surrounded by a capsule of dense fibrous connective tissue that is continuous with the periosteum of the articulating bones. This capsule is called the *joint* or *fibrous capsule.* It protects and, in some cases, strengthens the joint. Synovial joints are also characterized by a layer of hyaline cartilage, called *articular cartilage.* Articular cartilage covers the

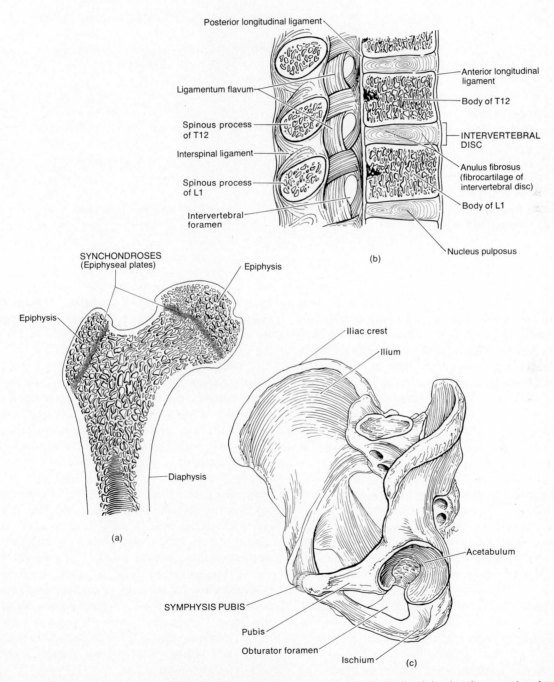

Figure 9-1 Cartilaginous joints. (a) Synchondrosis between the diaphysis and epiphysis of a growing femur. (b) Symphysis joint between the bodies of vertebrae seen in sagittal section. (c) Symphysis joint (symphysis pubis) between the coxal bones seen in an oblique view.

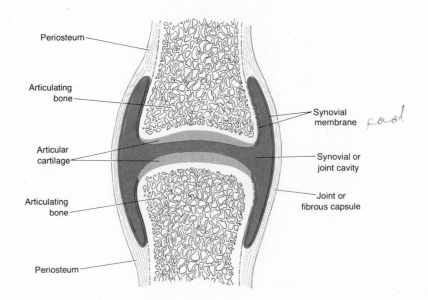

Figure 9—2   Structure of a synovial joint.

surfaces of the articulating bones but does not bind the bones together. These joints also have a *synovial membrane,* which lines the walls of the cavity and secretes synovial fluid to lubricate the joint. Synovial fluid consists of hyaluronic acid and interstitial fluid. When there is no joint movement, synovial fluid is viscous. But when there is joint movement, the fluid becomes less viscous. Synovial joints are held together by bands of collagenous fibers called *ligaments.* A ligament attaches to processes on one of the articulating bones, runs alongside the joint, and attaches to processes on the other bone.

Synovial joints are free of the limitations of fibrous and cartilaginous joints. In synovial joints, movement is determined by the location of the ligaments, by the muscles and their tendons, and by the presence of other bones that might restrict movement. The knee joint, one of the largest articulations in the body, illustrates the structure of a synovial joint and the limitations on its movement.

Examine Figure 9-3 and note the following relationships among the ligaments, tendons, and fibrocartilage in the knee joint:

1 Externally the joint is strengthened by muscles and tendons. These include the tendon of the quadriceps femoris muscle anteriorly and the gastrocnemius muscle posteriorly. The patella lies within the tendon of the quadriceps femoris.

2 The thickened portion of the quadriceps femoris tendon between the top of the patella and the tibia is called the patellar ligament. This ligament strengthens the anterior portion of the joint and prevents the lower leg from being flexed too far backward.

3 On either side of the joint are the fibular and tibial collateral ligaments. The fibular collateral ligament, between the femur and fibula, strengthens the lateral side of the joint. The tibial collateral ligament, between the femur and tibia, strengthens the medial side of the joint. Both ligaments prohibit side-to-side movement at the joint.

4 The oblique popliteal ligament is located on the posterior surface of the joint. It starts in a tendon that lies over the tibia and runs upward and laterally to the lateral side of the femur. It supports the back of the knee and prevents hyperextension – that is, bending the knee in the direction opposite to which it normally bends.

5 Internally, the joint is strengthened by the cruciate ligaments. The anterior cruciate ligament passes posteriorly and laterally from the tibia and attaches to the femur. The posterior cruciate ligament passes anteriorly and medially from the tibia and attaches to the femur. Both ligaments are believed to stabilize the knee joint during its movements.

6 Between the articular cartilage of the femur and tibia are **menisci**, concentric wedge-shaped pieces of fibrocartilage. These are called the lateral meniscus and the medial meniscus. The menisci provide support for

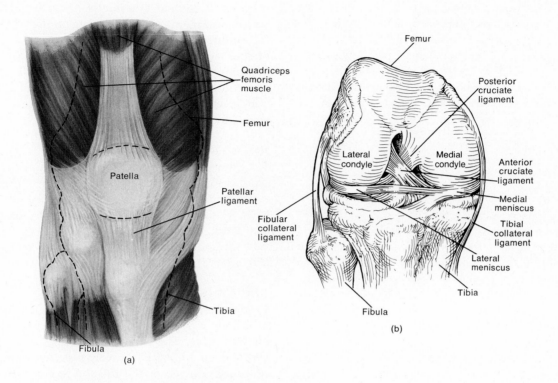

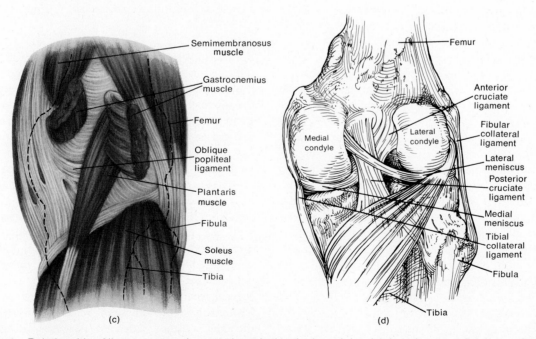

Figure 9–3 Relationship of ligaments, tendons, and menisci to the knee joint. (a) Anterior, superficial view. (b) Anterior view with many superficial structures removed. (c) Posterior, superficial view. (d) Posterior view with many superficial structures removed.

the continuous weight placed on the knee joint. Their surfaces also assist in rotation of the knee.

The various movements of the body could create friction between moving parts. To reduce this friction, saclike structures called **bursae** are situated in the body tissues. These sacs resemble joints in that their walls consist of connective tissue lined by a synovial membrane. They are also filled with synovial fluid. Bursae are located between the skin and bone in places where skin rubs over bone. They are also found between tendons and bones, muscles and bones, and ligaments and bones. As fluid-filled sacs, they cushion the movement of one part of the body over another.

Movements

GLIDING

A **gliding movement** is the simplest kind that can occur at a joint. One surface moves back and forth and from side to side over another surface without angular or rotary motion. Some joints that glide are those between the carpals and between the tarsals. The heads and tubercles of ribs glide on the bodies and transverse processes of vertebrae.

ANGULAR

Angular movements increase or decrease the angle between bones. Among the angular movements are flexion, extension, abduction, and adduction (Figure 9-4). **Flexion** usually involves a decrease in the angle between the anterior surfaces of the articulating bones. An exception to this definition is flexion of the knee and the toe joints in which there is a decrease in the angle between the posterior surfaces of the articulating bones. Examples of flexion include bending the head forward, where the joint is between the occipital bone and the atlas, bending the elbow, and bending the knee. Flexion of the foot at the ankle joint is called **dorsiflexion**. **Extension** involves an increase in the angle between the anterior surface of the articulating bones, with the two exceptions noted above. Extension restores a body part to its anatomical position after it has been flexed. Examples of extension are returning the head to the anatomical position after flexion, straightening the arm after flexion, and straightening the leg after flexion. Continuation of extension beyond the anatomical position, as in bending the head backward, is called **hyperextension**. Extension of the foot at the ankle joint is **plantar flexion**.

Abduction usually means movement of a bone *away from* the midline of the body. An example of abduction is moving the arms upward and away from

the body until they are held straight out at right angles to the chest. With the fingers and toes, however, the midline of the body is not used as the line of reference. Abduction of the fingers is a movement away from an imaginary line drawn through the middle finger – in other words, spreading the fingers. Abduction of the toes is relative to an imaginary line drawn through the second toe. **Adduction** is usually movement of a part *toward* the midline of the body. An example of adduction is returning the arms to the sides after abduction. As in abduction, adduction of the fingers is relative to the middle finger. Adduction of the toes is relative to the second toe.

ROTATION

Rotation is movement of a bone around its own axis. During rotation, no other motion is permitted. We rotate the atlas around the odontoid process of the axis when we shake the head no. Moving from the shoulder and turning the forearm up, then palm down, and then palm up again is an example of slight rotation of the arm (Figure 9-5a).

CIRCUMDUCTION

Circumduction is a movement in which the distal end of a bone moves in a circle while the proximal end remains stable. The bone describes a cone in the air. Circumduction typically involves flexion, abduction, adduction, extension, and rotation. It involves a 360° rotation. An example is moving the outstretched arm in a circle to wind up to pitch a ball (Figure 9-5b).

SPECIAL

Special movements are those found only at the joints indicated in Figure 9-6. **Inversion** is the movement of the sole of the foot inward at the ankle joint. **Eversion** is the movement of the sole outward at the ankle joint. **Protraction** is the movement of the mandible or clavicle forward on a plane parallel to the ground. Thrusting the jaw outward is protraction of the mandible. Bringing your arms forward until the elbows touch requires protraction of the clavicle. **Retraction** is the movement of a protracted part of the body backward on a plane parallel to the ground. Pulling the lower jaw back in line with the upper jaw is retraction of the mandible. **Supination** is a movement of the forearm in which the palm of the hand is turned forward (anterior). To demonstrate supination, flex your arm at the elbow to prevent rotation of the humerus in the shoulder joint. **Pronation** is a movement of the flexed forearm in which the palm is turned backward (posterior). **Elevation** is a movement in which a part of the body moves upward. You elevate your mandible when you close your mouth.

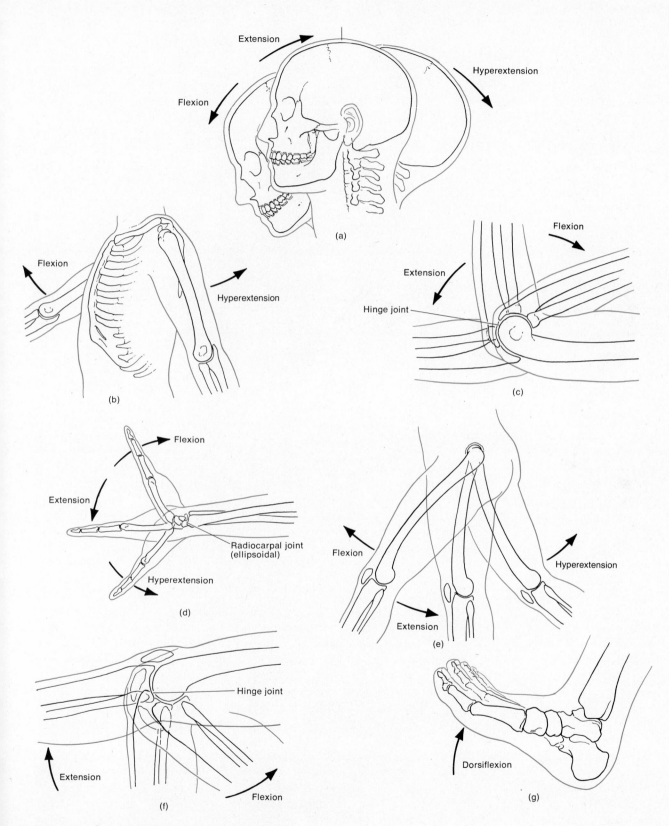

Figure 9–4 Angular movements at synovial joints.

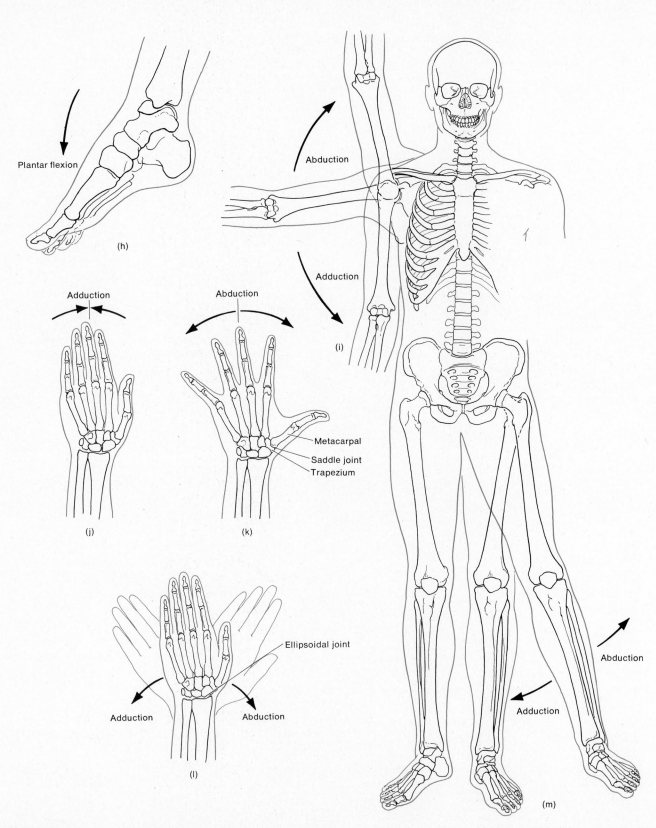

Figure 9–4 (cont.)   Angular movements at synovial joints.

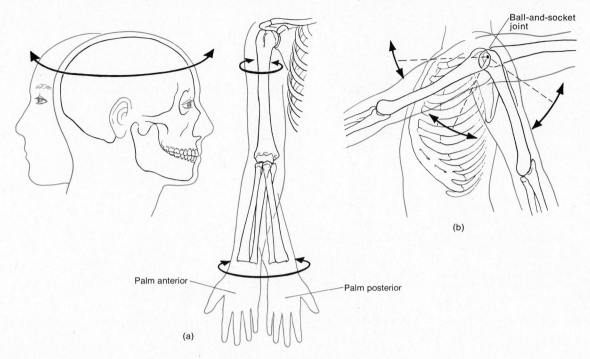

Figure 9—5 Rotation and circumduction. (a) Rotation at the atlas-axis joint (left) and rotation of the humerus (right). (b) Circumduction of the humerus at the shoulder joint.

Depression is a movement in which a part of the body moves downward. You depress your mandible when you open your mouth. The shoulders can also be elevated and depressed.

Types

GLIDING
The articulating surfaces of bones in **gliding joints (arthrodia)** are usually flat. Only side-to-side and back-and-forth movements are permitted. Since this joint allows movements in two planes, it is called *biaxial*. Twisting and rotation are inhibited at gliding joints generally because ligaments or adjacent bones restrict the range of movement. Examples are the joints between carpal bones, tarsal bones, the sternum and clavicle, and the scapula and clavicle.

HINGE
A **hinge joint (ginglymus)** is characterized by the convex surface of one bone that fits into the concave surface of another bone. Movement is primarily in a single plane and is usually flexion and extension. The joint is therefore known as *monaxial*. The motion is similar to that of a hinged door. Examples of hinge joints include the elbow, knee, ankle, and interphalangeal joints. The movement allowed by a hinge joint is illustrated by flexion and extension at the elbow and knee. (See Figure 9-4c, f.)

PIVOT
In a **pivot joint (trochoid)**, a rounded, pointed, or conical surface of one bone articulates with a shallow depression of another bone. The primary movement permitted is rotation, and the joint is therefore monaxial. Examples include the joints between the atlas and axis and between the proximal ends of the radius and ulna. Movement at a pivot joint is illustrated by supination and pronation of the palms and rotation of the head from side to side. (See Figure 9-5a.)

ELLIPSOIDAL
In an **ellipsoidal joint (condyloid)**, an oval-shaped condyle of one bone fits into an elliptical cavity of another bone. Since the joint permits side-to-side and back-and-forth movements, it is biaxial. The joint at the wrist between the radius and carpals is ellipsoidal. The movement permitted by such a joint is illustrated when you flex and extend and abduct and adduct the wrist. (See Figure 9-4d.)

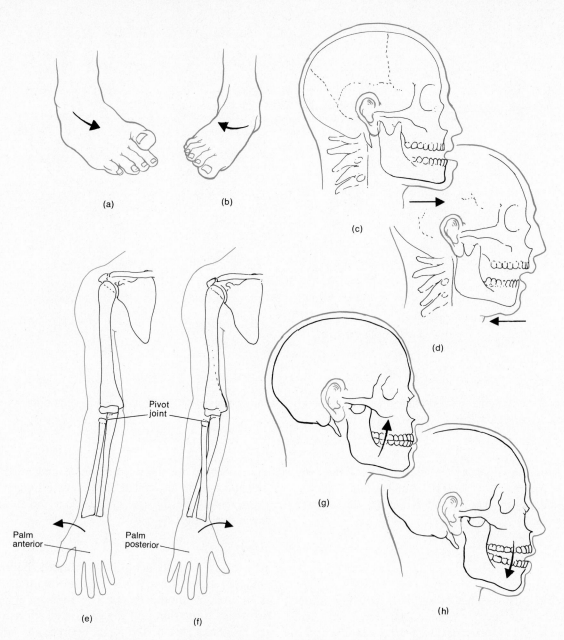

Figure 9–6   Special movements. (a) Inversion. (b) Eversion. (c) Protraction. (d) Retraction. (e) Supination. (f) Pronation. (g) Elevation. (h) Depression.

SADDLE

In a **saddle joint (sellaris),** the articular surfaces of both bones are saddle-shaped – in other words, concave in one direction and convex in the other. Essentially, the saddle joint is a modified ellipsoidal joint in which the movement is somewhat freer. Movements at a saddle joint are side to side and back and forth. Thus the joint is biaxial. The joint between the trapezium and metacarpal of the thumb is an example of a saddle joint. (See Figure 9-4k.)

BALL-AND-SOCKET

Ball-and-socket (spheroid) joints consist of a ball-like surface of one bone fitting into a cuplike depression of another bone. Such a joint permits *triaxial* movement. That is, there is movement in three planes of motion: flexion-extension, abduction-adduction, and rotation. Examples of ball-and-socket joints are the shoulder joint and hip joint. The range of movements at a ball-and-socket joint is illustrated by circumduction of the arm (Figure 9-5b).

Exhibit 9—1 JOINTS

Type	Description	Movement	Examples
Fibrous	No joint cavity; bones held together by a thin layer of fibrous tissue or dense fibrous tissue.		
Suture	Found only between bones of the skull; articulating bones separated by a thin layer of fibrous tissue.	None—synarthrotic.	Lambdoidal suture between occipital and parietal bones.
Syndesmosis	Articulating bones united by dense fibrous tissue.	Slight—amphiarthrotic.	Distal ends of tibia and fibula.
Cartilaginous	No joint cavity; articulating bones united by cartilage.		
Synchondrosis	Connecting material is hyaline cartilage.	None—synarthrotic.	Temporary joint between the diaphysis and epiphyses of a long bone.
Symphysis	Connecting material is a broad, flat disc of fibrocartilage.	Slight—amphiarthrotic.	Intervertebral joints and symphysis pubis.
Synovial	Joint cavity and articular cartilage present; synovial membrane lines cavity.	Freely movable— diarthrotic.	
Gliding	Articulating surfaces usually flat.	Biaxial (flexion-extension, abduction-adduction).	Intercarpal and intertarsal joints.
Hinge	Spool-like surface fits into a concave surface.	Monaxial (flexion-extension).	Elbow, knee, ankle, and interphalangeal joints.
Pivot	Rounded, pointed, or concave surface fits into a shallow depression.	Monaxial (rotation).	Atlas-axis and radioulnar joints.
Ellipsoidal	Oval-shaped condyle fits into an elliptical cavity.	Biaxial (flexion-extension, abduction-adduction).	Radiocarpal joint.
Saddle	Articular surfaces concave in one direction and convex in opposite direction.	Biaxial (flexion-extension, abduction-adduction).	Carpometacarpal joint of thumb.
Ball-and-socket	Ball-like surface fits into a cuplike depression.	Triaxial (flexion-extension, abduction-adduction, rotation).	Shoulder and hip joints.

The summary of joints presented in Exhibit 9-1 is based on the anatomy of the joints. Joints can also be classified according to movement. If we rearrange Exhibit 9-1 into a classification based on movement, we arrive at the following:

Synarthroses: immovable joints
1 Suture
2 Synchondrosis

Amphiarthroses: slightly movable joints
1 Symphysis
2 Syndesmosis

Diarthroses: freely movable joints
1 Gliding
2 Hinge
3 Pivot
4 Ellipsoidal
5 Saddle
6 Ball-and-socket

DISORDERS

ARTHRITIS

The term **arthritis** refers to at least 25 different diseases, the most common of which are rheumatoid arthritis, osteoarthritis, and gout. All these ailments are characterized by inflammation in one or more joints. Inflammation, pain, and stiffness may also be present in adjacent parts of the body, such as the muscles near the joint.

The causes of arthritis are unknown. In some cases, it has followed the stress of sprains, infections, and joint injury. Some researchers think that the cause is a bacterium or virus, whereas others suspect an allergy. Some believe the nervous system or hormones are involved, whereas others suspect a metabolic disorder. Still others believe that certain types of prolonged psychological stress, such as inhibited hostility, can upset homeostatic balance and bring on arthritic attacks.

RHEUMATISM

Rheumatism refers to any painful state of the supporting structures of the body, its bones, ligaments, joints, tendons, or muscles. Arthritis is a form of rheumatism in which the joints have become inflamed.

Rheumatism is not necessarily related to rheumatoid arthritis. About 25 percent of all Americans – most of them over age 40 – have rheumatism that is diagnosed as some form of osteoarthritis.

RHEUMATOID ARTHRITIS

Rheumatoid arthritis is the most common inflammatory form of arthritis. It involves inflammation of the joint, swelling, pain, and a loss of function. Usually this form occurs bilaterally – if your left knee is affected, your right knee may also be affected, although usually not to the same degree.

The disease may afflict at any age, but especially between 30 and 50 years old. The frequency of occurrence is approximately 57 percent in females and 43 percent in males. The symptoms in females are more severe than in males. One form of rheumatoid arthritis afflicts children shortly after birth – *juvenile rheumatoid arthritis* – and other forms can occur in later childhood.

The primary symptom of rheumatoid arthritis is inflammation of the synovial membrane. If it is completely untreated, the following sequential pathology may occur: The membrane thickens and synovial fluid accumulates. The resulting pressure causes pain and tenderness. The membrane then produces an abnormal tissue called pannus, which adheres to the surface of the articular cartilage. The pannus formation sometimes erodes the cartilage completely. When the cartilage is destroyed, fibrous tissue joins the exposed bone ends. The tissue ossifies and fuses the joint so that it is immovable – the ultimate crippling effect of rheumatoid arthritis. Most cases do not progress to this stage. But the range of motion of the joint is greatly inhibited by the severe inflammation and swelling.

Damaged joints may be surgically replaced, either partly or entirely, with artificial joints (Figure 9-7). The artificial parts are inserted after removal of the diseased portion of the articulating bone and its cartilage. The new metal or plastic joint is fixed in place with a special acrylic cement. When freshly mixed in the operating room, it hardens as strong as bone in minutes. These new parts function nearly as well as a normal joint – and much better than a diseased joint.

Joint surgery, especially hip and knee replacement, holds more promise than most other forms of surgery. The technique is being extended to other joints – wrist, elbow, shoulder, fingers, ankle – with much optimism.

OSTEOARTHRITIS

A degenerative joint disease far more common than rheumatoid arthritis, and usually less damaging, is **osteoarthritis**. It apparently results from a combination of aging, irritation of the joints, and wear and abrasion.

Degenerative joint disease is a noninflammatory, progressive disorder of movable joints, particularly weight-bearing joints. It is characterized pathologically by the deterioration of articular cartilage and by formation of new bone in the subchondral areas and at the margins of the joint. The cartilage slowly degenerates, and as the bone ends become exposed, they deposit small bumps, or *spurs,* of new osseous tissue. These spurs decrease the space of the joint cavity and restrict joint movement. Unlike rheumatoid arthritis, osteoarthritis usually affects only the articular cartilage. The synovial membrane is rarely destroyed, and other tissues are unaffected.

GOUTY ARTHRITIS

Uric acid is a waste product produced during the metabolism of the nucleic acid purine. Normally, all the acid is quickly excreted in the urine. In fact, it gives urine its name. The person who suffers from *gout* either produces excessive amounts of uric acid or is not able to excrete normal amounts. The result is a buildup of uric acid in the blood. This excess acid then reacts with sodium to form a salt called sodium urate. Crystals of this salt are deposited in soft tissues. Typical sites are the kidneys and the cartilage of the ears and joints.

In **gouty arthritis,** sodium urate crystals are deposited in the soft tissues of the joints. The crystals irritate the cartilage, causing inflammation, swelling, and acute pain. Eventually, the crystals destroy all the joint tissues. If the disorder is not treated, the ends of the articulating bones fuse and the joint becomes immovable.

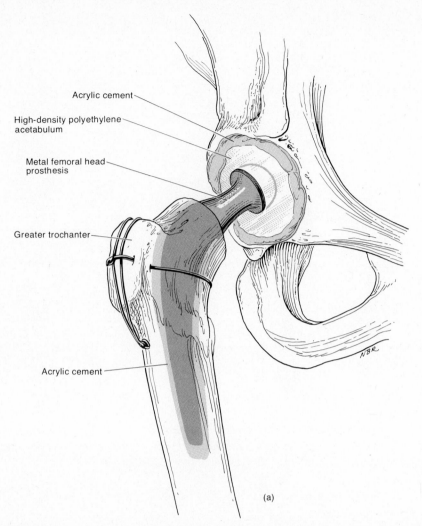

Acrylic cement

High-density polyethylene
acetabulum

Metal femoral head
prosthesis

Greater trochanter

Acrylic cement

(a)

Figure 9–7   Total hip and total knee replacement. (a) In the Charnley's technique and prosthesis for total hip replacement, the arthritic portions of the acetabulum and head of the femur are replaced by a prefabricated joint. The acetabulum is made of a polyethylene substance of high density, and the metallic femoral head is cemented into the femur with acrylic cement. The greater trochanter of the femur is reattached after surgery.

Gout occurs primarily in males of any age. It is believed to be the cause of 2 to 5 percent of all chronic joint diseases. Numerous studies indicate that gout is sometimes caused by an abnormal gene. The gene instructs the body to manufacture purine by a mechanism that produces unusually large amounts of uric acid. Diet and environmental factors such as stress and climate are also suspected causes of gout.

BURSITIS

An acute or chronic inflammation of a bursa is called **bursitis.** The condition may be caused by trauma, by an acute or chronic infection (including syphilis and tuberculosis), or by rheumatoid arthritis. Repeated excessive friction often results in a bursitis with local inflammation and the accumulation of fluid. Bunions are frequently associated with a friction bursitis over the head of the first metatarsal bone. Symptoms include pain, swelling, tenderness, and the limitation of motion involving the inflamed bursa.

TENDINITIS

Tendinitis or **tenosynovitis** frequently occurs as inflammation involving the tendon sheaths and synovial membrane surrounding certain joints. The wrists, shoulders, elbows (tennis elbow), finger joints (trigger finger), ankles, and associated tendons are most often affected. The affected sheaths may become visibly swollen because of fluid accumulation or they may remain dry. Local tenderness is variable and there may be disabling pain with movement of the body part. The condition often follows some form of trauma, strain, or excessive exercise. Treatment is with analgesic/anti-inflammatory drugs and, if warranted, cortisone injections.

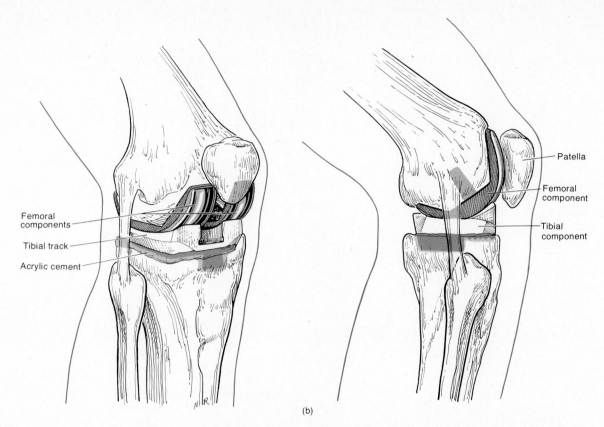

(b)

Figure 9–7 (cont.)   Total hip and total knee replacement. (b) Illustrated are two total knee replacement prostheses. In the technique on the left, called the polycentric type, there are two femoral and two tibial components. In the technique on the right, the Waldius type, there is one femoral component and one tibial component.

DISLOCATION

A **dislocation** or **luxation** is the displacement of a bone from a joint. The most common dislocations are those involving a finger, thumb, or shoulder. Those of the mandible, elbow, knee, or hip are less common. Symptoms include loss of motion, temporary paralysis of the involved joint, pain, swelling, and occasional shock. A partial or incomplete dislocation is called a **subluxation**. A dislocation is usually caused by a blow or fall, although unusual physical effort may lead to this condition.

SPRAIN

A **sprain** is the forcible wrenching or twisting of a joint with partial rupture or other injury to its attachments without luxation. There may be damage to the associated blood vessels, muscles, tendons, ligaments, or nerves. A sprain is more serious than a *strain,* which is simply the overstretching of a muscle without swelling. Severe sprains may be so painful that the joint cannot be moved.

SLIPPED DISC

Intervertebral discs are located between the bodies of adjacent vertebrae from the axis to the sacrum. Each disc is composed of an outer fibrous ring consisting of fibrocartilage called the *anulus fibrosus* and an inner soft, pulpy, highly elastic structure called the *nucleus pulposus.* (See Figure 9-1b.) The intervertebral discs absorb vertical shock. Under compression, they flatten, broaden, and bulge from their intervertebral spaces. The discs between the fourth and fifth lumbar vertebrae and between the fifth lumbar vertebra and sacrum are subject to great compressional forces. If the anterior and posterior ligaments of the discs become injured or weakened, the disc may become herniated – that is, the pressure developed in the nucleus pulposus is great enough to rupture the surrounding fibrocartilage. If this occurs, the nucleus pulposus may protrude posteriorly or into one of the adjacent vertebral bodies. This condition is called a **slipped disc.** Most often the nucleus pulposus slips posteriorly toward the spinal cord and spinal nerves. This movement exerts pressure on the spinal nerves causing considerable, sometimes very acute, pain. If the root of the sciatic nerve, which passes from the spinal cord to the foot, is pressured, the pain radiates down the back of the thigh, through the calf, and occasionally into the foot. If pressure is exerted on the spinal cord, nervous tissue may be destroyed.

MEDICAL TERMINOLOGY

Ankylosis (*agkyle* = stiff joint; *osis* = condition)   Severe or complete loss of movement at a joint.

Arthralgia (*algia* = pain)   Pain in a joint.

Arthrosis   A disease of a joint; also refers to an articulation, or joint.

Bursectomy (*ectomy* = removal of)   Removal of a bursa.

Chondritis   Inflammation of a cartilage.

Deterioration   The process or state of growing worse; disintegration or wearing away.

Detritus   Particulate matter produced by or remaining after the wearing away or disintegration of a substance or tissue.

Insidious   Hidden, not apparent – as a disease that does not exhibit distinct symptoms of its arrival.

Pyogenic   Producing suppuration (pus).

Reduce   To replace in normal position – as to reduce a fracture.

Rheumatology   The medical specialty devoted to arthritis.

Septic   Indicating the presence of microorganisms or their toxins.

Synovitis   Inflammation of a synovial membrane in a joint.

STUDY OUTLINE

CLASSIFICATION

1   A joint or articulation is a point of contact between two bones.

2   Closely fitting bones are strong but not freely movable. Loosely fitting joints are weaker but freely movable.

FIBROUS JOINTS

1   Bones held by fibrous connective tissue, with no joint cavity, are fibrous joints.

2   These joints include immovable sutures (found in the skull) and slightly movable syndesmoses (such as the tibiofibular articulation).

CARTILAGINOUS JOINTS

1   Bones held together by cartilage, with no joint cavity, are cartilaginous joints.

2   These joints include immovable synchondroses united by hyaline cartilage (temporary cartilage between diaphysis and epiphyses) and partially movable symphyses united by fibrocartilage (the symphysis pubis).

SYNOVIAL JOINTS

1   These joints contain a joint cavity, articular cartilage, and synovial membranes. They are held together by ligaments and tendons.

2   All synovial joints are freely movable.

3   Types of synovial joints include gliding joints (wrist bones), hinge joints (elbow), pivot joints (radioulnar), ellipsoidal joints (radiocarpal), saddle joints (carpometacarpal), and ball-and-socket joints (shoulder and hip).

4   Planes of movement at synovial joints include the monaxial, biaxial, and triaxial planes.

5   Types of movements at synovial joints include gliding movements, angular movements, rotation, circumduction, and special movements such as inversion, eversion, protraction, retraction, supination, pronation, elevation, and depression.

REVIEW QUESTIONS

1   Define an articulation. What factors determine the degree of movement at joints?

2   Distinguish among the three kinds of joints. List the subtypes. Be sure to include structure, degree of movement, and specific examples.

3   Using the knee as a typical joint, explain the components of a synovial joint. Indicate the relationship of ligaments and tendons to the strength of the joint and restrictions on movement.

4   What are bursae? What is their function?

5   Define the following principal movements: gliding, angular, rotation, circumduction, and special. Name a joint where each occurs.

6   Have your partner assume the anatomical position and execute for you each of the movements at joints discussed in the text. Reverse roles, and see if you can execute the same movements.

7   Contrast monaxial, biaxial, and triaxial planes of movement. Give examples of each, and name a joint at which each occurs.

Chapter 10
Muscle Tissue

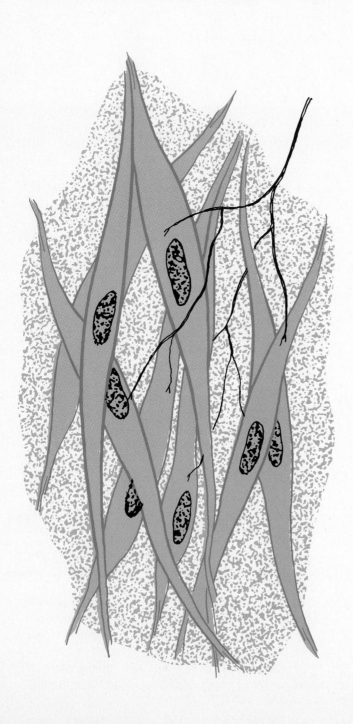

■ List the characteristics and functions of muscle tissue.

■ Compare the location, microscopic appearance, nervous control, and functions of the three kinds of muscle tissue.

■ Define fascia, epimysium, perimysium, endomysium, tendons, and aponeuroses. List their modes of attachment to muscles.

■ Describe the relationship of blood vessels and nerves to skeletal muscles.

■ Identify the histological characteristics of skeletal muscle.

■ Describe the physiology of contraction by listing the events associated with the sliding-filament theory.

■ Describe the physiological importance of the motor unit.

■ Describe the source of energy for muscular contraction.

■ Define the all-or-none principle of muscular contraction.

■ Contrast the normal contractions performed by skeletal muscles. Describe the phases of contraction in a typical myogram of a twitch contraction.

■ Contrast cardiac muscle tissue with smooth muscle tissue.

■ Compare oxygen debt and heat production as examples of muscle homeostasis.

■ Define such common muscular disorders as fibrosis, fibrositis, muscular dystrophy, and myasthenia gravis.

■ Compare spasms, cramps, convulsions, fibrillation, and tics as abnormal muscular contractions.

■ Define medical terminology associated with the muscular system.

Although bones and joints provide leverage and form the framework of the body, they are not capable of moving the body by themselves. Motion is an essential body function that results from the contraction of muscles. Muscle tissue constitutes about 40 to 50 percent of the total body weight and is composed of highly specialized cells with four striking characteristics.

CHARACTERISTICS

Irritability is the ability of muscle tissue to receive and respond to stimuli. A stimulus is a change in the internal or external environment strong enough to initiate a nerve impulse. A second characteristic of muscle is **contractility,** the ability to shorten and thicken, or contract, when a sufficient stimulus is received. Muscle tissue also exhibits **extensibility** – it can be stretched. Many skeletal muscles are arranged in opposing pairs. While one is contracting, the other is undergoing extension. Another characteristic of muscle tissue is **elasticity,** the ability of muscle to return to its original shape after contraction or extension.

FUNCTIONS

Through contraction, muscle performs three important functions: motion, maintenance of posture, and heat production.

The most obvious body motions are walking, running, and locomotion. Other movements, such as grasping a pencil or nodding the head, may be localized to certain parts of the body. These movements rely on the integrated functioning of the bones, joints, and muscles attached to the bones. Less noticeable kinds of motion produced by muscles are the beating of the heart, the churning of food in the stomach, the pushing of food through the intestines, the contraction of the gallbladder to release bile, and the contraction of the urinary bladder to expel urine.

In addition to the movement function, muscle tissue also enables the body to maintain posture. The contraction of skeletal muscles holds the body in stationary positions, such as standing and sitting.

The third function of muscle tissue is heat production. Skeletal muscle contractions produce heat and are thereby important in maintaining normal body temperature.

KINDS

Three kinds of muscle tissue are recognized: **skeletal, visceral,** and **cardiac.** These three types are further categorized by location, microscopic structure, and nervous control. Skeletal muscle tissue, which is named for its location, is attached to bones. It is *striated* muscle tissue because striations, or bandlike structures, are visible when the tissue is examined under a microscope. It is a *voluntary* muscle tissue because it can be made to contract by conscious control. Smooth, or *nonstriated,* muscle is located in the walls of hollow internal structures, such as blood vessels, the stomach, and the intestines. It is, therefore, described as *visceral* muscle tissue. It is *involuntary* muscle tissue because its contraction is usually not under conscious control. Cardiac muscle tissue forms the walls of the heart and is named for its location. Cardiac muscle tissue is also striated and is involuntary. Thus all muscle tissues are classified in the following ways: (1) skeletal, striated, voluntary muscle, (2) cardiac, striated, involuntary muscle, and (3) visceral, smooth, involuntary muscle tissue.

SKELETAL MUSCLE TISSUE

To understand the fundamental mechanisms of muscle movement, you will need some knowledge of its connective tissue components, its nerve and blood supply, and its histology, or microscopic structure.

Fascia

The term **fascia** is applied to a sheet or broad band of fibrous connective tissue beneath the skin or around muscles and other organs of the body. Fasciae may be divided into three types: superficial, deep, and subserous. The *superficial fascia,* or *subcutaneous layer,* is immediately deep to the skin. It covers the entire body and varies in thickness in different regions. On the back or dorsum of the hand it is quite thin, whereas over the inferior abdominal wall it is thick. The superficial fascia is composed of adipose tissue and loose connective tissue. The outer layer usually contains fat and varies considerably in thickness. The inner layer is thin and elastic. Between the two layers of superficial fascia are found arteries, veins, lymphatics, nerves, the mammary glands, and the facial muscles. Hair follicles, sweat glands, and sebaceous glands are embedded in the superficial fascia. The *deep fascia* is by far the most extensive of the three types. It is a dense connective tissue. Unlike the superficial fascia, the deep fascia does not contain fat. The deep fascia lines the body wall and extremities and holds muscles together, separating them into functioning groups.

The *subserous fascia* is located between the internal investing layer of deep fascia and a serous membrane. It covers the external surfaces of viscera in the thoracic and abdominal cavities.

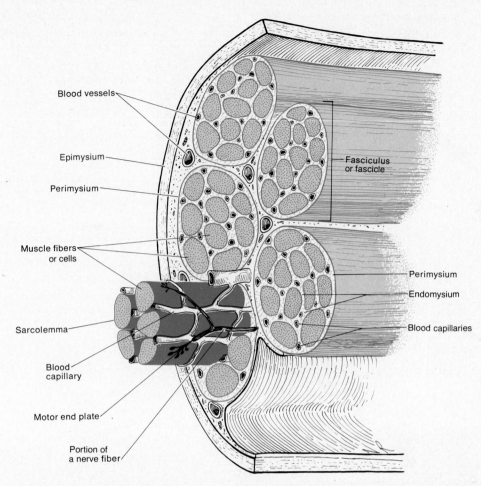

Blood vessels

Epimysium

Perimysium

Fasciculus
or fascicle

Muscle fibers
or cells

Perimysium

Endomysium

Sarcolemma

Blood capillaries

Blood
capillary

Motor end plate

Portion of
a nerve fiber

Figure 10–1   Relationships of connective tissue to skeletal muscle. Shown is a cross section of a skeletal muscle indicating the relative positions of the epimysium, perimysium and endomysium. Compare this figure with the photomicrograph in Figure 10–3b.

Connective Tissue Components

Skeletal muscles are further protected, strengthened, and attached to other structures by several connective tissue components. The entire muscle is usually wrapped with a substantial quantity of fibrous connective tissue called the epimysium (Figure 10-1). The epimysium is an extension of deep fascia. When the muscle is cut in cross section, invaginations of the epimysium are seen to divide the muscle into bundles called *fasciculi* or *fascicles*. These invaginations of the epimysium are called the perimysium. Perimysium, like epimysium, is an extension of deep fascia. In turn, invaginations of the perimysium, called endomysium, penetrate into the interior of each fascicle and separate each muscle cell. Endomysium is also an extension of deep fascia. The epimysium, perimysium, and endomysium are all continuous with the connective tissue that attaches the muscle to another structure, such as bone or other muscle. All three elements may be extended beyond the muscle cells as a *tendon* – a cord of connective tissue that

attaches a muscle to the periosteum of a bone. In other cases, the connective tissue elements may extend as a broad, flat band of tendons called an *aponeurosis*. This structure also attaches to the coverings of a bone or another muscle. When a muscle contracts, the tendon and its corresponding bone or muscle are pulled toward the contracting muscle. In this way skeletal muscles produce movement. Certain tendons, especially those of the wrist and ankle, are enclosed by tubes of fibrous connective tissue called *tendon sheaths*. They are lined by a synovial membrane that permits the tendon to slide easily within the sheath. The sheaths also prevent the tendons from slipping out of place.

Nerve and Blood Supply

Skeletal muscles are well supplied with nerves and blood vessels. This heavy infiltration of nervous and circulatory tissues is directly related to contraction, the chief characteristic of muscle. For a skeletal muscle cell to

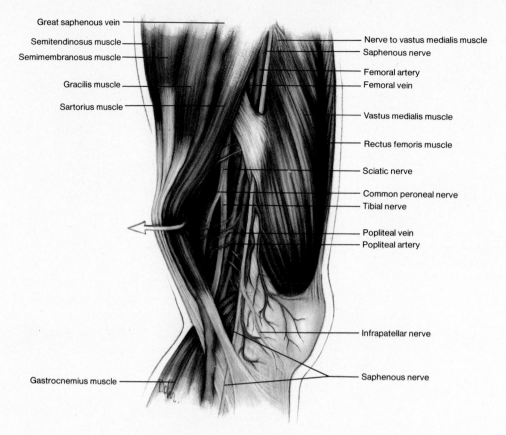

Figure 10–2   Relationship of blood vessels and nerves to skeletal muscles of the left thigh and knee in medial view.

contract, it must first be stimulated by an impulse from a nerve cell. Muscle contraction also requires a good deal of energy – meaning large amounts of nutrients and oxygen. Moreover, the waste products of these energy-producing reactions must be eliminated. Thus muscle action depends on the blood supply.

Generally, an artery and one or two veins accompany each nerve that penetrates a skeletal muscle. The larger branches of the blood vessel accompany the nerve branches through the connective tissue of the muscle (Figure 10-2). Microscopic blood vessels called capillaries are arranged in the endomysium. Each muscle cell is thus in close contact with one or more capillaries. Each skeletal muscle cell also makes contact with a portion of a nerve cell.

Histology

Muscle tissue attached to bones is generally termed **skeletal muscle tissue.** When a typical skeletal muscle is teased apart and viewed microscopically, it can be seen to consist of many elongated, cylindrical cells called **muscle fibers** (Figure 10-3a, b). These fibers lie parallel to each other and range from 10 to 100 μm in diameter.

Some fibers may reach lengths of 30 cm (12 inches) or more. Each muscle fiber is surrounded by a plasma membrane called the sarcolemma (*sarco* = flesh; *lemma* = sheath). The sarcolemma contains a quantity of cytoplasm called sarcoplasm. Within the sarcoplasm of a muscle fiber and lying close to the sarcolemma are many nuclei and a number of mitochondria. Also within a muscle fiber is the **sarcoplasmic reticulum,** a network of membrane-enclosed tubules comparable to smooth endoplasmic reticulum (Figure 10-3c). Running transversely through the fiber and perpendicularly to the sarcoplasmic reticulum are **T tubules.** The tubules open to the outside of the fiber. **A triad** consists of a T tubule and the segments of sarcoplasmic reticulum on either side.

A highly magnified view of skeletal muscle fibers reveals threadlike structures, about 1 or 2 μm in diameter, called **myofibrils** (Figure 10-3c, d, e). The prefix *myo* means muscle. The myofibrils run longitudinally through the muscle fiber and consist of two kinds of even smaller structures called **myofilaments.** The **thin myofilaments** are about 6 nm in diameter. The **thick myofilaments** are about 16 nm in diameter.

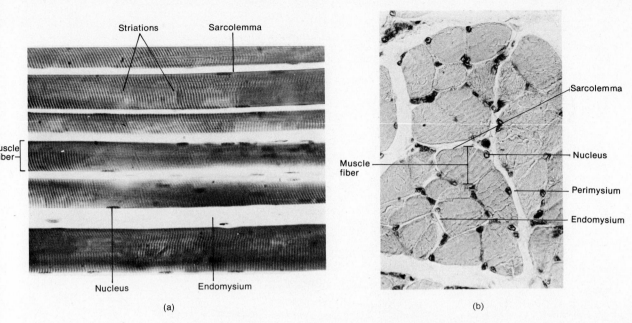

Striations Sarcolemma

Muscle
fiber

Nucleus Endomysium

(a)

Sarcolemma

Nucleus

Perimysium

Endomysium

Muscle
fiber

(b)

Figure 10–3 Histology of skeletal muscle tissue. (a) Photomicrograph of several muscle fibers in longitudinal section at a magnification of 640×. (b) Photomicrograph of several muscle fibers in cross section at a magnification of 640×. (Courtesy of Edward J. Reith, from *Atlas of Descriptive Histology,* by Edward J. Reith and Michael H. Ross, Harper & Row, Publishers, Inc., New York, 1970.)

The myofilaments of a myofibril do not extend the entire length of a muscle fiber – they are stacked in compartments called **sarcomeres.** Sarcomeres are partitioned by **Z lines,** which are narrow zones of dense material. Each sarcomere is about 2.6 μm long. In a relaxed muscle fiber – that is, one that is not contracting – the thin and thick myofilaments overlap and form a dark, dense band called the *anisotropic* or *A band.* Each A band is about 1.6 μm long. A light-colored, less dense area called the *isotropic* or *I band* is composed of thin myofilaments only. The I bands are about 1 μm long. This combination of alternating dark and light bands gives the muscle fiber its striped appearance. A narrow *H zone* contains thick myofilaments only. The H zone is about 0.5 μm long.

The thin myofilaments are composed mostly of a protein called *actin.* The actin molecules are arranged in a double-stranded coil that gives the thin myofilaments their characteristic shape (Figure 10-4a). Besides actin, the thin myofilaments contain two other molecules called *tropomyosin* and *troponin.* Together they are referred to as a *tropomyosin-troponin complex.*

The thick myofilaments are composed mostly of a protein called *myosin.* A myosin molecule is shaped like a rod with a round head. These rods form the long axis of the thick myofilaments, and the heads form projections called *cross bridges* (Figure 10-4b). The cross bridges are arranged in pairs and seem to spiral around the main axis of the thick myofilament. The relationship between thin and thick myofilaments is shown in Figure 10-4c.

CONTRACTION

Sliding-Filament Theory

During muscle contraction, the thin myofilaments slide inward toward the H zone, causing the sarcomere to shorten. However, the lengths of the thin and thick myofilaments do not change. The cross bridges of the thick myofilaments connect with portions of actin of the thin myofilaments. The myosin cross bridges move like the oars of a boat on the surface of the thin myofilaments – and the thin and thick myofilaments slide past each other (Figure 10-5). As the thin myofilaments move past the thick myofilaments, the width of the H zone between the ends of the thin myofilaments gets smaller and may even disappear when the thin myofilaments meet at the center of the sarcomere. In fact, the cross bridges may pull the thin myofilaments of each sarcomere so far inward that their ends overlap (Figure 10-6). As the thin myofilaments slide inward, the Z lines are drawn toward the A band and the sarcomere is shortened. The sliding of myofilaments and shortening of sarcomeres causes the shortening of the muscle fibers.

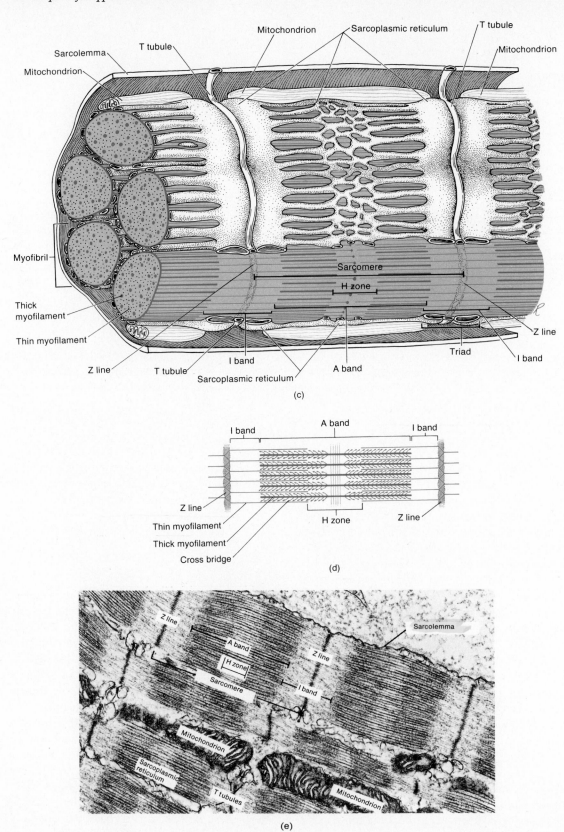

Figure 10–3 (cont.)   Histology of skeletal muscle tissue. (c) Enlarged aspect of several muscle fibers based on an electron micrograph. (d) Details of a sarcomere showing thin and thick myofilaments and various internal zones. (e) Electron micrograph of several sarcomeres at a magnification of 35,000X. (Photomicrographs courtesy of Edward J. Reith, from *Atlas of Descriptive Histology,* by Edward J. Reith and Michael H. Ross, Harper & Row, Publishers, Inc., New York, 1970. Electron micrograph courtesy of D. E. Kelly, from *Introduction to the Musculoskeletal System,* by Cornelius Rosse and D. Kay Clawson, Harper & Row, Publishers, Inc., New York, 1970.)

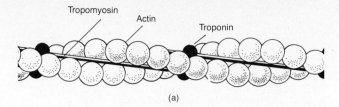

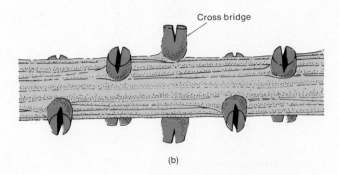

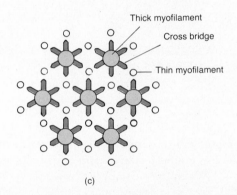

Figure 10–4 Detailed structure of myofilaments. (a) Thin myofilament. (b) Thick myofilament. (c) Cross section of several thin and thick myofilaments showing the arrangement of cross bridges. Note that each thick myofilament is surrounded by six thin myofilaments. (Parts (a) and (b) have been redrawn by permission from Arthur W. Ham, *Histology,* 7th ed., Philadelphia: J. B. Lippincott Company, 1974.)

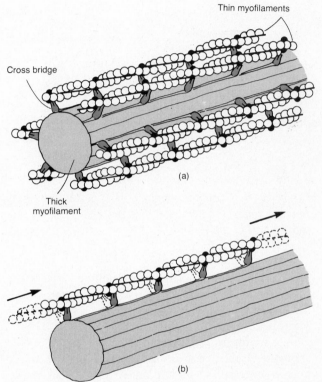

Figure 10–5 Movement of thin myofilaments past thick myofilaments. (a) Attachment of myosin cross bridges to actin of thin myofilaments. (b) Mechanism of movement of myosin cross bridges resulting in sliding of thin myofilaments toward H zone.

All these events associated with the movement of myofilaments are known as the **sliding-filament theory** of muscle contraction. Let's examine the theory in more detail.

Motor Unit

For skeletal muscle to contract, a stimulus must be applied to it. Such a stimulus is normally transmitted by nerve cells called neurons. A neuron has a threadlike process called a fiber, or axon, that may run 91 cm (3 ft) or more to a muscle. A bundle of such fibers from many different neurons composes a nerve. A neuron that transmits a stimulus to muscle tissue is called a motor neuron.

Upon entering a skeletal muscle, the axon of the motor neuron branches into fine endings that come into close approximation at grooves on the muscle membrane. The portion of the muscle membrane directly under the end of the axon is called a **motor end plate** (Figure 10-7). It can respond to a chemical called acetylcholine (ACh), which is released from the fine ending of the neuron. The area of contact between neuron and muscle fiber is called a **neuromuscular junction,** or **myoneural junction**. When a nerve impulse reaches a motor end plate, small vesicles in the terminal branches

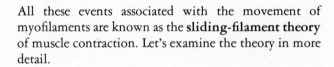

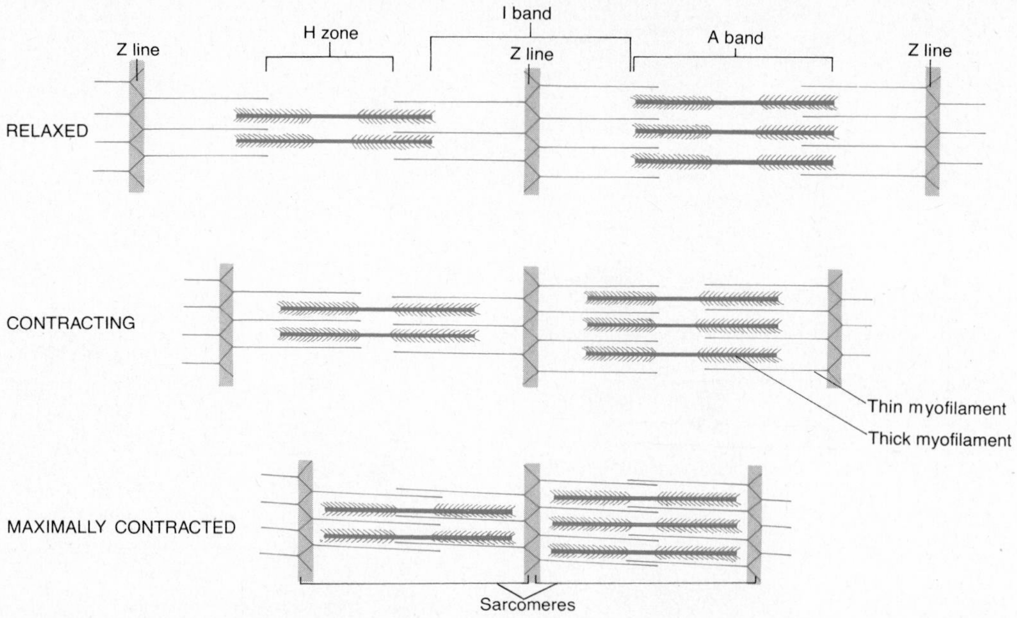

Figure 10–6   Sliding-filament theory of muscle contraction. Shown are the positions of the various parts of two sarcomeres in relaxed, contracting, and maximally contracted states. Note the movement of the thin myofilaments and the relative size of the H zone.

of the nerve fiber release acetylcholine. The ACh transmits the nerve impulse from the neuron, across the myoneural junction, to the muscle fiber, thus stimulating it to contract.

A motor neuron, together with the muscle cells it stimulates, is referred to as a **motor unit.** A single motor neuron innervates about 150 muscle fibers. This means that stimulation of one neuron will tend to cause the simultaneous contraction of about 150 muscle fibers. In addition, all the muscle fibers of a motor unit that are sufficiently stimulated will contract and relax together. Muscles that control precise movements, such as the eye muscles, have fewer than 10 muscle fibers to each motor unit. Some muscles have as few as one muscle fiber per motor unit. Muscles of the body that are responsible for gross movements may have as many as 500 muscle fibers in each motor unit.

Physiology

When a muscle fiber is relaxed, the concentration of calcium ions (Ca^{2+}) in the sarcoplasm is low – these ions are stored in the sarcoplasmic reticulum. Moreover, molecules of ATP are attached to the myosin cross bridges. The cross bridges are prevented from combining with actin of the thin myofilaments by the tropomyosin-troponin complex while the complex is attached to actin and the ATP is bound to the myosin cross bridges. In other words, the muscle fiber remains relaxed as long

as there are no calcium ions in the sarcoplasm, the tropomyosin-troponin complex is attached to actin, and ATP is bound to the myosin cross bridges.

When a nerve impulse reaches the motor end plate, the neuron releases acetylcholine, which causes an electrical change in the sarcolemma of the muscle fiber. This change travels over the surface of the sarcolemma and into the T tubules. When the impulse is conveyed from the T tubules to the sarcoplasmic reticulum, the reticulum releases the calcium ions from storage into the sarcoplasm surrounding the myofilaments. The calcium ions move to the myosin cross bridges and activate the myosin so that it can catalyze the breakdown of ATP into ADP + P. In the presence of calcium ions, myosin acts as an enzyme that breaks down ATP. Calcium ions also permit the tropomyosin-troponin complex to split from the thin myofilament so that the free receptor site of actin is now permitted to attach to the myosin cross bridge. The energy released from the breakdown of ATP is used for the attachment and movement of the myosin cross bridges and thus the sliding of the myofilaments. As the thin myofilaments slide past the thick myofilaments, the Z lines are drawn toward each other and the sarcomere shortens – the muscle contracts.

What happens when a muscle fiber goes from a contracted state back to a relaxed state? After the nerve impulse ends, the calcium ions return to the sarcoplasmic reticulum for storage. With their removal from the sarcoplasm, the enzymatic activity of myosin stops. The

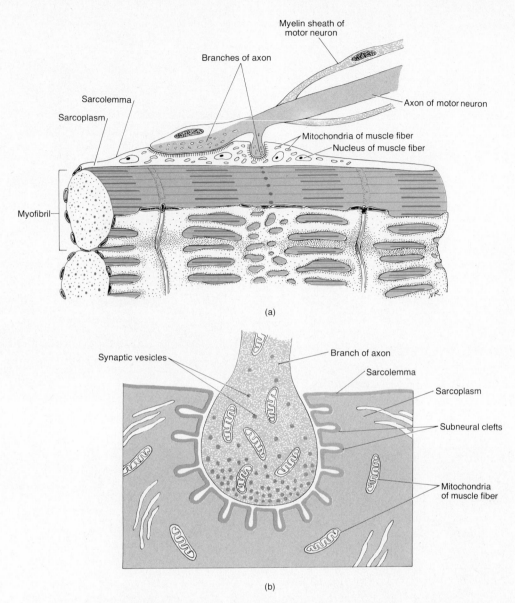

Myelin sheath of
motor neuron

Branches of axon

Sarcolemma

Sarcoplasm

Axon of motor neuron

Mitochondria of muscle fiber
Nucleus of muscle fiber

Myofibril

(a)

Synaptic vesicles

Branch of axon

Sarcolemma

Sarcoplasm

Subneural clefts

Mitochondria
of muscle fiber

(b)

Figure 10—7 Motor end plate. (a) Diagram as seen by a light microscope. (b) Enlarged aspect as seen by an electron microscope.

ADP is resynthesized into ATP, which again binds to the myosin cross bridges. The tropomyosin-troponin complex is reattached to the actin of the thin myofilaments. And the myosin cross bridges separate from the actin. Since the myosin cross bridges are broken, the thin myofilaments slip back to their resting position. The sarcomeres are thereby returned to their resting lengths and the muscle fiber resumes its resting state.

Energy

Contraction of a muscle requires energy. When a nerve impulse stimulates a muscle fiber, ATP, in the presence of ATPase (activated myosin), breaks down into ADP + P and energy is released. As far as we know, ATP is always the immediate source of energy for muscle contraction.

Like the other cells of the body, muscle cells synthesize ATP as follows:

$$\text{ADP} + \text{P} + \text{energy} \rightarrow \text{ATP}$$

The energy for replenishing ATP is derived from the breakdown of digested foods. However, unlike most other cells of the body, muscle fibers alternate between great activity and virtual inactivity. When a muscle is

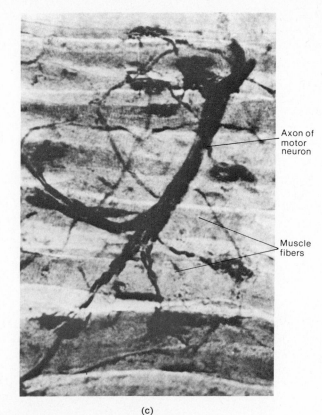

(c)

Figure 10–7 (cont.)   Motor end plate. (c) Photomicrograph at a magnification of 640X. (Courtesy of Fisher Scientific Company and S.T.E.M. Laboratories, Inc., Copyright 1975.)

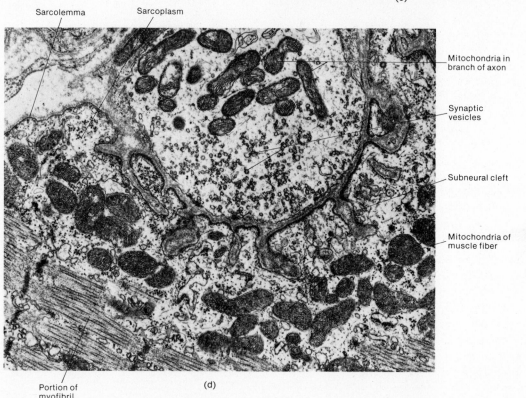

(d)

Figure 10–7 (cont.)   Motor end plate. (d) Electron micrograph at a magnification of 30,000X. (Courtesy of Cornelius Rosse and D. Kay Clawson, from *Introduction to the Musculoskeletal System,* Harper & Row, Publishers, Inc., New York, 1970.)

contracting, its energy requirements are high and the synthesis of ATP is accelerated. If the exercise is strenuous, ATP is used up even faster than it can be manufactured. Thus muscles must be able to build up a reserve supply of energy. They do this in two ways. A resting muscle needs little energy and produces much more ATP than it can use. At first, the muscle fiber stores the excess ATP on the thick myofilaments. When the fiber runs out of storage space for the ATP molecules, it combines the remainder of the ATP with a substance called *creatine*. Creatine, which is produced in the liver, can accept a high-energy phosphate from ATP to become the high-energy compound *creatine phosphate*. This reaction is anaerobic—it takes place without oxygen. It occurs as follows:

$$ATP + creatine \rightarrow creatine\ phosphate + ADP$$

Creatine phosphate is produced only when the muscle fibers are resting. During strenuous contraction, the reaction reverses itself. This reaction, also anaerobic, is shown in the following equation:

$$ADP + creatine\ phosphate \rightarrow ATP + creatine$$

All-or-None Principle

According to the **all-or-none principle,** the muscle fibers of a motor unit will contract to their fullest extent or will not contract at all, providing conditions remain constant. In other words, *muscle fibers* do not partly contract. The principle does not imply that the strength of the contraction is the same every time the fiber is stimulated. The strength of contraction may be decreased by fatigue, lack of nutrients, or lack of oxygen. The weakest stimulus from a neuron that can initiate a contraction is called a **threshold** or **liminal stimulus.** A stimulus of lesser intensity, or one that cannot initiate contraction, is referred to as a **subthreshold** or **subliminal stimulus.** Each muscle fiber in a motor unit has its own threshold level.

Kinds

The various skeletal muscles are capable of producing different contractions, depending on the stimulus.

TWITCH

The twitch contraction is a rapid, jerky response to a single stimulus. Twitch contractions can be artificially brought about in muscles of an animal. A recording of a twitch contraction is the classic way to illustrate the different phases of one single contraction. Figure 10-8 is

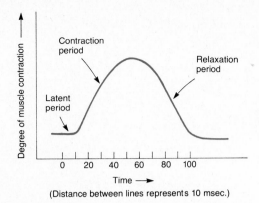

(Distance between lines represents 10 msec.)

Figure 10–8 Myogram of a twitch contraction.

a graph of a twitch contraction. This kind of record of a contraction is called a **myogram.** Note that a brief period exists between application of the stimulus and the beginning of contraction: the *latent period.* In frog muscle, it lasts about 10 milliseconds (0.01 second). The second phase, the *contraction period,* lasts about 40 milliseconds and is indicated by the upward tracing. The third phase, the *relaxation period,* lasts about 50 milliseconds and is indicated by the downward tracing. The duration of these periods depends on the muscle. The latent period, contraction period, and relaxation period for muscles that move the eyes are very short. Muscles that move the leg undergo longer periods.

If additional stimuli are applied to the muscle after the initial stimulus, other responses may be noted. For example, if two stimuli are applied one immediately after the other, the muscle will respond to the first stimulus but not to the second. When a muscle fiber receives enough stimulation to contract, it temporarily loses its irritability and cannot contract again until its responsiveness is regained. This period of lost irritability is the *refractory period.* Its duration also varies with the muscle involved. Skeletal muscle has a short refractory period. Cardiac muscle has a long refractory period.

TREPPE

The treppe contraction is the condition in which a skeletal muscle contracts more forcefully after it has contracted several times. It is demonstrated by stimulating an isolated muscle with a series of stimuli at the same frequency and voltage. Suppose a series of liminal stimuli are introduced into a muscle. Time must be allowed for the muscle to contract and relax, but this takes only 0.1 second in frog muscle. If stimuli are repeated at intervals of 0.5 second, the first few tracings on the myogram will show an increasing height with each contraction. This is treppe—the staircase phenomenon (Figure 10-9). It is the principle athletes use when

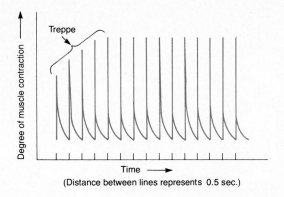

Figure 10–9 Myogram of a treppe contraction.

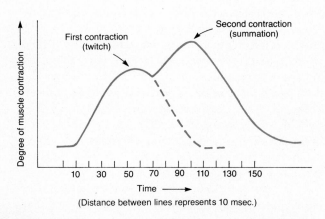

Figure 10–10 Myogram of summation of twitches. The second stimulus is applied at 70 msec., just before the muscle has finished contracting. The second contraction is even stronger than the first. The broken line represents the continuation of a twitch contraction.

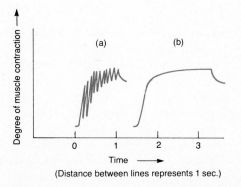

Figure 10–11 Myogram of incomplete and complete tetanus. (a) Incomplete. (b) Complete.

warming up. After the first few stimuli, the muscle reaches its peak of performance and undergoes its strongest contraction. After numerous contractions, the muscle does not relax completely, and the relaxation phase finally disappears.

TETANIC

When two stimuli are applied and the second is delayed until the refractory period is over, the skeletal muscle will respond to both stimuli. In fact, if the second stimulus is applied after the refractory period but before the muscle has finished contracting, the second contraction will be even stronger than the first. This phenomenon is **summation of twitches** or **wave summation** (Figure 10-10).

If a frog muscle is stimulated at a rate of 20 to 30 stimuli per second, the muscle can only partly relax between stimuli. As a result, the muscle maintains a sustained contraction called **incomplete tetanus** (Figure 10-11a). Stimulation at an increased rate (35 to 50 stimuli per second) results in **complete tetanus,** a sustained contraction that lacks even partial relaxation (Figure 10-11b). Essentially, both kinds of tetanus result from the rapid succession of separate twitches. Relaxation is either partial or does not occur at all. Voluntary contractions, such as contraction of the biceps in order to flex the forearm, are tetanic contractions.

TONIC

A sustained partial contraction of some of a skeletal muscle in response to stretch receptors is called **tone,** or **tonic contraction.** At any given time, some cells in a muscle are contracted while others are relaxed. This contraction tightens a muscle, but not enough fibers are contracting at any one time to produce movement. The same group of motor units does not function continuously. Instead, there is an asynchronous firing of alternating motor units that relieve one another so smoothly that the contraction can be sustained for long periods of time. Tone is essential for maintaining posture. When the muscles in the back of the neck are in tonic contraction, they keep the head in the anatomical position and prevent it from slumping forward onto the chest. In tonic contraction, these muscles do not apply enough force to pull the head back into hyperextension. The term **flaccid** is applied to muscles with less than normal tone. Such a loss of tone may be the result of damage or disease of the nerve that conducts a constant flow of impulses to the muscle. If the muscle does not receive impulses for an extended period of time, it may progress from flaccidity to **atrophy,** which is a state of wasting away. Muscles may also become flaccid and atrophied if they are not used. Bedridden individuals and

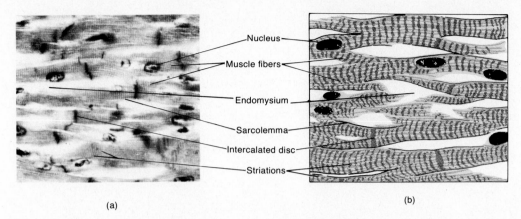

Nucleus

Muscle fibers

Endomysium

Sarcolemma

Intercalated disc

Striations

(a)

(b)

Figure 10–12   Histology of cardiac muscle tissue. (a) Photomicrograph of cardiac muscle tissue in longitudinal section at a magnification of 640X. (Courtesy of Edward J. Reith, from *Atlas of Descriptive Histology,* by Edward J. Reith and Michael H. Ross, Harper & Row, Publishers, Inc., New York, 1970.) (b) Diagram of the photomicrograph.

people with casts may experience atrophy because the flow of impulses to the inactive muscle is greatly reduced. On resumption of normal activities, the atrophied muscles recover.

ISOTONIC AND ISOMETRIC

Isotonic contractions are probably familiar to you. As the contraction occurs, the muscle shortens and pulls on another structure, such as a bone, to produce movement. During such a contraction, the tension remains constant (*iso* = equal; *tonos* = tension).

In an isometric contraction, there is a minimal shortening of the muscle. It remains nearly the same length, but the *tension* on the muscle increases greatly. Isometric contractions do not result in body movement. You can demonstrate such a contraction by carrying your books with your arm extended. The weight of the books pulls the arm downward, stretching the shoulder and arm muscles. The isometric contraction of the shoulder and arm muscles counteracts the stretch. The two forces – contraction and stretching – applied in opposite directions create the tension.

CARDIAC MUSCLE TISSUE

The principal constituent of the heart wall is **cardiac muscle tissue.** It has the same striated appearance as skeletal muscle tissue but it is involuntary. The cells of cardiac muscle tissue are roughly quadrangular and have only a single nucleus (Figure 10-12). The individual fibers are covered by a thin, poorly defined sarcolemma, and the internal myofibrils produce the characteristic striations. Cardiac muscle cells have the same basic arrangement of actin, myosin, and sarcoplasmic reticulum that is found in skeletal muscle cells. In addition,

they contain a system of transverse tubules similar to the T tubules of skeletal muscle. The nuclei in cardiac cells, however, are centrally located compared to the peripheral location of nuclei in skeletal muscle. While the fibers of skeletal muscle are arranged in a parallel fashion, those of cardiac muscle branch freely with other fibers to form two separate networks. The muscular walls and septum of the upper chambers of the heart (atria) compose one network. The muscular walls and septum of the lower chambers of the heart (ventricles) compose the other network. When a single fiber of either network is stimulated, all the fibers in the network become stimulated as well. Thus each network contracts as a functional unit. The fibers of each network were once thought to be fused together into a multinucleated mass called a syncytium. But it is now known that each fiber in a network is separated from the next fiber by an irregular transverse thickening of the sarcolemma called an **intercalated disc.** These discs strengthen the cardiac muscle tissue and aid in impulse conduction.

Under normal conditions, cardiac muscle tissue contracts rapidly, continuously, and rhythmically about 72 times a minute without stopping. This is a major physiological difference between cardiac and skeletal muscle tissue. Another difference is the source of stimulation. Skeletal muscle tissue ordinarily contracts only when stimulated by a nerve impulse. In contrast, cardiac muscle tissue can contract without nerve stimulation. Its source of stimulation is a conducting tissue of specialized muscle within the heart. About 72 times a minute, this tissue transmits electrical impulses that stimulate cardiac contraction. Nerve stimulation merely causes the conducting tissue to increase or decrease its rate of discharge. A third difference is that cardiac muscle tissue has a long refractory period that extends into part

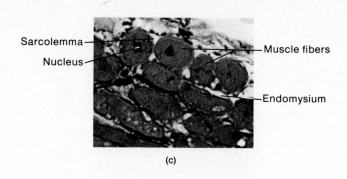

(c)

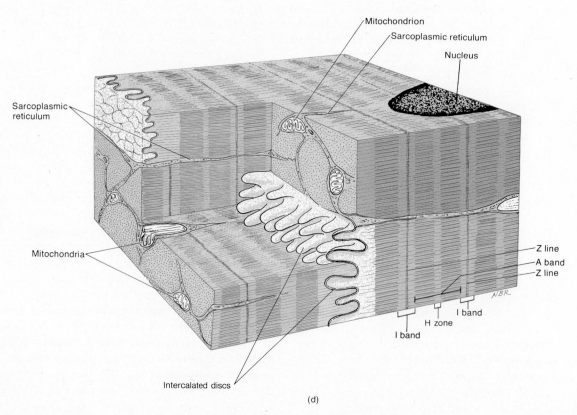

(d)

Figure 10–12 (cont.)   Histology of cardiac muscle tissue. (c) Photomicrograph of cardiac muscle tissue in cross section at a magnification of 100X. (Courtesy of Victor B. Eichler, Wichita State University.) (d) Diagram based on an electron micrograph.

of the relaxation period. As a result, even though the rate of the heartbeat can be drastically increased, the heart cannot undergo complete or incomplete tetanus.

SMOOTH MUSCLE TISSUE

Like cardiac muscle tissue, **smooth muscle tissue** is usually involuntary. However, it is nonstriated. A single fiber of smooth muscle tissue is about 5 to 10 mμ in diameter and 30 to 200 μm long. It is spindle-shaped,

and within the fiber is a single, oval, centrally located nucleus (Figure 10-13). Smooth muscle cells also contain actin and myosin filaments, but the filaments are not so orderly as in skeletal and cardiac muscle tissue. Thus well-differentiated striations do not occur.

Two kinds of smooth muscle tissue, visceral and multiunit, are recognized. The more common type is called *visceral muscle tissue.* It is found in wraparound sheets that form part of the walls of the hollow viscera such as the stomach, intestines, uterus, and urinary

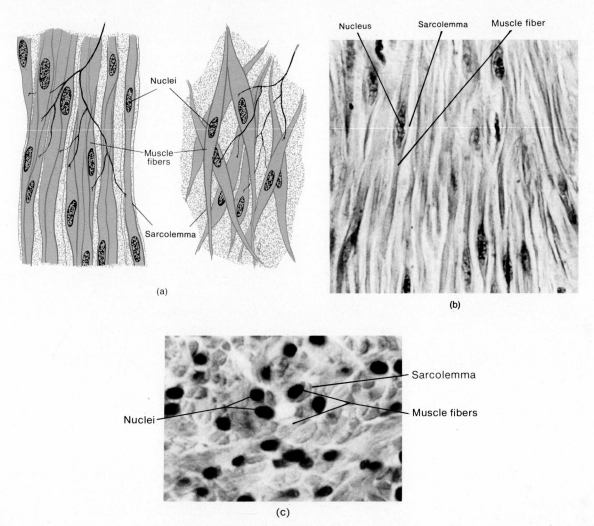

Figure 10–13 Histology of smooth muscle tissue. (a) Diagram of visceral smooth muscle tissue (left) and multi-unit smooth muscle tissue (right). (b) Photomicrograph of several smooth muscle fibers in longitudinal section at a magnification of 640X. (Courtesy of Edward J. Reith, from *Atlas of Descriptive Histology,* by Edward J. Reith and Michael H. Ross, Harper & Row, Publishers, Inc., New York, 1970.) (c) Photomicrograph of several smooth muscle fibers in cross section at a magnification of 200X. (Courtesy of Victor B. Eichler, Wichita State University.)

bladder. The terms *smooth muscle tissue* and *visceral muscle tissue* are sometimes used interchangeably. The fibers in visceral muscle tissue are tightly bound together to form a continuous network. When a neuron stimulates one fiber, the impulse travels over the other fibers so that contraction occurs in a wave over many adjacent fibers.

The second kind of smooth muscle tissue, *multiunit smooth muscle tissue,* consists of individual fibers each with its own motor-nerve endings. Whereas stimulation of a single visceral muscle fiber causes contraction of many adjacent fibers, stimulation of a single multiunit fiber causes contraction of only that fiber. In this respect, multiunit muscle tissue is like skeletal muscle tissue. Multiunit smooth muscle tissue is found in the walls of blood vessels, in the arrector pili

muscles that attach to hair follicles, and in the intrinsic muscles of the eye, such as the iris.

Both kinds of smooth muscle tissue contract more slowly and relax more slowly than skeletal muscle tissue. This characteristic is probably due to the poor arrangement of the thin and thick myofilaments of smooth muscle. Whereas skeletal muscle cells contract as individual units, visceral muscle cells contract in sequence as the impulse spreads from one to another.

HOMEOSTASIS

Muscle tissue has a vital role in maintaining the body's homeostasis. Two examples are the relationship of muscle tissue to oxygen and to heat production.

Oxygen Debt

The energy required to convert ADP into ATP is produced by the breakdown of digested foods. The primary nutrient that usually supplies this energy is the sugar glucose. The reaction proceeds as follows:

$$Glucose \longrightarrow pyruvic\ acid\ +\ energy$$

When the skeletal muscle is at rest, the breakdown of glucose is slow enough for the blood to supply sufficient oxygen to participate in the complete catabolism of pyruvic acid to the waste products carbon dioxide and water:

$$Pyruvic\ acid\ +\ O_2 \longrightarrow CO_2\ +\ H_2O\ +\ energy$$

Reactions such as this that involve oxygen are called *aerobic reactions.*

When a skeletal muscle is contracting and ATP production increases, the breakdown of glucose occurs too rapidly for the blood to supply oxygen for the pyruvic acid to be completely catabolized into carbon dioxide and water. The pyruvic acid is only partly catabolized to lactic acid. The conversion of pyruvic acid to lactic acid proceeds *anaerobically* (without oxygen) as follows:

$$Pyruvic\ acid \longrightarrow lactic\ acid$$

Most of the lactic acid diffuses from the skeletal muscles and is transported to the liver. There it is eventually resynthesized to glycogen or glucose, which can be reused later. Some of the lactic acid accumulates in the muscle tissue, however. Physiologists suspect that this acid is responsible for the feeling of muscle fatigue. Ultimately, the lactic acid in the skeletal muscle, as well as that which has diffused into the blood, must be catabolized to CO_2 and H_2O. And for this conversion additional oxygen is needed – the **oxygen debt.** When vigorous activity is over, labored breathing continues in order to pay back the debt. Thus the accumulation of lactic acid causes hard breathing and sufficient discomfort to stop muscle activity until homeostasis is restored.

Heat Production

The production of heat by skeletal muscle is an important homeostatic mechanism for maintaining normal body temperature. Of the total energy released during muscular contraction, only 20 to 30 percent is used for mechanical work (contraction). The rest is released as heat, which is utilized to help maintain a normal body temperature.

Heat production by muscles may be divided into two phases: (1) *initial heat,* which is produced by the contraction and relaxation of a muscle, and (2) *recovery heat,* which is produced after relaxation. Initial heat is independent of O_2 and is associated with ATP breakdown. Recovery heat is associated with the anaerobic breakdown of glucose to pyruvic acid and the aerobic conversion of lactic acid to CO_2 and H_2O.

DISORDERS

Disorders of the muscular system are related to disruptions of homeostasis. The disorders may involve a lack of nutrients, the accumulation of toxic products, disease, injury, disuse, or faulty nervous connections (innervations).

FIBROSIS

The formation of fibrous tissue in locations where it normally does not exist is called **fibrosis.** Skeletal and cardiac muscle fibers cannot undergo mitosis, and dead muscle fibers are normally replaced with fibrous connective tissue. Fibrosis, then, is often a consequence of muscle injury or degeneration.

FIBROSITIS

Fibrositis is an inflammation of fibrous tissue. If it occurs in the lumbar region, it is termed *lumbago.* Fibrositis is a common condition characterized by pain, stiffness, or soreness of fibrous tissue, especially in the muscle coverings. It is not destructive or progressive. It may persist for years or spontaneously disappear. Attacks of fibrositis may follow an injury, repeated muscular strain, or prolonged muscular tension. *Myositis* is an inflammation of muscle cells.

MUSCULAR DYSTROPHY

The term **muscular dystrophy** applies to a number of inherited myopathies, or muscle-destroying diseases. The word *dystrophy* means degeneration. The disease is characterized by degeneration of the individual muscle cells, which leads to a progressive atrophy (reduction in size) of the skeletal muscle. Usually the voluntary skeletal muscles are weakened equally on both sides of the body, whereas the internal muscles, such as the diaphragm, are not affected. Histologically, the changes that occur include the variation in muscle fiber size, degeneration of fibers, and deposition of fat.

MYASTHENIA GRAVIS

Myasthenia gravis is a weakness of the skeletal muscles. It is caused by an abnormality at the neuromuscular junction that prevents the muscle fibers from contracting. Recall that motor neurons stimulate the skeletal muscle fibers to contract by releasing acetylcholine. Myasthenia gravis is caused by failure of the neurons to release acetylcholine or by the release from the muscle fibers of an excess amount of cholinesterase, a chemical that interferes with the action of acetylcholine. As the disease progresses, more neuromuscular junctions become affected. The muscle becomes increasingly weaker and may eventually cease to function altogether.

The cause of myasthenia gravis is unknown. It is more common in females, occurring most frequently between the ages of 20 and 50. The muscles of the face and neck are most apt to be involved. Initial symptoms include a weakness of the eye muscles and difficulty in swallowing. Later, the individual has difficulty chewing and talking. Eventually, the muscles of the limbs may become involved. Death may result from paralysis of the respiratory muscles, but usually the disorder does not progress to this stage.

ABNORMAL CONTRACTIONS

One kind of abnormal contraction of a muscle is **spasm**: a sudden, involuntary contraction of short duration. A **cramp** is a painful spasmodic contraction of a muscle. It is an involuntary complete tetanic contraction. **Convulsions** are violent involuntary tetanic contractions of an entire group of muscles. Convulsions occur when motor neurons are stimulated by fever, poisons, hysteria, or changes in body chemistry due to withdrawal of certain drugs. The stimulated neurons send many bursts of seemingly disordered impulses to the muscle fibers. **Fibrillation** is the uncoordinated contraction of individual muscle fibers preventing the smooth contraction of the muscle. Cardiac muscle is particularly susceptible to this abnormality.

A **tic** is a spasmodic twitching made involuntarily by muscles that are ordinarily under voluntary control. Twitching of the eyelid and face muscles are examples. In general, tics are of psychological origin. They tend to develop in young individuals of nervous temperament.

MEDICAL TERMINOLOGY

Gangrene Death of tissue that results from almost complete interruption of its blood supply.

Myology Study of muscles.

Myomalacia (*malaco* = soft) Softening of a muscle.

Myopathy Any disease of muscle tissue.

Myosclerosis (*scler* = hard) Hardening of a muscle.

Myospasm Spasm of a muscle.

Myotonia Increased muscular irritability and contractility with decreased power of relaxation; tonic spasm of the muscle.

Trichinosis A myositis caused by the parasitic worm *Trichinella spiralis,* which may be found in the muscles of humans, rats, and pigs. People contract the disease by eating infected pork that is insufficiently cooked.

Volkmann's contracture (*contra* = against) Permanent contraction of a muscle due to replacement of destroyed muscle cells with fibrous tissue that lacks ability to stretch. Destruction of muscle cells may occur from interference with circulation caused by a tight bandage, a piece of elastic, or a cast.

Wryneck or torticollis Complete tetanus of one of the muscles of the neck; produces twisting of the neck and an unnatural position of the head.

CHARACTERISTICS

1 Irritability is the property of receiving and responding to stimuli.

2 Contractility is the ability to shorten and thicken, or contract.

3 Extensibility is the ability to be stretched or extended.

4 Elasticity is the ability to return to original shape after contraction or extension.

FUNCTIONS

1 Through contraction, muscle performs the three important functions of motion, maintenance of posture, and heat production.

KINDS

1 Skeletal muscle is attached to bones. It is striated and voluntary.

2 Smooth muscle is located in viscera. It is nonstriated and involuntary.

3 Cardiac muscle forms the walls of the heart. It is striated and involuntary.

SKELETAL MUSCLE TISSUE
FASCIA

1 The term *fascia* is applied to a sheet or broad band of fibrous connective tissue underneath the skin or around muscles and organs of the body.

2 Fascia is divided into three types: superficial, deep, and subserous.

CONNECTIVE TISSUE COMPONENTS

1 The entire muscle is covered by the epimysium. Fasciculi are covered by perimysium. Fibers are covered by endomysium.

2 Tendons and aponeuroses attach muscle to bone.

NERVE AND BLOOD SUPPLY

1 Nerves convey impulses, and blood provides nutrients and oxygen for contraction.

HISTOLOGY

1 The muscle consists of fibers covered by a sarcolemma. The fibers contain sarcoplasm, nuclei, sarcoplasmic reticulum, and T tubules.

2 Each fiber contains myofilaments (thin and thick). The myofilaments are compartmentalized into sarcomeres.

CONTRACTION
SLIDING-FILAMENT THEORY

1 A nerve impulse travels over the sarcolemma and enters the T tubules and sarcoplasmic reticulum.

2 The wave of depolarization leads to the release of calcium ions from the sarcoplasmic reticulum, triggering the contractile process.

3 Actual contraction is brought about when the thin myofilaments of one sarcomere slide toward each other.

MOTOR UNIT

1 A motor neuron transmits the stimulus to a skeletal muscle for contraction.

2 The region of the sarcolemma specialized to receive the stimulus is the motor end plate.

3 The area of contact between a motor neuron and muscle fiber is a neuromuscular or myoneural junction.

4 A motor neuron and the muscle fibers it stimulates form a motor unit.

PHYSIOLOGY

1 When a nerve impulse reaches the motor end plate, the neuron releases acetylcholine, which causes an electrical change.

2 This change releases calcium ions that activate the myosin, catalyzing the breakdown of ATP.

3 The energy released from the breakdown of ATP causes the sliding of the myofilaments.

ENERGY

1 The only direct source of energy is ATP.

2 When muscles are resting, ATP combines anaerobically with creatine to form creatine phosphate, which breaks down to produce ATP when muscles contract strenuously.

ALL-OR-NONE PRINCIPLE

1 Muscle fibers of a motor unit contract to their fullest extent or not at all.

2 The weakest stimulus capable of causing contraction is a threshold or liminal stimulus.

3 A stimulus not capable of inducing contraction is a subthreshold or subliminal stimulus.

KINDS

1 The various kinds of contractions are twitch, treppe, tetanic, tonic, isotonic, and isometric.

2 A record of a contraction is called a myogram. The refractory period is the time when a muscle has temporarily

lost irritability. Skeletal muscles have a short refractory period. Cardiac muscle has a long refractory period.

3 Summation is the stronger contraction of a muscle in response to additional stimuli.

CARDIAC MUSCLE TISSUE

1 This muscle is found only in the heart. It is striated and involuntary.

2 The cells are quadrangular and contain centrally placed nuclei.

3 The fibers form a continuous, branching network that contracts as a functional unit.

4 Intercalated discs provide strength and aid impulse conduction.

SMOOTH MUSCLE TISSUE

1 Smooth muscle is found in viscera. It is nonstriated and involuntary.

2 Visceral smooth muscle is found in the walls of viscera. The fibers are arranged in a network.

3 Multiunit smooth muscle is found in blood vessels and the eye. The fibers operate singly rather than as a unit.

HOMEOSTASIS

1 Oxygen debt is the amount of O_2 needed to convert accumulated lactic acid into CO_2 and H_2O. It occurs during strenuous exercise and is paid back by continuing to breathe rapidly after exercising. Until it is paid back, the homeostasis between muscular activity and oxygen requirements is not restored.

2 The heat given off during muscular contraction maintains the homeostasis of body temperature.

REVIEW QUESTIONS

1 How is the skeletal system related to the muscular system? What are the three basic functions of the muscular system?

2 What are the four characteristics of muscle tissue?

3 How can the three kinds of muscle tissue be distinguished?

4 What is fascia? What are the three different types of fascia and where are they found in the body?

5 Define epimysium, perimysium, endomysium, tendon, and aponeurosis. Describe the nerve and blood supply to a muscle.

6 Discuss the microscopic structure of skeletal muscle.

7 In considering the contraction of skeletal muscle, describe the following: sources of energy; motor unit; importance of calcium and troponin; sliding-filament theory.

8 What is the all-or-none principle? Relate it to a liminal and subliminal stimulus.

9 Define each of the following contractions and state the importance of each: isotonic, isometric, tonic, twitch, tetanic, treppe.

10 What is a myogram? How might a muscle become flaccid?

11 Describe the latent period, contraction period, and relaxation period of muscle contraction. Construct a diagram to substantiate your answer.

12 Define the refractory period. How does it differ between skeletal and cardiac muscle? What is summation?

13 Compare cardiac and smooth muscle with regard to microscopic structure, functions, and locations.

14 Discuss each of the following as examples of muscle homeostasis: steady state, oxygen debt, heat production.

15 What do you think might be the relationship between shivering (uncontrolled muscular contractions) and body temperature? Can you relate sweating (cooling of the skin) after strenuous exercise to the homeostasis of body temperature?

Chapter 11
The Muscular System

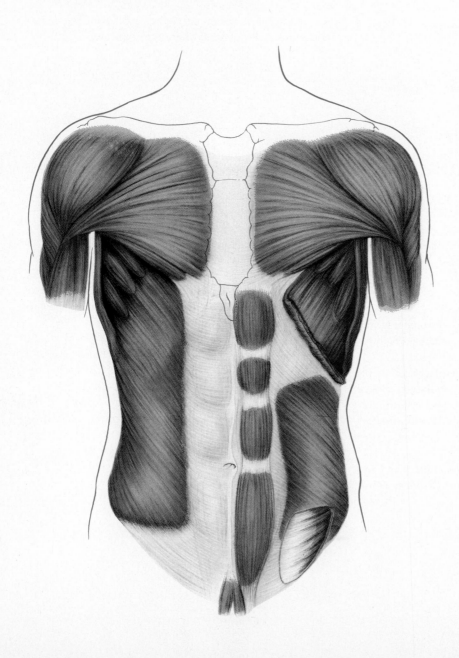

■ Describe the relationship between bones and skeletal muscles in producing body movements.

■ Define a lever and fulcrum and compare the three classes of levers on the basis of placement of the fulcrum, effort, and resistance.

■ Describe most body movements as activities of groups of muscles by explaining the roles of the prime mover, antagonist, and synergist.

■ Define the criteria employed in naming skeletal muscles.

■ Identify the principal skeletal muscles in different regions of the body by name, origin, insertion, action, and innervation.

■ Identify surface anatomy features of selected skeletal muscles.

■ Compare the common sites for intramuscular injections.

The term *muscle tissue* refers to all the contractile tissues of the body: skeletal, cardiac, and smooth muscle. The **muscular system,** however, refers to the *skeletal* muscle system: the skeletal muscle tissue and connective tissues that make up individual muscle organs such as the biceps. Cardiac and smooth muscle tissues are classified with other organ systems. For instance, cardiac muscle tissue is located in the heart, an organ of the circulatory system. Smooth muscle tissue of the intestine is part of the digestive system. Smooth muscle tissue of the urinary bladder is part of the urinary system. In this chapter, we discuss only the muscular system. We will see how skeletal muscles produce movement and describe the principal skeletal muscles.

HOW SKELETAL MUSCLES PRODUCE MOVEMENT

Origin and Insertion

Skeletal muscles produce movements by exerting force on tendons, which in turn pull on bones. Most muscles cross at least one joint and are attached to the articulating bones that form the joint (Figure 11-1a). When such a muscle contracts, it draws one articulating bone toward the other. The two articulating bones usually do not move equally in response to the contraction. One is held nearly in its original position because other muscles contract to pull it in the opposite direction or because its structure makes it less movable. Ordinarily, the attachment of a muscle tendon to the stationary bone is called the **origin.** The attachment of the other muscle tendon to the movable bone is the **insertion.** A good analogy is a spring on a door. The part of the spring attached to the door represents the insertion; the part attached to the frame is the origin. The fleshy portion of the muscle between the tendons of the origin and insertion is called the **belly,** or **gaster.** The origin is usually proximal and the insertion distal, especially in the appendages. In addition, muscles that move a body part generally do not cover the moving part. Figure 11-1a shows that although contraction of the biceps muscle moves the forearm, the belly of the muscle lies over the humerus.

Lever Systems

In producing a body movement, bones act as levers and joints function as fulcrums of these levers. A **lever** may be defined as a rigid rod that moves about on some fixed point called a **fulcrum.** A fulcrum may be symbolized as △. A lever is acted on at two different points by two different forces: the *resistance* ℝ and the *effort* (E). The resistance may be regarded as a force to be overcome, whereas the effort is exerted to overcome the resistance. The resistance may be the weight of the body part that is to be moved. The muscular effort (contraction) is applied to the bone at the insertion of the muscle and produces motion. Consider the biceps flexing the forearm at the elbow as a weight is lifted (Figure 11-1b). When the forearm is raised, the elbow is the fulcrum. The weight of the forearm plus the weight in the hand is the resistance. The shortening of the biceps pulling the forearm up is the effort.

Levers are categorized into three types by the position of the fulcrum, effort, and resistance.

1 In a **first-class lever,** the fulcrum is placed between the effort and resistance (Figure 11-2a). An example of a first-class lever is a seesaw. Examples of first-class levers in the body are not abundant, however. One is the head resting on the vertebral column. When the head is raised, the facial portion of the skull is the resistance. The joint between the atlas and occipital bone (atlanto-occipital joint) is the fulcrum. The muscles of the back in contraction represent the effort.

2 **Second-class levers** operate like a wheelbarrow. The fulcrum is at one end, the effort is at the opposite end, and the resistance is in between (Figure 11-2b). Most authorities agree that there are no examples of second-class levers in the body. Some, however, consider that raising the body on the toes (resistance) and utilizing the ball of the foot as the fulcrum is an example. Here, the calf muscles pull the heel upward as it shortens (effort).

3 **Third-class levers** are the most common levers in the body. They consist of the fulcrum at one end, the resistance at the opposite end, and the effort between them (Figure 11-2c). A common example is flexing the forearm at the elbow. The weight of the forearm is the resistance, the contraction of the biceps is the effort, and the elbow joint is the fulcrum.

Leverage

Leverage – the mechanical advantage gained by a lever – is largely responsible for a muscle's strength and range of movement. Consider strength first. Suppose we have two muscles of the same strength crossing and acting on a joint. Assume also that one is attached farther from the joint and one is closer. The muscle attached farther will produce the more powerful movement. Thus strength of movement depends on the placement of muscle attachments.

In considering the range of movement, again assume that we have two muscles of the same strength crossing and acting on a joint and that one is attached

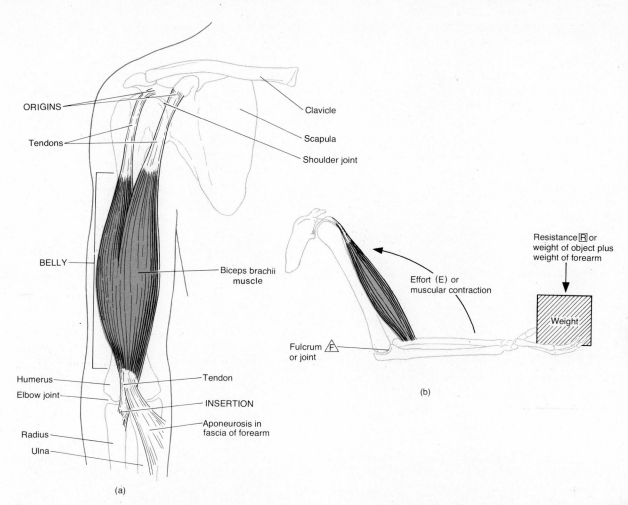

Figure 11—1   Relationship of muscles to bones. (a) Skeletal muscles produce movements by pulling on bones. (b) Bones serve as levers, and joints act as fulcrums for the levers. Here the lever-fulcrum principle is illustrated by the movement of the forearm lifting a weight. Note where the resistance and effort are applied in this example.

farther from the joint and one is closer. The muscle inserting closer to the joint will produce the greater range of movement. Thus range of movement depends on the placement of muscle attachments. Maximal strength and maximal range are therefore incompatible – strength and range vary inversely.

Group Actions

Most movements are coordinated by several skeletal muscles acting in groups rather than individually. As an example, take flexing the forearm at the elbow. A muscle that causes a desired action is referred to as the **agonist** or **prime mover.** In this instance, the biceps brachii is the agonist. Simultaneously with the contraction of the biceps brachii, another muscle, called the **antagonist,** is relaxing. In this movement, the triceps brachii serves as

the antagonist. The antagonist has an effect opposite to that of the agonist. That is, the antagonist relaxes and yields to the movement of the agonist. If we consider the extension of the forearm at the elbow, the triceps brachii would assume the role of the agonist and the biceps brachii would act as the antagonist. Most joints are operated by antagonistic groups of muscles. Still other muscles called **synergists,** or **fixators,** assist the agonist by reducing undesired action or unnecessary movements in the less mobile articulating bone. While flexing the forearm, the synergists, in this case the deltoid and pectoralis major muscles, hold the arm and shoulder in a suitable position for the flexing action. Whereas the deltoid abducts the humerus, the pectoralis major adducts and medially rotates the humerus. Essentially, synergists contract at the same time as the prime mover and help the prime mover produce an effective move-

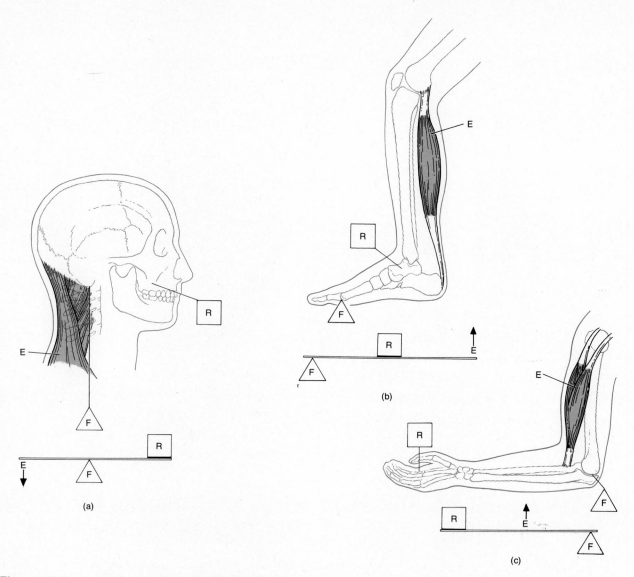

Figure 11–2 Classes of levers. (a) First-class lever. (b) Second-class lever. (c) Third-class lever. Each is defined on the basis of the placement of the fulcrum, effort, and resistance.

ment. Many muscles are, at various times, a prime mover, an antagonist, or a synergist, depending on the action.

NAMING SKELETAL MUSCLES

Most of the nearly 700 skeletal muscles are named on the basis of distinctive characteristics. If you understand them, you will remember the names of muscles. Some muscles are named on the basis of the *direction of the muscle fibers.* There are, for example, rectus (straight), transverse, and oblique muscles. Rectus fibers usually run parallel to the midline of the body. Transverse fibers run

perpendicular to the midline. Oblique fibers are diagonal to the midline. Muscles named according to these three directions include the rectus abdominis, transversus abdominis, and external oblique.

Another characteristic is *location.* The temporalis is named from its proximity to the temporal bone. The tibialis anterior is near the tibia. *Size* is another criterion. The term *maximus* means largest, and *minimus* means smallest. *Longus* means long; *brevis* means short. Examples include the gluteus maximus, gluteus minimus, adductor longus, and peroneus brevis.

Some muscles such as the biceps, triceps, and quadriceps are named for their *number of origins.* The

Exhibit 11—1   MUSCLES NAMED ACCORDING TO ACTION

Action	Definition	Example
Flexor	Usually decreases the anterior angle at a joint; some decrease the posterior angle.	Flexor carpi radialis
Extensor	Usually increases the anterior angle at a joint; some increase the posterior angle.	Extensor carpi ulnaris
Abductor	Moves a bone away from the midline.	Abductor hallucis longus
Adductor	Moves a bone closer to the midline.	Adductor longus
Levator	Produces an upward movement.	Levator scapulae
Depressor	Produces a downward movement.	Depressor labii inferioris
Supinator	Turns the palm upward or anteriorly.	Supinator
Pronator	Turns the palm downward or posteriorly.	Pronator teres
Dorsiflexor	Flexes the foot at the ankle joint.	Tibialis anterior
Plantar flexor	Extends the foot at the ankle joint.	Plantaris
Invertor	Turns the sole of the foot inward.	Tibialis anterior
Evertor	Turns the sole of the foot outward.	Peroneus tertius
Sphincter	Decreases the size of an opening.	Orbicularis oculi
Tensor	Makes a body part more rigid.	Tensor fasciae latae
Rotator	Moves a bone around its longitudinal axis.	Obturator

biceps has two origins. The triceps has three, and the quadriceps has four. Other muscles are named on the basis of *shape*. Common examples include the deltoid (meaning triangular) and trapezius (meaning trapezoid). Muscles may also be named after their *origin* and *insertion*. One example is the sternocleidomastoid, which originates on the sternum and clavicle and inserts at the mastoid process of the temporal bone. The stylohyoideus originates on the styloid process of the temporal bone and inserts at the hyoid bone.

Still another criterion used for naming muscles is *action*. Exhibit 11-1 lists the principal actions of muscles, their definitions, and examples of muscles that perform the actions. For convenience, the actions are grouped as antagonistic pairs where possible.

PRINCIPAL SKELETAL MUSCLES

Exhibits 11-2 through 11-16 list the muscles of the body in terms of their origins, insertions, actions, and innervations. Refer to Chapters 6 and 7 to review bone markings, since they serve as points of origin and insertion for muscles. By no means have all the muscles

of the body been included. Only the major ones are discussed.

Surface anatomy is the study of the form and markings of the body surface. A knowledge of surface anatomy will help you to identify certain superficial structures by visual inspection or palpation through the skin. *Palpation* means to feel with the hand. As you study the skeletal muscles of the body in this chapter, you will also be introduced to their surface anatomy by means of labeled photographs.

Let us now investigate the principal skeletal muscles of the body by examining Exhibits 11-2 through 11-16. Figure 11-3 shows general anterior and posterior views of the muscular system. Do not try to memorize all these muscles yet. As you study groups of muscles in subsequent Exhibits, refer to Figure 11-3 to see how each group is related to all others.

We have attempted to indicate whether the muscles are superficial or deep, anterior or posterior, medial or lateral. We have also tried to show the relationship of the muscles under consideration to other muscles in the area you are studying. If you have mastered the naming of muscles, their actions will have more meaning.

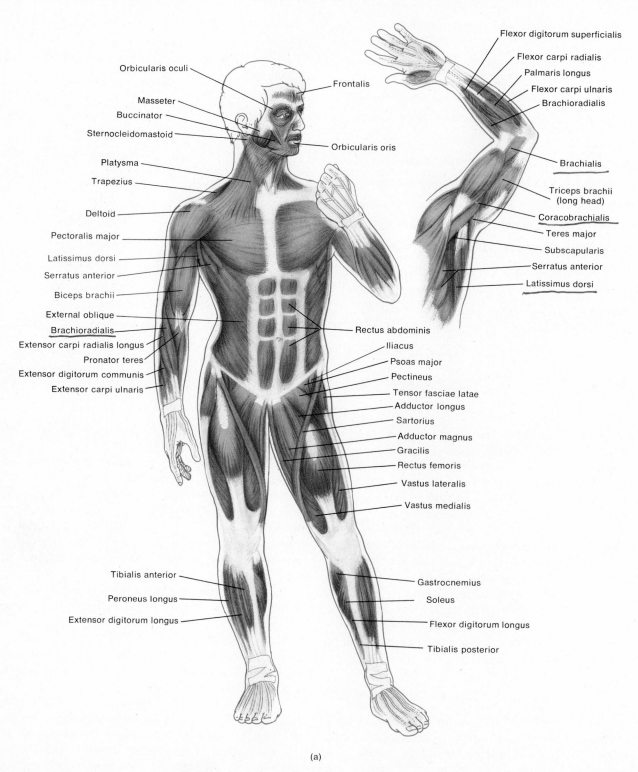

Orbicularis oculi

Masseter

Buccinator

Sternocleidomastoid

Platysma

Trapezius

Deltoid

Pectoralis major

Latissimus dorsi

Serratus anterior

Biceps brachii

External oblique

Brachioradialis

Extensor carpi radialis longus

Pronator teres

Extensor digitorum communis

Extensor carpi ulnaris

Frontalis

Orbicularis oris

Flexor digitorum superficialis

Flexor carpi radialis

Palmaris longus

Flexor carpi ulnaris

Brachioradialis

Brachialis

Triceps brachii (long head)

Coracobrachialis

Teres major

Subscapularis

Serratus anterior

Latissimus dorsi

Rectus abdominis

Iliacus

Psoas major

Pectineus

Tensor fasciae latae

Adductor longus

Sartorius

Adductor magnus

Gracilis

Rectus femoris

Vastus lateralis

Vastus medialis

Tibialis anterior

Peroneus longus

Extensor digitorum longus

Gastrocnemius

Soleus

Flexor digitorum longus

Tibialis posterior

(a)

Figure 11–3   Principal superficial muscles. (a) Anterior view.

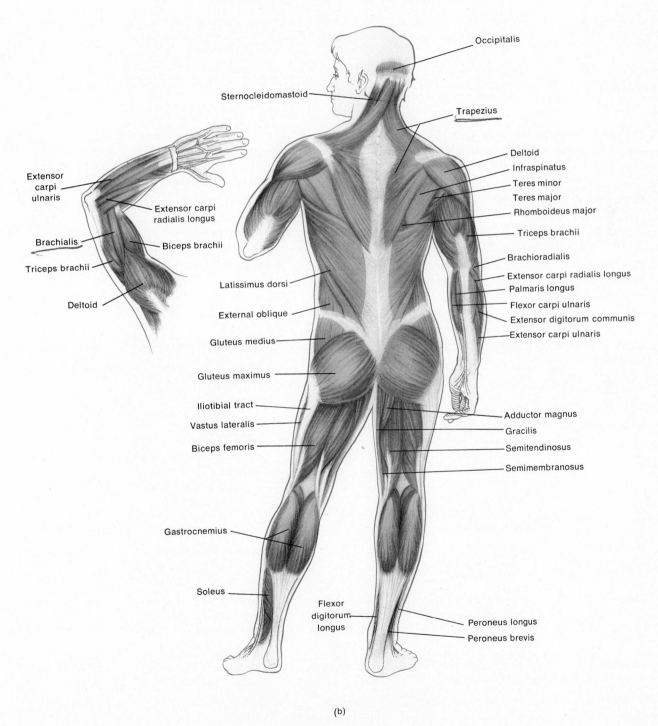

Occipitalis

Sternocleidomastoid

Trapezius

Deltoid
Infraspinatus
Teres minor
Teres major
Rhomboideus major

Triceps brachii

Brachioradialis

Extensor carpi radialis longus
Palmaris longus

Flexor carpi ulnaris
Extensor digitorum communis
Extensor carpi ulnaris

Extensor
carpi
ulnaris

Extensor carpi
radialis longus

Brachialis

Biceps brachii

Triceps brachii

Deltoid

Latissimus dorsi

External oblique

Gluteus medius

Gluteus maximus

Iliotibial tract

Vastus lateralis

Biceps femoris

Adductor magnus
Gracilis
Semitendinosus

Semimembranosus

Gastrocnemius

Soleus

Flexor
digitorum
longus

Peroneus longus
Peroneus brevis

(b)

Figure 11–3 (cont.)   Principal superficial muscles. (b) Posterior view.

Exhibit 11–2   MUSCLES OF FACIAL EXPRESSION (see Figure 11–4)

Muscle	Origin	Insertion	Action	Innervation
Epicranius (*epi* = over; *crani* = skull)	This muscle is divisible into two portions: the frontalis over the frontal bone and the occipitalis over the occipital bone. The two muscles are united by a strong aponeurosis, the galea aponeurotica, which covers the superior and lateral surfaces of the skull.			
Frontalis (*front* = forehead)	Galea aponeurotica	Skin superior to supraorbital line	Draws scalp forward, raises eyebrows, and wrinkles forehead horizontally.	Facial nerve (VII)
Occipitalis (*occipito* = base of skull)	Occipital bone and mastoid process of temporal bone	Galea aponeurotica	Draws scalp backward.	Facial nerve (VII)
Orbicularis oris (*orb* = circular; *or* = mouth)	Muscle fibers surrounding opening of mouth	Skin at corner of mouth	Closes lips, compresses lips against teeth, protrudes lips, and shapes lips during speech.	Facial nerve (VII)
Zygomaticus major (*zygomatic* = cheek bone; *major* = greater)	Zygomatic bone	Skin at angle of mouth and orbicularis oris	Draws angle of mouth upward and outward as in smiling or laughing.	Facial nerve (VII)
Levator labii superioris (*levator* = raises or elevates; *labii* = lip; *superioris* = upper)	Superior to infraorbital foramen of maxilla	Skin at angle of mouth and orbicularis oris	Elevates (raises) upper lip.	Facial nerve (VII)
Depressor labii inferioris (*depressor* = depresses or lowers; *inferioris* = lower)	Mandible	Skin of lower lip	Depresses (lowers) lower lip.	Facial nerve (VII)
Buccinator (*bucc* = cheek)	Alveolar processes of maxilla and mandible and pterygomandibular raphe (fibrous band extending from the pterygoid hamulus to the mandible)	Orbicularis oris	Major cheek muscle; compresses cheek as in blowing air out of mouth and causes the cheeks to cave in, producing the action of sucking.	Facial nerve (VII)
Mentalis (*mentum* = chin)	Mandible	Skin of chin	Elevates and protrudes lower lip and pulls skin of chin up as in pouting.	Facial nerve (VII)
Platysma (*platy* = flat, broad)	Fascia over deltoid and pectoralis major muscles	Mandible, muscles around angle of mouth, and skin of lower face	Draws outer part of lower lip downward and backward as in pouting; depresses mandible.	Facial nerve (VII)
Risorius (*risor* = laughter)	Fascia over parotid (salivary) gland	Skin at angle of mouth	Draws angle of mouth laterally as in tenseness.	Facial nerve (VII)
Orbicularis oculi (*ocul* = eye)	Medial wall of orbit	Circular path around orbit	Closes eye.	Facial nerve (VII)

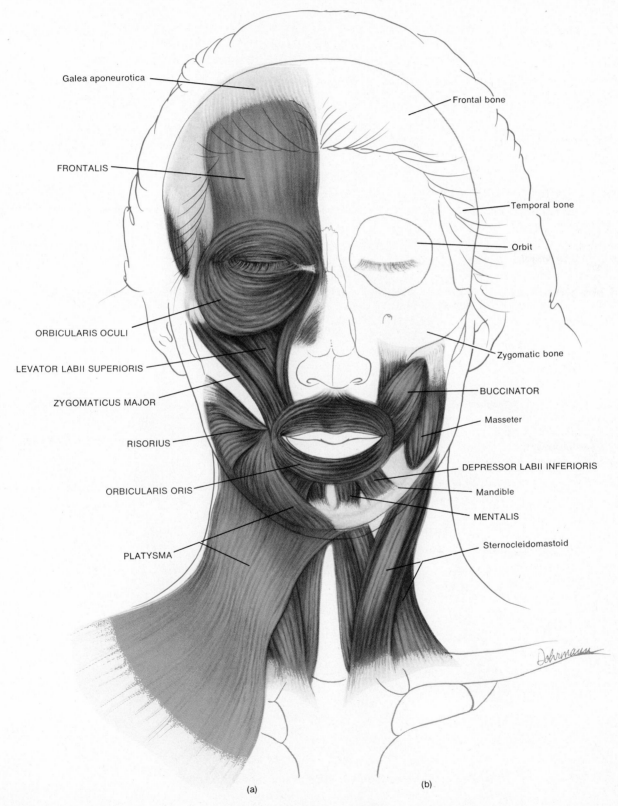

Figure 11—4 Muscles of facial expression. (a) Anterior superficial view. (b) Anterior deep view.

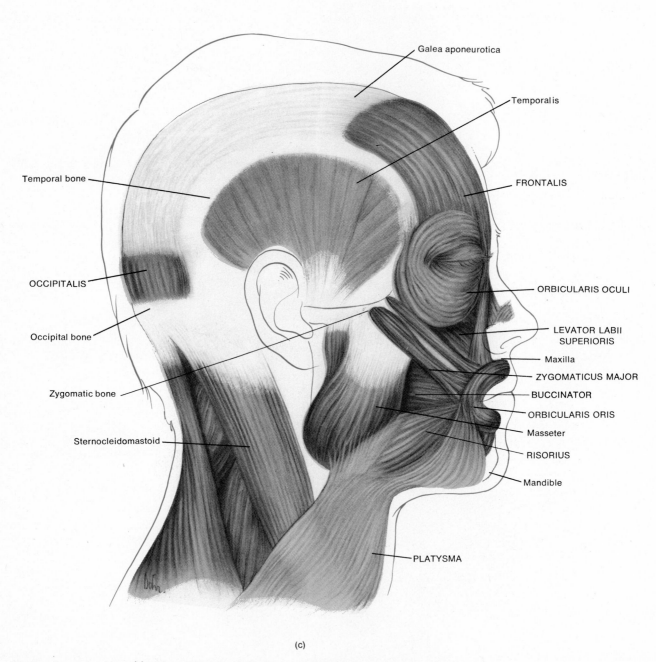

(c)

Figure 11–4 (cont.)   Muscles of facial expression. (c) Right lateral superficial view.

Exhibit 11—3 MUSCLES THAT MOVE THE LOWER JAW (see Figure 11—5)

Muscle	Origin	Insertion	Action	Innervation
Masseter (*maseter* = chewer)	Maxilla and zygomatic arch	Angle and ramus of mandible	Elevates mandible as in closing the mouth and protracts (protrudes) mandible.	Mandibular branch of trigeminal nerve (V)
Temporalis (*tempora* = temples)	Temporal bone	Coronoid process of mandible	Elevates and retracts mandible.	Temporal nerve from mandibular division of trigeminal nerve (V)
Medial pterygoid (*medial* = closer to midline; *pterygoid* = like a wing; pterygoid plate of sphenoid)	Medial surface of lateral pterygoid plate of sphenoid; maxilla	Angle and ramus of mandible	Elevates and protracts mandible and moves mandible from side to side.	Mandibular branch of trigeminal nerve (V)
Lateral pterygoid (*lateral* = farther from midline)	Greater wing and lateral surface of lateral pterygoid plate of sphenoid	Condyle of mandible; temporomandibular articulation	Protracts mandible, opens mouth, and moves mandible from side to side.	Mandibular branch of trigeminal nerve (V)

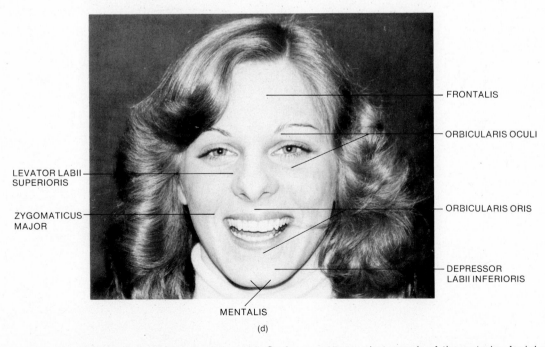

LEVATOR LABII SUPERIORIS

ZYGOMATICUS MAJOR

MENTALIS

FRONTALIS

ORBICULARIS OCULI

ORBICULARIS ORIS

DEPRESSOR LABII INFERIORIS

(d)

Figure 11—4 (cont.) Muscles of facial expression. (d) Surface anatomy photograph of the anterior facial muscles. (Courtesy of Donald Castellaro and Deborah Massimi.)

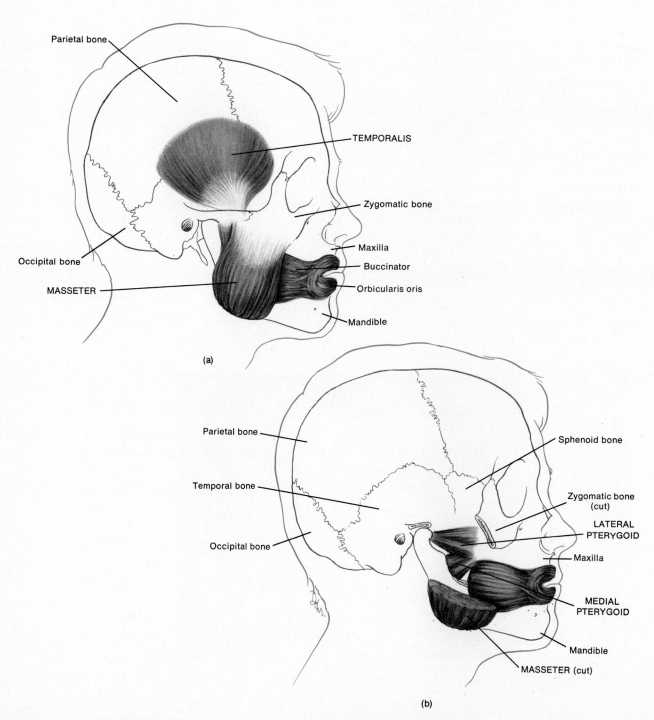

Figure 11–5 Muscles that move the lower jaw. (a) Superficial right lateral view. (b) Deep right lateral view.

Exhibit 11—4   MUSCLES THAT MOVE THE EYEBALLS— EXTRINSIC MUSCLES* (see Figure 11—6)

Muscle	Origin	Insertion	Action	Innervation
Superior rectus (*superior* = above; *rectus* = in this case, muscle fibers running parallel to long axis of eyeball)	Tendinous ring attached to bony orbit around optic foramen	Superior and central part of eyeball	Rolls eyeball upward.	Oculomotor nerve (III)
Inferior rectus (*inferior* = below)	Same as above	Inferior and central part of eyeball	Rolls eyeball downward.	Oculomotor nerve (III)
Lateral rectus	Same as above	Lateral side of eyeball	Rolls eyeball laterally.	Abducens nerve (VI)
Medial rectus	Same as above	Medial side of eyeball	Rolls eyeball medially.	Oculomotor nerve (III)
Superior oblique (*oblique* = in this case, muscle fibers running diagonally to long axis of eyeball)	Same as above	Eyeball between superior and lateral recti	Rotates eyeball on its axis; directs cornea downward and laterally; note that it moves through a ring of fibrocartilaginous tissue called the trochlea (*trochlea* = pulley).	Trochlear nerve (IV)
Inferior oblique	Maxilla (front of orbital cavity)	Eyeball between superior and lateral recti	Rotates eyeball on its axis; directs cornea upward and laterally.	Oculomotor nerve (III)

*Muscles situated on the outside of the eyeball.

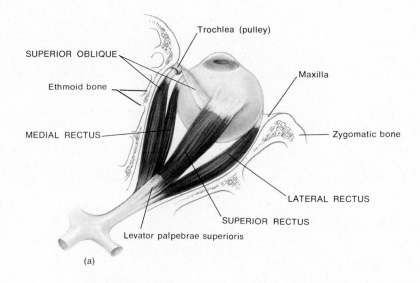

(a)

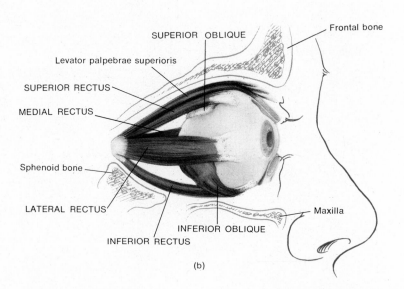

(b)

Figure 11–6   Extrinsic muscles of the eyeball. (a) Muscles of the right eyeball seen from above. (b) Right lateral view of muscles of the right eyeball.

Exhibit 11—5 MUSCLES THAT MOVE THE TONGUE (see Figure 11—7)

Muscle	Origin	Insertion	Action	Innervation
Genioglossus (*geneion* = chin; *glossus* = tongue)	Mandible	Undersurface of tongue and hyoid bone	Depresses and thrusts tongue forward (protraction).	Hypoglossal nerve (XII)
Styloglossus (*stylo* = stake or pole; styloid process of temporal)	Styloid process of temporal bone	Side and under-surface of tongue	Elevates tongue and draws it backward (retraction).	Hypoglossal nerve (XII)
Stylohyoid (*hyoeides* = U-shaped; pertaining to hyoid bone)	Styloid process of temporal bone	Body of hyoid bone	Elevates and retracts tongue.	Facial nerve (VII)
Hyoglossus	Body of hyoid bone	Side of tongue	Depresses tongue and draws down its sides.	Hypoglossal nerve (XII)

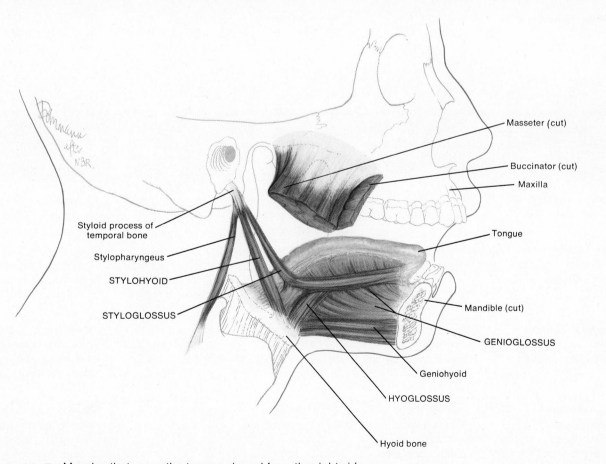

Figure 11—7 Muscles that move the tongue viewed from the right side.

Exhibit 11—6 MUSCLES THAT MOVE THE HEAD

Muscle	Origin	Insertion	Action	Innervation
Sternocleidomastoid (*sternum* = breastbone; *cleido* = clavicle; *mastoid* = mastoid process of temporal bone) (see Figure 11–10)	Sternum and clavicle	Mastoid process of temporal bone	Contraction of both muscles flexes the neck on the chest; contraction of one muscle rotates face toward side opposite contracting muscle.	Accessory nerve (XI); C2–C3
Semispinalis capitis (*semi* = half; *spine* = spinous process; *caput* = head) (see Figure 11–14)	Articular process of seventh cervical vertebra and transverse processes of first six thoracic vertebrae	Occipital bone	Both muscles extend head; contraction of one muscle rotates face toward same side as contracting muscle.	Dorsal rami of spinal nerves
Splenius capitis (*splenion* = bandage) (see Figure 11–14)	Ligamentum nuchae and spines of seventh cervical vertebra and first four thoracic vertebrae	Occipital bone and mastoid process of temporal bone	Both muscles extend head; contraction of one rotates it to same side as contracting muscle.	Dorsal rami of middle and lower cervical nerves
Longissimus capitis (*longissimus* = longest) (see Figure 11–14)	Transverse processes of last four cervical vertebrae	Mastoid process of temporal bone	Extends head and rotates face toward side opposite contracting muscle.	Dorsal rami of middle and lower cervical nerves

Exhibit 11—7 MUSCLES THAT ACT ON THE ANTERIOR ABDOMINAL WALL
(see Figure 11—8)

Muscle	Origin	Insertion	Action	Innervation
Rectus abdominis (*abdomino* = belly)	Pubic crest and symphysis pubis	Cartilage of fifth to seventh ribs and xiphoid process	Flexes vertebral column.	Branches of 7–12 intercostal nerves
External oblique (*external* = closer to the surface)	Lower eight ribs	Iliac crest; linea alba (midline aponeurosis)	Both muscles compress abdomen; one side alone bends vertebral column laterally.	Branches of 8–12 intercostal nerves, iliohypogastric and ilioinguinal nerves
Internal oblique (*internal* = farther from the surface)	Iliac crest, inguinal ligament, and thoracolumbar fascia	Cartilage of last three or four ribs	Compresses abdomen; one side alone bends vertebral column laterally.	Branches of 8–12 intercostal nerves, iliohypogastric, and ilioinguinal nerves
Transversus abdominis (*transverse* = muscle fibers run transversely to midline)	Iliac crest, inguinal ligament, lumbar fascia, and cartilages of last six ribs	Xiphoid process, linea alba, and pubis	Compresses abdomen.	Branches of 7–12 intercostal nerves, iliohypogastric, and ilioinguinal nerves

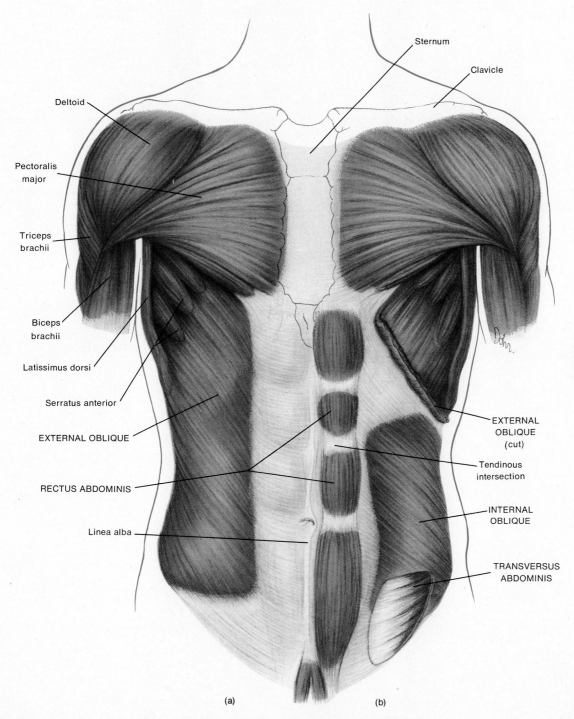

Figure 11—8 Muscles of the anterior abdominal wall. (a) Superficial view. (b) Deep view.

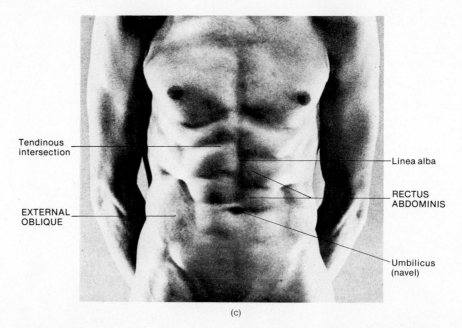

(c)

Figure 11–8 (cont.)   Muscles of the anterior abdominal wall. (c) Surface anatomy photograph. (Courtesy of R. D. Lockhart, *Living Anatomy,* Faber and Faber, 1974.)

Exhibit 11—8 MUSCLES USED IN BREATHING (see Figure 11—9)

Muscle	Origin	Insertion	Action	Innervation
Diaphragm (*dia* = across; *phragma* = wall)	Xiphoid process, costal cartilages of last six ribs, and lumbar vertebrae	Central tendon	Forms floor of thoracic cavity; contraction pulls central tendon downward and increases vertical length of thorax during inspiration.	Phrenic nerve
External intercostals (*inter* = between; *costa* = rib)	Inferior border of rib above	Superior border of rib below	Elevate ribs during inspiration and thus increase lateral and anteroposterior dimensions of thorax.	Intercostal nerves
Internal intercostals	Superior border of rib below	Inferior border of rib above	Draw adjacent ribs together during forced expiration and thus decrease the lateral and anteroposterior dimensions of thorax.	Intercostal nerves

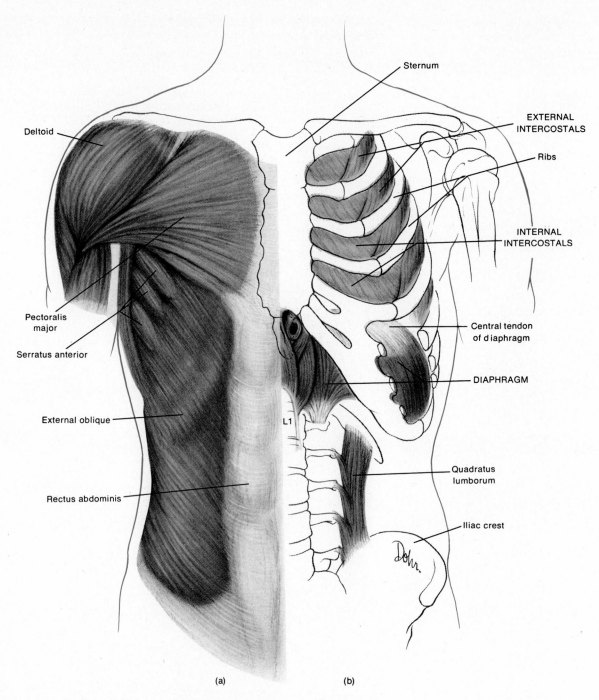

Figure 11–9   Muscles used in breathing. (a) Superficial view. (b) Deep view.

Exhibit 11—9 MUSCLES THAT MOVE THE SHOULDER GIRDLE
(see Figure 11—10)

Muscle	Origin	Insertion	Action	Innervation
Subclavius (*sub* = under; *clavius* = clavicle)	First rib	Clavicle	Depresses clavicle.	Nerve to subclavius
Pectoralis minor (*pectus* = breast, chest, thorax; *minor* = lesser)	Third through fifth ribs	Coracoid process of scapula	Depresses scapula, rotates shoulder joint anteriorly, and elevates third through fifth ribs during forced inspiration when scapula is fixed.	Medial pectoral nerve
Serratus anterior (*serratus* = serrated; *anterior* = front)	Upper eight or nine ribs	Vertebral border and inferior angle of scapula	Rotates scapula laterally and elevates ribs when scapula is fixed.	Long thoracic nerve
Trapezius (*trapezoides* = trapezoid-shaped)	Occipital bone, ligamentum nuchae, and spines of seventh cervical and all thoracic vertebrae	Acromion process of clavicle and spine of scapula	Elevates clavicle, adducts scapula, elevates or depresses scapula, and extends head.	Acessory nerve (XI); C3—C4
Levator scapulae (*levator* = raises; *scapulae* = scapula)	Upper four or five cervical vertebrae	Vertebral border of scapula	Elevates scapula.	Dorsal scapular nerve (C3—C5)
Rhomboideus major (*rhomboides* = rhomboid or diamond-shaped)	Spines of second to fifth thoracic vertebrae	Vertebral border of scapula	Adducts scapula and slightly rotates it upward.	Dorsal scapular nerve
Rhomboideus minor	Spines of seventh cervical and first thoracic vertebrae	Superior angle of scapula	Adducts scapula.	Dorsal scapular nerve

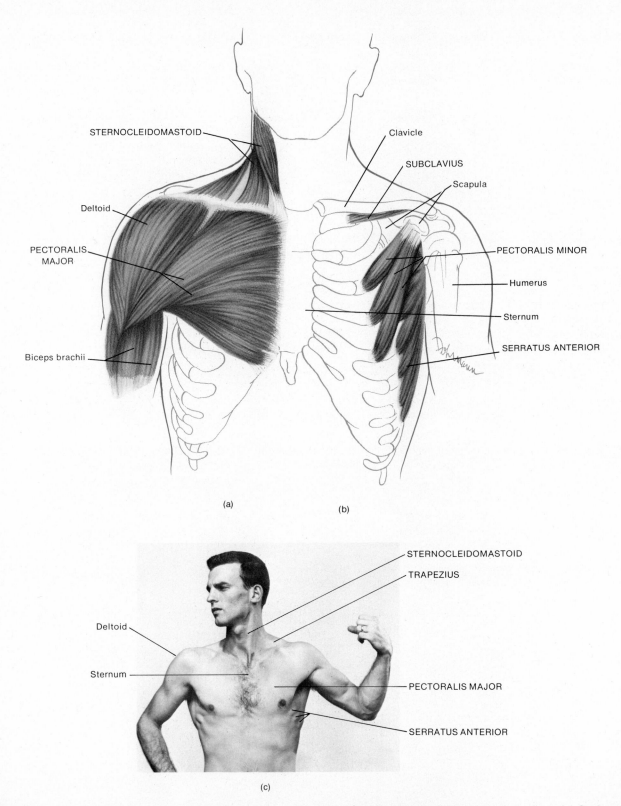

(a)   (b)

(c)

Figure 11–10   Muscles that move the shoulder girdle. (a) Anterior superficial view. (b) Anterior deep view. (c) Surface anatomy photograph of the neck and chest. (d) Posterior superficial view. (e) Posterior deep view. (f) Surface anatomy photograph of the back. (Photographs courtesy of Vincent P. Destro, Mayo foundation.)

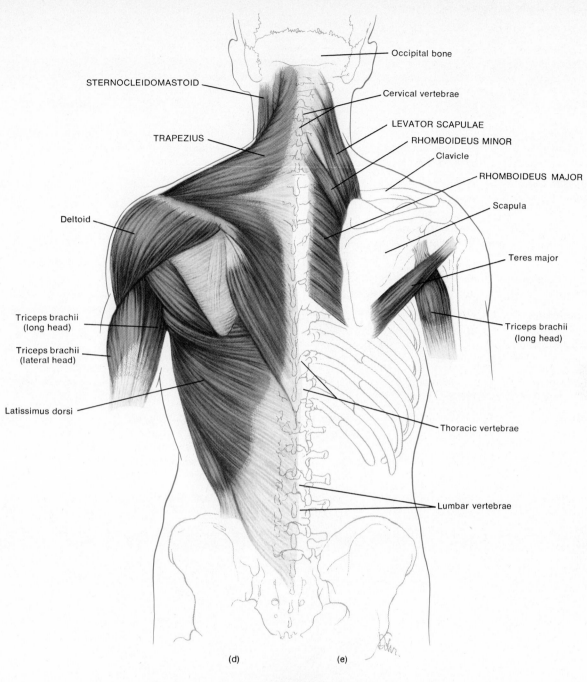

Occipital bone

STERNOCLEIDOMASTOID

Cervical vertebrae

LEVATOR SCAPULAE

RHOMBOIDEUS MINOR

Clavicle

RHOMBOIDEUS MAJOR

TRAPEZIUS

Scapula

Deltoid

Teres major

Triceps brachii
(long head)

Triceps brachii
(long head)

Triceps brachii
(lateral head)

Latissimus dorsi

Thoracic vertebrae

Lumbar vertebrae

(d)　　(e)

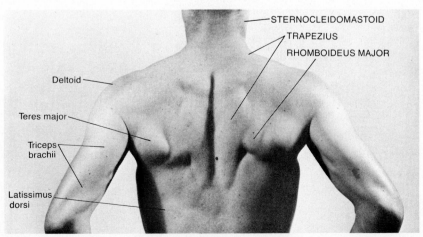

STERNOCLEIDOMASTOID

TRAPEZIUS

RHOMBOIDEUS MAJOR

Deltoid

Teres major

Triceps
brachii

Latissimus
dorsi

(f)

Exhibit 11—10   MUSCLES THAT MOVE THE ARM (see Figure 11—11)

Muscle	Origin	Insertion	Action	Innervation
Pectoralis major (see Figure 11—10)	Clavicle, sternum, cartilages of second to sixth ribs	Greater tubercle of humerus	Flexes, adducts, and rotates arm medially.	Medial and lateral pectoral nerve
Latissimus dorsi (*dorsum* = back)	Spines of lower six thoracic vertebrae, lumbar vertebrae, crests of sacrum and ilium, lower four ribs	Intertubercular groove of humerus	Extends, adducts, and rotates arm medially; draws shoulder downward and backward.	Thoracodorsal nerve
Deltoid (*delta* = triangular)	Clavicle and acromion process and spine of scapula	Deltoid tuberosity of humerus	Abducts arm.	Axillary nerve
Supraspinatus (*supra* = above; *spinatus* = spine of scapula)	Fossa superior to spine of scapula	Greater tubercle of humerus	Assists deltoid muscle in abducting arm.	Suprascapular nerve
Infraspinatus (*infra* = below)	Fossa inferior to spine of scapula	Greater tubercle of humerus	Rotates arm laterally.	Suprascapular nerve
Teres major (*teres* = long and round)	Inferior angle of scapula	Distal to lesser tubercle of humerus	Extends arm and draws it down; assists in adduction and medial rotation of arm.	Lower subscapular nerve
Teres minor	Axillary border of scapula	Greater tubercle of humerus	Rotates arm laterally.	Axillary nerve

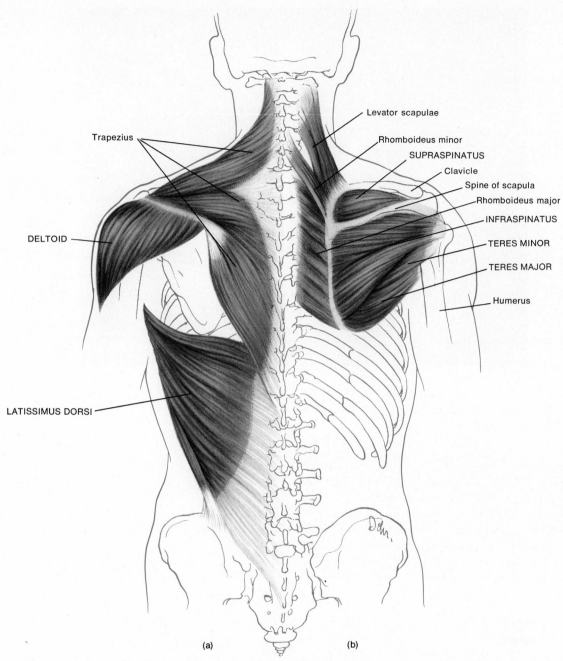

Figure 11–11   Muscles that move the arm. (a) Posterior superficial view. (b) Posterior deep view.

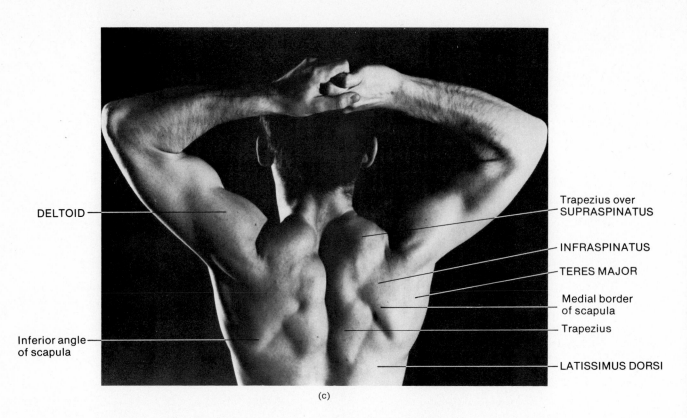

DELTOID

Inferior angle
of scapula

Trapezius over
SUPRASPINATUS

INFRASPINATUS

TERES MAJOR

Medial border
of scapula

Trapezius

LATISSIMUS DORSI

(c)

Figure 11–11 (cont.) Muscles that move the arm. (c) Surface anatomy photograph of the back. (Courtesy of J. Royce, *Surface Anatomy,* Davis, 1965.)

Before you move on to Exhibit 11-11 (Muscles That Move the Forearm), refer to Figure 1-4b. This figure is a cross section of the trunk through the heart and lungs. Several of the muscles you have studied up to this point are shown. Figure 1-4b will add to your understanding of how muscles are arranged with respect to each other from external to internal. It will also show you how muscles are oriented with regard to bones and viscera.

Exhibit 11–11 MUSCLES THAT MOVE THE FOREARM (see Figure 11–12)

Muscle	Origin	Insertion	Action	Innervation
Biceps brachii (*biceps* = two heads of origin; *brachion* = arm)	Long head originates from tuberosity above glenoid cavity and short head originates from coracoid process of scapula	Radial tuberosity and bicipital aponeurosis	Flexes and supinates forearm.	Musculocutaneous nerve
Brachialis	Anterior surface of humerus	Tuberosity and coronoid process of ulna	Flexes forearm.	Musculocutaneous, radial, and median nerves
Brachioradialis (*radialis* = radius) (see Figure 11–13 also)	Supracondyloid ridge of humerus	Superior to styloid process of radius	Flexes forearm.	Radial nerve
Triceps brachii (*triceps* = three heads of origin)	Long head originates from infraglenoid tuberosity of scapula, lateral head originates from lateral and posterior surface of humerus superior to radial groove, medial head originates from posterior surface of humerus inferior to radial groove	Olecranon of ulna	Extends forearm.	Radial nerve
Supinator (*supination* = turning palm upward or anteriorly)	Lateral epicondyle of humerus, ridge on ulna	Oblique line of radius	Supinates forearm.	Deep radial nerve
Pronator teres (*pronation* = turning palm downward or posteriorly)	Medial epicondyle of humerus, coronoid process of ulna	Midlateral surface of radius	Pronates forearm.	Median nerve

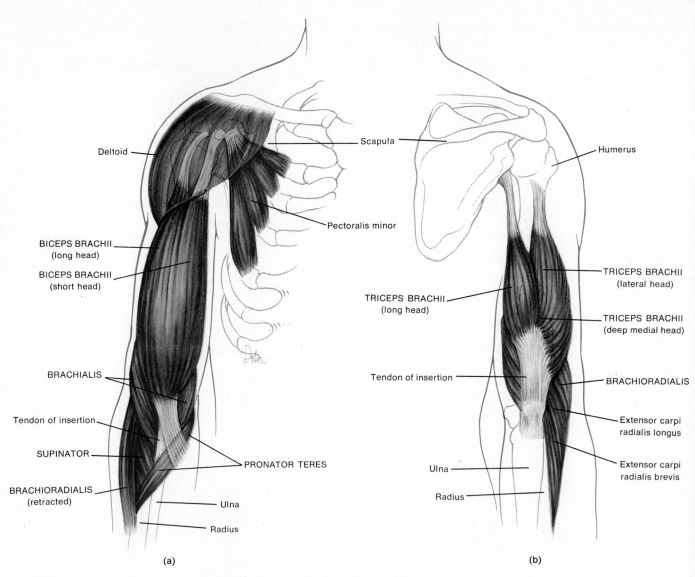

Figure 11-12 Muscles that move the forearm. (a) Anterior view. (b) Posterior view. (c) Surface anatomy photograph of the anterior arm and upper forearm. (Courtesy of Vincent P. Destro, Mayo Foundation.) (d) Surface anatomy photograph of the posterior arm. (Courtesy of Donald Castellaro and Richard Sollazzo.)

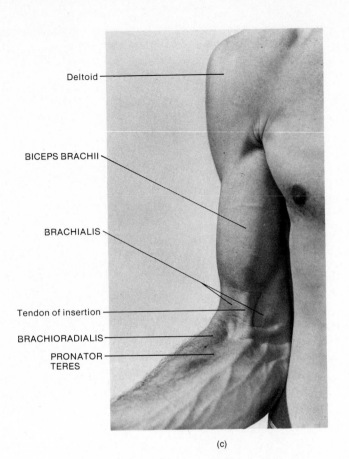

Deltoid

BICEPS BRACHII

BRACHIALIS

Tendon of insertion

BRACHIORADIALIS

PRONATOR
TERES

(c)

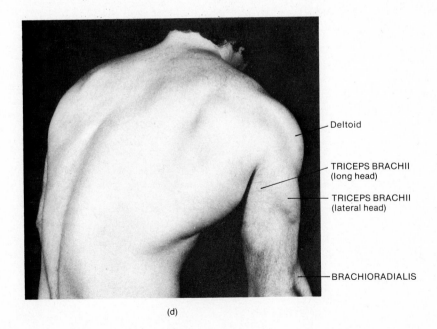

Deltoid

TRICEPS BRACHII
(long head)

TRICEPS BRACHII
(lateral head)

BRACHIORADIALIS

(d)

Exhibit 11—12 MUSCLES THAT MOVE THE WRIST AND FINGERS
(see Figure 11—13)

Muscle	Origin	Insertion	Action	Innervation
Flexor carpi radialis (*flexor* = decreases angle; *carpus* = wrist)	Medial epicondyle of humerus	Second and third metacarpals	Flexes and abducts wrist.	Median nerve
Flexor carpi ulnaris (*ulnaris* = ulna)	Medial epicondyle of humerus and upper dorsal border of ulna	Pisiform, hamate, and fifth metacarpal	Flexes and adducts wrist.	Ulnar nerve
Extensor carpi radialis longus (*extensor* = increases angle at a joint; *longus* = long)	Lateral epicondyle of humerus	Second metacarpal	Extends and abducts wrist.	Radial nerve
Extensor carpi ulnaris	Lateral epicondyle of humerus and dorsal border of ulna	Fifth metacarpal	Extends and adducts wrist.	Deep radial nerve
Flexor digitorum profundus (*digit* = finger or toe; *profundus* = deep)	Anterior medial surface of body of ulna	Bases of distal phalanges	Flexes distal phalanges of each finger.	Median and ulnar nerves
Flexor digitorum superficialis (*superficialis* = superficial)	Medial epicondyle of humerus, coronoid process of ulna, oblique line of radius	Middle phalanges	Flexes middle phalanges of each finger.	Median nerve
Extensor digitorum	Lateral epicondyle of humerus	Middle and distal phalanges of each finger	Extends phalanges.	Deep radial nerve
Extensor indicis (*indicis* = index)	Dorsal surface of ulna	Tendon of extensor digitorum of index finger	Extends index finger.	Deep radial nerve

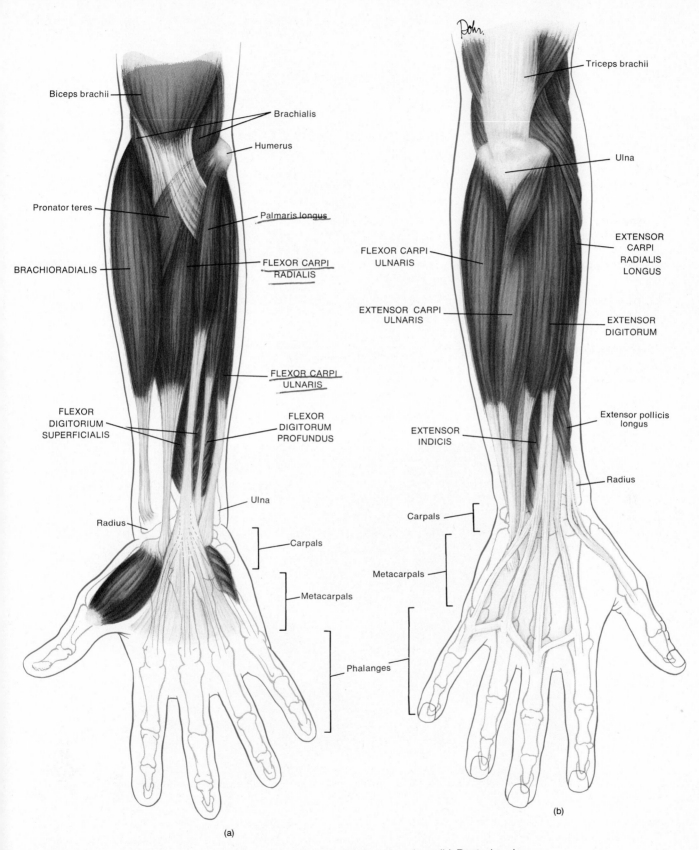

Figure 11–13 Muscles that move the wrist and fingers. (a) Anterior view. (b) Posterior view.

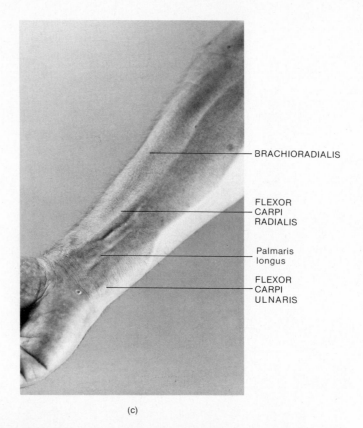

BRACHIORADIALIS

FLEXOR CARPI RADIALIS

Palmaris longus

FLEXOR CARPI ULNARIS

(c)

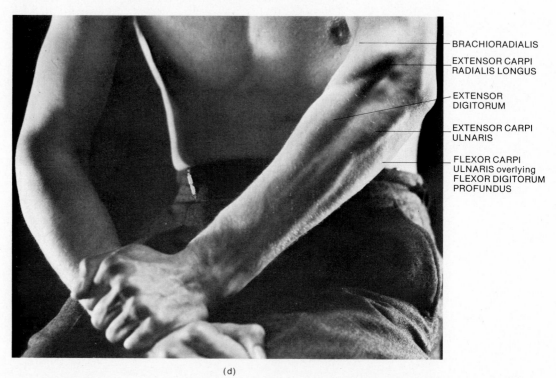

BRACHIORADIALIS

EXTENSOR CARPI RADIALIS LONGUS

EXTENSOR DIGITORUM

EXTENSOR CARPI ULNARIS

FLEXOR CARPI ULNARIS overlying FLEXOR DIGITORUM PROFUNDUS

(d)

Figure 11–13 (cont.) Muscles that move the wrist and fingers. (c) Surface anatomy photograph of the anterior forearm. (Courtesy of Vincent P. Destro, Mayo Foundation.) (d) Surface anatomy photograph of the posterolateral forearm. (Courtesy of R. D. Lockhart, *Living Anatomy,* Faber and Faber, 1974.)

Exhibit 11–13   MUSCLES THAT MOVE THE VERTEBRAL COLUMN
(see Figure 11–14)

Muscle	Origin	Insertion	Action	Innervation
Rectus abdominis (see Figure 11–8)	Body of pubis of coxal bone	Cartilages of fifth through seventh ribs	Flexes vertebral column at lumbar spine and compresses abdomen.	7–12 intercostal nerves
Quadratus lumborum (*quad* = four; *lumb* = lumbar region)	Iliac crest	Twelfth rib and upper four lumbar vertebrae	Flexes vertebral column laterally.	T12–L1
Sacrospinalis (erector spinae)				

This posterior muscle consists of three groupings: iliocostalis, longissimus, and spinalis. These groups, in turn, consist of a series of overlapping muscles. The iliocostalis group is laterally placed, the longissimus group is intermediate in placement, and the spinalis is medially placed.

Muscle	Origin	Insertion	Action	Innervation
Lateral				
Iliocostalis lumborum (*ilium* = flank; *lumbus* = loin)	Iliac crest	Lower six ribs	Extends lumbar region of vertebral column.	Dorsal rami of lumbar nerves
Iliocostalis thoracis (*thorax* = chest)	Lower six ribs	Upper six ribs	Maintains erect position of spine.	Dorsal rami of thoracic nerves
Iliocostalis cervicis (*cervix* = neck)	First six ribs	Transverse processes of fourth to sixth cervical vertebrae	Extends cervical region of vertebral column.	Dorsal rami of cervical nerves
Intermediate				
Longissimus thoracis	Transverse processes of lumbar vertebrae	Transverse processes of all thoracic and upper lumbar vertebrae and ninth and tenth ribs	Extends thoracic region of vertebral column.	Dorsal rami of spinal nerves
Longissimus cervicis	Transverse processes of fourth and fifth thoracic vertebrae	Transverse processes of second to sixth cervical vertebrae	Extends cervical region of vertebral column.	Dorsal rami of spinal nerves
Longissimus capitis	Transverse processes of upper four thoracic vertebrae	Mastoid process of temporal bone	Extends head and rotates it to opposite side.	Dorsal rami of middle and lower cervical nerves
Medial				
Spinalis thoracis	Spines of upper lumbar and lower thoracic vertebrae	Spines of upper thoracic vertebrae	Extends vertebral column.	Dorsal rami of spinal nerves

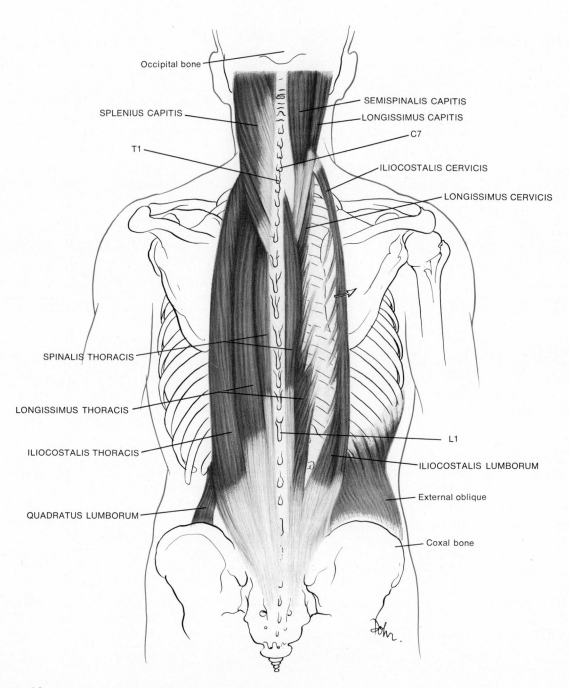

Occipital bone

SPLENIUS CAPITIS

SEMISPINALIS CAPITIS

LONGISSIMUS CAPITIS

C7

T1

ILIOCOSTALIS CERVICIS

LONGISSIMUS CERVICIS

SPINALIS THORACIS

LONGISSIMUS THORACIS

ILIOCOSTALIS THORACIS

L1

ILIOCOSTALIS LUMBORUM

External oblique

QUADRATUS LUMBORUM

Coxal bone

Figure 11–14 Muscles that move the vertebral column.

Exhibit 11—14 MUSCLES THAT MOVE THE THIGH (see Figure 11—15)

Muscle	Origin	Insertion	Action	Innervation
Psoas major (*psoa* = muscle of loin)	Transverse processes and bodies of lumbar vertebrae	Lesser trochanter of femur	Flexes and rotates thigh laterally; flexes vertebral column.	L2—L3
Iliacus (*iliac* = ilium) (Together the psoas major and iliacus are sometimes termed the iliopsoas muscle.)	Iliac fossa	Tendon of psoas major	Flexes and rotates thigh laterally; slight flexion of vertebral column.	Femoral nerve
Gluteus maximus (*gloutos* = buttock; (*maximus* = largest)	Iliac crest, sacrum, coccyx, and aponeurosis of sacrospinalis	Iliotibial tract of fascia lata and gluteal tuberosity of femur	Extends and rotates thigh laterally.	Inferior gluteal nerve
Gluteus medius (*media* = middle)	Ilium	Greater trochanter of femur	Abducts and rotates thigh medially.	Superior gluteal nerve
Gluteus minimus (*minimus* = small)	Ilium	Greater trochanter of femur	Abducts and rotates thigh laterally.	Superior gluteal nerve
Tensor fasciae latae (*tensor* = makes tense; *fascia* = band; *latus* = wide)	Iliac crest	Tibia by way of the iliotibial tract	Flexes and abducts thigh.	Superior gluteal nerve
Adductor longus (*adductor* = moves part closer to midline)	Pubic crest and symphysis pubis	Linea aspera of femur	Adducts, rotates, and flexes thigh.	Obturator nerve
Adductor brevis (*brevis* = short)	Inferior ramus of pubis	Linea aspera of femur	Adducts, rotates, and flexes thigh.	Obturator nerve
Adductor magnus (*magnus* = large)	Inferior ramus of pubis, ischium to ischial tuberosity	Linea aspera of femur	Adducts, flexes, and extends thigh (anterior part flexes, posterior part extends).	Obturator nerve
Piriformis (*pirum* = pear; *forma* = shape)	Sacrum	Greater trochanter of femur	Rotates thigh laterally and abducts it.	S2 or S1—S2
Obturator internus (*obturator* = closed because it arises over obturator foramen, which is closed by heavy membrane; *internus* = inside)	Margin of obturator foramen, pubis, and ischium	Greater trochanter of femur	Rotates thigh laterally and abducts it.	Obturator nerve
Pectineus (*pecten* = comb-shaped)	Fascia of pubis	Pectineal line of femur	Flexes, adducts, and rotates thigh laterally.	Femoral nerve

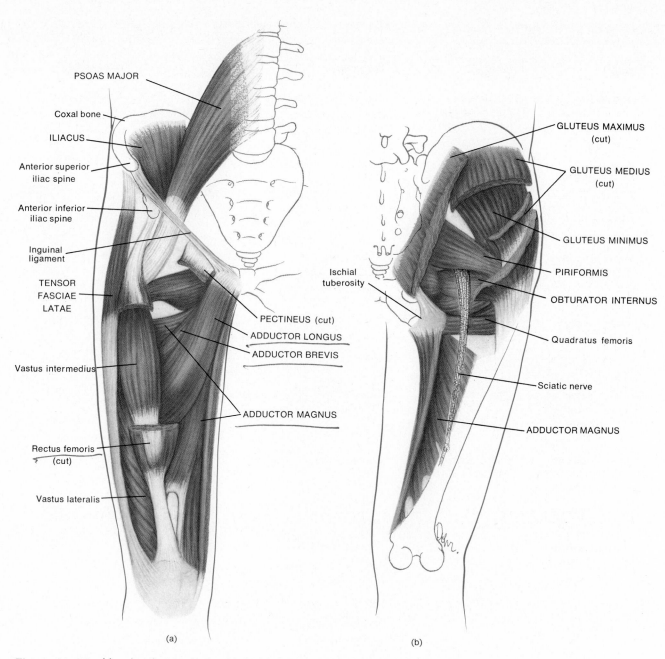

PSOAS MAJOR

Coxal bone

ILIACUS

Anterior superior
iliac spine

Anterior inferior
iliac spine

Inguinal
ligament

TENSOR
FASCIAE
LATAE

Vastus intermedius

Rectus femoris
(cut)

Vastus lateralis

PECTINEUS (cut)

ADDUCTOR LONGUS

ADDUCTOR BREVIS

ADDUCTOR MAGNUS

(a)

GLUTEUS MAXIMUS
(cut)

GLUTEUS MEDIUS
(cut)

GLUTEUS MINIMUS

PIRIFORMIS

OBTURATOR INTERNUS

Quadratus femoris

Sciatic nerve

ADDUCTOR MAGNUS

Ischial
tuberosity

(b)

Figure 11–15 Muscles that move the thigh. (a) Anterior view. (b) Posterior view.

Exhibit 11—15   MUSCLES THAT ACT ON THE LEG (see Figure 11—16)

Muscle	Origin	Insertion	Action	Innervation
Quadriceps femoris	A composite muscle that includes four distinct parts, usually described as four separate muscles. The common tendon that includes the patella and attaches to the tibial tuberosity is known as the patellar ligament.			
Rectus femoris (*rectus* = fibers parallel to midline; *femoris* = femur)	Anterior inferior iliac spine	Upper border of patella	All four heads extend legs; rectus portion alone also flexes thigh.	Femoral nerve
Vastus lateralis (*vastus* = large; *lateralis* = lateral)	Greater trochanter and linea aspera of femur	Upper border and sides of patella; tibial tuberosity through patellar ligament (tendon of quadriceps)		Femoral nerve
Vastus medialis (*medialis* = medial)	Linea aspera of femur			Femoral nerve
Vastus intermedius (*intermedius* = middle)	Anterior and lateral surfaces of body of femur			Femoral nerve
Hamstrings	A collective designation for three separate muscles.			
Biceps femoris	Long head arises from ischial tuberosity; short head arises from linea aspera of femur	Head of fibula and lateral condyle of tibia	Flexes leg and extends thigh.	Tibial nerve from sciatic nerve
Semitendinosus (*semi* = half; *tendo* = tendon)	Ischial tuberosity	Proximal part of medial surface of body of tibia	Flexes leg and extends thigh.	Tibial nerve from sciatic nerve
Semimembranosus (*membran* = membrane)	Ischial tuberosity	Medial condyle of tibia	Flexes leg and extends thigh.	Tibial nerve from sciatic nerve
Gracilis (*gracilis* = slender)	Symphysis pubis and pubic arch	Medial surface of body of tibia	Flexes leg and adducts thigh.	Obturator nerve
Sartorius (*sartor* = tailor; refers to cross-legged position of tailors)	Anterior superior spine of ilium	Medial surface of body of tibia	Flexes leg; flexes thigh and rotates it laterally, thus crossing leg.	Femoral nerve

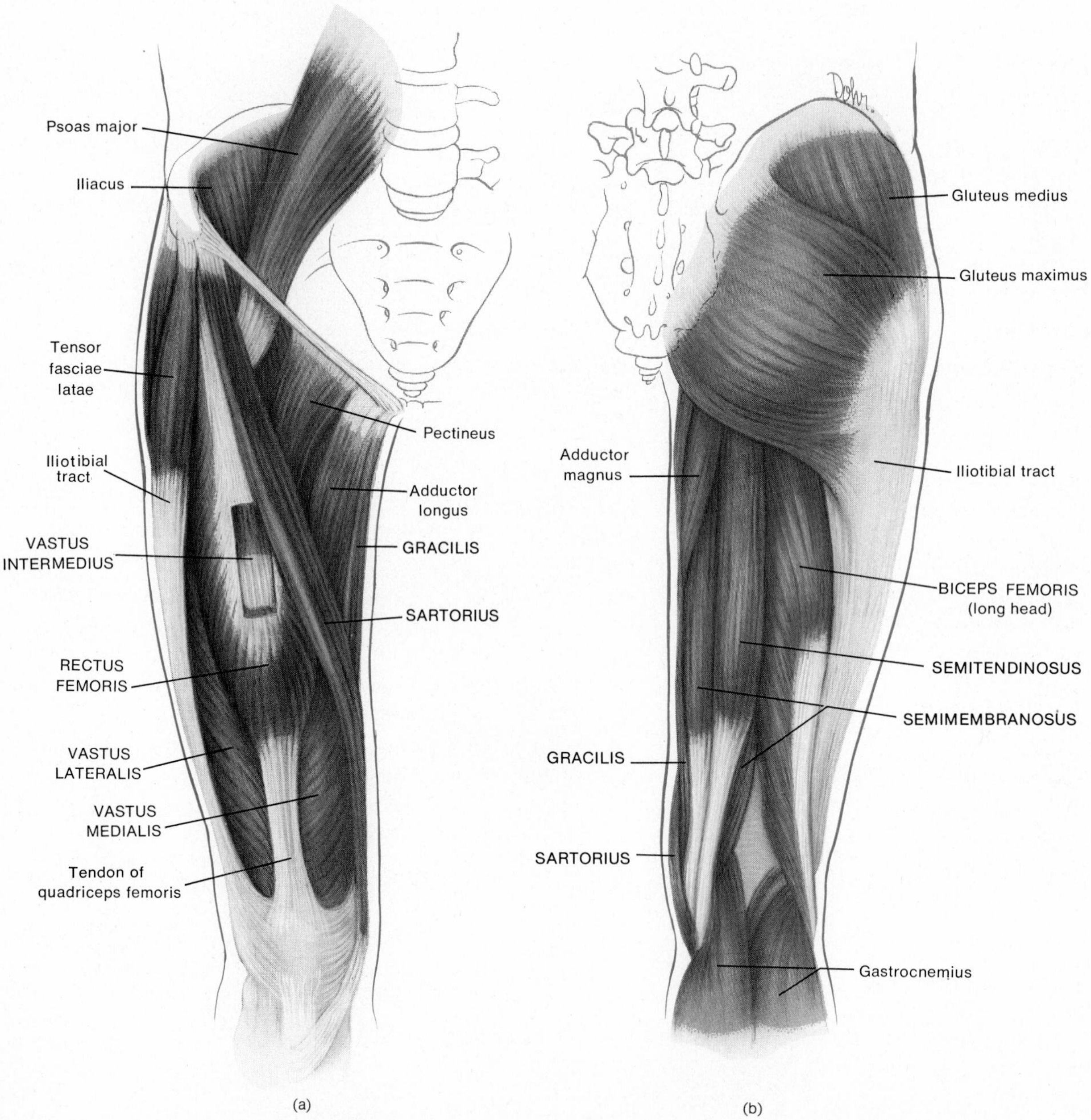

Figure 11–16   Muscles that act on the leg. (a) Anterior view. (b) Posterior view. (c) Surface anatomy photograph of the anterior thigh. (Courtesy of R. D. Lockhart, *Living Anatomy,* Faber and Faber, 1974.) (d) Surface anatomy photograph of the posterior thigh and leg. (Courtesy of Donald Castellaro and Richard Sollazzo.)

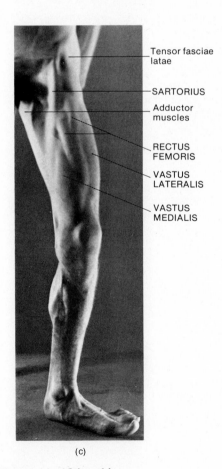

Tensor fasciae latae

SARTORIUS

Adductor muscles

RECTUS FEMORIS

VASTUS LATERALIS

VASTUS MEDIALIS

(c)

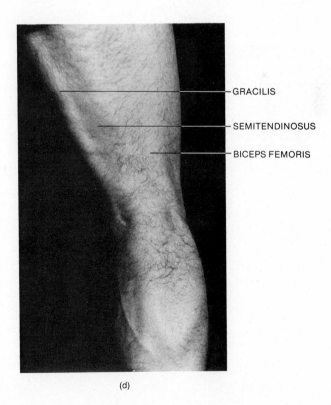

GRACILIS

SEMITENDINOSUS

BICEPS FEMORIS

(d)

Figure 11–16 (cont.)

Exhibit 11—16 MUSCLES THAT MOVE THE FOOT AND TOES
(see Figure 11—17)

Muscle	Origin	Insertion	Action	Innervation
Gastrocnemius (*gaster* = belly; *kneme* = leg)	Lateral and medial condyles of femur and capsule of knee	Calcaneus by way of calcaneal ("Achilles") tendon	Plantar flexes foot.	Tibial nerve
Soleus (*soleus* = sole of foot)	Head of fibula and medial border of tibia	Calcaneus by way of calcaneal ("Achilles") tendon	Plantar flexes foot.	Tibial nerve
Peroneus longus (*perone* = fibula)	Head and body of fibula and lateral condyle of tibia	First metatarsal and first cuneiform bone	Plantar flexes and everts foot.	Superficial peroneal nerve
Peroneus brevis	Body of fibula	Fifth metatarsal	Plantar flexes and everts foot.	Superficial peroneal nerve
Peroneus tertius (*tertius* = third)	Distal third of fibula	Fifth metatarsal	Dorsiflexes and everts foot.	Deep peroneal nerve
Tibialis anterior (*tibialis* = tibia)	Lateral condyle and body of tibia	First metatarsal and first cuneiform	Dorsiflexes and inverts foot.	Deep peroneal nerve
Tibialis posterior (*posterior* = back)	Interosseus membrane between tibia and fibula	Second, third, and fourth metatarsals; navicular; third cuneiform, cuboid	Plantar flexes and inverts foot.	Tibial nerve
Flexor digitorum longus (*digitorum* = digit, finger, or toe)	Tibia	Distal phalanges of four outer toes	Flexes toes and plantar flexes and inverts foot.	Tibial nerve
Extensor digitorum longus	Lateral condyle of tibia and anterior surface of fibula	Middle and distal phalanges of four outer toes	Extends toes and dorsiflexes and everts foot.	Deep peroneal nerve

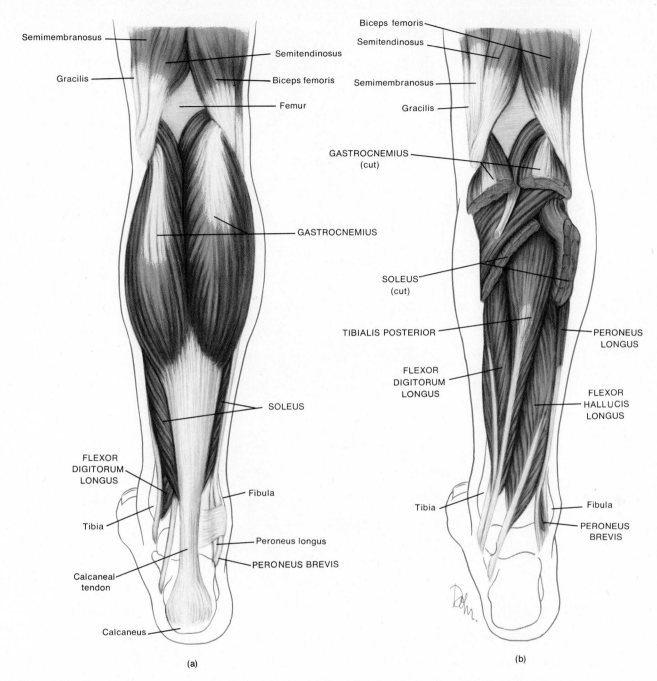

Figure 11–17 Muscles that move the foot and toes. (a) Superficial posterior view. (b) Deep posterior view.

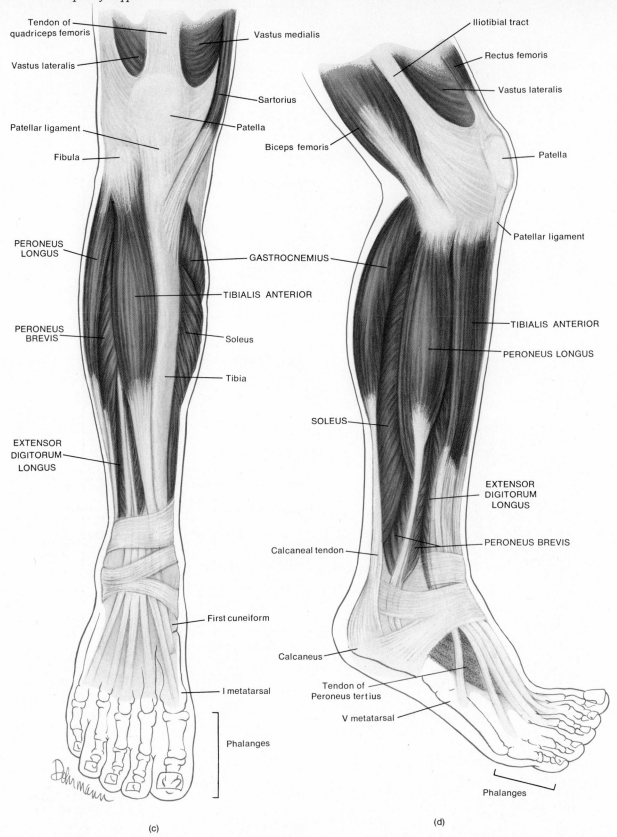

Tendon of quadriceps femoris

Vastus lateralis

Vastus medialis

Patellar ligament

Fibula

Sartorius

Patella

PERONEUS LONGUS

PERONEUS BREVIS

EXTENSOR DIGITORUM LONGUS

GASTROCNEMIUS

TIBIALIS ANTERIOR

Soleus

Tibia

First cuneiform

I metatarsal

Phalanges

(c)

Iliotibial tract

Rectus femoris

Vastus lateralis

Biceps femoris

Patella

Patellar ligament

GASTROCNEMIUS

TIBIALIS ANTERIOR

PERONEUS LONGUS

SOLEUS

EXTENSOR DIGITORUM LONGUS

PERONEUS BREVIS

Calcaneal tendon

Calcaneus

Tendon of Peroneus tertius

V metatarsal

Phalanges

(d)

Figure 11–17 (cont.) Muscles that move the foot and toes. (c) Superficial anterior view. (d) Superficial lateral view. (e) Surface anatomy photograph of the posterior leg. (Courtesy of Donald Castellaro and Richard Sollazzo.) (f) Surface anatomy photograph of the lateral leg. (Courtesy of R. D. Lockhart, *Living Anatomy,* Faber and Faber, 1974.)

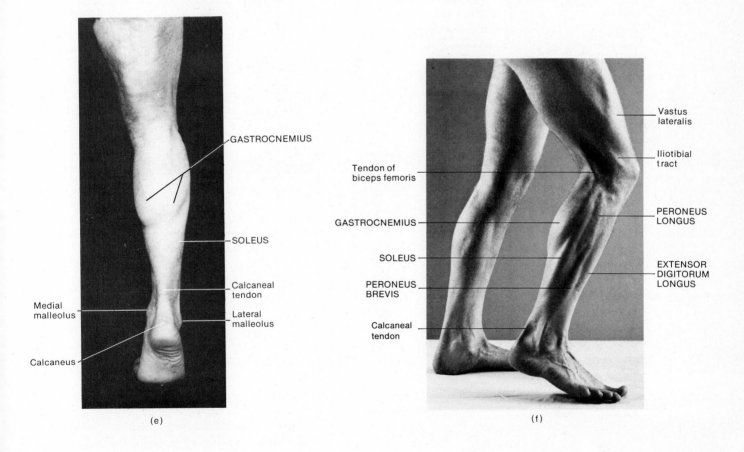

(e)

GASTROCNEMIUS

SOLEUS

Calcaneal
tendon

Medial
malleolus

Lateral
malleolus

Calcaneus

(f)

Tendon of
biceps femoris

GASTROCNEMIUS

SOLEUS

PERONEUS
BREVIS

Calcaneal
tendon

Vastus
lateralis

Iliotibial
tract

PERONEUS
LONGUS

EXTENSOR
DIGITORUM
LONGUS

INTRAMUSCULAR INJECTIONS

All methods of administering drugs with a needle are called **parenteral**. Among the parenteral routes of administration are:

1 Intradermal or **intracutaneous** The needle penetrates the epidermis and is inserted into the dermis.

2 Subcutaneous or **hypodermic** The needle is inserted into the subcutaneous layer.

3 Intramuscular The needle is inserted into muscle.

4 Intravenous The needle is inserted directly into a vein.

5 Intraspinal The needle is inserted into the vertebral canal.

At this point, we will discuss intramuscular injections only. Other parenteral routes are considered later.

An intramuscular injection penetrates the skin and subcutaneous tissue to enter the muscle itself. Intramuscular injections are preferred when prompt absorption is desired, when larger doses than can be given cutaneously are indicated, or when the drug is too irritating to give subcutaneously. The common sites for intramuscular injections include the buttock, lateral side of the thigh, and the deltoid region of the arm. Muscles in these areas, especially the gluteal muscles in the buttock, are fairly thick. Because of the large number of muscle fibers and extensive fascia, the drug has a large surface area for absorption. Absorption is further promoted by the extensive blood supply to muscles. Ideally, intramuscular injections should be given deep within the muscle and away from major nerves and blood vessels.

For many intramuscular injections, the preferred site is the buttock. The buttock should be divided into four quadrants and the upper outer quadrant used as the injection site. The iliac crest serves as a landmark for this quadrant. The spot for injection should be about 5 to 7½ cm (2 to 3 inches) below the iliac crest. The upper outer quadrant is chosen because, in this area, the muscle is quite thick with few nerves. Thus there is less chance of injuring the sciatic nerve. Injury to the nerve can cause paralysis of the lower extremity. The probability of injecting the drug into a blood vessel is also remote in this area. After the needle is inserted, the plunger should be pulled up for a few seconds. If the syringe fills with blood, the needle is in a blood vessel and a different injection site on the opposite buttock should be chosen.

Injections given in the lateral side of the thigh are inserted into the midportion of the vastus lateralis muscle. This site is determined by using the knee and greater trochanter of the femur as landmarks. The midportion of the muscle is located by measuring a handbreadth above the knee and a handbreadth below the greater trochanter.

The deltoid injection is given in the midportion of the muscle about two to three fingerbreadths below the acromion of the scapula and lateral to the axilla.

STUDY OUTLINE

HOW SKELETAL MUSCLES PRODUCE MOVEMENT

1 Skeletal muscles produce movement by pulling on bones.

2 The stationary attachment is the origin. The movable attachment is the insertion.

3 Bones serve as levers and joints as fulcrums.

4 Levers are acted on by two different forces: resistance and effort. There are first-class, second-class, and third-class levers.

5 The agonist or prime mover produces the desired action. The antagonist produces an opposite action. The synergist assists the agonist by reducing unnecessary movement.

NAMING SKELETAL MUSCLES

1 Skeletal muscles are named on the basis of distinctive criteria: direction of fibers, location, size, number of heads, shape, origin-insertion, and action.

PRINCIPAL SKELETAL MUSCLES

1 Surface anatomy is the study of the form and markings of the body surface.

2 A knowledge of surface anatomy will help you identify certain superficial structures by visual inspection or palpation through the skin.

INTRAMUSCULAR INJECTIONS

1 Parenteral routes may be intradermal, subcutaneous, intramuscular, intravenous, or intraspinal.

2 Advantages of intramuscular injections are prompt absorption, use of larger doses than can be given cutaneously, and minimal irritation.

3 Common sites for intramuscular injections are the buttock, lateral side of the thigh, and deltoid region of the shoulder.

1 What is meant by the muscular system? Explain fully.

2 Using the terms *origin, insertion,* and *belly,* explain how skeletal muscles produce body movement by pulling on bones.

3 What is a lever? Fulcrum? Apply these terms to the body, and indicate the nature of the forces that act on levers.

4 Define the role of the agonist (prime mover), antagonist, and synergist in producing body movement.

5 Select at random the names of some muscles presented in Exhibits 11-2 through 11-16, and see if you can determine the criteria employed for naming them.

6 Discuss the muscles and their actions involved in facial expression.

7 What muscles would you use to: (*a*) frown; (*b*) pout; (*c*) show surprise; (*d*) show your upper teeth; (*e*) pucker your lips; (*f*) squint; (*g*) blow up a balloon, (*h*) smile?

8 What are the principal muscles that move the mandible? Give the function of each.

9 What would happen if you lost tone in the masseter and temporal muscles?

10 What muscles move the eyeballs? In which direction does each muscle move the eyeball?

11 Describe the action of each muscle acting on the tongue.

12 What tongue, facial, and mandibular muscles would you use when chewing a piece of gum?

13 What muscles are responsible for moving the head? How do they do it?

14 Which muscles would you use to signify "yes" and "no" by moving your head?

15 What muscles compress the anterior abdominal wall?

16 What are the principal muscles involved in breathing? What are their actions?

17 In what directions is the shoulder girdle drawn? What muscles accomplish these movements?

18 What muscles are used to (*a*) raise your shoulders, (*b*) lower your shoulders, (*c*) join your hands behind your back, (*d*) join your hands in front of your chest?

19 What movements are possible at the shoulder joint? What muscles accomplish these movements?

20 What muscles move the forearm? In which directions do these movements occur?

21 What muscles move the forearm and what actions are used when drinking a glass of water?

22 Discuss the various movements possible at the wrist and fingers. What muscles accomplish these movements?

23 List the muscles and actions of the wrist and fingers that are used when writing.

24 Discuss the various muscles and movements of the vertebral column.

25 Perform an exercise involving each muscle listed in Exhibit 11-13.

26 What muscles accomplish movements of the femur? What actions are produced by these muscles?

27 Which muscles are used to flex and laterally rotate, abduct and medially rotate, and adduct the thigh?

28 What muscles act at the knee joint? What movements do these muscles perform?

29 Which muscles flex and extend the leg?

30 Discuss the muscles that plantar flex, evert, pronate, dorsiflex, and supinate the foot.

31 In which directions are the toes moved? What muscles bring about these movements?

32 What are the advantages of intramuscular injections?

Unit III
Control Systems of the Human Body

This unit will show you the significance of the nerve impulse in making rapid adjustments for maintaining homeostasis. You will learn how the nervous system detects changes in the environment, decides on a course of action, and responds to the change. We will also investigate the role of hormones in maintaining long-term homeostasis.

Chapter 12
Nervous Tissue

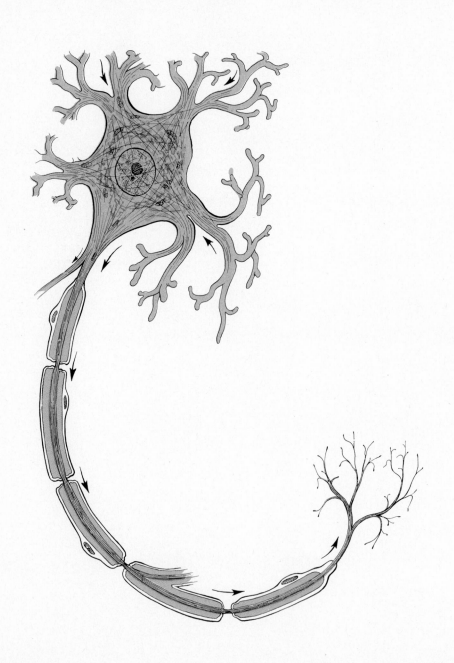

The **nervous system** is the body's control center and communications network. In human beings it performs two broad functions. First, it stimulates movements. Second, it shares the maintenance of homeostasis with the endocrine system. Human life cannot exist without a functioning nervous system. Skeletal and most smooth muscle cells cannot contract until stimulated by a nerve impulse. If the intercostal muscles and diaphragm do not contract, we cannot breathe. If the digestive glands are not stimulated to release their secretions, we cannot digest our food. It is obvious, then, that our cells cannot receive nutrients unless the digestive system is connected to a functioning nervous system. But suppose our muscles and glands could stimulate themselves. Even then, we could not live very long without our nervous system. The nervous system *senses* changes within the body and in the outside environment. It then interprets these changes and may initiate action to maintain homeostasis.

ORGANIZATION

The nervous system may be divided into two principal portions: the central nervous system and the peripheral nervous system (Figure 12-1).

The **central nervous system (CNS)** is the control center for the entire system and consists of the brain and spinal cord. All body sensations must be relayed from receptors to the central nervous system if they are to be interpreted and acted on. All the nerve impulses that stimulate muscles to contract and glands to secrete must also pass from the central system.

The various nerve processes that connect the brain and spinal cord with receptors, muscles, and glands comprise the **peripheral nervous system (PNS)**. The peripheral nervous system may be classified into an afferent system and an efferent system. The *afferent system* consists of nerve cells that convey information from receptors in the periphery of the body to the central nervous system. These nerve cells, called afferent (sensory) neurons, are the first cells to pick up incoming information. The *efferent system* consists of nerve cells that convey information from the central nervous system to muscles and glands. These nerve cells are called efferent (motor) neurons. The efferent system is subdivided into a somatic nervous system and an autonomic nervous system.

The *somatic nervous system* or *SNS* (*soma* = body) consists of efferent neurons that conduct impulses from the central nervous system to skeletal muscle tissue. Since the somatic nervous system produces movement only in skeletal muscle tissue, it is under conscious control and therefore voluntary.

The *autonomic nervous system* or *ANS,* by contrast, contains efferent neurons that convey impulses from the central nervous system to smooth muscle tissue, cardiac muscle tissue, and glands. The autonomic system produces responses only in involuntary muscles and glands. Thus the autonomic system is usually considered to be involuntary. With few exceptions, the viscera receive nerve fibers from the two divisions of the autonomic nervous system: the sympathetic division and the parasympathetic division.

HISTOLOGY

Despite the organizational complexity of the nervous system, it consists of only two principal kinds of cells. The first of these, the neurons, make up the nervous tissue that forms the structural and functional portion of the system. Neurons are highly specialized for impulse conduction and for all special functions attributed to the nervous system: thinking, controlling muscle activity, regulating glands. The second type of cell, the neuroglia, serves as a special supporting and protective component of the nervous system.

Neuroglia

The cells of the nervous system that perform the functions of support and protection are called **neuroglia** or **glial cells** (*neuro* = nerve; *glia* = glue). Many of the glial cells form a supporting network by twining around the nerve cells in the brain and spinal cord. Other glial cells bind nervous tissue to supporting structures and attach the neurons to their blood vessels. A few glial cells also serve specialized functions. For example, many nerve fibers are coated with a thick, fatty sheath produced by a particular type of neuroglia. Certain small glial cells are phagocytotic. They protect the central nervous system from disease by engulfing invading microbes and clearing away debris. These various functions performed by the neuroglia are divided among several different kinds of cells. Neuroglia are of clinical interest because they are a common source of tumors of the nervous system. Exhibit 12-1 lists the neuroglial cells and summarizes their functions.

Neurons

Nerve cells, called **neurons,** are responsible for conducting impulses from one part of the body to another.

STRUCTURE

A neuron consists of three distinct portions: (1) the cell body, (2) dendrites, and (3) an axon (Figure 12-2a). The

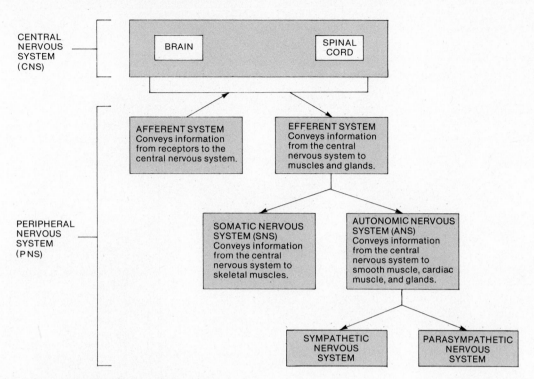

Figure 12-1 Organization of the nervous system.

cell body or **perikaryon** contains a well-defined nucleus and nucleolus surrounded by a granular cytoplasm. Within the cytoplasm are typical organelles such as mitochondria and a Golgi apparatus. Also located in the cytoplasm are structures characteristic of neurons: Nissl bodies and neurofibrils. *Nissl bodies* are orderly arrangements of granular (rough) ER and free ribosomes whose function is protein synthesis. Newly synthesized proteins pass from the perikaryon into the neuronal processes, mainly the axon, at the rate of about 1 mm (0.0394 inch)/day. These proteins replace those lost during metabolism and are used for growth of neurons and regeneration of peripheral nerve fibers. *Neurofibrils* are long, thin fibrils composed of microtubules. They may assume a function in support.

The cytoplasmic processes of neurons depend on the direction in which they conduct impulses. The processes are of two kinds: dendrites and axons. **Dendrites** are highly branched extensions of the cytoplasm of the cell body. A neuron usually has several main dendrites. Dendrites typically contain Nissl bodies and mitochondria. The function of dendrites is to conduct an impulse toward the cell body. The second type of cytoplasmic process, called an **axon** or **axis cylinder,** is a single, highly specialized, and long process that conducts impulses away from the cell body to another neuron or tissue.

An axon usually originates from the cell body as a small conical elevation called the *axon hillock.* An axon contains mitochondria and neurofibrils but no Nissl bodies. Its cytoplasm, called *axoplasm,* is surrounded by a plasma membrane known as the *axolemma.* Axons vary in length from a few millimeters (1 mm = 0.0394 inch) in the brain to a meter (3.28 ft) or more between the spinal cord and toes. Along the course of an axon, there may be side branches called *axon collaterals.* The axon and its collaterals terminate by branching into many fine filaments called *telodendria.*

The term **nerve fiber** is applied to an axon and its sheaths. Figure 12-2b shows a cross section of a nerve fiber of the peripheral nervous system. Many axons, especially large, peripheral axons, are surrounded by a white, phospholipid, segmented covering called the *myelin sheath.* Axons containing such a covering are *myelinated,* while those without it are *unmyelinated.* Myelin is responsible for the color of the white matter in the nerves, brain, and spinal cord. The myelin sheath is produced by flattened cells, called *Schwann cells,* located along the axon. They are neuroglial cells of the peripheral nervous system. In this process, a developing Schwann cell encircles the axon until its ends meet and overlap (Figure 12-3). The cell then winds around the axon several times and, in doing so, the cytoplasm and nucleus are pushed to the outside layer. The inner

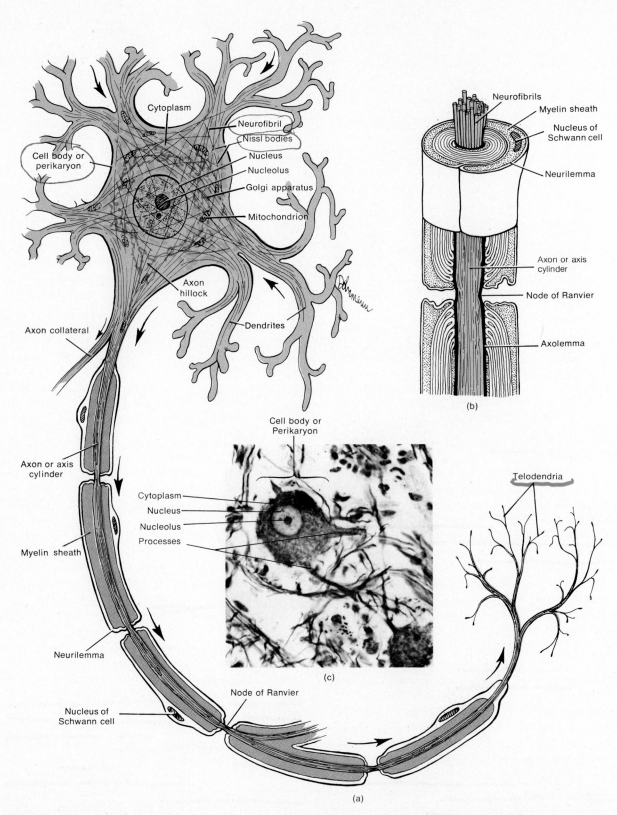

Figure 12-2 Structure of a neuron. (a) Shown in an entire multipolar neuron. The arrows indicate the direction in which a nerve impulse passes. (b) Cross section through a myelinated fiber. (c) Photomicrograph of a multipolar neuron from a sympathetic ganglion at a magnification of 640×. (Courtesy of Edward J. Reith, from *Atlas of Descriptive Histology,* by Edward J. Reith and Michael H. Ross, Harper & Row, Publishers, Inc., New York, 1970.)

Exhibit 12–1 NEUROGLIA OF CENTRAL NERVOUS SYSTEM

Type	Description	Microscopic Appearance	Function
Astrocytes (*astro* = star; *cyte* = cell)	Star-shaped cells with numerous processes.		Twine around nerve cells to form supporting network in brain and spinal cord; attach neurons to their blood vessels.
Oligodendrocytes (*oligo* = few; *dendro* = tree)	Resemble astrocytes in some ways, but processes are fewer and shorter.		Give support by forming semirigid connective tissue rows between neurons in brain and spinal cord; produce a thick, fatty myelin sheath on neurons of central nervous system.
Microglia (*micro* = small)	Small cells with few processes; normally stationary; if nervous tissue is damaged, they may migrate to injured area.		Engulf and destroy microbes and cellular debris.

portion, consisting of several layers of Schwann cell membrane, is the myelin sheath. The function of the myelin sheath is to insulate and maintain the axon. The *neurilemma* or *sheath of Schwann* is the peripheral nucleated cytoplasmic layer of the Schwann cell that encloses the myelin sheath (*lemma* = sheath). The neurilemma is found only around fibers of the peripheral nervous system. Its function is to assist in the regeneration of injured axons. Between the segments of the myelin sheath are unmyelinated gaps called *nodes of Ranvier*. Unmyelinated fibers are also enclosed by Schwann cells, but without multiple wrappings.

Nerve fibers of the central nervous system may be myelinated or unmyelinated. Myelination of central nervous system axons is accomplished by oligodendrocytes in somewhat the same manner that Schwann cells myelinate peripheral nervous system axons. Myelinated axons of the central nervous system also contain nodes of Ranvier, but they are not so numerous.

CLASSIFICATION

The different neurons in the body may be classified by structure and function.

The structural classification is based on the number of processes extending from the cell body. **Multipolar neurons** have several dendrites and one axon. Most neurons in the brain and spinal cord are of this type. **Bipolar neurons** have one dendrite and one axon and are found in the retina of the eye, the inner ear, and the olfactory area. The third structural type of neuron is the **unipolar neuron**. It has only one process extending from the cell body. The single process divides into a central branch, which functions as an axon, and a peripheral branch, which functions as a dendrite. Unipolar neurons originate in the embryo as bipolar neurons. During development, the axon and dendrite fuse into a single process.

The functional classification of neurons is based on the direction in which they transmit impulses. **Sensory neurons,** called **afferent neurons,** transmit impulses from receptors in the skin and sense organs to the brain and spinal cord. Sensory neurons also transmit impulses from receptors in the viscera. Sensory neurons are usually unipolar. **Motor neurons,** called **efferent neurons,** convey impulses from the brain and spinal cord to effectors, which may be either muscles or glands. Other neurons, called **association (connecting** or **internun-**

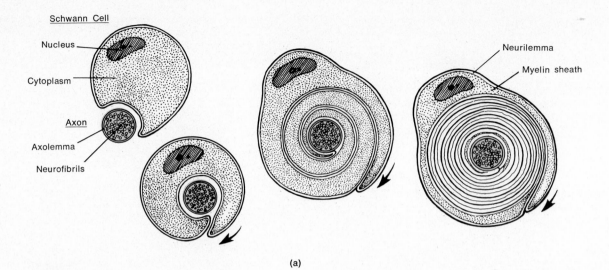

Schwann Cell

Nucleus

Cytoplasm

Axon

Axolemma

Neurofibrils

Neurilemma

Myelin sheath

(a)

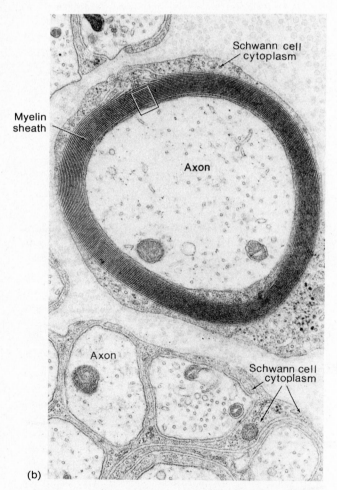

Schwann cell cytoplasm

Myelin sheath

Axon

Axon

Schwann cell cytoplasm

(b)

Figure 12-3 Myelin sheath. (a) Stages in the formation of a myelin sheath by a Schwann cell. (b) Electron micrograph of a nerve in cross section showing a myelinated nerve axon and several unmyelinated nerve axons at a magnification of 12,000×. (Courtesy of William Bloom and Don W. Fawcett, *A Textbook of Histology*, W. B. Saunders, Philadelphia, 1968.)

cial) neurons, carry impulses from sensory neurons to motor neurons and are located in the brain and spinal cord.

The processes of afferent and efferent neurons are arranged into bundles called *nerves*. Since nerves lie outside the central nervous system, they belong to the peripheral nervous system. The functional components of nerves are the nerve fibers, which may be grouped according to the following scheme:

1 General somatic afferent These fibers belong to the somatic portion of the peripheral nervous system. They conduct impulses from the skin, skeletal muscles, and joints to the central nervous system.

2 General somatic efferent These fibers also belong to the somatic portion of the peripheral nervous system. They conduct impulses from the central nervous system to skeletal muscles. Impulses over these fibers cause the contraction of skeletal muscles.

3 General visceral afferent These fibers convey impulses from the viscera and blood vessels to the central nervous system.

4 General visceral efferent These fibers belong to the autonomic nervous system. Also called *autonomic fibers*, they convey impulses from the central nervous system to cause contractions of smooth muscle, cardiac muscle, and glands. Impulses passing over these fibers cause contractions of smooth and cardiac muscle and secretion by glands.

STRUCTURAL VARIATION

Although all neurons conform to the general plan previously described, there are considerable differences in structure. For example, cell bodies range in size from 5 μm for the smallest cells to 135 μm for large motor neurons. The pattern of dendritic branching is also varied and distinctive for neurons in different parts of the body. Moreover, the axons of very small neurons are only a fraction of a millimeter in length and lack a myelin sheath, while some axons of large neurons are over a meter long and are usually enclosed in a myelin sheath.

A few patterns of diversity are shown in Figure 12-4. Note the structure of a typical afferent (sensory) neuron. Compare it to the typical efferent (motor) neuron. What structural differences do you observe? A few examples of association neurons are also shown: a *stellate cell, cell of Martinotti,* and a *horizontal cell of Cajal.* All are found in the cerebral cortex, the outer layer of the cerebrum. Note the *granule cell,* an association neuron in the cortex of the cerebellum.

PHYSIOLOGY

Two striking features of nervous tissue are its limited ability to regenerate and its highly developed ability to transmit electrical messages called nerve impulses.

Regeneration

Unlike the cells of epithelial tissue, neurons have but limited powers for regeneration. Around the time of birth, the cell bodies of most developing nerve cells lose their mitotic apparatus (centrioles and spindles) and thus their ability to reproduce. Thus the neuron cannot be replaced by other reproducing cells in the tissue. A neuron destroyed is permanently lost. However, myelinated axons in the peripheral nervous system can often be repaired if the cell body remains intact. Most nerves outside the brain and spinal cord have many myelinated axons. A person who injures a nerve in the arm thus has a good chance of regaining nerve function. Axons in the brain and spinal cord are myelinated by oligodendroglial cells, however. Scar tissue is formed by other neuroglial cells faster than the axon can repair itself. Hence injury to the brain or spinal cord is permanent because axonal regeneration is blocked.

Nerve Impulse

At this point, we will consider the nerve impulse – your body's quickest way of controlling and maintaining homeostasis.

MEMBRANE POTENTIALS

Studies of cell membranes, especially in nerve and muscle cells, indicate that when a cell is at rest there is a considerable difference between the ion concentration outside and inside the plasma membrane. In a resting neuron (one that is not conducting an impulse), there is a difference in electrical charges on either side of the membrane. This difference, called a *potential difference,* is partly the result of an unequal distribution of potassium (K^+) ions and sodium (Na^+) ions on either side of the membrane. In neurons, the K^+ ion concentration inside the cell is about 28 to 30 times greater than it is outside. The Na^+ ion concentration is about 14 times greater outside than inside. Another significant factor is the presence of large nondiffusible negatively charged ions trapped in the cell. Most of them are proteins. What causes the outside of the nerve cell membrane to differ from the inside?

Even when a nerve cell is not conducting an impulse, it is transporting ions across its membrane. Na^+ ions do not easily diffuse into the cell, and a few ions that do enter the membrane are actively transported out. This transportation mechanism is called the "sodium pump" (Figure 12-5a). The cell also has a "potassium pump" that actively transports K^+ ions inward so that many more K^+ ions are located inside the cell than outside. Neurons also contain a large number of negative ions on the inside that cannot diffuse outside or diffuse very poorly. Since Na^+ ions are positive and are actively transported outside the cell, a positive charge develops outside the membrane. Even though K^+ ions are positive and are actively transported to the inside of the cell, there are insufficient K^+ ions to equalize the even larger number of nondiffusible negative ions trapped in the cell. Thus the inside of the membrane has a negative charge. This difference in charge on either side of the membrane of a resting neuron is the **membrane potential** or **resting potential.** Such a membrane is said to be **polarized.**

Electrical measurements of a polarized membrane indicate a voltage of about 70 millivolts (mV). This means that the inside of the membrane is 70 mV less than the outside – that is, membrane potential is -70 mV. In subsequent discussions of membrane potentials, we will use the mV value that refers to the inside of the membrane, -70 mV.

The events associated with the generation of nerve impulses will now be examined. The ability of nerve cells to respond to stimuli and convert them into nerve impulses is called **irritability**. A stimulus is a change in the environment of sufficient strength to initiate an impulse.

IRRITABILITY

If a stimulus of adequate strength is applied to a polarized membrane, the membrane's permeability to Na^+ ions is greatly increased at the point of stimulation (Figure 12-6a). Thus Na^+ ions can now diffuse across the membrane into the cell. And, since there are more Na^+ ions entering than leaving, the electrical potential of the membrane begins to change. At first, the potential inside the membrane changes from -70 mV toward

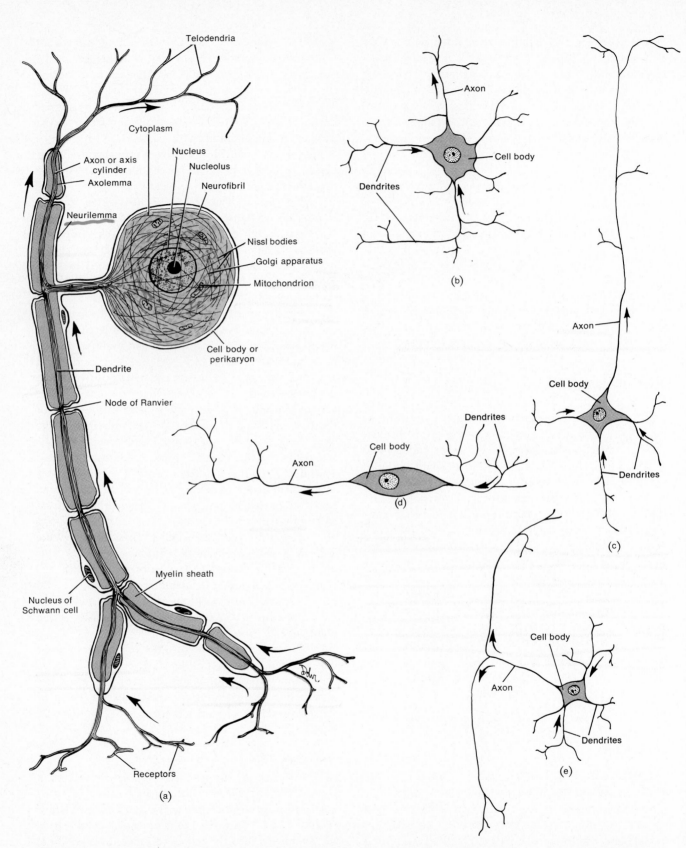

Figure 12-4   Varieties of neurons. (a) Typical afferent neuron. (b-e) Representative association neurons. (b) Stellate cell. (c) Cell of Martinotti. (d) Horizontal cell of Cajal. (e) Granule cell. Arrows indicate the direction of impulse conduction.

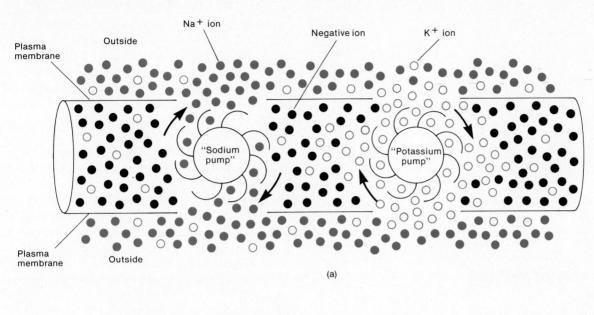

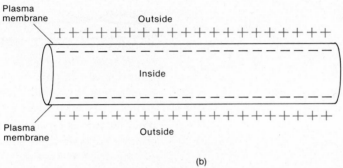

Figure 12-5 Development of the membrane or resting potential. (a) Schematic representation of the "sodium pump," "potassium pump," and distribution of negative ions. Note that the large population of Na+ ions outside the membrane results in an external positive charge. Although there are more K+ ions inside the cell membrane than outside, there are many more negative ions inside the membrane. This results in a net internal negative charge. (b) Simplified representation of a polarized membrane.

zero. At 0 mV the membrane is said to be **depolarized.** (Depolarization begins at the **threshold level**: about −60 mV.) Even after depolarization, the Na+ ions continue to rush inside and another membrane potential develops. This time the membrane potential is *reversed*— the inside of the membrane becomes positive and the outside negative. Electrical measurements indicate that the inside of the membrane is now +30 mV with respect to the outside. Thus the potential inside the membrane changes from −70 mV to zero to +30 mV.

A sensory neuron is generally stimulated at its distal end by a receptor, a structure sensitive to changes in the environment. Association and motor neurons are usually stimulated at their dendrites or cell bodies by another neuron. However, if a neuron's plasma membrane is depolarized at some point other than the usual one, the impulse will travel in both directions over the cell membrane of the entire neuron.

Once the events of depolarization have occurred, we say that an **action potential** or **nerve impulse** is initiated. It lasts about 1 millisecond. The stimulated, negatively charged point on the outside of the membrane sends out an electrical current to the positive point (still polarized) adjacent to it. This local current causes the adjacent inner part of the membrane to reverse its potential from −70 mV to +30 mV. The reversal repeats itself over and over until the nerve impulse travels the length of the neuron. The nerve impulse is essentially a wave of negativity that travels along the outside of a neuron cell membrane. Depolarization and reversal of potential require only about 0.5 millisecond. Of all the cells of the body, only muscle and nerve cells

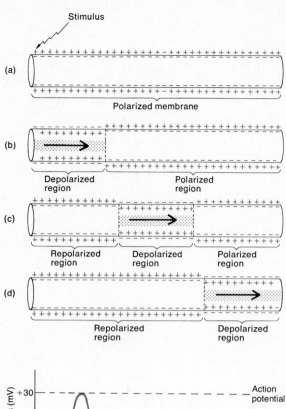

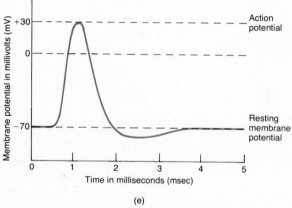

Figure 12-6 Initiation and transmission of a nerve impulse. (a–d) The colored area represents the region of the membrane that has initiated and is transmitting the nerve impulse. (e) Record of potential changes of a nerve impulse.

produce action potentials. Their ability to do this is called irritability.

By the time the impulse has traveled from one point on the membrane to the next, the previous point becomes **repolarized** – its resting potential is restored. Repolarization results from a new series of changes in membrane permeability. The membrane now becomes more permeable to K^+ ions than it was at its resting

potential level and is relatively impermeable again to Na^+ ions. The outward movement of K^+ ions causes the outer surface of the membrane to become electrically positive. The heavy loss of positive ions leaves the inner surface of the membrane negative again. Finally, any ions that have moved into or out of the nerve cell are restored to their original sites. Thus Na^+ ions are actively transported outside and K^+ ions are moved back into the cell. The repolarization period returns the cell to its resting potential, from $+30$ mV to -70 mV. The neuron is now prepared to receive another stimulus and transmit it in the same manner. In fact, until repolarization occurs, the neuron cannot transmit another impulse. The period of time during which the membrane recovers is called the refractory period.

A record of the electrical changes associated with a nerve impulse is illustrated in Figure 12-6e.

ALL-OR-NONE PRINCIPLE

Any stimulus strong enough to initiate an impulse is referred to as a *threshold* or *liminal stimulus*. When a stimulus is of threshold strength, we say the neuron has reached its threshold of stimulation. A nerve cell transmits an impulse according to the **all-or-none principle**: If a stimulus is strong enough to generate an action potential, the impulse is transmitted along the entire neuron at a constant and maximum strength for the existing conditions. The transmission is independent of any further intensity of the stimulus. However, transmission may be altered by conditions such as toxic materials in cells, fatigue, and malaise. Any stimulus weaker than a threshold stimulus is termed a *subthreshold* or *subliminal stimulus*. Such a stimulus is incapable of initiating a nerve impulse. If, however, a second stimulus or a series of subthreshold stimuli is quickly applied to the neuron, the cumulative effect may be sufficient to initiate an impulse. This phenomenon is called *summation*. If the second stimulus follows the original stimulus too closely, however, no response will occur because the nerve fiber needs sufficient time to recover from the passage of the first stimulus. This period of time, the **absolute refractory period,** depends on the fiber's diameter. Large fibers repolarize in about $\frac{1}{2,500}$ second. Their absolute refractory period is 0.4 millisecond. Thus a second nerve impulse can be transmitted $\frac{1}{2,500}$ second after the first – a total of up to 2,500 impulses per second. Small fibers, on the other hand, require as much as $\frac{1}{250}$ second to repolarize. Their absolute refractory period is 4 milliseconds. Thus they can transmit only 250 impulses per second. Under normal body conditions, the frequency of transmission may range between 10 and 500 impulses per second.

SALTATORY TRANSMISSION

Thus far we have considered nerve impulse transmission via unmyelinated fibers. In myelinated fibers, transmission is somewhat different. The myelin sheath surrounding a fiber contains a lipoprotein substance that does not conduct an electric current. It thus forms an insulating layer around the fiber. The myelin sheath is interrupted at various intervals called nodes of Ranvier. At these nodes, membrane depolarization can occur. But beneath the myelin sheath depolarization is impossible. When an impulse is transmitted along a myelinated fiber, it depolarizes the membrane around the first node of Ranvier, spreads around the outside of the myelin sheath to the next node, and so on. Thus the impulse jumps from node to node. This type of impulse transmission, characteristic of myelinated fibers, is called **saltatory transmission** (*saltare* = leaping).

Saltatory transmission is a valuable asset to your body. Since the impulse jumps long intervals as it moves from one node to the next, the speed of transmission is greatly increased. The impulse travels much faster than in the step-by-step depolarization process involved in an unmyelinated fiber of equal diameter.

SPEED OF NERVE IMPULSES

The speed of a nerve impulse is independent of stimulus strength. Once a neuron reaches its threshold of stimulation, the speed of the nerve impulse is normally determined by the size, type, and physiological condition of the fiber. Fibers with large diameters transmit impulses faster than those with small ones. Fibers with the greatest diameter are called *A fibers* and are all myelinated. The A fibers have a brief absolute refractory period and are capable of saltatory conduction. They transmit impulses at speeds up to 100 m/second. The A fibers are located in the axons of large sensory nerves that relay impulses associated with touch, pressure, position of joints, heat, and cold. They are also found in all motor nerves that convey impulses to the skeletal muscles. Sensory A fibers generally connect the brain and spinal cord with sensors that detect danger in the outside environment. Motor A fibers innervate the muscles that can do something about the situation. If you touch a hot object, information about the heat passes over sensory A fibers to the spinal cord. There it is relayed to motor A fibers that stimulate the muscles of the hand to withdraw instantaneously. The A fibers are located where split-second reaction may mean survival. Other fibers, called B and C fibers, transmit impulses more slowly and are generally found where instantaneous response is not a life-and-death matter. *B fibers* have a middle-sized diameter and a somewhat longer absolute

refractory period than A fibers. They are also myelinated and therefore capable of saltatory conduction. They transmit impulses at speeds of about 10 m/second. B fibers are found in nerves that transmit impulses from the skin and viscera to the brain and spinal cord. They also comprise all the axons of the visceral efferent neurons located in the motor nerves that leave the lower part of the brain and spinal cord and terminate in relay stations called *ganglia*. The ganglia ultimately link with other fibers that stimulate the smooth muscle and glands of the viscera. *C fibers* have the smallest diameter and the longest absolute refractory periods. They transmit impulses at the rate of about 0.5m/second. C fibers are unmyelinated and incapable of saltatory conduction. They are located in nerves that transmit impulses from the skin and in visceral nerves. These fibers transmit impulses for pain and perhaps impulses for touch, pressure, heat, and cold from the skin and pain receptors from the viscera. C fibers are located in all motor nerves that lead from the ganglia and stimulate the smooth muscle and glands of the viscera. Examples of the motor functions of B and C fibers are constricting and dilating the pupils, increasing and decreasing the heart rate, and contracting and relaxing the urinary bladder — functions of the autonomic nervous system.

Conduction Across Synapses

In addition to irritability, a neuron is also capable of **conductivity** — the ability to transmit an impulse to another neuron or to an effector such as a muscle or gland.

Impulses are transmitted from one neuron to another across a **synapse** — the junction between two neurons. The term *synapsis* means a connection. The synapse is essential for homeostasis because of its ability to transmit certain impulses and inhibit others. Figure 12-7c shows that within a synapse is an exceedingly minute gap about 200 Å across: the *synaptic cleft*. A *presynaptic neuron* is a neuron located before a synapse. A *postsynaptic neuron* is located after a synapse.

In Figure 12-7c the telodendrium of an axon terminates in rounded or oval expansions called *synaptic knobs* or *end feet*. The synaptic knobs of a presynaptic neuron may synapse with the following parts of a postsynaptic neuron: dendrites, cell body, or axon hillock. Accordingly, synapses may be classified as *axodendritic, axosomatic,* and *axoaxonic.* The synaptic knobs may all arise from a single presynaptic neuron, in which the knobs synapse with separate postsynaptic neurons. Such an arrangement is called **divergence** and

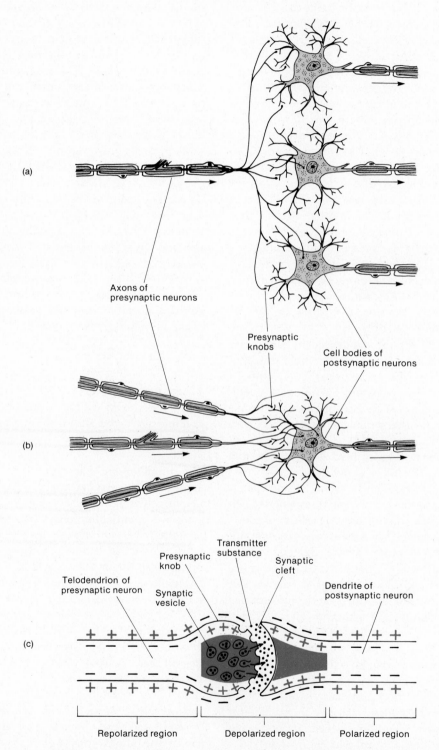

Figure 12-7 Impulse conduction at synapses. (a) Divergence of synapses. (b) Convergence of synapses. (c) Impulse conduction from a presynaptic knob to a postsynaptic dendrite across a synapse.

permits a single presynaptic neuron to influence several postsynaptic neurons or several muscle or gland cells at the same time. In another arrangement, called **convergence,** the synaptic knobs of several presynaptic neurons synapse with a single postsynaptic neuron. This arrangement permits stimulation or inhibition of the postsynaptic neuron.

Electron micrographic studies reveal that the synaptic knobs contain numerous granular structures called *synaptic vesicles.* These vesicles contain *chemical transmitter substances.* The transmitter's effect depends on the location of the synapse in the nervous system. When an impulse reaches a presynaptic knob of a presynaptic neuron, it causes the synaptic vesicles to rupture and release the transmitter. The transmitter then diffuses across the synaptic cleft and acts on the postsynaptic neuron or other tissue (muscle or gland). Depending on the chemical nature of the transmitter, several things may happen.

Excitatory Transmission

An **excitatory transmitter** is one that is released by an excitatory neuron. It can lower the postsynaptic neuron's membrane potential so that a new impulse can be generated across the synapse. If the potential is lowered enough, the membrane becomes depolarized, the potential inside the membrane becomes positive and the potential outside becomes negative, and a nerve impulse is initiated. Generally, the release of an excitatory transmitter by a single presynaptic knob is not sufficient to develop an action potential in a postsynaptic neuron. However, its release does bring the resting membrane potential closer to threshold level as Na^+ ions move into the cell. This change from the resting membrane potential level in the direction of the threshold level is called the **excitatory postsynaptic potential (EPSP)** (Figure 12-8a). The EPSP is always lower than the resting membrane potential of the neuron, but higher than its threshold level. The EPSP lasts only a few milliseconds. If several presynaptic knobs release their excitatory transmitters at the same time, however, the combined effect may initiate a nerve impulse – this effect is **summation.** If hundreds of presynaptic knobs release their excitatory transmitters simultaneously, they increase the chance of initiating a nerve impulse in the postsynaptic neuron. The greater the summation, the greater the probability an impulse will be initiated. When the summation is the result of the accumulation of transmitter substance from presynaptic knobs, it is called *spatial summation.* When summation is the result of the accumulation of transmitter substance from a single presynaptic knob firing two or more times in rapid

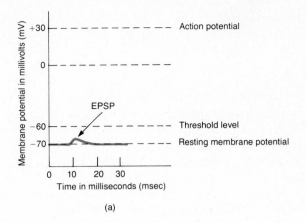

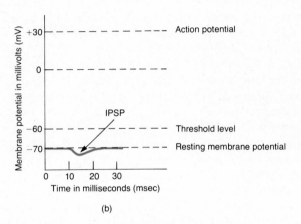

Figure 12-8 Comparison between excitatory and inhibitory postsynaptic potentials. (a) EPSP. (b) IPSP.

succession, it is called *temporal summation.* Since the EPSP lasts only about 15 milliseconds, the second firing must follow quickly if temporal summation is to occur. The time required for the impulse to cross a synapse – the *synaptic delay* – is about 0.5 millisecond. This delay is caused by the liberation of the transmitter substance, its passage across the synapse, stimulation of the postsynaptic neuron to become more permeable to Na^+ ions, and the inward movement of Na^+ ions that initiates the action potential in the postsynaptic neuron.

The release of an excitatory transmitter by a single presynaptic knob is not sufficient to initiate an action potential in a postsynaptic neuron. Nevertheless, the postsynaptic neuron does become more excitable to impulses from presynaptic neurons. This effect is called **facilitation.** If a postsynaptic neuron requires excitatory transmitter substance from 10 presynaptic knobs to initiate an action potential and only 7 presynaptic knobs fire, the postsynaptic membrane will not depolarize. But once these 7 presynaptic knobs have fired, they facilitate the postsynaptic neuron. That is, they prepare it for weaker subsequent firings of the other 3 presynaptic knobs. The result is impulse initiation.

Inhibitory Transmission

An **inhibitory transmitter** is one that is released by an inhibitory neuron and can inhibit an impulse at a synapse. Whereas excitatory transmitters make the postsynaptic neuron's resting membrane potential less negative, inhibitory transmitters make the resting membrane potential more negative. Inhibitory neurons are thought to release a transmitter that *hyperpolarizes* the postsynaptic neuron. That is, the cell interior becomes even more negative than the outside when it is at rest, making it even more difficult for the neuron to generate an action potential. The alteration of the postsynaptic membrane in which the resting membrane potential is made more negative is called the **inhibitory postsynaptic potential (IPSP)** (Figure 12-8b). The inhibitory transmitter probably alters the postsynaptic membrane's permeability to K^+ ions but not to Na^+ ions. As a result, the increased movement of K^+ ions out of the neuron results in an increased internal negativity. Just as the EPSP is less negative (lower) than the resting membrane potential of a neuron, the IPSP is more negative (higher) than the resting membrane potential.

Transmitter Substances

The exact chemical nature of excitatory and inhibitory transmitters is not known. It is thought, however, that the excitatory transmitter in a major portion of the central nervous system is **acetylcholine (ACh)**. This is the same transmitter discussed in Chapter 10 in relation to impulse transmission from a motor neuron to a muscle fiber through the neuromuscular junction. Acetylcholine lowers the postsynaptic neuron or muscle fiber membrane potential by increasing the permeability to Na^+ ions. Once this occurs, the membrane becomes depolarized, the potential is reversed, and a nerve impulse is initiated. As long as acetylcholine is present in the neuromuscular junction or synapse, it can stimulate a muscle fiber or postsynaptic neuron almost indefinitely. The transmission of a continuous succession of impulses by acetylcholine is normally prevented by an enzyme called **acetylcholinesterase (AChE)** or simply **cholinesterase.** Cholinesterase is released into the synaptic cleft by the postsynaptic neuron and into the neuromuscular junction by the muscle cell. Within $1/500$ second, cholinesterase inactivates acetylcholine. This action permits the membrane of the muscle fiber or postsynaptic neuron to repolarize almost immediately so that another impulse may be generated. When the next impulse comes

through, the synaptic knobs release more acetylcholine, the impulse is conducted, and cholinesterase again inactivates acetylcholine. This cycle is repeated over and over. Other suspected excitatory transmitters include **norepinephrine (noradrenalin** or **sympathin)**, which is found in high concentrations in certain neurons (postganglionic) of the sympathetic division of the autonomic nervous system and parts of the central nervous system; **serotonin,** which is found in high concentrations in the central nervous system; and **dopamine.**

There is considerable evidence that a substance called **gamma aminobutyric acid (GABA)** acts as an inhibitory transmitter in the central nervous system. It probably exerts its effect by hyperpolarizing the postsynaptic membrane according to the mechanism previously described.

Certain drugs have pronounced effects on synaptic transmission. Hypnotics and anesthetics, for example, depress synaptic transmission. Stimulants like caffeine and benzedrine facilitate synaptic transmission. Pressure has an effect on impulse transmission and conduction also. If excessive or prolonged pressure is applied to a nerve, impulse transmission is interrupted and part of the body may "go to sleep." Removal of the pressure results in a prickly sensation. This sensation is caused by an accumulation of waste products and a depressed circulation of blood.

One-Way Impulse Conduction

At a synapse there is only *one-way impulse conduction*—from a presynaptic axon to a postsynaptic dendrite, cell body, or axon hillock. Impulses must move forward over their pathway. They cannot back up into another presynaptic neuron. Such a mechanism is crucial in preventing impulse conduction along improper pathways. Imagine the result if impulses transmitted along the motor neurons that move your hand could move back and stimulate the sensory neuron which relays information about heat. You would feel heat, cry in pain, and go through all the emotions of being burned every time you simply wanted to move your hand.

Synaptic Fatigue

Since synaptic transmission depends on producing and liberating an excitatory transmitter, frequent or prolonged stimulation of the presynaptic neuron can exhaust the transmitter. Then the postsynaptic neuron is no longer stimulated—a condition called *synaptic fatigue.* The nature and degreee of fatigue depends on the synapse.

Integration at Synapses

A single postsynaptic neuron synapses with many presynaptic neurons. Some presynaptic knobs release excitatory transmitters and some release inhibitory transmitters. The sum of all the effects, excitatory and inhibitory, determines the effect on the postsynaptic neuron. Thus the postsynaptic neuron is an *integrator:* It receives signals, integrates them, and then responds accordingly. The postsynaptic neuron may respond in the following ways:

1 If the excitatory effect is greater than the inhibitory effect, but higher than the threshold level of stimulation, the result is facilitation.

2 If the excitatory effect is greater than the inhibitory effect, but equal to or lower than the threshold level of stimulation, the result is initiation of an impulse.

3 If the inhibitory effect is greater than the excitatory effect, the result is inhibition of the impulse.

Organization of Neuronal Synapses

The central nervous system contains millions of neurons. Their arrangement, however, is not haphazard. They are organized into definite patterns called *neuronal pools.* Each pool differs from all others and has its own role in regulating homeostasis.

A neuronal pool may contain thousands of neurons or even millions. To illustrate the composition of a neuronal pool, a simplified version is given in Figure 12-9. This example contains only five postsynaptic neurons: A, B, C, D, and E. It also contains two incoming presynaptic neurons: 1 and 2. The postsynaptic neurons are subject to stimulation by the incoming presynaptic neurons. The postsynaptic neurons in the pool may be stimulated by one or several presynaptic knobs. Moreover, the incoming presynaptic knobs may produce facilitation, excitation, or inhibition. A principal feature of a neuronal pool is the location of the presynaptic neurons in relation to the postsynaptic neurons. Compare the location of presynaptic axon 1 with that of postsynaptic neuron B. Since they are aligned, more presynaptic knobs of axon number 1 synapse with postsynaptic neuron B than with postsynaptic neurons A or C. We say that postsynaptic neuron B is in the center of the field of presynaptic axon 1. Consequently, postsynaptic neuron B usually receives sufficient presynaptic knobs from axon 1 to generate an impulse. This region where the neuron in the pool fires is called the *discharge zone.*

Now look outside the field of presynaptic axon 1. Note its relation to postsynaptic neuron A. Here postsynaptic neuron A in the pool is receiving few presynaptic knobs from the axon supplying postsynaptic neuron B. Thus there are insufficient presynaptic knobs to fire postsynaptic neuron A, but enough to cause facilitation. We therefore call this area the *facilitated zone.* Presynaptic axon 1, when stimulated, will cause excitation of postsynaptic neuron B and facilitation of postsynaptic neuron A.

Neuronal pools in the central nervous system are arranged in patterns over which the impulses are conducted. These are termed *circuits.* Simple circuits are arranged so that a presynaptic neuron stimulates a single neuron in a pool. The single neuron then stimulates another and so on. In other words, the impulse is relayed from one neuron to another in succession as a new impulse is generated at each synapse.

Most circuits, however, are more complex. In a *diverging circuit,* the impulse from a single presynaptic neuron causes the stimulation of increasing numbers of cells along the circuit (Figure 12-10a). An example of such a circuit is a single motor neuron in the brain stimulating numerous other motor neurons in the spinal cord that, in turn, leave the spinal cord where each stimulates many skeletal muscle fibers. Thus a single impulse may result in the contraction of several skeletal muscle fibers. In another kind of diverging circuit, impulses from one pathway are relayed to other pathways so the same information travels in various directions at the same time. This circuit is common along sensory pathways of the nervous system.

Another kind of circuit is called a *converging circuit* (Figure 12-10b). Unlike the diverging circuits, a postsynaptic neuron receives impulses from different fibers of the same source or different fibers from several sources. In the first pattern, the postsynaptic neuron receives impulses from several fibers of the same source. Here there is the possibility of strong excitation or inhibition. In the second pattern, the postsynaptic neuron receives impulses from several different sources. Here there is a possibility of reacting the same way to different stimuli. Suppose your reaction to vomit is distinctly unpleasant. The smell of vomit (one kind of stimulus), the sight of vomit (another kind), or just reading about vomit (still another kind) might all have the same effect on you – an unpleasant one.

Some circuits in your body are constructed so that once the presynaptic cell is stimulated, it will cause the postsynaptic cell to transmit a series of impulses. One such circuit is called a *reverberating circuit* (Figure 12-11a). In this pattern, the incoming impulse stimulates the first neuron, which stimulates the second, which stimulates the third, and so on. Branches from the second and third neurons synapse with the first, however, sending the impulse back through the circuit again and again. Thus the postsynaptic neuron generates

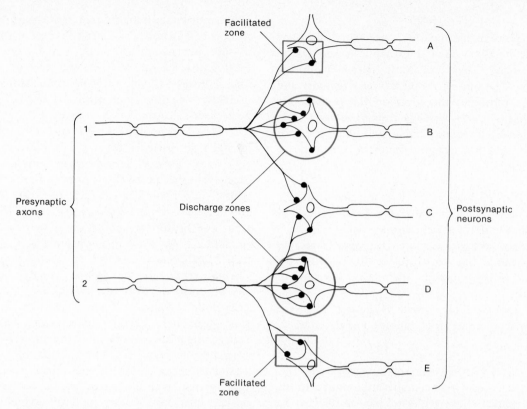

Figure 12-9 Relative positions of discharge and facilitated zones in a very simplified version of a neuronal pool.

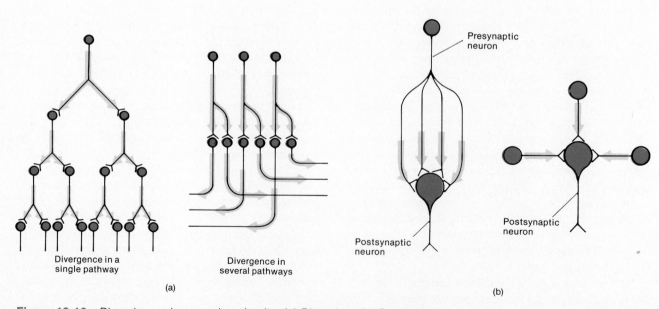

Figure 12-10 Diverging and converging circuits. (a) Diverging. (b) Converging.

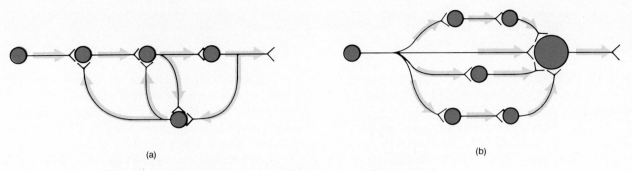

Figure 12-11   Reverberating and parallel circuits. (a) Reverberating. (b) Parallel.

continuous impulses. Generally, the circuit is broken by fatigue. A central feature of the reverberating circuit is that once fired, the output signal may last from a few seconds to many hours. The duration depends on the number and arrangement of neurons in the circuit. Among the body responses thought to be the result of output signals from reverberating circuits are the rate of breathing, coordinated muscular activities, waking up, and sleeping (when reverberation stops). Some scientists think reverberating circuits are related to short-term memory.

A final circuit worth consideration is the *parallel circuit* (Figure 12-11b). Like a reverberating circuit, a parallel circuit is constructed so the postsynaptic cell transmits a series of impulses. In a parallel circuit, a single presynaptic cell stimulates a group of neurons, each of which synapses with a common postsynaptic cell. The advantage of this circuit is that the postsynaptic neuron can send out a stream of impulses in succession as they are received. The impulses leave the postsynaptic neuron once every $\frac{1}{2,000}$ second. This circuit has no feedback system. Once all the neurons in the circuit have transmitted their impulses to the postsynaptic neuron, the circuit is broken. It is thought that the parallel circuit is employed for precise activities like mathematical calculations.

STUDY OUTLINE

ORGANIZATION

1 The nervous system controls and integrates all body activities by sensing changes, interpreting them, and reacting to them.

2 The central nervous system consists of the brain and spinal cord.

3 The peripheral nervous system is classified into an afferent system and an efferent system.

4 The efferent system is subdivided into a somatic nervous system and an autonomic nervous system.

5 The somatic nervous system consists of efferent neurons that conduct impulses from the central nervous system to skeletal muscle tissue.

6 The autonomic nervous system contains efferent neurons that convey impulses from the central nervous system to smooth muscle tissue, cardiac muscle tissue, and glands.

HISTOLOGY
NEUROGLIA

1 Neuroglia are specialized tissue cells that support, attach neurons to blood vessels, produce the myelin sheath, and carry out phagocytosis.

NEURONS

1 Neurons, or nerve cells, consist of a perikaryon or cell body, dendrites that pick up stimuli and convey impulses to the cell body, and usually a single axon. The axon transmits impulses from the neuron to the dendrites or cell body of another neuron or to an effector organ of the body.

2 On the basis of structure, neurons are multipolar, bipolar, and unipolar.

3 On the basis of function, sensory (afferent) neurons transmit impulses to the central nervous system; association neurons transmit impulses to other neurons including motor neurons and motor (efferent) neurons transmit impulses to effectors.

PHYSIOLOGY
REGENERATION

1 Around the time of birth, the cell body loses its mitotic apparatus and is no longer able to divide.

2 Nerve fibers (axis cylinders) that have a neurilemma are capable of regeneration.

NERVE IMPULSE

1 The nerve impulse represents the body's quickest way of controlling and maintaining homeostasis.

2 The membrane of a nonconducting neuron is positive on the outside and negative inside. This difference in charge is called a resting potential, and the membrane is said to be polarized.

3 When a stimulus causes the inside of the cell membrane to become positive and the outside negative, the membrane is said to have an action potential, which travels from point to point along the membrane. The traveling action potential is a nerve impulse.

CONDUCTION ACROSS SYNAPSES

1 An excitatory transmitter is one that is released by an excitatory nerve. It can lower the postsynaptic neuron's membrane potential so that a new impulse can be generated across the synapse.

2 An inhibitory transmitter is one that is released by an inhibitory neuron and can inhibit an impulse at a synapse.

3 The exact chemical nature of excitatory and inhibitory transmitters is not known. It is thought, however, that the excitatory transmitter in a major portion of the central nervous system is acetylcholine. An enzyme called cholinesterase inactivates acetylcholine.

4 At a synapse there is only one-way impulse conduction from a presynaptic axon to a postsynaptic dendrite, cell body, or axon hillock.

5 Since synaptic transmission depends on producing and liberating an excitatory transmitter, frequent or prolonged stimulation of the presynaptic neuron can exhaust the transmitter.

6 The postsynaptic neuron is an integrator. It receives signals, integrates them, and then responds accordingly.

7 The arrangement of neurons in the central nervous system is not haphazard. They are organized into definite patterns called neuronal pools. Each pool differs from all others and has its own role in regulating homeostasis.

REVIEW QUESTIONS

1 How does the nervous system maintain homeostasis? Distinguish between the central and peripheral nervous systems. Relate the terms *voluntary* and *involuntary* to the nervous system.

2 What are neuroglia? List their principal functions.

3 Define a neuron. Diagram and label a neuron. Next to each part list its function.

4 Discuss how neurons are classified by structure and function.

5 What determines neuron regeneration?

6 Define irritability and conductivity.

7 Outline the principal steps in the origin and transmission of a nerve impulse. What determines the speed of a nerve impulse?

8 Define resting potential, polarized membrane, action potential, depolarized membrane, and repolarized membrane.

9 What is the all-or-none principle? Relate it to threshold stimulus, subthreshold stimulus, and summation.

10 What events are involved in the transmission of a nerve impulse across a synapse?

11 How are nerve impulses inhibited? What advantage is this to the body?

12 What is synaptic fatigue?

13 Why is the postsynaptic neuron called an integrator?

14 Explain the different functional units into which synapses are organized.

15 Support this statement: "The nerve impulse is the body's best means for rapid correction of a deviation that tends to disrupt homeostasis."

Chapter 13
The Spinal Cord
and the
Spinal Nerves

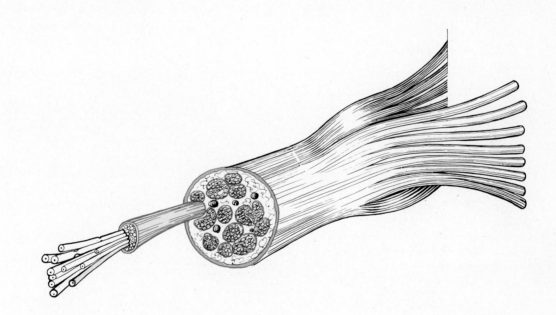

STUDENT OBJECTIVES

■ Define white matter, gray matter, nerves, ganglia, tracts, and nuclei.

■ Describe the principal structural features of the spinal cord.

■ Identify the factors responsible for maintaining and protecting the spinal cord.

■ Define spinal puncture. Discuss its location, the general technique, its purpose, and its significance.

■ Identify the conducting and reflex activities of the spinal cord.

■ Contrast the functions of ascending and descending tracts of the spinal cord.

■ Discuss the components of a reflex arc and its relationship to homeostasis.

■ Contrast the operation of a two-neuron and multi-neuron reflex arc.

■ Define a reflex.

■ Identify the relationship between reflexes and the maintenance of homeostasis.

■ Classify reflexes on the basis of organs stimulated and location of receptors.

■ List several clinically important reflexes.

■ List the distribution of the 31 pairs of spinal nerves.

■ Describe the structure of a typical spinal nerve.

■ Define a plexus.

■ Note the name, composition, and functions of the principal plexuses.

■ Define a dermatome.

■ Define spinal cord injury and list the immediate and long-range effects.

■ Describe peripheral nerve damage and conditions necessary for its regeneration.

■ Identify sciatica and neuritis.

■ Define medical terminology associated with the spinal cord and spinal nerves.

In this chapter, our main concern will be a study of the structure and function of the spinal cord and the nerves that originate from it.

GROUPING OF NEURAL TISSUE

The term **white matter** refers to aggregations of myelinated axons from many neurons supported by neuroglia. The lipid substance, myelin, has a whitish color that gives white matter its name. The gray areas of the nervous system are called **gray matter.** They contain either nerve cell bodies and dendrites or bundles of unmyelinated axons and neuroglia.

A **nerve** is a bundle of fibers located outside the central nervous system. Since the dendrites of somatic afferent neurons and axons of somatic efferent neurons of the peripheral nervous system are myelinated, most nerves are white matter. Nerve cell bodies that lie outside the central nervous system are generally grouped with other nerve cell bodies to form **ganglia** (*ganglion* = knot). Ganglia, since they are made up principally of unmyelinated nerve cell bodies, are masses of gray matter.

A **tract** is a bundle of fibers in the central nervous system. Tracts may run long distances up and down the spinal cord. Tracts also exist in the brain and connect parts of the brain with each other and with the spinal cord. The chief spinal tracts that conduct impulses up the cord are concerned with sensory impulses and are called *ascending tracts.* By contrast, spinal tracts that carry impulses down the cord are motor tracts and are called *descending tracts.* The major tracts consist of myelinated fibers and are therefore white matter. A **nucleus** is a mass of nerve cell bodies and dendrites in the central nervous system. It consists of gray matter. **Horns** or **columns** are the chief areas of gray matter in the spinal cord. The term *horn* describes the two-dimensional appearance of the organization of gray matter in the spinal cord as seen in cross section. The term *column* describes the three-dimensional appearance of the gray matter in longitudinal columns. Since the white matter of the spinal cord is also arranged in columns, we will refer to the gray matter as being arranged in horns.

SPINAL CORD

General Features

The **spinal cord** is a cylindrical structure that is slightly flattened anteriorly and posteriorly. It begins as a continuation of the medulla oblongata, the inferior part of the brain stem, and extends from the foramen magnum of the occipital bone to the level of the second lumbar vertebra (Figure 13-1). The length of the adult spinal cord ranges from 42 to 45 cm (16 to 18 inches). The diameter of the cord varies at different levels.

When the cord is viewed externally, two conspicuous enlargements can be seen. The superior enlargement, the *cervical enlargement,* extends from the fourth cervical to the first thoracic vertebra. Nerves that supply the upper extremities arise from the cervical enlargement. The inferior enlargement, called the *lumbar enlargement,* extends from the ninth to twelfth thoracic vertebra. Nerves that supply the lower extremities arise from the lumbar enlargement.

Below the lumbar enlargement, the spinal cord tapers to a conical portion known as the *conus medullaris.* The conus medullaris lies at about the level of the first or second lumbar vertebra. Arising from the conus medullaris is the *filum terminale,* a nonnervous fibrous tissue of the spinal cord that extends inferiorly to the coccyx. The filum terminale consists mostly of pia mater, the innermost of three membranes that cover and protect the spinal cord and brain. Some nerves that arise from the lower portion of the cord do not leave the vertebral column immediately. They angle inferiorly in the vertebral canal like wisps of coarse hair flowing from the end of the cord. They are appropriately named the *cauda equina* (horse's tail).

The spinal cord is a series of 31 segments, each giving rise to a pair of spinal nerves. *Spinal segment* refers to a region of the spinal cord from which a pair of spinal nerves arises. Figure 13-3 shows that the cord is divided into right and left sides by two grooves. One of these, the *anterior median fissure,* is a deep, wide groove on the anterior (ventral) surface. The other is the *posterior median sulcus,* a shallower, narrow groove on the posterior (dorsal) surface.

Protection and Coverings

The spinal cord is located in the vertebral canal of the vertebral column. The vertebral canal is formed by the foramina of all the vertebrae arranged on top of each other. Since the wall of the vertebral canal is essentially a ring of bone surrounding the spinal cord, the cord is well protected. A certain degree of protection is also provided by the meninges, the cerebrospinal fluid, and the vertebral ligaments.

The *meninges* are coverings that run continuously around the spinal cord and brain. Those associated specifically with the cord are known as spinal meninges (Figure 13-2). The outer spinal meninx is called the *dura mater* (or tough mother) and forms a tube from the level of the second sacral vertebra, where it is fused with the filum terminale, to the foramen magnum, where it is

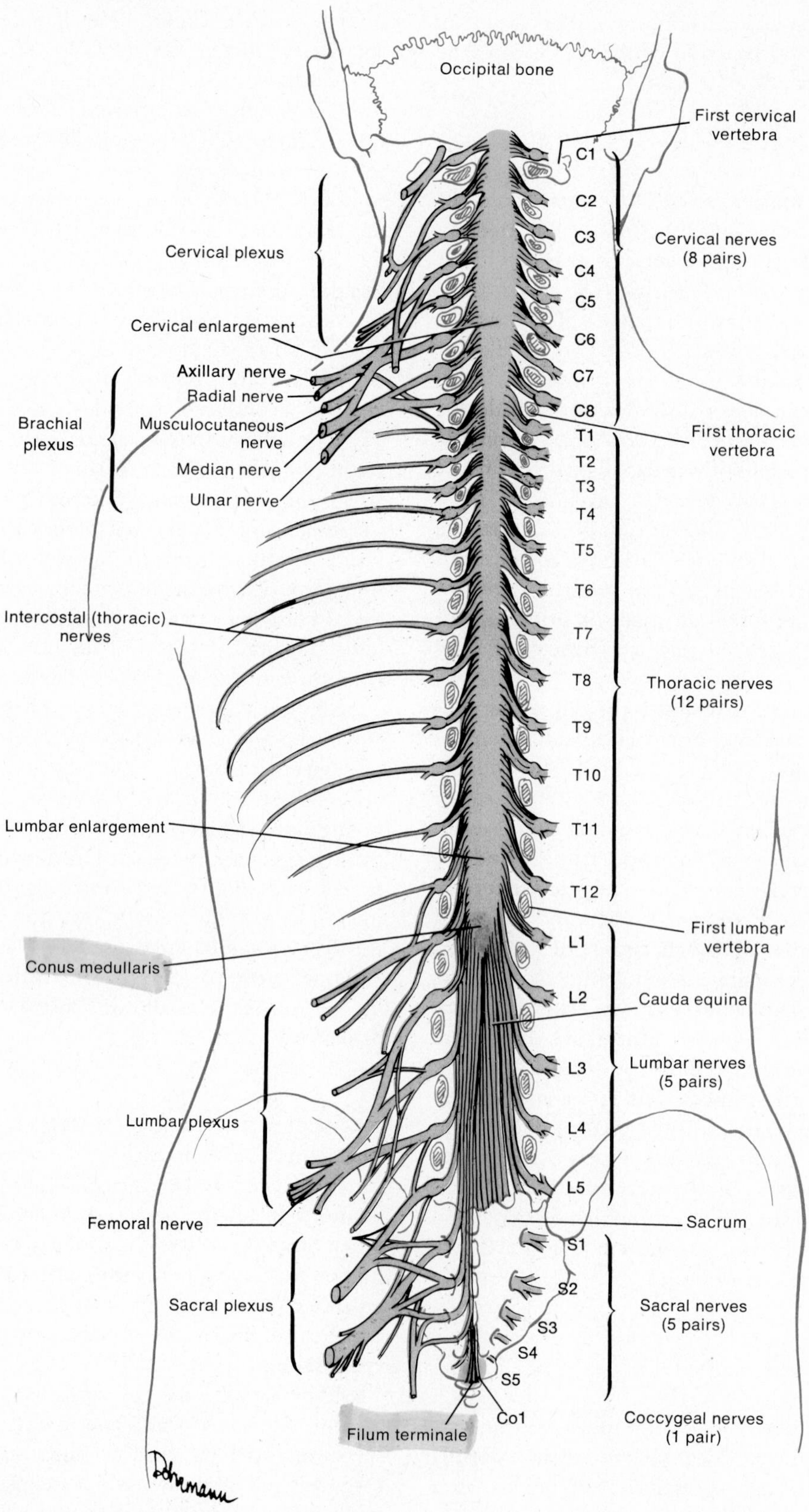

Figure 13-1   Spinal cord and spinal nerves in posterior view.

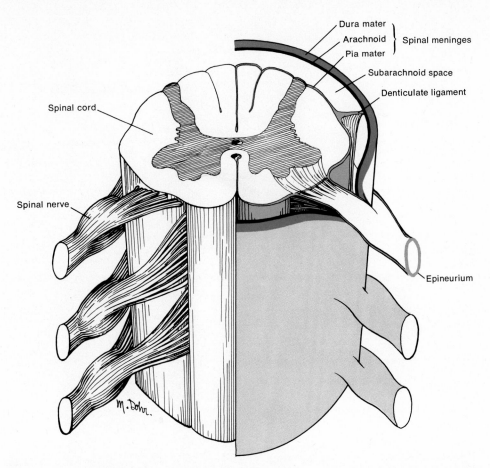

Figure 13-2 Location of the spinal meninges as seen on the left side in a cross section of the spinal cord.

continuous with the dura mater of the brain. The dura mater is composed of dense, fibrous connective tissue. Between the dura mater and the wall of the vertebral canal is the *epidural space,* which is filled with fat, connective tissue, and blood vessels. It serves as padding around the cord. The middle spinal meninx is called the *arachnoid* (or spider layer). It is a delicate connective tissue membrane that forms a tube inside the dura mater. It is also continuous with the arachnoid of the brain. Between the dura mater and the arachnoid is a space called the *subdural space,* which contains serous fluid. The inner meninx is known as the *pia mater* (or delicate mother). It is a transparent fibrous membrane that forms a tube around and adheres to the surface of the spinal cord and brain. It contains numerous blood vessels. Between the arachnoid and the pia mater is the *subarachnoid space,* where the cerebrospinal fluid circulates. All three spinal meninges cover the spinal nerves as they exit the spinal column through the intervertebral foramina. The spinal cord is suspended in the middle of its dural sheath by membranous extensions of the pia

mater. These extensions, called the *denticulate ligaments,* are attached laterally along the length of the cord between the ventral and dorsal nerve roots on either side. The ligaments protect the spinal cord against shock and sudden displacement.

Spinal Puncture

The removal of cerebrospinal fluid from the subarachnoid space in the interior lumbar region of the spinal cord is a *spinal (lumbar) puncture,* or *tap.* The procedure is normally performed between the third and fourth or fourth and fifth lumbar vertebrae. The spinous process of the fourth lumbar vertebra is easily located by drawing a line across the highest points of the iliac crests. This line will pass right through the spinous process of the fourth lumbar vertebra. A lumbar puncture is below the spinal cord and thus poses little danger to it. If the patient lies on one side, drawing the knees and chest together, the vertebrae separate slightly so that a needle can be conveniently inserted. Lumbar punctures are used to

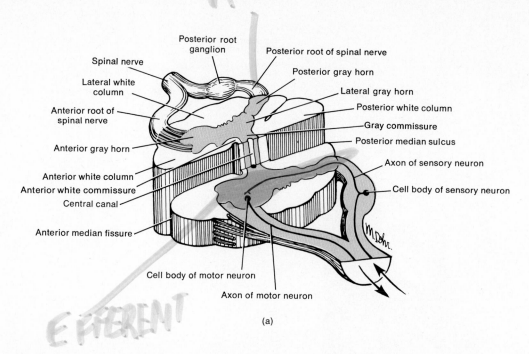

AFFERENT

EFFERENT

(a)

withdraw fluid for diagnostic purposes, to introduce antibiotics (as in the case of meningitis), and to administer anesthesia.

Structure in Cross Section

The spinal cord consists of both gray and white matter. Figure 13-3 shows that the gray matter lies in an area shaped like an H. The gray matter consists primarily of nerve cell bodies and unmyelinated axons and dendrites of association and motor neurons. The white matter surrounds the gray matter and consists of bundles of myelinated axons of motor and sensory neurons.

In the center of the gray matter is a cross bar of the H called the *gray commissure,* connecting the right and left portions of the H. In the center of the gray commissure is a small space called the *central canal.* This canal runs the length of the spinal cord and is continuous with the fourth ventricle of the medulla. It contains cerebrospinal fluid. Anterior to the gray commissure is the *anterior (ventral) white commissure,* which connects the white matter of the right and left sides of the spinal cord. The upright portions of the H are further subdivided into regions. Those closer to the front of the cord are called *anterior (ventral) gray horns.* They represent the motor part of the gray matter. The regions closer to the back of the cord are referred to as *posterior (dorsal) gray horns.* They represent the sensory part of the gray matter. The regions between the anterior and posterior gray horns are *lateral gray horns.* The lateral gray horns are most prominent in the thoracic and upper lumbar segments of the cord.

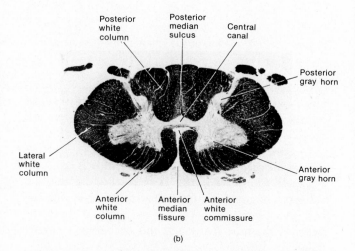

(b)

Figure 13-3 Spinal cord. (a) The organization of gray and white matter in the spinal cord as seen in a cross section of the spinal cord. The front of the figure has been sectioned at a lower level than the back so that you can see what is inside the posterior root ganglion, posterior root of the spinal nerve, anterior root of the spinal nerve, and the spinal nerve. (b) Photograph of the spinal cord at the seventh cervical segment, Weigert stain, at a magnification of 7×. (Courtesy of Murray L. Barr, *The Human Nervous System,* Harper & Row Publishers, Inc., New York, 1974.)

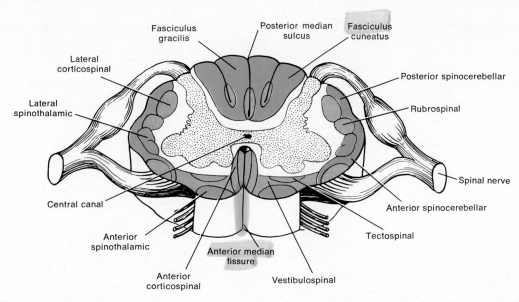

Figure 13-4 Selected tracts of the spinal cord. Ascending (sensory) tracts are indicated in color; descending (motor) tracts are shown in gray.

The gray matter of the cord also contains several nuclei that serve as relay stations for impulses and origins for certain nerves. Nuclei are clusters of nerve cell bodies and dendrites in the spinal cord and brain.

The white matter on each side of the cord, like the gray matter, is organized into regions. The anterior and posterior gray horns divide the white matter into three broad areas: *anterior (ventral) white column, posterior (dorsal) white column,* and *lateral white column.* Each column (or *funiculus*) in turn consists of distinct bundles of myelinated fibers that run the length of the cord. These bundles are called *fasciculi* or *tracts.* The longer *ascending tracts* consist of sensory axons that conduct impulses which enter the spinal cord and pass upward to the brain. The longer *descending tracts* consist of motor axons that conduct impulses from the brain downward through the spinal cord and out to muscles and glands. Thus the ascending tracts are sensory tracts and the descending tracts are motor tracts. Still other short tracts contain ascending or descending axons that convey impulses from one level of the cord to another.

Functions

A major function of the spinal cord is to convey sensory impulses from the periphery to the brain and to conduct motor impulses from the brain to the periphery. A second principal function is to provide reflexes.

CONDUCTION PATHWAY

The vital function of conveying sensory and motor information is carried out by the ascending and de-

scending tracts of the cord. The names of the tracts indicate the white column (funiculus) in which the tract travels, where the cell bodies of the tract originate, where the axons of the tract terminate, and the direction of impulse conduction within the tract. For example, the anterior spinothalamic tract is located in the anterior white column, it originates in the spinal cord, it terminates in the thalamus (a region of the brain), and it is an ascending (sensory) tract since it conveys impulses from the cord upward to the brain.

Important ascending and descending tracts are shown in Figure 13-4. Exhibit 13-1 summarizes the principal tracts.

REFLEX CENTER

The second principal function of the spinal cord is to provide reflexes. Spinal nerves are the paths of communication between the spinal cord tracts and the periphery. Figure 13-3 reveals that each pair of spinal nerves is connected to a segment of the cord by two points of attachment called roots. The **posterior** or **dorsal (sensory) root** contains sensory nerve fibers only and conducts impulses from the periphery to the spinal cord. These fibers extend into the posterior (dorsal) gray horn. Each dorsal root also has a swelling, the **posterior** or **dorsal (sensory) root ganglion,** which contains the cell bodies of the sensory neurons from the periphery. The other point of attachment of a spinal nerve to the cord is the **anterior** or **ventral (motor) root.** It contains motor nerve fibers only and conducts impulses from the spinal cord to the periphery. The cell bodies of the motor neurons are located in the gray matter of the cord. If the

Exhibit 13–1   SELECTED ASCENDING AND DESCENDING TRACTS OF SPINAL CORD

Tract	Location (White Column)	Origin	Termination	Function
Ascending tracts				
Anterior (ventral) spinothalamic	Anterior (ventral) column	Posterior (dorsal) gray horn on one side of cord but crosses to opposite side	Thalamus; impulse eventually conveyed to cerebral cortex	Conveys sensations for crude touch and pressure from one side of body to opposite side of thalamus. Eventually sensations reach cerebral cortex.
Lateral spinothalamic	Lateral column	Posterior (dorsal) gray horn on one side of cord but crosses to opposite side	Thalamus; impulse eventually conveyed to cerebral cortex	Conveys sensations for pain and temperature from one side of body to opposite side of thalamus Eventually sensations reach cerebral cortex.
Fasciculus gracilis and fasciculus cuneatus	Posterior (dorsal) column	Axons of afferent neurons from periphery that enter posterior (dorsal) column and rise to same side of medulla	Nucleus gracilis and nucleus cuneatus of medulla; impulse eventually conveyed to cerebral cortex	Convey sensations from one side of body to same side of medulla for fine touch; two-point discrimination (ability to distinguish that two points on skin are touched even though close together); proprioception (awareness of precise position of body parts and their direction of movement); stereognosis (ability to recognize size, shape, and texture of object); weight discrimination (ability to assess weight of an object); and vibrations. Eventually sensations may reach cerebral cortex.
Posterior (dorsal) spinocerebellar	Posterior (dorsal) portion of lateral column	Posterior (dorsal) gray horn on same side of cord	Cerebellum	Conveys sensations from one side of body to same side of cerebellum for subconscious proprioception.
Anterior (ventral) spinocerebellar	Anterior (ventral) portion of column	Posterior (dorsal) gray horn on one side of cord; tract contains both crossed and uncrossed fibers	Cerebellum	Conveys sensations from both sides of body to cerebellum for subconscious proprioception

motor impulse supplies a skeletal muscle, the cell bodies are located in the anterior (ventral) gray horn. If, however, the impulse supplies smooth muscle, cardiac muscle, or a gland through the autonomic nervous system, the cell bodies are located in the lateral gray horn.

■ *Reflex Arc and Homeostasis*   The path an impulse follows from its origin in the dendrites or cell body of a neuron in one part of the body to its termination elsewhere in the body is called a *conduction pathway.* All conduction pathways consist of circuits of neurons. One pathway is known as a **reflex arc,** the functional unit of the nervous system. A reflex arc contains two or more neurons over which impulses are conducted from a receptor to the brain or spinal cord and then to an effector. The basic components of a reflex arc are as follows:

Exhibit 13—1   (cont.)

Tract	Location (White Column)	Origin	Termination	Function
Descending tracts				
Lateral corticospinal	Lateral column	Cerebral cortex on one side of brain but crosses in base of medulla to opposite side of cord	Anterior (ventral) gray horn	Conveys motor impulses from one side of cortex to anterior gray horn of opposite side. Eventually impulses reach skeletal muscles on opposite side of body that coordinate precise, discrete movements.
Anterior (ventral) corticospinal	Anterior (ventral) column	Cerebral cortex on one side of brain, uncrossed in medulla, but crosses to opposite side of cord	Anterior (ventral) gray horn	Conveys motor impulses from one side of cortex to anterior gray horn of opposite side. Eventually impulses reach skeletal muscles on opposite side of body that coordinate precise, discrete movements.
Rubrospinal	Lateral column	Midbrain (red nucleus) but crosses to opposite side of cord	Anterior (ventral) gray horn	Conveys motor impulses from one side of midbrain to skeletal muscles on opposite side of body that are concerned with muscle tone and posture.
Tectospinal	Anterior (ventral) column	Midbrain but crosses to opposite side of cord	Anterior (ventral) gray horn	Conveys motor impulses from one side of midbrain to skeletal muscles on opposite side of body that control movements of head in response to auditory, visual, and cutaneous stimuli.
Vestibulospinal	Anterior (ventral) column	Medulla on one side of brain to same side of cord	Anterior (ventral) gray horn	Conveys motor impulses from one side of medulla to skeletal muscles on same side of body that regulate body tone in response to movements of head (equilibrium).

1 **Receptor**   The distal end of a dendrite or a sensory structure associated with the distal end of a dendrite. Its role in the reflex arc is to respond to a change in the internal or external environment by initiating a nerve impulse in a sensory neuron.

2 **Sensory neuron**   Once stimulated, the sensory neuron passes the impulse from the receptor to the central nervous system.

3 **Center**   A region, usually in the central nervous system, where an incoming sensory impulse generates an outgoing motor impulse. In the center, the impulse may be inhibited, transmitted, or rerouted. In the center of some reflex arcs, the sensory neuron directly generates the impulse in the motor neuron. The center

may also contain an association neuron between the sensory neuron and the motor neuron leading to a muscle or a gland.

4 **Motor neuron**   Transmits the impulse generated by the sensory or association neuron in the center to the organ of the body that will respond.

5 **Effector**   The organ of the body, either muscle or gland, that responds to the motor impulse. This response is called a *reflex action* or *reflex*.

Reflexes are fast responses to changes in the internal or external environment to maintain homeostasis. Reflexes carried out by the spinal cord alone are called *spinal reflexes*. Reflexes that result in the contrac-

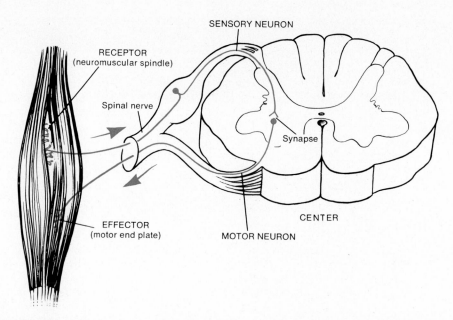

SENSORY NEURON

RECEPTOR
(neuromuscular spindle)

Spinal nerve

Synapse

EFFECTOR
(motor end plate)

MOTOR NEURON

CENTER

Figure 13-5   Stretch reflex. Notice that in a stretch reflex there are only two neurons involved and there is only one synapse in the pathway. Thus it is a monosynaptic reflex arc. Why is the reflex arc shown referred to as an ipsilateral reflex arc?

tion of skeletal muscles are known as *somatic reflexes.* Those that cause the contraction of smooth or cardiac muscle or secretion by glands are *visceral (autonomic) reflexes.* Our concern at this point is to examine a few somatic spinal reflexes: the stretch reflex, the flexor reflex, and the crossed extensor reflex.

■ *Stretch Reflex*   The **stretch reflex** is based on a *two-neuron,* or *monosynaptic, reflex arc.* Only two neurons are involved and there is only one synapse in the pathway (Figure 13-5). This reflex results in the contraction of a muscle when it is stretched. Slight stretching of a muscle stimulates receptors in the muscle called *neuromuscular spindles.* The spindles monitor changes in the length of the muscle. Once the spindle is stimulated, an impulse is sent along a sensory neuron to the spinal cord. The sensory neuron lies in the posterior root of a spinal nerve and synapses with a motor neuron in the anterior gray horn. The sensory neuron generates an impulse at the synapse that is transmitted along the motor neuron. The motor neuron lies in the anterior root of the spinal nerve and terminates in a skeletal muscle. Once the impulse reaches the stretched muscle, it contracts. Thus the stretch is counteracted by contraction. Since the sensory impulse enters the spinal cord on the same side that the motor impulse leaves the spinal cord, the reflex arc is called an *ipsilateral reflex arc.* All monosynaptic reflex arcs are ipsilateral. The stretch reflex is essential in

maintaining muscle tone. Moreover, it is the basis for several tests used in neurological examinations. One such reflex is the *knee jerk,* or *patellar reflex.* This reflex is illustrated in Figure 13-5. This reflex is tested by tapping the patellar ligament (stimulus). Neuromuscular spindles in the quadriceps femoris muscle attached to the ligament send the sensory impulse to the spinal cord and the returning motor impulse causes contraction of the muscle. The response is extension of the leg at the knee, or a knee jerk.

■ *Flexor Reflex and Crossed Extensor Reflex*   Reflexes other than stretch reflexes involve association neurons in addition to the sensory and motor neuron — they are *polysynaptic reflex arcs.* One example of a reflex based on a polysynaptic reflex arc is the **flexor reflex,** or **withdrawl reflex** (Figure 13-6). Suppose you step on a tack. As a result of the painful stimulus, you immediately withdraw your foot. What has happened? A sensory neuron transmits an impulse from the receptor to the spinal cord. A second impulse is generated in an association neuron, which generates a third impulse in a motor neuron. The motor neuron stimulates the muscles of your foot and you withdraw it. Thus a flexor reflex is protective. It moves an extremity to avoid pain.

This stretch reflex is also ipsilateral. The incoming and outgoing impulses are on the same side of the spinal cord. The stretch reflex also illustrates another feature of

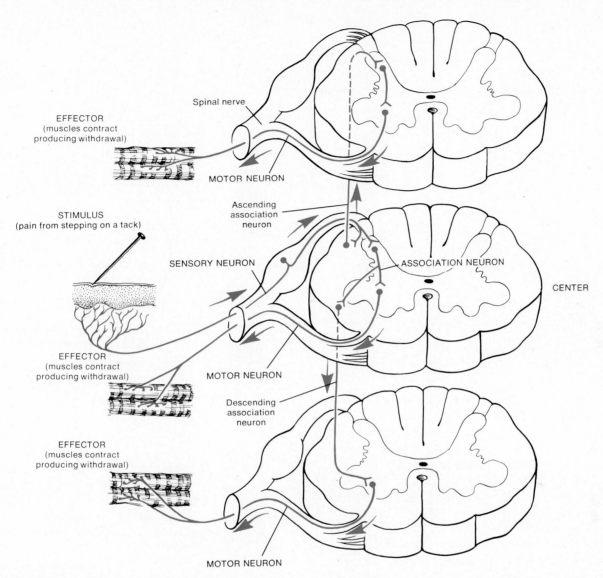

Figure 13-6 Flexor reflex. This reflex arc is a polysynaptic ipsilateral reflex arc because it involves more than one synapse. This is because it contains association neurons as well as sensory and motor neurons. Why is the reflex arc shown also an intersegmental reflex arc?

reflex arcs. In the monosynaptic stretch reflex, the returning motor impulse affects only the quadriceps muscle of the thigh. When you withdraw your entire lower or upper extremity from a noxious stimulus, more than one muscle is involved. Therefore several motor neurons are simultaneously returning impulses to several upper and lower extremity muscles at the same time. Thus a single sensory impulse causes several motor responses. This kind of reflex arc, in which a single sensory neuron splits into ascending and descending branches, each forming a synapse with association neurons at different segments of the cord, is called an

intersegmental reflex arc. Because of intersegmental reflex arcs, a single sensory neuron can activate several motor neurons and thereby cause stimulation of more than one effector.

Something else may happen when you step on a tack. You may lose your balance as your body weight shifts to the other foot. Then you do whatever you can to regain your balance so you do not fall. This means motor impulses are also sent to your unstimulated foot and both upper extremities. The motor impulses that travel to your unaffected foot cause extension at the knee so you can place your entire body weight on the foot.

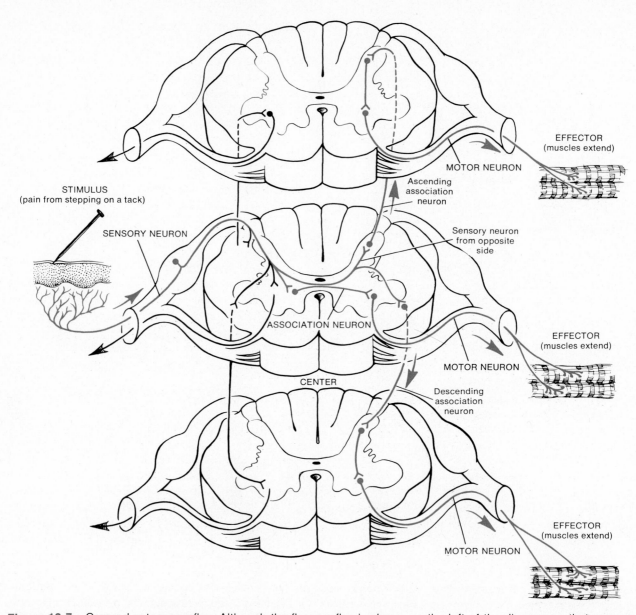

STIMULUS
(pain from stepping on a tack)

SENSORY NEURON

ASSOCIATION NEURON

CENTER

EFFECTOR
(muscles extend)

MOTOR NEURON

Ascending
association
neuron

Sensory neuron
from opposite
side

EFFECTOR
(muscles extend)

MOTOR NEURON

Descending
association
neuron

EFFECTOR
(muscles extend)

MOTOR NEURON

Figure 13-7   Crossed extensor reflex. Although the flexor reflex is shown on the left of the diagram so that you can correlate it with the crossed extensor reflex on the right, concentrate your attention on the crossed extensor reflex. Why is the crossed extensor reflex classified as a contralateral reflex arc?

These impulses cross the spinal cord as shown in Figure 13-7. The incoming sensory impulse not only initiates the flexor reflex that causes you to withdraw, it also initiates an extensor reflex. The incoming sensory impulse crosses to the opposite side of the spinal cord through association neurons at that level and several levels above and below the point of sensory stimulation. From these levels, the motor neurons cause extension of the knee, thus maintaining balance. Unlike the flexor reflex, which passes over an ipsilateral reflex arc, the extensor reflex passes over a *contralateral reflex arc*—the

impulse enters one side of the spinal cord and exits on the opposite side. The reflex just described in which extension of the muscles in one limb occurs as a result of flexion of the muscles of the opposite limb is simply called a **crossed extensor reflex.**

The flexor reflex and crossed extensor reflex also illustrate *reciprocal inhibition,* another feature of many reflexes. Reciprocal inhibition occurs when a reflex excites a muscle to cause its contraction and also inhibits another muscle to allow its extension. Thus, in this reflex, excitation and inhibition occur simultaneously.

In the flexor reflex, when the flexor muscles of your lower extremity are contracting, the extensor muscles of the same extremity are being extended. If both sets of muscles contracted at the same time, you would not be able to flex your limb because both sets of muscles would pull on the limb bones. But, because of reciprocal inhibition, one set of muscles contracts while the other is being extended.

In the crossed extensor reflex, reciprocal inhibition also occurs. While you are flexing the muscles of the limb that has been stimulated by the tack, the muscles of your other limb are producing extension to help maintain balance. Reciprocal inhibition is vital in coordinating body movements. In flexing the forearm at the elbow, there is a prime mover, an antagonist, and a synergist. The prime mover (biceps) contracts to cause flexion, the antagonist (triceps) extends to yield to the action of the prime mover, and the synergist (deltoid) helps the prime mover perform its role efficiently.

REFLEXES AND DIAGNOSIS

Reflexes are often used for diagnosing disorders of the nervous system and locating injured tissue. If a reflex ceases to function or functions abnormally, the physician may suspect that the damage lies somewhere along a particular conduction pathway. Visceral reflexes, however, are usually not practical tools for diagnosis. It is difficult to stimulate visceral receptors, since they are deep in the body. In contrast, many somatic reflexes can be tested simply by tapping or stroking the body.

Superficial reflexes are elicited by stroking the skin with a hard object, such as an applicator stick. The object is quickly passed over the skin once, depressing the skin but not producing a scratch.

Any skeletal muscle can normally be stimulated to contract by a slight sudden stretch of its tendon, which can be created by administering a light tap. Many muscle tendons are deeply buried, however, and cannot be readily tapped through the skin. Reflexes that involve a stretch stimulus to a tendon are called *deep tendon reflexes.* Deep tendon reflexes provide information about the integrity and function of the reflex arcs and spinal cord segments without involving the higher centers.

To obtain a substantial response when testing deep tendon reflexes, the muscle must be slightly stretched before the tap is administered. If it is stretched an appropriate amount, tapping the tendon elicits a muscle contraction.

If reflexes are weak or absent, *reinforcement* can be used. In this method, muscle groups other than those being tested are tensed voluntarily with isometric contractions to increase reflex activity in other parts of the body. For example, the person can be asked to hook the fingers together and then try to pull them apart. This action may increase the strength of reflexes involving other muscles. If the reflex can be demonstrated, it is certain that the sensory and motor nerve connections are intact between muscle and spinal cord.

Muscle reflexes can help determine the spinal cord's excitability. When a large number of facilitatory impulses are transmitted from the brain to the spinal cord, the muscle reflexes become so sensitive that simply tapping the knee tendon with the tip of one's finger may cause the leg to jump a considerable distance. On the other hand, the cord may be so intensely inhibited by other impulses from the brain that almost no degree of pounding on the muscles or tendons can elicit a response.

Neurological impairment can be evaluated by using a stopwatch to time the reflex response. Sensitivity of sensory end organs in a muscle is demonstrated by stretching it by as little as 0.05 mm and for as short a duration as $\frac{1}{20}$ second.

Among the reflexes of clinical significance are:

1 **Patellar reflex** (knee jerk) This reflex involves extension of the lower leg by contraction of the quadriceps femoris muscle in response to tapping the patellar ligament. The reflex is blocked by damaged afferent or efferent nerves to the muscle or reflex centers in the second, third, or fourth lumbar segments of the spinal cord. This reflex is also absent in people with chronic diabetes and neurosyphilis. The reflex is exaggerated in disease or injury involving the corticospinal tracts descending from cortex to spinal cord. This reflex may also be exaggerated by applying a second stimulus (a sudden loud noise) while tapping the patellar tendon.

2 **Achilles reflex** (ankle jerk) This reflex involves extension (plantar flexion) of the foot by contraction of the gastrocnemius and soleus muscles in response to tapping the Achilles tendon. Blockage of the ankle jerk indicates damage to the nerves supplying the posterior leg muscles or to the nerve cells in the lumbosacral region of the spinal cord. This reflex is also absent in people with chronic diabetes, neurosyphilis, alcoholism, and subarachnoid hemorrhages. An exaggerated Achilles reflex indicates cervical cord compression or a lesion of the motor tracts of the first or second sacral segments of the cord.

3 **Babinski reflex** This reflex results from light stimulation to the outer margin of the sole of the foot. The great toe is extended, with or without fanning of the other toes. This phenomenon occurs in normal children under $1\frac{1}{2}$ years of age and is due to incomplete development of the nervous system. The myelination of fibers in the corticospinal tract has not reached comple-

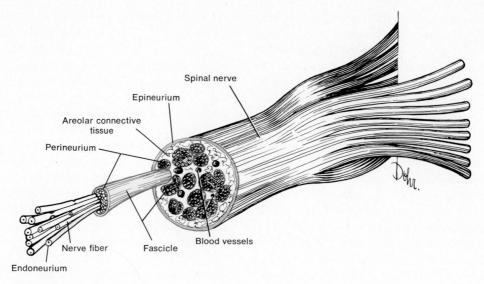

Figure 13-8   Coverings of a spinal nerve.

tion. A positive Babinski reflex after age 1½ is considered abnormal and indicates an interruption of the corticospinal tract as the result of a lesion of the tract, usually in the upper portion. The normal response after 1½ years of age is the *plantar reflex,* or negative Babinski – a curling under of all the toes accompanied by a slight turning in and flexion of the anterior part of the foot.

4   **Abdominal reflex**   This reflex compresses the abdominal wall in response to stroking the side of the abdomen. Two separate reflexes, the upper abdominal reflex and the lower abdominal reflex, are involved. The patient should be lying down and relaxed with arms at the sides and knees slightly flexed. The response is an abdominal muscle contraction that results in a lateral deviation of the umbilicus to the side opposite the stimulus. Abscence of this reflex is associated with lesions of the corticospinal system. It may also be absent because of lesions of the peripheral nerves, lesions of reflex centers in the thoracic part of the cord, and multiple sclerosis.

SPINAL NERVES
Composition and Coverings
A **spinal nerve** has two points of attachment to the cord: a posterior root and an anterior root. A short distance from the spinal cord the roots unite to form a spinal nerve. Since the posterior root contains sensory fibers and the anterior root contains motor fibers, a spinal nerve is a *mixed nerve.* The posterior (dorsal) root ganglion contains cell bodies of sensory neurons. The posterior and anterior roots unite to form the spinal

nerve at the intervertebral foramen.

In Figure 13-8, you can see that the nerve contains many fibers surrounded by different coverings. The individual fibers, whether myelinated or unmyelinated, are wrapped in a connective tissue called the *endoneurium.* Groups of fibers with their endoneurium are arranged in bundles called fascicles, and each bundle is wrapped in connective tissue called the *perineurium.* The outermost covering around the entire nerve is the *epineurium.* The spinal meninges fuse with the epineurium as the nerve exists from the vertebral canal.

Names
The 31 pairs of spinal nerves are named and numbered according to the region and level of the spinal cord from which they emerge (see Figure 13-1). The first cervical pair emerges between the atlas and the occipital bone. All other spinal nerves leave the backbone from the intervertebral foramina between adjoining vertebrae. There are 8 pairs of cervical nerves, 12 pairs of thoracic nerves, 5 pairs of lumbar nerves, 5 pairs of sacral nerves, and 1 pair of coccygeal nerves. During fetal life, the spinal cord and vertebral column grow at different rates, the cord growing more slowly. Thus not all the spinal cord segments are in line with their corresponding vertebrae. Remember that the spinal cord terminates near the level of the first or second lumbar vertebra. Thus the lower lumbar, sacral, and coccygeal nerves must descend more and more to reach their foramina before emerging from the vertebral column. This arrangement constitutes the cauda equina.

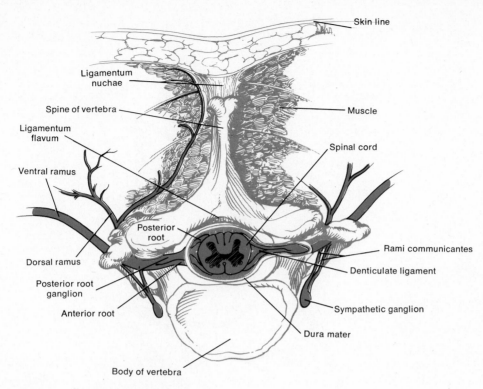

Figure 13-9 Branches of a typical spinal nerve.

Branches

Shortly after a spinal nerve leaves its intervertebral foramen, it divides into several branches (Figure 13-9). These branches are known as rami. The *dorsal ramus* innervates (supplies) the muscles and skin of the dorsal surface of the back. The *ventral ramus* of a spinal nerve innervates all the structures of the extremities and the lateral and ventral trunk. Except for thoracic nerves T2 to T12, the ventral rami of other spinal nerves enter into the formation of plexuses before supplying a part of the body. In addition to dorsal and ventral rami, spinal nerves also give off a *meningeal branch.* This branch supplies the vertebrae, vertebral ligaments, blood vessels of the spinal cord, and the meninges. Other branches of a spinal nerve are the *rami communicantes.*

Plexuses

The ventral rami of spinal nerves, except for T2 to T12, do not go directly to the structures of the body they supply. Instead, they form networks with adjacent nerves on either side of the body. Such networks are called **plexuses,** meaning braid. The principal plexuses are the cervical plexus, the brachial plexus, the lumbar plexus, and the sacral plexus. Emerging from the plexuses are nerves bearing names that are often descriptive of the general regions they supply or the course they take. Each of these nerves, in turn, may have several branches named for the specific structures they innervate.

CERVICAL PLEXUS

The **cervical plexus** is formed by the ventral rami of the first four cervical nerves (C1 to C4) with contributions from C5. There is one on each side of the neck alongside the first four cervical vertebrae (Figure 13-10). The *roots* of the plexus indicated in the diagram are simply continuations of the ventral rami. The cervical plexus supplies the skin and muscles of the head, neck, and upper part of the shoulders. Branches of the cervical plexus also connect with cranial nerves XI (accessory) and XII (hypoglossal). A major pair of nerves arising from the cervical plexuses are the phrenic nerves supplying motor fibers to the diaphragm. Damage to the spinal cord above the origin of the phrenic nerves results in paralysis of the diaphragm and death since the phrenic nerves no longer send impulses to the diaphragm. Contractions of the diaphragm are essential for breathing.

Exhibit 13-2 summarizes the nerves and distributions of the cervical plexus. The relationship of the cervical plexus to the other plexuses is shown in Figure 13-1.

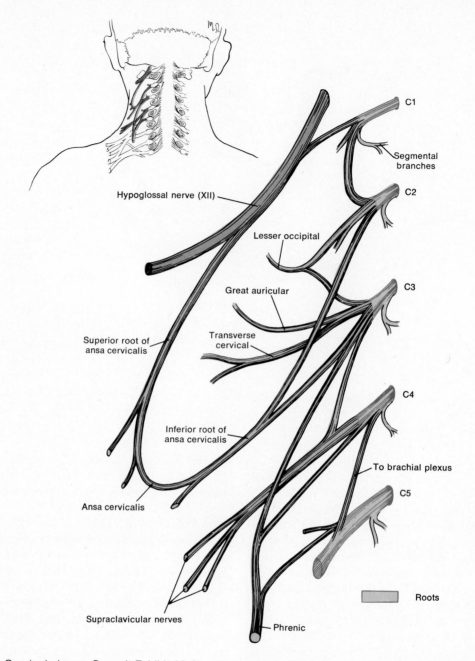

Figure 13-10 Cervical plexus. Consult Exhibit 13-2 so that you can determine the distribution of each of the nerves of the plexus.

BRACHIAL PLEXUS

The **brachial plexus** is formed by the ventral rami of spinal nerves C5 to C8 and T1 with contributions from C4 and T2. On either side of the last four cervical and first thoracic vertebrae, the brachial plexus extends downward and laterally, passes over the first rib behind the clavicle, and then enters the axilla (Figure 13-11). The brachial plexus constitutes the entire nerve supply for the upper extremities, as well as a number of neck and shoulder muscles.

The *roots* of the brachial plexus, like those of the cervical plexus, are continuations of the ventral rami of the spinal nerves. The roots of C5 and C6 unite to form the *upper trunk,* C7 becomes the *middle trunk,* and C8 to T1 form the *lower trunk.* Each trunk, in turn, divides into an *anterior division* and a *posterior division.* The divisions then unite to form cords. The *posterior cord* is formed by the union of the posterior divisions of the upper, middle, and lower trunks, and the *medial cord* is formed as a continuation of the anterior division of the lower trunk.

Exhibit 13–2 CERVICAL PLEXUS

Nerve	Origin	Distribution
Superficial or cutaneous branches		
Lesser occipital	C2–C3	Skin of scalp behind and above ear.
Greater auricular	C2–C3	Skin in front, below, and over ear.
Transverse cervical	C2–C3	Skin over anterior aspect of neck.
Supraclaviculars	C3–C4	Skin over upper portion of chest and shoulder.
Deep or largely motor branches		
Ansa cervicalis		This nerve is divided into a superior root and an inferior root.
Superior root	C1–C2	Infrahyoid, thyrohyoid, and geniohyoid muscles of neck.
Inferior root	C3–C4	Omohyoid, sternohyoid, and sternothyroid muscles of neck.
Phrenic	C3–C5	Diaphragm between chest and abdomen.
Segmental branches	C1–C5	Prevertebral (deep) muscles of neck, levator scapulae, sternocleidomastoid, and trapezius muscles.

The peripheral nerves arise from the cords. Thus the brachial plexus begins as roots that unite to form trunks. The trunks branch into divisions, the divisions form cords, and the cords give rise to the peripheral nerves.

A summary of the nerves and distributions of the brachial plexus is given in Exhibit 13-3. The relationship of the brachial plexus to the other plexuses is shown in Figure 13-1.

LUMBAR PLEXUS

The **lumbar plexus** is formed by the ventral rami of spinal nerves L1 to L4. It differs from the brachial plexus in that there is no intricate interlacing of fibers. It also consists of *roots* and an *anterior* and *posterior division*. On either side of the first four lumbar vertebrae, the lumbar plexus passes obliquely outward behind the psoas major muscle (posterior division) and anterior to the quadratus lumborum muscle (anterior division) and then gives rise to its peripheral nerves (Figure 13-12, page 304). The lumbar plexus supplies the anterolateral abdominal wall, external genitals, and part of the lower extremity. The largest nerve arising from the lumbar plexus is the femoral nerve. Injury to the nerve is indicated by an inability to extend the leg and by loss of sensation in the skin over the anteromedial aspect of the thigh.

A summary of the nerves and distributions of the lumbar plexus is presented in Exhibit 13-4 on page 305. The relationship of the lumbar plexus to the other plexuses is shown in Figure 13-1.

SACRAL PLEXUS

The **sacral plexus** is formed by the ventral rami of spinal nerves L4 to L5 and S1 to S4. It is situated largely in front of the sacrum (Figure 13-13, page 306). Like the lumbar plexus, it contains *roots* and an *anterior* and *posterior division*. The sacral plexus supplies the buttocks, perineum, and lower extremities. The largest nerve arising from the sacral plexus – and, in fact, the largest nerve in the body – is the sciatic nerve. This nerve may be injured because of a slipped disc, dislocated hip, pressure from the uterus during pregnancy, or an improperly given gluteal intramuscular injection. The sciatic nerve supplies the entire musculature of the leg and foot.

A summary of the nerves and distributions of the sacral plexus is given in Exhibit 13-5 on page 307. The relationship of the sacral plexus to the other plexuses is shown in Figure 13-1.

INTERCOSTAL (THORACIC) NERVES

Spinal nerves T2 to T12 do not enter into the formation of plexuses. Instead, these **intercostal** or **thoracic** nerves are distributed directly to the structures they supply (see Figure 13-1). After leaving the vertebral foramina, ventral rami of nerves T3 to T6 pass in the costal grooves of the ribs and are distributed to the intercostal muscles and skin of the anterior and lateral chest wall. Nerves T7 to T12 supply the abdominal muscles and overlying skin. T2 supplies the skin of the axilla and posteromedial aspect of the arm. The dorsal rami of the intercostal nerves supply the muscles and skin of the dorsal aspect of the thorax.

Dermatomes

The skin over the entire body is supplied segmentally by spinal nerves. This means that the spinal nerves innervate

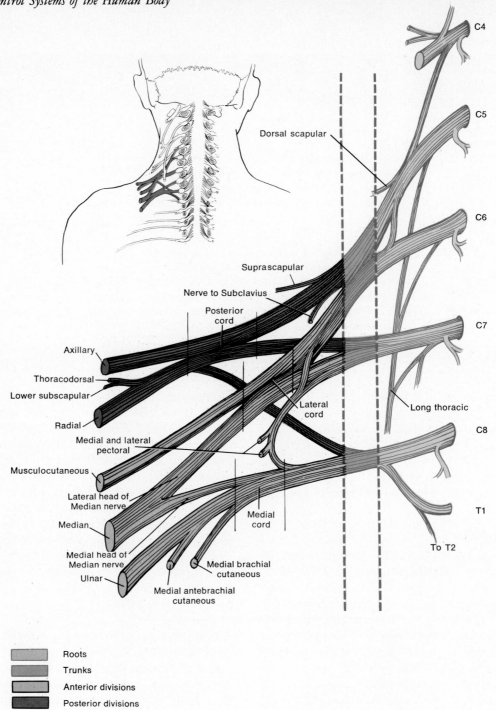

Figure 13-11 Brachial plexus. Consult Exhibit 13-3 so that you can determine the distribution of each of the nerves of the plexus.

specific, constant segments of the skin. With the exception of spinal nerve C1, all other spinal nerves supply branches to the skin. The skin segment supplied by a spinal nerve is a **dermatome** (Figure 13-14, page 308). In the neck and trunk, the dermatomes form consecutive bands of skin. In the trunk, there is an overlap of adjacent dermatome nerve supply. Thus there is little loss of sensation if only a single nerve supply to a dermatome is interrupted. Most of the skin of the face and scalp is supplied by cranial nerve V (trigeminal).

Since a physician knows that a particular dermatome is associated with a particular spinal nerve, it is possible to determine which segment of the spinal cord or spinal nerve is malfunctioning. If a dermatome is stimulated and the sensation is not perceived, it can be assumed that the nerve supplying the dermatome is involved.

Exhibit 13–3 BRACHIAL PLEXUS

Nerve	Origin	Distribution
Root nerves		
Dorsal scapular	C5	Levator scapulae, rhomboideus major, and rhomboideus minor muscles.
Long thoracic	C5–C7	Serratus anterior muscle.
Trunk nerves		
Nerve to subclavius	C5–C6	Subclavius muscle.
Suprascapular	C5–C6	Supraspinatus and infraspinatus muscles.
Lateral cord nerves		
Musculocutaneous	C5–C7	Coracobrachialis, biceps brachii, and brachialis muscles.
Median (lateral head)	C5–C7	See distribution for Median (medial head).
Lateral pectoral	C5–C7	Pectoralis major muscle.
Posterior cord nerves		
Upper subscapular	C5–C6	Subscapularis muscle.
Thoracodorsal	C6–C8	Latissimus dorsi muscle.
Lower subscapular	C5–C6	Subscapularis and teres major muscles.
Axillary (circumflex)	C5–C6	Deltoid and teres minor muscles; skin over deltoid and upper posterior aspect of arm.
Radial	C5–C8, T1	Extensor muscles of arm and forearm (triceps brachii, brachioradialis, extensor carpi radialis longus, extensor digitorum, extensor carpi ulnaris, extensor indicis); skin of posterior arm and forearm, lateral two-thirds of dorsum of hand, and fingers over proximal and middle phalanges.
Medial cord nerves		
Medial pectoral	C8–T1	Pectoralis major and pectoralis minor muscles.
Medial brachial cutaneous	C8–T1	Skin of medial and posterior aspects of lower third of arm.
Medial antebrachial cutaneous	C8–T1	Skin of medial and posterior aspects of forearm.
Median (medial head)	C5–C8, T1	Medial and lateral heads of median nerve form median nerve. Distributed to flexors of forearm (pronator teres, flexor carpi radialis, flexor digitorum superficialis) except flexor carpi ulnaris and flexor digitorum profundus; skin of lateral two-thirds of palm of hand and fingers.
Ulnar	C8–T1	Flexor carpi ulnaris and flexor digitorum profundus muscles; skin of medial side of hand, little finger, and medial half of ring finger.
Other cutaneous distributions		
Intercostobrachial	Second intercostal nerve	Skin over medial side of arm.
Upper lateral brachial cutaneous	Axillary	Skin over deltoid muscle and down to elbow.
Posterior brachial cutaneous	Radial	Skin over posterior aspect of arm.
Lower lateral brachial cutaneous	Radial	Skin over lateral aspect of elbow.
Lateral antebrachial cutaneous	Musculocutaneous	Skin over lateral aspect of forearm.
Posterior antebrachial cutaneous	Musculocutaneous	Skin over posterior aspect of forearm.

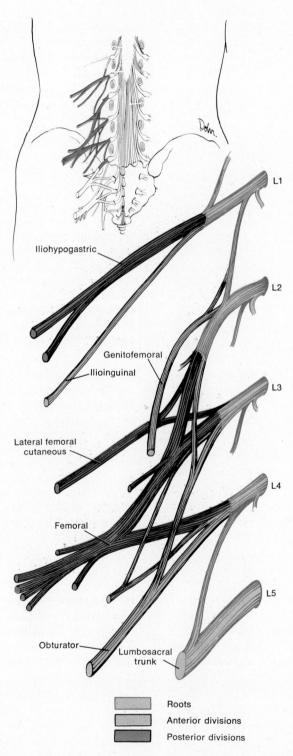

Figure 13-12   Lumbar plexus. Consult Exhibit 13-4 so that you can determine the distribution of each of the nerves of the plexus.

Exhibit 13–4 LUMBAR PLEXUS

Nerve	Origin	Distribution
Iliohypogastric	T12–L1	Muscles of anterolateral abdominal wall (external oblique, internal oblique, transversus abdominis); skin of lower abdomen and buttock.
Ilioinguinal	L1	Muscles of anterolateral abdominal wall as indicated above; skin of upper medial aspect of thigh, root of penis and scrotum in male, and labia majora and mons pubis in female.
Genitofemoral	L1–L2	Cremaster muscle; skin over middle anterior surface of thigh, scrotum in male, and labia majora in female.
Lateral femoral cutaneous	L2–L3	Skin over lateral, anterior, and posterior aspects of thigh.
Femoral	L2–L4	Flexor muscles of thigh (iliacus, psoas major, pectineus, rectus femoris, sartorius); extensor muscles of leg (rectus femoris, vastus lateralis, vastus medialis, vastus intermedius); skin on front and over medial aspect of thigh and medial side of leg and foot.
Obturator	L2–L4	Adductor muscles of leg (obturator externus, pectineus, adductor longus, adductor brevis, adductor magnus, gracilis); skin over medial aspect of thigh.
Saphenous	L2–L4	Skin over medial aspect of leg.

DISORDERS

SPINAL CORD INJURY

The spinal cord may be damaged by fracture or dislocation of the vertebrae enclosing it or by wounds. All can result in **transection** – partial or complete severing of the spinal cord. Complete transection means that all ascending and descending pathways are cut. It results in loss of all sensation and voluntary muscular movement below the level of transection. In fact, individuals with complete cervical transections close to the base of the skull usually die of asphyxiation before treatment can be administered. This happens because impulses from the phrenic nerves to the breathing muscles are interrupted. If the upper cervical cord is partially transected, both the upper and lower extremities are paralyzed and the patient is classified as *quadriplegic*. Partial transection between the cervical and lumbar enlargements results in paralysis of the lower extremities only, and the patient is classified as *paraplegic*.

In the case of partial transection, **spinal shock** lasts from a few days to several weeks. During this period, all reflex activity is abolished, a condition called *areflexia*. In time, however, there is a return of reflex activity. The first reflex to return is the knee jerk. Its reappearance may take several days. Next the flexion reflexes return. This may take up to several months. Then the crossed extensor reflexes return. Visceral reflexes such as erection and ejaculation are also affected by transection. Moreover, bladder and bowel function are no longer under voluntary control.

PERIPHERAL NERVE DAMAGE

If the cell body of a neuron is destroyed, the neuron cannot be replaced and its function is permanently lost. Axons that have a neurilemma can be repaired, however, as long as the cell body is intact and fibers are in association with Schwann cells. Most nerves that lie outside the brain and spinal cord consist of axons and dendrites that are covered with a neurilemma. A person who injures a nerve in the upper extremity, for example, has a good chance of regaining nerve function. Axons in the brain and spinal cord do not have a neurilemma. Injury there is permanent.

When there is damage to an axon (or to dendrites of somatic afferent neurons), there are usually changes in the cell body and always changes in the portions of the nerve processes distal to the site of damage. These changes associated with the cell body are referred to as the axon reaction or retrograde degeneration. Those associated with the distal portion of the cut fiber are called wallerian degeneration. The axon reaction occurs in essentially the same way, whether the damaged fiber is in the central or peripheral nervous system. The wallerian reaction, however, depends on whether the fiber is central or peripheral. We will consider the axon reaction first.

AXON REACTION

When there is damage to an axon of a central or peripheral neuron, certain structural changes occur in the cell body – the axon reaction. One of the most significant features of the axon

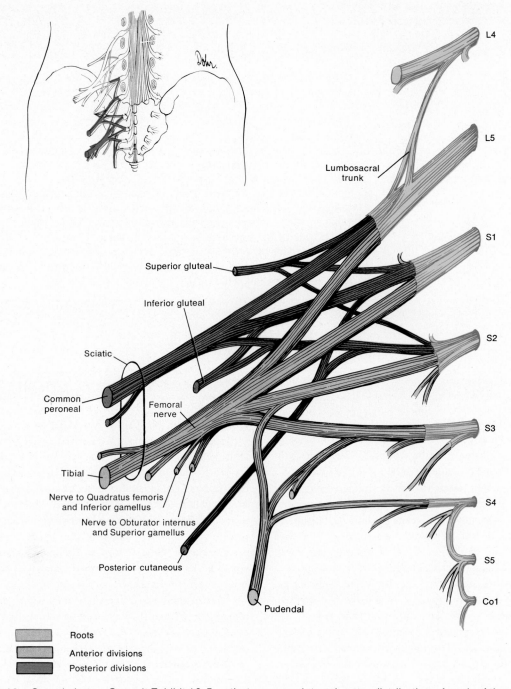

Superior gluteal

Inferior gluteal

Sciatic

Common peroneal

Femoral nerve

Tibial

Nerve to Quadratus femoris and Inferior gamellus

Nerve to Obturator internus and Superior gamellus

Posterior cutaneous

Pudendal

Lumbosacral trunk

L4

L5

S1

S2

S3

S4

S5

Co1

Roots

Anterior divisions

Posterior divisions

Figure 13-13   Sacral plexus. Consult Exhibit 13-5 so that you can determine the distribution of each of the nerves of the plexus.

reaction occurs 24–48 hours after damage. The Nissl bodies, arranged in an orderly fashion in an uninjured cell body, break down into finely granular masses. This alteration is called *chromatolysis*. It begins between the axon hillock and nucleus but spreads throughout the cell body. As a result of chromatolysis, the cell body swells and the swelling reaches its maximum between 10 and 20 days after injury. Chromatolysis results in a loss of ribosomes by the rough endoplasmic reticulum and an increase in the number of free ribosomes. Another sign of the axon reaction is the off-center position of the nucleus in the cell body. This change makes it possible to identify the cell bodies of damaged fibers through a micro-

Exhibit 13–5   SACRAL PLEXUS

Nerve	Origin	Distribution
Superior gluteal	L4–L5, S1	Gluteus minimus and gluteus medius muscles and tensor fasciae latae.
Inferior gluteal	L5–S2	Gluteus maximus muscle.
Nerve to piriformis	S1–S2	Piriformis muscle.
Nerve to quadratus femoris	L4–L5, S1	Inferior gemellus and quadratus femoris muscles.
Nerve to obturator internus	L5–S2	Superior gemellus and obturator internus muscles.
Perforating cutaneous	S2–S3	Skin over lower medial aspect of buttock.
Posterior cutaneous	S1–S3	Skin over anal region, upper posterior aspect of thigh, upper part of calf, scrotum in male, and labia majora in female.
Sciatic	L4–S3	Actually two nerves: tibial and common peroneal, bound together by common sheath of connective tissue. It splits into its two divisions, usually at knee. (See below for distributions.) As sciatic nerve descends through thigh, it sends branches to hamstring muscles (biceps femoris, semitendinosus, semimembranosus) and adductor magnus.
Tibial (medial popliteal)	L4–S3	Gastrocnemius, plantaris, soleus, popliteus, tibialis posterior, flexor digitorum, and hallucis muscles. Branches of tibial nerve in foot are medial plantar nerve and lateral plantar nerve.
Medial plantar		Abductor hallucis, flexor digitorum brevis, and flexor hallucis muscles; skin over medial two-thirds of plantar surface of foot.
Lateral plantar		Remaining muscles of foot not supplied by medial plantar nerve; skin over lateral third of plantar surface of foot.
Common peroneal (lateral popliteal)	L4–S2	Divides into a superficial peroneal and a deep peroneal branch. (Distributions described below.)
Superficial peroneal		Peroneus longus and peroneus brevis muscles; skin over distal third of anterior aspect of leg and dorsum of foot.
Deep peroneal		Tibialis anterior, extensor hallucis longus, peroneus tertius, and extensor digitorum brevis muscles; skin over great and second toes.
Pudendal	S2–S4	Muscles of perineum; skin of penis and scrotum in male and clitoris, labia majora, labia minora, and lower vagina in female.

scope. Following chromatolysis, there are signs of recovery in the cell body. There is an acceleration of RNA and protein synthesis, which favors regeneration of the axon. Recovery often takes several months and involves the restoration of normal levels of RNA and proteins and the Nissl bodies to their usual, uninjured patterns.

WALLERIAN DEGENERATION

The part of the axon distal to the damage becomes slightly swollen and then breaks up into fragments by the third to fifth day. The myelin sheath around the axon also undergoes degeneration. Degeneration of the distal portion of the axon and myelin sheath is called **wallerian degeneration.** Following degeneration, there is phagocytosis of the remains.

Even though there is degeneration of the axon and myelin sheath, the neurilemma of the Schwann cells remains. The Schwann cells in the proximal portion of the axon multiply by mitosis and grow toward the injured area. This growth results in the formation of a tube that extends the proximal portion of the axon into the injured area. This "tunnel" provides a means for new axons to grow from the proximal area, across the injured area, and into the injured area so that the severed connections of the axon can be reestablished.

REGENERATION

Accelerated protein synthesis is required for repair of the damaged axon. The proteins synthesized in the cell body pass into the axon at about the rate of 1 mm (0.039 inch)/day. The

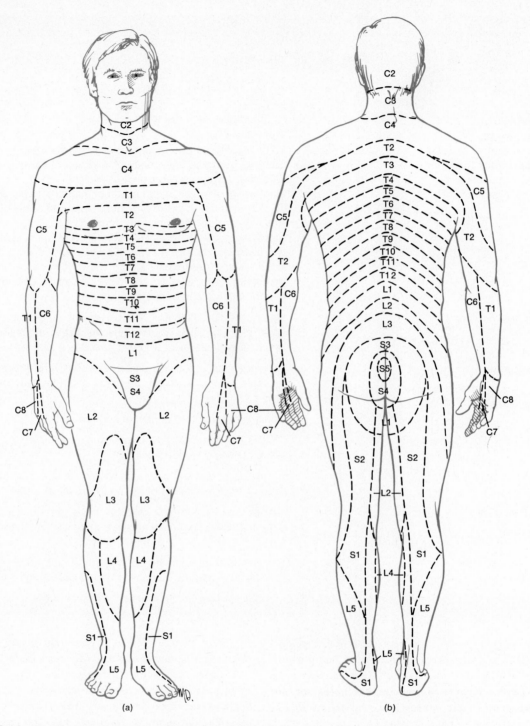

Figure 13-14 Distribution of spinal nerves to dermatomes. (a) Anterior view. (b) Posterior view.

proteins assist in regenerating the damaged axon. During the first few days following damage, regenerating axons begin to invade the tube formed by the Schwann cells. Axons from the proximal area grow at the rate of about 1.5 mm (0.059 inch)/day across the area of damage, find their way into the distal neurilemmal tubes, and grow toward the distally located receptors and effectors. Thus sensory and motor connections are reestablished. In time, a new myelin sheath is also produced by the Schwann cells. However, function is never completely restored after a nerve is severed.

SCIATICA

Sciatica is a type of neuritis characterized by severe pain along the path of the sciatic nerve or its branches. The term is commonly applied to a number of disorders affecting this nerve. Because of its length and size, the sciatic nerve is exposed to many kinds of injury. Inflammation of or injury to the nerve causes pain that passes from the back or thigh down its length into the leg, foot, and toes. Probably the most common cause of sciatica is a slipped or herniated disc. Other causes include irritation from osteoarthritis, back injuries, or pressure on the nerve from certain types of exertion. Other cases are idiopathic and sciatica may be associated with diabetes mellitus, gout, or vitamin deficiencies.

NEURITIS

Neuritis is inflammation of a single nerve, two or more nerves in separate areas, or many nerves simultaneously. It may result from irritation to the nerve produced by direct blows, bone fractures, contusions, or penetrating injuries. Additional causes include vitamin deficiency (usually thiamine) and poisons such as carbon monoxide, carbon tetrachloride, heavy metals, and some drugs. Neuritis exists in many forms and is usually considered a symptom rather than a disease. A thorough physical examination, along with laboratory studies, is necessary to discover its exact cause.

MEDICAL TERMINOLOGY

Analgesia (*an* = without; *algia* = painful condition)   Insensitivity to pain.

Anesthesia (*esthesia* = feeling)   Loss of feeling.

Coma   Abnormally deep unconsciousness with an absence of voluntary response to stimuli; varying degrees of reflex activity remain. A coma may be due to illness or injury.

Contusion   Injury to tissues without breakage of skin; a bruise.

Epidural   External to the dura mater.

Idiopathic   Self-originated; occurring without known cause or due to some other condition already present.

Lethargy   A condition of functional torpor or sluggishness.

Neuralgia (*neur* = nerve)   Attacks of pain along the entire length or branch of a peripheral sensory nerve.

Paralysis   Diminished or total loss of motor function resulting from damage to nervous tissue or muscle.

Stupor   Condition of unconsciousness, torpor, or lethargy with suppression of sense or feeling.

Torpor   Abnormal inactivity or no response to normal stimuli.

STUDY OUTLINE

GROUPING OF NEURAL TISSUE

1 White matter is an aggregation of myelinated axons and associated neuroglia.

2 Gray matter is a collection of nerve cell bodies and dendrites or unmyelinated axons along with associated neuroglia.

3 A nerve is a bundle of nerve fibers outside the central nervous system.

4 A ganglion is a collection of cell bodies outside the central nervous system.

5 A bundle of fibers of similar function in the central nervous system forms a tract.

6 A mass of nerve cell bodies and dendrites in the gray matter of the brain forms a nucleus.

7 A horn or column is an area of gray matter in the spinal cord.

SPINAL CORD
GENERAL FEATURES

1 The gray matter in the spinal cord is divided into horns and the white matter into funiculi or columns.

2 In the center of the spinal cord is the central canal, which runs the length of the spinal cord and contains cerebrospinal fluid.

3 There are ascending (sensory) tracts and descending (motor) tracts.

PROTECTION AND COVERINGS

1 The spinal cord is protected by the vertebral canal, the meninges, the cerebrospinal fluid, and the vertebral ligaments.

2 The meninges are three coverings that run continuously around the spinal cord and brain: dura mater, arachnoid, and pia mater.

3 Removal of cerebrospinal fluid from the subarachnoid space or ventricle is called a spinal (lumbar) puncture.

4 A spinal puncture is used to diagnose pathologies and to introduce antibiotics and anesthetics.

FUNCTIONS AND REFLEXES

1 A major function of the spinal cord is to convey sensory impulses from the periphery to the brain and to conduct major impulses from the brain to the periphery.

2 Another function is to provide reflexes.

3 A reflex arc is the shortest route that can be taken by an impulse from a receptor to an effector.

4 A two-neuron reflex arc contains one sensory and one motor neuron. Stretch reflexes such as the patellar reflex are all monosynaptic.

5 A multineuron reflex arc contains a sensory, association, and motor neuron. A withdrawal or flexor reflex such as pulling the hand away from a hot object is an example.

6 A reflex is a quick, involuntary response to a stimulus that passes along a reflex arc.

7 Reflexes represent the body's principal mechanisms for responding to changes in the internal and external environment to maintain homeostasis.

8 Two types of reflexes are the deep tendon reflexes and the superficial reflexes.

9 Among clinically important somatic reflexes are the patellar reflex, the Achilles reflex, the Babinski reflex, and the abdominal reflex.

SPINAL NERVES

1 Thirty-one pairs of spinal nerves originate from the spinal cord. They are attached to the cord by a dorsal and a ventral root. All are mixed nerves.

2 The principal branches of the spinal nerves are the dorsal and ventral rami and the visceral branches (rami communicantes).

3 In all but the thoracic level, the ventral branches form plexuses (networks of fibers) before they innervate various parts of the body.

4 A dermatome is the skin segment supplied by a spinal nerve.

5 Dermatomes are used to locate spinal nerve dysfunction.

REVIEW QUESTIONS

1 Define the following terms: white matter, gray matter, nerve, ganglion, tract, nucleus.

2 How is the central nervous system differentiated from the peripheral nervous system? What are the components of the peripheral nervous system?

3 Where is the spinal cord located? Describe its general external appearance.

4 Diagram and label a cross section of the spinal cord and explain the function of each area. What are the functions of the spinal cord?

5 Compare the ascending and descending tracts of the spinal cord with respect to location, origin, termination, and function.

6 What is a spinal (lumbar) puncture?

7 Define a reflex arc. What are its components? Distinguish between a monosynaptic and polysynaptic reflex arc.

8 Define a reflex. How are reflexes related to the maintenance of homeostasis?

9 Distinguish the following reflexes: somatic, visceral, intersegmental, ipsilateral, contralateral.

10 Indicate the clinical importance of the following reflexes: patellar, Achilles, Babinski, abdominal.

11 What is a spinal nerve? How is a spinal nerve related to the spinal cord? How are spinal nerves named?

12 What are the four branches of a spinal nerve? Describe the distribution of each.

13 Define a plexus. What are the principal plexuses? What are the general destinations of their peripheral nerves?

14 Define a dermatome.

Chapter 14
The Brain
and the
Cranial Nerves

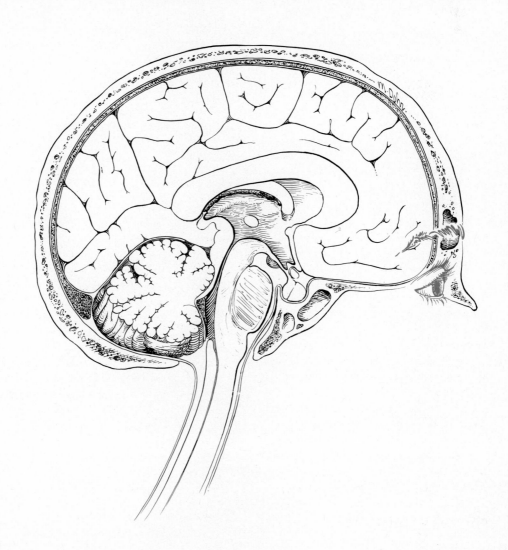

STUDENT OBJECTIVES

■ Identify the structures responsible for protecting the brain.

■ Describe the formation and circulation of cerebrospinal fluid.

■ Define the condition hydrocephalus and its treatment.

■ Describe the blood–brain barrier.

■ Compare the components of the brain stem with regard to structure and function.

■ Identify the structural features of the cerebrum.

■ Compare the motor, association, and sensory functions of the cerebrum.

■ Describe the principle of an electroencephalograph and its significance in the diagnosis of certain disorders.

■ Identify the anatomical characteristics and functions of the cerebellum.

■ Identify by number and name the 12 pairs of cranial nerves.

■ List the clinical symptoms of these disorders of the nervous system: poliomyelitis, syphilis, cerebral palsy, Parkinsonism, multiple sclerosis, cerebral vascular accidents, and dyslexia.

■ Describe Tay-Sachs disease.

■ Define medical terminology associated with the central nervous system.

Now we will consider how the brain is protected, what its principal parts are, how it is related to the spinal cord, and how it is related to the 12 pairs of cranial nerves.

BRAIN

Protection and Coverings

The **brain** of an average adult is one of the largest organs of the body, weighing about 1,300 g (3 lb). Figure 14-1 shows that the brain is mushroom-shaped. It is divided into four principal parts: brain stem, diencephalon, cerebrum, and cerebellum. The **brain stem,** the stalk of the mushroom, consists of the medulla oblongata, pons varolii, and midbrain. The lower end of the brain stem is a continuation of the spinal cord. Above the brain stem is the **diencephalon,** consisting primarily of the thalamus and hypothalamus. The **cerebrum** spreads over the diencephalon. The cerebrum constitutes about seven-eighths of the total weight of the brain and occupies most of the skull. Inferior to the cerebrum and posterior to the brain stem is the **cerebellum.**

The brain is protected by the cranial bones. Like the spinal cord, the brain is also protected by meninges. The *cranial meninges* surround the brain and are continuous with the spinal meninges. The cranial meninges have the same basic structure and bear the same names as the spinal meninges: the outermost *dura mater,* middle *arachnoid,* and innermost *pia mater* (Figure 14-2). The cranial dura mater consists of two layers. The thicker, outer layer (periosteal layer) lightly adheres to the cranial bones and serves as a periosteum. The thinner, inner layer (meningeal layer) includes a mesothelial layer on its smooth surface. The spinal dura mater corresponds to the meningeal layer of the cranial dura mater.

Cerebrospinal Fluid

The brain, as well as the rest of the central nervous system, is further protected against injury by **cerebrospinal fluid.** This fluid circulates through the subarachnoid space around the brain and spinal cord and through the ventricles of the brain. The subarachnoid space is the area between the arachnoid and pia mater. The **ventricles** are cavities in the brain that communicate with each other, with the central canal of the spinal cord, and with the subarachnoid space. Each of the two *lateral ventricles* is located in a hemisphere (side) of the cerebrum under the corpus callosum (Figure 14-2). The *third ventricle* is a slit between and inferior to the right and left halves of the thalamus and between the lateral ventricles. Each lateral ventricle communicates with the

third ventricle by a narrow, oval opening: the *interventricular foramen,* or *foramen of Monro.* The *fourth ventricle* lies between the interior brain stem and the cerebellum. It communicates with the third ventricle via the *cerebral aqueduct (aqueduct of Sylvius),* which passes through the midbrain. The roof of the fourth ventricle has three openings: a *median aperture (foramen of Magendie)* and two *lateral apertures (foramina of Luschka).* Through these openings, the fourth ventricle also communicates with the subarachnoid space of the brain and cord.

The entire central nervous system contains about 125 ml (4 oz) of cerebrospinal fluid. It is a clear, colorless fluid of watery consistency. Chemically, it contains proteins, glucose, urea, and salts. It also contains some white blood cells. The fluid serves as a shock absorber for the central nervous system. It also circulates nutritive substances filtered from the blood. Cerebrospinal fluid is formed primarily by filtration from networks of capillaries, called **choroid plexuses,** located in the ventricles. The fluid formed in the choroid plexuses of the lateral ventricles circulates through the interventricular foramen to the third ventricle, where more fluid is added by the choroid plexus in the third ventricle. It then flows through the cerebral aqueduct to the fourth ventricle. Here there are contributions from the choroid plexus in the fourth ventricle. The fluid then circulates through the apertures of the fourth ventricle into the subarachnoid space around the back of the brain. It also passes downward to the subarachnoid space around the posterior surface of the spinal cord, up the anterior surface of the spinal cord, and around the anterior part of the brain. From here it is gradually reabsorbed into veins. Some cerebrospinal fluid may be formed by ependymal (neuroglial) cells lining the central canal of the spinal cord. This small quantity of fluid ascends to reach the fourth ventricle. Most of the fluid is absorbed into the superior sagittal sinus. The absorption actually occurs through **arachnoid villi** – fingerlike projections of the arachnoid that push into the superior sagittal sinus. Normally, cerebrospinal fluid is absorbed as rapidly as it is formed.

If an obstruction, such as a tumor, arises in the brain and interferes with the drainage of fluid from the ventricles into the subarachnoid space, large amounts of fluid accumulate in the ventricles. Fluid pressure inside the brain increases, and, if the fontanels have not yet closed, the head bulges to relieve the pressure. This condition is called **internal hydrocephalus** (*hydro* = water; *cephalo* = head). If an obstruction interferes with drainage somewhere in the subarachnoid space and cerebrospinal fluid accumulates inside the space, the condition is termed **external hydrocephalus.**

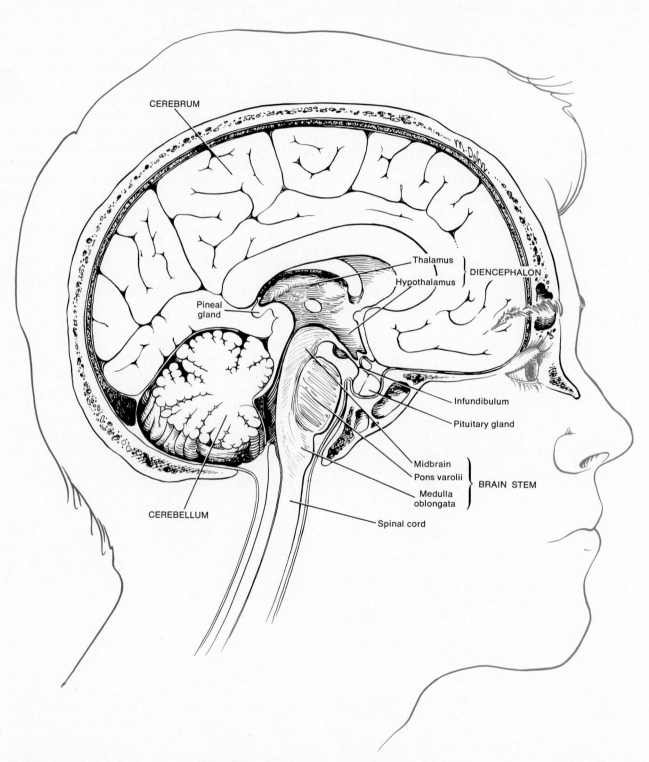

Figure 14-1 Brain. (a) Principal parts of the medial aspect of the brain seen in sagittal section. The infundibulum and pituitary gland are discussed in conjunction with the endocrine system.

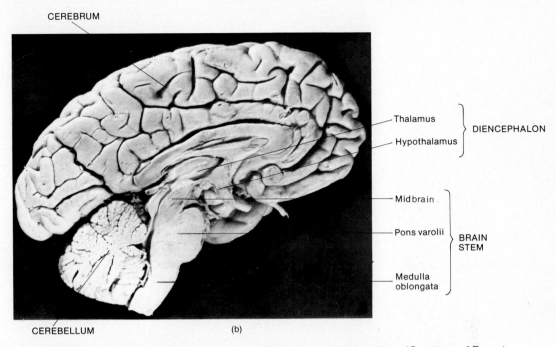

CEREBRUM

Thalamus
Hypothalamus
} DIENCEPHALON

Midbrain
Pons varolii
Medulla oblongata
} BRAIN STEM

CEREBELLUM

(b)

Figure 14-1 (cont.)   Brain. (b) Photograph of the medial aspect of the brain seen in sagittal section. (Courtesy of Ernest Gardner *et al., Anatomy: A Regional Study of Human Structure,* 4th ed., W. B. Saunders, Philadelphia, 1975.)

Blood Supply

The brain is well supplied with blood vessels, which supply oxygen and nutrients. Although the brain actually consumes less oxygen than most other organs of the body, it must receive a constant supply. If the blood flow to the brain is interrupted for even a few moments, unconsciousness may result. A 1- or 2-minute interruption may weaken the brain cells by starving them of oxygen. If the cells are totally deprived of oxygen for 4 minutes, many are permanently injured. Occasionally during childbirth the oxygen supply from the mother's blood is interrupted before the baby leaves the birth canal and can breathe. Often such babies are stillborn or suffer permanent brain damage that may result in mental retardation, epilepsy, and paralysis.

Blood supplying the brain also contains glucose, the principal source of energy for brain cells. Because carbohydrate storage in the brain is limited, the supply of glucose must be continuous. If blood entering the brain has a low glucose level, mental confusion, dizziness, convulsions, and even loss of consciousness may occur.

Glucose, oxygen, and certain ions pass rapidly from the circulating blood into brain cells. Other substances, such as creatinine, urea, chloride, insulin, and sucrose, enter quite slowly. Still other substances—proteins and most antibiotics—do not pass at all from the blood into brain cells. The differential rates of passage of certain

materials from the blood into the brain suggest a concept called the **blood–brain barrier.** Electron micrograph studies of the capillaries of the brain reveal that they differ from other capillaries. Brain capillaries either lack pores or contain fewer pores than other body capillaries. Brain capillaries are also constructed of more densely packed cells and are surrounded by large numbers of glial cells and a continuous basement membrane. These features form a barrier to the passage of certain materials. Thus substances that cross the barrier are either very small molecules or require the assistance of a carrier molecule to cross by active transport. The function of the blood–brain barrier is not known. It may protect brain cells from harmful substances.

Brain Stem

MEDULLA OBLONGATA

The **medulla oblongata,** or simply **medulla,** is a continuation of the upper portion of the spinal cord and forms the inferior part of the brain stem. Its position in relation to the other parts of the brain may be noted in Figure 14-1. It lies just superior to the level of the foramen magnum and extends upward to the inferior portion of the pons varolii. The medulla measures only 3 cm (about 1 inch) in length.

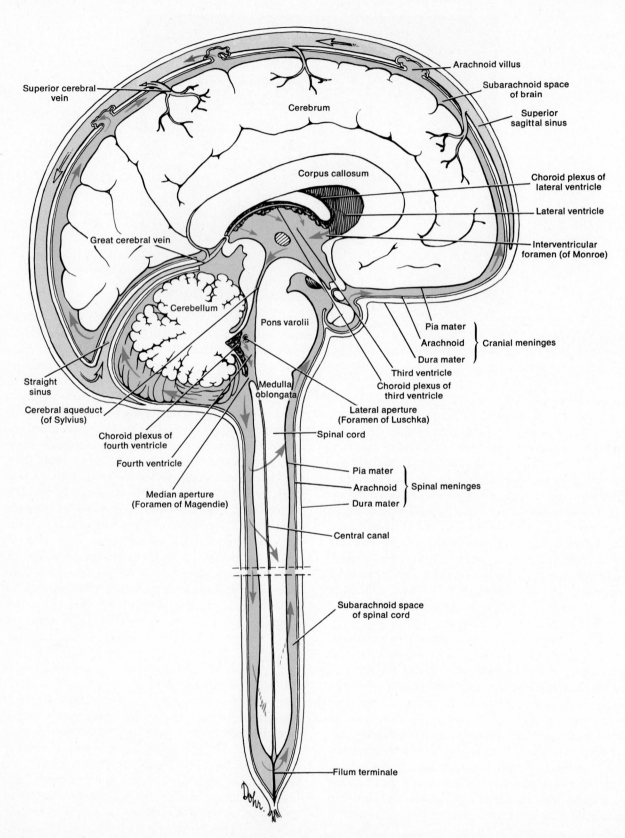

Figure 14-2 Brain and meninges seen in sagittal section. The direction of flow of cerebrospinal fluid is indicated by colored arrows.

The medulla contains all ascending and descending tracts that communicate between the spinal cord and various parts of the brain. These tracts constitute the white matter of the medulla. Some tracts cross as they pass through the medulla. Let us see how this crossing occurs and what it means.

On the ventral side of the medulla are two roughly triangular structures called *pyramids* (Figure 14-3). The pyramids are composed of the largest motor tracts that pass from the outer region of the cerebrum (cerebral cortex) to the spinal cord. Just above the junction of the medulla with the spinal cord, most of the fibers in the left pyramid cross to the right side, and most of the fibers in the right pyramid cross to the left. This crossing is called the **decussation of pyramids**. The adaptive value, if any, of this phenomenon is unknown. The principal motor fibers that undergo decussation belong to the lateral corticospinal tracts. These tracts originate in the cerebral cortex and pass inferiorly to the medulla. The fibers cross in the pyramids and descend in the lateral columns of the spinal cord, terminating in the anterior gray horns. Here synapses occur with motor neurons that terminate in skeletal muscles. As a result of the crossing, fibers that originate in the left cerebral cortex activate muscles on the right side of the body, and fibers that originate in the right cerebral cortex activate muscles on the left side. Decussation explains why motor areas of one side of the cerebral cortex control muscular movements on the opposite side of the body.

The dorsal side of the medulla contains two pairs of prominent nuclei: the right and left *nucleus gracilis* and *nucleus cuneatus.* These nuclei receive sensory fibers from ascending tracts (right and left fasciculus gracilis and fasciculus cuneatus) of the spinal cord and relay the sensory information to the opposite side of the medulla. This information is conveyed to the thalamus and then to the sensory areas of the cerebral cortex. Nearly all sensory impulses received on one side of the body cross in the medulla or spinal cord and are perceived in the opposite side of the cerebral cortex.

In addition to its function as a conduction pathway for motor and sensory impulses between the brain and spinal cord, the medulla also contains an area of dispersed gray matter containing some white fibers. This region is called the **reticular formation.** Actually, portions of the reticular formation are located in the spinal cord, pons, midbrain, and diencephalon. The reticular formation functions in consciousness and arousal. Within the medulla are three vital reflex centers of the reticular system. The *cardiac center* regulates heart beat; the *respiratory center* adjusts the rate and depth of breathing; the *vasoconstrictor center* regulates the diameter of blood vessels. Other centers in the medulla coordinate swallowing, vomiting, coughing, sneezing, and hiccuping.

The medulla also contains the nuclei of origin for four pairs of cranial nerves. These are the cochlear and vestibular branches of the vestibulocochlear nerves (VIII), which are concerned with hearing and equilibrium (there is also a nucleus for the vestibular branches in the pons); the glossopharyngeal nerves (IX), which relay impulses related to swallowing, salivation, and taste; the vagus nerves (X), which relay impulses to and from many thoracic and abdominal viscera; the accessory nerves (XI), which convey impulses related to head and shoulder movements (a part of this nerve also arises from the first five segments of the spinal cord); and the hypoglossal nerves (XII), which convey impulses that involve tongue movements.

On each lateral surface of the medulla is an oval projection called the *olive.* The olive contains an inferior olivary nucleus and two accessory olivary nuclei. The nuclei are connected to the cerebellum by fibers.

PONS VAROLII

The relationship of the **pons varolii** or **pons** to other parts of the brain can be seen in Figures 14–1 and 14–3a. The pons, which means bridge, lies directly above the medulla and anterior to the cerebellum. It measures about 2.5 cm (1 inch) in length. Like the medulla, the pons consists of white fibers scattered throughout with nuclei. As the name implies, the pons is a bridge connecting the spinal cord with the brain and parts of the brain with each other. These connections are provided by fibers that run in two principal directions. The transverse fibers connect with the cerebellum through the middle cerebellar peduncles. The longitudinal fibers of the pons belong to the motor and sensory tracts that connect the spinal cord or medulla with the upper parts of the brain stem.

The nuclei for certain paired cranial nerves are also contained in the pons. These include the trigeminal nerves (V), which relay impulses for chewing and for sensations of the head and face; the abducens nerves (VI), which regulate certain eyeball movements; the facial nerves (VII), which conduct impulses related to taste, salivation, and facial expression; and the vestibular branches of the vestibulocochlear nerves (VIII), which are concerned with equilibrium.

Another important nucleus in the reticular formation of the pons is the *pneumotaxic center.* Together with the respiratory center in the medulla, it helps control respiration.

MIDBRAIN

The **midbrain** or **mesencephalon** extends from the pons to the lower portion of the diencephalon (Figure 14-1). It is about 2.5 cm (1 inch) in length. The cerebral aqueduct passes through the midbrain and connects the third ventricle above with the fourth ventricle below.

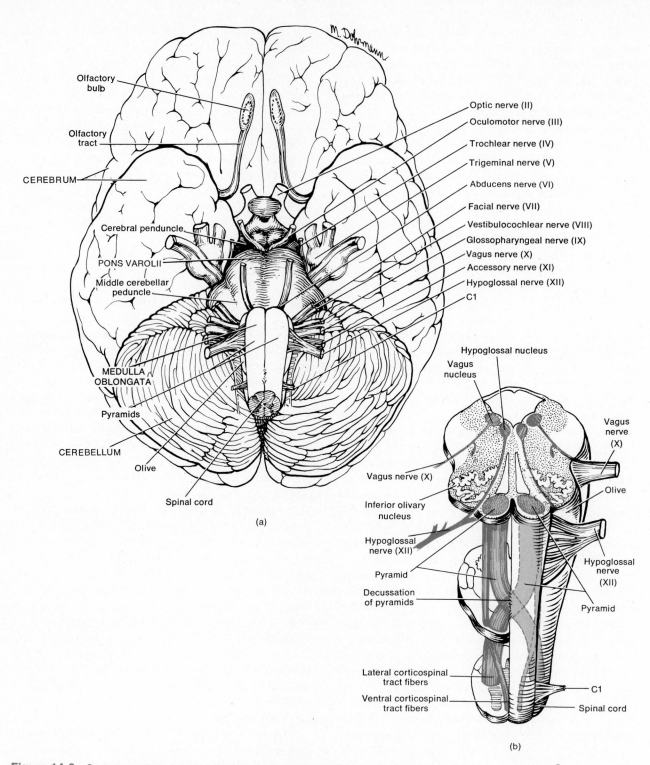

Figure 14-3   Structure of the brain. (a) Ventral view. (b) Details of the medulla.

The ventral portion of the midbrain contains a pair of fiber bundles referred to as *cerebral peduncles*. The cerebral peduncles contain many motor fibers that convey impulses from the cerebral cortex to the pons and spinal cord. They also contain sensory fibers that pass from the spinal cord to the thalamus. The cerebral peduncles constitute the main connection for tracts between upper parts of the brain and lower parts of the brain and the spinal cord.

The dorsal portion of the midbrain is called the *tectum* and contains four rounded eminences: the *corpora quadrigemina*. Two of the eminences are known as the *superior colliculi.* These serve as a reflex center for movements of the eyeballs and head in response to visual and other stimuli. The other two eminences are the *inferior colliculi.* They serve as reflex centers for movements of the head and trunk in response to auditory stimuli. The midbrain also contains the *substantia nigra,* a large, heavily pigmented nucleus near the cerebral peduncles.

A major nucleus in the reticular formation of the midbrain is the *red nucleus.* Fibers from the cerebellum and cerebral cortex terminate in the red nucleus. The red nucleus is also the origin of cell bodies of the descending rubrospinal tract. Other nuclei in the midbrain are associated with cranial nerves. These include the oculomotor nerves (III), which mediate some movements of the eyeballs and changes in pupil size and lens shape, and the trochlear nerves (IV), which conduct impulses that move the eyeballs.

A structure called the *medial lemniscus* is common to the medulla, pons, and midbrain. The medial lemniscus is a band of white fibers containing axons that convey impulses for fine touch, proprioception, and vibrations from the medulla to the thalamus.

Diencephalon

The **diencephalon** consists principally of the thalamus and hypothalamus. The relationship of these structures to the rest of the brain is shown in Figure 14-1.

THALAMUS

The **thalamus** is a large oval structure above the midbrain (Figure 14-4a). It consists of two masses of gray matter covered by a thin layer of white matter. It measures about 3 cm (1 inch) in length and constitutes four-fifths of the diencephalon. The thalamus contains numerous nuclei organized into masses (Figure 14-4c). Some nuclei in the thalamus serve as relay stations for all sensory impulses, except smell, to the cerebral cortex. These include the *medial geniculate nuclei* (hearing), the *lateral geniculate nuclei* (vision), and the *ventral posterior nuclei* (general sensations and taste). Other nuclei are centers for synapses in the somatic motor system. These include the *ventral lateral nuclei* (voluntary motor actions) and *ventral anterior nuclei* (voluntary motor actions and arousal). The thalamus is the principal relay station for sensory impulses that reach the cerebral cortex from the spinal cord, brain stem, cerebellum, and parts of the cerebrum.

The thalamus also functions as an interpretation center. That is, some sensory impulses that enter the thalamus are interpreted there. At the thalamic level, one can have conscious recognition of pain and temperature and some awareness of crude touch and pressure. The thalamus also contains a *reticular nucleus* in its reticular

formation and an *anterior nucleus* in the floor of the lateral ventricle.

HYPOTHALAMUS

The **hypothalamus** is a small portion of the diencephalon and its relationship to other parts of the brain is shown in Figures 14-1 and 14-4a. The hypothalamus forms the floor and part of the lateral walls of the third ventricle. The hypothalamus is protected by the sphenoid bone and indirectly by the sella turcica of the sphenoid bone. Despite its small size, nuclei in the hypothalamus control many body activities, most of them related to homeostasis. The chief functions of the hypothalamus are these:

1 It controls and integrates the autonomic nervous system, which stimulates smooth muscle, regulates the rate of contraction of cardiac muscle, and controls the secretions of many glands.. Through the autonomic nervous system, the hypothalamus is the main regulator of visceral activities. It regulates heart rate, movement of food through the digestive tract, and contraction of the urinary bladder.

2 It is involved in the reception of sensory impulses from the viscera.

3 It is the principal intermediary between the nervous system and endocrine system – the two major control systems of the body. The hypothalamus lies just above the pituitary, the main endocrine gland. When the hypothalamus detects certain changes in the body, it releases chemicals called regulating factors that stimulate or inhibit the anterior pituitary gland. The anterior pituitary then releases or holds back hormones that regulate carbohydrates, fats, proteins, certain ions, and sexual functions.

4 It is the center for the mind-over-body phenomenon. When the cerebral cortex interprets strong emotions, it often sends impulses along tracts that connect the cortex with the hypothalamus. The hypothalamus then directs impulses via the autonomic nervous system and also releases chemicals that stimulate the anterior pituitary gland. The result can be a wide range of changes in body activities. For instance, when you panic, impulses leave the hypothalamus to stimulate your heart to beat faster. Likewise, continued psychological stress can produce long-term abnormalities in body function that result in serious illness. These are so-called psychosomatic disorders. Psychosomatic disorders are real.

5 It is associated with feelings of rage and aggression.

6 It controls normal body temperature. Certain cells of the hypothalamus serve as a thermostat – a mechanism sensitive to changes in temperature. If blood flowing through the hypothalamus is above normal

Everything is in duplicate

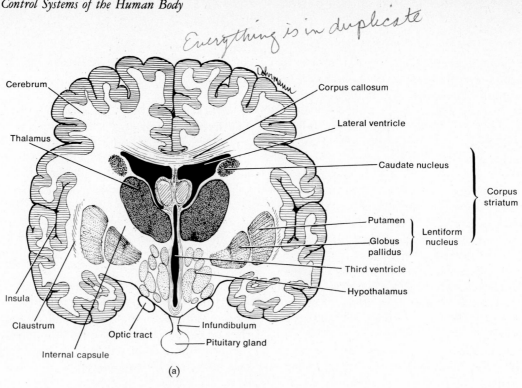

(a)

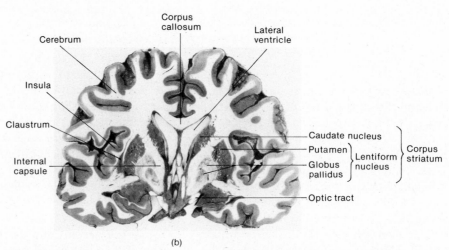

(b)

Figure 14-4   Thalamus. (a) Frontal section showing the thalamus and associated structures. (b) Photograph of a frontal section of the cerebrum anterior to the thalamus. (Courtesy of Murray L. Barr, *The Human Nervous System,* Harper & Row Publishers, Inc., 1974.)

temperature, the hypothalamus directs impulses along the autonomic nervous system to stimulate activities that promote heat loss. Heat can be lost through relaxation of the smooth muscle in the blood vessels and by sweating. Conversely, if the temperature of the blood is below normal, the hypothalamus generates impulses that promote heat retention. Heat can be retained through the contraction of cutaneous blood vessels, cessation of sweating, and shivering.

7 It regulates food intake through two centers.

The *feeding center* is stimulated by hunger sensations from an empty stomach. When sufficient food has been ingested, the *satiety center* is stimulated and sends out impulses that inhibit the feeding center.

8 It contains a *thirst center.* Certain cells in the hypothalamus are stimulated when the extracellular fluid volume is reduced. The stimulated cells produce the sensation of thirst in the hypothalamus.

9 It is one of the centers that maintain the waking state and sleep patterns.

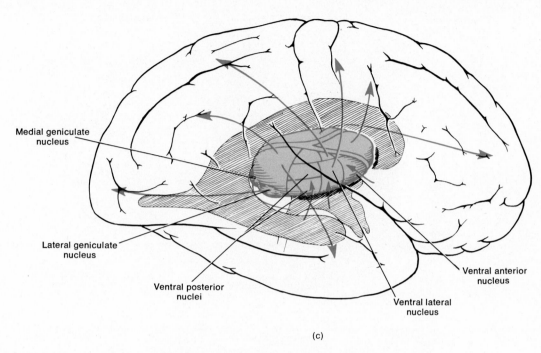

Medial geniculate
nucleus

Lateral geniculate
nucleus

Ventral posterior
nuclei

Ventral lateral
nucleus

Ventral anterior
nucleus

(c)

Figure 14-4 (cont.) Thalamus. (c) Right lateral view of the thalamic nuclei.

Cerebrum

Supported on the brain stem and forming the bulk of the brain is the **cerebrum** (Figure 14-1). The surface of the cerebrum is composed of gray matter 2 to 4 mm (0.08 to 0.16 inch) thick and is referred to as the *cerebral cortex* (*cortex* = rind or bark). The cortex, containing millions of cells, consists of six layers of nerve cell bodies. Beneath the cortex lies the cerebral white matter.

During embryonic development when there is a rapid increase in brain size, the gray matter of the cortex enlarges out of proportion to the underlying white matter. As a result, the cortical region rolls and folds upon itself. The upfolds are called *gyri* or *convolutions* (Figure 14-5b). The deep downfolds are referred to as *fissures;* the shallow downfolds are *sulci.* The most prominent fissure, the *longitudinal fissure,* nearly separates the cerebrum into right and left halves, or *hemispheres* (Figure 14-5b). The hemispheres, however, are connected internally by a large bundle of transverse fibers composed of white matter called the *corpus callosum.* Between the hemispheres in an extension of the cranial dura mater called the *falx cerebri.*

LOBES

Each cerebral hemisphere is further subdivided into four lobes by deep sulci or fissures (Figure 14-5). The *central*

sulcus, or *fissure of Rolando,* separates the *frontal lobe* from the *parietal lobe.* A major gyrus, the *precentral gyrus,* is located immediately anterior to the central sulcus. The *lateral cerebral sulcus,* or *fissure of Sylvius,* separates the *frontal lobe* from the *temporal lobe.* The *parietooccipital sulcus* separates the *parietal lobe* from the *occipital lobe.* Another prominent fissure, the *transverse fissure,* separates the cerebrum from the cerebellum. The frontal lobe, parietal lobe, temporal lobe, and occipital lobe are named after the bones that cover them. A fifth part of the cerebrum, the *insula (island of Reil),* lies deep within the lateral cerebral fissure, under the parietal, frontal, and temporal lobes. It cannot be seen in an external view of the brain.

WHITE MATTER

The white matter underlying the cortex consists of myelinated axons running in three principal directions:

1 Association fibers connect and transmit impulses between gyri in the same hemisphere.

2 Commissural fibers transmit impulses from the gyri in one cerebral hemisphere to the corresponding gyri in the opposite cerebral hemisphere. Three impor-

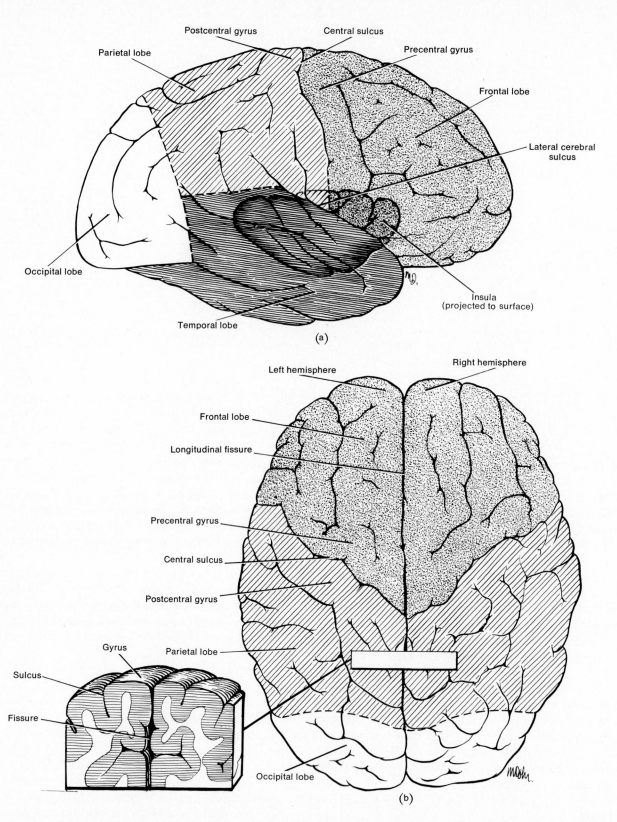

Figure 14-5 Lobes and fissures of the cerebrum. (a) Right lateral view. Since the insula cannot be seen externally, it has been projected to the surface. It can be seen in Figure 14-4a. (b) Superior view. The insert in (b) indicates the relative differences between a gyrus, sulcus, and fissure.

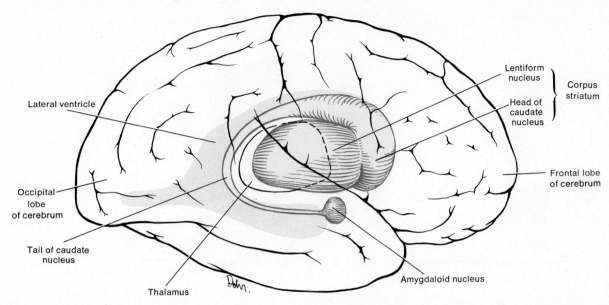

Figure 14-6 Basal ganglia. In this right lateral view of the cerebrum, the basal ganglia have been projected to the surface. Refer to Figure 14-4a to note the positions of the basal ganglia in the frontal section of the cerebrum.

tant groups of commissural fibers are the *corpus callosum, anterior commissure,* and *posterior commissure.*

3 Projection fibers form ascending and descending tracts that transmit impulses from the cerebrum to other parts of the brain and spinal cord.

BASAL GANGLIA

The **basal ganglia** or **cerebral nuclei** are paired masses of gray matter in each cerebral hemisphere (Figure 14-6). The largest of the basal ganglia of each hemisphere is the *corpus striatum.* It consists of the *caudate nucleus* and the *lentiform nucleus.* The lentiform nucleus, in turn, is subdivided into a lateral portion called the *putamen* and a medial portion called the *globus pallidus.* Figure 14-4a shows the two divisions of the lentiform nucleus and a structure called the *internal capsule.* It is made up of a group of sensory and motor white matter tracts that connect the cerebral cortex with the brain stem and spinal cord. The portion of the internal capsule passing between the lentiform nucleus and the caudate nucleus and between the lentiform nucleus and thalamus is sometimes considered part of the corpus striatum.

Other structures frequently considered part of the basal ganglia are the claustrum and amygdaloid nucleus. The *claustrum* is a thin sheet of gray matter lateral to the putamen. The *amygdaloid nucleus* is located at the tail end of the caudate nucleus. Some authorities also consider the *substantia nigra,* the *subthalamic nucleus,* and the *red nucleus* to be part of the basal ganglia. The substantia nigra is a large motor nucleus in the midbrain.

The subthalamic nucleus lies against the internal capsule. Its major connection is with the globus pallidus.

The basal ganglia are interconnected by many fibers. They are also connected to the cerebral cortex, thalamus, and hypothalamus. The caudate nucleus and the putamen control large subconscious movements of the skeletal muscles—such as swinging the arms while walking. Such gross movements are also consciously controlled by the cerebral cortex. The globus pallidus is concerned with the regulation of muscle tone required for specific body movements.

LIMBIC SYSTEM

Certain components of the cerebral hemispheres and diencephalon constitute the **limbic system.** It includes the following regions of gray matter:

1 Limbic lobe Formed by two gyri of the cerebral hemisphere: the cingulate gyrus and the hippocampal gyrus.

2 Hippocampus An extension of the hippocampal gyrus that extends into the floor of the lateral ventricle.

3 Amygdaloid nucleus Located at the tail end of the caudate nucleus.

4 Hypothalamus The regions of the hypothalamus that form part of the limbic system are the perifornical nuclei.

5 Anterior nucleus of the thalamus Located in the floor of the lateral ventricle.

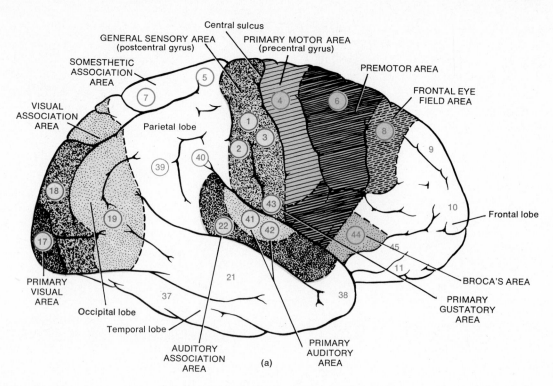

Figure 14-7 Functional areas of the cerebrum. (a) The lateral view indicates the sensory and motor areas. Although the right hemisphere is illustrated, Broca's area is in the left hemisphere of most people.

The limbic system functions in the emotional aspects of behavior related to survival. It also functions in memory. Although behavior is a function of the entire nervous system, the limbic system controls most of its involuntary aspects. Experiments on the limbic system of monkeys and other animals indicate that the amygdaloid nucleus assumes a major role in controlling the overall pattern of behavior – from heart rate to sexual activity.

Other experiments have shown that the limbic system is associated with pleasure and pain. When certain areas of the limbic system of the hypothalamus, thalamus, and midbrain are stimulated, experimental animals indicate they are experiencing intense punishment. When other areas are stimulated, the animals' reactions indicate they are experiencing extreme pleasure. In still other studies, stimulation of the perifornical nuclei of the hypothalamus result in a behavioral pattern called *rage*. The animal assumes a defensive posture – extending its claws, raising its tail, hissing, spitting, growling, and opening its eyes wide. Stimulating other areas of the limbic system results in an opposite behavioral pattern: docility, tameness, and affection.

FUNCTIONAL AREAS OF CEREBRAL CORTEX

The functions of the cerebrum are numerous and complex. In a general way, the cerebral cortex is divided into motor, sensory, and association areas. The **motor** **areas** control muscular movement. The **sensory areas** interpret sensory impulses. And the **association areas** are concerned with emotional and intellectual processes.

■ *Sensory Areas* The *general sensory area* or *somesthetic area* is located directly posterior to the central sulcus of the cerebrum on the postcentral gyrus. It extends from the longitudinal fissure on the top of the cerebrum to the lateral cerebral sulcus. In Figure 14-7a the general sensory area is designated by the areas numbered 1, 2, and 3.* The general sensory area receives sensations from cutaneous, muscular, and visceral receptors in various parts of the body. Each point of the general sensory area receives sensations from specific parts of the body. Essentially the entire body is spatially represented in the general sensory area. The portion of the sensory area receiving stimuli from body parts is not dependent on the size of the part but on the number of receptors. For example, a greater portion of the sensory area receives impulses from the lips than from the thorax. The major function of the general sensory area is to localize exactly the points of the body where the sensations originate. The thalamus is capable of localizing sensations in a

---

These numbers, as well as most of the others shown, are based on K. Brodmann's cytoarchitectural map of the cerebral cortex. His map, first published in 1909, is an attempt to correlate structure and function.

general way. That is, the thalamus receives sensations from large areas of the body but cannot distinguish between specific areas of stimulation. This ability is reserved to the general sensory area of the cortex.

Posterior to the general sensory area is the *somesthetic association area.* It corresponds to the areas numbered 5 and 7 in Figure 14-7a. The somesthetic association area receives input from the thalamus, other lower portions of the brain, and the general sensory area. The somesthetic association area integrates and interprets sensations. This area permits you to determine the exact shape and texture of an object without looking at it, to determine the orientation of one object to another as they are felt, and to sense the relationship of one body part to another. Another role of the somesthetic association area is the storage of memories of past sensory experiences. Thus you can compare sensations with previous experiences.

Other sensory areas of the cortex include:

1 **Primary visual area** (area 17)   Located on the medial surface of the occipital lobe and occasionally extends around to the lateral surface. It receives sensory impulses from the eyes and interprets shape and color.

2 **Visual association area** (areas 18 and 19)   Located in the occipital lobe and receives sensory signals from the primary visual area and the thalamus. It relates present to past visual experiences with recognition and evaluation of what is seen.

3 **Primary auditory area** (areas 41 and 42)   Located in the superior part of the temporal lobe near the lateral cerebral sulcus. It interprets the basic characteristics of sound such as pitch and rhythm.

4 **Auditory association area** (area 22)   Inferior to the primary auditory area in the temporal cortex. It determines if a sound is speech, music, or noise. It also interprets the meaning of speech by translating words into thoughts.

5 **Primary gustatory area** (area 43)   Located at the base of the postcentral gyrus above the lateral cerebral sulcus in the parietal cortex. It interprets sensations related to taste.

6 **Primary olfactory area**   Located in the temporal lobe on the medial aspect and interprets sensations related to smell.

7 **Gnostic area** (areas 5, 7, 39, and 40)   This *common integrative area* is located between the somesthetic, visual, and auditory association areas. The gnostic area receives impulses from these areas, as well as from the taste and smell areas, the thalamus, and lower portions of the brain stem. The gnostic area integrates all thoughts from the various sensory areas so that a common thought can be formed from the various sensory inputs. It then transmits signals to other parts of the brain to cause the appropriate response to the sensory signal.

■ *Motor Areas*   The *primary motor area* is located mainly in the precentral gyrus of the frontal lobe (Figure 14-7a). This region is also designated as area 4. Like the general sensory area, the primary motor area consists of points that control specific muscles or groups of muscles. Stimulation of a specific point of the primary motor area results in a muscular contraction, usually on the opposite side of the body.

The *premotor area* (area 6) is anterior to the primary motor area. It is concerned with learned motor activities of a complex and sequential nature. It generates impulses that cause a specific group of muscles to contract in a specific sequence. An example of this is writing. Thus the premotor area controls skilled movements.

The *frontal eye field area* (area 8) in the frontal cortex is sometimes included in the premotor area. This area controls voluntary scanning movements of the eyes—searching for a word in a dictionary, for instance.

The *language areas* are also significant parts of the motor cortex. When you listen to someone speaking, sounds are relayed to the primary auditory area of the cortex. The sounds are then interpreted as words in the auditory association area. The words are interpreted as thoughts in the gnostic area. Written words are interpreted by the visual association area and converted into thoughts by the gnostic area. Thus you can translate speech or written words into thoughts.

The translation of thoughts into speech involves *Broca's area* or the *motor speech area,* designated as area 44 and located in the frontal lobe just superior to the lateral cerebral sulcus. From this area, a sequence of signals is sent to the premotor regions that control the muscles of the larynx, throat, and mouth. The impulses from the premotor area to the muscles result in specific, coordinated contractions that enable you to speak. Simultaneously, impulses are sent from Broca's area to the primary motor area. From here, impulses reach your breathing muscles to regulate the proper flow of air past the vocal cords. The coordinated contractions of your speech and breathing muscles enable you to translate your thoughts into speech.

Broca's area is usually located in the left cerebral hemisphere of most individuals regardless of whether they are left-handed or right-handed. Injury to the sensory or motor speech areas results in *aphasia,* which is an inability to speak; *agraphia,* an inability to write; *word deafness,* an inability to understand spoken words; or *word blindness,* an inability to understand written words.

■ *Association Areas*   The *association areas* of the cerebrum are made up of association tracts that connect motor and sensory areas (see Figure 14-7b). The association region of the cortex occupies the greater portion of the lateral surfaces of the occipital, parietal, and temporal lobes and the frontal lobes anterior to the motor areas.

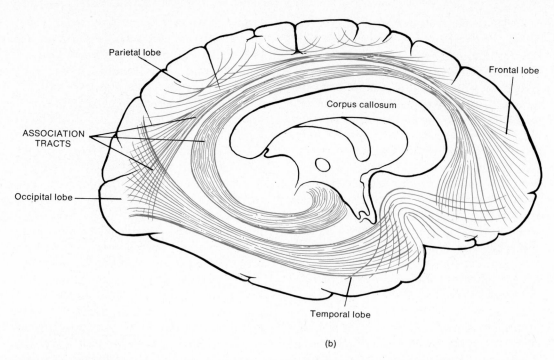

Figure 14-7 (cont.)   Functional areas of the cerebrum. (b) The sagittal section shows the association tracts.

The association areas are concerned with memory, emotions, reasoning, will, judgment, personality traits, and intelligence.

BRAIN WAVES

Brain cells can generate electrical activity as a result of literally millions of action potentials of individual neurons. These electrical potentials are called **brain waves** and indicate activity of the cerebral cortex. Brain waves pass easily through the skull and can be detected by sensors called electrodes. A record of such waves is called an **electroencephalogram (EEG).** An EEG is obtained by placing electrodes on the head and amplifying the waves with an electroencephalograph. As indicated in Figure 14-8, four kinds of waves are produced by normal individuals:

    1 Alpha waves   These rhythmic waves occur at a frequency of about 10 to 12 cycles/second. They are found in the EEGs of nearly all normal individuals when awake and in the resting state. These waves disappear entirely during sleep.

    2 Beta waves   The frequency of these waves is between 15 and 60 cycles/second. Beta waves generally appear when the nervous system is active – that is, during periods of sensory input and mental activity.

    3 Theta waves   These waves have frequencies of 5 to 8 cycles/second. Theta waves normally occur in children and in adults experiencing emotional stress.

    4 Delta waves   The frequency of these waves is between 1 to 5 cycles/second. Delta waves occur during sleep. They are normal in an awake infant. When produced by an awake adult, they indicate brain damage.

    Distinct EEG patterns appear in certain abnormalities. In fact, the EEG is used clinically in the diagnosis of epilepsy, infectious diseases, tumors, trauma, and hematomas. Electroencephalograms also furnish information regarding sleep and wakefulness.

Cerebellum

The **cerebellum** is the second-largest portion of the brain and occupies the inferior and posterior aspects of the cranial cavity. Specifically, it is below the posterior portion of the cerebrum and is separated from it by the *transverse fissure* (see Figure 14-1). The cerebellum is also separated from the cerebrum by an extension of the cranial dura mater called the *tentorium cerebelli.* The cerebellum is shaped somewhat like a butterfly. The central constricted area is the *vermis,* which means worm-shaped, and the lateral "wings" are referred to as *hemispheres* (Figure 14-9). Between the hemispheres is another extension of the cranial dura mater: the *falx cerebelli.* It passes only a short distance between the cerebellar hemispheres.

The surface of the cerebellum, called the *cortex,* consists of gray matter in a series of slender, parallel ridges called *gyri.* These gyri are less prominent than those located on the cerebral cortex. Beneath the gray matter are white matter tracts (*arbor vitae*) that resemble branches of a tree. Deep within the white matter are masses of gray matter: the *cerebellar nuclei.*

The cerebellum is attached to the brain stem by three paired bundles of fibers called *cerebellar peduncles.* These are as follows:

1 Inferior cerebellar peduncles, which connect the cerebellum with the medulla at the base of the brain stem and with the spinal cord.

2 Middle cerebellar peduncles, which connect the cerebellum with the pons.

3 Superior cerebellar peduncles, which connect the cerebellum with the midbrain.

The cerebellum is a motor area of the brain that produces certain subconscious movements in the skeletal muscles. These movements are required for coordination, for maintenance of posture, and for balance. The cerebellar peduncles are the fiber tracts that allow the cerebellum to perform its functions.

Let us now see how the cerebellum produces coordinated movement. Motor areas of the cerebral cortex voluntarily initiate muscle contraction. Once the movement has begun, the sensory areas of the cortex receive impulses from nerves in the joints. The impulses provide information about the extent of muscle contraction and the amount of joint movement. The term *proprioception* is applied to this sense of the position of one body part relative to another. The cerebral cortex uses the proprioceptive sensations to determine which muscles are required to contract next and what strength they are to contract in order to continue moving in the desired direction. Then a pattern of impulses is generated by the cerebral cortex along tracts to the pons and midbrain, which relay the impulses over the middle and superior cerebellar peduncles to the cerebellum. The cerebellum then generates subconscious motor impulses along the inferior cerebellar peduncles to the medulla and spinal cord. The impulses pass downward along the spinal cord and out the nerves that stimulate the prime movers and synergists to contract and that inhibit the contraction of the antagonists. The result is smooth, coordinated movement. A well-functioning cerebellum is essential for delicate movements such as playing the piano.

The cerebellum also transmits impulses that control postural muscles. That is, the cerebellum is required for maintaining normal muscle tone. The cerebellum

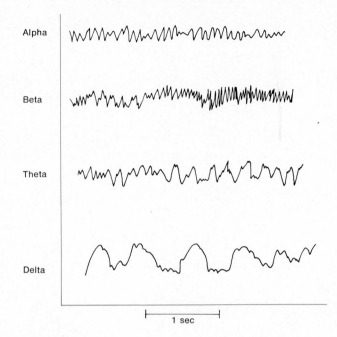

Figure 14-8 Kinds of waves recorded in an electroencephalograph.

also maintains body equilibrium. The inner ear contains structures that sense balance. Information such as whether the body is leaning to the left or right is transmitted from the inner ear to the cerebellum. The cerebellum then discharges impulses that cause the contraction of the muscles necessary for maintaining equilibrium.

Damage to the cerebellum through trauma or disease is characterized by certain symptoms involving skeletal muscles. There may be lack of muscle coordination, called *ataxia.* Blindfolded people with ataxia cannot touch the tip of their nose with a finger because they cannot coordinate movement with their sense of where a body part is located. Another sign of ataxia is a change in the speech pattern due to a lack of coordination of speech muscles. Cerebellar damage may also result in disturbances of gait in which the subject staggers or cannot coordinate normal walking movements.

CRANIAL NERVES

Of the 12 pairs of **cranial nerves,** 10 pairs originate from the brain stem, but all 12 pairs leave the skull through foramina in the base of the skull (see Figure 14-3a). The cranial nerves are designated in two ways—with Roman numerals and with names. The Roman numerals indicate the order in which the nerves arise

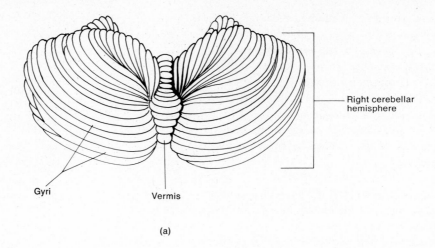

(a)

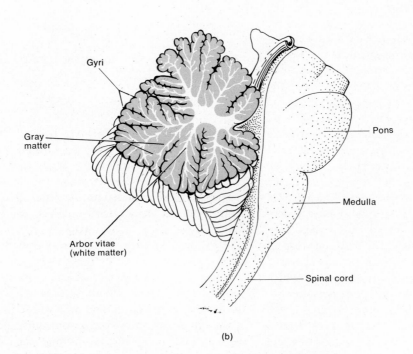

(b)

Figure 14-9   Cerebellum. (a) Viewed from below. (b) Viewed in sagittal section.

from the brain (front to back). The names indicate the distribution or function of the nerves. Some cranial nerves are termed *mixed nerves*. They contain both sensory and motor fibers. Other cranial nerves contain sensory fibers only. The cell bodies of sensory fibers are located in ganglia outside the brain. The cell bodies of motor fibers lie in nuclei within the brain.

Some motor fibers control subconscious movements, yet the somatic nervous system has been defined as a *conscious* system. The reason for this apparent contradiction is that some fibers of the autonomic nervous system leave the brain bundled together with somatic fibers of the cranial nerves. Therefore subconscious functions transmitted by the autonomic fibers are described along with the conscious functions of the somatic fibers of the cranial nerves. Although the cranial nerves are mentioned singly in this discussion, remember that they are paired structures.

I. Olfactory

(handwritten margin note: fibers arise in mucous membrane of upper nasal cavity.)

The **olfactory nerve** is entirely sensory and conveys impulses related to smell. It arises as bipolar neurons from the olfactory mucosa of the nasal cavity. The dendrites and cell bodies of these neurons are generally limited to the mucosa covering the superior nasal conchae and the adjacent nasal septum. Axons from the neurons pass through the cribriform plate of the ethmoid bone and synapse with other olfactory neurons in the *olfactory bulb,* an extension of the brain lying above the cribriform plate. The axons of these neurons make up the *olfactory tract.* The fibers from the tract terminate in the primary olfactory area in the cerebral cortex.

II. Optic

The **optic nerve** is also entirely sensory and conveys impulses related to vision. Impulses initiated by rods and cones of the retina are relayed by bipolar neurons to ganglion cells. Axons of the ganglion cells, the optic nerve fibers, enter the optic foramina where the two optic nerves unite to form the *optic chiasma.* Within the chiasma, fibers from the medial half of each retina cross to the opposite side; those from the lateral half remain on the same side. From the chiasma, the fibers pass posteriorly to the *optic tracts.* From the optic tracts, the majority of fibers terminate in a nucleus of the thalamus. They then synapse with neurons that pass to the visual areas of the cerebral cortex. Some fibers from the optic chiasma terminate in the superior colliculi of the midbrain. They synapse with neurons whose fibers terminate in the nuclei that convey impulses to the oculomotor (**III**), trochlear (**IV**), and abducens (**VI**) nerves—nerves that control the extrinsic (external) and intrinsic (internal) eye muscles. Through this relay, there are widespread motor responses to light stimuli.

III. Oculomotor

The **oculomotor nerve** is a mixed cranial nerve. It originates from neurons in a nucleus in the ventral portion of the midbrain. It runs forward, divides into a superior and inferior branch, and passes through the superior orbital fissure in the orbit. The superior branch is distributed to the superior rectus (an extrinsic eyeball muscle) and the levator palpebrae superioris (the muscle of the upper eyelid). The inferior branch is distributed to the medial rectus, inferior rectus, and inferior oblique muscles—all extrinsic eyeball muscles. These distributions to the levator palpebrae superioris and extrinsic eyeball muscles constitute the motor portion of the oculomotor nerve. Through these distributions, impulses are sent that control movements of the eyeball and upper eyelid.

The inferior branch of the oculomotor nerve also sends a branch to the *ciliary ganglion,* a relay center of the autonomic nervous system that connects a nucleus in the midbrain with the intrinsic eyeball muscles. These intrinsic muscles include the ciliary muscle of the eyeball and the sphincter muscle of the iris. Through the ciliary ganglion, the oculomotor nerve controls the smooth muscle (ciliary muscle) reponsible for accommodation of the lens for near vision and the smooth muscle (sphincter muscle of iris) responsible for constriction of the pupil.

The sensory portion of the oculomotor nerve consists of afferent fibers from proprioceptors in the eyeball muscles supplied by the nerve to the midbrain. These fibers convey impulses related to muscle sense (proprioception).

IV. Trochlear

The **trochlear nerve** is a mixed cranial nerve. It is the smallest of the 12 cranial nerves. The motor portion originates in a nucleus in the midbrain, and axons from the nucleus pass through the superior orbital fissure of the orbit. The motor fibers innervate the superior oblique muscle of the eyeball, another extrinsic eyeball muscle. It controls movement of the eyeball.

The sensory portion of the trochlear nerve consists of afferent fibers that run from proprioceptors in the superior oblique muscle to the nucleus of the nerve in the midbrain. The sensory portion is responsible for muscle sense.

V. Trigeminal

The **trigeminal nerve** is a mixed cranial nerve and the largest of the cranial nerves. As indicated by its name, the trigeminal nerve has three sensory branches: ophthalmic, maxillary, and mandibular. The trigeminal nerve contains two roots on the ventrolateral surface of the pons. The large sensory root has a swelling called the *semilunar (gasserian) ganglion* located in a fossa on the inner surface of the petrous portion of the temporal bone. From this ganglion, the *ophthalmic branch* enters the orbit via the superior orbital fissure, the *maxillary branch* enters the foramen rotundum, and the *mandibular branch* pierces the foramen ovale. The smaller motor root originates in a nucleus in the pons. The motor fibers join the mandibular branch and supply the muscles of mastication. These motor fibers, which control chewing movements, constitute the motor portion of the trigeminal nerve.

The sensory portion of the trigeminal nerve delivers impulses related to touch, pain, and temperature and consists of the ophthalmic, maxillary, and mandibular branches. The ophthalmic branch receives sensory fibers from the skin over the upper eyelid, eyeball, lacrimal glands, upper part of the nasal cavity, side of the nose, forehead, and anterior half of the scalp. The maxillary branch receives sensory fibers from the mucosa of the nose, palate, parts of the pharynx, upper teeth, upper lip, cheek, and lower eyelid. The mandibular branch transmits sensory fibers from the anterior two-thirds of the tongue (not taste), lower teeth, skin over the mandible and side of the head in front of the ear, and mucosa of the floor of the mouth. Sensory fibers from the three branches of the trigeminal nerve enter the semilunar ganglion and terminate in a nucleus in the pons. There are also sensory fibers from proprioceptors in the muscles of mastication.

VI. Abducens

The **abducens nerve** is a mixed cranial nerve that originates from a nucleus in the pons. The motor fibers extend from the nucleus to the lateral rectus muscle of the eyeball, an extrinsic eyeball muscle. Impulses over the fibers bring about movement of the eyeball. The sensory fibers run from proprioceptors in the lateral rectus muscle to the pons and mediate muscle sense. The abducens nerve reaches the lateral rectus muscle through the superior orbital fissure of the orbit.

VII. Facial

The **facial nerve** is a mixed nerve. Its motor fibers originate from a nucleus in the pons and enter the petrous portion of the temporal bone. The motor fibers are distributed to facial and scalp muscles. Impulses along these fibers cause contraction of the muscles of facial expression. Some motor fibers are also distributed to the sublingual and submandibular glands.

The sensory fibers extend from the taste buds of the anterior two-thirds of the tongue to the *geniculate ganglion,* a swelling of the facial nerve. From here, the fibers pass to a nucleus in the pons which sends fibers to the thalamus for relay to the gustatory area of the cerebral cortex. The sensory portion of the facial nerve conveys sensations related to taste. Proprioceptors are in the muscles of the face and scalp.

VIII. Vestibulocochlear

The **vestibulocochlear nerve** is a sensory cranial nerve. It consists of two branches: the cochlear (auditory)

branch and the vestibular branch. The *cochlear branch,* which conveys impulses associated with hearing, arises in the spiral organ of Corti in the cochlea of the internal ear. The cell bodies of the cochlear branch are in the *spiral ganglion* of the cochlea. From here the axons pass through a nucleus in the medulla and terminate in the thalamus. Ultimately, the fibers synapse with neurons that relay the impulses to the auditory areas of the cerebral cortex.

The *vestibular branch* arises in the semicircular canals, the saccule, and the utricle of the inner ear. Fibers from the semicircular canals, saccule, and utricle extend to the *vestibular ganglion,* where the cell bodies are contained. The cell bodies of the fibers synapse in the ganglion with fibers that extend to a nucleus in the medulla and pons and terminate in the thalamus. Some fibers also enter the cerebellum. The vestibular branch transmits impulses related to equilibrium.

IX. Glossopharyngeal

The **glossopharyngeal nerve** is a mixed cranial nerve. Its motor fibers originate in a nucleus in the medulla. The nerve exits the skull through the jugular foramen. The motor fibers are distributed to the swallowing muscles of the pharynx and the parotid gland to mediate swallowing movements and the secretion of saliva. The sensory fibers of the glossopharyngeal nerve supply the pharynx and taste buds of the posterior third of the tongue. Some sensory fibers also originate from receptors in the carotid sinus, which assumes a major role in blood pressure regulation. The sensory fibers of the glossopharyngeal nerve terminate in a nucleus in the thalamus. There are also sensory fibers from proprioceptors in the muscles innervated by this nerve.

X. Vagus

The **vagus nerve** is a mixed cranial nerve that is widely distributed from the head and neck into the thorax and abdomen. Its motor fibers originate in a nucleus of the medulla and terminate in the muscles of the pharynx, larynx, respiratory passageways, heart, esophagus, stomach, small intestine, most of the large intestine, and gallbladder. Impulses along the motor fibers generate visceral, cardiac, and skeletal muscle movement. Sensory fibers of the vagus nerve supply essentially the same structures as the motor fibers. They convey impulses for various sensations from the larynx and viscera. The fibers terminate in the medulla and pons. There are also sensory fibers from proprioceptors in the muscles supplied by this nerve.

XI. Accessory

The **accessory nerve** (formerly the spinal accessory nerve) is a mixed cranial nerve. It differs from all other cranial nerves in that it originates from both the brain stem and the spinal cord. The *bulbar (medullary) portion* originates from nuclei in the medulla, passes through the jugular foramen, and supplies the voluntary muscles of the pharynx, larynx, and soft palate that are used in swallowing. The *spinal portion* originates in the anterior gray horn of the first five segments of the cervical portion of the spinal cord. The fibers from the segments join, enter the foramen magnum, and exit through the jugular foramen along with the bulbar portion. The spinal portion conveys motor impulses to the sternocleidomastoid and trapezius muscles to coordinate head movements. The sensory fibers originate from proprioceptors in the muscles supplied by its motor neurons and terminate in upper cervical posterior root ganglia. They conduct impulses for proprioception.

XII. Hypoglossal

The **hypoglossal nerve** is a mixed nerve. The motor fibers originate in a nucleus in the medulla, pass through the hypoglossal canal, and supply the muscles of the tongue. These fibers conduct impulses related to speech and swallowing.

The sensory portion of the hypoglossal nerve consists of fibers originating from proprioceptors in the tongue muscles and terminating in the medulla. The sensory fibers conduct impulses for muscle sense.

DISORDERS

Many disorders can affect the central nervous system. Some are caused by viruses or bacteria. Others are caused by damage to the nervous system during birth. The origins of many conditions, however, are unknown. Here we discuss the origins and symptoms of some common central nervous system disorders.

POLIOMYELITIS

Poliomyelitis, also known as **infantile paralysis,** is a viral infection that is most common during childhood. Onset of the disease is marked by fever, severe headache, a stiff neck and back, deep muscle pain and weakness, and loss of certain somatic reflexes. The virus may spread via the respiratory passages and blood to the central nervous system where it destroys the motor nerve cell bodies, specifically those in the anterior horns of the spinal cord and in the nuclei of the cranial nerves. Injury to the spinal gray matter is the basis for the name of this disease (*polio* = gray matter; *myel* = spinal cord). Destruction of the anterior horns produces paralysis. The first sign of bulbar polio is difficulty in swallowing, breathing, and speaking. Poliomyelitis can cause death from respiratory or heart failure if the virus invades the brain cells of the vital medullary centers. In recent years, an immunization against the disease has been used.

SYPHILIS

Syphilis is a venereal disease caused by the *Treponema pallidum* bacterium. Venereal diseases are infectious disorders that can be spread through sexual contact. The disease progresses through several stages: primary, secondary, latent, and sometimes tertiary. During the *primary stage,* the chief symptom is an open sore, called a chancre, at the point of contact. The chancre eventually heals. About 6 weeks later, symptoms such as a skin rash, fever, and aches in the joints and muscles usher in the *secondary stage.* At this stage, syphilis can usually be treated with antibiotics. Even if individuals do not undergo treatment, their symptoms will eventually disappear. Within a few years, the disease will cease to be infectious. The symptoms of the disease disappear, but a blood test is generally positive. During this later "symptomless" period, called the *latent stage,* the bacteria may invade and slowly destroy body organs. Untreated syphilis is considered dangerous for this reason. When organ degeneration appears, the disease is said to be in the *tertiary stage.* If the syphilis bacteria attack the organs of the nervous system, the tertiary stage is called *neurosyphilis.* Neurosyphilis may take different forms, depending on the tissue involved. For instance, about 2 years after the onset of the disease, the bacteria may attack the meninges, producing meningitis. The blood vessels that supply the brain may also become infected. In this case, symptoms depend on the parts of the brain destroyed by oxygen and glucose starvation. Cerebellar damage is manifested by uncoordinated movements as in writing. As the motor areas become extensively damaged, victims may be unable to control urine and bowel movements. Eventually, they may become bedridden, unable even to feed themselves. Damage to the cerebral cortex produces memory loss and personality changes that range from irritability to hallucinations.

CEREBRAL PALSY

The term **cerebral palsy** refers to a group of motor disorders caused by damage to the motor areas of the brain during fetal life, birth, or infancy. One cause is infection of the mother with German measles during the first 3 months of pregnancy. During early pregnancy, certain cells in the fetus are dividing and differentiating in order to lay down the basic structures of the brain. These cells can be abnormally changed by toxin from the measles virus. Radiation during fetal life, temporary oxygen starvation during birth, and hydrocephalus during infancy may also damage brain cells.

Cases of cerebral palsy are categorized into three groups depending on whether the cortex, the basal ganglia of the cerebrum, or the cerebellum is affected most severely. Most cerebral palsy victims have at least some damage in all three areas. The location and extent of motor damage determine the symptoms. The victim may be deaf or partially blind. About 70 percent of cerebral palsy victims appear to be mentally retarded. The apparent mental slowness, however, is often due to the person's inability to speak or hear well. Such individuals are often more mentally acute than they appear.

Cerebral palsy is not a progressive disease. Thus it does not worsen as time elapses. Once the damage is done, however, it is irreversible.

PARKINSONISM

This disorder, also called **Parkinson's disease,** is a progressive malfunction of the basal ganglia of the cerebrum. The basal ganglia regulate subconscious contractions of skeletal muscles that aid activities desired by the motor areas of the cerebral cortex — swinging the arms when walking, for example. In Parkinsonism, the basal ganglia produce unnecessary skeletal movements that often interfere with voluntary movement. For instance, the muscles of the upper extremities may alternately contract and relax, causing the hands to shake. This shaking is called *tremor.* Other muscles may contract continuously, causing rigidity of the involved body part. *Rigidity* of the facial muscles gives the face a masklike appearance. The expression is characterized by a wide-eyed, unblinking stare and a slightly open mouth with uncontrolled drooling. Vision, hearing, and intelligence are unaffected by the disorder, indicating that Parkinsonism does not attack the cerebral cortex.

Parkinsonism seems to be caused by a malfunction at the neuron synapses. The motor neurons of the basal ganglia release the chemical transmitter acetylcholine. In normal people, the basal ganglia also produce a synaptic transmitter called dopamine, which quickly inactivates the acetylcholine and prevents continuous conduction across the synapse. People with Parkinsonism do not manufacture enough dopamine in their bodies. As a result, stimulated basal ganglia neurons do not easily stop conducting impulses. Injections of dopamine are useless; the blood–brain barrier stops it. However, symptoms are somewhat relieved by a drug developed a few years ago, levodopa, and its successors carbidopa and bromocriptine — none without distressing side effects.

MULTIPLE SCLEROSIS

Multiple sclerosis causes progressive destruction of the myelin sheaths of neurons in the central nervous system. The sheaths deteriorate to *scleroses,* which are hardened scars or plaques, in multiple regions — hence the name. The destruction of myelin sheaths interferes with the transmission of impulses from one neuron to another, literally short-circuiting conduction pathways. Multiple sclerosis is one of the most common disorders of the central nervous system. Usually the first symptoms occur between the ages of 20 and 40. Early symptoms are generally produced by the formation of a few plaques and are, consequently, mild. Plaque formation in the cerebellum may produce lack of coordination in one hand. The patient's handwriting becomes strained and irregular. A short-circuiting of pathways in the corticospinal tract may partially paralyze the leg muscles so that the patient drags a foot when walking. Other early symptoms include double vision and urinary tract infections. Following a period of remission during which the symptoms temporarily disappear, a new series of plaques develop and the victim suffers a second attack. One attack follows another over the years. Each time the plaques form, certain neurons are damaged by the hardening of their sheaths. Other neurons are uninjured by their plaques. The result is a progressive loss of function interspersed with remission periods during which the undamaged neurons regain their ability to transmit impulses.

The symptoms of multiple sclerosis depend on the areas of the central nervous system most heavily laden with plaques. Sclerosis of the white matter of the spinal cord is common. As the sheaths of the neurons in the corticospinal tract deteriorate, the patient loses the ability to contract skeletal muscles. Damage to the ascending tracts produces numbness and short-circuits impulses related to position of body parts and flexion of joints. Damage to either set of tracts also destroys spinal cord reflexes.

As the disease progresses, most voluntary motor control is eventually lost and the patient becomes bedridden. Death occurs anywhere from 7 to 30 years after the first symptoms appear. The usual cause of death is a severe infection resulting from the loss of motor acitivity. Without the constricting action of the urinary bladder wall, for example, the bladder never totally empties and stagnant urine provides an environment for bacterial growth. Bladder infection may then spread to the kidney, damaging kidney cells.

Multiple sclerosis may be caused by a virus. The occasional appearance of more than one case in a family suggests such an infectious agent, but the same circumstances also suggest a genetic predisposition. Like other demyelinating diseases, multiple sclerosis is incurable. Electrical stimulation of the spinal cord can improve function in certain patients, however.

CEREBRAL VASCULAR ACCIDENTS

The most common brain disorder is a **cerebrovascular accident (CVA),** also called a **stroke** or **cerebral apoplexy.**

A CVA is the destruction of brain tissue or infarction resulting from disorders in the vessels that supply the brain. Common causes of CVAs are intracerebral hemorrhage, embolism, and atherosclerosis of the cerebral arteries. An *intracerebral hemorrhage* is a rupture of a vessel in the pia mater or brain. Blood seeps into the brain and damages neurons by increasing intracranial fluid pressure. An *embolus* is a blood clot, air bubble, or bit of foreign material, most often debris from an inflammation, that becomes lodged in an artery and blocks circulation. *Atherosclerosis* is the formation of plaques in the artery walls. The plaques may slow down circulation by constricting the vessel. Both emboli and atherosclerosis cause brain damage by reducing the supply of oxygen and glucose needed by brain cells.

Many elderly people suffer mild CVAs as a result of short periods of reduced blood supply. Another cause is atherosclerosis. Another is *arteriosclerosis,* or hardening of the arteries, which occurs with aging. Damage is generally undetectable or very mild. During these mild CVAs the individual may have a short blackout, blurred vision, or dizziness and does not realize anything serious has occurred. A CVA can also cause sudden, massive damage, however. Severe CVAs cause about 21 percent of all deaths from cardiovascular disease. The person who recovers may suffer partial paralysis and mental disorders such as speech difficulty. The malfunction depends on the parts of the brain that were injured. Vascular disorders are more common after age 40.

DYSLEXIA

Dyslexia (*dys* = difficulty; *lexis* = words) is unrelated to basic intellectual capacity, but it causes a mysterious difficulty in handling words and symbols. Apparently some peculiarity in the brain's organizational pattern distorts the ability to read, write, and count. Letters in words seem transposed, reversed, or even topsy-turvy – *dog* becomes *god; b* changes identity with *d;* a sign saying "OIL" flip-flops into "710." Many dyslexics cannot orient themselves in the three dimensions of space and may show bodily awkwardness.

The cause of dyslexia is unknown, since it is unaccompanied by outward scars of detectable neurological damage and its symptoms vary from victim to victim. It occurs three times as often among boys as among girls. It has been variously attributed to defective vision, brain damage, lead in the air, physical trauma, or oxygen deprivation during birth. It remains an unsolved problem.

TAY-SACHS DISEASE

Tay-Sachs disease is a central nervous system affliction that brings death before age 5. The Tay-Sachs gene is carried by normal-appearing individuals descended from the Ashkenazi Jews of Eastern Europe. Approximately one in 3,600 of their offspring will be afflicted with Tay-Sachs disease. The disease is caused by the neuronal degeneration of the central nervous system because of excessive amounts of the sphingolipid known as ganglioside G_{m2} in the nerve cells of the brain. The afflicted child will develop normally until the age of 4 to 8 months. Then symptoms follow a course of progressive degeneration: paralysis, blindness, inability to eat, decubitus ulcers, and death from infection. There is no known cure.

MEDICAL TERMINOLOGY

Ambulant, ambulatory Walking or able to walk; not confined to bed.

Aphasia (*a* = without; *phasis* = speech) Diminished ability to comprehend or express spoken or written words due to damage to brain centers; the most common cause is a CVA.

Bacterial meningitis Acute inflammation of the meninges caused by bacteria.

Bradykinesia (*brady* = slow) Abnormal slowness of movement.

Contracture Abnormal shortening of muscle tissue rendering the muscle highly resistant to stretching. A contracture can lead to permanent disability.

Dyslexia Impaired ability to comprehend written language.

Familial Affecting members of the same family.

Lacrimation Secretion of tears.

Predisposition A special tendency toward some disease.

Recumbent Lying down.

Sinusitis Inflammation of paranasal sinuses. It often occurs during an upper respiratory infection, when infection in the nasal cavity spreads to the sinuses.

Spastic (*spas* = pull) Resembling spasms or convulsions.

Viral encephalitis An acute inflammation of the brain caused by a direct attack by various viruses or by an allergic reaction to any of the many viruses that are normally harmless to the central nervous system. If the virus affects the spinal cord as well, it is called *encephalomyelitis.*

BRAIN

PROTECTION AND COVERINGS

1 The brain is protected by the cranial bones, the meninges, and the cerebrospinal fluid.

CEREBROSPINAL FLUID

1 Cerebrospinal fluid protects by serving as a shock absorber. It also circulates nutritive substances from the blood.

2 The accumulation of cerebrospinal fluid in the head is called hydrocephalus. If the fluid accumulates in the ventricles, it is called internal hydrocephalus. If it accumulates in the subarachnoid space, it is called external hydrocephalus.

BLOOD SUPPLY

1 Interruption of the oxygen supply to the brain can result in paralysis, mental retardation, epilepsy, or death.

2 Glucose deficiency may produce dizziness, convulsions, and unconsciousness.

3 The blood–brain barrier is a concept that explains the differential rates of passage of certain materials from the blood into the brain.

BRAIN STEM

1 The medulla oblongata is continuous with the upper part of the spinal cord. It contains nuclei that are reflex centers for regulation of heart rate, respiratory rate, vasoconstriction, swallowing, coughing, vomiting, sneezing, and hiccuping.

2 The pons is superior to the medulla. It connects the spinal cord with the brain and links parts of the brain with each other. It relays impulses from the cerebral cortex to the cerebellum related to voluntary skeletal movements. It contains the nuclei for certain cranial nerves. The reticular formation of the pons contains the pneumotaxic center, which helps control respiration.

3 The midbrain connects the pons and the diencephalon. It conveys motor impulses from the cerebrum to the cerebellum and cord and conveys sensory impulses from cord to thalamus. It regulates auditory and visual reflexes.

DIENCEPHALON

1 The thalamus is superior to the midbrain. It contains nuclei that serve as relay stations for all sensory impulses except smell. These include hearing, vision, general sensation,

and taste. It also contains nuclei that are centers for synapses in the somatic motor system, such as voluntary motor actions and arousal. Conscious recognition of pain and temperature and general awareness of touch and pressure are also located in the thalamus.

2 The hypothalamus is anterior and inferior to the thalamus. It controls the autonomic nervous system, body temperature, food and fluid intake, waking state, and sleep.

CEREBRUM

1 The cerebrum is the largest brain portion. Its surface (cortex) contains gyri, fissures, and sulci.

2 The functions of the cerebral cortex are motor (voluntary muscular movement), sensory (interpreting sensory impulses), and associational (emotional and intellectual processes). It contains the basal ganglia or cerebral nuclei in each cerebral hemisphere. Certain parts of these hemispheres constitute the limbic system—sometimes called the "visceral" or "emotional" brain.

3 Brain waves generated by the cerebral cortex may be recorded on an EEG. They are used to diagnose epilepsy, infections, and tumors.

CEREBELLUM

1 The cerebellum is the second largest portion of the brain. It is posterior to the pons and medulla and inferior to the cerebrum. The cerebellum is the motor area of the brain that produces certain subconscious movements in the skeletal muscles. These movements are required for coordination, posture, and balance.

CRANIAL NERVES

1 Twelve pairs of cranial nerves originate from the brain.

2 The pairs are named primarily on the basis of distribution and numbered by order of attachment to the brain.

 Internal capsule

REVIEW QUESTIONS

1 How is the brain protected?

2 Describe the composition, formation, and circulation of cerebrospinal fluid.

3 What is the difference between internal and external hydrocephalus?

4 Discuss the importance of constant supply of oxygen and glucose to brain functioning.

5 What is the brain stem? Compare the medulla oblongata, pons varolii, and midbrain with regard to structure and function.

6 Define decussation. Give an example.

7 What are the functions of the thalamus?

8 Summarize the major functions of the hypothalamus.

9 What are the principal sulci and fissures of the cerebrum? How are they related to the lobes of the cerebrum? What is the cerebral cortex?

10 How are the names of the cranial bones related to the lobes of the cerebrum?

11 Compare the fibers in the white matter of the cerebrum with regard to direction and function.

12 What are basal ganglia? List two and describe their functions.

13 Discuss the motor, sensory, and association functions of the cerebrum. What is aphasia? Agraphia? Word deafness? Word blindness?

14 What is an electroencephalograph? What is its diagnostic value?

15 Describe the structure and functions of the cerebellum. What are cerebellar peduncles? With which body system is the cerebellum most closely associated in terms of coordination?

16 How would damage to the cerebellum affect skeletal muscles?

17 What is ataxia? List several signs of ataxia.

18 What is a cranial nerve? Why are some cranial nerves classified as mixed nerves? How are cranial nerves named and numbered?

19 List the location and function of each cranial nerve.

Chapter 15
The Sensory, Motor, and Integrative Systems

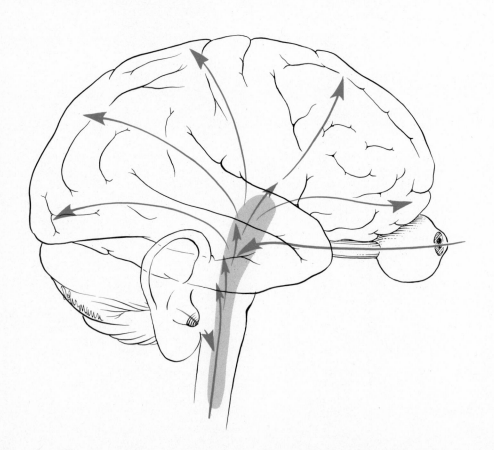

Now that you have examined the structure of the nervous system and its activities, we will see how its different parts cooperate in performing its three essential functions: (1) receiving sensory information; (2) transmitting motor impulses that result in movement or secretion; and (3) integration, an activity that deals with memory, sleep, and emotions. We will deal first with the input of sensory information.

SENSATIONS

Your ability to sense stimuli is vital to your survival. If pain could not be sensed, burns would be common. An inflamed appendix or stomach ulcer would progress unnoticed. A lack of sight would increase the risk of injury from unseen obstacles, a loss of smell would allow a harmful gas to be inhaled, a loss of hearing would prevent recognition of automobile horns, and a lack of taste would allow toxic substances to be ingested. In short, if you could not "sense" your environment and make the necessary homeostatic adjustments, you could not survive on your own.

Definition

In its broadest context, **sensation** refers to a state of awareness of external or internal conditions of the body. **Perception** refers to the conscious registration of a sensory stimulus. For a sensation to occur, four prerequisites must be fulfilled:

1 A **stimulus,** or change in the environment, capable of initiating a response by the nervous system must be present.

2 A **receptor** or **sense organ** must pick up the stimulus and convert it to a nerve impulse. A sense receptor or sense organ may be viewed as specialized nervous tissue that is extremely sensitive to internal or external conditions.

3 The impulse must be **conducted** along a nervous pathway from the receptor or sense organ to the brain.

4 A region of the brain must **translate** the impulse into a sensation.

Receptors are capable of converting a specific stimulus into a nerve impulse. The stimulus may be light, heat, pressure, mechanical energy, or chemical energy. Each stimulus is capable of causing the membrane of the receptor to depolarize. This depolarization is called a **generator potential (receptor potential).**

The generator potential is a graded response – within limits, the magnitude increases with stimulus strength and frequency. When the generator potential reaches the threshold level, it initiates an action potential (nerve impulse). Once initiated, the action potential is propagated along the nerve fiber. Whereas a generator potential is a local, graded response, an action potential is propagated and obeys the all-or-none principle. The function of a generator potential is to initiate an action potential by transducing a stimulus into a nerve impulse.

A receptor may be quite simple. It may consist of the dendrites of a single neuron in the skin that are sensitive to pain stimuli. Or it may be contained in a complex organ such as the eye. Regardless of complexity, all sense receptors contain the dendrites of sensory neurons. The dendrites occur either alone or in close association with specialized cells of other tissues. Receptors are very excitable and their threshold stimulus is low. Except for pain receptors, each is specialized by having a low threshold to a specific stimulus and a high threshold to all others.

Many sensory impulses are conducted to the sensory areas of the cerebral cortex. Only in this region can a stimulus produce conscious sensations. Sensory impulses that terminate in the spinal cord or brain stem can initiate motor activities, but they can never produce conscious sensations. Once a stimulus is received by a receptor and converted into an impulse, the impulse is conducted along an afferent pathway that enters either the spinal cord or the brain.

Characteristics

Conscious sensations or perceptions occur in the cortical regions of the brain. In other words, you see, hear, and feel pain in the brain. You seem to see with your eyes, hear with your ears, and feel pain in an injured part of your body only because the cortex interprets the sensation as coming from the stimulated sense receptor. The term **projection** describes this process by which the brain refers sensations to their point of stimulation.

A second characteristic of many sensations is **adaptation.** The perception of a sensation may disappear even though a stimulus is still being applied. When you get into a tub of hot water, you might feel a burning sensation. But soon the sensation decreases to one of comfortable warmth, even though the stimulus (hot water) is still present. This is due to fatigue at synapses.

Sensations may also be characterized by **afterimages.** That is, some sensations persist even though the stimulus has been removed. This phenomenon is the reverse of adaptation. One common example of afterimage occurs when you look at a bright light and then look away or close your eyes. You still see the light for several seconds or minutes afterward.

Another characteristic of sensations is **modality:** the specific sensation felt. The sensation may be one of pain, pressure, touch, body position, equilibrium, hearing, vision, smell, or taste. In other words, the distinct property by which one sensation may be distinguished from another is its modality.

Classification

One convenient method of classifying sensations is by location of the receptor. Receptors may be classified as exteroceptors, visceroceptors, and proprioceptors. **Exteroceptors** provide information about the external environment. They are sensitive to stimuli outside the body and transmit sensations of hearing, sight, smell, touch, pressure, temperature, and pain. Exteroceptors are located near the surface of the body.

Visceroceptors or **enteroceptors** provide information about the internal environment. These sensations arise from within the body and may be felt as pain, pressure, taste, fatigue, hunger, thirst, and nausea. Visceroceptors are located in blood vessels and viscera.

Proprioceptors allow us to feel position and movement. Such sensations give us information about muscle tension, the position and tension of our joints, and equilibrium. These receptors are located in muscles, tendons, joints, and the internal ear. They provide information about body position and movement.

Sensations may also be classified according to the simplicity or complexity of the receptor and the neural pathway involved. *General senses* involve a simple receptor and neural pathway. The receptors for general sensations are numerous and widespread. Examples include cutaneous sensations such as touch, pressure, heat, cold, and pain. *Special senses,* by contrast, involve complex receptors and neural pathways. The receptors for each special sense are found in only one or two specific areas of the body. Among the special senses are smell, taste, sight, and hearing.

GENERAL SENSES

Skin contains the receptor organs for many general senses. Receptor organs are also located in muscles, tendons, joints, subcutaneous tissue, and viscera.

Cutaneous Sensations

Cutaneous sensations include touch, pressure, cold, heat, and pain. The receptors for these sensations are in the skin, connective tissue, and the ends of the gastrointestinal tract. The cutaneous receptors are randomly distributed over the body surface so that certain parts of the skin are densely populated with receptors and other parts contain only a few. Areas of the body that have few cutaneous receptors are insensitive; those containing

many are very sensitive. This fact can be demonstrated by using the *two-point discrimination test* for touch. A compass is applied to the skin, and the distance in millimeters between the two points of the compass is varied. The subject then indicates when two points are felt and when only one is felt.

The compass may be placed on the tip of the tongue, an area where receptors are very densely packed. The distance between the two points can then be narrowed to 1.4 mm (0.06 inch). At this distance, the points are able to stimulate two different receptors and the subject feels touched by two objects. If the distance is less than 1.4 mm, the subject feels only one point, even though both points are touching the tongue. This is because the points are so close together that they reach only one receptor. The compass can then be placed on the back of the neck, where receptors are few and far apart. In this area of the skin, the subject feels two distinctly different points only if the distance between them is 36.2 mm (1.43 inch) or greater.

The results of this test indicate that the more sensitive the area, the closer the compass points may be placed and still be felt separately. The following order for these receptors, from greatest sensitivity to least, has been established: tip of tongue, tip of finger, side of nose, back of hand, back of neck.

Cutaneous receptors have simple structures. They consist of the dendrites of sensory neurons that may or may not be enclosed in a capsule of epithelial or connective tissue. Impulses received by cutaneous touch receptors pass along somatic afferent neurons in spinal and cranial nerves, through the thalamus, to the general sensory area of the parietal lobe of the cortex.

FINE TOUCH

Cutaneous receptors for **fine** or **light touch** include Meissner's corpuscles, Merkel's discs, and root hair plexuses (Figure 15-1). *Meissner's corpuscles* are egg-shaped receptors containing a mass of dendrites enclosed by connective tissue. They are located in the papillae of the skin and enable us to detect two points touched to the skin. Meissner's corpuscles are most numerous in the fingertips, palms of the hand, and soles of the feet. They are also abundant in the eyelids, tip of the tongue, lips, nipples, clitoris, and tip of the penis. *Merkel's discs* are receptors for touch that consist of disclike formations of dendrites attached to deeper layers of epidermal cells. They are distributed in many of the same locations as Meissner's corpuscles. *Root hair plexuses* are dendrites arranged in networks around the roots of hairs. If a hair shaft is moved, the dendrites are stimulated. Since the root hair plexuses are not surrounded by supportive or protective structures, they are called *free,* or *naked, nerve endings.*

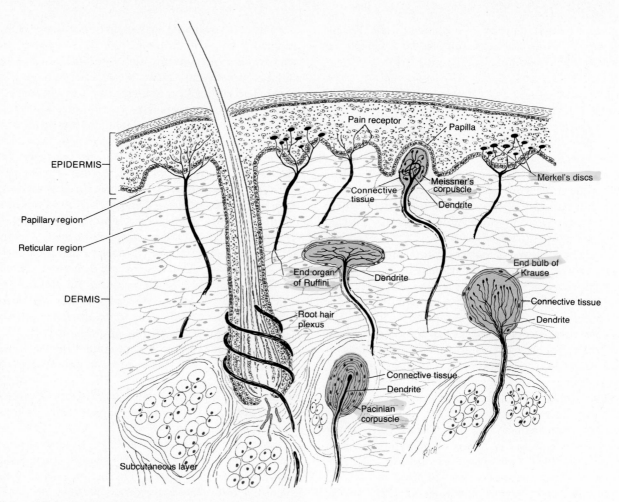

Figure 15-1 Structure and location of cutaneous receptors.

DEEP PRESSURE

Sensations of **deep pressure** are longer lasting and have less variation in intensity than do sensations of touch. Moreover, light touch is felt in a small, "pinprick" area whereas deep pressure is felt over a much larger area. The deep pressure receptors are oval structures called *Pacinian corpuscles* (Figure 15-1). They are composed of a capsule resembling an onion and consisting of connective tissue layers enclosing dendrites. Pacinian corpuscles are located in the subcutaneous tissue under the skin, the deep subcutaneous tissues that lie under mucous membranes, in serous membranes, around joints and tendons, in the perimysium of muscles, in the mammary glands, in the external genitalia of both sexes, and in certain viscera.

COLD

The cutaneous receptors for **cold** are not known but were once thought to be *end bulbs of Krause* (Figure 15-1). The commonest form of these receptors is an oval connective tissue capsule containing dendrites. They are widely distributed in the dermis and subcutaneous connective tissue and are also located in the conjunctiva of the eye, the tip of the tongue, and external genitals.

HEAT

The cutaneous receptors for **heat** are also not known but were once thought to be *end organs of Ruffini* (Figure 15-1). These receptors are deeply embedded in the dermis and are less abundant than end bulbs of Krause.

Pain Sensations

The receptors for **pain** are simply the branching ends of the dendrites of certain sensory neurons. Pain receptors are found in practically every tissue of the body. They may be stimulated by any type of stimulus. Excessive stimulation of a sense organ causes pain. When stimuli for other sensations such as touch, pressure, heat, and cold reach a certain threshold, they stimulate the sensation of pain as well. Additional stimuli for pain receptors include excessive distension or dilation of an organ, prolonged muscular contractions, muscle spasms, inadequate blood flow to an organ, or the presence of certain chemical substances. Pain receptors, because of their sensitivity to all stimuli, perform a protective function by identifying changes that may endanger the body. Pain receptors adapt only slightly or not at all. If there were adaptation to pain, it would be ignored and irreparable damage could result.

Sensory impulses for pain are conducted to the central nervous system along spinal and cranial nerves. The lateral spinothalamic tracts of the spinal cord relay impulses to the thalamus. From here the impulses may be relayed to the postcentral gyrus of the parietal lobe. Recognition of the kind and intensity of most pain is ultimately localized in the cerebral cortex. Some awareness of pain occurs at subcortical levels.

Pain may be divided into two types: somatic and visceral. **Somatic pain** arises from stimulation of the skin receptors. In this case, it is called *superficial somatic pain*. It may also arise from stimulation of receptors in skeletal muscles, joints, tendons, and fascia and is thus called *deep somatic pain*. **Visceral pain** results from stimulation of receptors in the viscera.

The ability of the cerebral cortex to locate the origin of pain is related to past experience. In most instances of somatic pain and in some instances of visceral pain, the cortex accurately projects the pain back to the stimulated area. If you burn your finger, you feel the pain in your finger. If the lining of your pleural cavity is inflamed, you experience pain there. In most instances of visceral pain, however, the sensation is not projected back to the point of stimulation. Rather, the pain may be felt in or just under the skin that overlies the stimulated organ. The pain may also be felt in a surface area far from the stimulated organ. This phenomenon is called **referred pain**. In general, the area to which the pain is referred and the visceral organ involved receive their innervation from the same segment of the spinal cord. Consider the following example. Afferent fibers from the heart as well as from the skin over the heart and along the left arm enter spinal cord segments T1 to T4. Thus the pain of a heart attack is typically felt in the skin over the heart and along the left arm. Figure 15-2 illustrates cutaneous regions to which visceral pain may be referred.

A kind of pain frequently experienced by amputees is called **phantom pain** – pain in a limb that has been amputated. Suppose a foot has been amputated. A sensory nerve that originally terminated in the foot is severed during the operation but repairs itself and returns to function in the remaining leg. From past experience the brain has projected the stimulation of these neurons back to the distal end of the limb. Thus when the distal ends of the severed neurons are now stimulated, the brain continues to project the sensation back to the missing part. Even though a foot has been amputated, the patient still "feels" pain in the toes.

Proprioceptive Sensations

An awareness of the activities of muscles, tendons, and joints is provided by the **proprioceptive** or **kinesthetic sense**. It informs us of the degree to which muscles are contracted and the amount of tension created in the tendons. The proprioceptive sense enables us to recognize the location and rate of movement of one body part in relation to others. It also allows us to estimate weight and determine the muscular work necessary to perform a task. With the proprioceptive sense, we can judge the position and movements of our limbs without using our eyes when we walk, type, or dress in the dark.

Proprioceptive receptors are located in muscles, tendons, joints, and the internal ear. The receptors for proprioception are of three types. The *joint kinesthetic receptors* are located in the capsules of joints. These receptors provide feedback information on the degree and rate of angulation (change of position) of a joint. The other two receptors for proprioception, neuromuscular spindles and Golgi tendon organs, provide feedback information from muscles. *Neuromuscular spindles* consist of the endings of sensory neurons that are wrapped around specialized muscle fibers. They are located in nearly all skeletal muscles and are numerous in the muscles of the extremities. Neuromuscular spindles provide feedback information on the degree of muscle stretch. This information is relayed to the central nervous system to assist in the coordination and efficiency of muscle contraction. Neuromuscular spindles are involved in the stretch and extensor reflexes. *Golgi tendon organs* are also proprioceptive receptors that provide information about skeletal muscles. These organs are located at the junction of muscle and tendon. They function by sensing the tension applied to a tendon. The degree of tension is related to the degree of muscle contraction and is translated by the central nervous system.

Proprioceptors adapt only slightly. This feature is advantageous since the brain must be apprised of the status of different parts of the body at all times so adjustments can be made to ensure coordination.

The afferent pathway for muscle sense consists of impulses generated by proprioceptors via cranial and spinal nerves to the central nervous system. Impulses for

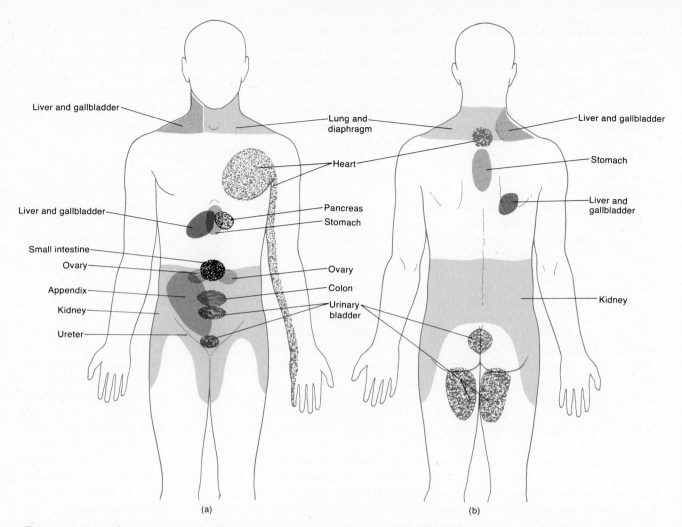

Liver and gallbladder

Lung and diaphragm

Liver and gallbladder

Heart

Stomach

Pancreas

Liver and gallbladder

Liver and gallbladder

Stomach

Small intestine

Ovary

Appendix

Kidney

Ureter

Ovary

Colon

Urinary bladder

Kidney

(a)

(b)

Figure 15-2   Referred pain. (a) Anterior view. (b) Posterior view. The shaded parts of the diagram indicate cutaneous areas to which visceral pain is referred.

conscious proprioception pass along ascending tracts in the cord, where they are relayed to the thalamus and cerebral cortex. The sensation is registered in the general sensory area in the parietal lobe of the cortex posterior to the central fissure. Proprioceptive impulses that have resulted in reflex action pass to the cerebellum along spinocerebellar tracts.

Levels of Sensation

Sensory fibers terminating in the spinal cord can generate spinal reflexes without immediate action by the brain. Sensory fibers terminating in the lower brain stem bring about far more complex motor reactions than simple spinal reflexes. When sensory impulses reach the lower brain stem, they cause subconscious motor reactions. Sensory impulses that reach the thalamus can be lo-

calized crudely in the body. In fact, at the thalamic level sensations are sorted by modality – that is, when sensory information reaches the cerebral cortex, we experience precise localization. It is at this level that memories of previous sensory information are stored and the perception of sensation occurs on the basis of past experience. Let us now examine how sensory information is transmitted from receptors to the central nervous system.

SENSORY PATHWAYS

Sensory information transmitted from the spinal cord to the brain is conducted along two general pathways: the posterior column pathway and the spinothalamic pathway.

In the **posterior column pathway** to the cerebral cortex there are three separate sensory neurons. The *first-order neuron* connects the receptor with the spinal cord and medulla on the same side of the body. The cell body of the first-order neuron is in the posterior root ganglion of a spinal or cranial nerve. The first-order neuron synapses with a *second-order neuron*. The second-order neuron passes from the medulla upward to the thalamus. The cell body of the second-order neuron is located in the nuclei cuneatus and gracilis of the medulla. Before passing into the thalamus, the second-order neuron crosses to the opposite side of the medulla and enters the medial lemniscus, a projection tract that terminates at the thalamus. In the thalamus, the second-order neuron synapses with a *third-order neuron*. The third-order neuron terminates in the somesthetic sensory area of the cerebral cortex. The posterior column pathway conducts impulses related to proprioception, fine touch, two-point discrimination, and vibrations.

The **spinothalamic pathway** is composed of three orders of sensory neurons also. The first-order neuron connects a receptor of the neck, trunk, and extremities with the spinal cord. The cell body of the first-order neuron is in the posterior root ganglion also. The first-order neuron synapses with the second-order neuron, which has its cell body in the posterior gray horn of the spinal cord. The fiber of the second-order neuron crosses to the opposite side of the spinal cord and passes upward to the brain stem in the lateral spinothalamic tract or ventral spinothalamic tract. The fibers from the second-order neuron terminate in the thalamus. There the second-order neuron synapses with a third-order neuron. The third-order neuron terminates in the somesthetic sensory area of the cerebral cortex. The spinothalamic pathway conveys sensory impulses for pain and temperature as well as crude touch and pressure.

The second-order neurons of the spinothalamic pathway enter the medulla, pons, and midbrain. Thus the spinothalamic pathway conducts sensory signals that result in subconscious motor reactions. By contrast, second-order neurons of the posterior column pathway have a direct connection with the thalamus and cerebral cortex. Thus the posterior column pathway conducts sensory information primarily into the conscious areas of the brain.

Now we can examine the anatomy of specific sensory pathways—for pain and temperature, for crude touch and pressure, and for fine touch, proprioception, and vibration.

Pain and Temperature

The sensory pathway for pain and temperature is called the **lateral spinothalamic pathway** (Figure 15-3). The first-order neuron conveys the impulse for pain or temperature from the appropriate receptor to the posterior gray horn on the same side of the spinal cord. In the horn, the first-order neuron synapses with a second-order neuron. The axon of the second-order neuron crosses to the opposite side of the cord. Here it becomes a component of the *lateral spinothalamic tract* in the lateral white column. The second-order neuron passes upward in the tract through the brain stem to a nucleus in the thalamus called the ventral posterolateral nucleus. In the thalamus, conscious recognition of pain and temperature occurs. The sensory impulse is then conveyed from the thalamus through the internal capsule to the somesthetic area of the cortex by a third-order neuron. The cortex analyzes the sensory information for the precise source, severity, and quality of the pain and heat stimuli.

Crude Touch and Pressure

The neural pathway that conducts impulses for crude touch and pressure is the **anterior (ventral) spinothalamic pathway** (Figure 15-4). By crude touch and pressure is meant the ability to perceive that something has touched the skin although its exact location, shape, size, or texture cannot be determined. The first-order neuron conveys the impulse from a crude touch or pressure receptor to the posterior gray horn on the same side of the spinal cord. In the horn, the first-order neuron synapses with a second-order neuron. The axon of the second-order neuron crosses to the opposite side of the cord and becomes a component of the *anterior spinothalamic tract* in the anterior white column. The second-order neuron passes upward in the tract through the brain stem to the ventral posterolateral nucleus of the thalamus. The sensory impulse is then relayed from the thalamus through the internal capsule to the somesthetic area of the cerebral cortex by a third-order neuron. Although there is some awareness of crude touch and pressure at the thalamic level, it is not fully perceived until the impulses reach the cortex.

Fine Touch, Proprioception, Vibration

The neural pathway for fine touch, proprioception, and vibration is called the **posterior column pathway** (Figure 15-5). This pathway conducts impulses that give rise to several discriminating senses:

1 Fine touch: the ability to recognize the exact location of stimulation and to distinguish that two points are touched, even though they are close together (two-point discrimination).

2 Stereognosis: recognizing the size, shape, and texture of an object.

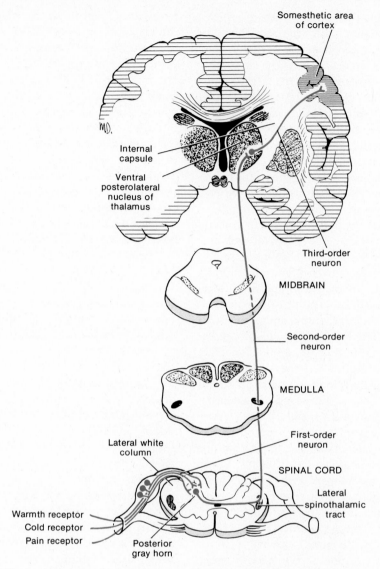

Figure 15-3   Sensory pathway for pain and temperature—the lateral spinothalamic pathway.

3 Weight discrimination: the ability to assess the weight of an object.

4 Proprioception: the awareness of the precise position of body parts and directions of movement.

5 The ability to sense vibrations.

First-order neurons for the discriminating senses just noted follow a pathway different from those for pain and temperature and crude touch and pressure. Instead of terminating in the posterior gray horn, the first-order neurons from appropriate receptors pass upward in the fasciculus gracilis or fasciculus cuneatus in the posterior white column of the cord. From here the first-order neurons enter either the nucleus gracilis or nucleus cuneatus in the medulla where they synapse with second-order neurons. The axons of the second-order neurons cross to the opposite side of the medulla and

ascend to the thalamus through the medial lemniscus, a projection tract of white fibers passing through the medulla, pons, and midbrain. The second-order neuron axons synapse with third-order neurons in the ventral posterior nucleus in the thalamus. In the thalamus, there is no conscious awareness of the discriminating senses, except for a possible crude awareness of vibrations. The third-order neurons convey the sensory impulses to the somesthetic area of the cerebral cortex. It is here that you perceive your sense of position and movement and fine touch.

Cerebellar Tracts

The *posterior spinocerebellar tract* is an uncrossed tract that conveys impulses concerned with subconscious muscle sense. That is, the tract assumes a role in reflex

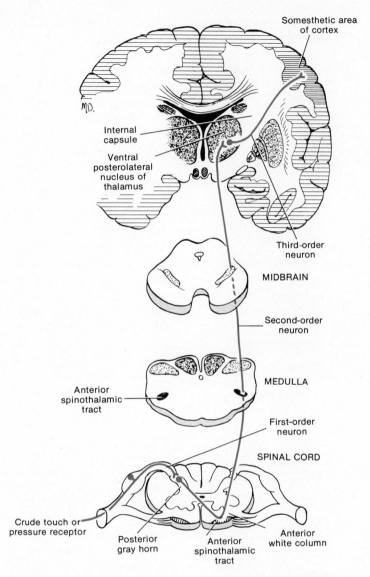

Figure 15-4 Sensory pathway for crude touch and pressure—the anterior spinothalamic pathway.

adjustments for posture and muscle tone. The nerve impulses originate in neurons that run between proprioceptors in muscles, tendons, and joints and the posterior gray horn of the spinal cord. Here the neurons synapse with afferent neurons that pass to the ipsilateral lateral white column of the cord to enter the posterior cerebellar tract. The tract enters the inferior cerebellar peduncles from the medulla and ends at the cerebellar cortex. In the cerebellum, synapses are made that ultimately result in the transmission of impulses back to the spinal cord to the anterior gray horn to synapse with the lower motor neurons leading to skeletal muscles.

The *anterior spinocerebellar tract* also conveys impulses for subconscious muscle sense. It, however, is made up of both crossed and uncrossed nerve fibers. Sensory neurons deliver impulses from proprioceptors to the posterior gray horn of the spinal cord. Here a synapse

occurs with neurons that make up the anterior spinocerebellar tracts. Some fibers cross to the opposite side of the spinal cord in the anterior white commissure. Others pass laterally to the ipsilateral anterior spinocerebellar tract and move upward, through the brain stem, to the pons to enter the cerebellum through the superior cerebellar peduncles. Here again the impulses for subconscious muscle sense are registered.

MOTOR PATHWAYS

The principal parts of the brain concerned with skeletal muscle control are the cerebral motor cortex, basal ganglia, reticular formation, and cerebellum. The motor cortex assumes the major role for controlling precise, discrete muscular movements. The basal ganglia largely integrate semivoluntary movements like walking, swim-

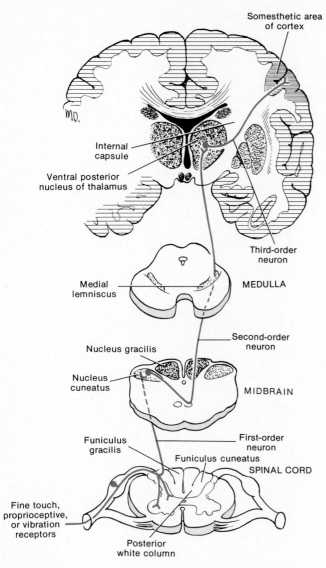

Somesthetic area
of cortex

Internal
capsule

Ventral posterior
nucleus of thalamus

Third-order
neuron

MEDULLA

Medial
lemniscus

Second-order
neuron

Nucleus gracilis

Nucleus
cuneatus

MIDBRAIN

Funiculus
gracilis

First-order
neuron

Funiculus cuneatus

SPINAL CORD

Fine touch,
proprioceptive,
or vibration
receptors

Posterior
white column

Figure 15-5 Sensory pathway for fine touch, proprioception, and vibration—the posterior column pathway.

ming, and laughing. The cerebellum, although not a control center, assists the motor cortex and basal ganglia by making body movements smooth and coordinated. Voluntary motor impulses are conveyed from the brain through the spinal cord by way of two major pathways: the pyramidal pathways and the extrapyramidal pathways.

Pyramidal Pathways

Voluntary motor impulses are conveyed from the motor areas of the brain to somatic efferent neurons leading to skeletal muscles via the **pyramidal pathways**. Most pyramidal fibers originate from cell bodies in the precentral gyrus. They descend through the internal capsule of the cerebrum and cross to the opposite side of the brain. They terminate in nuclei of cranial nerves that innervate voluntary muscles or in the anterior gray horn of the spinal cord. A short connecting neuron probably completes the connection of the pyramidal fibers with the motor neurons that activate voluntary muscles.

The pathways over which the impulses travel from the motor cortex to skeletal muscles have two components: *upper motor neurons (pyramidal fibers)* and *lower motor neurons (peripheral fibers)*. Here we consider three tracts of the pyramidal system:

1 Lateral corticospinal tract *(pyramidal tract proper)*   This tract begins in the motor cortex, descends through the internal capsule of the cerebrum, the cerebral peduncle of the midbrain, and then the pons on the same side as the point of origin (Figure 15-6). In the medulla, the fibers decussate to the opposite side to descend through the spinal cord in the lateral white column in the lateral corticospinal tract. Thus the motor cortex of the right side of the brain controls muscles on the left side of the body and vice versa. The upper motor neurons of the lateral corticospinal tract probably synapse with short association neurons in the anterior gray horn of the cord. These then synapse in the anterior gray horn with lower motor neurons that exit all levels of the cord via the ventral roots of spinal nerves. The lower motor neurons terminate in skeletal muscles.

2 Anterior corticospinal tract   About 15 percent of the upper motor neurons from the motor cortex do not cross in the medulla. These pass through the medulla and continue to descend on the same side to the anterior white column to become part of the *anterior (straight* or *uncrossed) corticospinal tract*. The fibers of these upper motor neurons decussate and probably synapse with association neurons in the anterior gray horn of the spinal cord of the side opposite the origin of the anterior corticospinal tract. The association neurons in the horn synapse with lower motor neurons that exit the cervical and upper thoracic segments of the cord via the ventral roots of spinal nerves. The lower motor neurons terminate in skeletal muscles that control muscles of the neck and part of the trunk.

3 Corticobulbar tract   The fibers of this tract begin in upper motor neurons in the motor cortex. They accompany the corticospinal tracts through the internal capsule to the brain stem, where they decussate and terminate in the nuclei of cranial nerves in the pons and medulla. These cranial nerves include the trigeminal (V), abducens (VI), facial (VII), glossopharyngeal (IX), vagus (X), accessory (XI), and hypoglossal (XII). The corticobulbar tract conveys impulses that largely control voluntary movements of the head and neck.

The various tracts of the pyramidal system convey impulses from the cortex that result in precise muscular movements.

Extrapyramidal Pathways

The **extrapyramidal pathways** include all descending tracts other than the pyramidal tracts. Generally, these include tracts that begin in the basal ganglia and reticular formation. The main extrapyramidal tracts are as follows:

1 Rubrospinal tract This tract originates in the red nucleus of the midbrain (after receiving fibers from the cerebellum), crosses over to descend in the lateral white column of the opposite side, and terminates in the anterior gray horns of the cervical and upper thoracic segments of the cord. The tract transmits impulses to skeletal muscles concerned with tone and posture.

2 Tectospinal tract This tract originates in the superior colliculus of the midbrain, crosses to the opposite side, descends in the anterior white column, and enters the anterior gray horns in the cervical segments of the cord. Its function is to transmit impulses that control movements of the head in response to auditory, visual, and cutaneous stimuli.

3 Vestibulospinal tract This tract originates in the vestibular nucleus of the medulla, descends on the same side of the cord in the anterior white column, and terminates in the anterior gray horns, mostly in the cervical and lumbosacral segments of the cord. It conveys impulses that regulate muscle tone in response to movements of the head. This tract, therefore, plays a major role in equilibrium.

Only one motor neuron carries the impulse from the cerebral cortex to the cranial nerve nuclei or spinal cord: an *upper motor neuron.* Only one motor neuron in the pathway actually terminates in a skeletal muscle: the *lower motor neuron.* This neuron, a somatic efferent neuron, always extends from the central nervous system to the skeletal muscle. Since it is the final transmitting neuron in the pathway, it is also called the *final common pathway.* This neuron is important clinically. If it is damaged or diseased, there is neither voluntary nor reflex action of the muscle it innervated and the muscle remains in a relaxed state—a condition called *flaccid paralysis.* Injury or disease of upper motor neurons in a motor pathway is characterized by varying degrees of continued contraction of the muscle (*spasticity*) and exaggerated reflexes. Another characteristic is the sign of Babinski; stroking the plantar surface along the outer border of the foot produces a slow dorsiflexion of the great toe accompanied by fanning of the lateral toes. The normal response is plantar flexion of the toes.

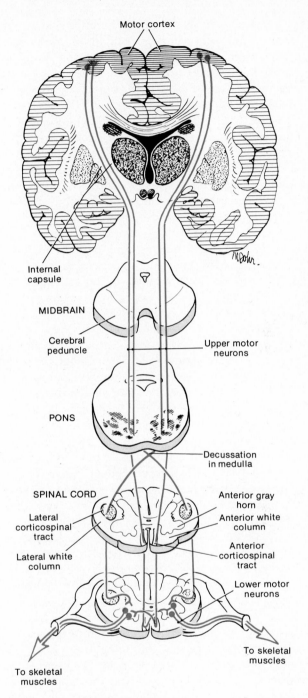

Figure 15-6 Pyramidal pathways.

Lower motor neurons are subjected to stimulation by many other presynaptic neurons. Some signals are excitatory; others are inhibitory. The algebraic sum of the opposing signals determines the final response of the lower motor neuron. It is not just a simple matter of the brain sending an impulse and the muscle always contracting.

Association neurons are of considerable importance in the motor pathways. Impulses from the brain are conveyed to association neurons before being received by lower motor neurons. These association neurons are integrators. They integrate the pattern of muscle contraction.

The basal ganglia have many connections with other parts of the brain. Through these connections, they help to control subconscious movements. The caudate nucleus controls gross intentional movements. The caudate nucleus and putamen, together with the cortex, control patterns of movement. The globus pallidus controls positioning of the body for performing a complex movement. The subthalamic nucleus is thought to control walking and possibly rhythmic movements. Many potential functions of the basal ganglia are held in check by the cerebrum. Thus if the cerebral cortex is damaged early in life, a person can still perform many gross muscular movements.

The role of the cerebellum is significant also. The cerebellum is connected to other parts of the brain that are concerned with movement. The vestibulocerebellar tract transmits impulses from the equilibrium apparatus in the ear to the cerebellum. The olivocerebellar tract transmits impulses from the basal ganglia to the cerebellum. The corticopontocerebellar tract conveys impulses from the cerebrum to the cerebellum. The spinocerebellar tracts relay proprioceptive information to the cerebellum. Thus the cerebellum receives considerable information regarding the overall physical status of the body. Using this information, the cerebellum generates impulses that integrate body responses.

Take tennis, for example. To make a good serve, you must bring your racket forward just far enough to make solid contact. How do you stop at the exact point without swinging too far? This is where the cerebellum comes in. It receives information about your body status while you are serving. Before you even hit the ball, the cerebellum has already sent information to the cerebral cortex and basal ganglia informing them that your swing must stop at an exact point. In response to cerebellar stimulation, the cortex and basal ganglia transmit motor impulses to your opposing body muscles to stop the swing. The cerebellar function of stopping overshoot when you want to zero in on a target is called its *damping function*. The cerebellum also helps you to coordinate different body parts while walking, running, and swimming. Finally, the cerebellum helps you maintain equilibrium.

INTEGRATIVE FUNCTIONS

We turn now to a fascinating, though poorly understood, function of the cerebrum: integration. The **integrative functions** include cerebral activities such as memory, sleep and wakefulness, and emotions. The role of the limbic system in emotional behavior was discussed in Chapter 14.

Memory

Memory may be defined as the ability to recall thoughts originally established by incoming sensory impulses. Memory may be classified into two kinds: short-term and long-term.

Short-term memory is the ability to recall bits of information. One example is finding a number in the phone book and then dialing it. If the number has no special significance, it is usually forgotten in a few seconds. Short-term memories leave no permanent imprint on the brain. One theory of short-term memory claims that memories may be caused by reverberating neuronal circuits—an incoming impulse stimulates the first neuron, which stimulates the second, which stimulates the third, and so on. Branches from the second and third neurons synapse with the first, sending the impulse back through the circuit again and again. Thus the output neuron generates continuous impulses. Once fired, the output signal may last from a few seconds to many hours, depending on the arrangement of neurons in the circuit. If this pattern is applied to short-term memory, an incoming thought—the phone number—continues in the brain even after the initial stimulus is gone. Thus you can recall the memory only for as long as the reverberation continues.

The concept of a reverberating circuit does not, however, explain long-term memory. *Long-term memory* is the persistence of an incoming impulse for years. One theory explains long-term memory on the basis of another principle: facilitation at synapses. When an incoming thought enters a neuronal circuit, the synapses in the circuit become facilitated for the passage of a similar signal later on. Thus an incoming signal facilitates the synapses in the circuit used for that signal over and over and you recall the memory. Such a neuronal circuit is called an *engram*.

The first incoming thought leading to long-term memory lasts only a brief time. How then does a short-term memory result in a long-term memory? One explanation is that the reverberating circuit of a short-term memory may persist for up to an hour after the initial thought. This reverberation establishes the engram. Later another incoming thought can cause facilitation of the neurons in the engram and the result is long-term memory. Another theory suggests that long-term memory is related to protein synthesis by RNA. According to this notion, the storage of memories results from the production of RNA that synthesizes specific proteins for each memory stored. The theory also

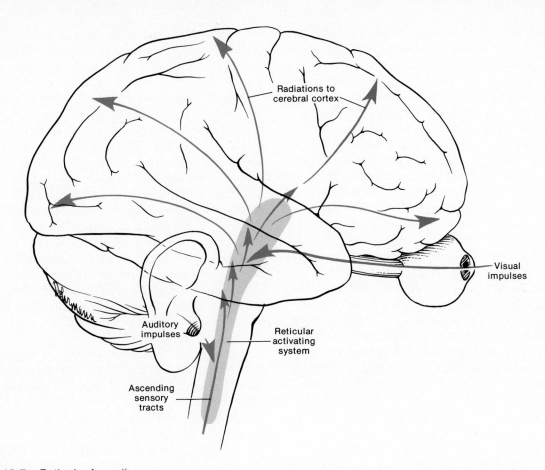

Figure 15-7   Reticular formation.

suggests that when a memory is recalled, the protein is destroyed but the RNA template remains so the memory is not lost.

The portions of the brain thought to be associated with memory include (1) the association cortex of the frontal, parietal, occipital, and temporal lobes and (2) parts of the limbic system, especially the hippocampus.

Sleep and Wakefulness

Humans sleep and awaken in a fairly constant 24-hour rhythm called a *circadian rhythm.* This cycle occurs even when there is total light or total darkness. Since neuronal fatigue precedes sleep and the signs of fatigue disappear after sleep, fatigue is apparently one cause of sleep. Moreover, EEG recordings indicate that during wakefulness the cerebral cortex is very active, sending impulses continuously through the body. During sleep, however, fewer impulses are transmitted by the cerebral cortex. The activity of the cerebral cortex is thought to be related to the reticular formation.

The reticular formation has numerous connections with the cerebral cortex (Figure 15-7). Stimulation of

portions of the reticular formation results in increased cortical activity. Thus the reticular formation is also known as the **reticular activating system (RAS).** One part of the system, the mesencephalic part, is composed of areas of gray matter of the pons and midbrain. When this area is stimulated, many impulses pass upward into the thalamus and disperse to widespread areas of the cerebral cortex. The effect is a generalized increase in cortical activity. The other part of the RAS, the thalamic part, consists of gray matter in the thalamus. When the thalamic part is stimulated, signals from specific parts of the thalamus cause activity in specific parts of the cerebral cortex. Apparently the mesencephalic part of the RAS causes general wakefulness (consciousness) and the thalamic part causes *arousal* – that is, awakening from deep sleep.

For the *arousal reaction* to occur, the RAS must be stimulated by input signals. Almost any sensory input can activate the RAS: pain stimuli, proprioceptive signals, bright light, an alarm clock. Once the RAS is activated, the cerebral cortex is also activated and you experience arousal. Signals from the cerebral cortex can also stimulate the RAS. Such signals may originate in the somesthetic cortex, the motor cortex, or the limbic system. When the signals activate the RAS, the RAS activates the cerebral cortex and arousal occurs.

Following arousal, the RAS and cerebral cortex continue to activate each other through a feedback system consisting of many circuits. The RAS also has a feedback system with the spinal cord that is composed of many circuits. Impulses from the activated RAS are transmitted down the spinal cord and then to skeletal muscles. Muscle activation causes proprioceptors to return impulses that activate the RAS. The two feedback systems maintain activation of the RAS, which in turn maintains activation of the cerebral cortex. The result is a state of wakefulness called *consciousness*. Since humans experience different levels of consciousness (alertness, attentiveness, relaxation, nonattentiveness), it is assumed that the level of consciousness depends on the number of feedback circuits operating at the time. In fact, consciousness may be altered by various factors. Amphetamines probably activate the RAS to produce a state of wakefulness and alertness. Meditation produces a lack of consciousness. Anesthetics produce a state of consciousness called anesthesia. Damage to the nervous system, as well as disease, can produce a lack of consciousness called coma. And drugs such as LSD can alter consciousness also.

If this theory of wakefulness is accepted, how then does sleep occur if the activating feedback systems are in continual operation? One explanation is that the synapses in the feedback circuits eventually undergo fatigue. The feedback system slows tremendously or is inhibited. Inactivation of the RAS produces a state known as *sleep*.

Just as there are different levels of consciousness, there are different levels of sleep. Normal sleep consists of two types: nonrapid eye movement sleep (non-REM) and rapid eye movement sleep (REM).

Non-REM sleep consists of four stages. Each has been identified by EEG recordings:

Stage 1. The person is relaxing with eyes closed. During this time, respirations are regular, pulse is even, and the person has fleeting thoughts. If awakened during stage 1, the person will frequently say he has not been sleeping. During stage 1, alpha waves appear on the EEG.

Stage 2. It is a little harder to awaken the person. Fragments of dreams may be experienced, and the eyes may slowly roll from side to side. During stage 2, the EEG shows *sleep spindles:* sudden, short bursts of sharply pointed alpha waves that occur 14–16 cycles/second.

Stage 3. The person is very relaxed. Body temperature begins to fall and blood pressure decreases. During this stage, it is difficult to awaken the person and the EEG shows a mixture of sleep spindles and delta waves. This stage occurs about 20 minutes after falling asleep.

Stage 4. Deep sleep occurs. The person is very relaxed and responds slowly if awakened. Bed-wetting and sleepwalking may occur during this stage. The EEG of stage 4 is dominated by delta waves.

In a typical 7 or 8 hour sleep period, a person goes from stages 1 to 4 of non-REM sleep. Then the person ascends to stages 3 and 2 and then to REM sleep within 50 to 90 minutes.

In **REM sleep,** the EEG readings are similar to those of stage 1 of non-REM sleep. There are significant physiological differences, however. During REM sleep, respirations and pulse rate increase and are irregular. Blood pressure also fluctuates considerably. It is during REM sleep that most dreaming occurs. Following REM sleep, which lasts about 10 to 20 minutes, the person descends again to stages 3 and 4 of non-REM sleep. This cycle repeats itself from three to five times during the entire sleep period. Each time the person returns to REM sleep, however, more time is spent at this stage. As much as 50 percent of an infant's sleep is REM sleep as contrasted with 20 percent for adults. Most sedatives significantly decrease REM sleep.

STUDY OUTLINE

SENSATIONS
DEFINITION
1 Sensation is awareness of conditions and changes in these conditions inside and outside the body.

2 The prerequisites for sensation are receiving a stimulus, converting it into an impulse, conducting the impulse to the brain, and translating the impulse into a sensation.

3 A receptor generates a stimulus for a sensation.

CHARACTERISTICS
1 Projection occurs when the brain refers a sensation to the point of stimulation.

2 Adaptation is the loss of sensation even though the stimulus is still applied.

3 An afterimage is the persistence of a sensation even though the stimulus is removed.

4 The modality is the property by which one sensation is distinguished from another.

CLASSIFICATION
1 Exteroceptors receive stimuli from the external environment.

2 Visceroceptors receive stimuli from blood vessels and viscera.

3 Proprioceptors receive stimuli from muscles, tendons, and joints for body position and movement.

GENERAL SENSES
CUTANEOUS SENSATIONS
1 Cutaneous sensations include touch, pressure, cold, heat, and pain. Receptors for these sensations are located in the skin, connective tissue, and the ends of the gastrointestinal tract.

2 Receptors for these sensations are Meissner's corpuscles, Merkel's discs, root hair plexuses, Pacinian corpuscles, Krause's end bulbs, and Ruffini's end organs.

PAIN SENSATIONS
1 Receptors are located in nearly every body tissue.

2 Two kinds of pain recognized in the parietal lobe of the cortex are somatic and visceral.

3 Referred pain is felt in the skin near or away from the organ sending pain impulses.

4 With phantom pain, a person "feels" pain in a limb that has been amputated.

PROPRIOCEPTIVE SENSATIONS
1 Receptors located in muscles, tendons, and joints convey impulses related to muscle tone, movement of body parts, and body position.

LEVELS OF SENSATION
1 Sensory fibers terminating in the lower brain stem bring about far more complex motor reactions than simple spinal reflexes.

2 When sensory impulses reach the lower brain stem, they cause subconscious motor reactions.

3 Sensory impulses that reach the thalamus can be localized crudely in the body.

4 When sensory impulses reach the cerebral cortex, we experience precise localization.

SENSORY PATHWAYS
1 In the posterior column pathway and the spinothalamic pathway there are first-order, second-order, and third-order neurons.

2 The sensory pathway for pain and temperature is the lateral spinothalamic pathway.

3 The neural pathway that conducts impulses for crude touch and pressure is the anterior spinothalamic pathway.

4 The neural pathway for fine touch, proprioception, and vibration is the posterior column pathway.

5 The pathways to the cerebellum are the anterior and posterior spinocerebellar tracts.

MOTOR PATHWAYS
1 Voluntary motor impulses are conveyed from the brain through the spinal cord along the corticospinal pathway and the extracorticospinal pathway.

2 Major extracorticospinal tracts are the rubrospinal, tectospinal, and vestibulospinal tracts.

INTEGRATIVE FUNCTIONS
1 Memory is defined as the ability to recall thoughts originally established by incoming sensory impulses.

2 Sleep and wakefulness are integrative functions.

3 Non-REM sleep consists of four stages identified by EEG recordings.

1 Define a sensation and a sense receptor. What prerequisities are necessary for the perception of a sensation?

2 Describe the following characteristics of a sensation: projection, adaptation, afterimage, modality.

3 Name some examples of adaptation not discussed in the text.

4 Compare the location and function of exteroceptors, visceroceptors, and proprioceptors.

5 Distinguish between a general sense and a special sense.

6 What is a cutaneous sensation? How are cutaneous receptors distributed over the body? Relate your response to the two-point discrimination test.

7 For each of the following cutaneous sensations, describe the receptor involved in terms of structure, function, and location: touch, pressure, cold, heat.

8 How do cutaneous sensations help maintain homeostasis?

9 Why are pain receptors important? Differentiate somatic pain, visceral pain, referred pain, and phantom pain.

10 Why is the concept of referred pain useful to the physician diagnosing internal disorders?

11 What is the proprioceptive sense? Where are the receptors for this sense located?

12 Relate proprioception to the maintenance of homeostasis.

13 What is the sensory pathway for pain and temperature?

14 Which pathway controls crude touch and pressure?

15 Which pathway is responsible for fine touch, proprioception, and vibrations?

16 Which pathways control voluntary motor impulses from the brain through the spinal cord?

17 Define memory. What are the two kinds of memory?

18 What are the four stages of non-REM sleep?

Chapter 16
The Autonomic Nervous System

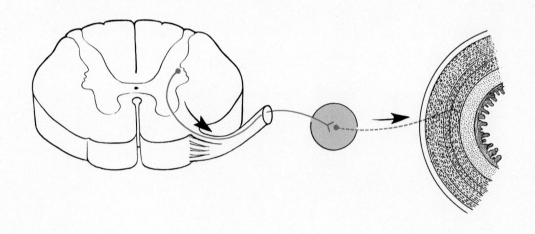

STUDENT OBJECTIVES

■ Compare the structural and functional differences between the somatic efferent and autonomic portions of the nervous system.

■ Identify the structural features of the autonomic nervous system.

■ Compare the sympathetic and parasympathetic divisions of the autonomic nervous system in terms of structure, physiology, and chemical transmitters released.

■ Describe a visceral autonomic reflex and its components.

■ Explain the role of the hypothalamus and its relationship to the sympathetic and parasympathetic division.

■ Explain the relationship between biofeedback and the autonomic nervous system.

■ Describe the relationship between meditation and the autonomic nervous system.

The portion of the nervous system that regulates the activities of smooth muscle, cardiac muscle, and glands is the **autonomic nervous system.** Structurally, the system consists of visceral efferent neurons organized into nerves, ganglia, and plexuses. Functionally, it usually operates without conscious control. Physiologists originally thought the system functioned autonomously with no control from the central nervous system—hence its name, the autonomic nervous system. In truth, the autonomic system is neither structurally nor functionally independent of the central nervous system. It is regulated by centers in the brain, in particular by the cerebral cortex, hypothalamus, and medulla oblongata. However, the autonomic nervous system does differ from the somatic efferent in some ways. For convenience of study the two are separated.

SOMATIC EFFERENT AND AUTONOMIC NERVOUS SYSTEMS

Whereas the somatic efferent nervous system produces conscious movement in skeletal muscles, the autonomic nervous system (visceral efferent nervous system) regulates visceral activities. And it generally does so involuntarily and automatically. Examples of visceral activities regulated by the autonomic nervous system are changes in the size of the pupil, accommodation for near vision, dilatation and constriction of blood vessels, adjustment of the rate and force of the heartbeat, emptying of the urinary bladder by contraction of its smooth muscle, movements of the gastrointestinal tract, formation of gooseflesh, and secretion by most glands. These activities usually lie beyond conscious control. They are automatic.

The autonomic nervous system is entirely motor. All its axons are efferent fibers, which transmit impulses from the central nervous system to visceral effectors. Autonomic fibers are called **visceral efferent fibers.** **Visceral effectors** include cardiac muscle, smooth muscle, and glandular epithelium. This does not mean there are no afferent (sensory) impulses from visceral effectors, however. Impulses that give rise to visceral sensations pass over visceral afferent neurons that have cell bodies located outside but close to the central nervous system. Some functions of these afferent neurons were described with the cranial and spinal nerves. However, the hypothalamus, which largely controls the autonomic nervous system, also receives impulses from the visceral sensory fibers.

The autonomic nervous system consists of two principal divisions: the **sympathetic** and the **parasympathetic.** Many organs innervated by the autonomic nervous system receive visceral efferent neurons from both components of the autonomic system—one set from the sympathetic division, another from the parasympathetic division. In general, impulses transmitted by the fibers of one division stimulate the organ to start or increase activity, whereas impulses from the other division decrease or halt the organ's activity. Organs that receive impulses from both sympathetic and parasympathetic fibers have *dual innervation.* In the somatic efferent nervous system, only one kind of motor neuron innervates an organ, which is always a skeletal muscle. When the somatic neurons stimulate the cells of the skeletal muscle, the muscle becomes active. When the neuron ceases to stimulate the muscle, contraction stops altogether. Skeletal muscle cells of each motor unit contract only when stimulated by their motor neuron. When the impulse stops, contraction stops.

STRUCTURE

The sympathetic and parasympathetic divisions of the autonomic nervous system are also referred to as the **thoracolumbar** and **craniosacral** divisions, respectively. Let us see what this means by discussing the general features applicable to both divisions.

Visceral Efferent Neurons

Autonomic visceral efferent pathways always consist of two neurons. One runs from the central nervous system to a ganglion. The other runs directly from the ganglion to the effector.

The first of these visceral efferent neurons in an autonomic pathway is called a **preganglionic neuron** (Figure 16-1). Preganglionic neurons have their cell bodies in the brain or spinal cord. Their myelinated axons, called **preganglionic fibers,** pass out of the central nervous system as part of a cranial or spinal nerve. At some point, they leave these nerves and run to autonomic ganglia where they synapse with the dendrites or cell bodies of postganglionic neurons.

Postganglionic neurons, the second visceral efferent neurons in an autonomic pathway, lie entirely outside the central nervous system. Their cell bodies and dendrites (if they have dendrites) are located in the autonomic ganglia, where the synapse with the preganglionic fibers occurs. The axons of postganglionic neurons are called **postganglionic fibers.** Postganglionic fibers are nonmyelinated, and they terminate in visceral effectors.

Thus preganglionic neurons convey efferent impulses from the central nervous system to autonomic ganglia. Postganglionic neurons relay the impulses from the autonomic ganglia to visceral effectors.

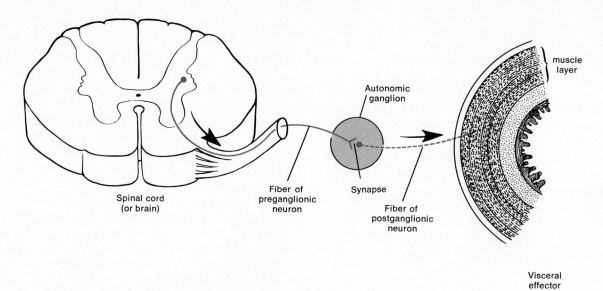

Figure 16-1   Relationship between preganglionic and postganglionic neurons.

Preganglionic Neurons

In the sympathetic division, the preganglionic neurons have their cell bodies in the lateral gray horns of the thoracic segments and first two lumbar segments of the spinal cord (Figure 16-2). It is for this reason that the sympathetic division is also called the **thoracolumbar division** and the fibers of the sympathetic preganglionic neurons are known as the **thoracolumbar outflow.**

The cell bodies of the preganglionic neurons of the parasympathetic division are located in nuclei in the brain stem and in the lateral gray horns of the second through fourth sacral segments of the spinal cord (Figure 16-2) – hence the synonymous term **craniosacral division.** The fibers of the parasympathetic preganglionic neurons are referred to as the **craniosacral outflow.**

Autonomic Ganglia

Autonomic pathways also include **autonomic ganglia,** where synapses between visceral efferent neurons occur. Autonomic ganglia differ from posterior root ganglia. The latter contain cell bodies of sensory neurons and no synapses occur in them. The autonomic ganglia may be divided into three general groups (Figure 16-3). The *sympathetic trunk* or *vertebral chain ganglia* are a series of ganglia that lie in a vertical row on either side of the vertebral column, extending from the base of the skull to the coccyx. They are also known as *paravertebral* or *lateral ganglia.* They receive preganglionic fibers only from the thoracolumbar (sympathetic) division (Figure 16-2).

A second kind of ganglion of the sympathetic division of the autonomic nervous system is called a *prevertebral* or *collateral ganglion* (Figure 16-3). The ganglia of this group lie anterior to the spinal column and close to the large abdominal arteries from which their names are derived. Examples of prevertebral ganglia so named are the celiac ganglion, on either side of the celiac artery just below the diaphragm; the superior mesenteric ganglion, near the beginning of the superior mesenteric artery in the upper abdomen; and the inferior mesenteric ganglion, located near the beginning of the inferior mesenteric artery in the middle of the abdomen (Figure 16-2). Prevertebral ganglia receive preganglionic fibers from the thoracolumbar (sympathetic) division.

The third kind of autonomic ganglion belongs to the parasympathetic division and is called a *terminal* or *intramural ganglion.* The ganglia of this group are located at the end of a visceral efferent pathway very close to visceral effectors or within the walls of visceral effectors. Terminal ganglia receive preganglionic fibers from the craniosacral (parasympathetic) division. The preganglionic fibers do not pass through sympathetic trunk ganglia (Figure 16-2).

In addition to autonomic ganglia, the autonomic nervous system also contains **autonomic plexuses.** Slender nerve fibers from ganglia containing postganglionic nerve cell bodies arranged in a branching network constitute an autonomic plexus.

Postganglionic Neurons

Axons from preganglionic neurons of the sympathetic division pass to ganglia of the sympathetic trunk. They can either synapse in the sympathetic chain ganglia with postganglionic sympathetics or they can continue,

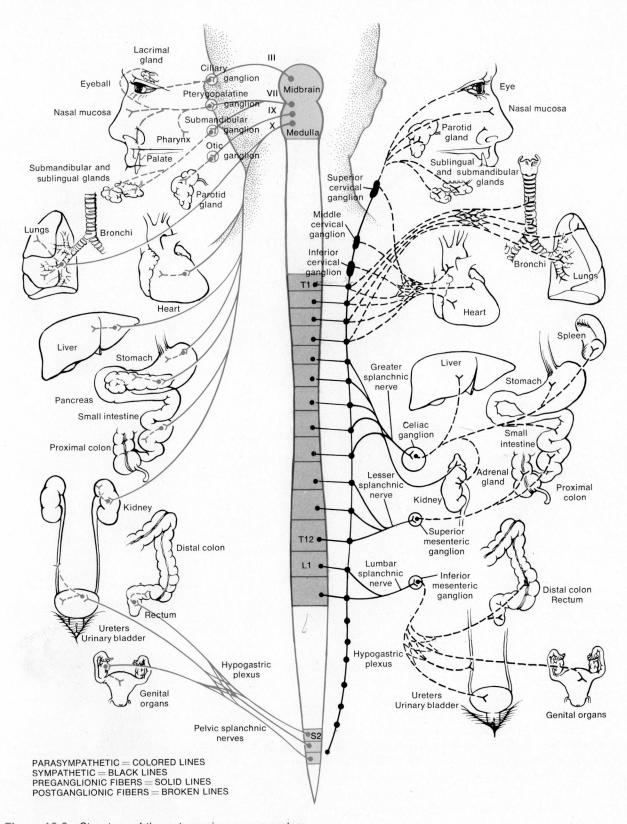

Figure 16-2 Structure of the autonomic nervous system.

Labels within the figure:

Lacrimal gland
Ciliary ganglion
Eyeball
Pterygopalatine ganglion
Nasal mucosa
Submandibular ganglion
Pharynx
Otic ganglion
Palate
Submandibular and sublingual glands
Parotid gland
Lungs
Bronchi
Heart
Liver
Stomach
Pancreas
Small intestine
Proximal colon
Kidney
Distal colon
Rectum
Ureters
Urinary bladder
Hypogastric plexus
Genital organs
Pelvic splanchnic nerves

III
VII
IX
X
Midbrain
Medulla

Eye
Nasal mucosa
Parotid gland
Sublingual and submandibular glands
Superior cervical ganglion
Middle cervical ganglion
Inferior cervical ganglion
T1
Bronchi
Lungs
Heart
Spleen
Greater splanchnic nerve
Liver
Stomach
Celiac ganglion
Small intestine
Lesser splanchnic nerve
Adrenal gland
Kidney
Proximal colon
T12
Superior mesenteric ganglion
L1
Lumbar splanchnic nerve
Inferior mesenteric ganglion
Distal colon Rectum
Hypogastric plexus
Ureters
Urinary bladder
Genital organs
S2

PARASYMPATHETIC = COLORED LINES
SYMPATHETIC = BLACK LINES
PREGANGLIONIC FIBERS = SOLID LINES
POSTGANGLIONIC FIBERS = BROKEN LINES

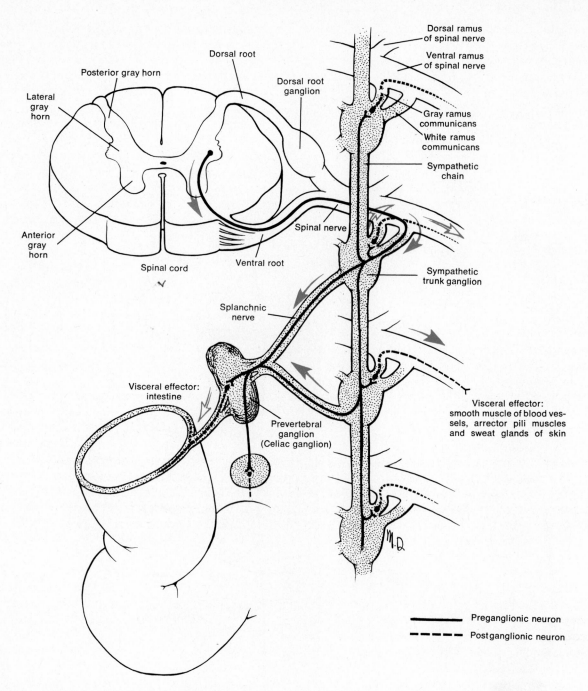

Figure 16-3   Ganglia and rami communicantes of the sympathetic autonomic nervous system.

without synapsing, through the chain ganglia to end at a prevertebral ganglion where synapses with the postganglionic sympathetics can take place. Each sympathetic preganglionic fiber synapses with several postganglionic fibers in the ganglion, and the postganglionic fibers pass to several visceral effectors. Upon exiting their

ganglia, the postsynaptic fibers innervate their visceral effectors.

Axons from preganglionic neurons of the parasympathetic division pass to terminal ganglia near or within a visceral effector. In the ganglion, the presynaptic neuron usually synapses with only four or five postsynaptic neurons to a single visceral effector. Upon exiting their ganglia, the postsynaptic fibers supply their visceral effectors.

With this background in mind, we can now examine some specific structural features of the sympathetic and parasympathetic divisions of the autonomic nervous system.

Sympathetic Division

The preganglionic fibers of the sympathetic division have their cell bodies located in the lateral gray horn of the spinal cord in the thoracic and first two lumbar segments (Figure 16-2). The preganglionic fibers are myelinated and leave the spinal cord through the ventral root of a spinal nerve along with the somatic efferent fibers of the same segmental levels. After exiting through the intervertebral foramina, the preganglionic sympathetic fibers enter a white ramus to pass to the nearest sympathetic trunk ganglion on the same side. Collectively, the white rami are called the **white rami communicantes.** Their name indicates that they contain myelinated fibers. Only thoracic and upper lumbar nerves have white rami communicantes. The white rami communicantes connect the ventral ramus of the spinal nerve with the ganglia of the sympathetic trunk.

When a preganglionic fiber of a white ramus communicans enters the sympathetic trunk, it may terminate (synapse) in several ways. Some fibers synapse in the first ganglion at the level of entry. Others pass up or down the sympathetic trunk for a variable distance to form the fibers on which the ganglia are strung. These fibers, known as *sympathetic chains* (Figure 16-3), may not synapse until they reach a ganglion in the cervical or sacral area. Some postganglionic fibers leaving the sympathetic trunk ganglia pass directly to visceral effectors of the head, neck, chest, and abdomen. Most, however, rejoin the spinal nerves before supplying peripheral visceral effectors such as sweat glands and the smooth muscle in blood vessels and around hair follicles. The **gray ramus communicans** is the structure containing the postganglionic fibers that run from the ganglion of the sympathetic trunk to the spinal nerve (Figure 16-3). The term *gray* refers to the fact that the fiber is unmyelinated. All spinal nerves have gray rami communicantes. Gray rami communicantes outnumber the white rami since there is a gray ramus leading to each of the 31 pairs of spinal nerves.

In most cases, a sympathetic preganglionic fiber terminates by synapsing with a large number of postganglionic cell bodies in a ganglion, usually 20 or more. Often the postganglionic fibers then terminate in widely separated organs of the body. Thus an impulse that starts in a single preganglionic neuron may affect several visceral effectors. For this reason, most sympathetic responses have widespread effects on the body.

The sympathetic trunks are two in number, situated anterolaterally to the spinal cord, one on either side. Each consists of a series of ganglia arranged more or less segmentally. The divisions of the sympathetic trunk are named on the basis of location. Typically, there are 22 ganglia in each chain: 3 cervical, 11 thoracic, 4 lumbar, and 4 sacral. Although the trunk extends downward from the neck, thorax, and abdomen to the coccyx, it receives preganglionic fibers only from the thoracic and lumbar segments of the spinal cord (Figure 16-2).

The cervical portion of each sympathetic trunk is located in the neck anterior to the prevertebral muscles. It is subdivided into a superior, middle, and inferior ganglion (Figure 16-2). The *superior cervical ganglion* is behind the internal carotid artery on a level with the second or third cervical vertebra. Postganglionic fibers leaving the ganglion serve the head where they are distributed to the eye, nasal mucosa, and the submandibular, sublingual, and parotid salivary glands. The *middle cervical ganglion* is situated at the level of the sixth cervical vertebra. Postganglionic fibers from it innervate the heart. The *inferior cervical ganglion* is located near the first rib. Its postganglionic fibers also supply the heart.

The thoracic portion of each sympathetic trunk usually consists of 11 segmentally arranged ganglia, lying ventral to the necks of the corresponding ribs. This portion of the sympathetic trunk receives most of the sympathetic preganglionic fibers. Postganglionic fibers from the thoracic sympathetic trunk innervate heart, lungs, bronchi, and other thoracic viscera.

The lumbar portion of each sympathetic trunk is found on either side of the corresponding lumbar vertebrae. The sacral portion of the sympathetic trunk lies in the pelvic cavity on the medial side of the sacral foramina.

Some preganglionic fibers pass through the sympathetic trunk without terminating in the trunk. Beyond the trunk, they form nerves known as *splanchnic nerves* (Figure 16-2). After passing through the trunk of ganglia, the splanchnic nerves terminate in the *celiac (solar) plexus*. In the plexus, the preganglionic fibers synapse in ganglia with postganglionic cell bodies. These ganglia are prevertebral ganglia. The greater splanchnic nerve passes to the celiac ganglion of the celiac plexus. From here, postganglionic fibers are distributed to the stomach, spleen, liver, kidney, and small intestine. The lesser splanchnic nerve passes through the celiac plexus to the superior mesenteric ganglion of the superior mesenteric plexus. Postganglionic fibers from this ganglion innervate the small intestine and colon. The lowest splanchnic nerve, not always present, enters the renal plexus. Postganglionics supply the renal artery and ureter. The lumbar splanchnic nerve enters the inferior mesenteric plexus. In the plexus, the preganglionic fibers synapse with postganglionic fibers in the inferior mesenteric ganglion. These fibers pass through the hypogastric plexus and supply the distal colon and rectum, urinary bladder, and genital organs. As noted earlier, the postganglionic fibers leaving the prevertebral ganglia follow the course of various arteries to abdominal and pelvic visceral effectors.

Parasympathetic Division

The preganglionic cell bodies of the parasympathetic division are found in nuclei in the brain stem and the lateral gray horn of the second through fourth sacral segments of the spinal cord (Figure 16-2). Their fibers emerge as part of a cranial nerve or as part of the ventral root of a spinal nerve. The **cranial parasympathetic outflow** consists of preganglionic fibers that leave the brain stem by way of the oculomotor nerves **(III)**, facial nerves **(VII)**, glossopharyngeal nerves **(IX)**, and vagus nerves **(X)**. The **sacral parasympathetic outflow** consists of preganglionic fibers that leave the ventral roots of the second through fourth sacral nerves. The preganglionic fibers of both the cranial and sacral outflows end in terminal ganglia where they synapse with postganglionic neurons. We will first look at the cranial outflow.

Four pairs of cranial parasympathetic ganglia innervate structures in the head and are located close to the organs they innervate. The *ciliary ganglion* is near the back of an orbit lateral to each optic nerve. Preganglionic fibers pass with the oculomotor nerve **(III)** to the ciliary ganglion. Postganglionic fibers from the ganglion innervate smooth muscle cells in the eyeball. Each *pterygopalatine ganglion* is situated lateral to a sphenopalatine foramen. It receives preganglionic fibers from the facial nerve **(VII)** and transmits postganglionic fibers to the nasal mucosa, palate, pharynx, and lacrimal gland. Each *submandibular ganglion* is found near the duct of a submandibular salivary gland. It receives preganglionic fibers from the facial nerve **(VII)** and transmits postganglionic fibers that innervate the submandibular and sublingual salivary glands. The *otic ganglia* are situated just below each foramen ovale. The otic ganglion receives preganglionic fibers from the glossopharyngeal nerve **(IX)** and transmits postganglionic fibers that innervate the parotid salivary gland. Ganglia associated

with the cranial outflow are classified as terminal ganglia. Since the terminal ganglia are close to their visceral effectors, postganglionic parasympathetic fibers are short. Postganglionic sympathetic fibers are relatively long.

The last component of the cranial outflow, the preganglionic fibers that leave the brain via the vagus nerves **(X)**, has the most extensive distribution of the parasympathetic fibers. Each vagus nerve enters into the formation of several plexuses in the thorax and abdomen. As it passes through the thorax, it sends fibers to the *superficial cardiac plexus* in the arch of the aorta and the *deep cervical plexus* anterior to the branching of the trachea. These plexuses contain terminal ganglia, and the postganglionic parasympathetic fibers emerging from them supply the heart. Also in the thorax is the *pulmonary plexus,* in front and behind the roots of the lungs and within the lungs themselves. It receives preganglionic fibers from the vagus and transmits postganglionic parasympathetic fibers to the lungs and bronchi. Other plexuses associated with the vagus nerve are described in later chapters in conjunction with the appropriate thoracic, abdominal, and pelvic viscera. Postganglionic fibers from these plexuses innervate viscera such as the liver, pancreas, stomach, kidney, small intestine, and part of the colon.

The sacral parasympathetic outflow consists of preganglionic fibers from the ventral roots of the second through fourth sacral nerves. Collectively, they form the *pelvic splanchnic nerves.* They pass into the hypogastric plexus. From ganglia in the plexus, parasympathetic postganglionic fibers are distributed to the colon, ureters, urinary bladder, and reproductive organs.

The salient structural features of the sympathetic and parasympathetic divisions are compared in Exhibit 16-1.

PHYSIOLOGY

Chemical Transmitters

Autonomic fibers, like other axons of the nervous system, release chemical transmitters at synapses as well as at points of contact between autonomic fibers and visceral effectors. These latter points are called **neuroeffector junctions.** On the basis of the chemical transmitter produced, autonomic fibers may be classified as either cholinergic or adrenergic. **Cholinergic fibers** release *acetylcholine* and include the following: (1) all sympathetic and parasympathetic preganglionic axons, (2) all parasympathetic postganglionic axons, and (3) some sympathetic postganglionic axons. The cholinergic sympathetic postganglionic axons include those to

Exhibit 16—1 STRUCTURAL FEATURES OF SYMPATHETIC AND PARASYMPATHETIC DIVISIONS

Sympathetic	Parasympathetic
Forms thoracolumbar outflow.	Forms craniosacral outflow.
Contains sympathetic trunk and prevertebral ganglia.	Contains terminal ganglia.
Ganglia are close to the CNS and distant from visceral effectors.	Ganglia are near or within visceral effectors.
Each preganglionic fiber synapses with many postganglionic neurons that pass to many visceral effectors.	Each preganglionic fiber usually synapses with four or five postganglionic neurons that pass to a single visceral effector.
Distributed throughout the body, including the skin.	Distribution limited primarily to head and viscera of thorax, abdomen, and pelvis.

sweat glands and those to blood vessels in skeletal muscles and the external genitalia. Since acetylcholine is quickly inactivated by the enzyme cholinesterase, the effects of cholinergic fibers are shortlived and local. **Adrenergic fibers** produce the chemical transmitter *norepinephrine,* or *noradrenalin,* also called *sympathin.* Most sympathetic postganglionic axons are adrenergic. Since norepinephrine is inactivated much more slowly than acetylcholine and norepinephrine may enter the bloodstream, the effects of sympathetic stimulation are longer lasting and more widespread than parasympathetic stimulation.

Activities

Most visceral effectors have dual innervation. That is, they receive fibers from both the sympathetic and the parasympathetic divisions. In these cases, impulses from one division stimulate the organ's activities, whereas impulses from the other division inhibit the organ's activities. The stimulating division may be either the sympathetic or the parasympathetic, depending on the organ. For example, sympathetic impulses increase heart activity whereas parasympathetic impulses decrease heart activity. On the other hand, parasympathetic impulses increase digestive activities whereas sympathetic impulses inhibit them. A summary of the activities of the autonomic system is presented in Exhibit 16-2.

The parasympathetic division is primarily concerned with activities that restore and conserve body energy. It is a rest-repose system. Under normal body conditions, for instance, parasympathetic impulses to the digestive glands and the smooth muscle of the digestive system dominate over sympathetic impulses. Thus energy-supplying foods can be digested and absorbed by the body.

The sympathetic division, by contrast, is primarily concerned with processes involving the expenditure of energy. When the body is in homeostasis, the main function of the sympathetic division is to counteract the parasympathetic effects just enough to carry out normal processes requiring energy. During extreme stress, however, the sympathetic dominates the parasympathetic. When people are confronted with a dangerous situation, for example, their bodies become alert and they sometimes perform feats of unusual strength. Fear stimulates the sympathetic division. Activation of the sympathetic division sets into operation a series of physiological responses collectively called the *fight-or-flight response.* It produces the following effects: (1) The pupils of the eyes dilate. (2) The heart rate increases. (3) The blood vessels of the skin and viscera constrict. (4) The remainder of the blood vessels dilate. This reaction causes a rise in blood pressure and a faster flow of blood into the dilated blood vessels of skeletal muscles, cardiac muscle, lungs, and brain – organs involved in fighting off danger. Rapid breathing occurs as the bronchioles dilate to allow faster movement of air in and out of the lungs. Blood sugar level rises as liver glycogen is converted to glucose to supply the body's additional energy needs. The sympathetic divison also stimulates the medulla of the adrenal gland to produce epinephrine and norepinephrine, hormones that intensify and prolong the sympathetic effects noted above. During this period of stress, the sympathetic effects inhibit other processes that are not essential for meeting the situation. Muscular movements of the gastrointestinal tract and digestive secretions are slowed down or even stopped.

VISCERAL AUTONOMIC REFLEXES

A **visceral autonomic reflex** adjusts the activity of a visceral effector. In other words, it results in the contraction of smooth or cardiac muscle or secretion by a gland. Such reflexes assume a key role in activities such as

No question

Exhibit 16—2 ACTIVITIES OF AUTONOMIC NERVOUS SYSTEM

Visceral Effector	Effect of Sympathetic Stimulation	Effect of Parasympathetic Stimulation
Eye		
Iris	Contracts dilator muscle of iris and brings about dilatation of pupil.	Contracts sphincter muscle of iris and brings about constriction of pupil.
Ciliary muscle	No innervation.	Contracts ciliary muscle and accommodates lens for near vision.
Glands		
Sweat	Stimulates secretion.	No innervation.
Lacrimal (tear)	No innervation.	Normal or excessive secretion.
Salivary	Vasoconstriction which decreases salivary secretion.	Stimulation of salivary secretion and vasodilation.
Gastric	No known effect.	Secretion stimulated.
Intestinal	No known effect.	Secretion stimulated.
Adrenal medulla	Promotes epinephrine and norepinephrine secretion.	No innervation.
Lungs (bronchial tubes)	Dilatation.	Constriction.
Heart	Increases rate and strength of contraction; dilates coronary vessels that supply blood to heart muscle cells.	Decreases rate and strength of contraction; constricts coronary vessels.
Blood vessels		
Skin	Constriction.	No innervation for most.
Skeletal muscle	Dilatation.	No innervation.
Visceral organs (except heart and lungs)	Constriction.	No innervation for most.
Liver	Promotes glycogenolysis; decreases bile secretion.	Promotes glycogenesis; increases bile secretion.
Stomach	Decreases motility.	Increases motility.
Intestines	Decreases motility.	Increases motility.
Kidney	Constriction of blood vessels that results in decreased urine volume.	No effect.
Pancreas	Inhibits secretion.	Promotes secretion.
Spleen	Contraction and discharge of stored blood into general circulation.	No innervation.
Urinary bladder	Relaxes muscular wall.	Contracts muscular wall.
Arrector pili of hair follicles	Contraction results in erection of hairs ("goose pimples").	No innervation.
Uterus	Inhibits contraction if nonpregnant; stimulates contraction if pregnant.	Minimal effect.
Sex organs	Vasoconstriction of ductus deferens, seminal vesicle, prostate; results in ejaculation.	Vasodilation and erection.

regulating heart action, blood pressure, respiration, digestion, defecation, and urinary bladder functions.

A visceral autonomic reflex arc consists of the following components:

1 **Receptor** The receptor is the distal end of an afferent neuron in an exteroceptor or enteroceptor.

2 **Afferent neuron** This neuron, either a somatic afferent or visceral afferent neuron, conducts the sensory impulse to the spinal cord or brain.

3 **Association neurons** These neurons are found in the central nervous system.

4 **Visceral efferent preganglionic neuron** In the thoracic and abdominal regions, this neuron is in the lateral gray horn of the spinal cord. The axon passes through the ventral root of the spinal nerve, the spinal nerve, and the white ramus communicans. It then enters a sympathetic trunk or prevertebral ganglion, where it synapses with a postganglionic neuron. In the cranial and sacral regions, the visceral efferent preganglionic axon leaves the central nervous system and passes to a terminal ganglion, where it synapses with a postganglionic neuron. The role of the visceral efferent preganglionic neuron is to convey a motor impulse from the brain or spinal cord to an autonomic ganglion.

5 **Visceral efferent postganglionic neuron** This neuron conducts a motor impulse from a visceral efferent preganglionic neuron to the visceral effector.

6 **Visceral effector** A visceral effector is smooth muscle, cardiac muscle, or a gland.

The basic difference between a somatic reflex arc and a visceral autonomic reflex arc is that in a somatic reflex, only one efferent neuron is involved. In a visceral autonomic reflex arc, two efferent neurons are involved.

Visceral sensations do not always reach the cerebral cortex. Therefore they remain at subconscious levels. Under normal conditions, you are not aware of muscular contractions of the digestive organs, heartbeat, changes in the diameter of blood vessels, and pupil dilation and constriction. When your body is making adjustments in such visceral activities, they are handled by visceral reflex arcs whose centers are in the spinal cord or lower regions of the brain. Among such centers are the cardiac, respiratory, vasomotor, swallowing, and vomiting centers in the medulla and the temperature control center in the hypothalamus. Thus stimuli delivered by somatic or visceral afferent neurons synapse in these centers, and the returning motor impulses conducted by visceral efferent neurons bring about an adjustment in the visceral effector without conscious recognition. The impulses are interpreted and acted on subconsciously.

Some visceral sensations do give rise to conscious recognition: hunger, nausea, and fullness of the urinary bladder and rectum.

CONTROL BY HIGHER CENTERS

The autonomic nervous system is not a separate nervous system. Axons from many parts of the central nervous system are connected to both the sympathetic and the parasympathetic divisions of the autonomic nervous system and thus exert considerable control over it. Autonomic centers in the cerebral cortex are connected to autonomic centers of the thalamus, for example. These, in turn, are connected to the hypothalamus. In this hierarchy of command, the thalamus sorts incoming impulses before they reach the cerebral cortex. The cerebral cortex then turns over control and integration of visceral activities to the hypothalamus. It is at the level of the hypothalamus that the major control and integration of the autonomic nervous system is exerted.

The hypothalamus is connected to both the sympathetic and the parasympathetic divisions of the autonomic nervous system. The posterior and lateral portions of the hypothalamus appear to control the sympathetic division. When these areas are stimulated, there is an increase in visceral activities – an increase in heart rate, a rise in blood pressure due to vasoconstriction of blood vessels, an increase in the rate and depth of respiration, dilatation of the pupils, and inhibition of the digestive tract. On the other hand, the anterior and medial portions of the hypothalamus seem to control the parasympathetic division. Stimulation of these areas results in a decrease in heart rate, lowering of blood pressure, constriction of the pupils, and increased motility of the digestive tract.

Control of the autonomic nervous system by the cerebral cortex occurs primarily during emotional stress. In extreme anxiety, the cerebral cortex can stimulate the hypothalamus. This stimulation, in turn, increases heart rate and blood pressure. You may have experienced this very reaction before taking an examination. If the cortex is stimulated by an extremely unpleasant sight, the stimulation causes vasodilation of blood vessels, a lowering of blood pressure, and fainting.

Biofeedback

In its simplest terms, **biofeedback** is a process in which people get constant signals, or feedback, on visceral body functions such as blood pressure, heart rate, and muscle tension. By using special monitoring devices, they can control these visceral functions consciously.

Suppose you are connected to a monitor that informs you of your heartbeat by means of lights. A red light indicates a fast heart beat, an amber light a normal rate, and a green light a slow rate. When you see the red light flash, you know your heart is beating too fast. You have been informed of a visceral response. This is the biofeedback. According to some researchers, you can be taught to slow down the heart rate by thinking of something pleasant and thus relaxing the body. The green light flashes and your reward is a slower heart rate. In similar experiments, some individuals have learned to control heart rhythm.

Researchers estimate that between 5 and 10 percent of the American population suffers from migraine headaches. Moreover, no effective treatment has been developed that does not have significant side effects and serious risks. One approach to alleviating migraine headaches is biofeedback.* The first clue that biofeedback could be used in this way came when a patient in the voluntary control laboratory of the Menninger Foundation demonstrated that with a 10°F rise in hand skin temperature (as a result of vasodilation and increased blood flow) she could spontaneously recover from a migraine headache.

In a study conducted at the Menninger Foundation, subjects suffering from migraine headaches received instructions in the use of a monitor that registers the skin temperature of the right index finger. Subjects were also given a typewritten sheet containing two sets of phrases. The first set was designed to help them relax the entire body. The second set was designed to bring about an increased flow of blood in the hands. The subjects practiced raising their skin temperature at home for 5 to 15 minutes a day. When skin temperature increased, the monitor emitted a high-pitched sound. In time, the monitor was abandoned.

Once the subjects learned how to vasodilate their blood vessels, the migraine headaches lessened. Since migraine headaches are believed to involve a distension of blood vessels in the head, the shunting of blood from head to hands relieved the distension and thus the pain.

Other experiments have shown that biofeedback can be applied to childbirth. Women were given monitors hooked up to their fingers and arms to measure electrical conductivity of the skin and skeletal muscle tension. Both conductivity and tension increase with nervousness and make labor difficult. Muscle tension was recorded as a sirenlike sound that became louder with nervousness. Skin conductivity was recorded as a crack-ling noise that also increased with nervousness. The monitors kept the women informed of their nervousness. This was the biofeedback. Having pleasant thoughts, reduced the sound levels. The reward was less nervousness. The results of the study indicate that the women needed less medication during labor and labor time itself was shortened.

There is no way to determine where biofeedback will lead. Perhaps the outstanding contribution of biofeedback research has been to demonstrate that the autonomic nervous system is not autonomous. Visceral responses can be controlled. Strong supporters of biofeedback point out that its possible applications in medicine are legion. They envision the use of biofeedback in lowering blood pressure in patients with hypertension, altering heart rates and rhythms, relieving pain from migraine headaches, making delivery easier, and controlling anxiety related to a host of illnesses that may be linked to stress. One researcher concludes that "an important trend is beginning to take place in the areas of psychosomatic disorders and medicine. This is the increasing involvement of the patient in his own treatment. The traditional doctor-patient relationship is giving way slowly to a shared responsibility."

Meditation

Yoga, which literally means union, is defined as a higher consciousness achieved through a fully rested and relaxed body and a fully awake and relaxed mind. One widely practiced technique for achieving higher consciousness is called **transcendental meditation.** One sits in a comfortable position with the eyes closed and concentrates on a suitable sound or thought.

Research indicates that transcendental meditation can alter physiological responses. Oxygen consumption decreases drastically along with carbon dioxide elimination. Subjects have experienced a reduction in metabolic rate and blood pressure. Researchers also observed a decrease in heart rate, an increase in the intensity of alpha brain waves, a sharp decrease in the amount of lactic acid in the blood, and an increase in the skin's electrical resistance. These last four responses are characteristic of a highly relaxed state of mind. Alpha waves are found in the EEGs of almost all individuals awake and in a resting state, but they disappear during sleep.

These responses have been called an **integrated response** – essentially, a hypometabolic state due to inactivation of the sympathetic division of the autonomic nervous system. This state is exactly the opposite of the fight-or-flight response described earlier – which is a hyperactive state of the sympathetic division. The integrated response suggests that the central nervous system does exert control over the autonomic nervous system.

*Much of the following discussion of the use of biofeedback for the treatment of migraine headaches is based upon the information provided by Dr. Joseph D. Sargent of the Menninger Foundation, Topeka, Kansas.

SOMATIC EFFERENT AND AUTONOMIC NERVOUS SYSTEMS

1 The autonomic nervous system automatically regulates the activities of smooth muscle, cardiac muscle, and glands.

2 It usually operates without conscious control.

3 It is regulated by centers in the brain, in particular by the cerebral cortex, the hypothalamus, and the medulla oblongata.

4 The somatic efferent nervous system produces conscious movement in skeletal muscles. The autonomic nervous system (visceral efferent nervous system) regulates visceral activities.

STRUCTURE

1 The autonomic nervous system consists of visceral efferent neurons organized into nerves, ganglia, and plexuses.

2 It is entirely motor. All autonomic axons are efferent fibers.

3 Efferent neurons are preganglionic (with myelinated axons) and postganglionic (with unmyelinated axons).

4 The autonomic system consists of two principal divisions: the sympathetic and the parasympathetic (also called the thoracolumbar and craniosacral divisions).

5 Autonomic ganglia are classified as sympathetic trunk ganglia (on sides of spinal column), prevertebral ganglia (anterior to spinal column), and terminal ganglia (near or inside visceral effectors).

6 Sympathetic responses are widespread and, in general, concerned with energy expenditure. Parasympathetic responses are restricted and are typically concerned with energy restoration and conservation.

PHYSIOLOGY

1 Autonomic fibers release chemical transmitters at synapses. On the basis of the transmitter produced, these fibers may be classified as cholinergic or adrenergic.

2 Cholinergic fibers release acetylcholine. Adrenergic fibers produce norepinephrine (noradrenalin), also called sympathin.

VISCERAL AUTONOMIC REFLEXES

1 A visceral autonomic reflex adjusts the activity of a visceral effector.

2 A visceral autonomic reflex arc consists of a receptor, afferent neuron, association neuron, visceral efferent preganglionic neuron, visceral efferent postganglionic neuron, and visceral effector.

CONTROL BY HIGHER CENTERS

1 The hypothalamus controls and integrates the autonomic nervous system.

2 The hypothalamus is connected to both the sympathetic and the parasympathetic divisions.

3 Biofeedback is a process by which people get constant signals on visceral body functions. It has been used to control heart rate and other functions.

4 Yoga is a higher consciousness achieved through a fully rested and relaxed body and a fully awake and relaxed mind.

5 Transcendental meditation produces the following physiological responses: decreased oxygen consumption and carbon dioxide elimination, reduced metabolic rate, decrease in heart rate, increase in the intensity of alpha brain waves, a sharp decrease in the amount of lactic acid in the blood, and an increase in the skin's electrical resistance.

1 What are the principal components of the autonomic nervous system? What is its general function? Why is it called involuntary?

2 What is the principal anatomical difference between the voluntary nervous system and the autonomic nervous system?

3 Relate the role of visceral efferent fibers and visceral effectors to the autonomic nervous system.

4 Distinguish the following with respect to location and function: preganglionic neurons and postganglionic neurons.

5 What is an autonomic ganglion? Describe the location and function of the three types of autonomic ganglia. Define white and gray ramus communicantes.

6 On what basis are the sympathetic and parasympathetic divisions of the autonomic nervous system differentiated anatomically and functionally?

7 Discuss the distinction between cholinergic and adrenergic fibers of the autonomic nervous system.

8 Give examples of the antagonistic effects of the sympathetic and parasympathetic divisions of the autonomic nervous system.

9 Summarize the principal functional differences between the voluntary nervous system and the autonomic nervous system.

10 Below is a diagram of the human body as it might appear in a fear situation. Specific parts of the body have been labeled. Write the *sympathetic response* for each labeled part.

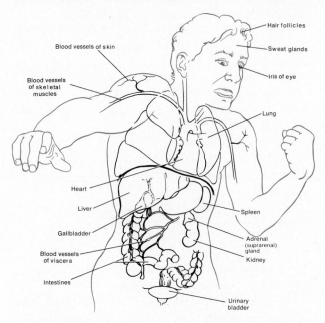

11 Define a visceral autonomic reflex and give three examples.

12 Describe a complete visceral autonomic reflex in proper sequence.

13 Describe how the hypothalamus controls and integrates the autonomic nervous system.

14 Define biofeedback. Explain how it could be useful.

15 What is transcendental meditation? How is the integrated response related to the autonomic nervous system?

Chapter 17
The Special Senses

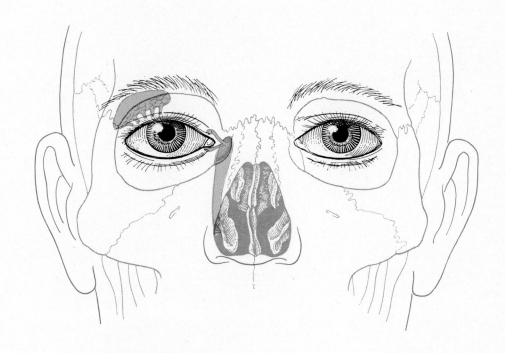

STUDENT OBJECTIVES

■ Locate the receptors for olfaction and describe the neural pathway for smell.

■ Identify the gustatory receptors and describe the neural pathway for taste.

■ Describe the structure and physiology of the accessory visual organs.

■ List the structural divisions of the eye.

■ Discuss retinal image formation by describing refraction, accommodation, constriction of the pupil, convergence, and inverted image formation.

■ Define emmetropia, myopia, hypermetropia, and astigmatism.

■ Diagram and discuss the rhodopsin cycle responsible for light sensitivity of rods.

■ Describe the afferent pathway of light impulses to the brain.

■ Define the anatomical subdivisions of the ear.

■ List the principal events in the physiology of hearing.

■ Identify the receptor organs for equilibrium.

■ Discuss the receptor organs' roles in maintaining static and dynamic equilibrium.

■ Contrast the causes and symptoms of cataracts, glaucoma, conjunctivitis, trachoma, Ménière's disease, and impacted cerumen.

■ Define medical terminology associated with the sense organs.

In contrast to the general senses, the special senses of smell, taste, sight, hearing, and equilibrium have receptor organs that are highly complex. Like the general senses, however, the special senses allow us to detect changes in our environment.

OLFACTORY SENSATIONS

The receptors for the **olfactory sense** are located in the nasal epithelium in the superior portion of the nasal fossae (left and right chambers of the nasal cavity) on either side of the nasal septum (Figure 17-1). The nasal epithelium consists of two principal kinds of cells: supporting and olfactory. The *supporting cells* are columnar epithelial cells of the mucous membrane lining the nose. Olfactory glands in the mucosa keep the mucous membrane moist. The *olfactory cells* are bipolar neurons whose cell bodies lie between the supporting cells. The distal (free) end of each olfactory cell contains six to eight dendrites, called *olfactory hairs.* The unmyelinated axons of the olfactory cells unite to form the *olfactory nerves,* which pass through foramina in the cribriform plate of the ethmoid bone. The olfactory nerves terminate in paired masses of gray matter called the *olfactory bulbs.* The olfactory bulbs lie beneath the frontal lobes of the cerebrum on either side of the crista galli of the ethmoid bone. The first synapse of the olfactory neural pathway occurs in the olfactory bulbs between the axons of the olfactory nerves and the dendrites of neurons inside the olfactory bulbs. Axons of these neurons run posteriorly to form the *olfactory tract.* From here, impulses are conveyed to the olfactory portion of the cortex. In the cortex, the impulses are interpreted as odor and give rise to the sensation of smell. A few of the surface anatomy features of the nose are indicated in Figure 23-3.

The mechanism by which the stimulus for smell is converted to a nerve impulse is explained by three widely accepted theories. One theory holds that substances capable of producing odors emit gaseous particles. On entering the nasal fossae, these particles become dissolved in the mucus of the nasal membrane. This fluid then acts chemically on the olfactory hairs to create a nerve impulse. According to the second theory, radiant energy given off by the molecules of the stimulating substance is the stimulus rather than the molecules themselves. The third theory purports that substances detected by smell are usually soluble in fat. Since the membrane of an olfactory hair is largely fat, it is assumed that molecules of substances to be smelled are dissolved in the membrane where they initiate a nerve impulse.

The sensation of smell happens quickly, but adaptation to odors also occurs rapidly. We become accustomed to some odors and are able to endure others. Rapid adaptation also accounts for the failure of a person to detect a gas that accumulates slowly in a room. The cortex stores memories of odors quite well. Once you have smelled a substance, you generally recognize its odor if you smell it again.

Both the supporting cells of the nasal epithelium and tear glands are innervated by branches of the trigeminal nerve (V). The nerve receives stimuli of pain, cold, heat, tickling, and pressure. Olfactory stimuli such as pepper, ammonia, and chloroform are irritating and may cause tearing because they stimulate the lacrimal and nasal mucosal receptors of the trigeminal nerve as well as the olfactory neurons.

GUSTATORY SENSATIONS

The receptors for **gustatory sensations,** or sensations of taste, are located in the taste buds (Figure 17-2). Taste buds are most numerous on the tongue, but they are also found on the soft palate and in the throat. The *taste buds* are oval bodies consisting of two kinds of cells. The *supporting cells* are a specialized epithelium that forms a capsule. Inside each capsule are 4 to 20 *gustatory cells.* Each gustatory cell contains a hairlike process (*gustatory hair*) that projects to the external surface through an opening in the taste bud called the *taste pore.* Gustatory cells make contact with taste stimuli through the taste pore.

Taste buds are found in some connective tissue elevations on the tongue called **papillae.** The papillae give the upper surface of the tongue its rough appearance. *Circumvallate papillae,* the largest type, are circular and form an inverted V-shaped row at the posterior portion of the tongue. *Fungiform* (mushroom-shaped) *papillae* are knoblike elevations found primarily on the tip and sides of the tongue. All circumvallate and most fungiform papillae contain taste buds. *Filiform papillae* are threadlike structures that cover the anterior two-thirds of the tongue.

For gustatory cells to be stimulated, the substances we taste must be in solution in the saliva so they can enter the taste pores. Despite the many substances we seem to taste, there are basically only four taste sensations: sour, salt, bitter, and sweet. All other "tastes," such as chocolate, pepper, and coffee, are actually odors. Each of the four tastes is due to a different response to different chemicals. Certain regions of the tongue react more strongly than others to certain taste sensations. Although the tip of the tongue reacts to all four taste sensations, it is highly sensitive to sweet and salty substances. The posterior portion of the tongue is highly sensitive to bitter substances. The lateral edges of the tongue are more sensitive to sour substances.

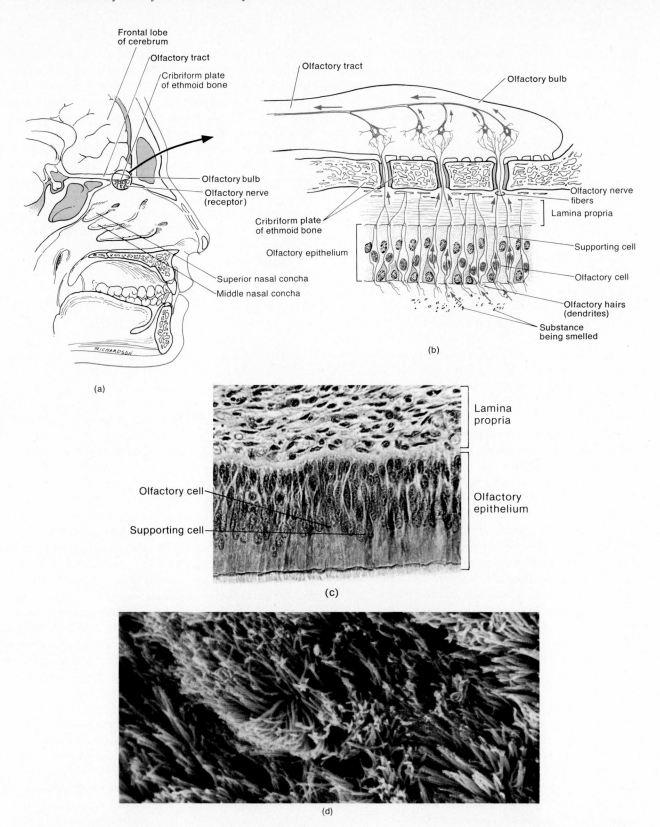

Figure 17-1 Olfactory receptors. (a) Location of receptors in right nasal fossa. (b) Enlarged aspect of olfactory receptors. (c) Photomicrograph of the olfactory mucosa at a magnification of 400×. (Courtesy of Donald I. Patt, from *Comparative Vertebrate Histology,* by Donald I. Patt and Gail R. Patt, Harper & Row, Publishers, Inc., New York, 1969.) (d) Scanning electron micrograph of the ciliated nasal mucosa at a magnification of 5,000×. (Courtesy of Fisher Scientific Company and S.T.E.M. Laboratories, Inc., Copyright 1975.)

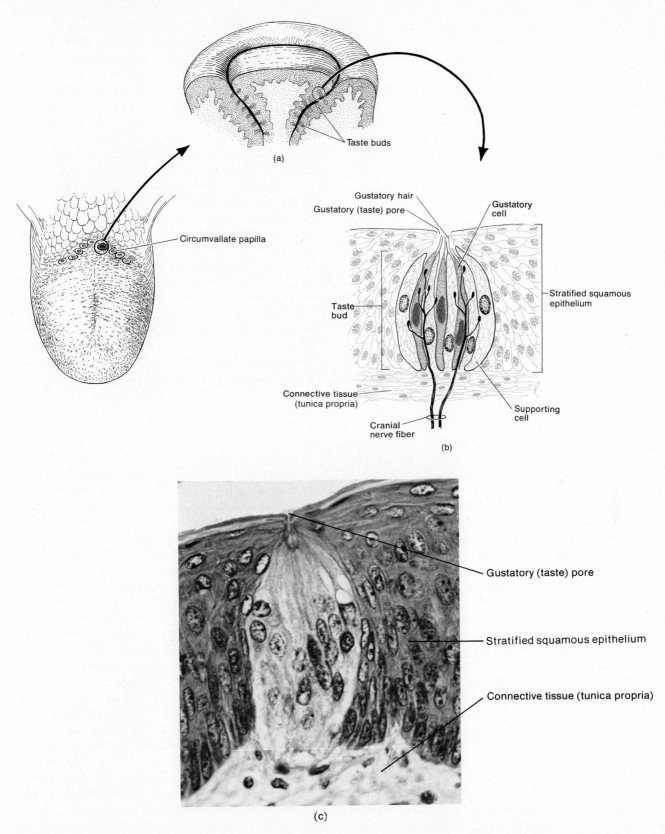

Figure 17-2 Gustatory receptors. (a) Location of taste buds relative to papilla. (b) Structure of a taste bud. (c) Photomicrograph of a taste bud at a magnification of 575×. (Courtesy of Edward J. Reith, from *Atlas of Descriptive Histology,* by Edward J. Reith and Michael H. Ross, Harper & Row, Publishers, Inc., New York, 1970.)

The cranial nerves that supply afferent fibers to taste buds are the facial nerve (VII), which supplies the anterior two-thirds of the tongue; the glossopharyngeal (IX), which supplies the posterior one-third of the tongue; and the vagus (X), which supplies the area of the throat around the epiglottis. Taste impulses are conveyed from the gustatory cells in taste buds along the three nerves just cited to the medulla and then to the thalamus. They terminate in the parietal lobe of the cortex.

VISUAL SENSATIONS

The structures related to vision are the eyeball, the optic nerve, the brain, and a number of accessory structures.

Accessory Structures of Eye

Among the accessory organs are the eyebrows, eyelids, eyelashes, and the lacrimal apparatus, which produces tears (Figure 17-3). The *eyebrows* form a transverse arch at the junction of the upper eyelid and forehead. Structurally, they resemble the hairy scalp. The skin of the eyebrows is richly supplied with sebaceous glands. The hairs are generally coarse and directed laterally. Deep to the skin of the eyebrows are the fibers of the orbicularis oculi muscles. The eyebrows protect the eyeballs from falling objects, perspiration, and the direct rays of the sun.

The upper and lower *eyelids,* or *palpebrae,* consist of dense folds of skin. They shade the eyes during sleep, protect the eyes from excessive light and foreign objects, and spread lubricating secretions over the eyeballs. The upper eyelid is more movable than the lower and contains in its superior region a special levator muscle known as the *levator palpebrae superioris.* The space between the upper and lower eyelids that exposes the eyeball is called the *palpebral fissure.* Its angles are known as the *lateral canthus,* which is narrower and closer to the temporal bone, and the *medial canthus,* which is broader and nearer the nasal bone. In the medial canthus there is a small reddish elevation, the *lacrimal caruncle,* containing sebaceous and sudoriferous glands. A whitish material secreted by the caruncle collects in the medial canthus.

From superficial to deep, each eyelid consists of epidermis, dermis, subcutaneous areolar connective tissue, fibers of the orbicularis oculi muscle, a tarsal plate, tarsal glands, and a conjunctiva. The *tarsal plate* is a thick fold of connective tissue that gives form and support to the eyelid. Embedded in grooves on the deep surface of

each tarsal plate is a row of elongated tarsal glands known as *Meibomian glands.* These are modified sebaceous glands, and their oily secretion helps keep the eyelids from adhering to each other. Infection of these glands produces a tumor or cyst on the eyelid called a *chalazion.* The *conjunctiva* is a thin mucous membrane called the *palpebral conjunctiva* when it lines the inner aspect of the eyelids. It is called the *bulbar* or *ocular conjunctiva* when it is reflected from the eyelids on to the eyeball to the periphery of the cornea.

Projecting from the border of each eyelid, anterior to the Meibomian glands, is a row of short, thick hairs: the *eyelashes.* In the upper lid, they are long and turn upward; in the lower lid, they are short and turn downward.

Sebaceous glands at the base of the hair follicles of the eyelashes called *glands of Zeis* pour a lubricating fluid into the follicles. Infection of these glands is called a *sty.*

The *lacrimal apparatus* is a term used for a group of structures that manufactures and drains away tears. These structures are the lacrimal glands, the excretory lacrimal ducts, the lacrimal canals, the lacrimal sacs, and the nasolacrimal ducts. A *lacrimal gland* is a compound tubuloacinar gland located at the superior lateral portion of each orbit. Each is about the size and shape of an almond. Leading from the lacrimal glands are 6 to 12 *excretory lacrimal ducts* that empty lacrimal fluid, or tears, onto the surface of the conjunctiva of the upper lid. From here the lacrimal fluid passes medially and enters two small openings *(puncta lacrimalia)* that appear as two small dots, one in each papilla of the eyelid, at the medial canthus of the eye. The lacrimal secretion then passes into two ducts, the *lacrimal canals,* and is next conveyed into the lacrimal sac. The lacrimal canals are located in the lacrimal grooves of the lacrimal bones. The *lacrimal sac* is the superior expanded portion of the nasolacrimal duct, a canal that transports the lacrimal secretion into the inferior meatus of the nose.

The *lacrimal secretion* is a watery solution containing salts, some mucus, and a bactericidal enzyme called lysozyme. It cleans, lubricates, and moistens the eyeball. After being secreted by the lacrimal glands, it is spread over the surface of the eyeball by the blinking of the eyelids. Usually, 1 ml per day is produced. Normally, the secretion is carried away by evaporation or by passing into the lacrimal canals and then into the nasal cavities as fast as it is produced. If, however, an irritating substance makes contact with the conjunctiva, the lacrimal glands are stimulated to oversecrete. Tears then accumulate more rapidly than they can be carried away. This is a protective mechanism since the tears dilute and wash away the irritating substance. "Watery" eyes also occur

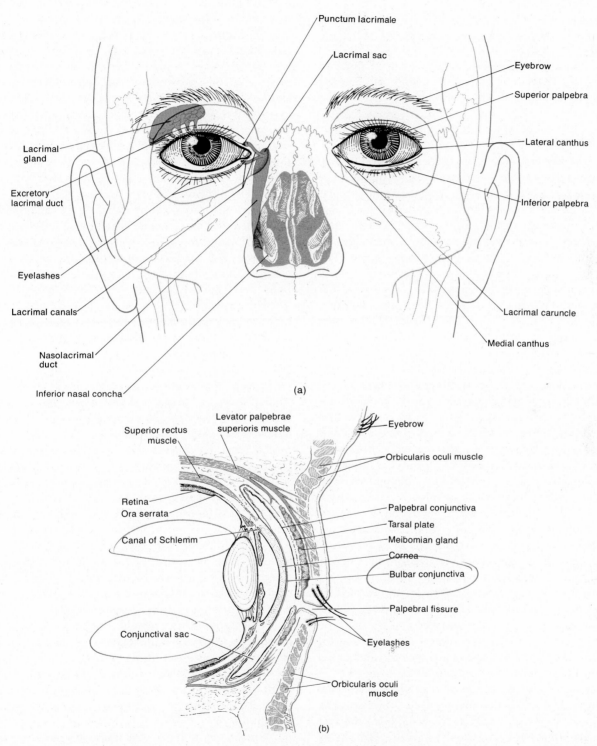

Figure 17-3 Accessory structures of the eye. (a) Anterior view. (b) Sagittal section of the eyelids and anterior portion of the eyeball.

when an inflammation of the nasal mucosa, such as a cold, obstructs the nasolacrimal ducts so that drainage of tears is blocked.

Structure of Eyeball

The adult **eyeball** measures about 2.5 cm (1 inch) in diameter. Of its total surface area, only the anterior one-sixth is exposed. The remainder is recessed and protected by the orbit into which it fits. Anatomically, the eyeball can be divided into three layers: fibrous tunic, vascular tunic, and retina (Figure 17-4).

The **fibrous tunic** is the outer coat of the eyeball. It can be divided into two regions: the posterior sclera and the anterior cornea. The *sclera,* the "white of the eye," is a white coat of fibrous tissue that covers all the eyeball except the anterior colored portion. The sclera gives shape to the eyeball and protects its inner parts. Its posterior surface is pierced by the optic nerve. The anterior portion of the fibrous tunic is called the *cornea.* It is a nonvascular, nervous, transparent fibrous coat that covers the iris, the colored part of the eye. The outer surface of the cornea is covered by an epithelial layer that is continuous with the epithelium of the bulbar conjunctiva. At the junction of the sclera and cornea is a venous sinus known as the *canal of Schlemm.*

The **vascular tunic** is the middle layer of the eyeball and is composed of three portions: the posterior choroid, the anterior ciliary body, and the iris. Collectively, these three structures are called the *uvea.* The *choroid* is a thin, dark-brown membrane that lines most of the internal surface of the sclera. It contains numerous blood vessels and a large amount of pigment. The choroid absorbs light rays so they are not reflected back out of the eyeball. Through its blood supply, it nourishes the retina. The optic nerve also pierces the choroid at the back of the eyeball. The anterior portion of the choroid becomes the *ciliary body.* It is the thickest portion of the vascular tunic. It extends from the *ora serrata* of the retina (inner tunic) to a point just behind the sclerocorneal junction. The ora serrata is simply the jagged margin of the retina. This second division of the vascular tunic contains the *ciliary muscle*–a smooth muscle that alters the shape of the lens for near or far vision. The *iris* is the third portion of the vascular tunic. It consists of circular and radial smooth muscle fibers arranged to form a doughnut-shaped structure. The black hole in the center of the iris is the *pupil,* the area through which light enters the eyeball. The iris is suspended between the cornea and the lens and is attached at its outer margin to the ciliary body. A principal function of the iris is to regulate the amount of light entering the eyeball. When the eye is stimulated by bright light, the circular muscles of the iris contract and decrease the size of the pupil. When the eye must adjust to dim light, the radial muscles of the iris contract and increase the pupil's size.

The third and inner coat of the eye, the **retina,** lies only in the posterior portion of the eye. Its primary function is image formation. It consists of a nervous tissue layer and a pigmented layer. The retina covers the choroid. At the edge of the ciliary body, it appears to end in a scalloped border called the ora serrata. This is where the outer nervous layer or visual portion of the retina ends. The inner pigmented layer extends anteriorly over the back of the ciliary body and the iris as the nonvisual portion of the retina. The inner nervous layer contains three zones of neurons. These three zones, named in the order in which they conduct impulses, are the photoreceptor neurons, bipolar neurons, and ganglion neurons.

The dendrites of the photoreceptor neurons are called rods and cones because of their shapes. They are visual receptors highly specialized for stimulation by light rays. **Rods** are specialized for vision in dim light. They also allow us to discriminate between different shades of dark and light and permit us to see shapes and movement. **Cones,** by contrast, are specialized for color vision and sharpness of vision *(visual acuity).* Cones are stimulated only by bright light. This is why we cannot see color by moonlight. It is estimated that there are 7 million cones and somewhere between 10 and 20 times as many rods. Cones are most densely concentrated in the *central fovea,* a small depression in the center of the macula lutea. The *macula lutea,* or yellow spot, is in the exact center of the retina. The fovea is the area of sharpest vision because of the high concentration of cones. Rods are absent from the fovea and macula, but they increase in density toward the periphery of the retina.

When impulses for sight have passed through the photoreceptor neurons, they are conducted across synapses to the bipolar neurons in the intermediate zone of the nervous layer of the retina. From here the impulses are passed to the ganglion neurons.

The axons of the ganglion neurons extend posteriorly to a small area of the retina called the *optic disc,* or *blind spot.* This region contains openings through which the fibers of the ganglion neurons exit as the optic nerve. Since this area contains neither rods nor cones, and only nerve fibers, no image is formed on it. Thus it is called the blind spot.

In addition to the fibrous tunic, vascular tunic, and retina, the eyeball itself contains the lens, just behind the pupil and iris. The *lens* is constructed of numerous layers of protein fibers arranged like the layers of an onion.

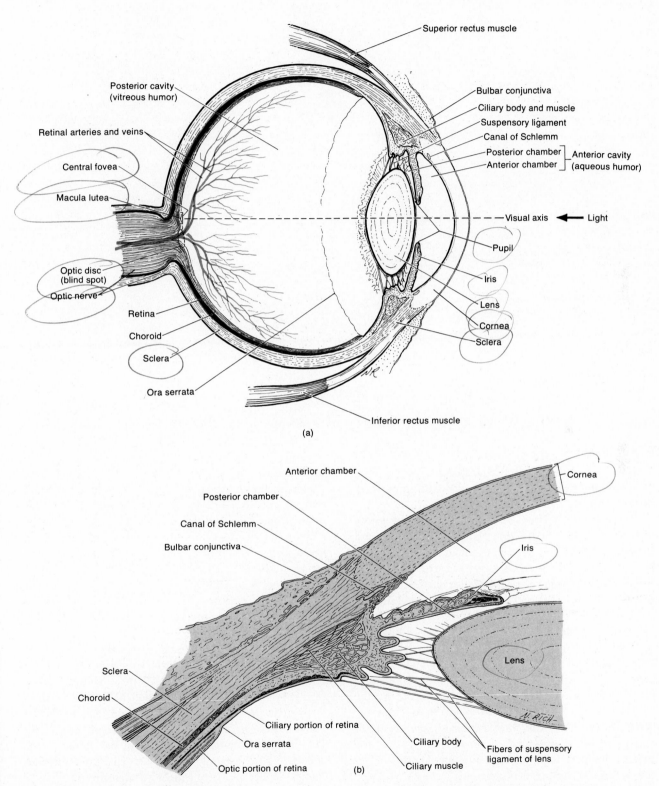

Figure 17-4 Structure of the eyeball. (a) Gross anatomy in sagittal section. (b) Section through the anterior part of the eyeball at the sclerocorneal junction.

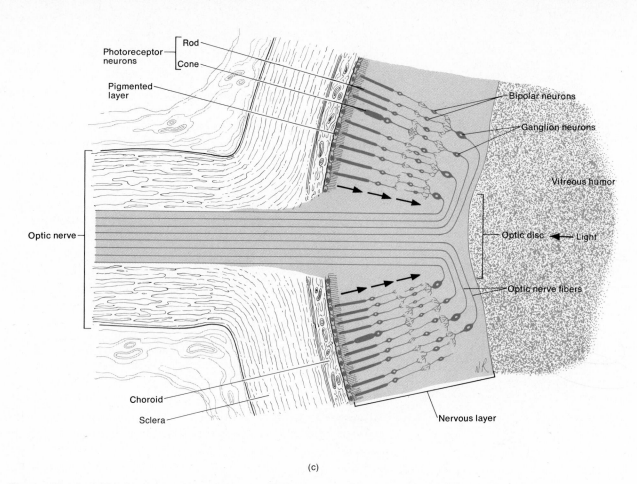

Photoreceptor neurons — Rod, Cone
Pigmented layer
Optic nerve
Choroid
Sclera
Bipolar neurons
Ganglion neurons
Vitreous humor
Optic disc ← Light
Optic nerve fibers
Nervous layer

(c)

Figure 17-4 (cont.)　Structure of the eyeball. (c) Diagram of the microscopic structure of the retina.

Normally, the lens is perfectly transparent. It is enclosed by a clear connective tissue capsule and held in position by the *suspensory ligament.*

The interior of the eyeball contains a large cavity divided into two smaller ones: the anterior cavity and the posterior cavity. They are separated from each other by the lens. The *anterior cavity,* in turn, has two subdivisions referred to as the anterior chamber and the posterior chamber. The *anterior chamber* lies behind the cornea and in front of the iris. The *posterior chamber* lies behind the iris and in front of the suspensory ligament and lens. The anterior cavity is filled with a watery fluid called the *aqueous humor.* The fluid is believed to be secreted into the posterior chamber by choroid plexuses of the ciliary processes of the ciliary bodies behind the iris. It is very similar to cerebrospinal fluid. From the posterior chamber, the fluid permeates the posterior cavity and then passes forward between the iris and the lens, through the pupil, into the anterior chamber. From the anterior chamber, the aqueous humor, which is continually produced, is drained off into the *canal of Schlemm* and then into the blood. The anterior chamber

thus serves a function similar to the subarachnoid space around the brain and spinal cord. The canal of Schlemm is analogous to a venous sinus of the dura mater. The pressure in the eye, called *intraocular pressure,* is produced mainly by the aqueous humor. The intraocular pressure keeps the retina smoothly applied to the choroid so the retina may form clear images. Besides maintaining normal intraocular pressure, the aqueous humor is also the principal link between the circulatory system and the lens and cornea. Neither the lens nor the cornea has blood vessels.

The second, and larger, cavity of the eyeball is the *posterior cavity.* It lies between the lens and the retina and contains a jellylike substance called the *vitreous humor.* This substance contributes to intraocular pressure, helps to prevent the eyeball from collapsing, and holds the retina flush against the internal portions of the eyeball. The vitreous humor, unlike the aqueous humor, does not undergo constant replacement. It is formed during embryonic life and is not replaced thereafter.

Exhibit 17-1 (pages 380–381) presents some surface features of the eye.

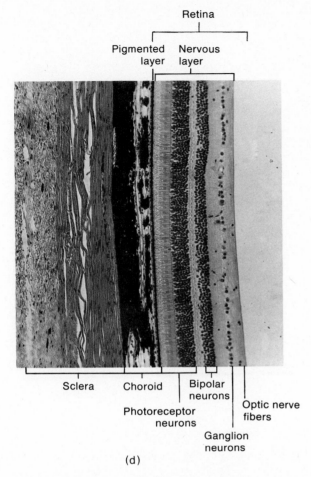

Retina
Pigmented layer | Nervous layer

Sclera Choroid Bipolar neurons
Photoreceptor neurons
Ganglion neurons
Optic nerve fibers

(d)

Figure 17-4 (cont.) Structure of the eyeball. (d) Photomicrograph of the posterior wall of the eyeball, showing the layers of the retina at a magnification of 100×. (Courtesy of Donald I. Patt, from *Comparative Vertebrate Histology,* by Donald I. Patt and Gail R. Patt, Harper & Row, Publishers, Inc., New York, 1969.)

Physiology of Vision

Before light can reach the rods and cones of the retina, it must pass through the cornea, aqueous humor, pupil, lens, and vitreous humor. Moreover, for vision to occur, light reaching the rods and cones must form an image on the retina. The resulting nerve impulses must then be conducted to the visual areas of the cerebral cortex. In discussing the physiology of vision, let us first consider retinal image formation.

RETINAL IMAGE FORMATION

The formation of an image on the retina requires four basic processes, all concerned with focusing light rays:

1 Refraction of light rays.
2 Accommodation of the lens.
3 Constriction of the pupil.
4 Convergence of the eyes.

Accommodation and pupil size are functions of the smooth muscle cells of the ciliary muscle and the dilator and sphincter muscles of the iris. They are termed *intrinsic eye muscles* since they are inside the eyeball. Convergence is a function of the voluntary muscles attached to the outside of the eyeball called the *extrinsic eye muscles.*

■ *Refraction and Accommodation* When light rays traveling through a transparent medium (such as air) pass into a second transparent medium with a different density (such as water), they bend at the surface of the two media. This is *refraction* (Figure 17-5). The eye has four such media of refraction: cornea, aqueous humor, lens, and vitreous humor. Light rays entering the eye from the air are refracted at the following points: (1) the anterior surface of the cornea as they pass from the lighter air into the denser cornea; (2) the posterior surface of the cornea as they pass into the less dense aqueous humor; (3) the anterior surface of the lens as they pass from the aqueous humor into the denser lens; and (4) the posterior surface of the lens as they pass from the lens into the less dense vitreous humor.

When an object is 6 m (20 ft) or more away from the viewer, the light rays reflected from the object are nearly parallel to one another. The degree of refraction that takes place at each surface in the eye is very precise. Therefore the parallel rays are sufficiently bent to fall exactly on the central fovea, where vision is sharpest. However, light rays that are reflected from close-by objects are divergent rather than parallel. As a result, they must be refracted toward each other to a greater extent. This change in refraction is brought about by the lens of the eye.

If the surfaces of a lens curve outward, as in a convex lens, the lens will refract the rays toward each other so they eventually intersect. The more the lens curves outward, the more acutely it bends the rays toward each other. Conversely, when the surfaces of a lens curve inward, as in a concave lens, the rays bend away from each other. The lens of the eye is biconvex. Furthermore, it has the unique ability to change the focusing power of the eye by becoming moderately curved at one moment and greatly curved the next. When the eye is focusing on a close object, the lens curves greatly in order to bend the rays toward the central fovea. This increase in the curvature of the lens is called *accommodation* (Figure 17-6). The ciliary muscle

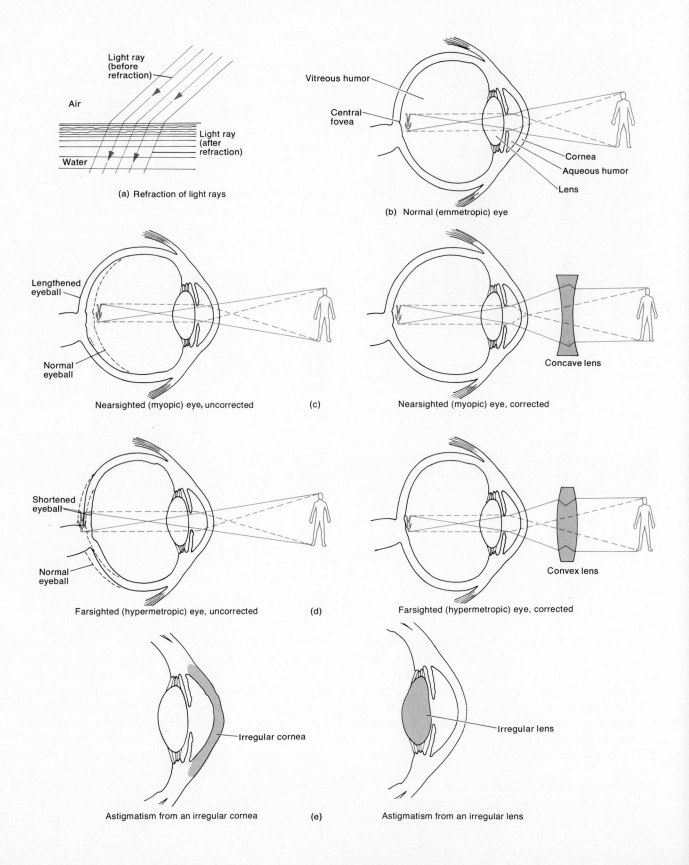

(a) Refraction of light rays

(b) Normal (emmetropic) eye

Nearsighted (myopic) eye, uncorrected

(c)

Nearsighted (myopic) eye, corrected

Farsighted (hypermetropic) eye, uncorrected

(d)

Farsighted (hypermetropic) eye, corrected

Astigmatism from an irregular cornea

(e)

Astigmatism from an irregular lens

Figure 17-5 Normal and abnormal refraction in the eyeball. (a) Refraction of light rays passing from air into water. (b) In the normal or emmetropic eye, light rays from an object are bent sufficiently by the four refracting media and converged on the central fovea. A clear image is formed. (c) In the nearsighted or myopic eye, the image is focused in front of the retina. The condition may result from an elongated eyeball or a thickened lens. Correction is by use of a concave lens which diverges entering light rays. (d) In the farsighted or hypermetropic eye, the image is focused behind the retina. The condition may be the result of the eyeball being too short or the lens being too thin. Correction is by a convex lens which refracts entering light rays. (e) Astigmatism. In this condition, the curvature of the cornea or lens is uneven. As a result, horizontal and vertical rays are focused at two different points on the retina. Suitable glasses correct the refraction of an astigmatic eye. On the left, astigmatism resulting from an irregular cornea. On the right, astigmatism resulting from an irregular lens. The image is not focused on the area of sharpest vision of the retina. This results in blurred or distorted vision.

contracts, pulling the ciliary body and choroid forward toward the lens. This action releases the tension on the lens and suspensory ligament. Due to its elasticity, the lens shortens, thickens, and bulges. In near vision, the ciliary muscle is contracted and the lens is bulging. In far vision, the ciliary muscle is relaxed and the lens is flatter. With aging, the lens loses elasticity and, therefore, its ability to accommodate.

The normal eye, known as an *emmetropic eye,* can sufficiently refract light rays from an object 6 m (20 ft) away to focus a clear object on the retina. Many individuals, however, do not have this ability because of abnormalities related to improper refraction. Among

these abnormalities are *myopia* (nearsightedness), *hypermetropia* (farsightedness), and *astigmatism* (irregularities in the surface of the lens or cornea). The conditions are illustrated and explained in Figure 17-5c–e.

■ *Constriction of Pupil* The muscles of the iris also assume a function in the formation of clear retinal images. Part of the accommodation mechanism consists of the contraction of the circular muscle fibers of the iris to constrict the pupil. Constricting the pupil means narrowing the diameter of the hole through which light enters the eye. This action occurs simultaneously with accommodation of the lens and prevents light rays from entering the eye through the periphery of the lens. Light rays entering at the periphery would not be brought to focus on the retina and would result in blurred vision. The pupil, as noted earlier, also constricts in bright light to protect the retina from sudden or intense stimulation.

■ *Convergence* Birds see a set of objects off to the left through one eye and an entirely different set off to the right through the other. This characteristic doubles their field of vision and allows them to detect predators behind them. In human beings, both eyes focus on only one set of objects—a characteristic called *single binocular vision.*

Single binocular vision occurs when light rays from an object are directed toward corresponding points on the two retinas. When we stare straight ahead at a distant object, the incoming light rays are aimed directly at both pupils and are refracted to identical spots on the retinas of both eyes. But as we move close to the object, our eyes must rotate medially—that is, become "crossed"—for the light rays from the object to hit the same points on both retinas. The term *convergence* refers to this medial movement of the two eyeballs so they are both directed toward the object being viewed. The

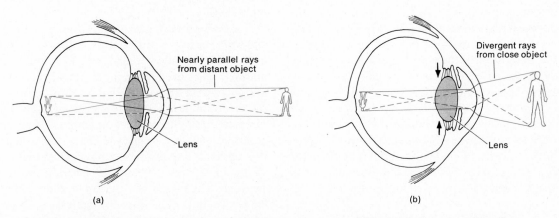

Figure 17-6 Accommodation. (a) For objects 6 m (20 ft) or more away. (b) For objects nearer than 6 m.

Exhibit 17–1   SURFACE ANATOMY OF EYE

Anterior view of right eye

Photograph courtesy of Donald Castellaro and Deborah Massimi.

nearer the object, the greater the degree of convergence necessary to maintain single binocular vision. Convergence is brought about by the coordinated action of the extrinsic eye muscles.

■ *Inverted Image*   Images are focused upside down on the retina. They also undergo mirror reversal. That is, light reflected from the right side of an object hits the left side of the retina and vice versa. Note in Figure 17-5b that reflected light from the top of the object crosses light from the bottom of the object and strikes the retina below the central fovea. Reflected light from the bottom of the object crosses light from the top of the object and strikes the retina above the central fovea. The reason why we do not see a topsy-turvy world is that the brain learns early in life to coordinate visual images with the exact locations of objects. The brain stores memories of reaching and touching objects and automatically turns visual images right-side-up and right-side-around.

STIMULATION OF PHOTORECEPTORS
After an image is formed on the retina by refraction, accommodation, constriction of the pupil, and conver-

gence, light impulses must be converted into nerve impulses by the rods and cones. Rods contain a reddish purple pigmented compound called *rhodopsin,* or *visual purple.* This substance consists of the protein scotopsin plus retinene, a derivative of vitamin A. When light rays strike a rod, rhodopsin rapidly breaks down. This chemical breakdown stimulates impulse conduction by the rods (Figure 17-7).

Rhodopsin is highly light sensitive – even the light rays from the moon or a candle will break down some of it and thereby allow us to see. The rods, then, are uniquely specialized for night vision. However, they are of only limited help for daylight vision. In bright light, the rhodopsin is destroyed faster than it can be manufactured. In dim light, production is able to keep pace with a slower rate of breakdown. These characteristics of rhodopsin are responsible for the experience of having to adjust to a dark room after walking in from the sunshine. The period of adjustment is the time it takes for the completely destroyed rhodopsin to reform. The adjustment period is normal. *Night blindness* is the lack of normal night vision following the adjustment period. It is most often caused by vitamin A deficiency.

Exhibit 17—1 (cont.)

1 Pupil Opening of center of iris of eyeball for light transmission.

2 Iris Circular pigmented muscular membrane behind cornea.

3 Sclera "White" of eye, a coat of fibrous tissue that covers entire eyeball except for cornea.

4 Conjunctiva Membrane that covers exposed surface of eyeball and lines eyelids.

5 Eyelids Folds of skin and muscle lined by conjunctiva.

6 Palpebral fissure Space between eyelids when they are open.

7 Medial canthus Site of union and upper and lower eyelids near nose.

8 Lateral canthus Site of union of upper and lower eyelids away from nose.

9 Lacrimal caruncle Fleshy, yellowish projection of medial commissure that contains modified sweat and sebaceous glands.

10 Eyelashes Hairs on margins of eyelids, usually arranged in two or three rows.

11 Eyebrows Several rows of hairs superior to upper eyelids.

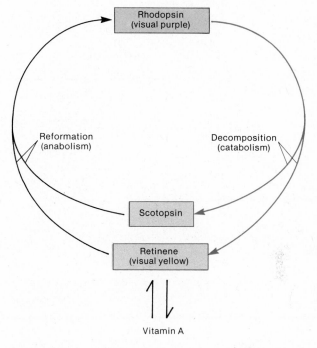

Figure 17-7 Rhodopsin cycle. Color indicates breakdown reactions in light. Black indicates reformation of rhodopsin in darkness.

Cones, which are the receptors for daylight and color, contain photosensitive chemicals that require bright light for their breakdown. Unlike rhodopsin, the photosensitive chemicals of the cones reform quickly. It is believed that there are three types of cones and that each contains a different visual pigment. Each pigment has a different maximum absorption of light of a different color so that each responds best to light of a given wavelength (color). One type of cone responds best to red light, the second to green light, and the third to blue light. Just as an artist can obtain almost any color by mixing primary colors, it is believed that cones can perceive any color by differential stimulation. Stimulation of a cone by two or more colors may produce any combination of colors.

AFFERENT PATHWAY TO BRAIN

From the rods and cones, impulses are transmitted through bipolar neurons to ganglion cells. The cell bodies of the ganglion cells lie in the retina, and their axons leave the eyeball via the optic nerve (Figure 17-8). The axons pass through the *optic chiasma,* a crossing point of the optic nerves. Some fibers cross to the opposite side. Others remain uncrossed. On passing through the optic chiasma, the fibers, now part of the *optic tract,* enter the brain and terminate in the thalamus. Here the fibers synapse with third-order neurons whose axons pass to the visual centers located in the occipital lobes of the cerebral cortex.

Analysis of the afferent pathway to the brain reveals that the visual field of each eye is divided into two regions: the *medial* (or *nasal*) *half* and the *lateral* (or *temporal*) *half.* For each eye, light rays from an object in the nasal half of the visual field fall on the temporal half of the retina. Light rays from an object in the temporal half of the vision field fall on the nasal half of the retina. Note that in the optic chiasma nerve fibers from the nasal halves of the retinas cross and continue on to the thalamus. Note also that nerve fibers from the temporal halves of the retinas do not cross but continue directly on to the thalamus. As a result, the visual center in the cortex of the right occipital lobe "sees" the left side of an object via impulses from the temporal half of the retina of the right eye and the nasal half of the retina of the left eye. The cortex of the left occipital lobe interprets visual sensations from the right side of an object via impulses

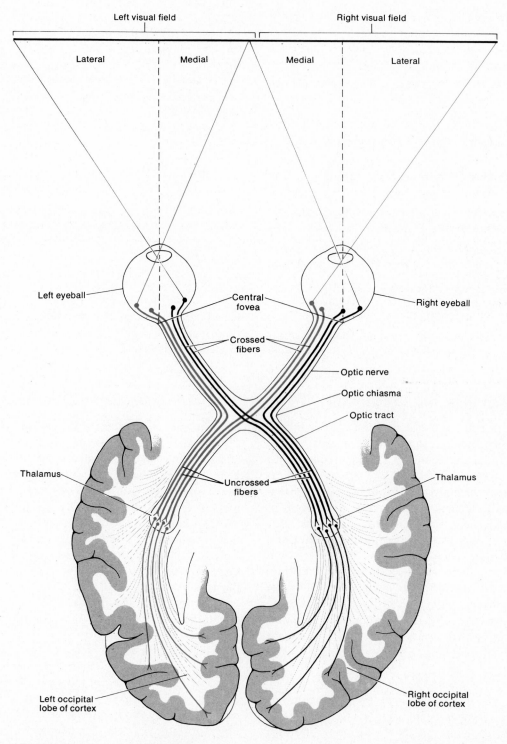

Figure 17-8   Afferent pathway for visual impulses.

from the nasal half of the right eye and the temporal half of the left eye.

AUDITORY SENSATIONS AND EQUILIBRIUM

In addition to containing receptors for sound waves, the **ear** also contains receptors for equilibrium. Anatomically, the ear is divided into three principal regions: the external or outer ear, the middle ear, and the internal or inner ear.

External or Outer Ear

The *external* or *outer ear* is structurally designed to collect sound waves and direct them inward (Figure 17-9a). It consists of the pinna, the external auditory canal, and the tympanic membrane, also called the eardrum. The *pinna,* or *auricle,* is a trumpet-shaped flap of elastic cartilage covered by thick skin. The rim of the pinna is called the helix; the inferior portion is the lobe. (See Exhibit 17-2 on page 386 for surface features of the external ear.) The pinna is attached to the head by ligaments and muscles. The *external auditory canal* or *meatus* is a tube about 2.5 cm (1 inch) in length that lies in the external auditory meatus of the temporal bone. It leads from the pinna to the eardrum. The walls of the canal consist of bone lined with cartilage that is continuous with the cartilage of the pinna. The cartilage in the external auditory canal is covered with thin, highly sensitive skin. Near the exterior opening, the canal contains a few hairs and specialized sebaceous glands called *ceruminous glands,* which secrete *cerumen* (earwax). The combination of hairs and cerumen prevents foreign objects from entering the ear. The *tympanic membrane* or *eardrum* is a thin, semitransparent partition of fibrous connective tissue between the external auditory meatus and the middle ear. Its external surface is concave and covered with skin. Its internal surface is convex and covered with a mucous membrane.

Middle Ear

Also called the *tympanic cavity,* the *middle ear* is a small, epithelial-lined, air-filled cavity hollowed out of the temporal bone (Figure 17-9a, b). The cavity is separated from the external ear by the eardrum and from the internal ear by a thin bony partition that contains two small openings: the oval window and the round window. The posterior wall of the cavity communicates with the mastoid cells of the temporal bone through a chamber called the *tympanic antrum.* This anatomical fact explains why a middle ear infection may spread to the

temporal bone, causing mastoiditis, or even to the brain. The anterior wall of the cavity contains an opening that leads directly into the *Eustachian tube,* also called the *auditory tube* or *meatus.* The Eustachian tube connects the middle ear with the nose and nasopharynx of the throat. Through this passageway, infections may travel from the throat and nose to the ear. The function of the tube is to equalize air pressure on both sides of the tympanic membrane. Abrupt changes in external or internal air pressure might otherwise cause the eardrum to rupture. Since the tube opens during swallowing and yawning, these activities allow atmospheric air to enter or leave the middle ear until the internal pressure equals the external pressure. Any sudden pressure changes against the eardrum may be equalized by deliberately swallowing.

Extending across the middle ear are three exceedingly small bones called **auditory ossicles.** These are called the malleus, incus, and stapes. According to their shapes, they are commonly named the hammer, anvil, and stirrup, respectively. The "handle" of the **malleus** is attached to the internal surface of the tympanic membrane. Its head articulates with the base of the incus. The **incus** is the intermediate bone in the series and articulates with the stapes. The base or footplate of the **stapes** fits into a small opening between the middle and inner ear called the *fenestra vestibuli,* or *oval window.* Directly below the oval window is another opening, the *fenestra cochlea,* or *round window.* This opening, which also separates the middle and inner ears, is enclosed by a membrane called the secondary tympanic membrane. The auditory ossicles are attached to the tympanic membrane, to each other, and to the oval window by means of ligaments and muscles.

Internal or Inner Ear

The *internal or inner ear* is also called the *labyrinth* (Figure 17-10) because of its complicated series of canals. Structurally, it consists of two main divisions: (1) a bony labyrinth and (2) a membranous labyrinth that fits in the bony labyrinth. The *bony labyrinth* is a series of cavities in the petrous portion of the temporal bone. It can be divided into three areas named on the basis of shape: the vestibule, cochlea, and semicircular canals. The bony labyrinth is lined with periosteum and contains a fluid called *perilymph.* This fluid surrounds the *membranous labyrinth,* a series of sacs and tubes lying inside and having the same general form as the bony labyrinth. Epithelium lines the membranous labyrinth and contains a fluid called *endolymph.*

The *vestibule* constitutes the oval central portion of the bony labyrinth. The membranous labyrinth in the

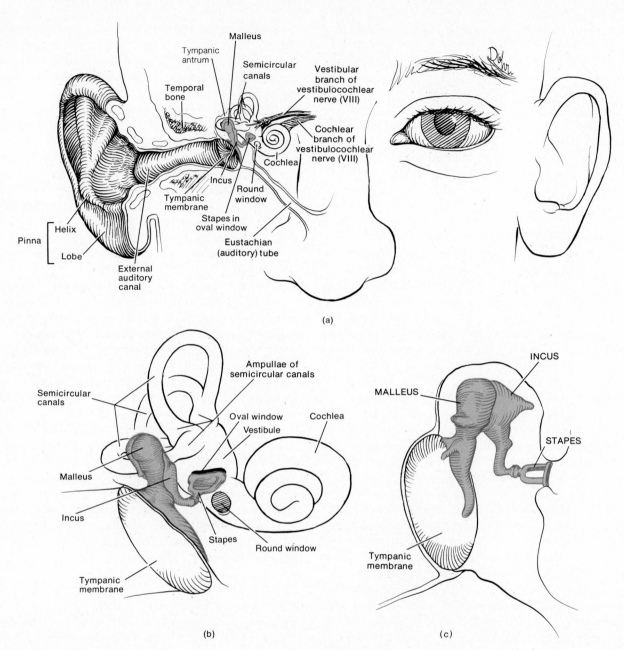

Figure 17-9   Structure of the auditory apparatus. (a) Divisions of the right ear into external, middle, and internal portions seen in a coronal section through the right side of the skull. The middle ear is shown in color. (b) Details of the middle ear and the bony labyrinth of the internal ear. (c) Ossicles of the middle ear.

vestibule consists of two sacs called the *utricle* and *saccule.* These sacs are connected to each other by a small duct.

Projecting upward and posteriorly from the vestibule are the three bony *semicircular canals.* Each is arranged at approximately right angles to the other two. On the basis of their positions, they are called the superior, posterior, and lateral canals. One end of each

canal enlarges into a swelling called the *ampulla.* Inside the bony semicircular canals lie portions of the membranous labyrinth: the *semicircular ducts* or *membranous semicircular canals.* These structures are almost identical in shape to the bony semicircular canals and communicate with the utricle of the vestibule.

Lying in front of the vestibule is the *cochlea,* so designated because of its resemblance to a snail's shell.

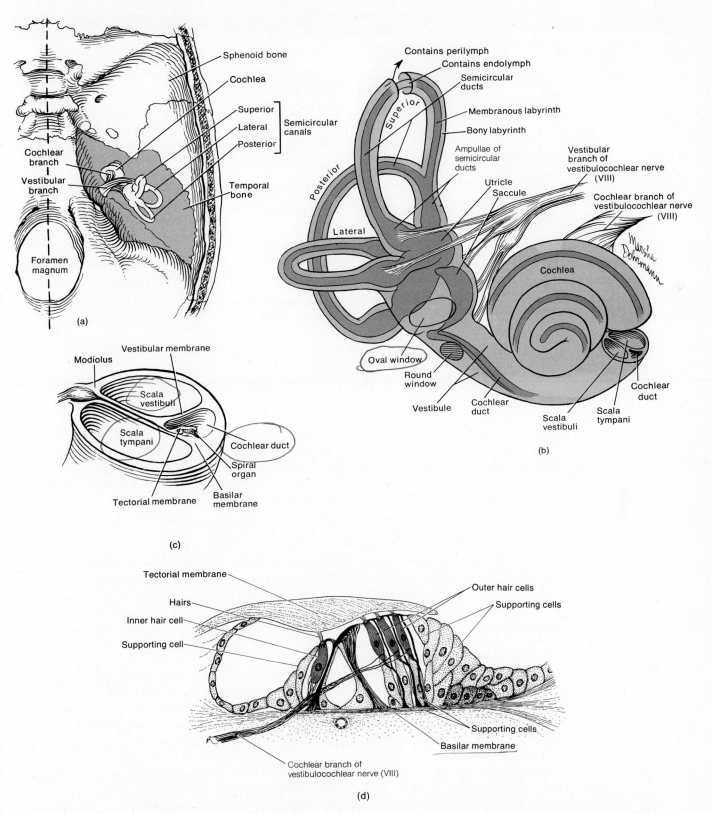

Figure 17-10 Details of the internal ear. (a) Relative position of the bony labyrinth projected to the inner surface of the floor of the skull. (b) The outer, gray area belongs to the bony labyrinth. The inner, colored area belongs to the membranous labyrinth. (c) Cross section through the cochlea. (d) Enlargement of the spiral organ or organ of Corti.

Exhibit 17–2   SURFACE ANATOMY OF EAR

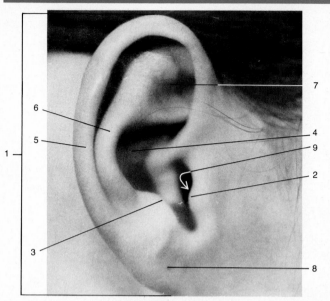

Lateral view of right ear
Photograph courtesy of Donald Castellaro and Deborah Massimi.

1 Auricle   Portion of external ear not contained in head, also called pinna or trumpet.

2 Tragus   Cartilaginous projection anterior to external opening to ear.

3 Antitragus   Cartilaginous projection opposite tragus.

4 Concha   Hollow of auricle.

5 Helix   Superior and posterior free margin of auricle.

6 Antihelix   Semicircular ridge posterior and superior to concha.

7 Triangular fossa   Depression in superior portion of antihelix.

8 Lobule   Interior portion of auricle devoid of cartilage.

9 External auditory canal   Canal extending from external ear to eardrum.

The cochlea consists of a bony spiral canal that makes about 2¾ turns around a central bony core called the *modiolus.* A cross section through the cochlea shows the canal is divided by partitions into three separate channels resembling the letter Y on its side. The stem of the Y is a bony shelf that protrudes into the canal. The wings of the Y are composed of the bony labyrinth. The channel above the bony partition is the *scala vestibuli;* the channel below is the *scala tympani.* The cochlea adjoins the wall of the vestibule, into which the scala vestibuli opens. The scala tympani terminates at the round window. The perilymph of the vestibule is continuous with that of the scala vestibuli. The third channel (between the wings of the Y) is the membranous labyrinth: the *cochlear duct.* The cochlear duct is separated from the scala vestibuli by the *vestibular membrane,* also called *Reissner's membrane.* It is separated from the scala tympani by the *basilar membrane.* Resting on the basilar membrane is the *spiral organ,* or *organ of Corti,* the organ of hearing. The organ of Corti is a series of epithelial cells on the inner surface of the basilar membrane. It consists of a number of supporting cells and hair cells, which are the receptors for auditory sensations. The inner hair cells are medially placed in a single row and extend the entire length of the cochlea. The outer hair cells are arranged in several rows throughout the cochlea. The hair cells have long hairlike processes at their free ends that extend into the endolymph of the cochlear duct. The basal ends of the hair cells are in contact with fibers of the cochlear branch of the vestibulocochlear nerve (VIII). Projecting over and in contact with the hair cells of the organ of Corti is the *tectorial membrane,* a delicate and flexible gelatinous membrane.

Physiology of Hearing

Sound waves result from the alternate compression and decompression of air. They originate from a vibrating object and travel through air much as waves travel on water. The events involved in the physiology of hearing sound waves are illustrated in Figure 17-11 and listed below:

1 Sound waves that reach the ear are directed by the pinna into the external auditory canal.

2 When the waves strike the tympanic membrane, the alternate compression and decompression of the air cause the membrane to vibrate.

3 The central area of the tympanic membrane is connected to the malleus, which also starts to vibrate. The vibration is then picked up by the incus, which transmits the vibration to the stapes.

4 As the stapes moves back and forth, it pushes the oval window in and out.

5 The movement of the oval window sets up waves in the perilymph.

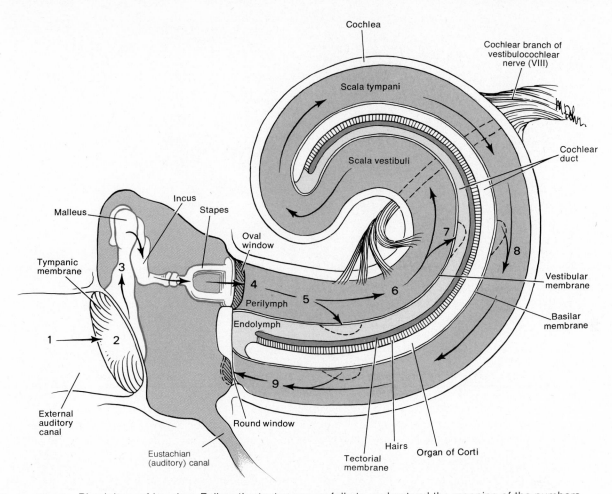

Figure 17-11 Physiology of hearing. Follow the text very carefully to understand the meaning of the numbers.

6 As the window bulges inward, it pushes the perilymph of the scala vestibuli up into the cochlea.

7 This pressure pushes the vestibular membrane inward and increases the pressure of the endolymph inside the cochlear duct.

8 The basilar membrane gives under the pressure and bulges out into the scala tympani.

9 The sudden pressure in the scala tympani pushes the perilymph toward the round window, causing it to bulge back into the middle ear. Conversely, as the sound wave subsides, the stapes moves backward and the procedure is reversed. That is, the fluid moves in the opposite direction along the same pathway and the basilar membrane bulges into the cochlear duct.

10 When the basilar membrane vibrates, the hair cells of the organ of Corti are moved against the tectorial membrane. In some unknown manner, the movement of the hairs stimulates the dendrites of neurons at their base and sound waves are converted into nerve impulses.

11 The impulses are then passed on to the cochlear branch of the vestibulocochlear nerve (Figure 17-12) and

the medulla. Here some impulses cross to the opposite side and finally travel to the auditory area of the temporal lobe of the cerebral cortex.

If sound waves passed directly to the oval window without passing through the tympanic membrane and auditory bones, hearing would be inadequate. A minimal amount of sound energy is required to transmit sound waves through the perilymph of the cochlea. Since the tympanic membrane has a surface area about 22 times larger than that of the oval window, it can collect about 22 times more sound energy. This energy is sufficient to transmit sound waves through the perilymph.

Physiology of Equilibrium

The term *equilibrium* has two meanings. One kind of equilibrium, called *static equilibrium,* refers to the orientation of the body (mainly the head) relative to the ground. The second kind, *dynamic equilibrium,* is the

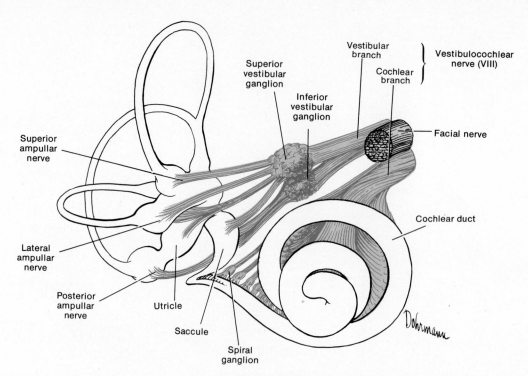

Figure 17-12 Principal nerves of the internal ear. Shown here is the right membranous labyrinth with the bony labyrinth removed. Nerve branches from each of the ampullae of the semicircular ducts and from the utricle and saccule form the vestibular branch of the vestibulocochlear nerve. The vestibular ganglia contain the cell bodies of the vestibular branch. The cochlear branch of the vestibulocochlear nerve arises in the spiral organ or organ of Corti. The spiral ganglion contains the cell bodies of the cochlear branch.

maintenance of body position (mainly the head) in response to sudden movements. The receptor organs for equilibrium are the saccule, utricle, and semicircular ducts.

The utricle and saccule each contain sensory hair cells that project into the cavity of the membranous labyrinth. The hairs are coated with a gelatinous layer in which particles of calcium carbonate, called *otoliths,* are embedded. When the head tips downward, the otoliths slide with gravity in the direction of the ground. As the particles move, they exert a downward pull on the gelatinous mass, which in turn exerts a downward pull on the hairs and makes them bend. The movement of the hairs stimulates the dendrites at the bases of the hair cells. The impulse is then transmitted to the temporal lobe of the brain through the vestibular branch of the vestibulocochlear nerve (see Figure 17-12). The utricle and saccule are considered to be sense organs of static equilibrium. They provide information regarding the

orientation of the head in space and are essential for maintaining posture.

The three semicircular ducts maintain dynamic equilibrium. They are positioned at right angles to each other in three planes: frontal (the superior duct), sagittal (the posterior duct), and lateral (the lateral duct). This positioning permits correction of an imbalance in three planes. In the ampulla, the dilated portion of each duct, there is a small elevation called the *crista* (Figure 17-13a). Each crista is composed of a group of hair cells covered by a mass of gelatinous material called the *cupula* (Figure 17-13b). When the head moves, the endolymph in the semicircular ducts flows over the hairs and bends them. The movement of the hairs stimulates sensory neurons, and the impulses pass over the vestibular branch of the vestibulocochlear nerve. The impulses then reach the temporal lobe of the cerebrum and are sent to the muscles that must contract in order to maintain body balance in the new position.

DISORDERS

The special sense organs can be altered or damaged by numerous disorders. The causes of disorder can range from congenital origins to the effects of old age. Here we discuss a few common disorders of the eyes and ears.

CATARACT

The most prevalent disorder resulting in blindness is **cataract** formation. This disorder causes the lens or its capsule to lose its transparency. Cataracts can occur at any age, but we will

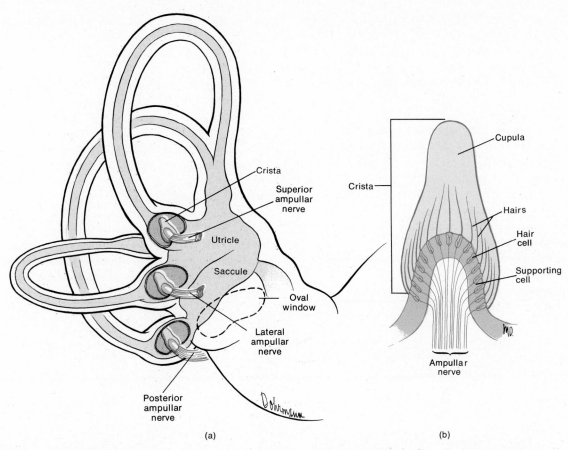

Figure 17-13 Semicircular ducts and dynamic equilibrium. (a) Position of the cristae relative to the membranous ampullae. (b) Enlarged aspect of a crista.

discuss the type that develops with old age. As a person gets older the cells in the lenses may degenerate and be replaced with nontransparent fibrous protein. Or the lenses may start to manufacture nontransparent protein. The main symptom of cataract is a progressive, painless loss of vision. The degree of loss depends on the location and extent of the opacity. If vision loss is gradual, frequent changes in glasses may help maintain useful vision for a while. Eventually, though, the changes may be so extensive that light rays are blocked out altogether. At this point, surgery is indicated. Essentially, the surgical procedure consists of removing the opaque lens and substituting an artificial lens by means of eyeglasses or by implanting a plastic lens inside the eyeball to replace the natural one.

GLAUCOMA

The second most common cause of blindness, especially in the elderly, is **glaucoma.** This disorder is characterized by an abnormally high pressure of fluid inside the eyeball. The aqueous humor does not return into the bloodstream through the canal of Schlemm as quickly as it is formed. The fluid accumulates and, by compressing the lens into the vitreous humor, puts pressure on the neurons of the retina. If the pressure continues over a long period of time, it destroys the

neurons and brings about blindness. It can affect a person of any age, but 95 percent of the victims are over 40. Glaucoma affects the eyesight of more than 1 million people in this country.

CONJUNCTIVITIS

Many different eye inflammations exist, but the most common type is **conjunctivitis (pinkeye)** – an inflammation of the membrane that lines the insides of the eyelids and covers the cornea. Conjunctivitis can be caused by microorganisms – most often the pneumococci or staphylococci bacteria. In such cases, the inflammation is very contagious. It can also be caused by a number of irritants, in which case the inflammation is not contagious. Irritants include dust, smoke, wind, air pollution, and excessive glare. The condition may be acute or chronic. The epidemic type in children is extremely contagious, but normally it is not serious.

TRACHOMA

This chronic contagious conjunctivitis is caused by an organism called the TRIC agent. The organism has characteristics of both viruses and bacteria. **Trachoma** is characterized by many granulations or fleshy projections on the eyelids. If untreated, these projections can irritate and inflame the cornea and reduce vision. The disease produces an excessive growth of subconjunctival tissue and the invasion of blood vessels into the upper half of the front of the cornea. The disease progresses until it covers the entire cornea, bringing about a loss of vision because of corneal opacity.

MÉNIÈRE'S DISEASE

An important cause of deafness and loss of equilibrium in adults is **Ménière's disease** of the inner ear. It is a disturbance or malfunction of any part of the inner ear. It can be the result of many causes:

1 Infection of the middle ear.
2 Trauma from brain concussion producing hemorrhage or splitting of the labyrinth.
3 Cardiovascular diseases, such as arteriosclerosis and blood vessel disturbances.
4 Congenital malformation of the labyrinth.
5 Excessive formation of endolymph.
6 Allergy.

The last two causes can produce an increase in pressure in the cochlear duct and vestibular system. This pressure, in turn, causes a progressive atrophy of the hair cells of the cochlear or semicircular ducts.

If the cochlear duct is injured, typical symptoms are hissing, roaring, or ringing in the ears and deafness. If the semicircular ducts are involved, the person feels dizzy and nauseous. The dizzy spells can last from a few minutes to several days. Ménière's disease is a chronic disorder. It affects both sexes equally and usually begins in late middle life.

IMPACTED CERUMEN

Some people produce an abnormal amount of cerumen, or earwax, in the external auditory canal. Here it becomes impacted and prevents sound waves from reaching the tympanic membrane. The treatment for **impacted cerumen** is usually periodic ear irrigation or removal of wax with a blunt instrument.

MEDICAL TERMINOLOGY

Achromatopsia (*a* = without; *chrom* = color)   Complete color blindness.

Ametropia (*ametro* = disproportionate; *ops* = eye)   Refractive defect of the eye resulting in an inability to focus images properly on the retina.

Blepharitis (*blepharo* = eyelid)   An eyelid inflammation.

Eustachitis   Eustachian tube infection or inflammation.

Keratitis (*keralo* = cornea)   An inflammation or infection of the cornea.

Keratoplasty (*plasty* = form)   Corneal graft or transplant: an opaque cornea is removed and replaced with a normal transparent cornea to restore vision.

Labyrinthitis   Inner ear or labyrinth inflammation.

Myringitis (*myringa* = eardrum)   Inflammation of the eardrum; also called tympanitis.

Nystagmus   A constant, rapid involuntary eyeball movement, possibly caused by a disease of the central nervous system.

Otalgia (*oto* = ear; *algia* = pain)   Earache.

Otitis   Inflammation of the ear.

Presbyopia (*presby* = old)   Inability to focus on nearby objects due to loss of elasticity of the crystalline lens. The loss is usually caused by aging.

Ptosis (*ptosis* = fall)   Falling or drooping of the eyelid. (This expression is also used for the slipping of any organ below its normal position.)

Retinoblastoma (*blast* = bud; *oma* = tumor)   A common tumor arising from immature retinal cells and accounting for 2 percent of childhood malignancies.

Strabismus   An eye muscle disorder, commonly called "crossed eyes." The eyeballs do not move in unison. It may be caused by lack of coordination of the extrinsic eye muscles.

Tinnitus   A ringing in the ears.

OLFACTORY SENSATIONS

1 Receptor cells in the nasal epithelium send impulses to the olfactory bulbs, to olfactory tracts, to the cortex.

GUSTATORY SENSATIONS

1 Receptors in the taste buds send impulses to the cranial nerves, thalamus, and cortex.

VISUAL SENSATIONS

1 The eye is constructed of three coats: *(a)* fibrous tunic (sclera and cornea); *(b)* vascular tunic (choroid, ciliary body, and iris); and *(c)* retina, which contains rods and cones.

2 The anterior cavity contains aqueous humor. The posterior cavity contains vitreous humor.

3 The refractive media of the eye are the cornea, aqueous humor, lens, and vitreous humor.

4 Retinal image formation involves refraction of light, accommodation of lens, constriction of pupil, convergence, and inverted image formation.

5 Improper refraction may result from myopia (nearsightedness), hypermetropia (farsightedness), and astigmatism (corneal or lens abnormalities).

6 Rods and cones convert light rays into visual nerve impulses. Rhodopsin is necessary for the conversion in rod cells.

7 Impulses from rods and cones are conveyed through the retina to the optic nerve, the optic chiasma, the optic tract, the thalamus, and the cortex.

AUDITORY SENSATIONS AND EQUILIBRIUM

1 The ear consists of three anatomical subdivisions: *(a)* the outer ear (pinna, external auditory canal, and tympanic membrane); *(b)* the middle ear (ossicles, oval window, round window, and opening into the Eustachian tube); and *(c)* the inner ear (bony labyrinth and membranous labyrinth).

2 Sound waves are caused by the alternate compression and decompression of air.

3 Waves enter the external auditory canal, strike the tympanic membrane, pass through the ossicles, strike the oval window, set up waves in the perilymph, strike the vestibular membrane and scala tympani, increase pressure in the endolymph, strike the basilar membrane, and stimulate hairs on the spiral organ. A sound impulse is then initiated.

4 Static equilibrium is balance relative to the pull of gravity.

5 Dynamic equilibrium is balance in response to body movement.

1 Discuss the origin and path of an impulse that results in smelling.

2 How are papillae related to taste buds? Describe the structure and location of the papillae. Discuss how an impulse for taste travels from a taste bud to the brain.

3 Describe the structure and importance of the following accessory structures of the eye: eyelids, eyelashes, eyebrows, and lacrimal apparatus.

4 By means of a labeled diagram, indicate the principal anatomical structures of the eye. How is the retina adapted to its function?

5 Distinguish a sty from a chalazion.

6 How do extrinsic eye muscles differ from intrinsic eye muscles?

7 Descibe the location and contents of the chambers of the eye. What is intraocular pressure? How is the canal of Schlemm related to this pressure?

8 Explain how each of the following events is related to the physiology of vision: *(a)* refraction of light, *(b)* accommodation, *(c)* constriction of the pupil, *(d)* convergence, and *(e)* inverted image formation.

9 Distinguish emmetropia, myopia, hypermetropia, and astigmatism by means of a diagram.

10 How is a light stimulus converted into an impulse? Relate your discussion to the rhodopsin cycle by means of a diagram.

11 What is night blindness? What causes it?

12 Describe the path of a visual impulse from the optic nerve to the brain.

13 Define visual field. Relate the visual field to image formation on the retina.

14 Diagram the principal parts of the outer, middle, and inner ear. Describe the function of each part labeled.

15 Explain the events involved in the transmission of sound from the pinna to the spiral organ.

16 What is the afferent pathway for sound impulses from the vestibulocochlear nerve to the brain?

17 Compare the function of the saccule and utricle in maintaining static equilibrium with the role of the semicircular ducts in maintaining dynamic equilibrium.

Chapter 18
The Endocrine System

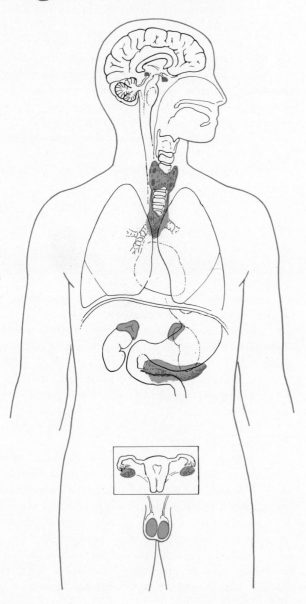

- Discuss the function of the endocrine system in maintaining homeostasis.

- Define an endocrine gland.

- Identify the relationship between an endocrine gland, a target organ, and cyclic AMP.

- Define the anatomical and physiological relationship between the pituitary gland and hypothalamus.

- List the seven hormones of the adenohypophysis, their target organs, and functions.

- Define the source of hormones stored by the neurohypophysis, their target organs, and functions.

- Define a negative feedback mechanism.

- Relate a negative feedback mechanism to the regulation of hormones secreted by the adenohypophysis.

- Discuss how thyroxin is synthesized and stored by thyroid follicles and transported.

- Identify the physiological effects and regulation of secretion of thyroxin and thyrocalcitonin.

- Describe the physiological effects and regulation of the parathyroid hormone.

- Identify the principal effects of abnormal secretion of the parathyroid hormone on calcium metabolism.

- Distinguish the effects of adrenal cortical mineralocorticoids, glucocorticoids, and gonadocorticoids on physiological activities.

- Identify the function of the adrenal medullary secretions as supplements of sympathetic responses.

- Compare the roles of glucagon and insulin in the control of blood sugar level.

- Identify the physiological effects of the hormones secreted by the pineal gland.

- Define the general role of the thymus in antibody production.

- Define the general adaptation syndrome and compare homeostatic responses and stress responses.

- Identify the body reactions during the alarm, resistance, and exhaustion stages of stress.

- Describe the disorders associated with hypersecretion or hyposecretion of hormones produced by the different endocrine glands.

- Define medical terminology associated with the endocrine system.

The nervous system controls the body through electrical impulses delivered over neurons. The body's other control system, the endocrine system, affects bodily activities by releasing chemical messengers, called hormones, into the bloodstream. Obviously, the body could not function if the two great control systems were to pull in opposite directions. The nervous and endocrine systems therefore coordinate their activities like an interlocking supersystem. Certain parts of the nervous system routinely stimulate or inhibit the release of hormones. The hormones, in turn, are quite capable of stimulating or inhibiting the flow of nerve impulses. In this chapter, you will study the endocrine glands—the body's means of chemical control.

ENDOCRINE GLANDS

The body contains two kinds of glands: exocrine and endocrine. **Exocrine glands** secrete their products into ducts. The ducts then carry the secretions into body cavities or to the body's surface. They include sweat, sebaceous, mucous, and digestive glands. **Endocrine glands,** by contrast, secrete their products into the extracellular space around the secretory cells. Since they secrete internally, the term *endo,* meaning within, is used. The secretion passes into the capillaries to be transported in the blood. Since they have no ducts, endocrine glands are also called *ductless glands.* The endocrine glands of the body are the pituitary (hypophysis), thyroid, parathyroids, adrenals (suprarenals), pancreas, ovaries, testes, pineal, and thymus. The placenta or "afterbirth" is, in some ways, a temporary endocrine gland. Moreover, the kidneys, stomach, and small intestine also produce hormones. The endocrine glands make up the **endocrine system.** The location of many organs of the endocrine system is illustrated in Figure 18-1.

The secretions of endocrine glands are called **hormones** (*hormone* = set in motion). A hormone may be a protein, an amine, or a steroid. Amines, like proteins, contain carbon, hydrogen, and nitrogen. Unlike proteins, they lack oxygen and contain no peptide bonds. A steroid is a lipid with a cholesterol-type nucleus. The one thing all hormones have in common—whether protein, amine, or steroid—is the function of maintaining homeostasis by changing the physiological activities of cells. A hormone may stimulate changes in the cells of an organ or in groups of organs: the *target organs.* Or the hormone may directly affect the activities of all the cells in the body.

HORMONES AND CYCLIC AMP

The general effects of most hormones are fairly well known. Insulin and glucagon regulate blood sugar level.

Prolactin stimulates milk secretion by the mammary glands. The thyroid hormones, thyroxin (T_4) and triiodothyronine (T_3), help control metabolic rate. Progesterone helps prepare the uterus for implantation. However, the manner in which specific hormones affect different cells of the body is not entirely clear.

We have learned that one chemical substance seems to be involved in many reactions involving hormones: cyclic AMP (see Figure 2-15). Cyclic AMP is synthesized from ATP, the principal energy-storing chemical in cells. This synthesis requires the enzyme adenyl cyclase. Current speculation is that a hormone circulating in the blood reaches a target cell and brings a specific message for that cell. The hormone is called the first messenger. To give the cell its message, the hormone must attach to a specific receptor site on the plasma membrane (Figure 18-2). This attachment increases the activity of adenyl cyclase in the plasma membrane. In the presence of the enzyme, ATP is converted into cyclic AMP in the cell. The cyclic AMP then diffuses throughout the cell and acts as a second messenger and performs a specific function according to the message indicated by the hormone.

In the cell, cyclic AMP activates the appropriate enzymes to get a specific job done. In liver or skeletal muscle cells, cyclic AMP activates certain enzymes that change glucose into glycogen to lower blood sugar level according to a message from insulin but raises blood sugar level by activating other enzymes that change glycogen into glucose according to a message from glucagon. In cells of the thyroid gland, cyclic AMP stimulates the secretion of thyroid hormones in response to a message from another hormone called thyroid-stimulating hormone (TSH). In other words, cyclic AMP induces a specific target cell to perform a specific function based on the message it receives from the hormone that attaches to the plasma membrane. Thus target cells attract only specific hormones to the receptor sites of their membranes as the hormones circulate through the blood. In this way, different target cells accomplish different functions—a very efficient way for all your body cells to cooperate in maintaining homeostasis. It is also a very efficient way of ensuring that target cells function properly. Keep in mind that hormones are not enzymes; they alter enzymatic activity.

High levels of cyclic AMP persist only briefly because it is degraded by an enzyme called cAMP phosphodiesterase. Prostaglandins may influence the formation of cyclic AMP in the cell membrane. Thus prostaglandins may help regulate hormonal action.

Hormones known to use cyclic AMP as a second messenger include human growth hormone, epinephrine, norepinephrine, glucagon, adrenocorticotropic

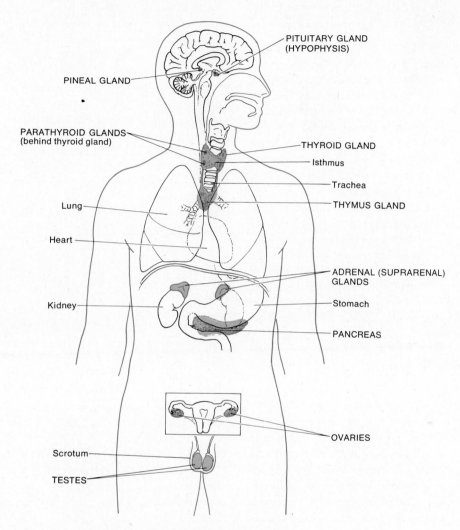

Figure 18-1 Location of endocrine glands and associated structures.

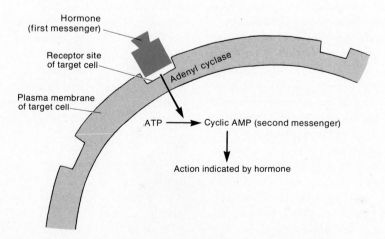

Figure 18-2 Proposed relationship between a hormone, target cell, and cyclic AMP.

hormone, luteinizing hormone, angiotensin, antidiuretic hormone, thyroid-stimulating hormone, melanocyte-stimulating hormone, thyroxin, gastrin, and serotonin.

PITUITARY (HYPOPHYSIS)

The hormones of the **pituitary gland,** also called the **hypophysis,** regulate so many body activities that the pituitary has been nicknamed the "master gland." Surprisingly, the hypophysis is a small round structure measuring about 1.3 cm (0.5 inch) in diameter. It lies in the sella turcica of the sphenoid bone and is attached to the hypothalamus of the brain via a stalklike structure. This structure is called the *infundibulum.*

The pituitary is divided structurally and functionally into an anterior lobe and a posterior lobe. Both are connected to the hypothalamus. The *anterior lobe* contains many glandular epithelium cells and forms the glandular part of the pituitary. A system of blood vessels connects the anterior lobe with the hypothalamus. The *posterior lobe* contains axonic ends of neurons whose cell bodies are located in the hypothalamus. The nerve fibers that terminate in the posterior lobe are supported by neuroglial cells called pituicytes. Other nerve fibers connect the neurohypophysis directly with the hypothalamus.

Adenohypophysis

The anterior lobe of the pituitary is also called the **adenohypophysis.** It releases hormones that regulate a whole range of bodily activities–from growth to reproduction. However, the release of these hormones is either stimulated or inhibited by chemical secretions from the hypothalamus of the brain. Such chemicals are called **regulating factors.** They constitute a link between the nervous system and the endocrine system. The hypothalamic releasing factors are delivered to the adenohypophysis in the following way. The blood supply to the adenohypophysis and infundibulum is derived from several *superior hypophyseal arteries.* These arteries are branches of the internal carotid and posterior communicating arteries (Figure 18-3). The superior hypophyseal arteries form a network or plexus of capillaries, the *primary plexus,* in the infundibulum near the inferior portion of the hypothalamus. Regulating factors from the hypothalamus diffuse into this plexus. This plexus drains into veins, known as the *hypophyseal portal veins,* that pass down the infundibulum. At the inferior portion of the infundibulum, the veins form a *secondary plexus* in this adenohypophysis. From this plexus, hormones of the adenohypophysis pass into the anterior hypophyseal veins for distribution to tissue cells.

When the anterior lobe receives the proper chemical stimulation from the hypothalamus, its glandular cells secrete any one of seven hormones. The glandular cells themselves are called acidophils, basophils, or chromophobes, depending on the way their cytoplasm reacts to laboratory stains. The acidophils, which stain pink, secrete two hormones: human growth hormone, which controls body growth; and prolactin, which initiates milk secretion from the breasts. The basophils stain darkly and release the other five hormones. These are thyroid-stimulating hormone, which controls the thyroid gland; adrenocorticotropic hormone, which regulates the cortical regions of the adrenal glands; follicle-stimulating hormone, which stimulates the production of egg and sperm in the reproductive organs; luteinizing hormone, which stimulates other sexual and reproductive activities; and melanocyte-stimulating hormone, which is related to skin pigmentation. The chromophobes may also be involved in the secretion of the adrenocorticotropic hormone.

Except for the growth hormone and melanocyte-stimulating hormone, all the secretions are referred to as *tropic hormones,* which means that their target organs are other endocrine glands. Prolactin, follicle-stimulating hormone, and luteinizing hormone are also called *gonadotropic hormones* because they regulate the functions of the gonads. The gonads (ovaries and testes) are the endocrine glands that produce sex hormones.

HUMAN GROWTH HORMONE (HGH)

The *human growth hormone (HGH)* is also known as *somatotropin* and the *somatotropic hormone (STH).* The word root *soma* means body, whereas *trop* means nourishment. Its principal function is to act on the hard and soft tissues to increase their rate of growth and maintain their size once growth is attained. HGH causes cells to grow and multiply by directly increasing the rate at which amino acids enter cells and are built up into proteins. This process is accomplished through cyclic AMP. The building processes are called *anabolism.* Thus HGH is considered to be a hormone of protein anabolism since it increases the rate of protein synthesis. HGH also causes cells to switch from burning carbohydrates to burning fats for energy. For example, it stimulates adipose tissue to release fat. And it stimulates other cells to break down the released fat molecules. When chemical bonds are broken, energy is released. Since energy-releasing processes are referred to as *catabolism,* we can say that HGH promotes fat catabolism. At the same time, HGH accelerates the rate at which glycogen stored in the liver is converted into glucose and released into the blood. Since the cells are using fats for energy, however, they do not consume much glucose.

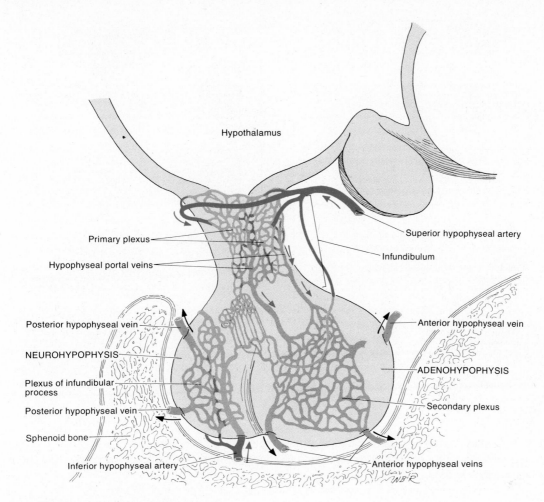

Figure 18-3 Blood supply of the pituitary gland.

The result is an increase in blood sugar level, a condition called *hyperglycemia.* Excessive amounts of the hormone may lead to diabetes. This process is called the *diabetogenic effect.*

The control of HGH secretion is not yet clearly established. Among the factors that increase secretion are low blood sugar level, increased intake of amino acids, stress, and exercise. By contrast, a high blood sugar level decreases HGH secretion (Figure 18-4). Apparently, a lower than normal amount of glucose in the blood stimulates the hypothalamus to secrete a regulating factor called the *human growth hormone releasing factor (HGHRF).* When HGHRF is released into the bloodstream, it circulates to the adenohypophysis and stimulates the lobe to secrete HGH. As soon as blood sugar level returns to normal, HGHRF secretion shuts off. It is believed that an abnormally high level of blood sugar causes the hypothalamus to secrete another regulating factor called the *human growth hormone inhibiting factor (HGHIF)* or *somatostatin.* This regulating factor inhibits

the release of HGHRF and thus the secretion of HGH.

The regulation of HGH illustrates two phenomena that are typical of all the secretions of the adenohypophysis. First, each hormone is controlled by its own regulating factor from the hypothalamus. In some cases the regulating factor stimulates hormone secretion; in other cases, it inhibits secretion. Second, secretion is generally regulated through negative feedback systems. Since hormones are chemical regulators of homeostasis, these feedback systems are hardly surprising. Continued heavy secretion of a hormone would overshoot the goal and send the body out of balance in the opposite direction.

THYROID-STIMULATING HORMONE (TSH)

This hormone is also called *thyrotropin* and *TSH.* It stimulates the synthesis and secretion of the hormones produced by the thyroid gland. Secretion is controlled by a regulating factor produced by the hypothalamus called

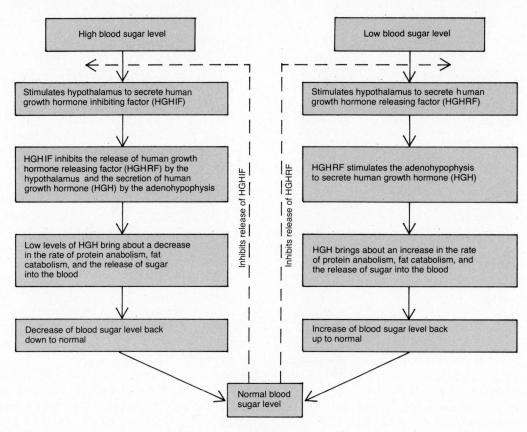

Figure 18-4 Regulation of the secretion of human growth hormone (HGH). Like other hormones of the adenohypophysis, the secretion of HGH is controlled by regulating factors. Like most hormones of the body, HGH secretion and inhibition involve negative feedback systems.

thyrotropin-releasing factor (TRF). Release of TRF depends on blood levels of thyroxin and the body's metabolic rate and operates according to a negative feedback system.

ADRENOCORTICOTROPIC HORMONE (ACTH)
This hormone, also called *adrenocorticotropin* and *ACTH,* has a dual function. Its tropic function is to control the production and secretion of certain adrenal cortex hormones. It also increases fat mobilization, brings about hyperglycemia, and helps provide resistance to stress. Secretion of ACTH is governed by a regulating factor produced by the hypothalamus called *adrenocorticotropic hormone-releasing factor (ACTHRF)*. Release of ACTHRF depends on a number of stimuli and hormones and operates as a negative feedback system.

FOLLICLE-STIMULATING HORMONE (FSH)
In the female, *follicle-stimulating hormone (FSH)* is transported from the adenohypophysis by the blood to the ovaries, where it stimulates the development of ova each month. FSH also stimulates cells in the ovaries to secrete estrogens, or female sex hormones. In the male,

FSH stimulates the testes to produce sperm and secrete testosterone, a male sex hormone. Secretion of FSH is subject to a regulating factor produced by the hypothalamus called *follicle-stimulating hormone releasing factor (FSHRF)*. FSHRF is released in response to estrogens and possibly progesterone in the female and testosterone in the male and involves a negative feedback system.

LUTEINIZING HORMONE (LH OR ICSH)
The *luteinizing hormone* is called *luteotropin (LH)* in the female and *interstitial cell stimulating hormone (ICSH)* in the male. In the female, together with estrogens, LH stimulates the ovary to release the developed ovum and prepares the uterus for implantation of a fertilized ovum. It also stimulates the secretion of progesterone (another female sex hormone) and readies the mammary glands for milk secretion. In the male, ICSH stimulates the interstitial cells in the testes to develop and secrete testosterone. Secretion of LH or ICSH is controlled by a regulating factor produced by the hypothalamus called *luteotropic releasing factor (LRF)*. Release of LRF is governed by a negative feedback system involving estrogens, progesterone, and testosterone.

PROLACTIN (PR)

Prolactin (PR) or the *lactogenic hormone,* together with other hormones, initiates and maintains milk secretion by the mammary glands, a process called lactation. Prolactin acts directly on tissues. By itself, it has little effect – it requires preparation by estrogens, progesterone, corticosteroids, and insulin. When the mammary glands have been primed by these hormones, prolactin brings about milk secretion.

Prolactin has both an inhibitory and an excitatory negative control system. During menstrual cycles, a *prolactin inhibiting factor (PIF),* a regulating factor from the hypothalamus, inhibits the release of prolactin from the anterior pituitary. As the levels of estrogens and progesterone fall during the late secretory phase of the menstrual cycle, the secretion of PIF diminishes and the blood level of prolactin rises. However, its rising level does not last long enough to have much effect on the breasts, which may be tender because of the presence of prolactin just before menstruation. As the menstrual cycle starts up again and the level of estrogens again rises, PIF is again secreted and the prolactin level drops.

Prolactin levels rise during pregnancy. Apparently a regulating factor from the hypothalamus, called *prolactin releasing factor (PRF),* stimulates prolactin secretion after long periods of inhibition. Prolactin levels fall after delivery and rise again during breast feeding. Suckling reduces hypothalamic secretion of PIF. Mechanical stimulation of the female breast brings about massive secretion of prolactin; stimulation of the male breast does not.

MELANOCYTE-STIMULATING HORMONE (MSH)

The *melanocyte-stimulating hormone (MSH)* increases skin pigmentation by stimulating the dispersion of melanin granules in melanocytes. In the absence of the hormone, the skin may be pallid. An excess of MSH may cause darkening of the skin. Secretion of MSH is stimulated by a hypothalamic regulating factor called *melanocyte-stimulating hormone releasing factor (MSHRF).* It is inhibited by a *melanocyte-stimulating hormone inhibiting factor (MSHIF).*

Neurohypophysis

In a strict sense, the posterior lobe, or **neurohypophysis,** is not an endocrine gland since it does not synthesize hormones. The posterior lobe consists of supporting cells called *pituicytes,* which are similar in appearance to the neuroglia of the nervous system. It also contains neuron fibers that establish an important connection with the hypothalamus (Figure 18-5). The cell bodies of the neurons originate in nuclei in the hypothalamus. The fibers project from the hypothalamus, form the *hypothalamic-hypophyseal tract,* and terminate on blood capillaries in the neurohypophysis. The cell bodies of the neurons produce two hormones: oxytocin and the antidiuretic hormone. Following their production, the hormones are transported in the neuron fibers into the neurohypophysis and stored in the axon terminals. Later, when the hypothalamus is properly stimulated, it sends impulses over the neurons. The impulses cause the release of the hormones from the axon terminals into the blood.

The blood supply to the neurohypophysis is from the *inferior hypophyseal arteries,* derived from the internal carotid arteries. In the neurohypophysis, the inferior hypophyseal arteries form a plexus of capillaries called the *plexus of the infundibular process.* From this plexus hormones stored in the neurohypophysis pass into the *posterior hypophyseal veins* for distribution to tissue cells.

OXYTOCIN

Oxytocin stimulates the contraction of the smooth muscle cells in the pregnant uterus and the contractile cells around the ducts of the mammary glands. It is released in large quantities just prior to giving birth. When labor begins, the uterus and vagina are distended. This distension initiates afferent impulses to the hypothalamus that stimulate the secretion of more oxytocin by the hypothalamus. The oxytocin migrates along the nerve fibers of the hypothalamus to the neurohypophysis. The impulses also cause the neurohypophysis to release oxytocin into the blood. It is then carried by the blood to the uterus to reinforce uterine contractions. Oxytocin also affects milk ejection. Milk formed by the glandular cells of the breasts is stored until the baby begins active sucking. From about 30 seconds to 1 minute after nursing begins, the baby receives no milk. During this latent period, nerve impulses from the nipple are transmitted to the hypothalamus. The hypothalamus sends impulses down the neurosecretory neurons that release oxytocin from their axonic ends in the neurohypophysis. Oxytocin then flows from the neurohypophysis via the blood to the breasts, where it stimulates smooth muscle cells to contract and eject milk out of the mammary glands. Oxytocin is inhibited by progesterone but works together with estrogens. Estrogens and progesterone inhibit, via PIF, the release of prolactin. Prolactin accumulates in the pituitary gland during gestation. The sucking stimulation that produces the release of oxytocin also inhibits the release of PIF.

ANTIDIURETIC HORMONE (ADH)

The principal physiological activity of ADH is its effect on urine volume. ADH causes the kidneys to remove water from newly forming urine and return it to the

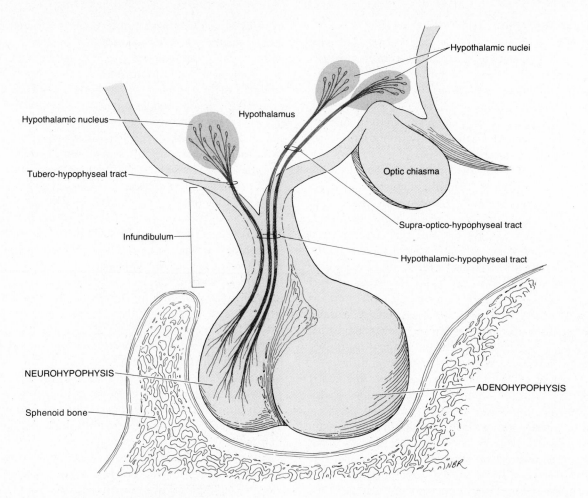

Hypothalamic nuclei

Hypothalamic nucleus

Hypothalamus

Tubero-hypophyseal tract

Optic chiasma

Supra-optico-hypophyseal tract

Hypothalamic-hypophyseal tract

Infundibulum

NEUROHYPOPHYSIS

ADENOHYPOPHYSIS

Sphenoid bone

Figure 18-5 Hypothalamic-hypophyseal tract.

bloodstream—thus decreasing urine volume. In the absence of ADH, urine output may be increased tenfold. An **antidiuretic** is any chemical substance that prevents excessive urine production.

ADH can raise blood pressure by bringing about constriction of arterioles. This effect is noted if large quantities of the purified hormone are injected. Only rarely, however, does the body secrete enough hormone to affect blood pressure significantly.

The amount of ADH normally secreted varies with the body's needs. When the body is dehydrated, the concentration of water in the blood falls below normal limits as the salt-to-water ratio changes. Receptors in the hypothalamus detect the low water concentration in the plasma and stimulate the hypothalamus to produce ADH. The hormone travels down neuron fibers to the neurohypophysis. It is then released into the bloodstream and transported to the kidneys. The kidneys respond by decreasing urine output, and water is conserved. ADH also decreases the rate at which

perspiration is produced during dehydration. By contrast, if the blood contains a higher than normal water concentration, the receptors detect the increase and hormone secretion is stopped. The kidneys can then release large quantities of urine, and the volume of body fluid is brought down to normal.

Secretion of ADH can also be altered by a number of special conditions. Pain, stress, acetylcholine, and nicotine all stimulate secretion of the hormone. Alcohol inhibits secretion and thereby increases urine output. This is why thirst is one symptom of a hangover.

THYROID

The endocrine organ located just below the larynx is called the **thyroid gland.** The right and left lateral lobes lie one on either side of the trachea and are connected by a mass of tissue called an *isthmus* that lies in front of the trachea just below the cricoid cartilage (Figure 18-6). The *pyramidal lobe,* when present, extends upward. The

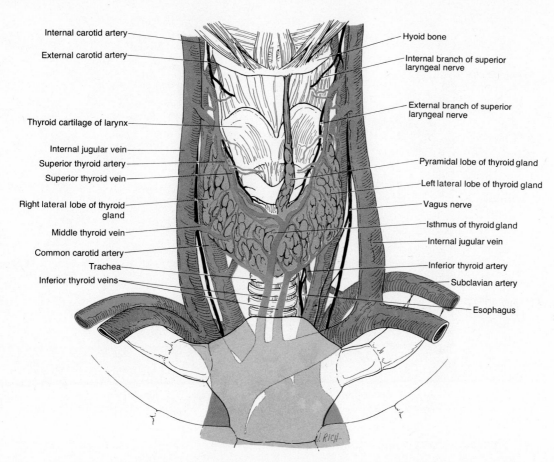

Internal carotid artery
External carotid artery

Thyroid cartilage of larynx

Internal jugular vein
Superior thyroid artery
Superior thyroid vein

Right lateral lobe of thyroid gland
Middle thyroid vein
Common carotid artery
Trachea
Inferior thyroid veins

Hyoid bone
Internal branch of superior laryngeal nerve

External branch of superior laryngeal nerve

Pyramidal lobe of thyroid gland
Left lateral lobe of thyroid gland
Vagus nerve
Isthmus of thyroid gland
Internal jugular vein
Inferior thyroid artery
Subclavian artery
Esophagus

Figure 18-6 Location and blood supply of the thyroid gland in anterior view. The pyramidal lobe of the thyroid is not constant and, when present, it may be attached to the hyoid bone by a muscle.

gland weighs about 25 g (almost 1 oz) and has a rich blood supply, receiving 80 to 120 ml of blood per minute. Histologically, the thyroid is composed of spherical sacs called *thyroid follicles* (Figure 18-7). The walls of each follicle consist of cells that reach the surface of the lumen of the follicle (principal cells) and cells that do not reach the lumen (parafollicular cells). The principal cells manufacture the hormones thyroxin (T_4 or tetraiodothyronine) and triiodothyronine (T_3). Together these hormones are referred to as the thyroid hormones. Thyroxin is considered to be the major hormone produced by the principal cells. The parafollicular cells produce the hormone thyrocalcitonin. Each thyroid follicle is filled with *thyroid colloid,* a stored form of the thyroid hormones.

Physiology

One of the thyroid gland's unique features is its ability to store hormones and release them in a steady flow over a long period of time. The principal hormone, *thyroxin,* is synthesized from iodine and an amino acid called tyrosine. Synthesis usually occurs on a continuous basis. Thyroxin combines with a protein in the gland called thyroglobulin, the main constituent of the thyroid colloid. Thyroxin-thyroglobulin is stored in the thyroid follicles until thyroxin is needed by the body (Figure 18-8a). Then thyroxin splits apart from thyroglobulin and is released into the blood. In the blood, thyroxin combines with plasma proteins for transportation to target tissue cells. Most thyroxin combines with a plasma protein called *thyroxin-binding globulin (TBG).* The thyroxin is released from the TBG as it enters tissue cells. Triiodothyronine is also synthesized from iodine and tyrosine and is stored in combination with thyroglobulin. When released from the follicles into the blood, triiodothyronine is also transported by TBG to tissue cells.

Thyroxin

Thyroxin is produced by the principal cells of the thyroid gland. Its major function is to control metabo-

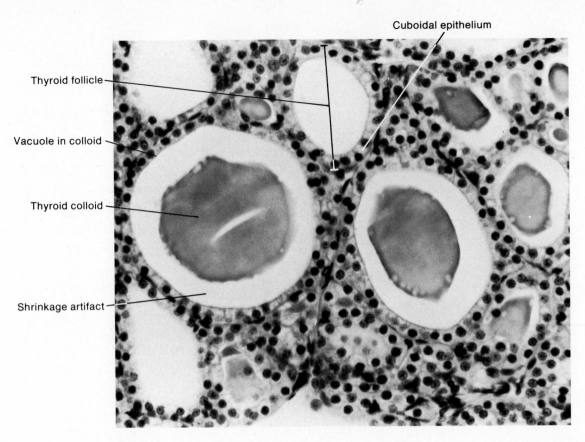

Thyroid follicle

Vacuole in colloid

Thyroid colloid

Shrinkage artifact

Cuboidal epithelium

Figure 18-7   Histology of the thyroid gland at a magnification of 90×. (Courtesy of Victor B. Eichler, Wichita State University.)

lism by regulating the catabolic or energy-releasing processes and the anabolic or building-up processes. Thyroxin increases the rate at which carbohydrates are burned. And it stimulates cells to break down proteins for energy instead of using them for building processes. At the same time, thyroxin decreases the breakdown of fats. The overall effect, though, is to increase catabolism. It thus produces energy and raises body temperature as heat energy is given off. This process is called the *calorigenic effect*. Thyroxin also helps regulate tissue growth and development. It works with HGH to accelerate body growth, especially the growth of nervous tissue. Thyroxin deficiency during fetal development can result in fewer and smaller neurons, defective myelination of nerve fibers, and mental retardation. During the early years of life it can cause small stature and prevent certain organs from developing. Finally, thyroxin increases the reactivity of the nervous system, which causes an increase in heart rate and motility of the gastrointestinal tract.

The secretion of thyroxin seems to be brought on by several factors (Figure 18-8b). If thyroxin in the blood falls below the normal level or the metabolic rate decreases, chemical sensors in the hypothalamus detect the change in blood chemistry and stimulate the hypothalamus to secrete a regulating factor called *thyrotropin-releasing factor (TRF)*. TRF stimulates the adenohypophysis to secrete thyroid-stimulating hormone (TSH). Then TSH stimulates the thyroid to release thyroxin until the metabolic rate returns to normal. Conditions that increase the body's need for energy – a cold environment, high altitude, pregnancy – also trigger this feedback system and increase the secretion of thyroxin.

Thyroid activity can be inhibited by a number of other factors. When large amounts of certain sex hormones (estrogens and androgens) are circulating with the blood, for example, TSH secretion diminishes. Aging slows down the activities of most glands. Thus thyroid production decreases as you get older.

Thyrocalcitonin (TCT)

The thyroid hormone produced by the parafollicular cells of the thyroid gland is *thyrocalcitonin* or *calcitonin* or

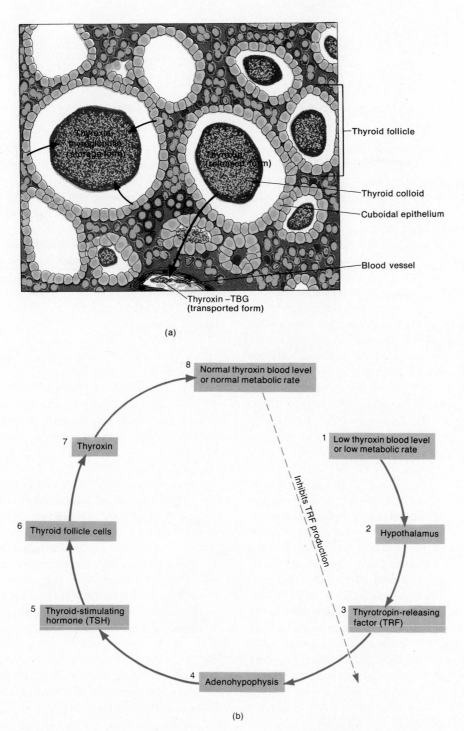

Figure 18-8 Thyroid gland. (a) Storage, release, and transportation of thyroxin. (b) Regulation of thyroxin secretion.

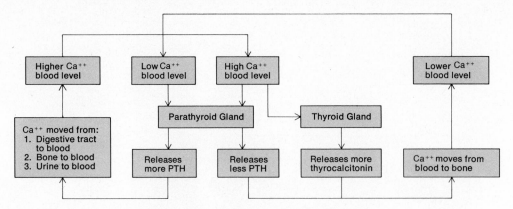

Figure 18-9   Regulation of the secretion of the parathyroid hormone and thyrocalcitonin.

TCT. It is involved in the homeostasis of blood calcium level. Thyrocalcitonin lowers the amount of calcium in the blood by inhibiting bone breakdown and accelerating the absorption of calcium by the bones. It appears to exert its influence by antagonizing a number of bone resorptive agents such as vitamin D, vitamin A, and the parathyroid hormone (PTH). If thyrocalcitonin is administered to a person with a normal level of blood calcium, it causes *hypocalcemia* (low blood calcium level). Hypocalcemia is also a complication of magnesium deficiency. If thyrocalcitonin is given to a person with *hypercalcemia* (high blood calcium level), the level returns to normal. It is suspected that the blood calcium level directly controls the secretion of thyrocalcitonin according to a negative feedback system that does not involve the pituitary gland (Figure 18-9).

PARATHYROIDS

Structure

Embedded on the posterior surfaces of the lateral lobes of the thyroid are small, round masses of tissue called the **parathyroid glands.** Typically two parathyroids, superior and inferior, are attached to each lateral thyroid lobe (Figure 18-10). They measure about 3 to 8 mm (0.1 to 0.3 inch) in length, 2 to 5 mm (0.07 to 0.2 inch) in width, and 0.5 to 2 mm (0.02 to 0.07 inch) in thickness.

Histologically, the parathyroids contain two kinds of epithelial cells. The first kind, called *principal* or *chief cells,* is believed to be the major synthesizer of parathyroid hormone. Some researchers believe that the other kind of cell, called an *oxyphil cell,* synthesizes a reserve capacity of hormone.

Parathyroid Hormone (PTH)

Parathyroid hormone (PTH) or *parathormone* controls the homeostasis of ions in the blood, especially calcium and phosphate ions. If adequate amounts of vitamin D are present, PTH increases the rate of calcium absorption from the intestine into the blood. PTH also increases the absorption of some magnesium and phosphate ions from the intestine into the blood. Moreover, PTH increases the number and activity of osteoclasts, or bone-destroying cells. As a result, bone tissue is broken down and calcium and phosphate ions are released into the blood. Recall that thyrocalcitonin secreted by the thyroid has the opposite effect. PTH also produces two changes in the kidneys:

1 It increases the rate at which the kidneys remove calcium and magnesium ions from the urine and return them to the blood.

2 It accelerates the transportation of phosphate ions from the blood into the urine for elimination. More phosphate is lost through the urine than is gained from the bones.

The overall effect of PTH, then, is to decrease blood phosphate level and increase blood calcium level. As far as blood calcium level is concerned, PTH and thyrocalcitonin are antagonists.

PTH secretion is not controlled by the pituitary gland. When the calcium ion level of the blood falls, more PTH is released (see Figure 18-9). Conversely, when the calcium ion level of the blood rises, less PTH (and more thyrocalcitonin) is secreted. This is another example of a negative feedback control system.

ADRENALS (SUPRARENALS)

The body has two **adrenal (suprarenal) glands** superior to each kidney (Figure 18-11). Each adrenal gland is structurally and functionally differentiated into two sections: the outer *adrenal cortex,* which makes up the bulk of the gland, and the inner *adrenal medulla* (Figure

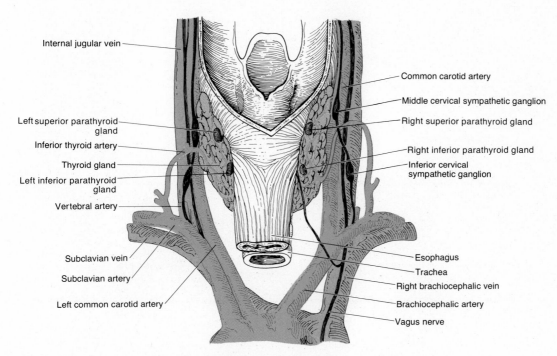

Figure 18-10 Location and blood supply of the parathyroid glands in posterior view.

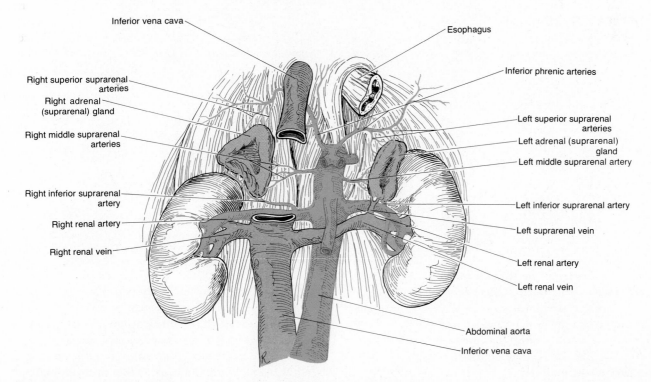

Figure 18-11 Location and blood supply of the adrenal (suprarenal) glands.

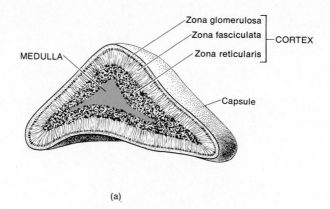

(a)

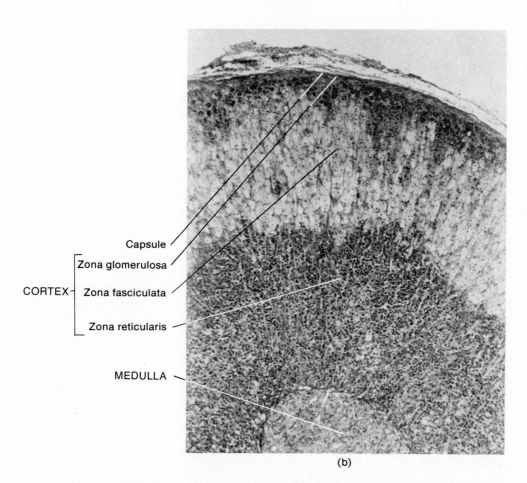

(b)

Figure 18-12 Subdivisions and histology of the adrenal (suprarenal) glands. (a) Diagram of the subdivisions. (b) Histology of the subdivisions at a magnification of 15×. (Courtesy of Victor B. Eichler, Wichita State University.)

18-12a). Covering the gland are a thick layer of fatty connective tissue and an outer, thin fibrous capsule.

The average dimensions of the adult adrenal gland are about 50 mm (2 inches) in length, 30 mm (1.1 inch) in width, and 10 mm (0.4 inch) in thickness. The adrenals, like the thyroid, are among the most vascular organs of the body.

Adrenal Cortex

Histologically, the cortex is subdivided into three zones (Figure 18-12). Each zone has a different cellular arrangement and secretes different hormones. The outer zone, directly underneath the connective tissue covering, is referred to as the *zona glomerulosa*. Its cells are arranged in arched loops or round balls, and they primarily secrete a group of hormones called mineralocorticoids. The middle zone, or *zona fasciculata,* is the widest of the three zones and consists of cells arranged in long, straight cords. The zona fasciculata secretes mainly glucocorticoid hormones. The inner zone, the *zona reticularis,* contains cords of cells that branch freely. This zone synthesizes mostly sex hormones, chiefly male hormones called androgens.

MINERALOCORTICOIDS

Mineralocorticoids help control electrolyte homeostasis, particularly the concentrations of sodium and potassium. Although the adrenal cortex secretes at least three different hormones classified as mineralocorticoids, one of these hormones is responsible for about 95 percent of the mineralocorticoid activity—*aldosterone.* Aldosterone acts on the tubule cells in the kidneys and causes them to increase their reabsorption of sodium. As a result, sodium ions are removed from the urine and returned to the blood. In this manner, aldosterone prevents rapid depletion of sodium from the body. On the other hand, aldosterone decreases reabsorption of potassium. Large amounts of potassium are moved from the blood into the urine.

These two basic functions—conservation of sodium and elimination of potassium—cause a number of secondary effects. For example, a large proportion of the sodium reabsorption occurs through an exchange reaction whereby positive hydrogen ions pass into the urine to replace the positive sodium ions. Since this mechanism removes hydrogen ions, it makes the blood less acidic and prevents acidosis. The movement of Na^+ ions also sets up a positively charged field in the blood vessels around the kidney tubules. As a result, negatively charged chloride and bicarbonate ions are drawn out of the urine and back into the blood. Finally, the increase in sodium-ion concentration in the blood vessels causes water to move by osmosis from the urine into the blood. In summary, aldosterone causes potassium excretion and sodium reabsorption. The sodium reabsorption leads to the elimination of H^+ ions; the retention of Na^+, Cl^-, and HCO_3^-; and the retention of water.

The control of aldosterone secretion is complex. Apparently, several mechanisms operate (Figure 18-13)—one of these is the *renin-angiotensin pathway.* A decrease in blood volume from dehydration or Na^+ deficiency brings about a drop in blood pressure. The low blood pressure stimulates certain kidney cells, called juxtaglomerular cells, to secrete into the blood an enzyme called *renin.* In this pathway, renin converts *angiotensinogen,* a plasma protein produced by the liver, into *angiotensin I,* which is then converted into *angiotensin II.* Angiotensin II stimulates the adrenal cortex to produce more aldosterone. Aldosterone brings about increased Na^+ and water reabsorption. This reabsorption leads to an increase in extracellular fluid volume and a restoration of blood pressure to normal. A second mechanism for the control of aldosterone involves K^+ concentration. An increased K^+ concentration in extracellular fluid directly stimulates aldosterone secretion by the adrenal cortex and causes the elimination of excess K^+ by the kidneys. A decreased K^+ concentration in extracellular fluid decreases aldosterone production, and thus less K^+ than usual is eliminated by the kidneys. Hemorrhage also stimulates the adrenal cortex to produce aldosterone.

GLUCOCORTICOIDS

The *glucocorticoids* are a group of hormones concerned with normal metabolism and resistance to stress. Three glucocorticoids are *hydrocortisone (cortisol), corticosterone,* and *cortisone.* Of the three, hydrocortisone is the most abundant. The glucocorticoids have the following effects on the body:

1 Glucocorticoids work with other hormones in promoting normal metabolism. Their role is to make sure enough energy is provided. They increase the rate at which amino acids are removed from cells and transported to the liver. The amino acids may be synthesized into new proteins, such as the enzymes needed for metabolic reactions. If the body's reserves of glycogen and fat are low, the liver may convert the amino acids to glucose. This conversion of another substance into glucose is called gluconeogenesis. Glucocorticoids help the body to store fat.

2 Glucocorticoids work in many ways to provide resistance to stress. A sudden increase in available glucose by way of gluconeogenesis from amino acids makes the body more alert. Additional glucose gives the body energy for combating a range of stressors: fright, temperature extremes, high altitude, bleeding, infection. Glucocorticoids also make the blood vessels more sensitive to vessel-constricting chemicals. They thereby raise blood pressure. This effect is advantageous if the stressor happens to be blood loss, which causes a drop in blood pressure.

3 Glucocorticoids decrease the blood vessel dilatation and edema associated with inflammations. They are thus anti-inflammatory compounds. Unfortunately, they also decrease connective-tissue regeneration and are thereby responsible for slow wound healing.

The control of glucocorticoid secretion is a typical negative feedback mechanism (Figure 18-14). The two stimuli are stress and low blood level of glucocorticoids.

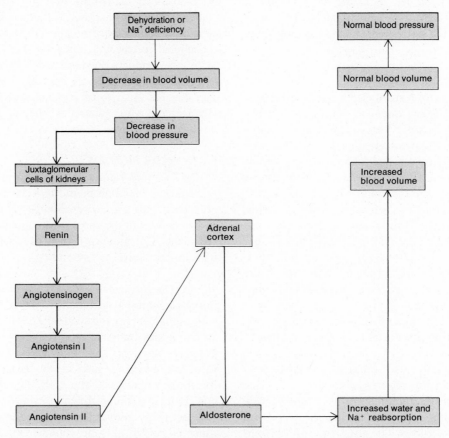

Figure 18-13   Proposed mechanism for the regulation of the secretion of aldosterone by the renin-angiotensin pathway.

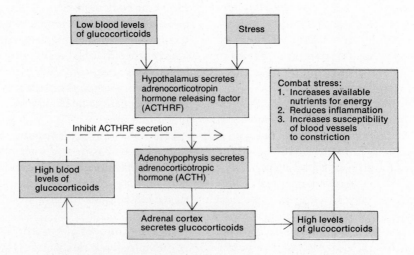

Figure 18-14   Regulation of the secretion of glucocorticoids.

Such stress could be emotional pressure or physical damage caused by contusions, broken bones, disease, or tissue destruction. The stress may directly stimulate the hypothalamus. For example, the hypothalamus could be stimulated by low blood pressure resulting from excessive bleeding. Or the stimulus may be relayed to the hypothalamus from other parts of the nervous system. In any case, either stress or an abnormally low level of glucocorticoids stimulates the hypothalamus to secrete a regulating factor called *adrenocorticotropin hormone releasing factor (ACTHRF)*. This secretion initiates the release of ACTH from the anterior lobe of the pituitary. ACTH is carried through the blood to the adrenal cortex, where it then stimulates glucocorticoid secretion.

GONADOCORTICOIDS
The adrenal cortices secrete both male and female *gonadocorticoids,* or *sex hormones*. These are estrogens and androgens, usually secreted in minute amounts.

Adrenal Medulla
The adrenal medulla consists of hormone-producing cells, called *chromaffin cells,* which surround large blood-containing sinuses. Chromaffin cells develop from the same source as the postganglionic cells of the sympathetic division of the nervous system. They are directly innervated by preganglionic cells of the sympathetic division of the autonomic nervous system and may be regarded as postganglionic cells that are specialized to secrete. In all other visceral effectors, preganglionic sympathetic fibers first synapse with postganglionic neurons before innervating the effector. In the adrenal medulla, however, the preganglionic fibers pass directly into the chromaffin cells of the gland. The secretion of hormones from the chromaffin cells is directly controlled by the autonomic nervous system, and innervation by the preganglionic fibers allows the gland to respond rapidly to a stimulus.

ACTION OF HORMONES
The two principal hormones synthesized by the adrenal medulla are epinephrine and norepinephrine. Epinephrine constitutes about 80 percent of the total secretion of the gland and is more potent in its action than norepinephrine. Both hormones are *sympathomimetic.* That is, they produce effects similar to those brought about by the sympathetic division of the autonomic nervous system. And, to a large extent, they are responsible for the fight-or-flight response. Like the glucocorticoids of the adrenal cortices, these hormones help the body resist stress. However, unlike the cortical hormones, the medullary hormones are not essential for life.

CONTROL OF HORMONES
Under stress, impulses received by the hy[conveyed to sympathetic preganglionic ? cause the chromaffin cells to increase ? epinephrine and norepinephrine. Epine? blood pressure by increasing heart rate ? the blood vessels. It accelerates the rate of respiration, dilates respiratory passageways, decreases the rate of digestion, increases the efficiency of muscular contractions, increases blood sugar level, and stimulates cellular metabolism. Hypoglycemia may also stimulate medullary secretion of epinephrine and norepinephrine.

PANCREAS
Because of its functions, the **pancreas** can be classified as both an endocrine and an exocrine gland. The pancreas is a flattened organ located posterior and slightly inferior to the stomach (Figure 18-15). The adult pancreas consists of a head, body, and tail. Its average length is 12 to 15 cm (5 to 6 inches). The endocrine portion of the pancreas consists of clusters of cells called *islets of Langerhans* (Figure 18-16). Two kinds of cells are found in these clusters: (1) *Alpha cells* constitute about 25 percent of the islet cells and secrete the hormone glucagon; (2) *beta cells* constitute about 75 percent of the islet cells and secrete the hormone insulin. The islets are surrounded by blood capillaries and the cells that form the exocrine part of the gland.

The endocrine secretions of the pancreas – glucagon and insulin – are concerned with control of the blood sugar level. Let us now examine how this regulation takes place.

Glucagon
The product of the alpha cells is *glucagon,* a hormone whose principal physiological activity is to increase the blood glucose level (Figure 18-17). Glucagon does this by accelerating the conversion of liver glycogen into glucose. The liver then releases the glucose into the blood, and the blood sugar level rises. Secretion of glucagon is directly controlled by the level of blood sugar via a negative feedback system. When the blood sugar level falls below normal, chemical sensors in the alpha cells of the islets stimulate the cells to secrete glucagon. When blood sugar rises, the cells are no longer stimulated and production slackens. If for some reason the self-regulating device fails and the alpha cells secrete glucagon continuously, hyperglycemia may result. Glucagon secretion is also inhibited by GIF or somatostatin.

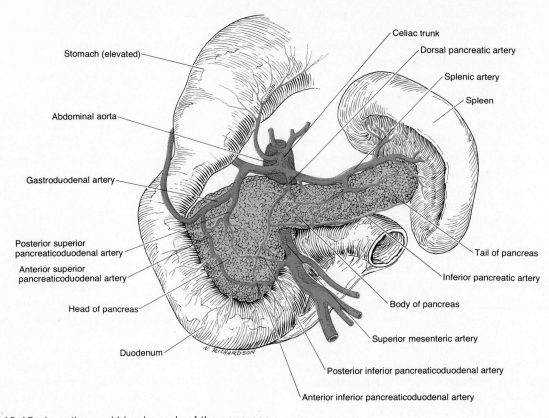

Stomach (elevated)

Celiac trunk

Dorsal pancreatic artery

Splenic artery

Spleen

Abdominal aorta

Gastroduodenal artery

Posterior superior
pancreaticoduodenal artery

Anterior superior
pancreaticoduodenal artery

Head of pancreas

Tail of pancreas

Inferior pancreatic artery

Body of pancreas

Superior mesenteric artery

Duodenum

Posterior inferior pancreaticoduodenal artery

Anterior inferior pancreaticoduodenal artery

N. RICHARDSON

Figure 18-15 Location and blood supply of the pancreas.

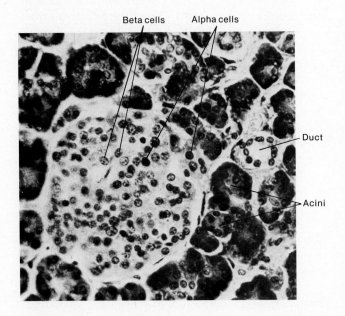

Beta cells

Alpha cells

Duct

Acini

Figure 18-16 Histology of the pancreas. Enlarged aspect of a single islet of Langerhans and surrounding acini at a magnification of 100×. (Courtesy of Victor B. Eichler, Wichita State University.)

Insulin

The beta cells of the islets produce a hormone called *insulin.* This hormone increases the build-up of proteins in cells, but its chief physiological action is opposite that of glucagon. Insulin decreases blood sugar level in two ways (Figure 18-17). It accelerates the transport of glucose from the blood into cells, especially skeletal muscle cells. It also accelerates the conversion of glucose into glycogen. The regulation of insulin secretion, like that of glucagon secretion, is directly determined by the level of sugar in the blood and is based on a negative feedback system. Other hormones can indirectly affect insulin production, however. For instance, HGH raises blood glucose level, and the rise in glucose level triggers insulin secretion. ACTH, by stimulating the secretion of glucocorticoids, brings about hyperglycemia and also indirectly stimulates the release of insulin. GIF or somatostatin inhibits the secretion of insulin.

OVARIES AND TESTES

The female gonads, called the **ovaries,** are paired oval bodies located in the pelvic cavity. The ovaries produce female sex hormones responsible for the development

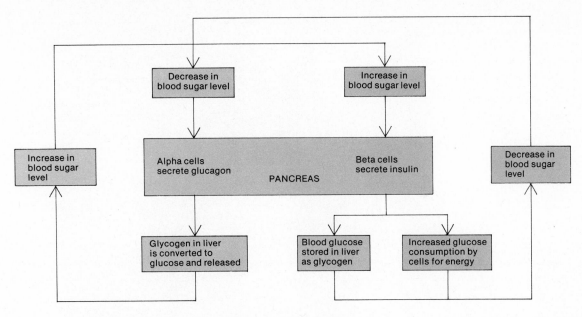

Figure 18-17 Regulation of the secretion of glucagon and insulin.

and maintenance of the female sexual characteristics. Along with the gonadotropic hormones of the pituitary, the sex hormones also regulate the menstrual cycle, maintain pregnancy, and ready the mammary glands for lactation. The male has two oval glands, called **testes,** that lie in the scrotum. The testes produce the male sex hormones that stimulate the development and maintenance of the male sexual characteristics.

PINEAL

The cone-shaped gland located in the roof of the third ventricle is known as the **pineal gland,** or **epiphysis cerebri** (see Figure 18-1). It is 5 to 8 mm (0.2 to 0.3 inch) long and 9 mm wide. It is covered by a capsule formed by the pia mater and consists of masses of parenchymal and glial cells. Around the cells are scattered preganglionic sympathetic fibers. The pineal gland starts to degenerate at about age 7, and in the adult it is largely fibrous tissue.

Although many anatomical facts concerning the pineal gland have been known for years, its physiology is still somewhat obscure. One hormone secreted by the pineal gland is *melatonin,* which appears to affect the secretion of hormones by the ovaries. It has been known for years that light stimulates the sexual endocrine glands. Researchers have also discovered that blood levels of melatonin are low during the day and high at night. Putting these observations together, some investigators now believe that melatonin inhibits the activi-

ties of the ovaries. During daylight hours, light entering the eye stimulates neurons to transmit impulses to the pineal that inhibit melatonin secretion. Without melatonin interference, the ovaries are free to step up their hormone production. But at night, the pineal gland is able to release melatonin, and ovarian function is slowed down. One function of the pineal gland might very well be regulating the activities of the sexual endocrine glands, particularly the menstrual cycle.

The pineal may also secrete a hormone called *adrenoglomerulotropin.* This hormone may stimulate the zona glomerulosa of the adrenal cortex to secrete aldosterone. Still other functions attributed to the pineal gland are secretion of a growth-inhibiting factor and secretion of a hormone called *serotonin* that is involved in normal brain physiology.

THYMUS

Usually a bilobed organ, the **thymus gland** is located in the upper mediastinum posterior to the sternum and between the lungs (Figure 18-18). The gland is conspicuous in the infant, and during puberty it reaches its absolute maximum size. After puberty, the thymic tissue, which consists primarily of lymphocytes, is replaced by fat. By the time the person reaches maturity, the gland has atrophied. The thymus also plays an essential role in the maturation of a population of lymphocytes called T cells (thymus-dependent cells).

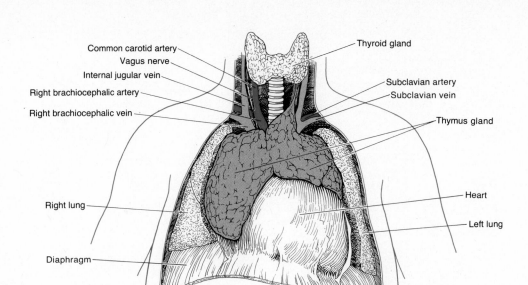

Common carotid artery
Vagus nerve
Internal jugular vein
Right brachiocephalic artery
Right brachiocephalic vein

Thyroid gland
Subclavian artery
Subclavian vein
Thymus gland

Right lung

Heart
Left lung

Diaphragm

Figure 18-18   Location of the thymus gland in a young child.

STRESS AND HOMEOSTASIS

Homeostasis may be viewed as specific responses of the body to specific stimuli. When blood calcium goes up, the rise stimulates the thyroid gland to release thyrocalcitonin. When blood calcium falls, thyrocalcitonin secretion is inhibited and parathyroid hormone secretion is stimulated. Homeostatic mechanisms "fine tune" the body. If the mechanisms are successful, the internal environment maintains a uniform chemistry, temperature, and pressure.

General Adaptation Syndrome

Homeostatic mechanisms are geared toward counteracting the everyday stresses of living. If a stress is extreme or unusual, however, the normal ways of keeping the body in balance may not be sufficient. In this case, the stress triggers a wide-ranging set of bodily changes called the **general adaptation syndrome.** Unlike the homeostatic mechanisms, the general adaptation syndrome does not maintain a constant internal environment. In fact, it does just the opposite. For instance, blood pressure and blood sugar level are raised above normal. The purpose of these changes in the internal environment is to gear up the body to meet an emergency.

Stressors

The hypothalamus can be called the body's watchdog. It has sensors that detect changes in the chemistry, temperature, and pressure of the blood. It is informed of emotions through tracts that connect it with the emotional centers of the cerebral cortex. When the hypothalamus senses stress, it initiates a chain of reactions that produce the general adaptation syndrome. The stresses that produce the syndrome are called **stressors.** A stressor may be almost any disturbance— heat or cold, environmental poisons, poisons given off by bacteria during a raging infection, heavy bleeding from a wound or surgery, or a strong emotional reaction.

When a stressor appears, it stimulates the hypothalamus to initiate the syndrome through two pathways. The first pathway is stimulation of the sympathetic nervous system and adrenal medulla. This stimulation produces an immediate set of responses called the alarm reaction. The second pathway, called the resistance reaction, involves the anterior pituitary gland and adrenal cortex. The resistance reaction is slower to start, but its effects last longer.

Alarm Reaction

The **alarm reaction,** or "fight-or-flight" response, is the body's initial reaction to a stressor (Figure 18-19a). It is actually a complex of reactions initiated by the hypothalamic stimulation of the sympathetic nervous system and the adrenal medulla. The responses of the visceral effectors are immediate and short-lived. They are designed to counteract a danger by mobilizing the body's resources for immediate physical activity. In essence, the alarm reaction brings tremendous amounts of glucose and oxygen to the organs that are most active in warding off danger. These are the brain, which must become highly alert; the skeletal muscles, which may have to

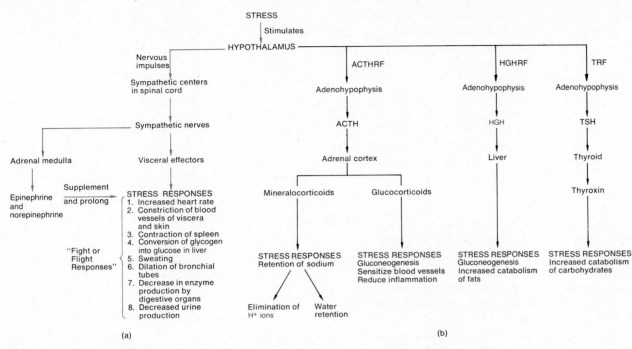

Figure 18-19 Stress responses. (a) During the alarm stage. (b) During the resistance stage. Colored arrows indicate immediate reactions. Black arrows indicate long-term reactions.

fight off an attacker; and the heart, which must work furiously to pump enough materials to the brain and muscles. The hyperglycemia associated with sympathetic activity is produced by:

1 Epinephrine and norepinephrine from the adrenal medulla, stimulating liver glycogenolysis.

2 Fat mobilization and glucose-sparing by HGH release.

3 Protein mobilization by glucocorticoids.

The effect of HGH and glucocorticoids is to stimulate liver gluconeogenesis—HGH by way of fat mobilization and glucocorticoids by way of protein mobilization.

Among the stress responses that characterize the alarm stage are the following:

1 The heart rate and the strength of cardiac muscle contraction increase. This response circulates substances in the blood very quickly to areas where they are needed to combat the stress.

2 Blood vessels supplying the skin and viscera, except the heart and lungs, undergo constriction. At the same time, blood vessels supplying the skeletal muscles and brain undergo dilation. These responses route more blood to organs active in the stress responses while decreasing blood supply to organs that do not assume an immediate, active role.

3 The spleen contracts and discharges stored blood into the general circulation to provide additional blood. Moreover, red blood cell production is accelerated and the ability of the blood to clot is increased. These preparations are made to combat bleeding.

4 The liver transforms large amounts of stored glycogen into glucose and releases it into the bloodstream. The glucose is broken down by the active cells to provide the energy needed to meet the stressor.

5 Sweat production increases. This response helps lower body temperature, which is elevated as circulation increases and body catabolism increases. Profuse sweating also helps to eliminate wastes produced as a result of accelerated catabolism.

6 The rate of breathing increases, and the respiratory passageways widen to accommodate more air. This response enables the body to acquire more oxygen, which is needed in the decomposition reactions of catabolism. It also allows the body to eliminate more carbon dioxide, which is produced as a side product during catabolism.

7 Production of saliva, stomach enzymes, and intestinal enzymes decreases. This reaction takes place since digestive activity is not essential for counteracting the stress.

8 Sympathetic impulses to the adrenal medulla increase its secretion of epinephrine and norepinephrine. These hormones supplement and prolong many sympa-

thetic responses—increasing heart rate and strength, constricting blood vessels, accelerating the rate of breathing, widening respiratory passageways, increasing the rate of catabolism, decreasing the rate of digestion, and increasing blood sugar level.

If you group the stress responses of the alarm stage by function, you will note that they are designed to increase circulation rapidly, promote catabolism for energy production, and decrease nonessential activities.

Resistance Reaction

The second stage in the stress responses is the **resistance reaction** (Figure 18-19b). Unlike the short-lived alarm reaction that is initiated by nervous impulses from the hypothalamus, the resistance reaction is initiated by regulating factors secreted by the hypothalamus and is a long-term reaction. The regulating factors are ACTHRF, HGHRF, and TRF.

ACTHRF stimulates the adenohypophysis to increase its secretion of ACTH. ACTH stimulates the adrenal cortex to secrete some of its hormones. The adrenal cortex is also indirectly stimulated by the alarm reaction. During the alarm reaction, kidney activity is cut back because of decreased blood circulation to the kidneys since it is not essential for meeting sudden danger. The resultant decrease in urine production stimulates the secretion of the mineralocorticoids.

The mineralocorticoids secreted by the adrenal cortex bring about the conservation of sodium ions by the body. A secondary effect of sodium conservation is the elimination of H^+ ions. The H^+ ions build up in high concentrations as a result of increased catabolism and tend to make the blood more acidic. Thus, during stress, a lowering of body pH is prevented. Sodium retention also leads to water retention—thus maintaining the high blood pressure that is typical of the alarm reaction. It also helps make up for fluid lost through severe bleeding.

The glucocorticoids, which are produced in high concentrations during stress, bring about the following reactions:

1 The glucocorticoids accelerate protein catabolism and the conversion of amino acids into glucose so that the body has a large supply of energy long after the immediate stores of glucose have been used up. The glucocorticoids also stimulate the removal of proteins from cell structures and stimulate the liver to break them down into amino acids. The amino acids can then be rebuilt into enzymes that are needed to catalyze the increased chemical activities of the cells or converted to glucose.

2 The glucocorticoids make blood vessels more sensitive to stimuli that bring about their constriction. This response counteracts a drop in blood pressure caused by bleeding.

3 The glucocorticoids also inhibit the production of fibroblasts, which develop into connective tissue cells. Injured fibroblasts release chemicals that play a role in stimulating the inflammatory response. Thus the glucocorticoids reduce inflammation and prevent it from becoming disruptive rather than protective. Unfortunately, through their effect on fibroblasts, the glucocorticoids also discourage connective tissue formation. Wound healing is therefore slow during a prolonged resistance stage.

Two other regulating factors are secreted by the hypothalamus in response to a stressor: TRF and HGHRF. TRF causes the adenohypophysis to secrete TSH. HGHRF causes it to secrete HGH. TSH stimulates the thyroid to secrete thyroxin, which increases the catabolism of carbohydrates. HGH stimulates the catabolism of fats and the conversion of glycogen to glucose. The combined actions of TSH and HGH increase catabolism and thereby supply additional energy for the body.

The resistance stage of the general adaptation syndrome allows the body to continue fighting a stressor long after the effects of the alarm reaction have dissipated. It increases the rate at which life processes occur. It also provides the energy, functional proteins, and circulatory changes required for meeting emotional crises, performing strenuous tasks, fighting infection, or resisting the threat of bleeding to death. During the resistance stage, blood chemistry returns to nearly normal. The cells use glucose at the same rate it enters the bloodstream. Thus blood sugar level returns to normal. Blood pH is brought under control by the kidneys as they excrete more hydrogen ions. However, blood pressure remains abnormally high because the retention of water increases the volume of blood.

All of us are confronted by stressors from time to time, and we have all experienced the resistance stage. Generally, this stage is successful in seeing us through a stressful situation, and our bodies then return to normal. Occasionally, the resistance stage fails to combat the stressor, however, and the body "gives up." In this case, the general adaptation syndrome moves into the stage of exhaustion.

Exhaustion

A major cause of exhaustion is loss of potassium ions. When the mineralocorticoids stimulate the kidney to retain sodium ions, potassium and hydrogen ions are traded off for sodium ions and secreted in the urine. As the chief positive ion in cells, potassium is partly responsible for controlling the water concentration of the cytoplasm. As the cells lose more and more potassium, they function less and less effectively. Finally they start to die. This condition is called the **stage of exhaustion.** Unless it is rapidly reversed, vital organs cease functioning and the person dies. Another cause of exhaustion is depletion of the adrenal glucocorticoids.

In this case, blood glucose level suddenly falls and the cells do not receive enough nutrients. A final cause is a weak organ. A long-term or strong resistance reaction puts heavy demands on the body, particularly on the heart, blood vessels, and adrenal cortex. They may not be up to handling the demands, or they may suddenly fail under the strain. In this respect, ability to handle stressors is determined to a large degree by general health.

DISORDERS

Disorders of the endocrine system, in general, are based on underproduction or overproduction of hormones. The term **hyposecretion** describes an underproduction; **hypersecretion** means an oversecretion.

PITUITARY

The anterior pituitary gland produces many hormones. All these hormones, with the exception of HGH and MSH, directly control the activities of other endocrine glands. Thus hyposecretion or hypersecretion of an anterior pituitary hormone produces widespread and complicated abnormalities.

Among the clinically interesting disorders related to the adenohypophysis are those involving HGH. This hormone builds up cells, particularly those of bone tissue. If HGH is hyposecreted during the growth years, bone growth is slow and the epiphyseal plates close before normal height is reached. This condition is called **pituitary dwarfism.** Other organs of the body also fail to grow, and the pituitary dwarf is childlike in many physical respects. Treatment requires administration of HGH during childhood before the epiphyseal plates close.

If HGH secretion is normal during childhood but low in adult life, a rare condition called **pituitary cachexia (Simmond's disease)** occurs. The tissues waste away (atrophy). The victim becomes quite thin and shows signs of premature aging. The atrophy occurs because the person is not receiving enough HGH to stimulate the protein-building activities required for replacing cells and cell parts.

Hypersecretion of HGH produces completely different disorders. Hypersecretion during childhood results in **giantism,** an abnormal increase in the length of long bones. Hypersecretion during adulthood is called **acromegaly.** Acromegaly cannot produce further lengthening of the long bones because the epiphyseal plates are already closed. Instead, the bones of the hands, feet, cheeks, and jaws thicken. Other tissues also grow. The eyelids, lips, tongue, and nose enlarge and the skin thickens and furrows, especially on the forehead and soles of the feet.

The principal abnormality associated with dysfunction of the neurohypophysis is **diabetes** ("overflow") **insipidus** ("tasteless"). This disorder should not be confused with diabetes mellitus ("sugar"), a disorder of the pancreas characterized by sugar in the urine. Diabetes insipidus is the result of a hyposecretion of ADH, usually caused by damage to the neurohypophysis or the hypothalamus. Symptoms include excretion of large amounts of urine and subsequent thirst. Diabetes insipidus is treated by administering ADH.

THYROID

Hyposecretion of thyroxin during the growth years results in **cretinism.** Two outstanding clinical symptoms of the cretin are dwarfism and mental retardation. The first is caused by failure of the skeleton to grow and mature. The second is caused by failure of the brain to develop fully. Recall that one function of thyroid hormone is to control tissue growth and development. Cretins also exhibit retarded sexual development and a yellowish skin color. Flat pads of fat develop, giving the cretin the characteristic round face and thick nose; a large, thick, protruding tongue; and protruding abdomen. Because the energy-producing metabolic reactions are slow, the cretin has a low body temperature and general lethargy. Carbohydrates are stored rather than utilized. Heart rate is also slow. If the condition is diagnosed early, the symptoms can be reversed by administering thyroid hormone.

Hypothyroidism during the adult years produces **myxedema.** Lack of thyroxin causes the body to retain water. A hallmark of this disorder is an edema that causes the facial tissues to swell and look puffy. The retention of water also brings an increase in blood volume that frequently causes high blood pressure. Like the cretin, the person with myxedema suffers from slow heart rate, low body temperature, muscular weakness, general lethargy, and a tendency to gain weight easily. The long-term combination of a slow heart rate and high blood pressure may overwork the heart muscles, causing the heart to enlarge. Because the brain has already reached maturity, the person with myxedema does not experience mental retardation. However, in moderately severe cases, nerve reactivity may be dulled so that the person lacks mental alertness. Myxedema occurs eight times more frequently in females than in males. Its symptoms are abolished by the administration of thyroxin.

Hypersecretion of thyroxin gives rise to **exophthalmic goiter.** This disease, like myxedema, is also more frequent in females. One of its primary symptoms is an enlarged thyroid, called a *goiter,* which may be two to three times its original size. Two other symptoms are an edema behind the eye, which causes the eye to "pop out" (**exophthalmos**), and an abnormally high metabolic rate. The high metabolic rate produces a range of effects that are generally opposite to those of myxedema—increased pulse, high body temperature, and moist, flushed skin. The person loses weight and is usually full of "nervous" energy. The thyroxin also increases the responsiveness of the nervous system, causing the person to become irritable and exhibit tremors of the extended fingers. Hyperthyroidism is usually treated by administering drugs that suppress thyroxin synthesis or surgically removing part of the gland.

The term **goiter** simply means an enlargement of the thyroid gland. It is a symptom of many thyroid disorders. It may also occur if the gland does not receive enough iodine to produce sufficient thyroxin for the body's needs. The follicular cells then enlarge in a futile attempt to produce more thyroxin, and they secrete large quantities of colloid. This condition is called *simple goiter.* Simple goiter is most often caused by a lower than average amount of iodine in the diet. It may also develop if iodine intake is not increased during certain conditions that put a high demand on the body for thyroxin—such as frequent exposure to cold and high fat and protein diets.

PARATHYROIDS

A normal amount of calcium in the extracellular fluid is necessary to maintain the resting state of neurons. A deficiency of calcium caused by **hypoparathyroidism** causes neurons to depolarize without the usual stimulus. As a result, nervous impulses increase and result in muscle twitches, spasms, and convulsions. This condition is called **tetany.** The effects of hypocalcemic tetany are observed in the **Trousseau** and **Chvostek signs.** Trousseau sign is observed when the binding of a blood pressure cuff around the upper arm produces contraction of the fingers and inability to open the hand. The Chvostek sign is a contracture of the facial muscles elicited by tapping the facial nerves at the angle of the jaw. Hypoparathyroidism results from surgical removal of the parathyroids or from parathyroid damage caused by parathyroid disease, infection, hemorrhage, or mechanical injury.

Hyperparathyroidism causes demineralization of bone. This condition is called **osteitis fibrosa cystica** because the areas of destroyed bone tissue are replaced by cavities that fill with fibrous tissue. The bones thus become deformed and are highly susceptible to fracture. Hyperparathyroidism is usually caused by a tumor in the parathyroids.

ADRENALS

Hypersecretion of the mineralocorticoid aldosterone results in a decrease in the body's potassium concentration. If potassium depletion is great, neurons cannot depolarize and muscular paralysis results. Hypersecretion also brings about excessive retention of sodium and water. The water increases the volume of the blood and causes high blood pressure. It also increases the volume of the interstitial fluid, producing edema.

Hyposecretion of glucocorticoids results in the condition called **Addison's disease.** Clinical symptoms include hypoglycemia, which leads to muscular weakness, mental lethargy, and weight loss. Increased potassium and decreased sodium lead to low blood pressure and dehydration. **Cushing's syndrome** is a hypersecretion of glucocorticoids, especially hydrocortisone and cortisone. The condition is characterized by the redistribution of fat. The result is spindly legs accompanied by a characteristic "moon face," "buffalo hump" on the back, and pendulous abdomen. Facial skin is flushed, and the skin covering the abdomen develops stretch marks. The individual also bruises easily, and wound healing is poor.

The **adrenogenital syndrome** results from overproduction of sex hormones, particularly the male androgens, by the adrenal cortex. Hypersecretion in male infants and young male children results in an enlarged penis. In young boys, it also causes premature development of male sexual characteristics. Hypersecretion in adult males is characterized by overgrowth of body hair, enlargement of the penis, and increased sexual drive. Hypersecretion in young girls results in premature sexual development. Hypersecretion in girls and women usually produces a receding hairline, baldness, an increase in body hair, deepening of the voice, muscular arms and legs, small breasts, and an enlarged clitoris.

PANCREAS

Hyposecretion of insulin results in a number of clinical symptoms referred to as **diabetes mellitus.** Typically an inherited disease, diabetes mellitus is caused by the destruction or malfunction of the beta cells. Among the symptoms are hyperglycemia and excretion of glucose in the urine as hyperglycemia increases. There is also an inability to reabsorb water, resulting in increased urine production, dehydration, loss of sodium, and thirst. Although the cells need glucose for energy-releasing reactions, glucose cannot enter the cells without the help of insulin. The cells start breaking down large quantities of fats and proteins into glucose. When the fats are decomposed, excessive amounts of organic acids called ketone bodies are formed and blood pH falls, causing a form of acidosis called **ketosis.** The catabolism of stored fats and proteins also causes weight loss. As lipids are transported by the blood from storage depots to hungry cells, lipid particles are deposited on the walls of blood vessels. The deposition leads to atherosclerosis and a multitude of circulatory problems.

MEDICAL TERMINOLOGY

Aldosteronism A disorder caused by hypersecretion of adrenal mineralocorticoids; potassium depletion occurs, sometimes causing paralysis; sodium and water are retained, causing high blood pressure and edema.

Antidiuretic Any chemical substance that prevents excessive urine production.

Feminizing adenoma Malignant tumors of the adrenal gland that secrete abnormally high amounts of female sex hormones and produce female secondary sexual characteristics in the male.

Hyperplasia Excessive development of tissue.

Hypoplasia Defective development of tissue.

Neuroblastoma Malignant tumor arising from the adrenal medulla associated with metastases to bones.

Thyroid storm An aggravation of all symptoms of hyperthyroidism resulting from trauma, surgery, unusual emotional stress, or labor.

Virilism Masculinization.

Virilizing adenoma Malignant tumors of the adrenal gland that secrete high amounts of male sex hormones and produce male secondary sexual characteristics in the female.

STUDY OUTLINE

ENDOCRINE GLANDS

1 Exocrine glands (sweat, sebaceous, digestive) secrete their products through ducts into body cavities or onto body surfaces.

2 Endocrine glands are ductless and secrete hormones into the blood.

HORMONES AND CYCLIC AMP

1 Hormones are proteins, amines, or steroids that change the physiological activities of cells in order to maintain homeostasis.

2 The hormone is called the first messenger, and cyclic AMP functions as a second messenger inside the cell, and performs a specific function according to the message indicated by the hormone.

3 Organs that exhibit changes in response to hormones are called target organs.

PITUITARY (HYPOPHYSIS)

1 This gland is differentiated into the adenohypophysis (the anterior lobe and glandular portion) and the neurohypophysis (the posterior lobe and nervous portion).

2 The adenohypophysis secretes tropic hormones and gonadotropic hormones. These hormones are regulated by neurohumors and, like most hormones, are involved in negative feedback systems.

3 Hormones of the adenohypophysis are *(a)* human growth hormone (regulates growth and is controlled by HGHRF); *(b)* thyroid-stimulating hormone (regulates activities of thyroid and is controlled by TRF); *(c)* adrenocorticotropic hormone (regulates adrenal cortex and is controlled by ACTHRF); *(d)* follicle-stimulating hormone (regulates ovaries and testes and is controlled by FSHRF); *(e)* luteinizing hormone (regulates female and male reproductive activities and is controlled by LRF); *(f)* prolactin (initiates milk secretion and is controlled by PRF and PIF); and *(g)* melanocyte-stimulating hormone (increases skin pigmentation and is controlled by MSHRF).

4 Hormones of the neurohypophysis are oxytocin (stimulates contraction of uterus and ejection of milk) and ADH (stimulates water reabsorption by the kidneys).

THYROID

1 This gland synthesizes thyroxin, which controls the rate of metabolism by increasing the catabolism of carbohydrates and proteins. It also produces triiodothyronine.

2 Thyrocalcitonin inhibits the activity of osteoclasts and stimulates activity of osteoblasts to reduce hypercalcemia.

PARATHYROIDS

1 Parathyroid hormone regulates the homeostasis of calcium and phosphate by stimulating osteoclasts in response to hypocalcemia.

ADRENALS (SUPRARENALS)

1 These glands consist of an outer cortex and inner medulla.

2 Cortical secretions are mineralocorticoids such as aldosterone that regulate sodium reabsorption and potassium excretion; glucocorticoids, which are essential to normal metabolism and resistance to stress; and gonadocorticoids (male and female sex hormones).

3 Medullary secretions are epinephrine and norepinephrine, which produce effects similar to sympathetic responses.

PANCREAS

1 Alpha cells of the pancreas secrete glucagon, which increases blood glucose level.

2 Beta cells secrete insulin, which decreases blood glucose level.

OVARIES AND TESTES

1 Ovaries are located in the pelvic cavity and produce sex hormones related to development and maintenance of female sexual characteristics.

2 Testes lie in the scrotum and produce sex hormones related to development and maintenance of male sexual characteristics.

PINEAL

1 This gland secretes melatonin (possibly regulates menstrual cycle), adrenoglomerulotropin (may stimulate adrenal cortex), and serotonin (involved in normal brain physiology).

THYMUS

1 This gland is necessary for the maturation of the thymus-dependent lymphocytes of the immune system.

STRESS AND HOMEOSTASIS

1 A stress is a condition of the body produced in response to extreme stimuli.

GENERAL ADAPTATION SYNDROME

1 If the stress is extreme or unusual, it triggers a wide-ranging set of bodily changes called the general adaptation syndrome.

2 Unlike the homeostatic mechanisms, this syndrome does not maintain a constant internal environment.

STRESSORS

1 The stresses that produce the general adaptation syndrome are called stressors.

2 Stressors include surgical operations, poisons, infections, fever, and strong emotional responses.

ALARM REACTION

1 The alarm reaction is initiated by nerve impulses from the hypothalamus to the sympathetic division of the autonomic nervous system and adrenal medulla. Responses are immediate and short-lived. They are "fight-or-flight" responses that increase circulation, promote catabolism for energy production, and decrease nonessential activities.

RESISTANCE REACTION

1 The resistance reaction is initiated by regulating factors secreted by the hypothalamus.

2 The regulating factors are ACTHRF, HGHRF, and TRF.

3 ACTHRF stimulates the adenohypophysis to increase its secretion of ACTH, which in turn stimulates the adrenal cortex to secrete hormones.

4 Resistance reactions are long-term and accelerate catabolism to provide energy to counteract stress.

5 Glucocorticoids are produced in high concentrations during stress. They create many distinct physiological effects.

EXHAUSTION

1 The stage of exhaustion results from dramatic changes during alarm and resistance reactions.

2 If stress is too great, exhaustion may lead to death.

REVIEW QUESTIONS

1 Distinguish between an endocrine gland and an exocrine gland. What is the relationship between an endocrine gland and a target organ and cyclic AMP?

2 What is a hormone? Distinguish between tropic and gonadotropic hormones.

3 In what respect is the pituitary gland actually two glands? Describe the histology of the adenohypophysis. Why does the anterior lobe of the gland have such an abundant blood supply?

4 What hormones are produced by the adenohypophysis? What are their functions?

5 Relate the importance of neurohumors to secretions of the adenohypophysis. How are negative feedback systems related to hormonal regulation?

6 Discuss the histology of the neurohypophysis and the function and regulation of its hormones.

7 Describe the location and histology of the thyroid gland. How are the thyroid hormones made, stored, and secreted?

8 Discuss the physiological effects of thyroxin and thyrocalcitonin. How are these hormones regulated?

9 Where are the parathyroids located? What is their histology? What are the functions of the parathyroid hormone (parathormone)? How is it regulated?

10 Compare the adrenal cortex and adrenal medulla with regard to location and histology.

11 Describe the hormones produced by the adrenal cortex in terms of type, normal function, and control.

12 What relationship does the adrenal medulla have to the autonomic nervous system? What is the action of adrenal medullary hormones?

13 Describe the location of the pancreas and the histology of the islets of Langerhans. What are the actions of glucagon and insulin?

14 Where is the pineal gland located? What are its assumed functions?

15 Describe the location of the thymus gland. What is its function?

16 Define the general adaptation syndrome. What is a stressor?

17 How do homeostatic responses differ from stress responses?

18 Outline the reactions of the body during the alarm stage, resistance stage, and stage of exhaustion when placed under stress. What is the central role of the hypothalamus during stress?

19 With the knowledge you have acquired in this chapter, respond to the following statement: "The combined activities of the nervous and endocrine systems are essential for maintaining homeostasis and overcoming stress."

Unit IV

Maintenance of the Human Body

This unit explains how the body maintains homeostasis on a day-to-day basis. In these chapters you will be studying the interrelations among the cardiovascular, lymphatic, respiratory, digestive, and urinary systems. You will also learn about metabolism, fluid and electrolyte balance, and acid–base dynamics.

Chapter 19
The Cardiovascular System: The Blood

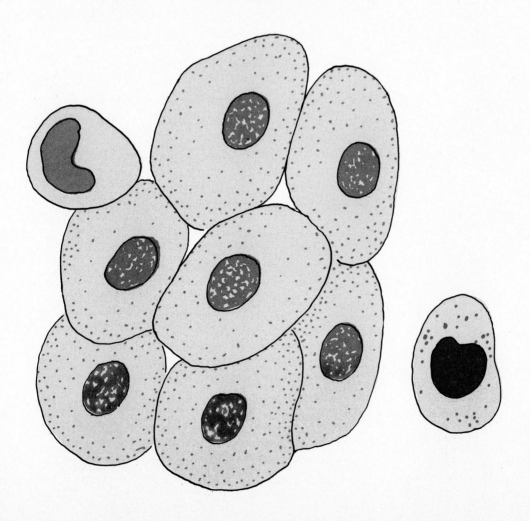

STUDENT OBJECTIVES

■ Contrast the general roles of blood, lymph, and interstitial fluid in maintaining homeostasis.

■ Define the principal physical characteristics of blood and its functions in the body.

■ Identify the plasma and formed element constituents of blood.

■ Compare the origins of the formed elements in blood and the reticuloendothelial cells.

■ Describe the structure of erythrocytes and their function in the transport of oxygen and carbon dioxide.

■ Define erythropoiesis and describe erythrocyte production and destruction.

■ Describe the importance of a reticulocyte count in the diagnosis of abnormal rates of erythrocyte production.

■ List the structural features and types of leucocytes.

■ Explain the significance of a differential count.

■ Discuss the role of leucocytes in phagocytosis and antibody production.

■ Discuss the structure of thrombocytes and explain their role in blood clotting.

■ List the components of plasma and explain their importance.

■ Describe the stages involved in blood clotting.

■ Name the factors that promote and inhibit blood clotting.

■ Contrast a thrombus and an embolus.

■ Define clotting time, bleeding time, and prothrombin time.

■ Explain ABO and Rh blood grouping.

■ Define the antigen-antibody reaction as the basis for ABO blood grouping.

■ Define the antigen-antibody reaction of the Rh blood grouping system.

■ Define erythroblastosis fetalis as a harmful antigen-antibody reaction.

■ Compare the location, composition, and function of interstitial fluid and lymph.

■ Contrast the causes of hemorrhagic, hemolytic, aplastic, and sickle cell anemia.

■ Compare the clinical symptoms of polycythemia and leukemia.

■ Identify the clinical symptoms of infectious mononucleosis.

■ Define medical terminology associated with blood and transfusion.

The more specialized a cell becomes, the less capable it is of carrying on an independent existence. For instance, a specialized cell is less capable of protecting itself from extreme temperatures, toxic chemicals, and changes in pH. Often it cannot go looking for food or devour whole bits of food. And, if it is firmly implanted in a tissue, it cannot move away from its own wastes. The substance that bathes the cell and carries out these vital functions for it is called interstitial fluid (also known as intercellular or tissue fluid).

The interstitial fluid, in turn, must be serviced by blood and lymph. The blood picks up oxygen from the lungs, nutrients from the digestive tract, hormones from the endocrine glands, and enzymes from still other parts of the body. The blood then transports these substances to all the tissues where they diffuse from the capillaries into the interstitial fluid. In the interstitial fluid, the substances are passed on to the cells and exchanged for wastes.

Since the blood must service all the tissues of the body, it can be an ideal medium for the transport of disease-causing organisms. To protect itself from disease spread, the body has a lymphatic system – a collection of vessels containing a fluid called lymph. The lymph picks up materials, including wastes, from the interstitial fluid, cleanses them of bacteria, and returns them to the blood. The blood then carries the wastes to the lungs, kidneys, and sweat glands, where they are eliminated from the body. The blood also takes wastes to the liver, where they are detoxified and recycled.

The blood, heart, and blood vessels constitute the **cardiovascular system.** The lymph, lymph vessels, and lymph glands make up the **lymphatic system.** Let us first take a look at the substance known as blood.

The red body fluid that flows through all the vessels except the lymph vessels is called **blood.** Blood is a viscous fluid – it is thicker and more adhesive than water. Water is considered to have a viscosity of 1. The viscosity of blood, by comparison, ranges from 4.5 to 5.5. This means that it flows $4\frac{1}{2}$ to $5\frac{1}{2}$ times more slowly than water. The adhesive quality of blood, or its stickiness, may be felt by touching it. Blood is also slightly heavier than water. Other physical characteristics of blood include a temperature of about 38°C (100.4°F), a pH range of 7.35 to 7.45 (slightly alkaline), and a 0.85 to 0.90 percent concentration of salt (NaCl). Blood constitutes about 8 percent of the total body weight. The blood volume of an average-sized man is between 5 and 6 liters (5 to 6 qt). An average-sized woman has 4 to 5 liters.

Despite its simple appearance, blood is a complex liquid that performs a number of critical functions:

1 It transports oxygen from the lungs to all cells of the body.

2 It transports carbon dioxide from the cells to the lungs.

3 It transports nutrients from the digestive organs to the cells.

4 It transports waste products from the cells to the kidneys, lungs, and sweat glands.

5 It transports hormones from endocrine glands to the cells.

6 It transports enzymes to various cells.

7 It regulates body pH through buffers and amino acids.

8 It regulates normal body temperature because it contains a large volume of water (an excellent heat absorber and coolant).

9 It regulates the water content of cells, principally through dissolved sodium ions.

10 It prevents body fluid loss through the clotting mechanism.

11 It protects against toxins and foreign microbes through special combat-unit cells.

Microscopically, blood is composed of two portions: plasma, which is a liquid containing dissolved substances, and formed elements, which are cells and cell-like bodies suspended in the plasma.

FORMED ELEMENTS

In clinical practice, the most common classification of the **formed elements** of the blood is the following:

Erythrocytes (red blood cells)
Leucocytes (white blood cells)
  Granular leucocytes (granulocytes)
    Neutrophils
    Eosinophils
    Basophils
  Agranular leucocytes (agranulocytes)
    Lymphocytes
    Monocytes
Thrombocytes (platelets)

Origin

The process by which blood cells are formed is called **hemopoiesis** or **hematopoiesis.** During embryonic and fetal life, there are no clear-cut centers for blood cell production. The yolk sac, liver, spleen, thymus gland, lymph nodes, and bone marrow all participate at various times in producing the formed elements. In the adult, however, we can pinpoint the production process. Red bone marrow (myeloid tissue) is responsible for producing red blood cells, granular leucocytes, and platelets. Lymphoid tissue – spleen, tonsils, lymph nodes – shares the responsibility with myeloid tissue for producing agranular leucocytes. Undifferentiated mesenchymal cells in red bone marrow are transformed into **hemocytoblasts,** immature cells that are eventually capable of developing into mature blood cells (Figure 19-1). For example, the hemocytoblasts develop into:

1 **Rubriblasts** *(proerythroblasts),* that go on to form mature red blood cells.

2 **Myeloblasts,** that go on to form mature neutrophils, eosinophils, and basophils.

3 **Megakaryoblasts,** that go on to form mature platelets.

4 **Lymphoblasts,** that eventually form lymphocytes.

5 **Monoblasts,** that eventually form monocytes.

Mature red blood cells do not live in the bloodstream very long. Their disintegrating bodies pose the danger of clogging small blood vessels, so certain cells clear away their bodies after they die. These cells are called *reticuloendothelial cells* and are also formed from primitive reticular cells. Reticuloendothelial cells enter the spleen, tonsils, lymph nodes, liver, lungs, and other organs and become highly specialized for phagocytosis. The reticuloendothelial cells in the lymph nodes are particularly active in destroying microbes and their toxins. The reticuloendothelial cells in the liver and spleen concentrate on ingesting dead blood cells.

Erythrocytes

Microscopically, **red blood cells,** or **erythrocytes,** appear as biconcave discs averaging about 7.7 μm in diameter (Figure 19-1). Mature red blood cells are quite simple in structure. They lack a nucleus and can neither reproduce nor carry on extensive metabolic activities. The cell contains a network of protein called the stroma; some cytoplasm; lipid substances, including cholesterol; and a red pigment called hemoglobin. *Hemoglobin,* which constitutes about 33 percent of the cell volume, is responsible for the red color of blood.

The erythrocytes combine with oxygen and carbon dioxide and transport them through the blood vessels. Red blood cells are highly specialized for this purpose. The hemoglobin molecule consists of a protein called globin and a pigment called heme, which contains iron. As the erythrocyte passes through the lungs, each of the four iron atoms in the hemoglobin molecule combines with a molecule of oxygen. The oxygen is transported in this state to other tissues of the body. In the tissues, the iron-oxygen reaction reverses, and the oxygen is released to diffuse into the interstitial fluid. On the return trip, the globin portion combines with a molecule of carbon dioxide from the interstitial fluid. This complex is transported and released in the lungs. Red blood cells

contain a great deal of hemoglobin molecules in order to increase their carrying capacity. One estimate is 280 million molecules of hemoglobin per erythrocyte. Hemoglobin is contained in the stroma of a red blood cell because hemoglobin molecules are small and, if they were free in plasma, they would leak through the endothelial membranes of blood vessels and be lost in the urine. The shape of a red blood cell increases its carrying capacity. A biconcave structure has a much greater surface area than, say, a sphere or cube. The erythrocyte thus provides the maximum surface area for the diffusion of gas molecules that pass through the membrane to combine with hemoglobin.

A red blood cell does not live long. Its cell membrane becomes fragile, and the cell is nonfunctional in about 120 days. The main reason for its short life is its inability to replace enzymes, particularly carbonic anhydrase. This enzyme is involved in the transportation of carbon dioxide by red blood cells. A healthy male has about 5.4 million red blood cells per cubic millimeter of blood, however, and a healthy female has about 4.8 million. The higher value in the male is because of his higher rate of metabolism. To maintain normal quantities of erythrocytes, the body must produce new mature cells at the astonishing rate of 2 million per second. In the adult, production takes place in the red bone marrow in the spongy bone of the cranium, ribs, sternum, bodies of vertebrae, and proximal epiphyses of the humerus and femur. The process by which erythrocytes are formed is called **erythropoiesis.**

Erythropoiesis starts with the transformation of a hemocytoblast into a rubriblast. The *rubriblast* (proerythroblast) gives rise to a *prorubricyte* (early erythroblast), which then develops into a *rubricyte* (intermediate erythroblast), the first cell in the sequence that begins to synthesize hemoglobin. The rubricyte next develops into a *metarubricyte* (late erythroblast). In the metarubricyte, hemoglobin synthesis is at a maximum and the nucleus is lost by extrusion. In the next stage, the metarubricyte develops into a *reticulocyte,* which in turn becomes an *erythrocyte,* or mature red blood cell. Once the erythrocyte is formed, it circulates through the blood vessels. Aged erythrocytes are destroyed by reticuloendothelial cells in the liver and spleen. The hemoglobin molecules are split apart—the iron is reused and the rest of the molecule is converted into other substances for reuse or elimination.

Normally erythropoiesis and red cell destruction proceed at the same pace. But if the body suddenly needs more erythrocytes, or if erythropoiesis is not keeping up with red blood cell destruction, a homeostatic mechanism steps up erythrocyte production. The stimulus for this mechanism is oxygen deficiency in the kidney cells. This mechanism is not surprising because the chief function of the erythrocytes is to deliver oxygen. As

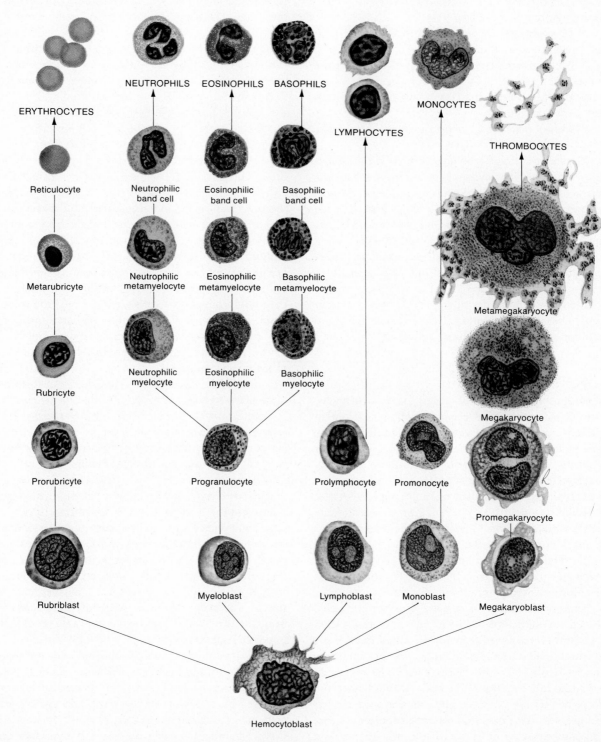

Figure 19-1   Origin, development, and structure of blood cells.

soon as the kidney cells become oxygen-deficient, they release a hormone called **erythropoietin.** This hormone circulates through the blood to the red bone marrow, where it stimulates hemocytoblasts to develop into red blood cells.

A diagnostic test that informs the physician about the rate of erythropoiesis is the **reticulocyte count.** Some reticulocytes are normally released into the bloodstream before they become mature red blood cells. If the number of reticulocytes in a sample of blood is less than 0.5 percent of the number of mature red blood cells in the sample, erythropoiesis is occurring too slowly. A low reticulocyte count might confirm a diagnosis of nutritional or pernicious anemia. Or it might indicate a kidney disease that prevents the kidney cells from producing erythropoietin. If the reticulocytes number more than 1.5 percent of the mature red blood cells, erythropoiesis is abnormally rapid. Any number of problems may be responsible for a high reticulocyte count: anemia, oxygen deficiency, and uncontrolled red blood cell production caused by a cancer in the bone marrow. If the individual has been suffering from a nutritional or pernicious anemia, the high count may indicate that treatment has been effective and the bone marrow is making up for lost time.

Leucocytes

Unlike red blood cells, **leucocytes,** or **white blood cells,** have nuclei and do not contain hemoglobin (Figure 19-1). They are far less numerous, averaging from 5,000 to 9,000 cells per cubic millimeter of blood. Red blood cells, therefore, outnumber white blood cells about 700 to 1. Leucocytes fall into two major groups. The first group contains the *granular leucocytes.* These develop from red bone marrow. They have granules in the cytoplasm and possess lobed nuclei. Three kinds of granular leucocytes exist: *neutrophils (polymorphs), eosinophils,* and *basophils.* The second principal group of leucocytes is called the *agranular leucocytes.* They develop from lymphoid and myeloid tissue. No cytoplasmic granules can be seen under a light microscope. Their nuclei are usually spherical. The two kinds of agranular leucocytes are *lymphocytes* and *monocytes.*

The general function of the leucocytes is to combat inflammation and infection. Some leucocytes are actively **phagocytotic** – they can ingest bacteria and dispose of dead matter. Most leucocytes also possess, to some degree, the ability to crawl through the minute spaces between the cells that form the walls of capillaries, the smallest blood vessels, and through connective and epithelial tissue. This movement, like that of amoebas, is called **diapedesis.** First, part of the cell membrane stretches out like an arm. Then the cytoplasm and

nucleus flow into the projection. Finally, the rest of the membrane snaps up into place. Another projection is made, and so on, until the cell has crawled to its destination.

Diagnosis of an injury or infection may involve a **differential count** – calculation of the number of each kind of white cell in 100 white blood cells. A normal differential count might appear as follows:

Neutrophils	60–70%
Eosinophils	2–4%
Basophils	0.5–1%
Lymphocytes	20–25%
Monocytes	3–8%
	100%

Particular attention is paid to the neutrophils in a differential count. The neutrophils are the most active white cells in response to tissue destruction. Their major role is phagocytosis. They also release the enzyme lysozyme, which destroys certain bacteria. More often than not, a high neutrophil count indicates damage by invading bacteria. An increase in the number of monocytes generally indicates a chronic (of long duration) infection such as tuberculosis. Apparently monocytes take longer to reach the site of infection than do neutrophils, but once they arrive they do so in larger numbers and destroy more microbes. Monocytes, like neutrophils, are phagocytic. They clean up cellular debris following an infection. High eosinophil counts indicate allergic conditions, since eosinophils are believed to combat the allergens that cause allergies. Eosinophils leave the capillaries, enter the tissue fluid, and produce antihistamines that destroy antigen-antibody complexes. Basophils are also believed to be involved in allergic reactions. Basophils leave the capillaries, enter the tissues, and become the mast cells of the tissues, liberating heparin, histamine, and serotonin.

The term **leucocytosis** refers to an increase in the number of white blood cells. If the increase exceeds 10,000, a pathological condition is usually indicated. An abnormally low level of white blood cells is termed **leucopenia.**

Some leucocytes, called lymphocytes, are involved in the production of antibodies and are discussed in detail in Chapter 22.

Antibodies are special proteins that inactivate antigens. An **antigen** is any substance that will stimulate the production of antibodies capable of reacting specifically with the substance. Most antigens are proteins, and most are not synthesized by the body. Many of the proteins that make up the cell structures and enzymes of bacteria are antigens. The toxins released by

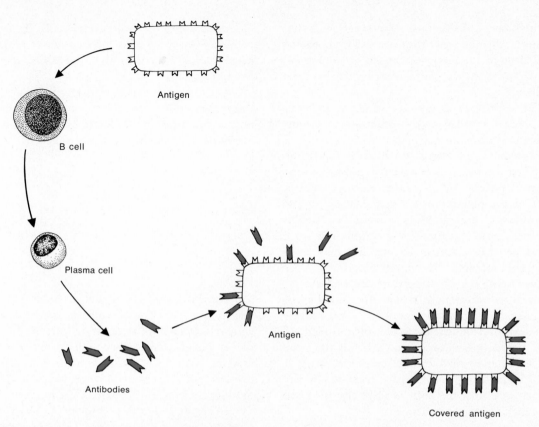

Figure 19-2 Antigen-antibody response. An antigen entering the body stimulates a B cell to develop into an antibody-producing plasma cell. The antibodies attach to the antigen, cover it, and render it harmless.

bacteria are also antigens. When antigens enter the body, they react chemically with substances in the lymphocytes and stimulate some lymphocytes, called B cells, to become *plasma cells* (Figure 19-2). The plasma cells then produce **antibodies:** globulin-type proteins that attach to antigens much as enzymes attach to substrates. Like enzymes, a specific antibody will generally attach only to a certain antigen. However, unlike enzymes, which enhance the reactivity of the substrate, antibodies "cover" their antigens so the antigens cannot come in contact with other chemicals in the body. In this way, bacterial poisons can be sealed up and rendered harmless. The bacteria themselves are destroyed by the antibodies. This process is called the **antigen-antibody response.** Eosinophils in tissues phagocytose the antigen-antibody complexes. The antigen-antibody response helps us to combat infection. It gives us immunity to some diseases. And it is responsible for blood types, allergies, and the body's rejection of organs transplanted from an individual with a different genetic makeup.

Foreign bacteria exist everywhere in the environment and have continuous access to the body through the mouth, nose, and pores of the skin. Furthermore, many cells, especially those of epithelial tissue, age and

die, and their remains must be disposed of daily. Even when the body is healthy, the leucocytes actively ingest bacteria and debris. However, a leucocyte can phagocytose only a certain number of substances before they interfere with the leucocyte's normal metabolic activities and bring on its death. Consequently, the life span of most leucocytes is very short. In a healthy body, some white blood cells will live a couple of days. During a period of infection they may live only a few hours.

Thrombocytes

If a hemocytoblast does not become an erythrocyte or granular leucocyte, it may develop into still another kind of cell, called a megakaryoblast, that is transformed into a megakaryocyte. Megakaryocytes are large cells whose cytoplasm breaks up into fragments. Each fragment becomes enclosed by a piece of the cell membrane and is called a **thrombocyte** or **platelet** (Figure 19-1). Platelets are disc-shaped cells without a nucleus. They average from 2 to 4 μm in diameter. Between 250,000 and 500,000 platelets appear in each cubic millimeter of blood.

The platelets prevent fluid loss by initiating a chain of reactions that results in blood clotting. Like the other

Exhibit 19—1 CHEMICAL COMPOSITION AND DESCRIPTION OF SUBSTANCES IN PLASMA

Constituent	Description	Constituent	Description
Water	Constitutes about 92 percent of plasma and is liquid portion of blood. Ninety percent of water is derived from absorption from digestive tract; 10 percent comes from cellular respiration. Water acts as solvent and suspending medium for solid components of blood and absorbs heat.	2. Nonprotein nitrogen (NPN) substances	Contain nitrogen but are not proteins. These substances include urea, uric acid, creatine, creatinine, ammonium salts. Represent breakdown products of protein metabolism and are carried by blood to organs of excretion.
Solutes		3. Food substances	Once foods are broken down in digestive tract, products of digestion are passed into blood for distribution to all body cells. These products include amino acids (from proteins), glucose (from carbohydrates), and fats (from lipids).
1. Proteins	Constitute 7 to 9 percent of solutes in plasma.		
Albumins	Constitute 55 to 64 percent of plasma proteins and are smallest plasma proteins. Produced by liver and provide blood with viscosity, a factor related to maintenance and regulation of blood pressure. Also exert considerable osmotic pressure to maintain water balance between blood and tissues and regulate blood volume.	4. Regulatory substances	Enzymes are produced by body cells and catalyze chemical reactions. Hormones, produced by endocrine glands, regulate growth and development in body.
		5. Respiratory gases	Oxygen and carbon dioxide are carried by blood. These gases are more closely associated with hemoglobin of red blood cells than plasma itself.
Globulins	Constitute about 15 percent of plasma proteins. Protein group to which antibodies produced by leucocytes belong. Gamma globulins attack measles and hepatitis viruses, tetanus bacteria, and possibly poliomyelitis virus.	6. Electrolytes	A number of ions constitute inorganic salts of plasma. Cations include Na^+, K^+, Ca^{2+}, Mg^{2+}. Anions include Cl^-, PO_4^{3-}, So_4^{2-}, HCO_3^-. Salts help maintain osmotic pressure, normal pH, physiological balance between tissues and blood.
Fibrinogen	Represents small fraction of plasma proteins (4 percent). Produced by liver and plays essential role in clotting.		

formed elements of the blood, platelets have a short life, probably only 1 week, because they are used up in clotting and are just too simple to carry on much metabolic activity.

Plasma

When the formed elements are removed from blood, a straw-colored liquid called **plasma** is left. Exhibit 19-1 outlines the chemical composition of plasma. Note that 7 to 9 percent of the solutes are proteins. Some of these proteins are also found elsewhere in the body, but in

blood they are called *plasma proteins*. Albumins, which constitute the majority of plasma proteins, are responsible for blood's viscosity. Along with the electrolytes, albumins also regulate blood volume by preventing all the water in the blood from diffusing into the interstitial fluid. Recall that water moves by osmosis from an area of low solute (high water) concentration to an area of high solute (low water) concentration. Globulins, which are antibody proteins released by plasma cells, form a small component of the plasma proteins. Gamma globulin is especially well known because it is able to form an antigen-antibody complex with the proteins of the hepatitis and measles viruses and the tetanus bacterium. Fibrinogen, a third plasma protein, takes part in the blood-clotting mechanism along with the platelets.

CLOTTING

There are several aspects of blood clotting that merit discussion here: clot formation, clot retraction and fibrinolysis, clot prevention, and clotting tests.

Formation

Normally, blood maintains its liquid state as long as it remains in the vessels. If it is drawn from the body, however, it thickens and forms a jelly. Eventually, the gel separates from the liquid. The straw-colored liquid, called **serum,** is simply plasma minus its clotting proteins. The gel is called a **clot** and consists of a network of insoluble fibers in which the cellular components of blood are trapped.

The process of clotting is called **hemostasis** or **coagulation.** Its purpose is to prevent blood loss when a blood vessel is ruptured. If the blood clots too easily, the result can be thrombosis – clotting in an unbroken blood vessel. If the blood takes too long to clot, a hemorrhage can result.

When a blood vessel is injured, two reactions occur to stop the bleeding. One reaction closes the vessel. The trauma to a blood vessel generates impulses that cause the smooth muscle in the blood vessel wall to contract, closing the vessel and thus preventing blood loss. The second reaction involves clot formation to plug the rupture by means of various chemicals known as **coagulation factors.** In plasma these factors are called *plasma coagulation factors.* A few *platelet coagulation factors* are released by platelets. One coagulation factor is released by damaged body tissues. Plasma coagulation factors are assigned a Roman numeral (see Exhibit 19-2).

Clotting is a complex process, but it can be described basically as a sequence of three stages. Stage 1 is concerned with the formation of a substance called thromboplastin. Stage 2 involves the conversion of prothrombin, a plasma protein, into thrombin, an enzyme. This stage requires the presence of thromboplastin and several other plasma coagulation factors. In stage 3, thrombin catalyzes the conversion of fibrinogen, another plasma protein, into fibrin. Fibrin forms the threads of the clot. A summary of these stages is as follows:

Stage 1:   Thromboplastin (formation or release)
Stage 2:   Prothrombin $\longrightarrow$ thrombin
Stage 3:   Fibrinogen $\longrightarrow$ fibrin

Depending on whether thromboplastin is released by damaged tissues or formed by platelet disintegration,

blood clotting may proceed along one of two routes: the extrinsic pathway or the intrinsic pathway.

EXTRINSIC PATHWAY

The **extrinsic pathway** of blood clotting begins when a blood vessel is ruptured (Figure 19-3). The damaged tissues surrounding the blood vessel or the ruptured area of the blood vessel itself release a lipoprotein called *tissue thromboplastin.* Tissue thromboplastin, in reaction with plasma coagulation factors IV, V, VII, and X, forms *extrinsic thromboplastin.* This process is stage 1 of the extrinsic pathway. In stage 2 prothrombin is converted to thrombin. This conversion requires extrinsic thromboplastin and several plasma coagulating factors such as IV, V, VII, and X. In stage 3 fibrinogen, which is soluble, is converted to fibrin, which is insoluble, by the thrombin formed in stage 2. This reaction requires plasma coagulation factors IV and XIII.

INTRINSIC PATHWAY

The **intrinsic pathway** of clotting begins with the rough surface created by a ruptured vessel (Figure 19-3). Under normal conditions the negatively charged platelet membrane and negatively charged endothelial lining of a blood vessel repel each other. Thus platelets do not adhere to the endothelial lining. However, a ruptured blood vessel results in the loss of the negative charge by the endothelial lining. Thus the platelets adhere to the ruptured area. This clumping together of platelets causes them to disintegrate and release their platelet coagulation factors into the plasma.

In stage 1 of the intrinsic pathway, four platelet coagulation factors (Pf_1, Pf_2, Pf_3, Pf_4), in reaction with seven plasma coagulation factors (IV, V, VIII, IX, X, XI, XII), form *intrinsic thromboplastin.* In stage 2 prothrombin is converted to thrombin by intrinsic thromboplastin and several plasma coagulating factors including IV, V, VII, and X. Stage 3 of the intrinsic pathway is the same as that for the extrinsic pathway. It involves the conversion of fibrinogen to fibrin in the presence of thrombin and plasma coagulation factors IV and XIII.

Besides helping to convert fibrinogen to fibrin, thrombin causes more platelets to adhere to each other, disintegrate, and release even more platelet coagulation factors. This cyclic feature of the intrinsic pathway ensures continual platelet disintegration until the clot is formed. Once the clot is formed, it plugs the ruptured area of the blood vessel and thus prevents hemorrhage (bleeding). Permanent repair of the blood vessel can then take place. In time, fibroblasts form connective tissue in the ruptured area and new endothelial cells repair the lining.

Exhibit 19—2 COAGULATION FACTORS*

Factor and Synonym	Comments	Factor and Synonym	Comments
I: Fibrinogen	Produced by liver. Important factor in stage 3 of clotting, in which it is converted to fibrin. Plasma minus fibrinogen is called serum.	X: Stuart factor or Stuart-Prower factor	Synthesized by liver. Formation dependent on vitamin K. Required for stages 1 and 2 of extrinsic and intrinsic pathways. Deficiency results in nosebleeds, bleeding into a joint, or bleeding into soft tissues.
II: Prothrombin	Produced by liver. Its synthesis requires vitamin K. Important in stage 2 of clotting, in which it is converted to thrombin.	XI: Plasma thromboplastin antecedent (PTA)	Synthesized by liver. Required for stage 1 of intrinsic pathway. Deficiency results in hemophilia C, a mild hemophilia of both males and females.
III: Thromboplastin	In extrinsic pathway it is known as extrinsic thromboplastin and is formed from tissue thromboplastin. In intrinsic pathway it is called intrinsic thromboplastin and is formed from platelet disintegration. Formation of thromboplastin signifies end of stage 1 of clotting.	XII: Hageman factor	Required for stage 1 of intrinsic pathway. Known to be activated by contact with glass and may assume role in initiating coagulation outside body.
IV: Calcium ions	Apparently involved in all three stages of clotting. Removal of calcium or its binding in plasma prevents coagulation.	XIII: Fibrin stabilizing factor (FSF)	Required for stage 3 of clotting.
V: Proaccelerin or labile factor	Produced in liver. Required for stages 1 and 2 of both extrinsic and intrinsic pathways.	Pf_1: Platelet factor 1 or platelet accelerator	Essentially same as plasma coagulation factor V.
VI:	No longer used in coagulation theory. Number has not been reassigned.	Pf_2: Platelet factor 2 or thrombin accelerator	Accelerates formation of thrombin in stage 1 of intrinsic pathway and conversion of fibrinogen to fibrin.
VII: Serum prothrombin conversion accelerator (SPCA) or stable factor	Synthesized in liver. Formation requires vitamin K. Required in stage 1 of extrinsic pathway.	Pf_3: Platelet factor 3 or platelet thromboplastic factor	In stage 1 of intrinsic pathway.
VIII: Antihemophilic factor	Synthesized by liver. Required for stage 1 of intrinsic pathway. Deficiency causes classic hemophilia or hemophilia A—an inherited disorder, primarily of males, in which bleeding may occur spontaneously or after only minor trauma.	Pf_4: Platelet factor 4	Binds heparin, an anticoagulant, during clotting.
IX: Christmas factor or plasma thromboplastin component (PTC)	Synthesized by liver. Formation requires vitamin K. Required for stage 1 of intrinsic pathway. Deficiency causes disorder called hemophilia B, a disease similar to hemophilia A.		

*Roman numerals I to XIII are plasma coagulation factors. Pf_1 to Pf_4 are platelet coagulation factors.

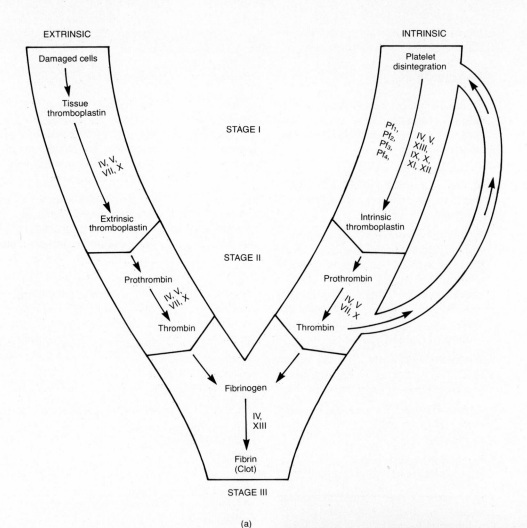

EXTRINSIC

Damaged cells

Tissue
thromboplastin

STAGE I

IV, V,
VII, X

Extrinsic
thromboplastin

STAGE II

Prothrombin

IV, V,
VII, X

Thrombin

Fibrinogen

IV,
XIII

Fibrin
(Clot)

STAGE III

INTRINSIC

Platelet
disintegration

Pf₁,
Pf₂,
Pf₃,
Pf₄,

IV, V,
XIII,
IX, X,
XI, XII

Intrinsic
thromboplastin

Prothrombin

IV, V
VII, X

Thrombin

(a)

Figure 19-3   Blood clotting. (a) Extrinsic and intrinsic
pathways. (b) Scanning electron micrograph of fibrin
threads and red blood cells at a magnification of 5,000×.
(Courtesy of Fisher Scientific Company and S.T.E.M.
Laboratories, Inc., Copyright 1975.)

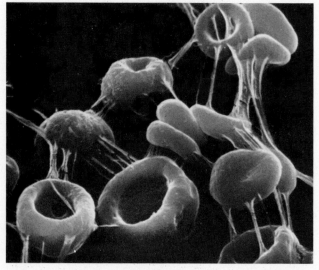

(b)

Retraction and Fibrinolysis

Normal coagulation involves two additional events after clot formation: clot retraction and fibrinolysis. **Clot retraction** or **syneresis** is the consolidation or tightening of the fibrin clot. The fibrin threads attached to the damaged surfaces of the blood vessel gradually contract. As the clot retracts, the ruptured area of the blood vessel gets smaller. Thus the risk of hemorrhage is further decreased. During retraction, a clear yellow fluid escapes between the fibrin threads – serum. Although some serum escapes, the formed elements in blood remain trapped in the fibrin threads. Normal clot retraction depends on an adequate number of platelets. Apparently, platelets in the clot disintegrate and, through the intrinsic pathway, form more fibrin.

The second event following clot formation – **fibrinolysis** – involves dissolution of the blood clot. Once the blood vessel is repaired, the clot dissolves. At the beginning of the coagulation pathway, certain enzymes are released by damaged tissues. These enzymes activate an inactive plasma enzyme called *plasminogen* into an active form called *plasmin*. Plasmin dissolves the fibrin clot.

Clot formation is a vital mechanism that prevents excessive loss of blood from the body. To form clots, the body needs calcium and vitamin K. Vitamin K is not involved in the actual clot formation, but it is required for the synthesis of prothrombin, which occurs in the liver. The vitamin is normally produced by bacteria that live in the intestine. It is also fat-soluble. It can thus be absorbed through the mucosa of the intestine and into the blood only if it is attached to fat. People suffering from disorders that prevent absorption of fat often experience uncontrolled bleeding. Clotting may be encouraged by applying a thrombin or fibrin spray, a rough surface such as gauze, or heat.

Prevention

Unwanted clotting may be brought on by the formation of cholesterol-containing masses called plaques in the walls of the blood vessels. They result in a rough surface that is perfect for the adhesion of platelets and is often the site of clotting. Clotting in an unbroken blood vessel is called **thrombosis.** The clot itself is a **thrombus.** A thrombus may dissolve. If it remains intact, it may damage tissues by cutting off the oxygen supply. Equally serious is the possibility that the thrombus will become dislodged and be carried with the blood to a smaller vessel. In a smaller vessel, the clot may block the circulation to a vital organ. A blood clot, bubble of air, or piece of debris transported by the bloodstream is called an **embolus.** When an embolus becomes lodged in a vessel and cuts off circulation, the condition is called an **embolism.**

No matter how healthy the body is, occasional rough spots appear on uncut vessel walls. In fact, it is believed that blood clotting is a continuous process inside blood vessels that is continually combated by clot-preventing and clot-dissolving mechanisms. Blood contains antithrombic substances – substances that prevent thrombin formation. One of these is *heparin,* which inhibits the conversion of prothrombin to thrombin and prevents most thrombus formation. Heparin is used in open-heart surgery to prevent clotting.

In general, any chemical substance that prevents clotting is an **anticoagulant.** Examples of anticoagulants are heparin, dicumarol, and the citrates and oxalates. Heparin is a quick-acting anticoagulant that blocks the clotting mechanism in five places. It is extracted from donated human blood. The pharmaceutical preparation *dicumarol* may be given to patients who are thrombosis-prone. Dicumarol acts as an antagonist to vitamin K and thus lowers the level of prothrombin. Dicumarol is slower acting than heparin and is used primarily as a preventative. The citrates and oxalates are used by laboratories and blood banks to prevent blood samples from clotting. These substances react with calcium to form insoluble compounds. In this way, the blood calcium is tied up and is no longer free to catalyze the conversion of prothrombin to thrombin. Citrated blood can be used for transfusions. The liver metabolizes the citrate, freeing the calcium ions. Oxalated blood cannot be used for transfusions.

Tests

The time required for blood to coagulate, usually from 5 to 15 minutes, is known as **clotting time.** This time is used as an index of a person's blood-clotting properties. One method for determining clotting time involves taking a sample of blood from a vein and placing 1 ml into each of three Pyrex tubes. The tubes are then submerged in a water bath at 37°C (98.6°F) and examined every 30 seconds for clot formation. Clotting is initiated when the platelets break up upon coming into contact with the glass. When the clot adheres to the tube, the end point is reached and the time is recorded. Blood taken from individuals with hemophilia clots very slowly or not at all.

Bleeding time is the time required for the cessation of bleeding from a small skin puncture. This time is usually measured by puncturing the ear lobe. As the droplets of blood escape, they are blotted by gently touching the wound with filter paper. When the paper is no longer stained, the bleeding has stopped. Normally,

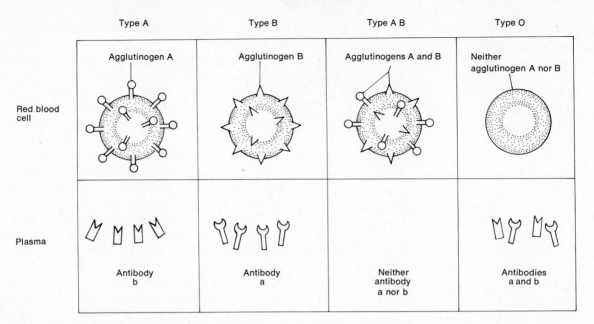

Figure 19-4   Agglutinogens (antigens) and agglutinins (antibodies) involved in the ABO blood grouping system.

bleeding time varies from 1 to 4 minutes. Unlike clotting time, which involves only the breakdown of platelets, bleeding time also involves constriction of injured blood vessels and all the steps of clot formation.

Prothrombin time is a test used to determine the amount of prothrombin in the blood. First the blood is treated with oxalate to tie up the calcium and make the blood incoagulable. Then calcium, thromboplastin, and plasma coagulation factors V and VII are all mixed with the blood sample. The length of time required for the blood to clot is the prothrombin time and depends on the amount of prothrombin in the sample. Normal prothrombin time is about 12 seconds.

GROUPING (TYPING)

The surfaces of erythrocytes contain genetically determined antigens called **agglutinogens.** These proteins are responsible for the two major blood group classifications: the ABO group and the Rh system.

ABO

The *ABO blood grouping* is based on two agglutinogens symbolized as A and B (Figure 19-4). Individuals whose erythrocytes manufacture only agglutinogen A are said to have blood type A. Those who manufacture only agglutinogen B are type B. Individuals who manufacture both A and B are typed AB. Others, who manufacture neither, are called type O.

Agglutinogen on Erythrocyte Membrane	*Blood Type*
A	A
B	B
AB	AB
Neither A nor B	O

These four blood types are not equally distributed. The incidence in the white population in the United States is as follows: type A (41 percent), type B (10 percent), type AB (4 percent), and type O (45 percent). Among blacks the frequencies are as follows: type A (27 percent), type B (20 percent), type AB (7 percent), and type O (46 percent).

The blood plasma of many people contains genetically determined antibodies referred to as **agglutinins.** These are antibody a (anti-A), which attacks agglutinogen A, and antibody b (anti-B), which attacks B. The antibodies formed by each of the four blood types are shown in Figure 19-4. You do not have antibodies that attack the agglutinogens of your own erythrocytes. A person with blood type A does not have antibody a. But you do have an antibody against any agglutinogen you yourself do not synthesize. Suppose type A blood is accidentally given to a person who does not have A agglutinogens. The person's body recognizes that the A protein is foreign and therefore treats it as an antigen. Antibody a's rush to the foreign erythrocytes, attack them, and cause them to *agglutinate* (clump) – hence the names agglutinogen and agglutinins. This reaction is another example of an antigen-antibody response.

Exhibit 19–3 INTERREACTIONS OF CELLS AND PLASMA OF ABO SYSTEM

Blood type	Agglutinogen	Agglutinin	Plasma causes agglutination of	Cells agglutinated by plasma of
A	A	b	B, AB	B, O
B	B	a	A, AB	A, O
AB	AB	None	None	A, B, O
O	None	a, b	A, B, AB	None

When blood is given a patient, care must be taken to ensure that the individual's antibodies will not agglutinate the donated erythrocytes and cause clumping. Destruction of the donated cells will not only undo the work of the transfusion, but the clumps can block vessels and may lead to death. Figure 19-4 shows that a person can receive blood from others in the same blood group.

Type A blood will agglutinate when mixed with types B and O. It may be given to individuals whose blood type is A or AB, but never to those whose blood type is B or O. Type B blood will agglutinate when mixed with types A and O. Thus it may be given to individuals whose blood type is B or AB, but never to individuals whose blood type is either A or O. Type AB blood will agglutinate when mixed with blood types A, B, or O. It may be given to individuals whose blood type is AB only. No other blood types may receive it. Type AB individuals requiring blood may be given types AB, A, B, and O—type AB blood has neither a nor b antibodies. Since type AB blood can theoretically receive all other blood types, it is called the *universal recipient*. Type O blood will not agglutinate with any of the other types. Theoretically, it may be given to individuals whose blood type is O, A, B, and AB. Since type O blood has no agglutinogens to serve as antigens in another person's body, it is known as the *universal donor*. Type O persons requiring blood may receive only type O blood. In practice, only matching blood types are used for transfusions.

Exhibit 19-3 summarizes the interreactions of the four blood types. Note that the degree of agglutination depends on the titer (strength or amount) of agglutinin (antibody) in the blood.

Rh

When blood is transfused, the technician must make sure the donor and recipient blood types are safely matched—not only for ABO group type but also for Rh

type. Otherwise, an ABO or an Rh incompatability could produce a severe, even fatal, reaction.

The *Rh system* is so named because it was first worked out in the blood of the Rhesus monkey. Like the ABO grouping, the Rh system is based on agglutinogens that lie on the surfaces of erythrocytes. Individuals whose erythrocytes have the Rh agglutinogens are designated *Rh⁺*. Those who lack Rh agglutinogens are designated *Rh⁻*. It is estimated that 85 percent of whites and 88 percent of blacks in the United States are Rh⁺, whereas 15 percent of whites and 12 percent of blacks are Rh⁻.

Under normal circumstances, human plasma does not contain anti-Rh antibodies. However, if an Rh⁻ person receives Rh⁺ blood, the body starts to make anti-Rh antibodies that will remain in the blood. If a second transfusion of Rh⁺ blood is given later, the previously formed anti-Rh antibodies will react against the donated blood and a severe reaction may occur. One of the most common problems with Rh incompatibility arises from pregnancy. During delivery, some of the fetus's blood may leak from the placenta (afterbirth) into the mother's bloodstream. If the fetus is Rh⁺ and the woman is Rh⁻, she, upon exposure to the Rh⁺ fetal cells, will make anti-Rh antibodies. If the woman becomes pregnant again, her anti-Rh antibodies will cross the placenta and make their way into the bloodstream of the baby. If the fetus is Rh⁻, no problem will occur since Rh⁻ blood does not have the Rh antigen. If the fetus is Rh⁺, an antigen-antibody response called *hemolysis* may occur in the fetal blood. Hemolysis means a breakage of erythrocytes resulting in the liberation of hemoglobin. The hemolysis brought on by fetal-maternal incompatibility is called **erythroblastosis fetalis**. When a baby is born with erythroblastosis, all the blood is slowly removed and replaced with antibody-free blood. It is even possible to transfuse blood into the unborn child if erythroblastosis is suspected. More important, though, is the fact that erythroblastosis can be prevented with RhoGAM, a drug administered to

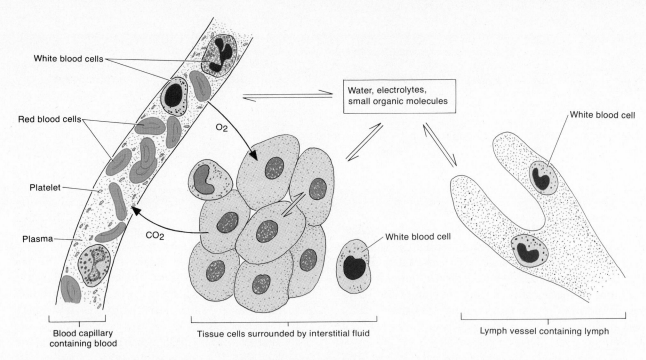

White blood cells

Red blood cells

Platelet

Plasma

O₂

CO₂

Water, electrolytes, small organic molecules

White blood cell

White blood cell

Blood capillary containing blood

Tissue cells surrounded by interstitial fluid

Lymph vessel containing lymph

Figure 19-5 Composition of blood, interstitial fluid, and lymph.

Rh⁻ mothers right after delivery or abortion. RhoGAM antibodies tie up the fetal agglutinogens so the mother cannot respond to the foreign antigens by producing antibodies. Thus the fetus of the next pregnancy is protected. In the case of an Rh⁺ mother and an Rh⁻ child, there are no complications since the fetus cannot make antibodies.

INTERSTITIAL FLUID AND LYMPH

For all practical purposes interstitial fluid and lymph are the same. The major difference between the two is location. When the fluid bathes the cells, it is called **interstitial fluid** or **intercellular fluid** or **tissue fluid**. When it flows through the lymphatic vessels, it is called **lymph** (Figure 19-5). Both fluids are similar in composition to plasma. The principal chemical difference is that they contain less protein because the larger protein

molecules are not easily filtered through the cells that form the walls of the capillaries. Keep in mind that whole blood does not flow into the tissue spaces; it remains in closed vessels. Certain constituents of the plasma do move, however, and once they move out of the blood, they are called interstitial fluid. The transfer of materials between blood and interstitial fluid occurs by osmosis, diffusion, and filtration across the cells that make up the capillary walls. Both interstitial fluid and lymph contain variable numbers of leucocytes. Leucocytes can enter the tissue fluid by diapedesis, and the lymphoid tissue itself is the site of nongranular leucocyte production. However, interstitial fluid and lymph both lack erythrocytes and platelets.

Other substances, especially organic molecules, in interstitial fluid and lymph vary in relation to the location of the sample analyzed. The lymph vessels that drain the organs of the digestive tract, for example, contain a great deal of lipid absorbed from food.

DISORDERS

ANEMIA

Anemia is a sign, not a diagnosis. Many kinds of anemia exist, all characterized by insufficient erythrocytes or hemoglobin. These conditions lead to fatigue and intolerance to cold, both of which are related to lack of oxygen needed for energy and heat production and paleness due to low hemoglobin content.

HEMORRHAGIC ANEMIA

An excessive loss of erythrocytes through bleeding is called **hemorrhagic anemia**. Common causes are large wounds, stomach ulcers, and heavy menstrual bleeding. If bleeding is extraordinarily heavy, the anemia is termed acute. Excessive blood loss can be fatal. Slow, prolonged bleeding is apt to produce a chronic anemia; chief symptom is fatigue.

HEMOLYTIC ANEMIA

If an erythrocyte cell membrane ruptures prematurely, the cell remains as a "ghost" and its hemoglobin pours out into the plasma. A characteristic sign of this condition is distortions in the shapes of erythrocytes that are progressing toward hemolysis. There may also be a sharp increase in the number of reticulocytes since the destruction of red blood cells stimulates erythropoiesis.

The premature destruction of red cells may result from inherent defects: hemoglobin defects, abnormal red cell enzymes, or defects of red cell membrane. Agents that may cause *hemolytic anemia* are parasites, toxins, and antibodies from incompatible blood (Rh mother and Rh fetus, for instance). Erythroblastosis fetalis of the newborn is an example of a hemolytic anemia.

APLASTIC ANEMIA

Destruction or inhibition of the red bone marrow results in **aplastic anemia.** Typically, the marrow is replaced by fatty tissue, fibrous tissue, or tumor cells. Toxins and certain medications are causes. Many of the medications inhibit the enzymes involved in hemopoiesis.

SICKLE CELL ANEMIA

The erythrocytes of a person with **sickle cell anemia** manufacture an abnormal kind of hemoglobin. When the erythrocyte gives up its oxygen to the interstitial fluid, its hemoglobin tends to lose its integrity in places of low oxygen tension and form long, stiff, rodlike structures that bend the erythrocyte into a sickle shape (Figure 19-6). The sickled cells rupture easily. Even though erythropoiesis is stimulated by the loss of the cells, it cannot keep pace with the hemolysis. The individual consequently suffers from a hemolytic anemia that reduces the amount of oxygen that can be supplied to the tissues. Prolonged oxygen reduction may eventually cause extensive tissue damage. Furthermore, because of the shape of the sickled cells, they tend to get stuck in blood vessels – and can cut off blood supply to an organ altogether.

Sickle cell anemia is inherited. The gene responsible for the tendency of the erythrocytes to sickle during hypoxia prevents erythrocytes from rupturing during a malarial crisis. Sickle cell genes are found primarily among populations, or descendents of populations, that live in the malaria belt around the world – including parts of Mediterranean Europe and subtropical Africa and Asia. A person with only one of the sickling genes is said to have sickle cell trait. Such an individual has a high resistance to malaria – a factor that may have tremendous survival value – but does not develop the anemia. Only people who inherit a sickling gene from both parents get sickle cell anemia.

POLYCYTHEMIA

The term **polycythemia** refers to an abnormal increase in the number of red blood cells. Increases of 2 to 3 million cells per

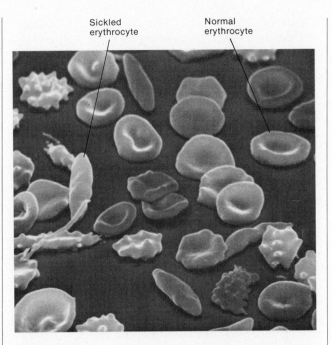

Sickled erythrocyte   Normal erythrocyte

Figure 19-6 Scanning electron micrograph of erythrocytes in sickle cell anemia at a magnification of 5,000×. (Courtesy of Fisher Scientific Company and S.T.E.M. Laboratories, Inc., Copyright 1975.)

cubic millimeter are considered to be polycythemic. The disorder is harmful because the blood's viscosity is greatly increased due to the extra red blood cells, and viscosity contributes to thrombosis and hemorrhage. It also causes a rise in blood pressure. The thrombosis results from too many red blood cells piling up as they try to enter smaller vessels. The hemorrhage is due to widespread hyperemia (unusually large amount of blood in an organ part).

A clinical test important in diagnosing polycythemia and anemia is the hematocrit. **Hematocrit** is the percentage of blood that is made up of red blood cells. It is determined by centrifuging blood and noting the ratio of red blood cells to whole blood. The average hematocrit for males is 47 percent. This means that in 100 ml of blood there are 47 ml of cells and 53 ml of plasma. The average hematocrit for females is 42 percent. Anemic blood may have a hematocrit of 15 percent; polycythemic blood may have a hematocrit of 65 percent.

INFECTIOUS MONONUCLEOSIS

Infectious mononucleosis is a contagious disease thought to be of viral origin that occurs mainly in children and young adults. Its trademark is an elevated white count with an abnormally high percentage of lymphocytes and mononucleocytes. An increase in the number of monocytes usually indicates a chronic infection. Symptoms include slight fever, sore throat, brilliant red throat and soft palate, stiff neck, cough, and malaise. The spleen may enlarge. Secondary complications involving the liver, heart, kidneys, and nervous system may develop. There is no cure for mononucleosis, and treatment consists of watching for and treating complications. Usually the disease runs its course in a few weeks, and the individual generally suffers no permanent ill effects.

LEUKEMIA

Also called "cancer of the blood," **leukemia** is an uncontrolled, greatly accelerated production of white cells. Many of the cells fail to reach maturity. As with most cancers, the symptoms result not so much from the cancer cells themselves as from their interference with normal body processes. The anemia and bleeding problems commonly seen in leukemia

result from the crowding out of normal bone marrow cells preventing normal production of red blood cells and platelets. The most common cause of death from leukemia is internal hemorrhaging, especially cerebral hemorrhage that destroys the vital centers in the brain. Another frequent cause of death is uncontrolled infection owing to lack of mature or normal white blood cells.

MEDICAL TERMINOLOGY

Blood plasma substitute   This substance mimics the characteristics of plasma. It is used to maintain blood volume during emergency conditions (such as hemorrhage) until blood can be matched or to prevent dehydration if a patient cannot swallow liquids. It is also used to replace fluid and electrolytes after loss of blood during surgery.

Citrated whole blood   Whole blood protected from coagulation by a citrate.

Direct transfusion (immediate)   Transfer of blood directly from one person to another without exposing the blood to air.

Exchange transfusion   Removing blood from the recipient while simultaneously replacing it with donor blood. This method is used for erythroblastosis fetalis and poisoning.

Fractionated blood (*fract* = break)   Blood that has been separated into its components. Only the part needed by the patient is given.

Hem, Hemo, Hema, Hemato (*heme* = iron)   Various combining forms meaning blood.

Hemolysis (laking)   A swelling and subsequent rupture of erythrocytes with the liberation of hemoglobin into the surrounding fluid.

Hemorrhage (*rrhage* = bursting forth)   Bleeding, either internal (from blood vessels into tissues) or external (from blood vessels directly to the surface of the body).

Heparinized whole blood   Whole blood in a heparin solution to prevent coagulation.

Indirect transfusion (mediate)   Transfer of blood from a donor to a container and then to the recipient, permitting blood to be stored for an emergency. The blood may be separated into its components so a patient receives only a needed part.

Normal plasma   Cell-free plasma contains normal concentrations of all solutes. It is used to bring blood volume up to normal when excessive numbers of blood cells have not been lost.

Platelet concentrates   Platelets are obtained from freshly drawn whole blood and are transfused for platelet-deficiency disorders such as hemophilia.

Reciprocal transfusion   Blood is transferred from a person who has recovered from a contagious infection into the vessels of a patient suffering with the same infection. An equal amount of blood is returned from the patient to the well person. This method allows the patient to receive antibody-bearing lymphocytes from the recovered person.

Septicemia   (*sep* = decay; *emia* = condition of blood) Toxins or disease-causing bacteria in the blood. Also called "blood poisoning."

Transfusion   The transfer of whole blood, blood components (red blood cells only or plasma only), or bone marrow directly into the bloodstream.

Venesection   Opening of a vein for withdrawal of blood.

Whole blood   Blood containing all formed elements, plasma, and plasma solutes in natural concentration.

STUDY OUTLINE

1 The principal function of blood is transportation of O_2, CO_2, nutrients, wastes, hormones, and enzymes. It regulates pH, body temperature, and water content of cells and protects against disease. Blood consists of plasma and formed elements.

FORMED ELEMENTS

1 Formed elements are erythrocytes, leucocytes, and thrombocytes. Wandering epithelial cells become hemocytoblasts. Hemocytoblasts in red bone marrow develop into erythrocytes, granular leucocytes, and thrombocytes. Hemocytoblasts entrapped in lymphatic tissue and found in red bone marrow develop into agranular leucocytes.

2 Erythrocytes, or red blood cells, are biconcave discs without nuclei and containing hemoglobin. Erythrocyte formation is called erythropoiesis and occurs in adult red marrow of certain bones. A reticulocyte count is a diagnostic test that indicates the rate of erythropoiesis.

3 Leucocytes, or white blood cells, are nucleated cells. Two principal types are granular (neutrophils, eosinophils, basophils) and agranular (lymphocytes and monocytes).

4 One function of leucocytes, especially neutrophils and monocytes, is to combat inflammation and infection through phagocytosis.

5 In response to the presence of foreign proteins called antigens, some lymphocytes are changed into tissue plasma cells. Plasma cells produce antibodies, which cover antigens and render them harmless. This antigen-antibody response combats infection and provides immunity.

6 Thrombocytes, or platelets, are disc-shaped structures without nuclei. They are formed from megakaryocytes and are involved in clotting.

7 The liquid portion of blood, called plasma, consists of 92 percent water and 8 percent solutes. Principal solutes include proteins (albumins, globulins, fibrinogen), foods, enzymes and hormones, gases, and electrolytes.

CLOTTING

1 A clot is a network of insoluble protein (fibrin) in which formed elements of blood are trapped.

2 The chemicals involved in clotting are known as coagulation factors.

3 There are two kinds of coagulation factors: plasma and platelet coagulation factors.

4 Blood clotting involves two pathways: the intrinsic and the extrinsic.

5 Clotting in a blood vessel is called thrombosis. A thrombus that moves from its site of origin is called an embolus.

6 Clinically important clotting tests are clotting time (time required for blood to coagulate), bleeding time (time required for the cessation of bleeding from a small skin puncture), and prothrombin time (time required for the blood to coagulate, which depends on the amount of prothrombin in the blood sample).

GROUPING (TYPING)

1 ABO and Rh systems are based on antigen-antibody responses.

2 In the ABO system, agglutinogens (antigens) A and B determine blood type. Plasma contains agglutinins (antibodies) that clump agglutinogens which are foreign to the individual.

3 In the Rh system, individuals whose erythrocytes have Rh agglutinogens are classified as Rh^+. Those who lack the antigen are Rh^-.

INTERSTITIAL FLUID AND LYMPH

1 Interstitial fluid bathes body cells, whereas lymph is found in lymphatic vessels.

2 These fluids are similar in chemical composition. They differ chemically from plasma in that both contain less protein, a variable number of leucocytes, and no platelets or erythrocytes.

REVIEW QUESTIONS

1 How are blood, interstitial fluid, and lymph related to the maintenance of homeostasis?

2 Distinguish between the cardiovascular system and lymphatic system.

3 Define the principal physical characteristics of blood. List the functions of blood and their relationship to other systems of the body.

4 Distinguish between plasma and formed elements. Where are the formed elements produced?

5 What are reticuloendothelial cells? Describe their function.

6 Describe the microscopic appearance of erythrocytes. What is the essential function of erythrocytes?

7 Define erythropoiesis. Relate erythropoiesis to the homeostasis of the red blood cell count. What factors accelerate and decelerate erythropoiesis?

8 What is a reticulocyte count? What is its diagnostic significance?

9 Describe the classification of leucocytes. What are their functions?

10 What is the importance of diapedesis and phagocytosis in fighting bacterial invasion?

11 What is a differential count? What is its significance?

12 Distinguish between leucocytosis and leucopenia.

13 Describe the antigen-antibody response. How is it protective?

14 What are the major chemicals in plasma? What do they do?

15 What is the difference between plasma and serum?

16 Briefly describe the process of clot formation. What is fibrinolysis? Why does blood usually not remain clotted in vessels?

17 List the various coagulation factors.

18 What are the pathways involved in blood clotting?

19 Define the following: thrombus, embolus, anticoagulant, clotting time, bleeding time, prothrombin time.

20 What is the basis for ABO blood grouping? What are agglutinogens and agglutinins?

21 What is the basis for the Rh system? How does erythroblastosis fetalis occur? How may it be prevented?

22 Compare interstitial fluid and lymph with regard to location, chemical composition, and function.

Chapter 20
The Cardiovascular System: The Heart

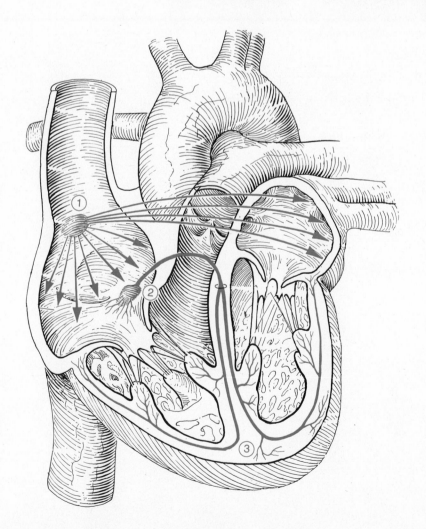

STUDENT OBJECTIVES

■ Describe the location of the heart in the mediastinum.

■ Distinguish between the structure and location of fibrous and serous pericardium.

■ Contrast the structure of the epicardium, myocardium, and endocardium.

■ Identify the blood vessels, chambers, and valves of the heart.

■ Describe the initiation and conduction of nerve impulses through the electrical conduction system of the heart.

■ Label and explain the deflection waves of a normal electrocardiogram.

■ Describe the route of blood in coronary circulation.

■ Compare angina pectoris and myocardial infarction as abnormalities of coronary circulation.

■ Define systole and diastole as the two principal events of the cardiac cycle.

■ Describe the pressure changes associated with blood flow through the heart.

■ Relate the events of the cardiac cycle to time.

■ Describe the sounds of the heart and their clinical significance.

■ Define cardiac output and explain what determines it.

■ Define Starling's law of the heart.

■ Contrast the effects of sympathetic and parasympathetic stimulation of the heart.

■ Define the role of pressoreceptors and chemoreceptors in controlling heart rate.

■ Define circulatory shock.

■ Describe the homeostatic mechanisms that compensate for circulatory shock.

■ Draw a shock cycle to illustrate the effects of severe shock on the circulatory organs.

■ Describe the use of hypothermia, the heart-lung bypass, artificial parts, and the artificial pacemaker.

■ List the six risk factors involved in heart attacks.

■ Explain why inadequate blood supply, anatomical disorders, and malfunctions of conduction are primary reasons for heart trouble.

■ Describe patent ductus arteriosus, septal defects, and valvular stenosis as congenital heart defects.

■ List the four abnormalities of the heart present in tetralogy of Fallot.

■ Define atrioventricular block, atrial flutter, atrial fibrillation, and ventricular fibrillation as abnormalities of the conduction system of the heart.

The **heart** is a hollow, muscular organ that pumps blood through the blood vessels. It is situated obliquely between the lungs in the mediastinum, and about two-thirds of its mass lies to the left of the body's midline (Figure 20-1a). The heart is shaped like a blunt cone about the size of a closed fist—12 cm (5 inches) long, 9 cm (3½ inches) wide at its broadest point, and 6 cm (2½ inches) thick. Its pointed end, the *apex,* projects downward, forward, and to the left and lies superior to the central depression of the diaphragm. Its broad end, or *base,* projects upward, backward, and to the right and lies just inferior to the second rib. The major parts of the heart to be considered here are the parietal pericardium, the wall and chambers, and the valves.

PARIETAL PERICARDIUM (PERICARDIAL SAC)

The heart is enclosed in a loose-fitting serous membrane called the **parietal pericardium** or **pericardial sac** (Figure 20-1b). It consists of two layers: the fibrous layer and the serous layer. The *fibrous layer* or *fibrous pericardium* is the outer layer and consists of a tough, fibrous connective tissue. The fibrous pericardium is attached to the large blood vessels entering and leaving the heart, to the diaphragm, and to the inside of the sternal wall of the thorax. It also adheres to the parietal pleurae. The fibrous pericardium prevents overdistension of the heart, provides a tough protective membrane around the heart, and anchors the heart in the mediastinum. The inner layer of the parietal pericardium is known as the *serous layer* or *serous pericardium.* This thin, more delicate membrane is continuous with the epicardium at the base of the heart and around the large blood vessels. The **epicardium** or **visceral pericardium** is the thin, transparent outer layer of the wall of the heart that is composed of serous tissue and mesothelium. Between the serous pericardium and the epicardium is a potential space called the *pericardial cavity.* The cavity contains a watery fluid, known as pericardial fluid, which prevents friction between the membranes as the heart moves.

WALL AND CHAMBERS

The wall of the heart is divided into three portions: the epicardium (external layer), the myocardium (middle layer), and the endocardium (inner layer). The **epicardium** is the same as the visceral pericardium. The **myocardium,** which is cardiac muscle tissue, comprises the bulk of the heart. Cardiac muscle fibers are involuntary, striated, and branched, and the tissue is arranged in interlacing bundles of fibers. The myocardium is responsible for the contraction of the heart. The **endocardium** is a thin layer of endothelium overlying a thin layer of connective tissue pierced by tiny blood vessels and bundles of smooth muscle. It lines the inside of the myocardium and covers the valves of the heart and the tendons that hold them open. It is continuous with the endothelial lining of the large blood vessels of the heart.

The interior of the heart is divided into four spaces or chambers, which receive the circulating blood (Figure 20-2). The two upper chambers are called the right and left **atria.** Each atrium has an appendage called an *auricle,* so named because of its resemblance to a dog's ear. The auricle increases the atrium's surface area. The lining of the atria is smooth, except for the anterior atrial walls and the lining of the auricles, which contain projecting muscle bundles that are parallel to each other and resemble the teeth of a comb: the *musculi pectinati.* These bundles give the lining of the auricles a ridged appearance. The atria are separated by a partition called the *interatrial septum.* On the posterior wall of the interatrial septum is an oval depression, the *fossa ovalis,* which corresponds to the site of the foramen ovale of the fetal heart. The two lower chambers, the right and left **ventricles,** are separated by an *interventricular septum.* The muscle tissue of the atria and ventricles is separated by connective tissue that also forms the valves. This "cardiac skeleton" effectively divides the myocardium into two separate muscle masses. Externally, a groove known as the *coronary sulcus* separates the atria from the ventricles. It encircles the heart and houses the coronary sinus and circumflex branch of the left coronary artery. The *anterior interventricular sulcus* and *posterior interventricular sulcus* separate the right and left ventricles externally. The sulci contain coronary blood vessels and a variable amount of fat.

The right atrium receives blood from all parts of the body except the lungs. It receives the blood through three veins. One of these veins is the *superior vena cava,* which brings blood from the upper portion of the body. Another is the *inferior vena cava,* which brings blood from the lower portions of the body. The third vein is the *coronary sinus,* which drains blood from most of the vessels supplying the walls of the heart. The right atrium then squeezes the blood into the right ventricle, which pumps it into the *pulmonary trunk.* The pulmonary trunk divides into a *right* and *left pulmonary artery,* each of which carries blood to the lungs. In the lungs, the blood releases its carbon dioxide and takes on oxygen. It returns to the heart via four *pulmonary veins* that empty into the left atrium. The blood is then squeezed into the left ventricle, which pumps the blood into the *ascending aorta.* From here aortic blood is passed into the *coronary arteries, arch of the aorta, descending thoracic aorta,* and *abdominal aorta.* These blood vessels transport the blood to all body parts except the lungs.

Figure 20-2 shows that the sizes of the four chambers vary according to function. The right atrium,

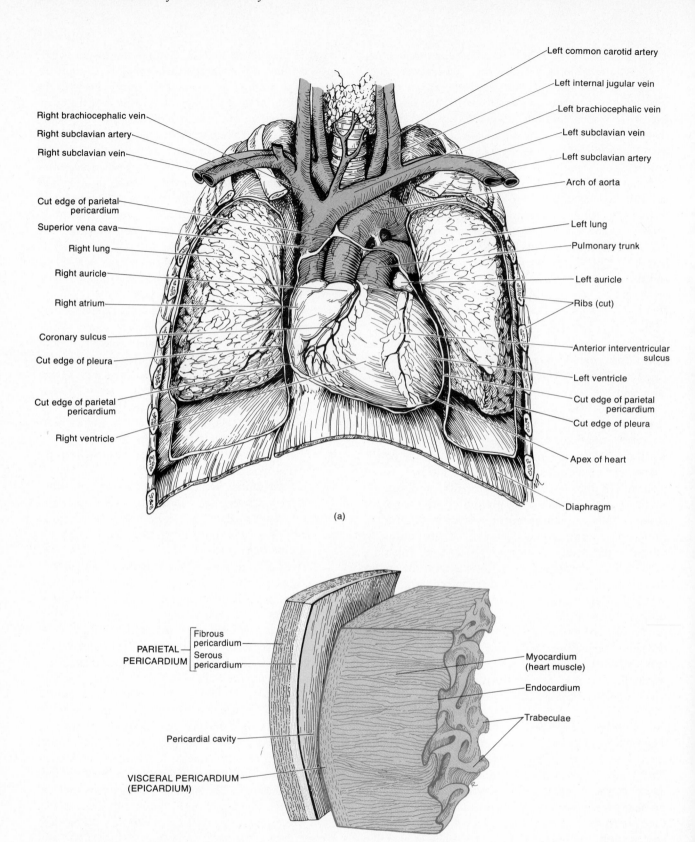

Left common carotid artery

Left internal jugular vein

Left brachiocephalic vein

Left subclavian vein

Left subclavian artery

Arch of aorta

Left lung

Pulmonary trunk

Left auricle

Ribs (cut)

Anterior interventricular sulcus

Left ventricle

Cut edge of parietal pericardium

Cut edge of pleura

Apex of heart

Diaphragm

Right brachiocephalic vein

Right subclavian artery

Right subclavian vein

Cut edge of parietal pericardium

Superior vena cava

Right lung

Right auricle

Right atrium

Coronary sulcus

Cut edge of pleura

Cut edge of parietal pericardium

Right ventricle

(a)

PARIETAL PERICARDIUM

Fibrous pericardium

Serous pericardium

Pericardial cavity

VISCERAL PERICARDIUM (EPICARDIUM)

Myocardium (heart muscle)

Endocardium

Trabeculae

(b)

Figure 20–1   Heart. (a) Position of the heart and associated blood vessels in the thoracic cavity. (b) Structure of the parietal pericardium and heart wall.

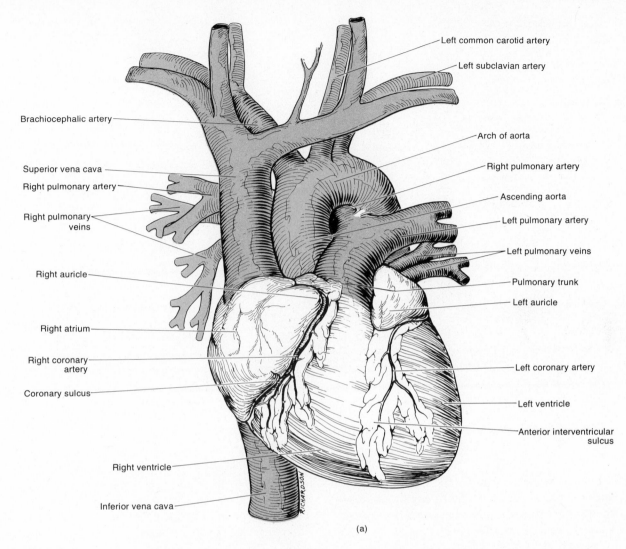

Left common carotid artery

Left subclavian artery

Brachiocephalic artery

Arch of aorta

Right pulmonary artery

Ascending aorta

Superior vena cava

Right pulmonary artery

Left pulmonary artery

Right pulmonary
veins

Left pulmonary veins

Right auricle

Pulmonary trunk

Left auricle

Right atrium

Left coronary artery

Right coronary
artery

Coronary sulcus

Left ventricle

Anterior interventricular
sulcus

Right ventricle

Inferior vena cava

(a)

Figure 20–2 Structure of the heart. (a) Anterior external view.

which must collect blood coming from almost all parts of the body, is slightly larger than the left atrium, which receives only the blood from the lungs. The thickness of the chamber walls varies too. The atria are thin-walled because they need only enough cardiac muscle tissue to squeeze the blood into the ventricles with the aid of a reduced pressure created by the expanding ventricles. The right ventricle has a much thicker layer of myocardium since it must send blood to the lungs and around back to the left atrium. The left ventricle has the thickest walls since it must pump blood through literally thousands of miles of vessels in the head, trunk, and extremities.

VALVES

As each chamber of the heart contracts, it pushes a portion of blood into a ventricle or out of the heart through an artery. But as the walls of the chambers relax, some structure must prevent the blood from flowing back into the chamber. That structure is a **valve**.

Atrioventricular (AV) valves lie between the atria and their ventricles. The atrioventricular valve between the right atrium and right ventricle is called the *tricuspid valve* because it consists of three flaps, or cusps. These flaps are fibrous tissues that grow out of the walls of the heart and are covered with endocardium. The pointed ends of the cusps project into the ventricle. Other names for this valve are the right atrioventricular or right AV valve. Cords called *chordae tendineae* connect the pointed ends to small conical projections—the *papillary muscles* (muscular columns)—located on the inner surface of the ventricles. The irregular surface of ridges and folds of the myocardium in the ventricles is known as the *trabeculae carneae*. The chordae tendineae and their muscles keep the flaps pointing in the direction

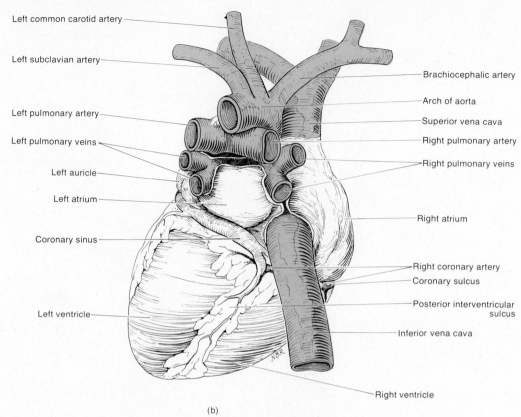

Left common carotid artery

Left subclavian artery

Left pulmonary artery

Left pulmonary veins

Left auricle

Left atrium

Coronary sinus

Left ventricle

Brachiocephalic artery

Arch of aorta

Superior vena cava

Right pulmonary artery

Right pulmonary veins

Right atrium

Right coronary artery

Coronary sulcus

Posterior interventricular sulcus

Inferior vena cava

Right ventricle

(b)

Figure 20–2 (cont.)   Structure of the heart. (b) Posterior external view.

of the blood flow. As the atrium relaxes and the ventricle squeezes the blood out of the heart, any blood driven back toward the atrium is pushed between the flaps and the ventricle walls (Figure 20-3a). This action drives the cusps upward until their edges meet and close the opening. At the same time, contraction of the papillary muscles prevents the valve from swinging upward into the atrium. The atrioventricular valve between the left atrium and left ventricle is called the *bicuspid* or *mitral valve.* It has two cusps that work in the same way as the cusps of the tricuspid valve. The bicuspid valve is also known as the left atrioventricular or left AV valve. Its cusps are also attached by way of the chordae tendineae to papillary muscles.

Each artery that leaves the heart has a valve that prevents blood from flowing back into the heart. These are the **semilunar valves**—*semilunar* meaning half-moon or crescent-shaped. The *pulmonary semilunar valve* lies in the opening where the pulmonary artery leaves the right ventricle. The *aortic semilunar valve* is situated at the opening between the left ventricle and the aorta. Both valves consist of three semilunar cusps. Each cusp is attached by its convex margin to the artery wall. The free borders of the cusps curve outward and project into the opening inside the blood vessel (Figure 20-3b). Like the atrioventricular valves, the semilunar valves permit blood to flow in only one direction—in this case, from the ventricles into the arteries.

CONDUCTION SYSTEM

The heart is innervated by the autonomic nervous system, but the autonomic neurons only increase or decrease the time it takes to complete a cardiac cycle. The chamber walls can go on contracting and relaxing, contracting and relaxing, without any direct stimulus from the nervous system. This action is possible because the heart has an intrinsic regulating system called the **conduction system.** The conduction system is composed of specialized muscle tissue that generates and distributes the electrical impulses which stimulate the cardiac muscle fibers to contract. These tissues are the sinu-atrial (sinoatrial) node, the atrioventricular node, the atrioventricular bundle, the bundle branches, and the Purkinje fibers. The cells of the conduction system develop during embryological life from certain cardiac muscle cells. These cells lose their ability to contract and become specialists in impulse transmission.

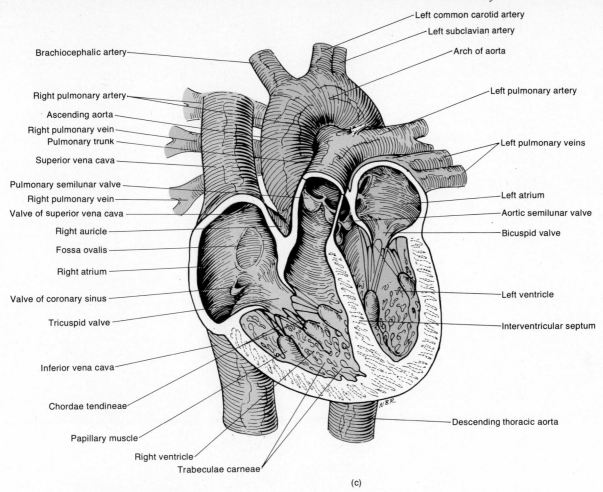

Left common carotid artery

Left subclavian artery

Arch of aorta

Brachiocephalic artery

Right pulmonary artery

Ascending aorta

Right pulmonary vein

Pulmonary trunk

Superior vena cava

Pulmonary semilunar valve

Right pulmonary vein

Valve of superior vena cava

Right auricle

Fossa ovalis

Right atrium

Valve of coronary sinus

Tricuspid valve

Inferior vena cava

Chordae tendineae

Papillary muscle

Right ventricle

Trabeculae carneae

Left pulmonary artery

Left pulmonary veins

Left atrium

Aortic semilunar valve

Bicuspid valve

Left ventricle

Interventricular septum

Descending thoracic aorta

(c)

Figure 20–2 (cont.) Structure of the heart. (c) Anterior internal view. (d) Path of blood through the heart.

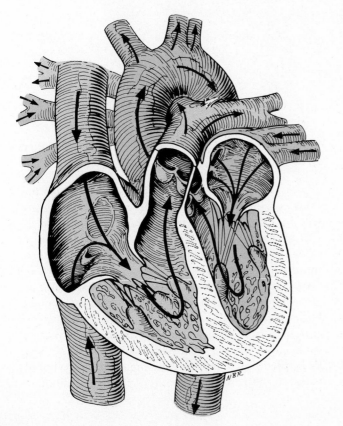

(d)

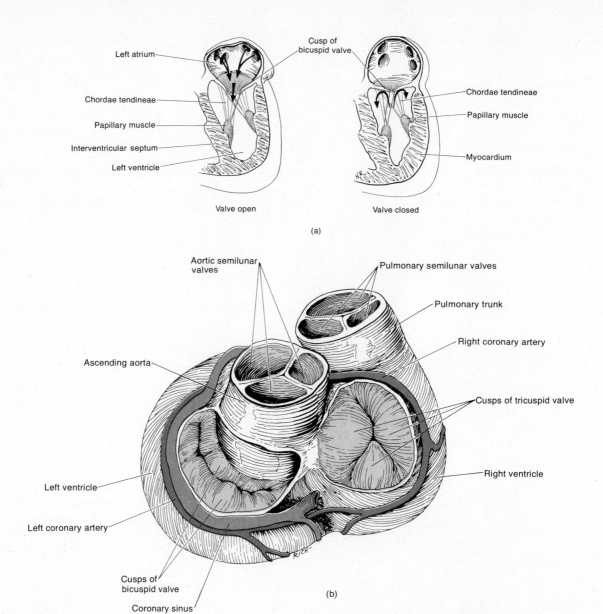

Left atrium

Cusp of bicuspid valve

Chordae tendineae

Papillary muscle

Interventricular septum

Left ventricle

Chordae tendineae

Papillary muscle

Myocardium

Valve open

Valve closed

(a)

Aortic semilunar valves

Pulmonary semilunar valves

Pulmonary trunk

Ascending aorta

Right coronary artery

Cusps of tricuspid valve

Left ventricle

Right ventricle

Left coronary artery

Cusps of bicuspid valve

Coronary sinus

(b)

Figure 20–3   Valves of the heart. (a) Structure and function of the bicuspid valve. (b) Valves of the heart viewed from above. The atria have been removed to expose the tricuspid and bicuspid valves.

A **node** of the conducting system is a compact mass of conducting cells. The **sinu-atrial (sinoatrial) node,** known as the **SA node** or **pacemaker,** is located in the right atrial wall inferior to the opening of the superior vena cava (Figure 20-4a). The SA node initiates each cardiac cycle, and thereby sets the basic pace for the heart rate—hence its common name, pacemaker. However, the rate set by the SA node may be altered by nervous impulses from the autonomic nervous system or by certain chemicals such as thyroid hormone, epinephrine, and acetylcholine. Once an electrical impulse is initiated by the SA node, the impulse spreads out over both atria, causing them to contract and at the same time

depolarizing the **atrioventricular (AV) node.** Because of its location near the inferior portion of the interatrial septum, the AV node is one of the last portions of the atria to be depolarized. From the AV node, a tract of conducting fibers called the **atrioventricular bundle** or **bundle of His** runs through the cardiac skeleton to the top of the interventricular septum. It then continues down both sides of the septum as the **right** and **left bundle branches.** The bundle of His distributes the charge over the medial surfaces of the ventricles. Actual contraction of the ventricles is stimulated by the **Purkinje fibers**—branches that emerge from the bundle branches and pass into the cells of the myocardium.

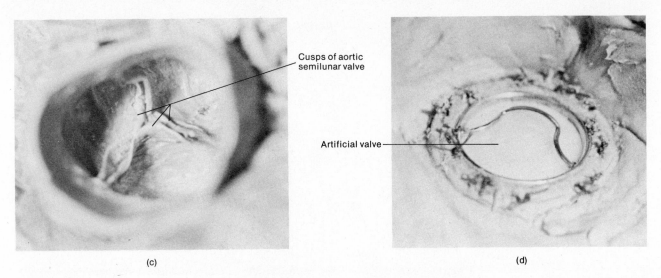

(c)   (d)

Figure 20–3 (cont.) Valves of the heart. (c) Photograph of a superior view of a semilunar valve. (d) Photograph of a superior view of an artificial valve. (Courtesy of John W. Eads, University of Arizona.)

ELECTROCARDIOGRAM

Impulse transmission through the conduction system generates electrical currents that may be detected on the body's surface. A recording of the electrical changes that accompany the cardiac cycle is called an **electrocardiogram (ECG)**. The instrument used to record the changes is an *electrocardiograph.*

Each portion of the cardiac cycle produces a different electrical impulse. These impulses are transmitted from the electrodes to a recording needle that graphs the impulses as a series of up-and-down waves called *deflection waves.* In a typical record (Figure 20-4b), three clearly recognizable waves accompany each cardiac cycle. The first wave, called the **P wave,** is a small upward wave. It indicates atrial depolarization – the spread of an impulse from the SA node through the muscle of the two atria. A fraction of a second after the P wave begins, the atria contract. Then there is a deflection wave called the **QRS wave (complex).** It begins as a downward deflection, continues as a large, upright, triangular wave, and ends as a downward wave at its base. This deflection represents atrial repolarization and ventricular depolarization – that is, the spread of the electrical impulse through the ventricles. The third recognizable deflection is a dome-shaped **T wave.** This wave indicates ventricular repolarization. There is no deflection to show atrial repolarization because the stronger QRS wave masks this event.

In reading an electrocardiogram, it is important to note the size of the deflection waves and certain time intervals. Enlargement of the P wave, for example, indicates enlargement of the atrium, as in mitral stenosis.

The *P-R interval* is measured from the beginning of the P wave to the beginning of the Q wave. It represents the conduction time from the beginning of atrial excitation to the beginning of ventricular excitation. The P-R interval is the time required for an impulse to travel through the atria and atrioventricular node to the remaining conducting tissues. The lengthening of this interval, as in arteriosclerotic heart disease and rheumatic fever, occurs because the heart tissue covered by the P-R interval, namely the atria and atrioventricular node, is scarred or inflamed. Thus the impulse must travel at a slower rate and the interval is lengthened. The normal P-R interval covers no more than 0.20 second.

An enlarged Q wave may indicate a myocardial infarction. An enlarged R wave generally indicates enlarged ventricles. The *S-T segment* begins at the end of the S wave and terminates at the beginning of the T wave. It represents the time between the end of the spread of the impulse through the ventricles and repolarization of the ventricles. The S-T segment is elevated in acute myocardial infarction and depressed when the heart muscle receives insufficient oxygen. The T wave represents ventricular repolarization. It is flat when the heart muscle is receiving insufficient oxygen, as in arteriosclerotic heart disease. It may be elevated when the body's potassium level is increased.

The ECG is invaluable in diagnosing abnormal cardiac rhythms and conduction patterns, detecting the presence of fetal life, determining the presence of several fetuses, and following the course of recovery from a heart attack.

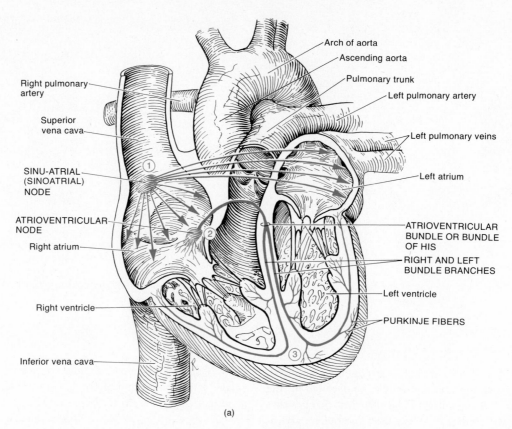

Right pulmonary artery

Superior vena cava

SINU-ATRIAL (SINOATRIAL) NODE

ATRIOVENTRICULAR NODE

Right atrium

Right ventricle

Inferior vena cava

Arch of aorta

Ascending aorta

Pulmonary trunk

Left pulmonary artery

Left pulmonary veins

Left atrium

ATRIOVENTRICULAR BUNDLE OR BUNDLE OF HIS

RIGHT AND LEFT BUNDLE BRANCHES

Left ventricle

PURKINJE FIBERS

(a)

Figure 20–4   Conduction system of the heart. (a) Location of the nodes and bundles of the conduction system.

BLOOD SUPPLY

The walls of the heart, like any other tissue, including large blood vessels, have their own blood vessels. Nutrients could not possibly diffuse through all the layers of cells that make up the heart tissue. And the blood in the left chambers of the heart would never supply enough oxygen. The flow of blood through the numerous vessels that pierce the myocardium is called the **coronary circulation** (Figure 20-5). The vessels that serve the myocardium include the *left coronary artery,* which originates as a branch of the ascending aorta. This artery runs under the left atrium and divides into the anterior interventricular and circumflex branches. The *anterior interventricular branch* follows the anterior interventricular sulcus and supplies oxygenated blood to the walls of both ventricles. The *circumflex branch* distributes oxygenated blood to the walls of the left ventricle and left atrium. The *right coronary artery* also originates as a branch of the ascending aorta. It runs under the right atrium and divides into the posterior interventricular and marginal branches. The *posterior interventricular branch* follows the posterior interventricular sulcus and supplies the walls of the two ventricles with oxygenated blood. The *marginal branch* transports oxygenated blood to the myocardium of the right ventricle and right

atrium. The left ventricle receives the most abundant blood supply because of the enormous work it must do.

As blood passes through the arterial system of the heart, it delivers oxygen and nutrients and collects carbon dioxide and wastes. Most of the deoxygenated blood, which carries the carbon dioxide and wastes, is collected by a large vein, the *coronary sinus,* which empties into the right atrium. A vascular sinus is a vein with a thin wall that has no smooth muscle to alter its diameter. The principal tributaries of the coronary sinus are the *great cardiac vein,* which drains the anterior aspect of the heart, and the *middle cardiac vein,* which drains the posterior aspect of the heart.

Most heart problems result from faulty coronary circulation. If a reduced oxygen supply weakens the cells but does not actually kill them, the condition is called **ischemia.** *Angina pectoris* ("chest pain") is ischemia of the myocardium. (Remember that pain impulses originating from most visceral muscles are referred to an area on the surface of the body.) Angina pectoris occurs when coronary circulation is somewhat reduced for some reason. Stress, which produces constriction of vessel walls, is a common cause. Equally common is strenuous exercise and a heavy meal. When any quantity of food enters the stomach, the body increases blood flow to the

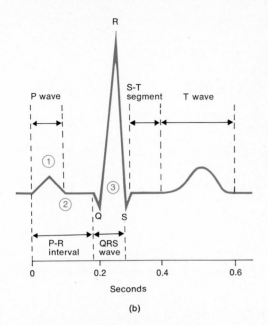

Figure 20-4 (cont.) Conduction system of the heart. (b) Recordings of a normal electrocardiogram.

digestive tract. As a consequence, some blood is diverted away from other organs, including the heart. Exercise, however, increases heart muscle activity and thus the heart's need for oxygen. Doing heavy work while food is in the stomach can therefore lead to oxygen deficiency in the myocardium. Angina pectoris weakens the heart muscle, but it does not produce a full-scale heart attack.

A much more serious problem is *myocardial infarction,* commonly called a "coronary" or "heart attack." **Infarction** means death of an area of tissue because of an interrupted blood supply. Myocardial infarction may result from a thrombus or embolus in one of the coronary arteries. The tissue distal to the obstruction dies, and the heart muscle loses at least some of its strength. The aftereffects depend partly on the size and location of the infarcted, or dead, area that is replaced by scar tissue.

CARDIAC CYCLE

In a normal heartbeat, the two atria contract simultaneously while the two ventricles relax. Then, when the two ventricles contract, the two atria relax. The term **systole** refers to the phase of contraction; **diastole** is the phase of relaxation. A **cardiac cycle,** or complete heartbeat, consists of the systole and diastole of both atria plus the systole and diastole of both ventricles.

Suppose we begin the cardiac cycle near the end of *atrial diastole* (Figure 20-6a, b). At this time, both atrioventricular valves are open and the semilunar valves are closed. Throughout most of atrial diastole, the ventricles receive blood from the atria. The atria are constantly filling with blood. The right atrium receives

deoxygenated blood from the superior and inferior venae cavae and coronary sinus. The left atrium receives oxygenated blood from the pulmonary veins. It is estimated that 75 percent of ventricular filling occurs during atrial diastole.

When the SA node fires, bringing about the end of atrial diastole, the atria depolarize and contract. This action produces the P wave on the ECG and signals the start of *atrial systole* (Figure 20-6c). Contraction of the atria forces the remaining blood into the ventricles. Deoxygenated blood from the right atrium passes into the right ventricle through the open tricuspid valve. Oxygenated blood passes from the left atrium into the left ventricle through the open mitral valve. While the atria are contracting, the ventricles are in diastole. During *ventricular diastole,* the ventricles are filling with blood and the semilunar valves in the aorta and pulmonary trunk are closed.

When atrial systole and ventricular diastole are completed, the events are reversed – the atria go into diastole and the ventricles go into systole. During *atrial diastole,* deoxygenated blood from the various parts of the body enters the right atrium. Simultaneously, oxygenated blood from the lungs enters the left atrium. During the first part of atrial diastole, the atrioventricular valves are closed since the ventricles are in systole. In *ventricular systole,* the ventricles contract and force blood into their respective vessels. Ventricular systole is initiated by the spread of the QRS wave through the ventricles. The right ventricle pumps deoxygenated blood to the lungs through the semilunar valve of the pulmonary trunk. The left ventricle pumps oxygenated blood through the open semilunar valve of the aorta. At the end of the ventricular systole, the semilunar valves close and both the atria and the ventricles relax. In a complete cardiac cycle the atria are in systole 0.1 second and in diastole 0.7 second. By contrast, the ventricles are in systole 0.3 second and in diastole 0.5 second.

Two phenomena control the movement of blood through the heart: the opening and closing of the valves and the contraction and relaxation of the myocardium. Both these activities occur without direct stimulation from the nervous system. The valves are controlled by pressure changes in each heart chamber. The contraction of the cardiac muscle is stimulated by its conduction system.

Pressure Changes

The pressure developed in a heart chamber is related primarily to the chamber's size and the volume of blood it contains. The greater the volume, the higher the pressure. As we discuss the pressure changes during the cardiac cycle, refer to Figure 20-6b. Here pressure

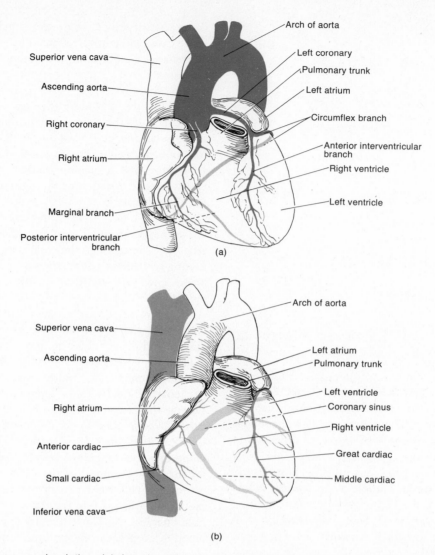

Superior vena cava

Ascending aorta

Right coronary

Right atrium

Marginal branch

Posterior interventricular branch

Arch of aorta

Left coronary

Pulmonary trunk

Left atrium

Circumflex branch

Anterior interventricular branch

Right ventricle

Left ventricle

(a)

Superior vena cava

Ascending aorta

Right atrium

Anterior cardiac

Small cardiac

Inferior vena cava

Arch of aorta

Left atrium

Pulmonary trunk

Left ventricle

Coronary sinus

Right ventricle

Great cardiac

Middle cardiac

(b)

Figure 20–5   Coronary circulation. (a) Anterior view of arterial distribution. (b) Anterior view of venous drainage.

changes associated with the left side of the heart are indicated. Although the pressures in the right side of the heart are somewhat lower, the same results are achieved. Blood flows from an area of higher pressure to an area of lower pressure.

The pressure in the atria is called *intraatrial pressure.* When the atria are in diastole, the pressure in them steadily increases because blood flows into them continuously from their vessels. As the ventricles go into diastole, ventricular pressure drops below atrial pressure and the atrioventricular valves open. Atrial blood drains into the ventricles, which are empty and consequently have a lower pressure. When atrial pressure builds up, the atria contract and send the remaining blood rushing into the ventricles.

As the ventricles fill with blood, their pressure, called *intraventricular pressure,* becomes greater than intraatrial pressure. At this point, a backflow of blood closes the atrioventricular valves. The ventricles start to

contract, but the semilunar valves remain closed from the previous cardiac cycle. No blood moves out into the aorta or pulmonary artery. The ventricles are closed.

Intraventricular pressure continues to rise until it is greater than the *intraarterial* pressure, the pressure in the aorta and pulmonary artery. When intraventricular pressure rises above intraarterial pressure, the semilunar valves are forced open and blood is ejected from the ventricles into the great vessels.

Once the ventricles eject their blood, intraventricular pressure falls below that in the great vessels. As a result, the semilunar valves are pushed closed by the backflow of blood created by the recoil of the elastic walls of the arteries. Once again, the ventricles are closed chambers. As the ventricles relax, intraventricular pressure decreases until it becomes less than intraatrial pressure. At that point, intraatrial pressure forces the tricuspid and mitral valves open, blood fills the ventricles, and another cycle begins.

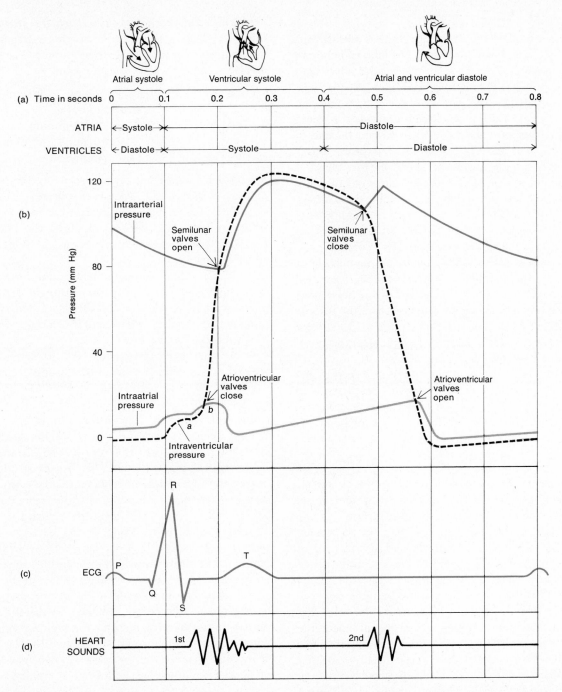

Figure 20–6 Cardiac cycle. (a) Systole and diastole of the atria and ventricles related to time. (b) Intraatrial, intraventricular, and intraarterial pressure changes along with the opening and closure of valves during the cardiac cycle. (c) Deflections in the ECG related to the cardiac cycle. (d) Heart sounds related to the cardiac cycle.

Timing

If we assume that the average heart beats 72 times per minute, then each cardiac cycle requires about 0.8 second (Figure 20-6a). During the first 0.1 second, the atria contract and the ventricles relax. The atrioventricular valves are open, and the semilunar valves are closed. For the next 0.3 second, the atria are relaxing and the ventricles are contracting. During the first part of this

period, all valves are closed. During the second part, the semilunars are open. The last 0.4 second of the cycle is the relaxation, or quiescent, period. All chambers are in diastole. And for the first part of the quiescent period, all valves are closed. During the latter part, the atrioventricular valves open and blood starts draining into the ventricles. When the heart beats faster than normal, the quiescent period is shortened accordingly.

Sounds

The first sound, which can be described as a **lubb** ($\overline{oo}$) sound, is a long, booming sound. The lubb is the sound created by the closure of the atrioventricular valves soon after ventricular systole begins (Figure 20-6d). The second sound, which is heard as a short, sharp sound, can be described as a **dupp** ($\breve{u}$) sound. Dupp is the sound created as the semilunar valves close toward the end of ventricular systole. A pause about two times longer comes between the second sound and the first sound of the next cycle. Thus the cardiac cycle can be heard as a lubb, dupp, pause; lubb, dupp, pause; lubb, dupp, pause. This is the sound of the heartbeat. But it comes primarily from the turbulence created by the closure of the valves and not from the contraction of the heart muscle.

Heart sounds provide valuable information about the valves. If the sounds are peculiar, they are called **murmurs.** Some murmurs are caused by the noise made by a little blood bubbling back up into an atrium because of improper closure of an atrioventricular valve. Murmurs do not always indicate a valve problem, however, and many have no clinical significance.

Although heart sounds are produced in part by the closure of valves, they are not necessarily heard best over these valves. Each sound tends to be clearest in a slightly different location closest to the surface of the body.

Surface Anatomy

The valves of the heart may be identified by surface projection (Figure 20-7). The pulmonary and aortic semilunar valves are represented on the surface by a line about 2.5 cm (1 inch) in length. The pulmonary semilunar valve lies horizontally behind the inner end of the left third costal cartilage and the adjoining part of the sternum. The aortic semilunar valve is placed obliquely behind the left side of the sternum, opposite the third intercostal space. The tricuspid valve lies behind the sternum, extending from the midline at the level of the fourth costal cartilage down toward the right sixth chondrosternal junction. The bicuspid lies behind the left side of the sternum obliquely at the level of the fourth costal cartilage. It is represented by a line about 3 cm in length.

CARDIAC OUTPUT

The amount of blood ejected from the left ventricle into the aorta per minute is called the **cardiac output,** or minute volume. **Cardiac output** is determined by two factors: (1) the amount of blood pumped by the left ventricle during each beat and (2) the number of heartbeats per minute. The amount of blood ejected by a ventricle during each systole is called the **stroke volume.** In a resting adult, stroke volume averages 70 ml and heart rate is about 72 beats per minute. The average cardiac volume, then, in a resting adult is

$$\begin{aligned}
\text{Cardiac output} &= \text{stroke volume} \times \text{ventricular} \\
&\quad \text{systole/minute} \\
&= 70 \text{ ml} \times 72/\text{minute} \\
&= 5{,}040 \text{ ml/minute or about} \\
&\quad 5 \text{ liter/minute}
\end{aligned}$$

Factors that increase heart rate or stroke volume tend to increase cardiac output. Factors that decrease heart rate or stroke volume tend to decrease cardiac output. If stroke volume falls dangerously low, the body can compensate to some extent by increasing the heartbeat and vice versa.

Stroke volume is determined by the force of the ventricular contraction. The more strongly the cardiac fibers contract, the more blood they eject. The strength of contraction is directly related to the amount of venous blood returned to the heart.

Within limits, contraction is more forceful when skeletal muscle fibers are stretched. During exercise, for example, a large amount of blood enters the heart and the increased diastolic filling stretches the fibers of the right ventricle. This increased length of the cardiac muscle fibers intensifies the force of the ventricular contraction—that is, the force of the beat. The increased incoming volume of blood is handled by an increased output through a more forceful ventricular contraction. As the increased amount of blood returns from the lungs to the left side of the heart, left ventricular stroke volume also increases. Thus, during exercise, cardiac output is increased. This phenomenon, by which the length of the cardiac muscle fiber determines the force of contraction, is referred to as **Starling's law of the heart.**

Normally, the amount of blood returning to the heart regulates the stroke volume and thereby affects cardiac output. During certain pathological conditions, however, stroke volume may fall dangerously low. If the ventricular myocardium is weak or damaged by an infarction, it cannot contract strongly. Or blood volume may be reduced by excessive bleeding. Stroke volume then falls because the cardiac fibers are not sufficiently stretched. In these cases, the body attempts to maintain a safe cardiac output by increasing the rate of contraction. The heart rate is regulated by the autonomic and endocrine systems.

REGULATION OF HEART RATE

Left to its own devices, the pacemaker would set a heart rate that never varied. Consequently, a number of

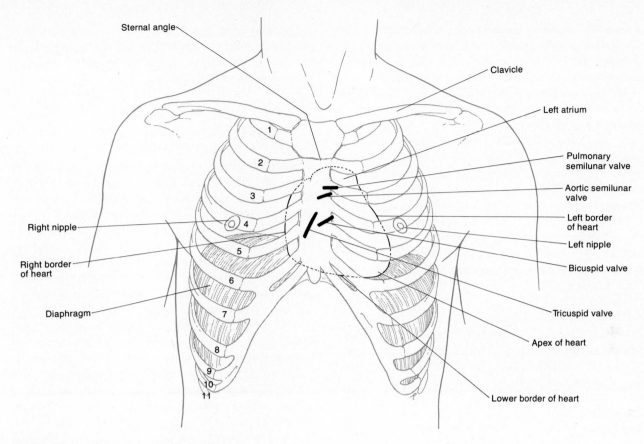

Figure 20–7 *Surface projection of the heart.*

reflexes exist to quicken heartbeat when tissues need more oxygen and to slow down the heart during periods of inactivity. The reflex arcs start in receptors located in blood vessels, pass to cardiac centers in the brain, and travel along autonomic nerves to the heart.

Autonomic Control

The pacemaker receives nerves from the parasympathetic and sympathetic divisions of the autonomic nervous system. Parasympathetic neurons in the vagus nerve travel to the heart. When stimulated by the **cardioinhibitory center** of the medulla, the parasympathetic fibers stimulate the pacemaker and the heart beat is decreased. The sympathetic pathway originates in the **cardioacceleratory center** of the medulla. It travels in a tract down the spinal cord and then passes over sympathetic nerves to the heart. Sympathetic stimulation counteracts parasympathetic stimulation and quickens the heartbeat. When neither cardiac center is stimulated by sensory neurons, the cardioacceleratory center tends to dominate. Sympathetic fibers then have a free rein to speed up heart rate until receptors intervene to stimulate the cardioinhibitory center.

PRESSORECEPTORS

Nerve cells capable of responding to changes in pressure are called **pressoreceptors**. Three of these pressoreceptors affect the rate of heartbeat: the carotid sinus reflex, the aortic reflex, and the right heart (atrial) reflex.

The *carotid sinus reflex* is concerned with maintaining normal blood pressure in the brain. The *carotid sinus* is a small widening of the internal carotid artery just above the point where it branches off from the common carotid artery. In the walls of the carotid sinus lie pressoreceptors. Any increase in blood pressure stretches the walls of the sinus, and the stretching stimulates the pressoreceptors. The impulses then travel from the pressoreceptors over afferent neurons in the glossopharyngeal (IX) nerves to the cardiac centers in the medulla. There the impulses stimulate the cardioinhibitory center and inhibit the cardioacceleratory center. Consequently, more impulses pass from the cardioinhibitory center via the vagus (X) nerves to the heart and fewer impulses pass from the cardioacceleratory center via sympathetic efferents to the heart. The result is a slower heartbeat. There is a subsequent decrease in cardiac output, a decrease in arterial blood volume, and restoration of blood pressure to normal. But if blood pressure falls, reflex acceleration of the heart takes place. The pressoreceptors in the carotid sinus do not stimulate the cardioinhibitory center, and the cardiac acceleratory center is free to dominate. The heart then beats faster to restore normal blood pressure. This inverse relationship between blood pressure and heart rate is referred to as **Marey's law of the heart.**

The *aortic reflex* is concerned with general systemic blood pressure. It is initiated by pressoreceptors in the walls of the aortic arch, and it operates like the carotid sinus reflex. The *right heart (atrial) reflex* responds to venous blood pressure. It is initiated by pressoreceptors in the superior and inferior venae cavae and in the right atrium. When venous pressure increases, the pressoreceptors send impulses that stimulate the cardioacceleratory center. This action causes the heart rate to increase. This mechanism is also called the **Bainbridge reflex.**

CHEMORECEPTORS

Structures that are sensitive to chemicals are called **chemoreceptors.** Certain chemoreceptors in the blood vessels are sensitive to oxygen. When blood oxygen is low, they become stimulated. Impulses are conveyed to the cardiac center, and the heart rate is accelerated. This action increases cardiac output, and more blood is pumped to the lungs and the cells of the body. Other chemoreceptors are sensitive to carbon dioxide. Any increase in the CO_2 concentration in the blood brings about inhibition of the cardioinhibitory center. The heart rate consequently increases.

Other Influences

Strong emotions such as fear, anger, and anxiety, along with a multitude of physiological stressors, increase heart rate through the general stress response. Mental states such as depression and grief tend to stimulate the cardioinhibitory center and decrease heart rate.

Certain chemicals produced by the body may act directly on the heart. Those that increase heart rate are also involved in the general stress response. Thyroxin, a hormone produced by the thyroid gland, increases overall body metabolism, including that of the heart cells. Epinephrine produced by the adrenal medulla imitates sympathetic effects and increases the force and frequency of the heartbeat. Norepinephrine, the chemical transmitter of postsynaptic sympathetic fibers, also stimulates the pacemaker to accelerate. Norepinephrine is another adrenal medullary hormone. By contrast, acetylcholine, the chemical transmitter of parasympathetic neurons, decreases heart rate.

Sex is another factor – the heartbeat is somewhat faster in females. Age is yet another factor – the heartbeat is fastest at birth, moderately fast in youth, average in adulthood, and below average in old age. Muscular exercise is one more factor – heartbeat increases in proportion to the work done.

CIRCULATORY SHOCK AND HOMEOSTASIS

When cardiac output or blood volume is reduced to the point where body tissues do not receive an adequate blood supply, **circulatory shock** results. Circulatory shock is caused by loss of blood volume through hemorrhage or through the release of histamine due to damage to body tissues (trauma). The characteristic symptoms of circulatory shock are a pale, clammy skin; cyanosis of ears and fingers; a feeble though rapid pulse; shallow and rapid breathing; lowered body temperature; and mental confusion or unconsciousness.

If the shock is mild, certain homeostatic mechanisms of the circulatory system compensate so that no serious damage results. Lowered blood pressure is compensated by constriction of blood vessels and water retention. Renin is secreted by the kidneys, aldosterone by the adrenal cortex, epinephrine by the adrenal medulla, and ADH by the posterior pituitary. Even though some blood is lost from circulation, blood return to the heart is normal and cardiac output remains essentially unchanged. Veins and many arterioles are constricted during compensation, but there is no constriction of arterioles supplying the heart and brain. As a consequence, blood flow to the heart and brain is normal or nearly so. Compensation is an effective homeostatic mechanism until about 900 ml of blood is lost.

If the shock is severe, death may occur. For instance, if the return of venous blood is greatly diminished by excessive blood loss, the compensatory mechanisms are insufficient. When the cardiac output decreases, the heart fails to pump enough blood to supply its own coronary vessels and the heart muscle weakens. In addition, prolonged vasoconstriction ultimately leads to tissue hypoxia, and vital organs such as the kidneys and liver are damaged. Essentially, the initial shock promotes more shock and a **circulatory shock cycle** is established. Once the shock reaches a certain level of severity, damage to the circulatory organs is so extensive that death ensues.

SURGERY

A number of surgical techniques now allow surgeons to perform new diagnostic procedures, correct heart defects, and make heart transplants.

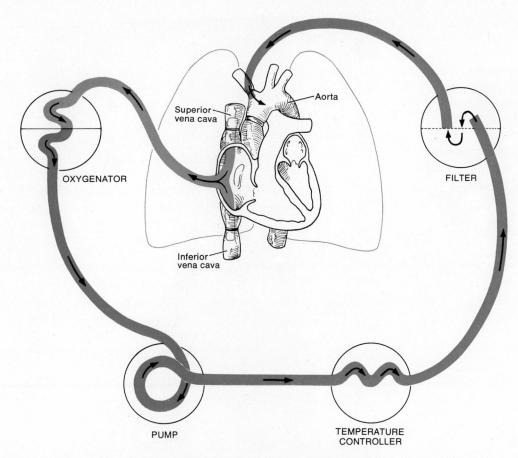

Figure 20–8 Principle of the heart-lung bypass. Blood drawn from the venae cavae is oxygenated, rewarmed to body temperature or cooled, filtered to remove air and emboli, and returned to the aorta and coronary arteries at the proper pressure. If necessary, drugs, anesthetics, and transfusions may be added to the circuit.

Open-Heart Surgery

Before surgeons can correct even the simplest heart defect, they have to be able to open up the heart and expose the chambers. But before open-heart surgery could be done, techniques had to be developed to capture the blood spurting out of the open chamber and pump it back into the vessels. One such life-support technique is extracorporeal circulation. Coupled with hypothermia, it has now made possible both heart surgery and heart transplants.

HYPOTHERMIA

Hypothermia, or body cooling, slows metabolism and reduces the oxygen needs of the tissues. Thus the heart and brain can withstand short periods of interrupted or reduced blood flow. Lost blood is then replaced by transfusion during and after the operation.

EXTRACORPOREAL CIRCULATION

To sustain the patient for long periods, surgeons turned to *extracorporeal circulation.* In this technique, blood bypasses the heart and lungs completely (Figure 20-8). It is pumped and oxygenated by a heart-lung machine outside the body. Modern heart-lung machines may also chill the blood to produce hypothermia as well.

ARTIFICIAL PARTS

The development of artificial blood vessels, heart valves, patches, and plugs made from synthetic materials has made it possible to repair many defects of the circulatory system. Replacing or bypassing diseased and damaged arteries with synthetic textile tubes has saved thousands of lives and limbs. The major development in treating coronary artery disease is the replacement of partially obstructed coronary vessels with synthetic tubes or sections of the saphenous veins. This procedure has saved many people from an invalid life. Severely damaged aortic, mitral, or tricuspid valves have been replaced by artificial ones. (See Figure 20-3d).

When major elements of the conduction system are disrupted, heart block of varying degrees may result. Complete heart block results most commonly from a

(a)

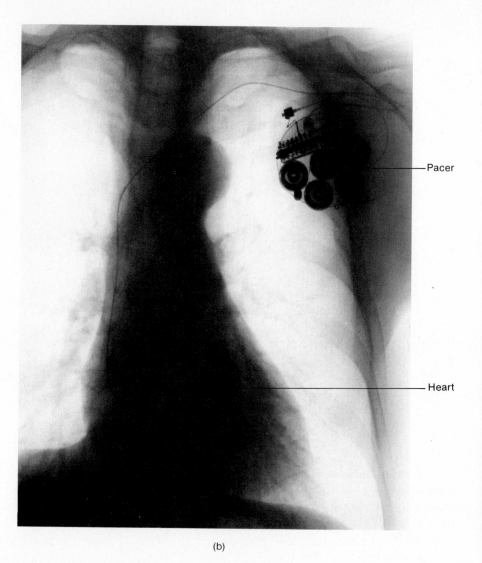

—Pacer

—Heart

(b)

Figure 20–9   Cardiac pacing system. (a) Cordis Omni-Stanicor pacer prior to implantation. (b) Anteroposterior projection of a Cordis Omni-Ventricor pacer. (Courtesy of Cordis Corporation, Miami, Florida.)

disturbance in AV conduction. The ventricles fail to receive atrial impulses, causing the ventricles and atria to beat independently of each other. In patients with heart block, normal heart rate can be restored and maintained with an *artificial pacemaker* (Figure 20-9). A pacemaker is a device that sends out small electrical charges that stimulate the heart. It consists of three basic parts: a *pulse generator,* which contains the battery cells; a *lead,* which is a flexible wire connected to the pulse generator; and an *electrode,* which delivers the charge to the heart.

Pulse generators are of two types. Demand (ventricular-inhibited) generators sense when the heart is sending out its own electrical charges and therefore will not send an electrical charge to the heart. But when the heart fails to beat, the demand generator senses the lack of activity and immediately sends an electrical charge to the heart. Demand generators are used for patients who still have some natural heart activity. The other pulse generator is called a fixed-rate (asynchronous) generator. This unit sends a series of electrical charges to the heart at a steady rate – usually 80 beats per minute. Unlike demand generators, fixed-rate generators supply all the stimulation since the natural activity of the heart is too slow.

Pacemaker leads are referred to as endocardial and myocardial. Endocardial leads are passed through a vein so the tip, called the electrode, makes contact with the endocardium. Myocardial leads are designed so that the electrode attaches directly to the outside of the heart.

DISORDERS

RISK FACTORS

It is estimated that one in every five persons who reaches 60 will have a **heart attack.** One in every four persons between 30 and 60 has the potential to be stricken. Heart disease is epidemic in this country, despite the fact that some of the causes can be foreseen and prevented.

The Framingham, Massachusetts, Heart Study, which began in 1950, is the longest and most famous study ever made of the susceptibility of a community to heart disease. Approximately 13,000 people in the town have participated by receiving examinations every 2 years since the study began. The results of this research indicate that people who develop combinations of certain risk factors eventually have heart attacks. These factors are high blood cholesterol level, high blood pressure, cigarette smoking, obesity, lack of exercise, and diabetes mellitus.

The first five risk factors all contribute to increasing the heart's workload. High blood cholesterol and hypertension will be discussed in the next chapter. Cigarette smoking, through the effects of nicotine, stimulates the adrenal gland to oversecrete aldosterone, epinephrine, and norepinephrine— powerful vasocontrictors. Overweight people develop miles of extra capillaries to nourish fat tissue. The heart has to work harder to pump the blood through more vessels. Without exercise, venous return gets less help from contracting skeletal muscles. In addition, regular exercise strengthens the smooth muscle of blood vessels and enables them to assist general circulation. Exercise also increases cardiac efficiency and output. In diabetes mellitus, fat metabolism dominates glucose metabolism. As a result, cholesterol levels get progressively higher and result in plaque formation, a situation that may lead to high blood pressure. The opposite is also true. High blood pressure drives fat into the vessel wall, encouraging atherosclerosis.

Generally, the immediate cause of heart trouble is one of the following: inadequate blood supply, anatomical disorders, or faulty conduction.

INADEQUATE BLOOD SUPPLY

Angina pectoris and myocardial infarction result from insufficient oxygen supply to the myocardium. Coronary artery disease kills about one in twelve of all Americans who die between the ages of 25 and 34. It claims almost one in four of all those who die between 35 and 44. It has been reported that 50 to 65 percent of all sudden deaths are due to coronary heart disease.

At least half the deaths from myocardial infarction occur before the patient reaches the hospital. These early deaths could result from an irregular heart rhythm—an *arrhythmia.* Sometimes this condition progresses to the stage called *cardiac*

arrest or ventricular fibrillation, in which the heart stops functioning. An arrhythmia is caused by disturbances in the conduction system. This abnormal rhythm of the heartbeat can result in cardiac arrest if the heart cannot supply the body's oxygen demands. Serious arrhythmias can be controlled, and the normal heart rhythm can be reestablished, if they are detected and treated early enough. Coronary care units have reduced hospital mortality rates from acute myocardial infarctions by about 30 to 20 percent or less by preventing or controlling serious arrhythmias.

ANATOMICAL DISORDERS

Less than 1 percent of all new babies have a **congenital,** or **inborn, heart defect.** Even so, the total number in this country each year is estimated to be 30,000 to 40,000. Some of these infants may live quite healthy and long lives without any need for repairing their hearts. But sometimes an inborn heart defect is so severe that an infant lives only a few hours. A common anatomical defect is patent ductus arteriosus. The connection between the aorta and the pulmonary artery remains open instead of closing completely after birth. This defect results in aortic blood flowing into the lower-pressure pulmonary trunk, thus increasing the pulmonary trunk blood pressure and overworking both ventricles and the heart.

A **septal defect** is an opening in the septum that separates the interior of the heart into a left and right side. *Atrial septal defect* is failure of the fetal foramen ovale to be closed off after birth, separating the two atria from one another. Because pressure in the right atrium is low, atrial septal defect generally allows a good deal of blood to flow from the left atrium to the right. This defect overloads the pulmonary circulation, produces fatigue and increases respiratory infections. If it occurs early in life, it inhibits growth because the systemic circulation may be deprived of a considerable portion of the blood destined for the organ and tissues of the body. *Ventricular septal defect* is caused by an abnormal development of the interventricular septum. Deoxygenated blood gets mixed with the oxygenated blood pumped into the systemic circulation. Consequently, the victim suffers *cyanosis,* a blue or dark purple discoloration of the skin. Cyanosis occurs whenever deoxygenated blood reaches the cells because of heart defect, lung defect, or suffocation. Septal openings can now be sewn shut or covered with synthetic patches.

Valvular stenosis is a narrowing, or *stenosis,* of one of the valves regulating blood flow in the heart. Narrowing may occur in the valve itself—most commonly in the mitral valve from rheumatic heart disease or in the aortic valve from sclerosis or rheumatic fever. Or it may occur near a valve. All stenoses are serious because they all place a severe workload on the heart by making it work harder to push the blood through the abnormally narrow valve openings. As a result of mitral stenosis, blood pressure is increased. Angina pectoris and heart failure may accompany this disorder. Most stenosed valves are totally replaced with artificial valves.

Tetralogy of Fallot is a combination of four defects causing a "blue baby" – a ventricular septal opening, an aorta that emerges from both ventricles instead of from the left ventricle, a stenosed pulmonary semilunar valve, and an enlarged right ventricle. Because of the ventricular septal defect, both oxygenated and unoxygenated blood are mixed in the ventricles. However, the body tissues are much more starved for oxygen than are those of a child with simple ventricular septal defect. Because the aorta also emerges from the right ventricle and the pulmonary artery is stenosed, very little blood ever gets to the lungs and pulmonary circulation is bypassed almost completely. Today it is possible to correct cases of tetralogy of Fallot when the patient is of proper age and condition. Open-heart operations are performed in which the narrowed pulmonary valve is cut open and the septal defect is sealed with a Dacron patch.

FAULTY CONDUCTION

Arrhythmia arises when electrical impulses through the heart are blocked at critical points in the conduction system. One such arrhythmia is called a **heart block.** Perhaps the most common blockage is in the atrioventricular node, which conducts impulses from the atria to the ventricles. This disturbance is called *atrioventricular (AV) block.* It usually indicates a myocardial infarction, arteriosclerosis, rheumatic heart disease, diphtheria, or syphilis. In a first-degree AV block, which can be detected only with an electrocardiograph, the transmission of impulses from the atria to the ventricles is delayed. Here the P-R interval is greater than it should be. In a second-degree AV block, impulses fail to reach the ventricles so the ventricular rate is about half that of the atrial rate. When ventricular contraction does not occur (dropped beat), oxygenated blood is not pumped efficiently to all parts of the body. The patient may feel faint or may collapse if there are many dropped ventricular beats. In a third-degree or complete AV block, practically no impulses reach the ventricles. Some never reach it at all. Atrial and ventricular rates get out of synchronization. The ventricles may go into systole at any time. This condition could occur when the atria are in systole or just before. Or the ventricles may rest for a few cardiac cycles.

With complete AV block, patients may have vertigo, unconsciousness, or convulsions. These symptoms result from a decreased cardiac output with diminished cerebral blood flow and cerebral hypoxia or lack of sufficient oxygen. Among the causes of AV block are excessive stimulation by the vagus nerves that depresses conductivity of the junctional fibers, destruction of the AV bundle as a result of coronary infarct, arteriosclerosis, myocarditis, or depression caused by various drugs. Other heart blocks include *intraatrial (IA) block, interventricular (IV) block,* and *bundle branch block (BBB).* In the latter condition, the ventricles do not contract together because of the delayed impulse in the blocked branch.

FLUTTER AND FIBRILLATION

Two rhythms that indicate heart trouble are atrial flutter and fibrillation. In **atrial flutter** the atrial rhythm averages between 240 and 360 beats per minute. The condition is essentially rapid atrial contractions accompanied by a second-degree AV block. It generally indicates severe damage to heart muscle. Atrial flutter usually becomes fibrillation after a few days or weeks. **Atrial fibrillation** is asynchronous contraction of the atrial muscles that causes the atria to contract irregularly and still faster. Atrial flutter and fibrillation occur in myocardial infarction, acute and chronic rheumatic heart disease, and hyperthyroidism. Atrial fibrillation results in complete uncoordination of atrial contraction so that atrial pumping ceases altogether. When the muscle fibrillates, the muscle fibers of the atrium quiver individually instead of contracting together. The quivering cancels out the pumping of the atrium. In a strong heart, atrial fibrillation reduces the pumping effectiveness of the heart by 25 to 30 percent.

Ventricular fibrillation is another rhythm that indicates heart trouble. It is characterized by asynchronous, haphazard ventricular muscle contractions. The rate may be rapid or slow. The impulse travels to the different parts of the ventricles at different rates. Thus part of the ventricle may be contracting while other parts are still unstimulated. Ventricular contraction becomes ineffective and circulatory failure and death occur immediately unless the arrhythmia is reversed quickly. Ventricular fibrillation may be caused by coronary occlusion. It sometimes occurs during surgical procedures on the heart or pericardium. It may be the cause of death in electrocution.

PARIETAL PERICARDIUM (PERICARDIAL SAC)

1 The pericardium, consisting of an outer fibrous layer and an inner serous layer, encloses the heart.

2 Between the outer fibrous layer and the inner serous layer is a space called the pericardial cavity.

3 The pericardial cavity contains fluid that prevents friction between the membranes.

WALL AND CHAMBERS

1 The wall of the heart has three layers: epicardium, myocardium, and endocardium.

2 The chambers include two upper atria and two lower ventricles.

3 The blood flows through the heart from the superior and inferior venae cavae and the coronary sinus to the right atrium. It then goes through the tricuspid valve to the right ventricle and through the pulmonary trunk to the lungs. It returns to the heart through the pulmonary veins into the left atrium, through the bicuspid valve to the left ventricle, and out through the aorta.

VALVES

1 Atrioventricular valves lie between the atria and their ventricles.

2 The atrioventricular valves are the tricuspid valve on the right side of the heart and the bicuspid valve on the left.

3 The chordae tendineae and their muscles keep the flaps of the valves pointing in the direction of blood flow.

4 The two arteries that leave the heart both have a semilunar valve.

5 All heart valves prevent the backflow of blood.

CONDUCTION SYSTEM

1 The conduction system consists of tissue specialized for impulse conduction.

2 Components of this system are the sinoatrial node (pacemaker), atrioventricular node, atrioventricular bundle, and Purkinje fibers.

ELECTROCARDIOGRAM

1 The record of electrical changes during each cardiac cycle is referred to as an electrocardiogram (ECG).

2 A normal ECG consists of a P wave (spread of impulse from SA node over atria), QRS wave (spread of impulse through ventricles), and T wave (ventricular repolarization). The P-R interval represents the conduction time from the beginning of atrial excitation to the beginning of ventricular excitation. The S-T segment represents the time between the end of the spread of the impulse through the ventricles and repolarization of the ventricles.

3 The ECG is invaluable in diagnosing abnormal cardiac rhythms and conduction patterns, detecting the presence of fetal life, determining the presence of several fetuses, and following the course of recovery from a heart attack.

BLOOD SUPPLY

1 The coronary circulation takes oxygenated blood through the arterial system of the myocardium.

2 Deoxygenated blood returns to the right atrium via the coronary sinus.

3 Complications of this system are angina pectoris and myocardial infarction.

CARDIAC CYCLE

1 This cycle consists of the systole (contraction) and diastole (relaxation) of both atria plus the systole and diastole of both ventricles followed by a short pause.

2 Two phenomena control the movement of blood through the heart. These are the opening and closing of the valves and the contraction and relaxation of the myocardium.

3 Blood flows through the heart from an area of higher pressure to an area of lower pressure.

4 With an average heartbeat of 72/minute, a complete cardiac cycle requires 0.8 second.

5 The first sound (lubb) represents the closing of the atrioventricular valves. The second sound (dupp) represents the closing of semilunar valves.

CARDIAC OUTPUT

1 Cardiac output is the amount of blood ejected by the left ventricle into the aorta per minute. It is calculated as follows: cardiac output = stroke volume × number of beats/minute.

2 Stroke volume is the amount of blood ejected by a ventricle per beat. According to Starling's law of the heart, the stretch of cardiac muscle fibers determines the force of contraction.

REGULATION OF HEART RATE

1 The pacemaker may be accelerated by sympathetic stimulation and slowed down by parasympathetic stimulation.

2 Pressoreceptors and chemoreceptors control heart rate.

3 Other influences on heart rate include strong emotions, depression, chemicals, sex, age, and muscular exercise.

CIRCULATORY SHOCK AND HOMEOSTASIS

1 Shock results when cardiac output is reduced or blood volume decreases to the point where body tissues become hypoxic.

2 Mild shock is compensated by vasoconstriction.

3 In severe shock, venous return is diminished and cardiac output decreases. The heart becomes hypoxic, prolonged vasoconstriction leads to hypoxia of other organs, and the shock cycle is intensified.

SURGERY

1 Hypothermia (body cooling) and extracorporeal circulation (heart-lung bypass) permit open-heart surgery.

2 Synthetic vessels, valves, and pacemakers are new life-saving devices.

REVIEW QUESTIONS

1 Describe the location of the heart in the mediastinum. Distinguish the subdivisions of the pericardium. What is the purpose of this structure?

2 Compare the three portions of the heart wall. Define atria and ventricles. What vessels enter or exit the atria and ventricles?

3 Discuss the principal valves in the heart and how they operate.

4 Describe the path of a nerve impulse through the heart's conducting system.

5 Define and label the deflection waves of a normal electrocardiogram. Why is the ECG an important diagnostic tool?

6 Describe the route of blood in the coronary circulation. Distinguish between angina pectoris and myocardial infarction.

7 Define systole and diastole and their relationship to the cardiac cycle.

8 Distinguish the principal events that occur during atrial systole, ventricular diastole, atrial diastole, and ventricular systole.

9 Describe the pressure changes associated with the movement of blood through the heart.

10 By means of a diagram, relate the events of the cardiac cycle to time. What is the quiescent period?

11 Describe the first and second heart sounds and indicate their clinical significance. Define a murmur.

12 What is cardiac output? How is it calculated? What factors alter cardiac output?

13 Define Starling's law of the heart. Why is it important?

14 Compare the effects of sympathetic and parasympathetic stimulation of the heart.

15 What is a pressoreceptor? Outline the operation of the carotid sinus reflex, the aortic reflex, and the right heart reflex.

16 Define a chemoreceptor. How do chemoreceptors operate? Cite two specific examples.

17 Define circulatory shock. What are its symptoms?

18 How is the body's homeostasis restored during mild shock?

19 Describe the effects of a severe shock on circulatory organs by drawing a shock cycle.

20 Explain the use of hypothermia and extracorporeal circulation in cardiovascular surgery.

21 Describe some artificial parts used in the treatment of cardiovascular disorders and explain why they might be needed.

Chapter 21
The Cardiovascular System: Vessels and Routes

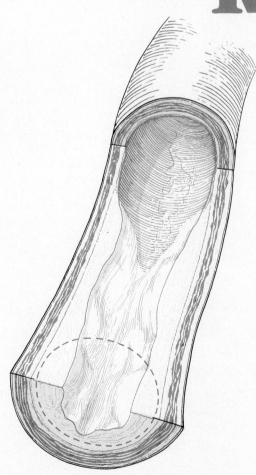

STUDENT OBJECTIVES

■ Contrast the structure and function of arteries, capillaries, and veins.

■ Compare systemic, hepatic portal, pulmonary, fetal circulation, and cerebral (circle of Willis).

■ Identify the principal arteries and veins of systemic circulation.

■ Describe the importance and route of blood involved in hepatic portal circulation.

■ Identify the major blood vessels of pulmonary circulation.

■ Contrast fetal and adult circulation.

■ Explain the fate of fetal circulation structures once postnatal circulation is established.

■ Define patent ductus arteriosus as an abnormality related to fetal circulation.

■ Describe how blood flows through vessels.

■ List the factors that resist the flow of blood through vessels.

■ List the factors that assist the return of venous blood to the heart.

■ Define pulse and identify the arteries where pulse may be felt.

■ Compare the several abnormal pulse rates.

■ Define blood pressure.

■ Describe one clinical method for recording systolic and diastolic pressure.

■ Contrast the clinical significance of systolic, diastolic, and pulse pressures.

■ List the causes and symptoms of aneurysms, atherosclerosis, and hypertension.

■ Describe the diagnosis of atherosclerosis by the use of angiography.

■ Define medical terminology associated with the cardiovascular system.

The blood vessels form a network of tubes that carry blood away from the heart, transport it to the tissues of the body, and then return it to the heart. Blood vessels are called either arteries, arterioles, capillaries, venules, or veins. **Arteries** are the vessels that carry blood from the heart to the tissues. Two large arteries leave the heart and divide into medium-sized vessels that head toward the various regions of the body. The medium-sized arteries, in turn, divide into small arteries which, in turn, divide into vessels called **arterioles.** As the arterioles enter a tissue, they branch into countless microscopic vessels called **capillaries.** Through the walls of the capillaries, substances are exchanged between the blood and body tissues. Before leaving the tissue, groups of capillaries reunite to form small veins called **venules.** These, in turn, merge to form progressively larger tubes – the veins themselves. **Veins,** in other words, are blood vessels that convey blood from the tissues back to the heart. Since blood vessels require oxygen and nutrients just like other tissues of the body, they also have blood vessels in their own walls called **vasa vasorum.**

BLOOD VESSELS

Arteries

Arteries and veins are fairly similar in construction (Figure 21-1a, d, e). Both have walls constructed of three coats or tunics and a hollow core, called a *lumen,* through which the blood flows. Arteries, however, are considerably thicker and stronger than veins. The pressure in an artery is always greater than in a vein. The inner coat of an arterial wall is called the *tunica interna* or *intima.* It is composed of a lining of endothelium (simple squamous epithelium) that is in contact with the blood. It also has an overlying layer of areolar connective tissue and an outer layer of elastic tissue called the internal elastic membrane. The middle coat, or *tunica media,* is usually the thickest layer. It consists of elastic fibers and smooth muscle. The outer coat, the *tunica externa* or *adventitia,* is composed principally of loose connective tissue. The tunica externa contains elastic and collagenous fibers and a few smooth muscle fibers. An external elastic membrane may separate the tunica media from the tunica externa.

As a result of the structure of the middle coat especially, arteries have two major properties: elasticity and contractility. When the ventricles of the heart contract and eject blood into the large arteries, the arteries expand to contain the extra blood. Then, as the ventricles relax, the elastic recoil of the arteries forces the blood onward. The contractility of an artery comes from its smooth muscle. The smooth muscle is arranged in rings around the lumen somewhat like a doughnut. As the muscle contracts, it squeezes the wall around the lumen and narrows the vessel. Such a decrease in the size of the lumen is called *vasoconstriction.* Conversely, if all the muscle fibers relax, the size of the arterial lumen increases. This increase is called *vasodilation.*

The contractility of arteries also serves a minor function in stopping bleeding. The blood flowing through an artery is under a great deal of pressure. Thus great quantities of blood can be quickly lost from a broken artery. When an artery is cut, its walls constrict so that blood does not escape quite so rapidly. However, there is a limit to how much vasoconstriction can help.

Most parts of the body receive branches from more than one artery. In such areas the distal ends of the vessels unite. The junction of two or more vessels supplying the same body region is called an **anastomosis.** Anastomoses may also occur between the origins of veins and between arterioles and venules. Anastomoses between arteries provide alternate routes by which blood can reach a tissue or organ. Thus if a vessel is occluded by disease, injury, or surgery, circulation to a part of the body is not necessarily stopped. The alternate route of blood to a body part through an anastomosis is known as **collateral circulation.** An alternate blood route may also be from nonanastomosing vessels that supply the same region of the body.

Capillaries

Capillaries are microscopic vessels measuring 4 to 12 μm in diameter. They usually connect arterioles with venules (Figure 21-1b) and permit the exchange of nutrients and gases between the blood and interstitial fluid. The structure of the capillaries is admirably suited for this purpose. First, the capillary walls are composed of only a single layer of cells (endothelium). Thus a substance in the blood must pass through the plasma membrane of just one cell to reach the interstitial fluid. This vital exchange of materials occurs only through capillary walls – the thick walls of arteries and veins present too great a barrier. Capillaries are also well suited to their function since they form a *capillary network* throughout the tissue. The network increases the surface area for passage and thereby allows a rapid exchange of large quantities of materials. Though capillaries lack the elastic connective fibers of arteries, their walls are still capable of distension. Thus they can adjust to the amount and force of blood flowing through them.

Microscopic blood vessels in certain parts of the body, such as the liver, are termed *sinusoids.* They are wider than capillaries and more tortuous. Also, instead

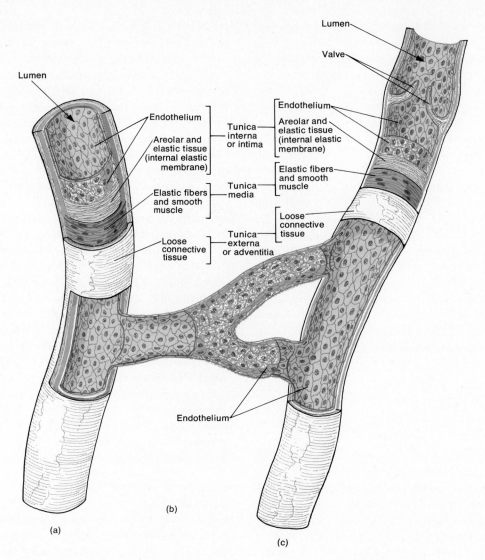

Figure 21–1 Blood vessels. (a-c) Structure of blood vessels. (a) Artery. (b) Capillary. (c) Vein. The relative size of the capillary is enlarged for emphasis.

of the usual endothelial lining, sinusoids are lined largely by phagocytic cells. In the liver, such cells are called Kupffer cells. Like capillaries, sinusoids convey blood from arterioles to venules. Other organs containing sinusoids include the spleen, adenohypophysis, and parathyroid glands.

Veins

Veins are composed of essentially the same three coats as arteries, but they have considerably less elastic tissue and smooth muscle (Figure 21-1c, f). However, veins do contain more white fibrous tissue. They are also distensible enough to adapt to variations in the volume and pressure of blood passing through them. Blood leaves a cut vein in an even flow rather than in the rapid spurts characteristic of arteries—by the time the blood leaves the capillaries and moves into the veins, it has lost a great

deal of pressure. Most of the structural differences between arteries and veins reflect this pressure difference. For example, veins do not need walls as strong as those of arteries. The low pressure in veins, however, has its disadvantages. When you stand, the pressure pushing blood up the veins in your lower extremities is barely enough to balance the force of gravity pushing it back down. For this reason, many veins, especially those in the limbs, contain valves that prevent backflow. Normal valves ensure the flow of blood toward the heart.

In people with weak valves, large quantities of blood are forced by gravity back down into distal parts of the vein. This pressure overloads the vein and pushes the walls outward. After repeated overloading, the walls lose their elasticity and become stretched and flabby. A vein damaged in this way is called a *varicose vein*. Because a varicosed wall is not able to exert a firm resistance against the blood, blood tends to accumulate in the

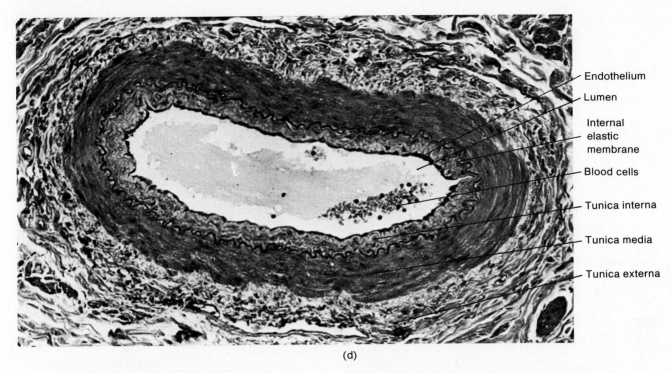

Endothelium

Lumen

Internal elastic membrane

Blood cells

Tunica interna

Tunica media

Tunica externa

(d)

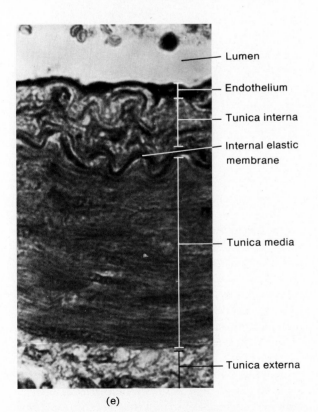

Lumen

Endothelium

Tunica interna

Internal elastic membrane

Tunica media

Tunica externa

(e)

Figure 21–1 (cont.) Blood vessels. (d-f) Histology of arteries and veins. (d) Cross section of an artery at a magnification of 50×. (e) Enlarged aspect of an artery at a magnification of 200×. (Courtesy of Victor B. Eichler, Wichita State University.)

pouched-out area of the vein, causing it to swell and forcing fluid into the surrounding tissue. Veins close to the surface of the legs are highly susceptible to varicosities. Veins that lie deeper are not so vulnerable because surrounding skeletal muscles prevent their walls from overstretching. Varicosities are also common in the veins that lie in the walls of the anal canal. These varicosities are called *hemorrhoids.*

Hemorrhoids may be caused in many people by constipation. Repeated straining during defecation forces blood down into the superior hemorrhoidal plexus, increasing pressure in these veins. Constipation is related to low-fiber diets, especially in this country. A recent hypothesis links increased intraabdominal pressure, caused by straining during evacuation of firm feces, directly to hemorrhoids or varicose veins and indirectly to thrombosis. A new development in the treatment of hemorrhoids is *cryosurgery (cryo = cold)* – external hemorrhoids can be destroyed by freezing them with a solution of nitrous oxide or liquid nitrogen.

Vascular (venous) sinuses, or simply *sinuses,* are veins with very thin walls – for example, the intracranial sinuses. These vessels consist of a wall of endothelium. They are supported by the dura mater, which replaces the tunica media and tunica externa. Intracranial vascular sinuses return deoxygenated blood from the brain to the heart. Another example of a vascular sinus is the coronary sinus of the heart.

CIRCULATORY ROUTES

Figure 21-2 shows a number of basic **circulatory routes** through which the blood travels. The largest route by far

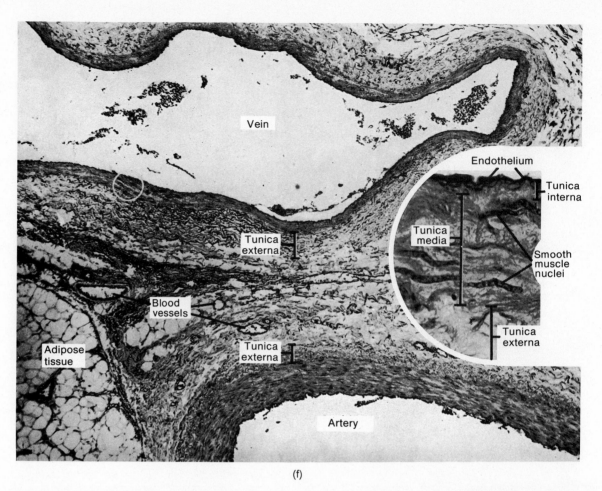

(f)

Figure 21–1 (cont.) Blood vessels. (f) Comparison of the structure of an artery and its accompanying vein at a magnification of 65×. The inset is an enlarged aspect of the layers of a vein at a magnification of 640×. (Courtesy of Edward J. Reith, from *Atlas of Descriptive Histology,* by Edward J. Reith and Michael H. Ross, Harper & Row, Publishers, Inc., New York, 1970.)

is the **systemic circulation.** This route includes all the oxygenated blood that leaves the left ventricle through the aorta and returns to the right atrium after traveling to all the organs including the nutrient arteries to the lungs. Two of the many subdivisions of the systemic circulation are the **coronary circulation,** which supplies the myocardium of the heart, and the **hepatic portal circulation,** which runs from the digestive tract to the liver. You may refer to Figure 20-5 for the details of coronary circulation. Blood leaving the aorta and traveling through the systemic arteries is a bright red color. As it moves through the capillaries, it loses its oxygen and takes on carbon dioxide, which gives the blood in the systemic veins its dark red color. When blood returns to the heart from the systemic route, it goes out of the right ventricle through the **pulmonary circulation** to the lungs. In the lungs, it loses its carbon dioxide and takes on oxygen. It is now bright red again. It returns to the left atrium of the heart and reenters the systemic

circulation. Another major route–the **fetal circulation**–exists only in the fetus and contains special structures that allow the developing human to exchange materials with its mother. **Cerebral circulation** (circle of Willis) is discussed in Exhibit 21-3.

Systemic Circulation

The flow of blood from the left ventricle to all parts of the body except the lungs and back to the right atrium is called the **systemic circulation.** The purpose of systemic circulation is to carry oxygen and nutrients to body tissues and to remove carbon dioxide and other wastes from the tissues. All systematic arteries branch from the *aorta,* which arises from the left ventricle of the heart.

As the aorta emerges from the left ventricle, it passes upward and deep to the pulmonary artery. At this point, it is called the *ascending aorta.* The ascending aorta gives off two coronary branches to the heart muscle.

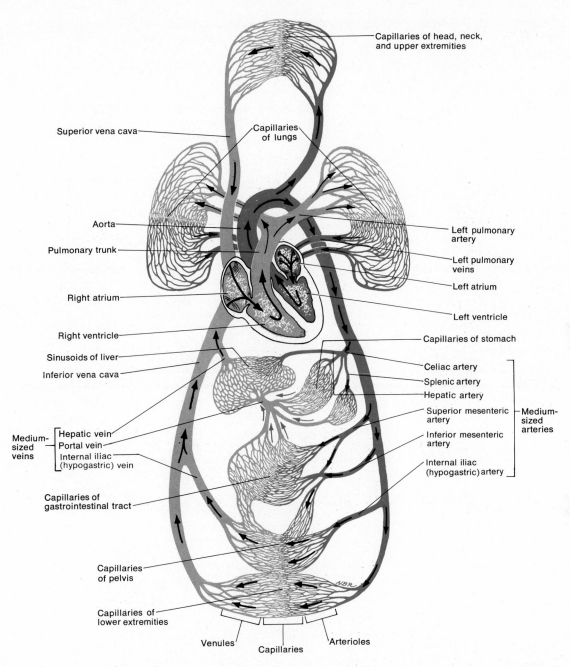

Figure 21—2   Circulatory routes. Systemic circulation is indicated by heavy black arrows; pulmonary circulation by thin black arrows; and hepatic portal circulation by thin colored arrows. Refer to Figure 20—5 for the details of coronary circulation and Figure 21—15 for the details of fetal circulation.

Then it turns to the left forming the *arch of the aorta* before descending to the level of the fourth lumbar vertebra as the *descending aorta.* The descending aorta lies close to the vertebral bodies, passes through the diaphragm, and terminates at the level of the fourth lumbar vertebra by dividing into two *common iliac arteries,* which carry blood to the lower extremities. The section of the descending aorta between the arch of aorta and the diaphragm is also referred to as the *thoracic aorta.* The section between the diaphragm and the common iliac

arteries is termed the *abdominal aorta.* Each section of the aorta gives off arteries that continue to branch into distributing arteries leading to organs and finally into the capillaries that pierce the tissues.

Blood is returned to the heart through the systemic veins. All the veins of the systemic circulation flow into either the *superior* or *inferior venae cavae* or the *coronary sinus.* They in turn empty into the right atrium. The principal arteries and veins are described and illustrated in Exhibits 21-1 to 21-12 and Figures 21-3 to 21-12.

Exhibit 21—1   AORTA AND ITS BRANCHES (see Figure 21-3)

Division of Aorta	Arterial Branch	Region Supplied
Ascending aorta	Right and left coronary	Heart
Arch of aorta	Brachiocephalic — Right common carotid	Right side of head and neck
	Brachiocephalic — Right subclavian	Right upper extremity
	Left common carotid	Left side of head and neck
	Left subclavian	Left upper extremity
Thoracic aorta	Intercostals	Intercostal and chest muscles, pleurae
	Superior phrenics	Posterior and superior surfaces of diaphragm
	Bronchials	Bronchi of lungs
	Esophageals	Esophagus
Abdominal aorta	Inferior phrenics	Inferior surface of diaphragm
	Celiac — Common hepatic	Liver
	Celiac — Left gastric	Stomach and esophagus
	Celiac — Splenic	Spleen, pancreas, stomach
	Superior mesenteric	Small intestine, cecum, ascending and transverse colons
	Suprarenals	Adrenal (suprarenal) glands
	Renals	Kidneys
	Gonadals — Testiculars	Testes
	Gonadals — Ovarians	Ovaries
	Inferior mesenteric	Transverse, descending, sigmoid colons; rectum
	Common iliacs — External iliacs	Lower extremities
	Common iliacs — Internal iliacs (hypogastrics)	Uterus, prostate, muscles of buttocks, urinary bladder

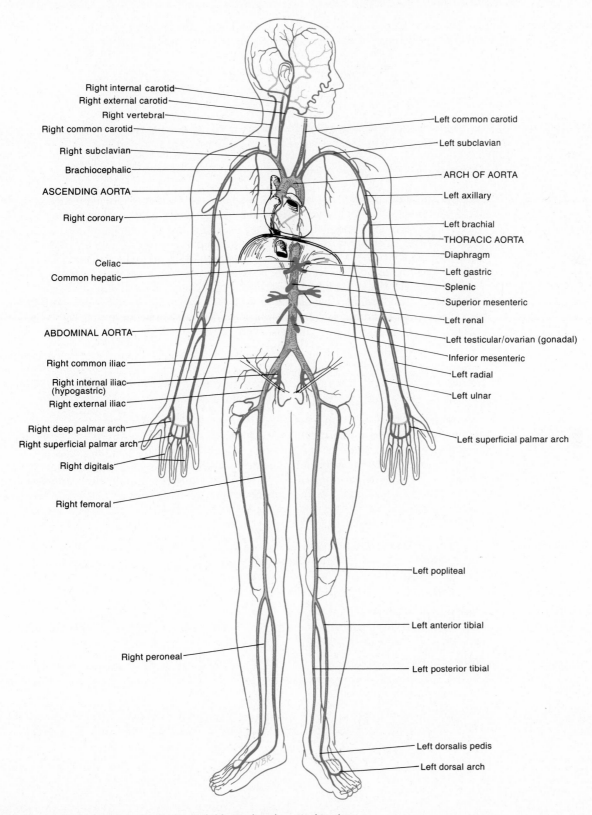

Figure 21–3 Aorta and its principal arterial branches in anterior view.

Exhibit 21—2   ASCENDING AORTA (see Figure 21-4)

Branch	Description and Region Supplied
Coronary arteries	These two branches arise from ascending aorta just superior to aortic semilunar valve. They form crown around heart giving off branches to atrial and ventricular myocardium.

Exhibit 21—3   ARCH OF AORTA (see Figure 21-5)

Branch	Description and Region Supplied
Brachiocephalic	**Brachiocephalic artery** is first branch off arch of aorta. Divides to form right subclavian artery and right common carotid artery. **Right subclavian artery** extends from brachiocephalic to first rib, passes into armpit, or axilla, and supplies arm, forearm, and hand. This artery is good example of giving same vessel different names as it passes through different regions. Continuation of right subclavian into axilla is called **axillary artery.** From here, it continues into upper arm as **brachial artery.** At bend of elbow, brachial artery divides into medial **ulnar** and lateral **radial arteries.** These vessels pass down to palm, one on each side of forearm. In palm, branches of two arteries anastomose to form two palmar arches—**superficial palmar arch** and **deep palmar arch.** From these arches arise **digital arteries,** which supply fingers and thumb.

Before passing into axilla, right subclavian gives off major branch to brain called **vertebral artery.** Right vertebral artery passes through foramina of transverse processes of cervical vertebrae and enters skull through foramen magnum to reach undersurface of brain. Here it unites with left vertebral artery to form **basilar artery.**

Anastomoses of left and right internal carotids along with basilar artery form arterial circle at base of brain called **circle of Willis.** From this anastomosis arise arteries supplying brain. Essentially, circle of Willis is formed by union of **anterior cerebral arteries** (branches of internal carotids) and **posterior cerebral arteries** (branches of basilar artery). Posterior cerebral arteries are connected with internal carotids by **posterior communicating arteries.** Anterior cerebral arteries are connected by **anterior communicating arteries.** Circle of Willis equalizes blood pressure to brain and provides alternate routes for blood to brain should arteries become damaged. |
Right common carotid	**Right common carotid artery** passes upward in neck. At upper level of larynx, it divides into **right external** and **right internal carotid arteries.** External carotid supplies right side of thyroid gland, tongue, throat, face, ear, scalp, and dura mater. Internal carotid supplies brain, right eye, and right sides of forehead and nose.
Left common carotid	**Left common carotid** branches directly from arch of aorta. Corresponding to right common carotid, it divides into basically same branches with same names—except that arteries are now labeled ''left'' instead of ''right.''
Left subclavian	**Left subclavian artery** is third branch off arch of aorta. It distributes blood to left vertebral artery and vessels of left upper extremity. Arteries branching from left subclavian are named like those of right subclavian.

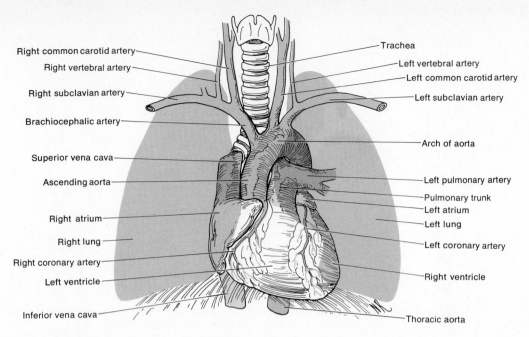

Figure 21–4 Ascending aorta and its arterial branches in anterior view.

Exhibit 21–4 THORACIC AORTA (see Figure 21-6)

Thoracic aorta runs from fourth to twelfth thoracic vertebrae. Along its course, it sends off numerous small arteries to viscera and skeletal muscles of chest. *Visceral branches* supply pericardium around heart, bronchial tubes that lead from windpipe to lungs, cells of lungs (but not areas of lungs that oxygenate blood), esophagus, and tissue lining mediastinum. *Parietal branches* supply chest muscles, diaphragm, and mammary glands.

Exhibit 21–5 ABDOMINAL AORTA (see Figure 21-6)

Branch	Description and Region Supplied
Visceral	
Celiac	**Celiac artery** (trunk) is first visceral aortic branch below diaphragm. It has three branches: (1) **common hepatic artery,** which supplies tissues of liver; (2) **left gastric artery,** which supplies stomach; and (3) **splenic artery,** which supplies spleen, pancreas, and stomach.
Superior mesenteric	**Superior mesenteric artery** distributes blood to small intestine and part of large intestine.
Suprarenals	Right and left **suprarenal arteries** supply blood to adrenal (suprarenal) glands.
Renals	Right and left **renal arteries** carry blood to kidneys.
Testiculars	Right and left **testicular arteries** extend into scrotum and terminate in testes.
Ovarians	Right and left **ovarian arteries** are distributed to ovaries.
Inferior mesenteric	**Inferior mesenteric artery** supplies major part of large intestine and rectum.
Parietal	
Inferior phrenics	**Inferior phrenic arteries** are distributed to undersurface of diaphragm.
Lumbars	**Lumbar arteries** supply spinal cord and its meninges and muscles and skin of lumbar region of back.
Middle sacral	**Middle sacral artery** supplies sacrum, coccyx, gluteus maximus muscles, and rectum.

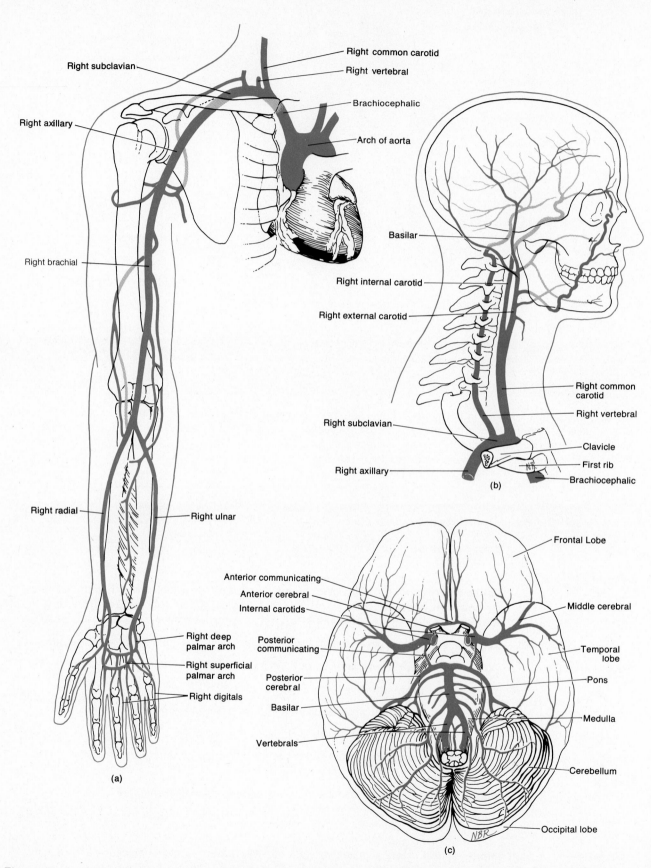

Figure 21–5 Arch of the aorta and its arterial branches. (a) Anterior view of the arteries of the right upper extremity. (b) Right lateral view of the arteries of the neck and head. (c) Arteries of the base of the brain.

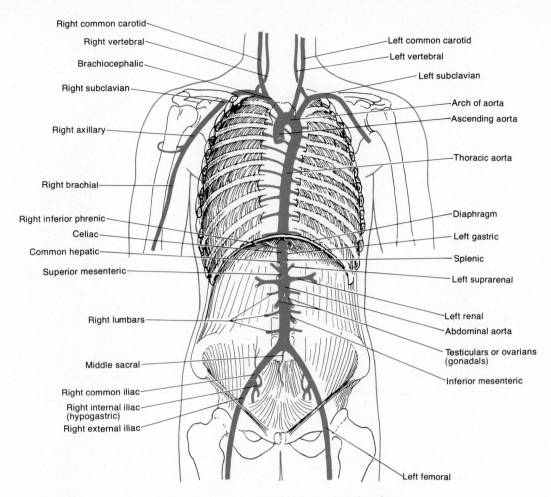

Right common carotid
Right vertebral
Brachiocephalic
Right subclavian
Right axillary
Right brachial
Right inferior phrenic
Celiac
Common hepatic
Superior mesenteric
Right lumbars
Middle sacral
Right common iliac
Right internal iliac (hypogastric)
Right external iliac

Left common carotid
Left vertebral
Left subclavian
Arch of aorta
Ascending aorta
Thoracic aorta
Diaphragm
Left gastric
Splenic
Left suprarenal
Left renal
Abdominal aorta
Testiculars or ovarians (gonadals)
Inferior mesenteric
Left femoral

Figure 21–6 Abdominal aorta and its principal arterial branches in anterior view.

Exhibit 21–6 ARTERIES OF PELVIS AND LOWER EXTREMITIES (see Figure 21-7)

Branch	Description and Region Supplied
	At about level of fourth lumbar vertebra, abdominal aorta divides into right and left **common iliac arteries.** Each passes downward about 5 cm (2 inches) and gives rise to two branches: internal iliac and external iliac.
Internal iliacs	**Internal iliac** or **hypogastric arteries** form branches that supply gluteal muscles, medial side of each thigh, urinary bladder, rectum, prostate gland, uterus, and vagina.
External iliacs	**External iliac arteries** diverge through pelvis, enter thighs, and here become right and left **femoral arteries.** Both femorals send branches back up to genitals and wall of abdomen. Other branches run to muscles of thigh. Femoral continues down medial and posterior side of thigh at back of knee joint, where it becomes **popliteal artery.** Between knee and ankle, popliteal runs down back of leg and is called **posterior tibial artery.** Below knee, **peroneal artery** branches off posterior tibial to supply structures on medial side of fibula and calcaneus. In calf, **anterior tibial artery** branches off popliteal and runs along front of leg. At ankle, it becomes **dorsalis pedis artery.** At ankle, posterior tibial divides into **medial** and **lateral plantar arteries.** These arteries anastomose with dorsalis pedis and supply blood to foot.

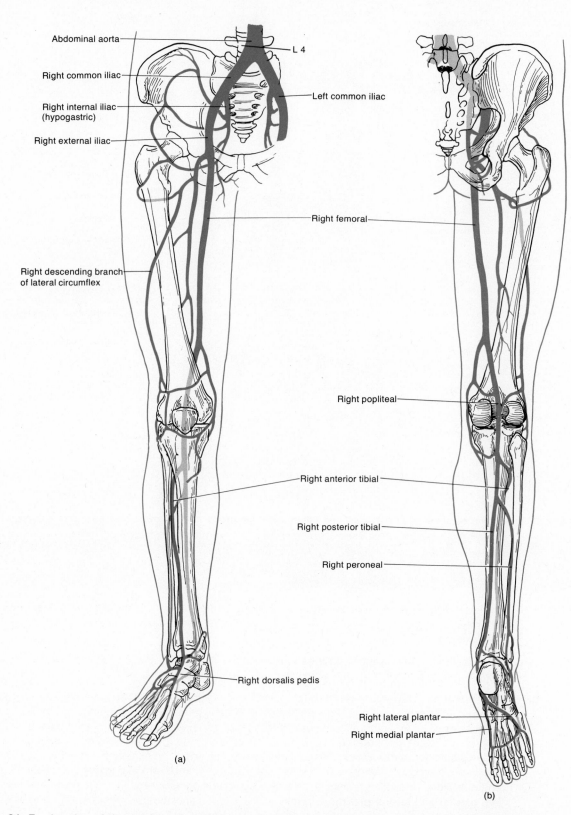

Abdominal aorta

L 4

Right common iliac

Left common iliac

Right internal iliac (hypogastric)

Right external iliac

Right femoral

Right descending branch of lateral circumflex

Right popliteal

Right anterior tibial

Right posterior tibial

Right peroneal

Right dorsalis pedis

Right lateral plantar

Right medial plantar

(a)

(b)

Figure 21–7   Arteries of the pelvis and right lower extremity. (a) Anterior view. (b) Posterior view.

Exhibit 21—7   VEINS OF SYSTEMIC CIRCULATION (see Figure 21-8)

Deep veins are located deep in body. They usually accompany arteries, and many have same names as corresponding arteries. **Superficial veins** are located just below skin and are visible.

All systemic veins return blood to right atrium of heart through one of three large vessels: coronary sinus, superior vena cava, and inferior vena cava. Return flow from coronary arteries is taken up by **cardiac veins,** which empty into large vein of heart called **coronary sinus.** From here, blood empties into right atrium of heart. Veins that empty into **superior vena cava** are veins of head and neck, upper extremities, thorax, and azygos veins. Veins that empty into **inferior vena cava** are veins of abdomen, pelvis, lower extremities, and azygos veins.

Exhibit 21—8   VEINS OF HEAD AND NECK (see Figure 21-9)

Vein	Description and Region Drained
Internal jugulars	Right and left **internal jugular veins** arise as continuation of **sigmoid sinuses** at base of skull. Intracranial vascular sinuses are located between layers of dura mater and receive blood from brain. Other sinuses that drain into internal jugular include **superior sagittal sinus, inferior sagittal sinus, straight sinus,** and **transverse (lateral) sinuses.** Internal jugulars descend on either side of neck. They receive blood from superior part of face and neck and pass behind clavicles, where they join with right and left **subclavian veins.** Unions of internal jugulars and subclavians form right and left **brachiocephalic veins.** From here blood flows into **superior vena cava.**
External jugulars	Left and right **external jugular veins** run down neck along outside of internal jugulars. They drain blood from parotid (salivary) glands, facial muscles, scalp, and other superficial structures into **subclavian veins.**

Exhibit 21—9   VEINS OF UPPER EXTREMITIES (see Figure 21-10)

Vein	Description and Region Drained
	Blood from each upper extremity is returned to heart by deep and superficial veins. Both sets of veins contain valves.
Deep veins	Deep veins run alongside arteries and are called **brachial, axillary,** and **subclavian veins.**
Superficial veins	Superficial veins anastomose extensively with each other and with deep veins.
Cephalics	**Cephalic vein** of each upper extremity begins in **dorsal arch** of hand and winds upward around radial border of forearm. Just below elbow, it unites with accessory cephalic vein to form cephalic vein of upper extremity. It eventually empties into axillary vein.
Basilics	**Basilic vein** of each upper extremity originates in ulnar part of dorsal arch. It extends along posterior surface of ulna to point below elbow where it joins **median cubital vein.** If a vein must be punctured for an injection, transfusion, or removal of a blood sample, median cubitals are preferred.
Axillaries	**Axillary vein** is continuation of basilic. It ends at about first rib, where it becomes subclavian.
Subclavians	Right and left **subclavian veins** unite with internal jugulars to form **brachiocephalic veins.** Thoracic duct of lymphatic system flows into left subclavian vein at junction with internal jugular. Right lymphatic duct enters right subclavian vein at corresponding junction.

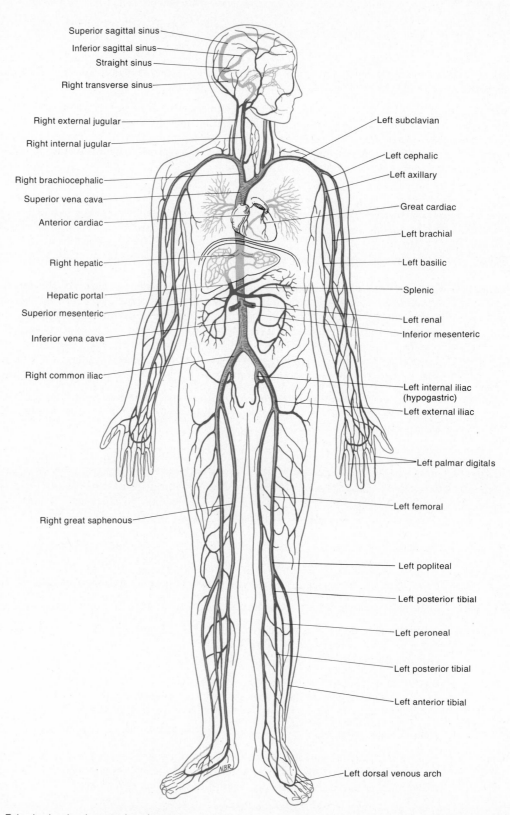

Superior sagittal sinus
Inferior sagittal sinus
Straight sinus
Right transverse sinus

Right external jugular
Right internal jugular

Right brachiocephalic
Superior vena cava
Anterior cardiac

Right hepatic

Hepatic portal
Superior mesenteric
Inferior vena cava

Right common iliac

Right great saphenous

Left subclavian
Left cephalic
Left axillary
Great cardiac
Left brachial
Left basilic
Splenic
Left renal
Inferior mesenteric
Left internal iliac (hypogastric)
Left external iliac
Left palmar digitals
Left femoral
Left popliteal
Left posterior tibial
Left peroneal
Left posterior tibial
Left anterior tibial
Left dorsal venous arch

Figure 21–8   Principal veins in anterior view.

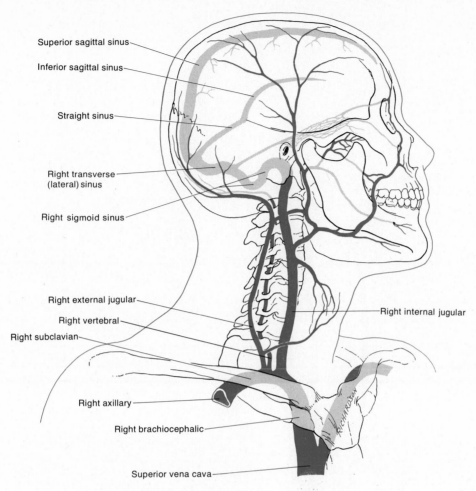

Superior sagittal sinus

Inferior sagittal sinus

Straight sinus

Right transverse (lateral) sinus

Right sigmoid sinus

Right external jugular

Right vertebral

Right subclavian

Right axillary

Right brachiocephalic

Superior vena cava

Right internal jugular

RICHARDSON

Figure 21–9   Veins of the neck and head in right lateral view.

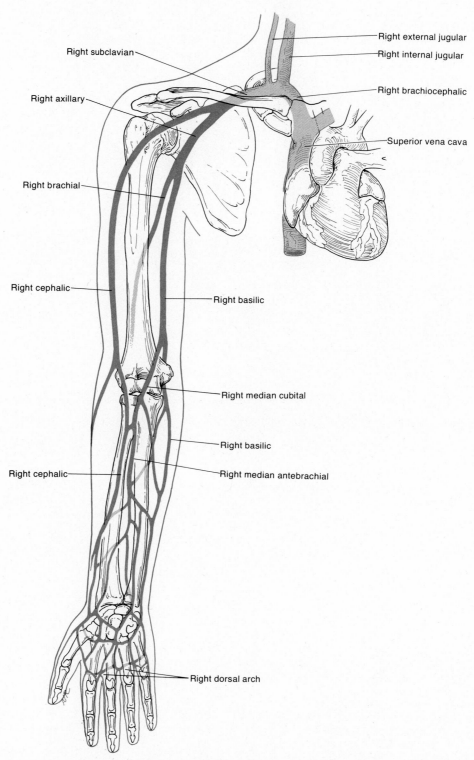

Figure 21–10   Veins of the right upper extremity in anterior view.

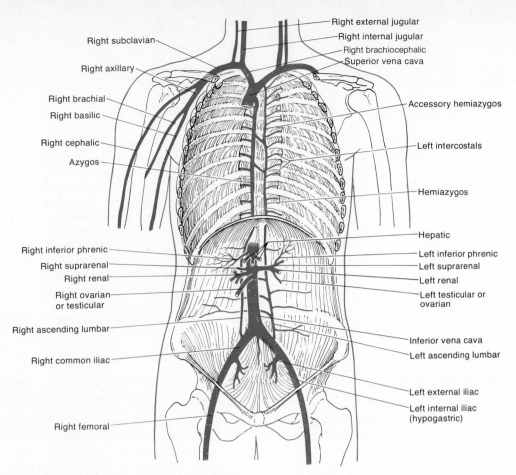

Right external jugular
Right internal jugular
Right brachiocephalic
Superior vena cava
Right subclavian
Right axillary
Right brachial
Right basilic
Right cephalic
Azygos
Accessory hemiazygos
Left intercostals
Hemiazygos
Hepatic
Right inferior phrenic
Right suprarenal
Right renal
Right ovarian or testicular
Right ascending lumbar
Right common iliac
Left inferior phrenic
Left suprarenal
Left renal
Left testicular or ovarian
Inferior vena cava
Left ascending lumbar
Left external iliac
Left internal iliac (hypogastric)
Right femoral

Figure 21–11   Veins of the thorax, abdomen, and pelvis in anterior view.

Exhibit 21—10 VEINS OF THORAX (see Figure 21-11)

Vein	Description and Region Drained
	Principal thoracic vessels that empty into superior vena cava are brachiocephalic and azygos veins.
Brachiocephalic	Right and left **brachiocephalic veins,** formed by union of subclavians and internal jugulars, drain blood from head, neck, upper extremities, mammary glands, and upper thorax. Brachiocephalics unite to form **superior vena cava.**
Azygos	**Azygos veins,** besides collecting blood from thorax, may serve as bypass for inferior vena cava that drains blood from lower body. Several small veins directly link azygos veins with inferior vena cava. And large veins that drain lower extremities and abdomen dump blood into azygos. If inferior vena cava or hepatic portal vein becomes obstructed, azygos veins can return blood from lower body to superior vena cava.
Azygos vein	**Azygos vein** lies in front of vertebral column, slightly right of midline. It begins as continuation of right ascending lumbar vein. It connects with inferior vena cava, right common iliac, and lumbar veins. Azygos receives blood from right intercostal veins that drain chest muscles; from hemi-azygos and accessory hemiazygos veins; from several esophageal, mediastinal, and pericardial veins; and from right bronchial vein. Vein ascends to fourth thoracic vertebra, arches over right lung, and empties into superior vena cava.
Hemiazygos vein	**Hemiazygos vein** is in front of vertebral column and slightly left of midline. It begins as continuation of left ascending lumbar vein. It receives blood from lower four or five intercostal veins and some esophageal and mediastinal veins. At level of ninth thoracic vertebra, it joins azygos vein.
Accessory hemiazygos vein	**Accessory hemiazygos vein** is also in front and to left of vertebral column. It receives blood from three or four intercostal veins and left bronchial vein. It joins azygos at level of eighth thoracic vertebra.

Exhibit 21—11   VEINS OF ABDOMEN AND PELVIS (see Figure 21-11)

Vein	Description and Region Drained
Inferior vena cava	**Inferior vena cava** is largest vein of body. It is formed by union of two **common iliac veins** that drain lower extremities and abdomen. It extends upward through abdomen and thorax to right atrium. Numerous small veins enter inferior vena cava. Most carry return flow from branches of abdominal aorta, and names correspond to names of arteries: left and right **renal veins** from kidneys; right **testicular vein** from testes (left testicular vein empties into left renal vein); right **ovarian vein** from ovaries (left ovarian vein also empties into left renal vein); right **suprarenal veins** from adrenal glands (left suprarenal vein empties into left renal vein); right **inferior phrenic vein** from diaphragm (left inferior phrenic vein sends tributary to left renal vein); and **hepatic veins** from liver. In addition, a series of parallel **lumbar veins** drain blood from both sides of posterior abdominal wall. Lumbars connect at right angles with right and left ascending lumbar veins, which form origin of corresponding azygos or hemiazygos vein. Lumbars drain blood into ascending lumbars and then run to inferior vena cava, where they release remainder of flow.

Exhibit 21—12   VEINS OF LOWER EXTREMITIES (see Figure 21-12)

Vein	Description and Region Drained
	Blood from each lower extremity is returned by superficial set and deep set of veins. Superficials are formed from extensive anastomoses close to surface. Deep veins follow large arterial trunks. Both sets have valves.
Superficial veins	Main superficial veins are great saphenous and small saphenous. Both, especially great saphenous, frequently become varicosed.
Great saphenous	**Great saphenous vein,** longest vein in body, begins at medial end of **dorsal venous arch** of foot. It passes in front of medial malleolus and then upward along medial aspect of leg and thigh. It receives tributaries from superficial tissues and connects with deep veins as well. It empties into femoral vein in groin.
Small saphenous	**Small saphenous vein** begins at lateral end of dorsal venous arch of foot. It passes behind lateral malleolus and ascends under skin of back of leg. It receives blood from foot and posterior portion of leg. It empties into popliteal vein behind knee.
Deep veins Posterior tibial	**Posterior tibial vein** is formed by union of **medial** and **lateral plantar veins** behind medial malleolus. It ascends deep in muscle at back of leg, receives blood from **peroneal vein,** and unites with anterior tibial vein just below knee.
Anterior tibial	**Anterior tibial vein** is upward continuation of **dorsalis pedis** veins in foot. It runs between tibia and fibula and unites with posterior tibial to form popliteal vein.
Popliteal	**Popliteal vein,** just behind knee, receives blood from anterior and posterior tibials and small saphenous vein.
Femoral	**Femoral vein** is upward continuation of popliteal just above knee. Femorals run up posterior of legs and drain deep structures of thighs. After receiving great saphenous veins in groin, they continue as right and left **external iliac veins.** Right and left **internal iliac veins** receive blood from pelvic wall and viscera, external genitals, buttocks, and medial aspect of thigh. Right and left **common iliac veins** are formed by union of internal and external iliacs. Common iliacs unite to form inferior vena cava.

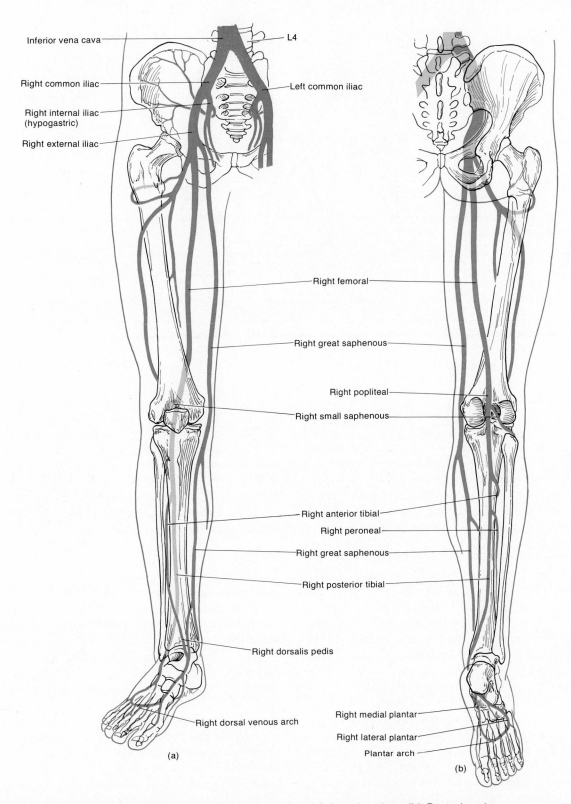

Inferior vena cava

L4

Right common iliac

Left common iliac

Right internal iliac
(hypogastric)

Right external iliac

Right femoral

Right great saphenous

Right popliteal

Right small saphenous

Right anterior tibial

Right peroneal

Right great saphenous

Right posterior tibial

Right dorsalis pedis

Right dorsal venous arch

Right medial plantar

Right lateral plantar

Plantar arch

(a)

(b)

Figure 21–12 Veins of the pelvis and right lower extremity. (a) Anterior view. (b) Posterior view.

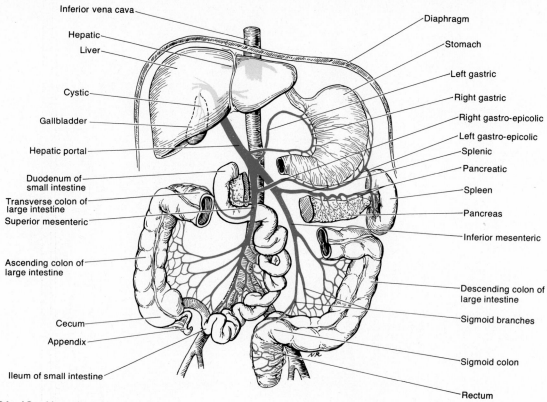

Figure 21–13 Hepatic portal circulation.

Hepatic Portal Circulation

Blood enters the liver from two sources. The hepatic artery delivers oxygenated blood from the systemic circulation; the hepatic portal vein delivers deoxygenated blood from the digestive organs. The term **hepatic portal circulation** refers to this flow of venous blood from the digestive organs to the liver before returning to the heart (Figure 21-13). Hepatic portal blood is rich with substances absorbed from the digestive tract. The liver monitors these substances before they pass into the general circulation. For example, the liver stores nutrients such as glucose. It modifies other digested substances so they may be used by cells. And it detoxifies harmful substances that have been absorbed by the digestive tract and destroys bacteria by phagocytosis.

The hepatic portal system includes veins that drain blood from the pancreas, spleen, stomach, intestines, and gallbladder and transport it to the portal vein of the liver. The *hepatic portal vein* is formed by the union of the superior mesenteric and splenic veins. The *superior mesenteric vein* drains blood from the small intestine and portions of the large intestine and stomach. The *splenic vein* drains the spleen and receives tributaries from the stomach, pancreas, and colon. The tributaries from the

stomach are the *coronary, pyloric,* and *gastro-epiploic veins.* The *pancreatic veins* come from the pancreas, and the *inferior mesenteric veins* come from portions of the colon. Before the hepatic portal vein enters the liver, it receives the *cystic vein* from the gallbladder. Ultimately, blood leaves the liver through the *hepatic veins,* which enter the inferior vena cava.

Pulmonary Circulation

The flow of deoxygenated blood from the right ventricle to the lungs and the return of oxygenated blood from the lungs to the left atrium is called the **pulmonary circulation** (Figure 21-14). The *pulmonary trunk* emerges from the right ventricle and passes upward, backward, and to the left. It then divides into two branches. The right pulmonary artery runs to the right lung; the left pulmonary artery goes to the left lung. On entering the lungs, the branches divide and subdivide. They get smaller and ultimately form capillaries around the alveoli in the lungs. Carbon dioxide is passed from the blood into the alveoli to be breathed out of the lungs. Oxygen breathed in by the lungs is passed from the alveoli into the blood. The capillaries then unite.

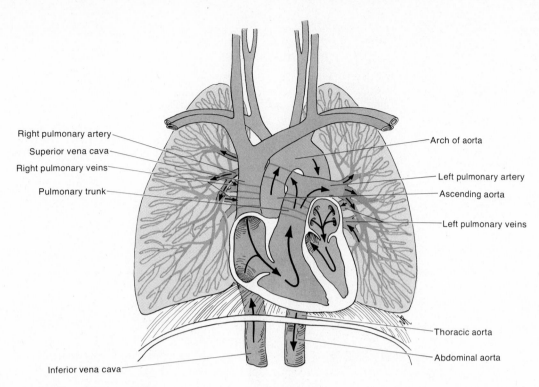

Figure 21–14 Pulmonary circulation.

They grow larger and become veins. Eventually, two *pulmonary veins* exit from each lung and transport the oxygenated blood to the left atrium. The pulmonary veins are the only postnatal veins that carry oxygenated blood. Contraction of the left ventricle then sends the blood into the systemic circulation.

Fetal Circulation

The circulatory system of a fetus, called **fetal circulation,** differs from an adult's because the lungs and digestive tract of a fetus are nonfunctional. The fetus derives its oxygen and nutrients and eliminates its carbon dioxide and wastes through the maternal blood (Figure 21-15).

The exchange of materials between fetal and maternal circulation occurs through a structure called the *placenta.* It is attached to the navel of the fetus by the umbilical cord, and it communicates with the mother through countless small blood vessels that emerge from the uterine wall. The umbilical cord contains blood vessels that branch into capillaries in the placenta. Wastes from the fetal blood diffuse out of the capillaries, into spaces containing maternal blood (intervillous spaces) in the placenta, and finally into the mother's uterine blood vessels. Nutrients travel the opposite route – from the maternal blood vessels to the intervillous spaces to

the fetal capillaries. There is no mixing of maternal and fetal blood since all exchanges occur through capillaries.

Blood passes from the fetus to the placenta via two *umbilical arteries.* These branches of the internal iliac arteries are included in the umbilical cord. At the placenta, the blood picks up oxygen and nutrients and eliminates carbon dioxide and wastes. The oxygenated blood returns from the placenta via the *umbilical vein.* This vein ascends to the liver of the fetus where it divides into two branches. Some blood flows through the branch that joins the hepatic portal vein and enters the liver. Although the fetal liver manufactures red blood cells, it does not function in digestion. Therefore, most of the blood flows into the second branch: the *ductus venosus.* The ductus venosus connects with the inferior vena cava.

In general, circulation through other portions of the fetus is not unlike the postnatal circulation. Deoxygenated blood returning from the lower regions is mingled with oxygenated blood from the ductus venosus in the inferior vena cava. This mixed blood then enters the right atrium. The circulation of blood through the upper portion of the fetus is also similar to the postnatal flow. Deoxygenated blood returning from the upper regions of the fetus is collected by the superior vena cava, and it also passes into the right atrium.

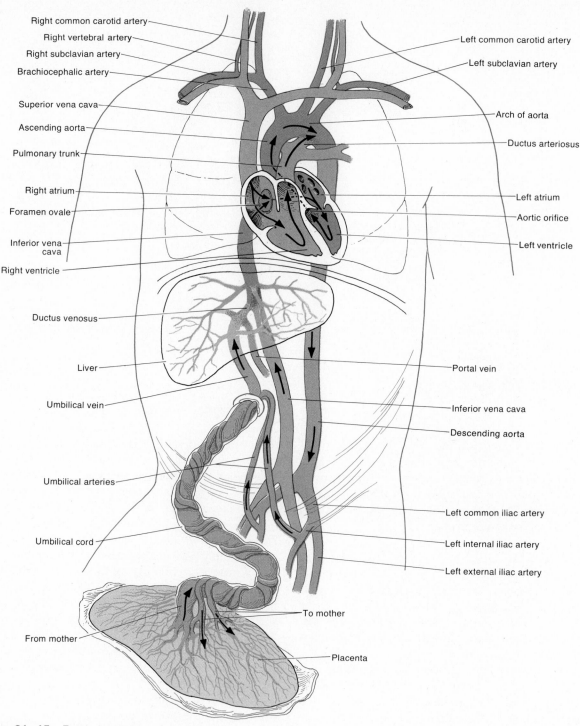

Figure 21–15   Fetal circulation.

Unlike postnatal circulation, most of the blood does not pass through the right ventricle to the lungs since the fetal lungs do not operate. In the fetus, an opening called the *foramen ovale* exists in the septum between the right and left atria. A valve in the inferior vena cava directs most of the blood through the foramen ovale so that it may be sent directly into the systemic circulation. The blood that does descend into the right ventricle is pumped into the pulmonary trunk, but little of this blood actually reaches the lungs. Most blood in the pulmonary trunk is sent through the *ductus arteriosus.* This small vessel connecting the pulmonary trunk with the aorta enables blood in excess of nutrient requirements to bypass the fetal lungs. The blood in the aorta is

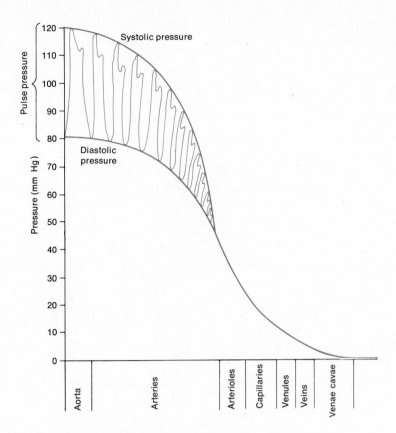

Figure 21—16   Blood pressure in various parts of the cardiovascular system.

carried to all parts of the fetus through its systemic branches. When the common iliac arteries branch into the external and internal iliacs, part of the blood flows into the internal iliacs. It then goes to the umbilical arteries and back to the placenta for another exchange of materials. The only vessel that carries fully oxygenated blood is the umbilical vein.

At birth, when lung, digestive, and liver functions are established, the special structures of fetal circulation are no longer needed. Thus:

1 The umbilical arteries atrophy to become the lateral umbilical ligaments.

2 The umbilical vein becomes the round ligament of the liver.

3 The placenta is passed by the mother as the "afterbirth."

4 The ductus venosus becomes the ligamentum venosum, a fibrous cord in the liver.

5 The foramen ovale normally closes shortly after birth to become the fossa ovalis, a depression in the interatrial septum.

6 The ductus arteriosus closes, atrophies, and becomes the ligamentum arteriosum.

Usually the ductus arteriosus closes shortly after birth. When it fails to close or closes imperfectly, blood shuttles uselessly back and forth between heart and lungs. This condition, *patent ductus arteriosus,* is easily remedied by surgery.

PHYSIOLOGY OF CIRCULATION

Blood flows through its system of closed vessels because of different pressures in various parts of the system. It always flows from regions of higher pressure to regions of lower pressure. The mean (average) pressure in the aorta is about 100 mm Hg (mercury). This pressure continually decreases from the aorta through the arteries to the arterioles (Figure 21-16). Capillary blood pressure is greater than the pressure in venules but less than the pressure in arterioles. Thus blood flows from arterioles into capillaries. In the venous system, the pressure falls slowly – from venules to veins to the venae cavae.

Thus, because of a continuous drop in pressure, blood flows from the aorta (100 mm Hg) to the arteries (100–40 mm Hg) to arterioles (40–25 mm Hg) to capillaries (25–12 mm Hg) to venules (12–8 mm Hg) to veins (10–5 mm Hg) to the venae cavae (2 mm Hg). Other mechanisms aid the flow of blood also. When blood leaves the capillaries, it enters the venules and veins, which are larger in diameter and thereby offer less resistance to flow. Contraction of skeletal muscles around the veins also helps drive blood toward the heart.

Pressure

Blood pressure is determined by two factors: the volume of blood delivered to a vessel in a certain amount of time and the resistance exerted against the flow. These two factors and their relationship to pressure can be summarized as follows:

$$\text{Pressure} = \text{flow (volume/minute)} \times \text{resistance}$$

Note that pressure increases if flow or resistance increases. When you narrow the diameter of the garden hose nozzle, the water comes out in a narrower stream under higher pressure. Decreasing the diameter of the opening increases the resistance to the flow of water.

In the circulatory system, pressure is produced by the amount of blood pumped out of the heart and by the resistance offered by the blood vessel walls. Either an increase in cardiac output or a narrowing of the vessel walls will increase pressure.

HEART ACTION AND BLOOD VOLUME

Cardiac output is the principal determinant of blood pressure. Without a delivery of blood to the vessels, there would be no pressure and no blood flow other than a slow oozing produced by the force of gravity. Cardiac output depends on the strength of ventricular contraction, on heart rate, and on the amount of blood available to be pumped. The normal volume of blood in a human body is about 5 liters (5 qt). Any decrease in this volume, as from hemorrhage, decreases the amount of blood that is circulated through the arteries each minute. As a result, blood pressure drops. Conversely, anything that increases blood volume, such as high salt intake and therefore water retention, increases blood pressure.

RESISTANCE

The force applied by the vessel walls against the flowing blood is *resistance.* Resistance can be exerted in a number of ways. And the body can increase or decrease resistance to adjust blood pressure. Let's examine some of the factors related to resistance.

■ *Arterial Walls*   During ventricular systole, blood is forced from the ventricles into filled arteries. To receive this extra supply of blood, the arterial walls, because of their *extensibility,* stretch outward. As soon as the force of systole is removed, the arterial walls, because of their *elasticity,* recoil inward to resume their normal shape. The pushing inward puts pressure (resistance) on the blood and forces it into the capillaries.

The resistance offered by the inward push of the arterial walls does not increase blood pressure because it is canceled out by the lack of resistance during disten-

sion. The extensibility and elasticity of arterial walls keep the blood pressure from fluctuating with every systole and diastole. As a spurt of blood leaves the ventricle, it is captured in the distended arterial walls. During diastole, the inward push of the walls keeps the blood flowing into the smaller arteries. Blood is thereby funneled into the capillaries at an even rate throughout the cardiac cycle. When arteries lose their extensibility and elasticity and become rigid, large amounts of blood spurt into the capillaries during systole and flow ceases during diastole. The spurting of blood into capillaries can easily damage their thin, fragile walls.

■ *Viscosity*   Another form of resistance is the *viscosity* of blood itself. In a thick fluid, molecules tend to be attracted to each other and resist flowing—molasses pours more slowly than water. Any condition that increases the thickness of blood also increases blood pressure. One such condition is an unusually high number of red blood cells. A depletion of the plasma protein albumin, which is also responsible for the thickness of blood, produces a decrease in blood pressure.

■ *Peripheral Resistance*   The walls of the blood vessels themselves offer the main resistance to blood flow. As blood is pushed into a vessel, it tends to spread out. The firm walls of the vessel offer *peripheral resistance* to spreading and channel the blood in a narrow stream. The smaller the diameter of a vessel, the more resistance it offers the blood. A major function of arterioles is to control peripheral resistance and, therefore, blood pressure by changing their diameters. The center for this regulation is the *vasomotor center* in the medulla (*vas* = vessel; *motor* = movement). Functionally, the vasomotor center consists of a vasoconstrictor and a vasodilator center.

Vasoconstrictor nerves conduct impulses from the medulla to the spinal cord and through sympathetic nerves to the arterioles. The impulses reduce the diameter of blood vessels by stimulating the smooth muscle to contract. Normally, the nerves continually send impulses to the smooth muscle of arterioles. As a result, the arterioles are kept in a state of tonic contraction. If there is an increase in blood pressure, however, the vasoconstrictor nerves are inhibited. The arteriole diameter passively increases and blood pressure falls because of the reduction in peripheral resistance.

Vasodilator nerves carry impulses that increase the diameter of vessels by inhibiting vasoconstriction. When blood pressure increases, the vasodilator nerves are stimulated, arteriole diameter increases, and blood pressure falls because of the decrease of peripheral resistance. Essentially, the vasomotor center, in conjunction with arterioles, adjusts blood pressure in response to varying needs of the body.

Production of Pressure Gradient

Blood always flows from an area of greater pressure to an area of lesser pressure. Although resistance increases pressure locally, the overall effect is a slow loss of pressure. The resistance of the artery walls against the blood produces friction, so that some kinetic energy of the flowing fluid is converted to heat energy and escapes. Thus, as the blood passes through the arteries to the capillaries and on to the veins, it continually loses pressure as it rubs against the vessel walls.

Factors Aiding Venous Return

The establishment of a pressure gradient is the primary reason why blood flows. But a number of other factors help the blood to return through the veins: an increased rate of flow, contractions of the skeletal muscles, and valves.

RATE OF FLOW

The less resistance offered the blood, the faster it flows:

$$\text{Flow} = \frac{\text{pressure}}{\text{resistance}}$$

A large vessel offers less peripheral resistance than a small one. Thus blood flows fastest through the aorta, which has the largest diameter and therefore the least peripheral resistance of any vessel in the body. The blood flows most slowly through the capillaries, which have minute diameters and offer a great deal of resistance. The slow flow through the capillaries allows time for the exchange of materials and is another example of how body structure is admirably suited to function. But as the blood reaches the larger venules and finally the very large veins, it picks up speed even though the pressure is decreasing—because the diameter of the veins increases as the blood nears the heart. The larger diameters of the veins offer much less resistance, which more than compensates for the progressive loss in blood pressure.

Blood flow is also related to the cross-sectional area of the blood vessels. Blood flows most rapidly where the cross-sectional area is least. The cross-sectional area of the aorta is 2.5 cm² and the velocity of blood there is 40 cm/second. Each time an artery branches, the cross-sectional area of the branches combined is greater than that of the original vessel. The cross-sectional area of arteries is 20 cm² and the velocity is 40–10 cm/second.

The cross-sectional area of arterioles is 40 cm² and the velocity is 10–0.1 cm/second. Capillaries have a cross-sectional area of 2,500 cm² and the velocity is less than 0.1 cm/second. This factor greatly aids the exchange of materials between capillary blood and body tissues. The cross-sectional area of venules is 250 cm² and the velocity of blood is 0.3 cm/second. As blood flows into veins, their cross-sectional area decreases to 80 cm² and the velocity increases to 0.3–5 cm/second. In the venae cavae, the cross-sectional area is 8 cm² and the velocity is 5–20 cm/second.

Thus the velocity of blood decreases as it flows from the aorta to arteries to arterioles to capillaries. The velocity of blood in the capillaries is slowest in the cardiovascular system. Thus there is adequate exchange (diffusion) time between the capillaries and adjacent tissues. As blood vessels leave capillaries and approach the heart, their cross-sectional area decreases. Therefore the velocity of blood increases as it flows from capillaries to venules to veins to the heart.

Blood flow can be measured to diagnose certain circulatory disorders. **Circulation time** is the time required for blood to pass from the right atrium, through pulmonary circulation, back to the left ventricle, through systemic circulation down to the foot, and back again to the right atrium. Such a trip usually takes about 23 seconds and requires about 28 heartbeats.

MUSCLES AND VALVES

The contraction of skeletal muscles is essential in returning blood to the heart. As the muscles contract, they tighten around the veins running through them. The pressure drives the blood toward the heart. This action is called milking. Individuals who are immobilized through injury or disease cannot take advantage of these contractions. As a result, the return of venous blood to the heart is slower and the heart has to work harder. Another important factor in maintaining venous circulation is breathing. During each inspiration, pressure in the thoracic cavity decreases and blood flows down a pressure gradient into the veins in the thoracic cavity. Each expiration would drive blood back away from the chest, but the veins have valves that further aid venous circulation. In many places of the body, especially the extremities, these valves offer little resistance to blood flowing toward the heart. Any pressure on the valves from blood moving backward closes them and stops the backflow.

CHECKING CIRCULATION

The pulse tells us about the rate of the heartbeat. The blood pressure measures the pressure driving the blood.

Pulse

The alternate expansion and elastic recoil of an artery with each systole of the left ventricle is called the **pulse**. Pulse is strongest in the arteries closest to the heart. It becomes weaker as it passes over the arterial system, and it disappears altogether in the capillaries. The pulse may be felt in any artery that lies near the surface of the body and over a bone or other firm tissue. The radial artery at the wrist is most commonly used. Other arteries that may be used for determining pulse are the:

1 Temporal artery, which is above and toward the outside of the eye.

2 Facial artery, which is at the lower jawbone on a line with the corners of the mouth.

3 Common carotid artery, which is on the side of the neck.

4 Brachial artery along the inner side of the biceps brachii muscle.

5 Femoral artery near the pelvic bone.

6 Popliteal artery behind the knee.

7 Posterior tibial artery behind the medial malleolus of the tibia.

8 Dorsalis pedis artery over the instep of the foot.

The pulse rate is the same as the heart rate and averages between 70 and 90 beats per minute in the resting state. The term *tachycardia* is applied to a rapid heart rate or pulse rate (*tachy* = fast). The term *bradycardia* (*brady* = slow) indicates a slow heart rate or pulse rate. Other factors should be noted besides pulse rate. For example, the intervals between beats should be equal in length. If a pulse is missed at intervals, the pulse is said to be irregular. Also, each pulse beat should be of equal strength. Irregularities in strength may indicate a lack of muscle tone in the heart or arteries.

Blood Pressure

Although the term **blood pressure** may be defined as the pressure exerted by the blood on the walls of any blood vessel, in the clinic it refers to the pressure only in the large arteries. Blood pressure is usually taken in the left brachial artery, and it is measured by a *sphygmoman-* *ometer* (*sphygmo* = pulse). A commonly used sphygmomanometer (Figure 21-17a) consists of a rubber cuff attached by a rubber tube to a compressible hand pump or bulb. Another tube attaches to the cuff and to a column of mercury marked off in millimeters. This column measures the pressure. The cuff is wrapped around the arm over the brachial artery and inflated by squeezing the bulb. The inflation creates a pressure on the artery (Figure 21-17b). The bulb is squeezed until the pressure in the cuff exceeds the pressure in the artery. At this point, the walls of the brachial artery are compressed tightly against each other, and no blood can flow through. Compression of the artery may be evidenced in two ways. First, if a stethoscope is placed below the cuff over the artery, no pulse can be heard. Second, no pulse can be felt by placing the fingers over the radial artery at the wrist.

Next the cuff is deflated gradually until the pressure in the cuff equals the maximal pressure in the brachial artery. At this point, the artery opens, a spurt of blood passes through, and the pulse may be heard through the stethoscope. As cuff pressure is further reduced, the sound suddenly becomes faint. Finally, the sound disappears altogether. When the first sound is heard, a reading on the mercury column is made. This sound corresponds to **systolic blood pressure** – the force with which blood is pushing against arterial walls during ventricular contraction. The pressure recorded on the mercury column when the sounds suddenly become faint is called **diastolic blood pressure**. It measures the force of blood in arteries during ventricular relaxation. Whereas systolic pressure indicates the force of the left ventricular contraction, diastolic pressure provides information about the resistance of blood vessels.

The average blood pressure of a young adult male is about 120 mm Hg systolic and 80 mm Hg diastolic. For convenience and brevity, these pressures are indicated as 120/80. In young adult females, the pressures are 8 to 10 mm Hg less. The difference between systolic and diastolic pressure is called **pulse pressure**. This pressure, which averages 40 mm Hg, provides information about the condition of the arteries. The higher the systolic pressure and the lower the diastolic pressure, the greater the pulse pressure. The normal ratio of systolic pressure to diastolic pressure to pulse pressure is about 3:2:1.

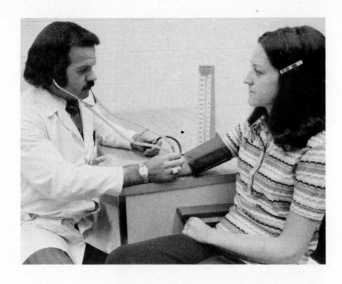

(a)

Figure 21—17 Measurement of blood pressure. (a) Use of a sphygmomanometer. (Courtesy of Lenny Patti.) (b) Pressure changes associated with the brachial artery.

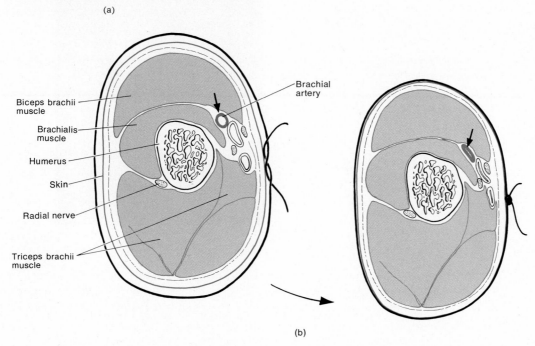

(b)

DISORDERS

ANEURYSM

A blood-filled sac formed by an outpouching in an arterial or venous wall is called an **aneurysm.** Aneurysms may occur in any major blood vessel and include the following types:

    1 **Berry**   A small aneurysm frequently in the cerebral artery. If it ruptures, it may cause a hemorrhage below the dura mater. Hemorrhaging is one cause of a stroke.

    2 **Ventricular**   A dilatation of a ventricle of the heart.

    3 **Aortic**   A dilatation of the aorta.

ATHEROSCLEROSIS

Atherosclerosis is a form of arteriosclerosis, which includes many diseases of the arterial wall. But atherosclerosis, the lipid-related arterial lesion, is the major disease responsible for the principal clinical complications. In this disorder, the tunica intima of an artery becomes thickened with soft fatty deposits called *atheromatous plaques* (Figure 21-18a).

    An *atheroma* is an abnormal mass of fatty or lipid material deposited in an arterial wall. Atheromas involve the abdominal aorta and major leg arteries more extensively than the

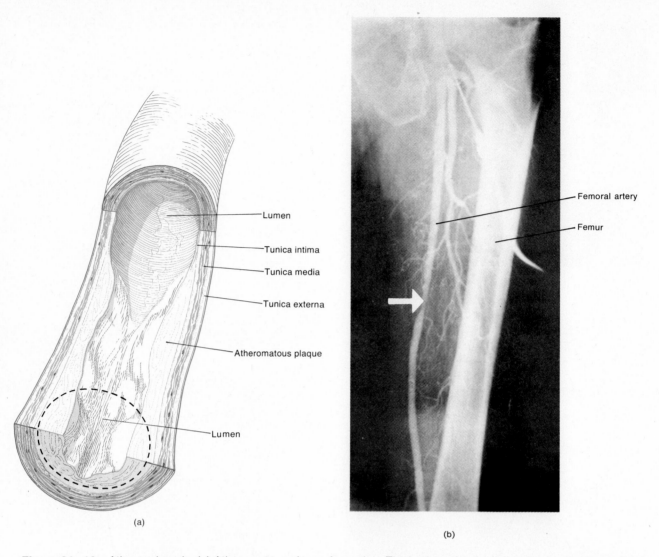

Labels in figure (a): Lumen, Tunica intima, Tunica media, Tunica externa, Atheromatous plaque, Lumen

Labels in figure (b): Femoral artery, Femur

(a)

(b)

Figure 21–18 Atherosclerosis. (a) Atheromatous plaque formation. The broken circular line indicates the approximate size of a normal lumen. (b) Femoral arteriogram showing an atheromatous plaque (arrow) in the middle third of the thigh. (Courtesy of Lester W. Paul and John H. Juhl, *The Essentials of Roentgen Interpretation*, 3d ed., Harper & Row, Publishers, Inc., New York, 1972.)

thoracic aorta. They may also be found in coronary, cerebral, and peripheral arteries. The atheroma looks like a pearly gray or yellow mound of tissue. As atheromas grow, they may impede blood flow in affected arteries and damage the tissues they supply. An additional danger is that the plaque may rupture and form a thrombus. If it breaks off and forms an embolus, it may obstruct small arteries, capillaries, and veins quite a distance from the site of formation. Moreover, the plaque may provide a roughened surface for clot formation.

The degree of arterial stenosis necessary to produce symptoms depends on the site of stenosis and the artery. The process causes a collateral circulation to develop around the stenotic block. These collaterals may prevent ischemic necrosis. When they do not, ischemia may progress to gangrene of the leg.

Atherosclerosis is generally a slow, progressive disease. It may start in childhood, and its development may produce absolutely no symptoms for 40 years or longer. Even if it reaches the advanced stages, the individual may feel no symptoms and the condition may be discovered only at postmortem examination. Diagnosis is possible by injecting radiopaque substances into the blood and then taking x rays of the arteries. This technique is called *angiography or arteriography*. The film is called an *arteriogram* (Figure 21-18b).

Animal experiments have given us considerable scientific information about the plaques. It is possible to produce the streaks in many animals by feeding them a diet high in fat and cholesterol. This diet raises the blood lipid levels – a condition called *hyperlipidemia* (*hyper* = above; *lipo* = fat). Hyperlipidemia increases the risk of atherosclerosis. Patients with a high blood level of cholesterol should be treated with diet and drug therapy.

HYPERTENSION

Hypertension, or high blood pressure, is the most common disease affecting the heart and blood vessels. Statistics from a recent National Health Survey indicate that hypertension afflicts at least 17 million American adults and perhaps as many as 22 million.

Primary hypertension, or essential hypertension, is a persistently elevated blood pressure that cannot be attributed to any particular organic cause. Specifically, the diastolic pressure continually exceeds 95 mm Hg. Approximately 85 percent of all hypertension cases fit this definition. The other 15 percent have *secondary hypertension.* Secondary hypertension is caused by disorders such as arteriosclerosis, kidney disease, and adrenal hypersecretion. Arteriosclerosis increases blood pressure by reducing the elasticity of the arterial walls and narrowing the space through which the blood can flow. Kidney disease and obstruction of blood flow may cause the kidney to release renin into the blood. This enzyme catalyzes the formation of angiotensin from a plasma protein. Angiotensin is a powerful blood-vessel constrictor—and the most potent agent known for raising blood pressure. Aldosteronism, the hypersecretion of aldosterone, may also cause an increase in blood pressure. Aldosterone is the adrenal cortex hormone that promotes the retention of salt and water by the kidneys. It thus tends to increase plasma volume. Pheochromocytoma is a tumor of the adrenal medulla. It produces and releases into the blood large quantities of norepinephrine and epinephrine. These hormones also raise blood pressure by stimulating the heart and constricting blood vessels.

The causes of primary hypertension are unknown. Medical science cannot cure it. However, almost all cases of hypertension, whether mild or severe, can be controlled by a variety of drugs that reduce blood pressure.

MEDICAL TERMINOLOGY

Arteriography   Recording of an image of arteries by x ray revealed by the direct injection of dyes.

Cardiac arrest   Complete stoppage of the heartbeat.

Cardiomegaly (*megalo* = great)   Heart enlargement.

Cyanosis   Slightly bluish, dark purple skin coloration due to oxygen deficiency in systemic blood.

Defibrillator   A mechanical device for applying electrical shock to the heart to terminate abnormal cardiac rhythms.

Epistaxis   A nosebleed.

Hematoma (hemangioma) (*hemo, hemato* = blood; *oma* = tumor)   Leakage of blood from a vessel, which clots to form a solid mass or swelling in tissue.

Occlusion   The closure or obstruction of the lumen of a structure such as a blood vessel.

Phlebitis   Inflammation of a vein.

Thrombophlebitis (*thrombo* = clot)   Inflammation of a vein with clot formation.

BLOOD VESSELS
ARTERIES
1 Arteries carry blood away from the heart. They are stronger and thicker than veins. Arteries consist of a tunica interna, tunica media (which maintains elasticity and contractility), and tunica externa.

2 Many arteries anastomose – the distal ends of two or more vessels unite. An alternate blood route from an anastomosis is called collateral circulation.

CAPILLARIES
1 Capillaries are microscopic blood vessels through which materials are exchanged between blood and interstitial fluid. They unite to form venules, which in turn form veins to carry blood back to the heart.

2 Capillaries branch to form an extensive capillary network throughout the tissue. This network increases the surface area, allowing a rapid exchange of large quantities of materials.

3 Microscopic blood vessels in the liver are called sinusoids.

VEINS
1 Veins have less elastic tissue and smooth muscle than arteries. They contain valves to prevent backflow of blood.

2 Weak valves can lead to varicose veins or hemorrhoids. A new development in the treatment of hemorrhoids is cryosurgery – destroying them by freezing with a solution of nitrous oxide or liquid nitrogen.

3 Vascular (venous) sinuses or simply sinuses are veins with very thin walls.

CIRCULATORY ROUTES
1 The largest circulatory route is the systemic circulation. Two of the many subdivisions of the systemic circulation are the coronary circulation and the hepatic portal circulation. Other routes include the cerebral, pulmonary, and fetal circulation.

SYSTEMIC CIRCULATION
1 The systemic circulation takes oxygenated blood from the left ventricle through the aorta to all parts of the body including lung tissue.

2 The aorta is divided into the ascending aorta, the arch of the aorta, and the descending aorta.

3 Blood is returned to the heart through the systemic veins. All the veins of the systemic circulation flow into either the superior or inferior venae cavae or the coronary sinus. They in turn empty into the right atrium.

HEPATIC PORTAL CIRCULATION
1 The hepatic portal circulation collects blood from the veins of the pancreas, spleen, stomach, intestines, and gallbladder and directs it into the hepatic portal vein of the liver.

2 This circulation enables the liver to utilize nutrients and detoxify harmful substances in the blood.

PULMONARY CIRCULATION
1 The pulmonary circulation takes deoxygenated blood from the right ventricle to the lungs and returns oxygenated blood from the lungs to the left atrium.

2 It allows blood to be oxygenated for systemic circulation.

FETAL CIRCULATION
1 The fetal circulation involves the exchange of materials between fetus and mother.

2 The fetus derives its oxygen and nutrients and eliminates its carbon dioxide and wastes through the maternal blood supply by means of a structure called the placenta.

3 At birth, when lung, digestive, and liver functions are established, the special structures of fetal circulation are no longer needed.

PHYSIOLOGY OF CIRCULATION
1 Blood flows from regions of higher to lower pressure.

2 The established pressure gradient is aorta (100 mm Hg) to arteries (100–40 mm Hg) to arterioles (40–25 mm Hg) to capillaries (25–12 mm Hg) to venules (12–8 mm Hg) to veins (10–5 mm Hg) to venae cavae (2 mm Hg) to right atrium (0 mm Hg).

PRESSURE
1 Pressure is determined by volume of blood, resistance against flow, cardiac output, and extensibility and elasticity of arteries.

2 Rate of flow is determined by the diameter of a blood vessel. The greater the diameter, the faster the flow.

3 Bloodflow is also related to the total cross-sectional area of the blood vessels. Blood flows most rapidly where the cross-sectional area is least.

PRODUCTION OF PRESSURE GRADIENT
1 As the blood passes through the arteries to the capillaries and on to the veins, it continually loses pressure as it rubs against vessel walls.

FACTORS AIDING VENOUS RETURN

1 The return of venous blood to the heart is assisted by increasing diameters of veins as they approach the heart, skeletal muscle contractions, respirations, and valves.

CHECKING CIRCULATION

1 The pulse tells us the rate of the heartbeat. The blood pressure measures the pressure the blood is under.

PULSE

1 Pulse is the alternate expansion and elastic recoil of an artery. It may be felt in any artery that lies near the surface or over a hard issue.

2 A normal rate is between 70 and 80 beats per minute.

BLOOD PRESSURE

1 Blood pressure is the pressure exerted by blood on the wall of an artery. It is measured by the use of a sphygmomanometer.

2 Systolic blood pressure is the force of blood recorded during ventricular contraction. Diastolic blood pressure is the force of blood recorded during ventricular relaxation. The average blood pressure is 120/80 mm Hg.

3 Pulse pressure is the difference between systolic and diastolic pressure. It averages 40 mm Hg and provides information about the condition of arteries.

REVIEW QUESTIONS

1 Describe the structural and functional differences among arteries, capillaries, and veins.

2 Discuss the importance of the elasticity and contractility of arteries. What is an anastomosis?

3 Define varicose veins and hemorrhoids.

4 What is meant by a circulatory route? Define systemic circulation.

5 Diagram the major divisions of the aorta, their principal arterial branches, and the regions supplied.

6 Trace a drop of blood from the arch of the aorta through its systemic circulatory route to the tip of the big toe on your left foot and back to the heart again. Remember that the major branches of the arch are the brachiocephalic artery, left common carotid artery, and left subclavian artery. Be sure to indicate which veins return the blood to the heart.

7 What is the circle of Willis? Why is it important?

8 What are visceral branches of an artery? Parietal branches?

9 What major organs are supplied by branches of the thoracic aorta? How is blood returned from these organs to the heart?

10 What organs are supplied by the celiac, superior mesenteric, renal, inferior mesenteric, inferior phrenic, and middle sacral arteries? How is blood returned to the heart?

11 Trace a drop of blood from the common iliac arteries through their branches to the big toe on your left foot and back to the heart again.

12 What is a deep vein? A superficial vein? Define a venous sinus in relation to blood vessels. What are the three major groups of systemic veins?

13 What is hepatic portal circulation? Describe the route by means of a diagram. Why is this route significant?

14 Define pulmonary circulation. Prepare a diagram to indicate the route. What is the purpose of the route?

15 Discuss in detail the anatomy and physiology of fetal circulation. Be sure to indicate the function of the umbilical arteries, umbilical vein, ductus venosus, foramen ovale, and ductus arteriosus.

16 What is the fate of the special structures involved in fetal circulation once postnatal circulation is established?

17 Describe the cause and treatment of patent ductus arteriosus.

18 How is the vasomotor center in the medulla related to peripheral resistance?

19 Why does blood flow faster in arteries and veins than in capillaries?

20 Identify the factors that assist the return of venous blood to the heart.

21 Define pulse. Where may pulse be felt?

22 Contrast the following: tachycardia, bradycardia, and irregular pulse.

23 What is blood pressure? Describe how systolic and diastolic blood pressure may be recorded by means of a sphygmomanometer.

24 Compare the clinical significance of systolic and diastolic pressure. How are these pressures written?

25 Define pulse pressure. What does this pressure indicate?

Chapter 22

The Lymphatic System

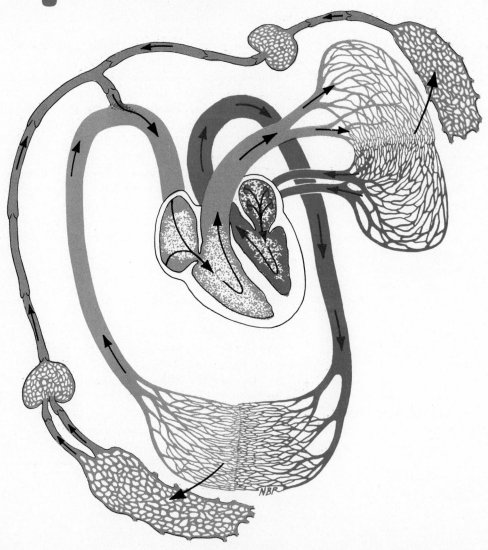

STUDENT OBJECTIVES

- Identify the components and functions of the lymphatic system.

- Compare the structure of veins and lymphatic vessels.

- Describe the structure and function of lymph nodes.

- Identify the clinically important groups of lymph nodes.

- Contrast the functions of the tonsils, spleen, and thymus gland as lymphatic organs.

- List the forces responsible for the circulation of lymph.

- Define immunity.

- Identify the nonspecific and specific defenses against infection.

- Describe the function of an antibody in the antigen-antibody reaction.

- Describe the role of the thymus gland in antibody production.

- Describe the role of lymphocytes and immunity.

- Differentiate between active and passive immunity.

- Describe the effects of an allergic reaction on the body.

- Explain the rejection phenomenon in transplantation.

- Contrast the various types of transplants.

- Discuss immunosuppressive techniques used to control rejection of transplants.

- Define implantation and autoimmune diseases.

- Define medical terminology associated with the lymphatic system.

Lymph, lymph vessels, a series of small masses of lymphoid tissue called lymph nodes, and three organs – tonsils, thymus, and spleen – make up the **lymphatic system.** The primary function of the lymphatic system is to drain, from the tissue spaces, protein-containing fluid that escapes from the blood capillaries. Such proteins cannot be directly reabsorbed. Other functions of the lymphatic system are to transport fats from the digestive tract to the blood, to produce lymphocytes, and to develop immunities.

LYMPHATIC VESSELS

Lymphatic vessels originate as blind-end tubes that begin in spaces between cells. The tubes are called **lymph capillaries.** Lymph capillaries originate in most parts of the body. They are, however, slightly larger and more permeable than blood capillaries. Just as blood capillaries converge to form venules and veins, lymph capillaries unite to form larger and larger lymph vessels called **lymphatics** (Figure 22-1). Lymphatics resemble veins in

structure but have thinner walls and more valves and contain lymph nodes at various intervals. Ultimately, lymphatics converge into two main channels – the thoracic duct and the right lymphatic duct.

The **thoracic duct,** or **left lymphatic duct,** begins as a dilation in front of the second lumbar vertebra. This dilation is called the *cisterna chyli.* The thoracic duct receives lymph from the left side of the head, neck, and chest, the left upper extremity, and the entire body below the ribs. It then empties the lymph into the left subclavian vein. A pair of valves at this junction prevents the passage of venous blood into the thoracic duct. The **right lymphatic duct** drains lymph from the upper right side of the body and empties it into the right subclavian vein at the junction of the right subclavian and jugular veins.

Lymphangiography is a procedure by which lymphatic vessels and lymph organs are filled with an opaque substance in order to be filmed. Such a film is called a *lymphangiogram.* Lymphangiograms are useful in detecting edema and carcinomas and in localizing lymph nodes

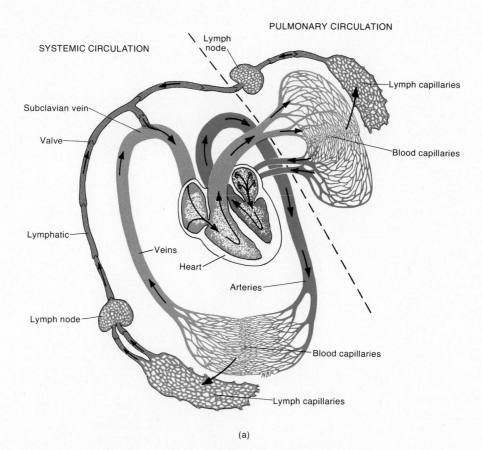

(a)

Figure 22–1 Lymphatic system. (a) Schematic representation of the relationship of the lymphatic system to the cardiovascular system.

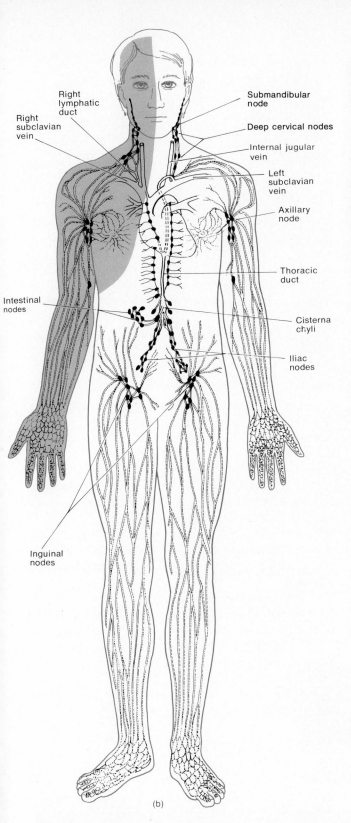

Right
lymphatic
duct

Right
subclavian
vein

Submandibular
node

Deep cervical nodes

Internal jugular
vein

Left
subclavian
vein

Axillary
node

Thoracic
duct

Intestinal
nodes

Cisterna
chyli

Iliac
nodes

Inguinal
nodes

(b)

Figure 22–1 (cont.) Lymphatic system. (b) Location of the principal lymphatics and lymph nodes. The colored area indicates those portions of the body drained by the right lymphatic duct. All other areas of the body are drained by the thoracic duct.

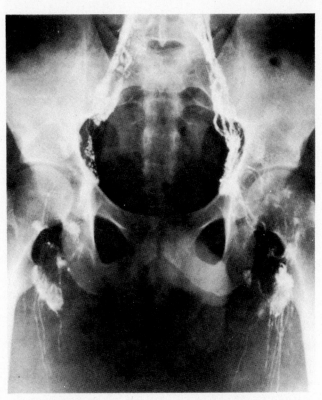

Figure 22–2 Normal lymphangiogram of the upper thighs and pelvis. Can you identify the lymphatics and lymph nodes? (Courtesy of Lester W. Paul and John H. Juhl, *The Essentials of Roentgen Interpretation,* 3d ed., Harper & Row, Publishers, Inc., New York, 1972.)

for surgical or radiotherapeutic treatment. A normal lymphangiogram of lymphatic vessels and a few nodes in the upper thighs and pelvis is shown in Figure 22-2.

LYMPH NODES

The oval or bean-shaped structures located along the length of lymphatics are called **lymph nodes** or **lymph glands.** They range from 1 to 25 mm (0.04 to 1 inch) in length. Structurally, a lymph node contains a slight depression on one side called a *hilum* where the blood vessels enter and leave the node (see Figure 22-3). Each node is covered by a *capsule* of fibrous connective tissue that extends into the node. The capsular extensions are called *trabeculae.* The capsule, trabeculae, and hilum constitute the stroma (framework) of a lymph node. The parenchyma of a lymph node is specialized into two regions. The outer *cortex* contains densely packed lymphocytes arranged in masses called *lymph nodules.* The nodules often contain lighter-staining central areas, the *germinal centers,* where lymphocytes are produced. The

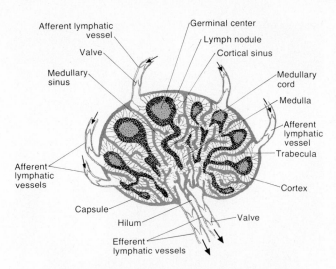

Figure 22–3 Structure of a lymph node. The path taken by circulating lymph is indicated by the arrows.

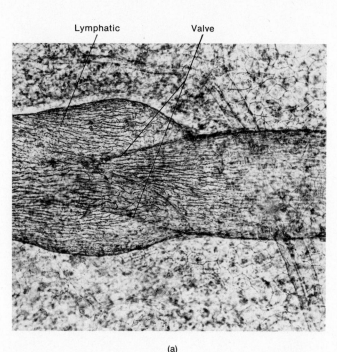

(a)

Figure 22–4 Histology of the lymphatics and lymph nodes. (a) Longitudinal section of a lymphatic at a magnification of $50\times$. (Courtesy of Carolina Biological Supply Company.)

inner region of a lymph node is called the *medulla.* In the medulla, the lymphocytes are arranged in strands called *medullary cords.*

The circulation of lymph through a node involves afferent lymphatic vessels, sinuses in the node, and efferent lymphatic vessels. *Afferent lymphatic vessels* enter the convex surface of the node at several points. They contain valves that open toward the node so that the circulation of lymph through the afferent lymphatic vessels is into the lymph node. Once inside the lymph node, the lymph enters the sinuses of the node, which are a series of irregular channels. Lymph from the afferent lymphatic vessels enters the *cortical sinuses* under the capsule. From here the lymph circulates to the *medullary sinuses* between the medullary cords. From these sinuses, the lymph circulates into the *efferent lymphatic vessels.* These vessels are located at the hilum of the lymph node. The efferent vessels are wider and fewer in number than the afferent vessels and contain valves that open away from the lymph node. Thus they convey lymph out of the node.

As the lymph circulates through the nodes, it is processed by fixed phagocytic cells of the reticuloendothelial system called macrophages that line the sinuses. Macrophages filter lymph of bacteria, dirt, and cell debris. Lymph nodes also give rise to lymphocytes and/or plasma cells. The plasma cells produce antibodies. At times the number of entering microbes is so great that the node itself may become infected. It then becomes enlarged and tender. Histological features of lymphatics and lymph nodes are shown in Figure 22-4.

Many lymph nodes appear in groups or chains in certain areas of the body. Among the larger and clinically important groups are the following:

1 **Deep cervical lymph nodes**   Located along the internal jugular veins; drain lymph from the head and neck.

2 **Submandibular lymph nodes**   Located superficial to the submandibular glands at the lower border of the mandible; drain the nose, lips, and teeth.

3 **Axillary lymph nodes**   Located in the underarm and chest region; drain the skin and muscles of the chest, including the breasts.

4 **Inguinal nodes**   Located in the groin; drain the external genitals and lower extremities.

LYMPHATIC ORGANS

Tonsils

Tonsils are basically masses of lymphoid tissue embedded in mucous membrane. The *pharyngeal tonsils* are embedded in the posterior wall of the nasopharynx (see Figure 23-4). When they become enlarged, they are called adenoids. The *palatine tonsils* are situated in the tonsillar fossae between the pharyngopalatine and glossopalatine arches (see Figure 24-4). These are the ones

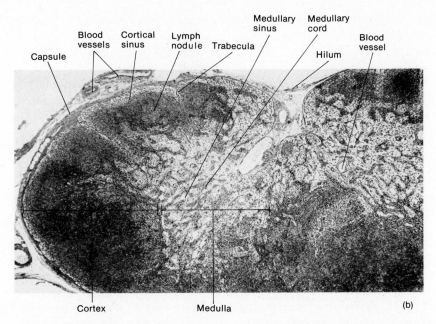

Capsule   Blood vessels   Cortical sinus   Lymph nodule   Trabecula   Medullary sinus   Medullary cord   Hilum   Blood vessel

Cortex   Medulla   (b)

Figure 22–4 (cont.) Histology of the lymphatics and lymph nodes. (b) Section through a lymph node at a magnification of 40×. (Courtesy of Edward J. Reith, from *Atlas of Descriptive Histology*, by Edward J. Reith and Michael H. Ross, Harper & Row, Publishers, Inc., New York, 1970.)

commonly removed by a tonsillectomy. The *lingual tonsil* is located at the base of the tongue and may also have to be removed by a tonsillectomy (see Figure 24-5). The tonsils are supplied with reticuloendothelial cells and filter lymph.

Spleen

The oval **spleen** is the largest mass of lymphatic tissue in the body, measuring about 12 cm (5 inches) in length. It is situated in the left hypochondriac region inferior to the diaphragm and posterolateral to the stomach. The spleen is surrounded by a capsule of fibroelastic tissue and scattered smooth muscle. The capsule, in turn, is covered by a serous membrane: the peritoneum. Like lymph nodes, the spleen contains trabeculae and a hilum. The capsule, trabeculae, and hilum constitute the stroma of the spleen.

The parenchyma of the spleen consists of two different kinds of tissue called white pulp and red pulp. *White pulp* is essentially lymphoid tissue arranged around arteries. The clusters of lymphocytes surrounding the arteries at intervals and the expansions are referred to as *splenic nodules* or *Malpighian corpuscles (bodies)*. The *red pulp* consists of venous sinuses filled with blood and cords of splenic tissue called *splenic* or *Billroth's cords*. Veins are closely associated with the red pulp.

The splenic artery and vein and the efferent lymphatics pass through the hilum. The spleen phago-

cytoses bacteria and worn-out red blood cells and platelets. It also produces lymphocytes and plasma cells. In addition, the spleen stores and releases blood in case of hemorrhage. The release seems to be purely sympathetic. The sympathetic impulses cause the smooth muscle of the spleen to contract.

Thymus Gland

The **thymus gland** is a bilobed mass of lymphatic tissue in the upper thoracic cavity. It is found along the trachea behind the sternum. The thymus is relatively large in children. It reaches its maximum size at puberty and then undergoes involution. Eventually, it is replaced by fat and connective tissue.

LYMPH CIRCULATION

The flow of lymph is from tissue spaces to the large lymphatic ducts to the subclavian veins. The flow is maintained primarily by the milking action of muscle tissue. Skeletal muscle contractions compress lymph vessels and force lymph toward the subclavian veins. Moreover, lymph vessels, like veins, contain valves, and the valves ensure the movement of lymph toward the subclavian veins. Another factor that maintains lymph flow is respiratory movements. These movements create a pressure gradient (difference) between the two ends of the lymphatic system. Lymph flows from the tissue spaces, where the pressure is higher, toward the thoracic region, where it is lower.

Edema, an excessive accumulation of lymph in tissue spaces, has several possible causes. One is an obstruction, such as an infected node, in the lymphatic channels between the lymphatic capillaries and the subclavian veins. Another is excessive lymph formation and increased permeability of blood capillary walls. A rise in capillary blood pressure, in which interstitial fluid is formed faster than it is passed into lymphatics, also may result in edema.

IMMUNITY

The human body continually attempts to maintain homeostasis by counteracting harmful stimuli in the environment. Frequently, these stimuli are disease-producing organisms, called *pathogens,* or their toxins. Your defenses against disease may be grouped into two broad areas: nonspecific and specific. These defenses provide you with your **immunity** – your ability to overcome the disease-producing effects of certain organisms. Nonspecific defenses represent a wide variety of body reactions against a wide range of pathogens. Specific defense is the production of a specific antibody against a specific pathogen or its toxin.

Nonspecific Defenses

A body reaction that deals with a variety of pathogens is a *nonspecific defense.* Nonspecific defenses are set up by the skin, mucous membranes, phagocytic leucocytes, and reticuloendothelial cells. The secretions of sweat and sebaceous glands are toxic to many microbes. The lacrimal glands of the eyes and the glands in the mucous membranes of the nose and mouth produce an enzyme called lysozyme, which is capable of destroying microbes. Many of the microbes swallowed with food are killed by the hydrochloric acid secretion of the stomach lining. Finally, throughout the respiratory tract dust particles that carry microbes are trapped in the sticky mucus. Cilia that line the tract move the mucus up to the mouth to be swallowed or spit out. If microbes penetrate the defenses of the skin and mucous membranes, the role of the leucocytes and reticuloendothelial cells is to kill them by phagocytosis.

Specific Defense

Although the nonspecific defenses of the body are generally effective against microbes, they cannot fight the battle alone. Nor can they combat the toxins produced by pathogens. The body's second line of defense, then, is a *specific defense* that involves antibody production. Antibodies are proteins that inactivate materials called antigens. The antigens are usually proteins themselves. Pathogens and their toxins are antigens. Tissues from another person, such as unmatched blood or a transplanted organ, can also be antigens. The antigen-antibody response is a specific defense – only a particular antibody can combat a particular antigen.

LYMPHOCYTES

Lymphoid tissue capable of responding to invading antigens is located in lymph nodes, the thymus gland, the spleen, and bone marrow. Within this lymphoid tissue are special lymphocytes that can respond to antigens in two ways. These lymphocytes are termed T cells and B cells.

T cells originate from the bone marrow and thymus gland. They are small lymphocytes that react with antigens and destroy them. They kill other cells (antigens) introduced by tissue transplantation, fungus cells, viruses, and bacteria associated with slow-developing diseases. When a T cell comes in contact with an antigen, a specific attachment occurs. Then the T cell secretes toxins and enzymes that dissolve the membrane of the antigen and digest its contents. Both the antigen and the T cell are destroyed in the process. T cells are the ones that bring about the rejection of transplants. They are responsible for what is called *cellular immunity.* Once produced, T cells circulate in the blood, spread throughout the body, and lodge in lymph nodes and other lymphoid tissues.

When a T cell is sensitized – that is, committed against a specific antigen – it is capable of forming large numbers of daughter cells that, in turn, divide many times and release thousands of T cells into circulation. Even after the antigen is removed from the body, T cells remain in lymphoid tissue and continue to form new T cells for months or years.

B cells originate from the bone marrow, spleen, and lymph nodes. When exposed to a specific antigen, they are transformed into *plasma cells.* Plasma cells produce antibodies, which are modified gamma globulins. The antibodies so produced are released into circulation. Once committed, some B cells remain in lymph nodes for years. If the same antigen is introduced into the body at a later date, they are transformed into plasma cells that produce antibodies. The immunity produced by B cells is known as *humoral immunity* since the antibodies circulate freely in the blood and tissue spaces. Humoral immunity protects mainly against acute diseases – pneumonia, staphylococcal infection, and streptococcal infection.

THYMUS GLAND

The thymus gland itself does not produce antibodies. However, it plays a crucial role in immunity. Long ago it was noticed that children born without a thymus gland cannot fight disease. These children usually die young from a serious infection. From fetal life to several years after birth, the thymus gland seems to set up the body's antibody system. Apparently, the thymus gland programs lymphocytes by altering their DNA. These cells develop into T cells. Once programmed, the T cells become established in lymph nodes and other lymphoid tissues.

The thymus gland is also believed to secrete a hormone, called *thymosin,* that enables B cells to develop into plasma cells. The plasma cells then produce antibodies against antigens. The overall function of the thymus gland is to prepare lymphocytes for immune responses. Once the thymus gland has performed its functions, it starts to degenerate.

ACTIVE AND PASSIVE IMMUNITY

Whether or not a microbe makes you sick often depends on whether you have been exposed to it before. The first time a lymphocyte is exposed to an antigen, its response is a little slow. The cell needs time to adjust its protein assembly lines or time to become sensitized. During this time, the microbes are free to multiply and produce toxins and other symptoms of the disease. If the same kind of microbe invades your body again in the future, your T cells and B cells may still be geared to recognizing the original antigen. In this case, the antigen-antibody response may occur before the antigens have a chance to bring on the symptoms. Such protection against future sickness is called *active immunity.* Active immunity may also be acquired through vaccinations with dead pathogens, attenuated (weakened) organisms, or low doses of their toxins. The proteins in the dead or attenuated pathogens are capable of stimulating an antigen-antibody response, but they cannot hurt you. Toxins are given in doses just high enough to stimulate an antigen-antibody response. The doses are too low, however, to cause disease.

Another form of immunity is called *passive immunity.* In this case antibodies or sensitized T cells are injected from an animal or person previously exposed to a disease. The globulin antibodies, for example, are effective against hepatitis and measles. Passive immunity gives you only temporary protection. It is generally used when a person has been exposed to a disease to which he or she is not immune and does not have enough time to manufacture antibodies.

The antigen-antibody response is essential to survival. Under certain circumstances, however, it may create problems. Two such problems are related to allergies and tissue rejection.

Allergy

A person who is overly reactive to an antigen is said to be *allergic* or *hypersensitive.* Whenever an allergic reaction occurs, there is tissue injury.

The antigens that induce an allergic reaction are called *allergens.* Examples of allergens include certain foods, antibiotics such as penicillin, cosmetics, chemicals in plants such as poison ivy, pollens, and even microbes. In the first stage of an allergic reaction, the cells of the body become sensitized by antibodies. That is, they become subject to injury if the allergen enters the body later. Sensitivity occurs when a person receives an initial dose of the allergen. When the allergen enters the body again in the second stage of an allergic reaction, tissue damage results.

In the allergic reaction, antibodies formed against the first dose of allergens remain attached to the cells that produced them. At this point, the cells become sensitized. When the antibodies react again with allergens introduced a second time, the antigen-antibody reaction destroys the cells as well as the allergens. Destruction of the cells triggers off several physiological responses:

1 Injured cells release histamine. Large quantities of histamine cause tissue inflammation and contraction of smooth muscle fibers – especially in the breathing tubes and blood vessels, causing them to constrict. Histamine also increases the permeability of blood vessels so that fluid moves from the vessels into the interstitial spaces, causing edema.

2 In a severe allergic reaction, *anaphylactic shock* may result from the prolonged effects of histamine. The respiratory tubes are continuously constricted, and the occurrence of edema is accelerated. This condition lowers the blood volume even more. If anaphylactic shock is not counteracted, death may result. The effects of anaphylaxis may be reversed by administering epinephrine or antihistamines, drugs that inactivate histamine.

Tissue Rejection

Transplantation involves the replacement of an injured or diseased tissue or organ. Usually, the body considers the proteins in the transplanted tissue or organ to be foreign and produces antibodies against them. This phenomenon is known as *tissue rejection.* Rejection can be somewhat reduced by administering drugs that inhibit the body's ability to form antibodies.

TYPES OF TRANSPLANTS

The closer the relationship of donor and recipient, the more successful the transplant. *Isografts* are transplants in which the donor and recipient have identical genetic backgrounds. They include transplants between identical twins and the transplantation of tissues from one part of the body to another. This type of transplant has been most successful.

An *allograft* is a transplant between individuals of the same species—but with different genetic backgrounds. The success of this type of transplant has been moderate. Frequently, it is used as a temporary measure until the damaged or diseased tissue is able to repair itself. Skin transplants from other individuals and blood transfusions might properly be considered allografts.

A *xenograft* is a transplant between animals of different species. This type of transplantation is used primarily as a physiological dressing over severe burns. Xenografts are presently restricted to laboratory animals.

The one organ allograft that has been quite successful is the thymus. Children born without a thymus can now receive the gland from an aborted fetus. The thymus-deficient child cannot produce antibodies and thus cannot reject the transplant. Rejection later on indicates that the child is manufacturing antibodies and no longer needs the organ.

IMMUNOSUPPRESSIVE THERAPY

Scientists are looking for *immunosuppressive drugs* that can stop antibody destruction of the transplant without destroying the antibody responses that protect us against microbes. To date, no drug has been entirely successful. It has been found, however, that horses produce antibodies that react with lymphocytes of other species but not with other foreign tissue. When this antibody was used with other drugs, it increased the first-year survival rate of kidney transplant patients to 90 percent.

IMPLANTATION

The replacement of a tissue or organ with an artificial device is known as *implantation*. Artificial replacements are constructed of materials, such as plastic, that do not stimulate antibody production. In many cases, the artificial device functions quite well. Plastics are widely used to replace blood vessels, valves, and bones. The major problem has been to develop devices that can duplicate complex physiological activities of organs and yet remain small enough to be implanted. Artificial pacemakers, which control heartbeat, have been implanted for years. However, the artificial kidney, which duplicates the activities of a natural one, is much too large to be placed inside the body. Perhaps someday the machine will be miniaturized to a point where it too can be implanted.

Autoimmune Diseases

Under normal conditions, the body's immune mechanism is able to recognize its own tissues and chemicals. Thus your immune mechanism normally produces neither T cells nor B cells against your own antigens. Such a recognition is called *tolerance*. There is no satisfactory explanation of tolerance. At times, however, tolerance breaks down and the body starts to produce T cells and B cells against its own antigens. This *autoimmune disease* seems to arise from a change in a natural body chemical that becomes antigenic and begins to stimulate an antigen-antibody response. Many persons develop autoimmune diseases from immunological damage to specific body tissues. In rheumatic fever, the body becomes immunized against tissues of the joints and heart. In myasthenia gravis, the person develops an immunity against muscles. As a result, nerve impulses are not transmitted to the muscles and paralysis results.

MEDICAL TERMINOLOGY

Adenitis (*adeno* = gland)   Enlarged, tender, and inflamed lymph nodes resulting from an infection.

Elephantiasis   Great enlargement of a limb (especially lower limbs) and scrotum resulting from obstruction of lymph glands or vessels caused by a tiny parasitic worm.

Hypersplenism   Abnormal splenic activity involving highly increased blood cell destruction.

Lymphadenectomy (*ectomy* = removal)   Removal of a lymph node.

Lymphadenopathy (*patho* = disease)   Enlarged, sometimes tender lymph glands.

Lymphangitis   Inflammation of the lymphatic vessels.

Lymphedema   Accumulation of lymph fluid producing subcutaneous tissue swelling.

Lymphoma   Any tumor composed of lymph tissue. Malignancy of reticuloendothelial cells of lymph nodes is called Hodgkin's disease.

Lymphangioma   A benign tumor of the lymph vessels.

Lymphostasis (*stasis* = halt)   A lymph flow stoppage.

LYMPHATIC VESSELS

1 Lymphatic vessels are similar in structure to veins. All lymphatics deliver lymph to the thoracic duct or right lymphatic duct.

LYMPH NODES

1 Lymph nodes are oval structures located along lymphatics. Lymph passing through the nodes is filtered, and it picks up antibodies and agranular leucocytes.

LYMPHATIC ORGANS

1 Lymph organs that filter lymph and add white blood cells and antibodies are the tonsils, spleen, and thymus gland.

LYMPH CIRCULATION

1 Lymph flows as a result of skeletal muscle contractions and respiratory movements.

IMMUNITY

1 The ability to overcome the harmful effects of disease-producing organisms is provided by nonspecific and specific defenses.

2 Nonspecific defenses include the skin and mucous membranes and their secretions; the inflammatory response and phagocytosis; and the reticuloendothelial system.

3 Specific defenses are the production of antibodies against a specific pathogen or its toxin. Antibody-producing cells are programmed by the thymus.

4 Within lymphoid tissue are special lymphocytes called T cells and B cells.

5 There are two types of immunity: active and passive.

6 An allergy is an overreaction to an antigen called an allergen. In allergic reactions, histamines are released by damaged tissues and bring about physiological responses that could cause anaphylactic shock.

7 Tissue rejection is the inactivation of foreign antigens (transplants) by antibodies. It is counteracted by immuno-suppressive drugs.

8 Types of transplants include isografts, allografts, and xenografts.

9 Examples of autoimmune diseases are rheumatic fever and myasthenia gravis.

REVIEW QUESTIONS

1 Identify the components and functions of the lymphatic system.

2 How do lymphatic vessels originate? Compare veins and lymphatics with regard to structure.

3 Construct a diagram to indicate the role of the thoracic duct and right lymphatic duct in draining lymph from different body regions.

4 What is a lymphangiogram? What is its diagnostic value?

5 Describe the structure of a lymph node. What functions do lymph nodes serve?

6 Identify four groups of clinically important lymph nodes. Compare the functions of the tonsils, spleen, and thymus gland as lymphatic organs.

7 Describe how lymph circulates.

8 Define immunity. Distinguish between nonspecific and specific defenses that provide immunity.

9 Contrast the functions of the skin and mucous membranes, blood, and reticuloendothelial system in providing nonspecific defenses against pathogens.

10 Distinguish between active and passive immunity.

11 Describe the role of the thymus gland in antibody production.

12 Describe in detail the allergic reaction.

13 Define the various transplants.

14 Explain the term *rejection* as it applies to transplants. What immunosuppressive techniques are used to overcome rejection?

15 Define autoimmune. Give two examples of autoimmune diseases.

Chapter 23
The Respiratory System

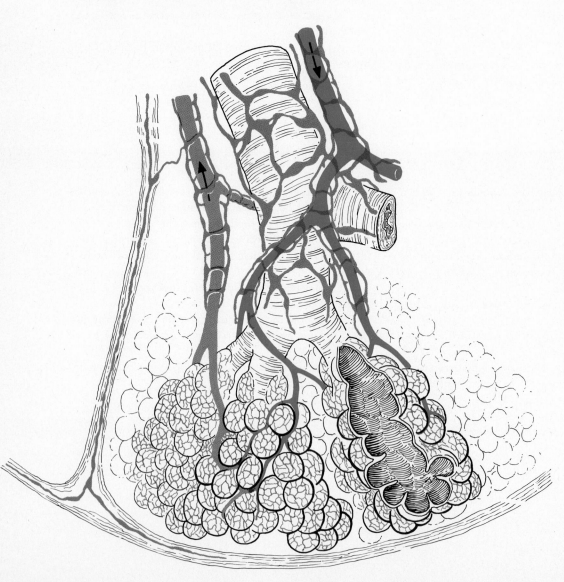

STUDENT OBJECTIVES

■ Identify the organs of the respiratory system.

■ Compare the structure of the external and internal nose.

■ Contrast the functions of the external and internal nose in filtering, warming, and moistening air.

■ Differentiate the three regions of the pharynx and describe their roles in respiration.

■ Identify the anatomical features of the larynx related to respiration and voice production.

■ Describe the tubes that form the bronchial tree with regard to structure and location.

■ Contrast tracheotomy and intubation as alternative methods for clearing air passageways.

■ Identify the coverings of the lungs and the gross anatomical features of the lungs.

■ Describe the structure of a lobule of the lung.

■ Describe the role of alveoli in the diffusion of respiratory gases.

■ List the sequence of pressure changes involved in inspiration and expiration.

■ Compare the volumes and capacities of air exchanged in respiration.

■ Define the partial pressure of a gas, Boyle's law, Charles' law, Dalton's law, and Henry's law.

■ Describe the mechanisms of external and internal respiration based on differences in partial pressure of O_2 and CO_2.

■ Describe how O_2 and CO_2 are carried by the blood.

■ Describe the parts of the nervous system that control respiration.

■ Compare the roles of the Hering-Breuer reflex and the pneumotaxic center in controlling respiration.

■ Describe the effects of chemical stimuli and pressure in determining the rate of respiration.

■ List the basic steps involved in heart-lung resuscitation including the Heimlich maneuver.

■ Describe the effects of pollutants on the epithelium of the respiratory system.

■ Describe the administration of medication by nebulization.

■ Define hay fever, bronchial asthma, emphysema, pneumonia, and hyaline membrane disease as disorders of the respiratory system.

■ Define medical terminology associated with the respiratory system.

Cells need a continuous supply of oxygen to carry out the activities that are vital to their survival. Many of these activities release quantities of carbon dioxide. Since an excessive amount of carbon dioxide is poisonous to cells, the gas must be eliminated quickly and efficiently. The two systems that supply oxygen and eliminate carbon dioxide are the cardiovascular system and the respiratory system. The **respiratory system** consists of organs that exchange gases between the atmosphere and blood. These organs are the nose, pharynx, larynx, trachea, bronchi, and lungs (Figure 23-1). In turn, the blood transports gases between the lungs and the cells. The overall exchange of gases between the atmosphere, the blood, and the cells is **respiration**. The respiratory and cardiovascular systems participate equally in respiration. Failure of either system has the same effect on the body: rapid death of cells from oxygen starvation and disruption of homeostasis.

ORGANS

Nose

The **nose** has an external portion and an internal portion inside the skull. Externally, the nose consists of a supporting framework of bone and cartilage covered with skin and lined with mucous membrane. The bridge of the nose is formed by the nasal bones, which hold it in a fixed position. Because it has a framework of pliable cartilage, the rest of the external nose is quite flexible. On the undersurface of the external nose are two openings called the *nostrils* or *external nares* (singular *naris;* see Figure 23-2). The surface anatomy of the nose is shown in Figure 23-3.

The internal region of the nose is a large cavity in the skull that lies below the cranium and above the mouth. Anteriorly, the internal nose merges with the external nose, and posteriorly it communicates with the throat (pharynx) through two openings called the *internal nares.* Four paranasal sinuses (frontal, sphenoidal, maxillary, and ethmoidal) and the nasolacrimal ducts also open into the internal nose. The lateral walls of the internal nose are formed by the ethmoid, maxillae, and inferior conchae bones. The ethmoid forms the roof. The floor is formed by the palatine bones and the maxilla of the hard palate. Occasionally the palatine and maxillary bones fail to fuse during embryonic life, and a child is born with a crack in the bony wall that separates the internal nose from the mouth. This condition is called *cleft palate.*

The inside of both the external and internal nose, called the *nasal cavity,* is divided into right and left chambers, the *nasal fossae,* by a vertical partition called the *nasal septum.* Cartilage is the primary material making up the anterior portion of the septum. The remainder is formed by the vomer and the perpendicular plate of the ethmoid. The anterior portions of the nasal fossae, which are just inside the nostrils, are called the *vestibules.* The vestibules are surrounded by cartilage as opposed to bone of the upper nasal cavity. The interior structures of the nose are specialized for three functions: incoming air is warmed, moistened, and filtered; olfactory stimuli are received; and large hollow resonating chambers are provided for speech sounds.

When air enters the nostrils, it passes first through the vestibule. The vestibule is lined by skin containing coarse hairs that filter out large dust particles. The air then passes into the rest of the fossa. Three shelves formed by projections of the superior, middle, and inferior conchae or turbinates extend out of the lateral wall of the fossa. The conchae, almost reaching the septum, subdivide each nasal fossa into a series of groovelike passageways—the *superior, middle,* and *inferior meati.* Mucous membrane lines the fossa and its shelves. The olfactory receptors lie in the membrane lining the upper portion of the fossa, also called the olfactory region. Below the olfactory region, the membrane contains pseudostratified ciliated columnar cells with many goblet cells and capillaries. As the air whirls around the turbinates and meati, it is warmed by the capillaries. Mucus secreted by the goblet cells moistens the air and traps dust particles. Drainage from the

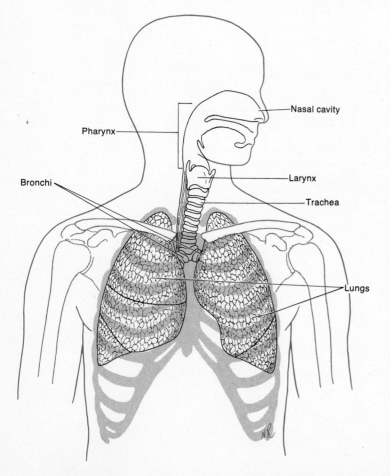

Figure 23–1   Organs of the respiratory system.

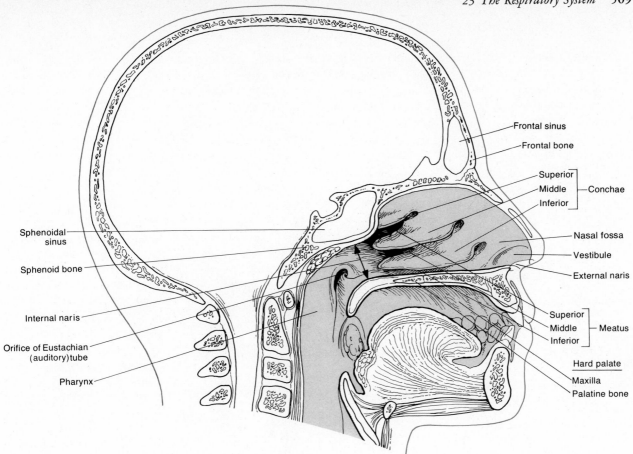

Figure 23–2 Right nasal fossa seen in sagittal section.

lacrimal ducts, and perhaps secretions from the paranasal sinuses, also help moisten the air. The cilia move the mucus-dust packages along to the throat so they can be eliminated from the body.

Pharynx

The **pharynx**, or throat, is a tube about 13 cm (5 inches) long that starts at the internal nares and runs partway down the neck (Figure 23-4). It lies just in back of the nasal cavity and oral cavity and just in front of the

1 Root Superior attachment of nose at forehead located between eyes.

2 Apex Tip of nose.

3 Dorsum nasi Rounded anterior border connecting root and apex; in profile, may be straight, convex, concave, or wavy.

4 Nasofacial angle Point at which side of nose blends with tissues of face.

5 Ala Convex flared portion of inferior lateral surface; unites with upper lip.

6 External nares External openings into nose.

7 Bridge Superior portion of dorsum nasi, superficial to nasal bones.

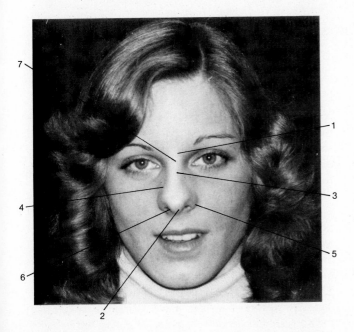

Figure 23–3 Surface anatomy of the nose in anterior view. (Courtesy of Donald Castellaro and Deborah Massimi.)

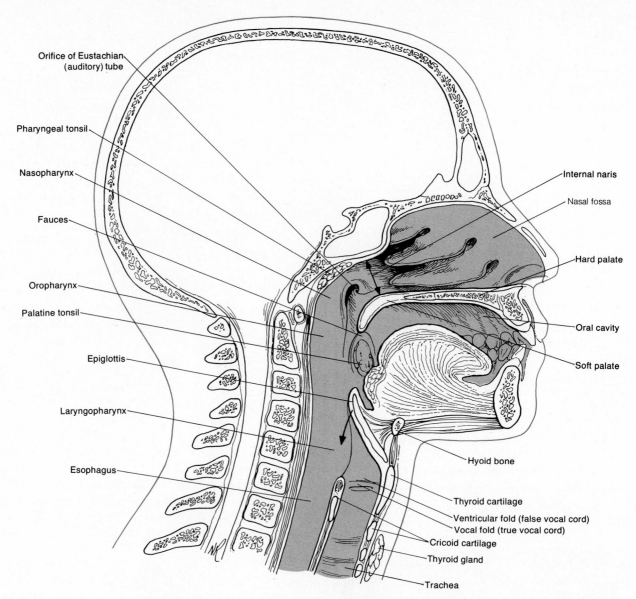

Orifice of Eustachian (auditory) tube

Pharyngeal tonsil

Nasopharynx

Fauces

Oropharynx

Palatine tonsil

Epiglottis

Laryngopharynx

Esophagus

Internal naris

Nasal fossa

Hard palate

Oral cavity

Soft palate

Hyoid bone

Thyroid cartilage

Ventricular fold (false vocal cord)

Vocal fold (true vocal cord)

Cricoid cartilage

Thyroid gland

Trachea

Figure 23–4   Head, neck, and upper chest seen in sagittal section.

cervical vertebrae. Its walls are composed of skeletal muscles and are lined with mucous membrane. The functions of the pharynx are limited to serving as a passageway for air and food and providing a resonating chamber for speech sounds.

The uppermost portion of the pharynx is called the *nasopharynx.* This part lies behind the internal nasal cavity and extends over the soft palate. There are four openings in its walls: two internal nares plus two openings that lead into the Eustachian (auditory) tubes. The posterior wall of the nasopharynx also contains the pharyngeal tonsil, or adenoid. Through the internal nares the nasopharynx exchanges air with the nasal cavities and receives the packages of dust-laden mucus. The nasopharynx has a lining of pseudostratified ciliated

epithelium. Cilia in the walls of the nasopharynx move the mucus down toward the mouth. The nasopharynx also exchanges small amounts of air with the auditory canal so that the air pressure inside the middle ear equals the pressure of the atmospheric air flowing through the nose and pharynx.

The second portion of the pharynx, the *oropharynx,* lies behind the oral cavity and extends from the soft palate down to the level of the hyoid bone. It receives only one opening: the *fauces,* or opening from the mouth. It is lined by stratified squamous epithelium. This portion of the pharynx is both respiratory and digestive in function since it is a common passageway for both air and food. Two pairs of tonsils, the palatine tonsils and the lingual tonsils, are found in the oropharynx. The lingual tonsils lie at the base of the tongue (see Figure 24-5).

The lowest portion of the pharynx is called the *laryngopharynx.* The laryngopharynx extends downward from the hyoid bone and empties into the esophagus (food tube) posteriorly and into the larynx (voice box) anteriorly. Like the oropharynx, the laryngopharynx is both a respiratory and a digestive pathway in function and is lined by stratified squamous epithelium.

Larynx

The **larynx,** or voice box, is a short passageway that connects the pharynx with the trachea. It lies in the midline of the neck anterior to the fourth through sixth cervical vertebrae. The walls of the larynx are supported by nine pieces of cartilage. Three are single and three are paired. The three single pieces are the large thyroid cartilage and the smaller epiglottic and cricoid cartilage. Of the paired cartilages, the arytenoid cartilages are the most important. The paired corniculate and cuneiform cartilages are of lesser significance. The *thyroid cartilage,* or Adam's apple, consists of two fused plates that form the anterior wall of the larynx and give it its triangular shape (Figure 23-5). In males the thyroid cartilage is bigger than it is in females.

The *epiglottis* is a large, leaf-shaped piece of cartilage lying on top of the larynx. The "stem" of the epiglottis is attached to the thyroid cartilage, but the "leaf" portion is unattached and free to move up and down like a trapdoor. During swallowing, the free edge of the epiglottis forms a lid over the larynx. In this way, the larynx is closed off and liquids and foods are routed into the esophagus and kept out of the trachea. If anything but air passes into the larynx, a cough reflex attempts to expel the material.

The *cricoid cartilage* is a ring of cartilage forming the lower walls of the larynx. It is attached to the first ring of cartilage of the trachea.

The paired *arytenoid cartilages* are pyramidal in shape and located at the superior border of the cricoid cartilage. They attach to the vocal folds and pharyngeal muscles and by their action can move the vocal cords. The *corniculate cartilages* are paired, cone-shaped cartilages. Each is located at the apex of each arytenoid cartilage. The paired *cuneiform cartilages* are rod-shaped cartilages in the mucous membrane fold that connects the epiglottis to the arytenoid cartilages.

Like the other respiratory passageways, the larynx is lined with a ciliated mucous membrane that traps dust not removed in the upper passages.

The mucous membrane of the larynx is arranged into two pairs of folds—an upper pair called the *ventricular folds* (or *false vocal cords*) and a lower pair

called simply the *vocal folds* (or *true vocal cords*). The air passageway between the folds is called the *glottis.* Under the mucous membrane of the true vocal cords lie bands of elastic ligaments stretched between pieces of rigid cartilage like the strings on a guitar. Skeletal muscles of the larynx, called intrinsic muscles, are attached internally to the pieces of rigid cartilage and to the vocal folds themselves. When the muscles contract, they pull the strings of elastic ligaments tight and stretch the cords out into the air passageways so that the glottis is narrowed. If air is directed against the vocal folds, they vibrate and set up sound waves in the column of air in the pharynx, nose, and mouth. The greater the pressure of air, the louder the sound.

Pitch is controlled by the tension on the true vocal cords. If the cords are pulled taut by the muscles, they vibrate more rapidly and a higher pitch results. Lower sounds are produced by decreasing the muscular tension on the cords. Vocal cords are usually thicker and longer in males than in females and vibrate more slowly. Thus men have a lower range of pitch than women.

Sound originates from the vibration of the true vocal cords. But other structures are necessary for converting the sound into recognizable speech. The pharynx, mouth, nasal cavities, and paranasal sinuses all act as resonating chambers that give the voice its human and individual quality. By constricting and relaxing the muscles in the walls of the pharynx we produce the vowel sounds. Muscles of the face, tongue, and lips help us to enunciate words.

Laryngitis is an inflammation of the larynx that is most often caused by a respiratory infection or irritants such as cigarette smoke. Inflammation of the vocal folds themselves causes hoarseness or loss of voice by interfering with the contraction of the cords or by causing them to swell to the point where they cannot vibrate freely. Many long-term smokers acquire a permanent hoarseness from the damage done by chronic inflammation.

Trachea

The **trachea,** or windpipe, is a tubular passageway for air about 12 cm (4 ½ inches) in length and 2.5 cm (1 inch) in diameter. It is located in front of the esophagus and extends from the larynx to the fifth thoracic vertebra, where it divides into right and left primary bronchi.

The tracheal epithelium is pseudostratified. It consists of ciliated columnar cells, goblet cells, and basal cells. The epithelium provides the same protection against dust as the membrane lining the larynx (Figure 23-6). The walls of the trachea are composed of smooth muscle and elastic connective tissue. They are encircled by a series of horizontal incomplete rings of cartilage that look like a series of letter C's stacked one on top of

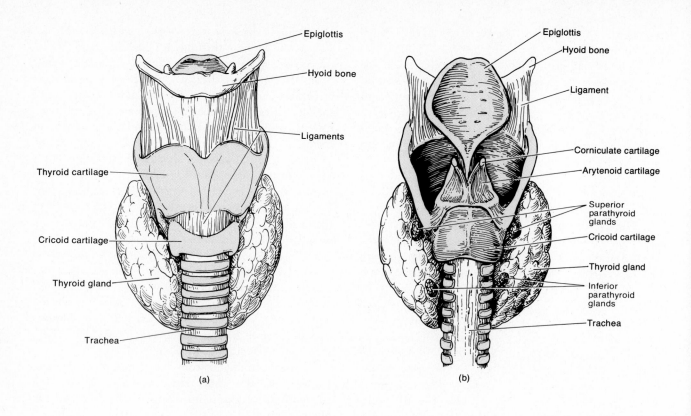

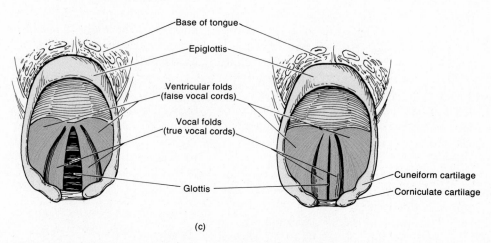

Figure 23–5 Larynx. (a) Anterior view. (b) Posterior view. (c) Viewed from above. In the figure on the left the true vocal cords are relaxed. In the figure on the right the true vocal cords are pulled taut.

the other. The open parts of the C's face the esophagus and permit it to expand into the trachea during swallowing. The solid parts of the C's provide a rigid support so the tracheal walls do not collapse inward and obstruct the air passageway.

Occasionally the respiratory passageways are unable to protect themselves from obstruction. The rings of cartilage may be accidentally crushed, or the mucous membrane may become inflamed and swell so much that

it closes off the air space. Inflamed membranes also secrete a great deal of mucus that may clog the lower respiratory passageways. Or a large object may be breathed in (aspirated) while the glottis is open. In any case, the passageways must be cleared quickly. If the obstruction is above the level of the chest, a *tracheotomy* may be performed. The first step in a tracheotomy is to make an incision in the neck and into the part of the trachea below the obstructed area. The patient breathes

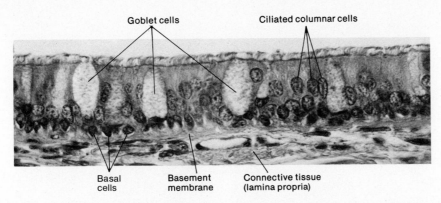

Goblet cells   Ciliated columnar cells

Basal cells   Basement membrane   Connective tissue (lamina propria)

Figure 23–6 Histology of the trachea. Shown here is an enlarged aspect of the tracheal epithelium at a magnification of 640×. (Courtesy of Edward J. Reith, from *Atlas of Descriptive Histology,* by Edward J. Reith and Michael H. Ross, Harper & Row, Publishers, Inc., New York, 1970.)

through a tube inserted through the incision. Another method is *intubation.* A tube is inserted into the mouth and passed down through the larynx and trachea. The firm walls of the tube push back any flexible obstruction, and the inside of the tube provides a passageway for air. If mucus is clogging the airways, it can be suctioned up through the tube.

Bronchi

The trachea terminates in the chest by dividing into a **right primary bronchus,** which goes to the right lung, and a **left primary bronchus,** which goes to the left lung (Figure 23-7a). The right primary bronchus is more vertical, shorter, and wider than the left. As a result, foreign objects that enter the air passageways frequently lodge in it. Like the trachea, the primary bronchi contain incomplete rings of cartilage and are lined by a ciliated columnar epithelium.

Upon entering the lungs, the primary bronchi divide to form smaller bronchi – the *secondary* or *lobar bronchi,* one for each lobe of the lung. (The right lung has three lobes; the left lung has two.) The secondary bronchi continue to branch, forming still smaller bronchi, called *tertiary* or *segmental bronchi,* which divide into *bronchioles.* Bronchioles, in turn, branch into even smaller tubes called *terminal bronchioles.* The continuous branching of the trachea into primary bronchi, secondary bronchi, bronchioles, and terminal bronchioles resembles a tree trunk with its branches and is commonly referred to as the *bronchial tree.* As the branching becomes more extensive in the bronchial tree, several structural changes may be noted. First, rings of cartilage are replaced by plates of cartilage that finally disappear in the bronchioles. Second, as the cartilage decreases, the amount of smooth muscle increases. In addition, the epithelium changes from ciliated columnar to simple

cuboidal in the terminal bronchioles. The fact that the walls of the bronchioles contain a great deal of smooth muscle but no cartilage is clinically significant. During an asthma attack the muscles go into spasm. Because there is no supporting cartilage, the spasms can close off the air passageways.

Bronchography is a technique for examining the bronchial tree. The patient breathes in air that contains a safe dosage of a radiopaque material. The element gives off rays that penetrate the chest walls and expose a film. The developed film, a *bronchogram,* provides a picture of the tree (Figure 23-7b).

Lungs

The **lungs** are paired, cone-shaped organs lying in the thoracic cavity (Figure 23-8). They are separated from each other by the heart and other structures in the mediastinum. Two layers of serous membrane, collectively called the *pleural membrane,* enclose and protect each lung. The outer layer is attached to the walls of the pleural cavity and is called the *parietal pleura.* The inner layer, the *visceral pleura,* covers the lungs themselves. Between the visceral and parietal pleura is a small potential space, the *pleural cavity,* which contains a lubricating fluid secreted by the membranes. This fluid prevents friction between the membranes and allows them to move easily on one another during breathing. Inflammation of the pleural membrane, or *pleurisy,* causes friction during breathing that can be quite painful when the swollen membranes rub against each other.

The lungs extend from the diaphragm to a point just above the clavicles and lie against the ribs in front and back. The broad inferior portion of the lung, the *base,* is concave and fits over the convex area of the diaphragm. The narrow superior portion of the lung is termed the *apex.* The surface of the lung lying against

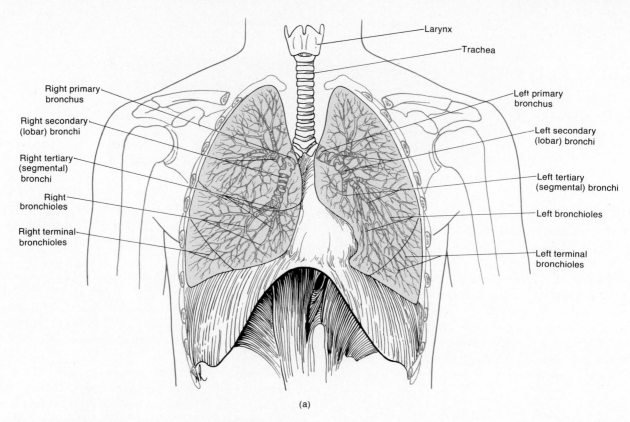

Larynx

Trachea

Right primary
bronchus

Right secondary
(lobar) bronchi

Right tertiary
(segmental)
bronchi

Right
bronchioles

Right terminal
bronchioles

Left primary
bronchus

Left secondary
(lobar) bronchi

Left tertiary
(segmental) bronchi

Left bronchioles

Left terminal
bronchioles

(a)

Figure 23–7   Air passageways to the lungs. (a) Diagram of the bronchial tree in relationship to the lungs.

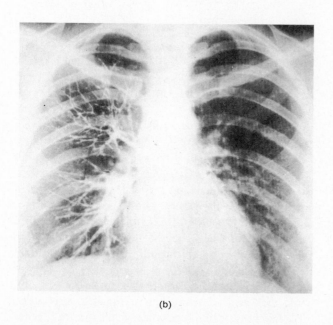

(b)

Figure 23–7 (cont.)   Air passageways to the lungs. (b) Bronchogram of the lungs. (Courtesy of Lester W. Paul and John H. Juhl, *The Essentials of Roetgen Interpretation,* 3d ed., Harper & Row, Publishers, Inc., New York, 1972.)

the ribs, the *costal surface,* is rounded to match the curvature of the ribs. The *mediastinal (medial) surface* of each lung contains a vertical slit, the *hilum,* through which bronchi, pulmonary vessels, and nerves enter and exit. The blood vessels, bronchi, and nerves are held together by the pleura and connective tissue, and they constitute the *root* of the lung. Medially, the left lung also contains a concavity, the *cardiac notch,* in which the heart lies.

The right lung is thicker and broader than the left. It is also somewhat shorter than the left because the diaphragm is higher on the right side to accommodate the liver that lies below it. The left lung is thinner, narrower, and longer than the right.

Each lung is divided into lobes by one or more fissures. Both lungs have an *oblique fissure,* which extends downward and forward. The right lung also has a *horizontal fissure.* The oblique fissure in the left lung separates the superior from the inferior lobe. The upper part of the oblique fissure of the right lung separates the superior lobe from the inferior lobe, whereas the lower part of the oblique fissure separates the inferior lobe from the middle lobe. The horizontal fissure of the right lung separates the superior lobe from the middle lobe.

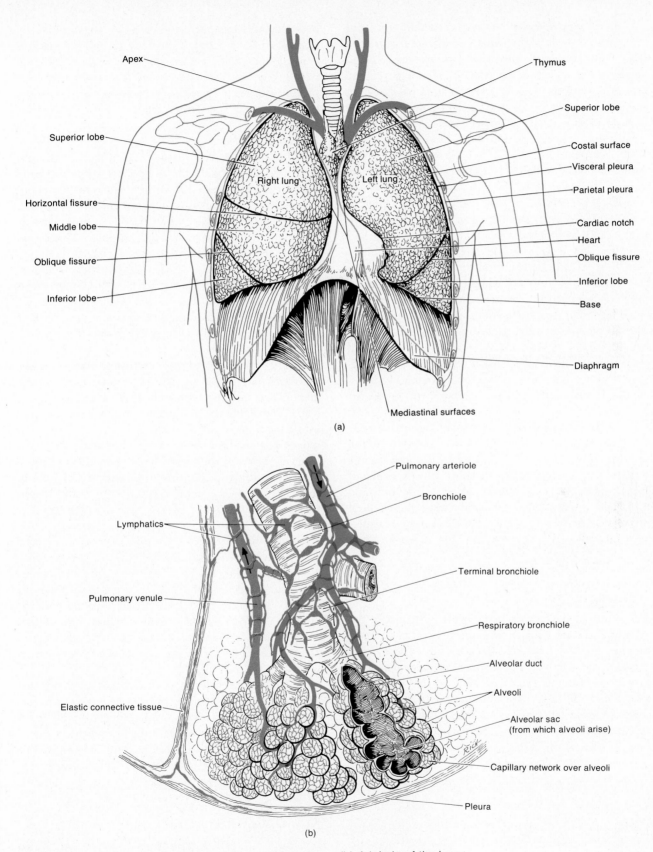

Figure 23–8 Lungs. (a) Coverings and external anatomy. (b) A lobule of the lung.

Each lobe receives its own secondary or lobar bronchus. Thus the right primary bronchus gives rise to three lobar bronchi called the *superior, middle,* and *inferior lobar* or *secondary bronchi.* The left primary bronchus gives rise to a *superior* and an *inferior lobar* or *secondary bronchus.* Within the substance of the lung, the lobar bronchi give rise to branches that are constant in both origin and distribution. Such branches are called *tertiary* or *segmental bronchi,* and the segment of lung tissue that each supplies is called a *bronchopulmonary segment.*

Each bronchopulmonary segment of the lungs is broken up into many small compartments called *lobules.* Every lobule is wrapped in elastic connective tissue and contains a lymphatic vessel, an arteriole, a venule, and a branch from a terminal bronchiole. Terminal bronchioles subdivide into microscopic branches called *respiratory bronchioles.* In the repiratory bronchioles, the epithelial lining changes from cuboidal to squamous as they become more distal. In addition, respiratory bronchioles contain alveoli. An **alveolus** is a cup-shaped outpouching lined by squamous epithelium and supported by a thin elastic membrane. Respiratory bronchioles, in turn, subdivide into several (2 to 11) *alveolar ducts* or *atria.* Around the circumference of the alveolar ducts are numerous alveoli and alveolar sacs. An *alveolar sac* or air sac is a cluster of alveoli that share a common opening. Over the alveoli, the arteriole and venule disperse into a network of capillaries. The exchange of gases between lungs and blood takes place by diffusion across the alveoli and the capillary walls. It has been estimated that each lung contains 150 million alveoli — providing an immense surface area for the exchange of gases.

RESPIRATION

The principal purpose of **respiration** is to supply the cells of the body with oxygen and remove the carbon dioxide produced by cellular activities. Three basic processes are involved. The first process is *ventilation,* or breathing — the movement of air between the atmosphere and the lungs. The second and third processes involve the exchange of gases within the body. *External respiration* is the exchange of gases between the lungs and blood. *Internal respiration* is the exchange of gases between the blood and the cells.

Ventilation

Ventilation or breathing is the process by which atmospheric gases are drawn down into the lungs and waste gases that have diffused into the lungs are expelled back up through the respiratory passageways. Air flows between the atmosphere and lungs for the same reason that blood flows through the body — a pressure gradient exists. We breathe in when the pressure inside the lungs is less than the air pressure in the atmosphere. We breathe out when the pressure inside the lungs is greater than the pressure in the atmosphere.

INSPIRATION

Breathing in is called **inspiration** or inhalation. Just before each inspiration the air pressure inside the lungs equals the pressure of the atmosphere, which is about 760 mm Hg at standard conditions. For air to flow into the lungs, the pressure inside the lungs must become lower than the pressure in the atmosphere. This condition is achieved by increasing the volume of the lungs. As you have observed, perhaps unknowingly, the pressure of a gas is inversely proportional to the volume of its container. If the size of a closed container is increased, the pressure of the air inside the container decreases. If the size of the container is decreased, then the pressure inside it increases.

The first step toward increasing lung volume involves contraction of the respiratory muscles — the diaphragm and external intercostal muscles (Figure 23-9). The diaphragm is the sheet of skeletal muscle that forms the floor of the thoracic cavity. As it contracts it moves downward, thereby increasing the depth of the thoracic cavity. At the same time, the external intercostal muscles contract, pulling the ribs upward and turning them slightly so the sternum is pushed forward. In this way the circumference of the thoracic cavity also is increased.

The overall increase in the size of the thoracic cavity causes its pressure, called *intrathoracic* or *intrapleural pressure,* to fall far below the pressure of the air inside the lungs. Consequently, the walls of the lungs are sucked outward by the partial vacuum. Expansion of the lungs is aided by the pleural membranes. The parietal pleura lining the chest cavity tends to stick to the visceral pleura around the lungs and to pull the visceral pleura with it.

When the volume of the lungs increases, the pressure inside the lungs, called the *intrapulmonic* or *intraalveolar pressure,* drops from 760 to 758 mm Hg. A pressure gradient is thus established between the atmosphere and the alveoli. Air rushes from the atmosphere into the lungs, and an inspiration takes place. Inspiration is frequently referred to as an active process because it is initiated by muscle contraction.

EXPIRATION

Breathing out, called **expiration** or exhalation, is also achieved by a pressure gradient. But this time the gradient is reversed so that the pressure in the lungs is

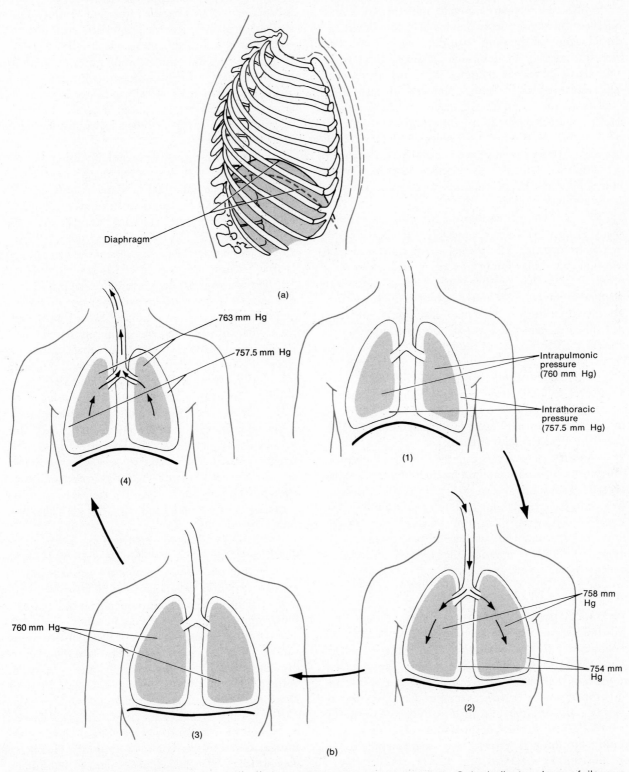

Figure 23—9 Breathing. (a) Changes in size of rib cage. Black indicates relaxed cage. Color indicates shape of rib cage during inspiration. (b) Pressure changes. (1) Lungs and pleural cavity just before inspiration. (2) Chest expanded and intrathoracic pressure decreased; lungs pulled outward and intrapulmonic pressure decreased. Air moves into lungs until intrapulmonic pressure equals atmospheric pressure (3). (4) Chest relaxes, intrathoracic pressure rises, and lungs snap inward. Intrapulmonic pressure raised, forcing air out until intrapulmonic pressure equals atmospheric pressure (1).

greater than the pressure of the atmosphere. Expiration starts when the respiratory muscles relax and the size of the chest cavity decreases in depth and circumference. As the intrathoracic pressure returns to its preinspiration level, the walls of the lungs are no longer sucked out. The elastic basement membranes of the alveoli and very elastic fibers in bronchioles and alveolar ducts snap back into their relaxed shape, and lung volume decreases. Intrapulmonic pressure increases, and air moves from the area of higher pressure (the alveoli) to the area of lower pressure (the atmosphere). Expiration is basically a passive process since no muscular contraction is required. However, the internal intercostals do aid in expiration.

Figure 23-9b shows that the intrathoracic pressure is always a little less than the pressure inside the lungs or in the atmosphere. The pleural cavities are sealed off from the outside environment and cannot equalize their pressure with that of the atmosphere. Nor can the diaphragm and rib cage move inward enough to bring the intrathoracic pressure up to atmospheric pressure. Actually, maintenance of a low intrathoracic pressure is vital to the functioning of the lungs. The alveoli are so elastic that at the end of an expiration they attempt to snap inward and collapse on themselves like the walls of a deflated balloon. Such a collapse, called *atelectasis,* which would obstruct the movement of air, is prevented by the slightly lower pressure in the pleural cavities that keeps the alveoli slightly inflated.

Another factor preventing the collapse of alveoli is the presence of a phospholipid produced by the alveolar cells. This substance, called *surfactant,* decreases surface tension in the lungs. That is, it forms a thin lining of the alveoli and prevents them from sticking together following expiration. Thus, as alveoli become smaller, for example following expiration, the tendency of alveoli to collapse is minimized because the surface tension does not increase.

Air Volumes Exchanged

In clinical practice the word *respiration* means one inspiration plus one expiration. The average healthy adult has 14 to 18 respirations a minute. During each respiration the lungs exchange volumes of air with the atmosphere. A lower than normal exchange volume is usually a sign of pulmonary malfunction. The apparatus commonly used to measure the amount of air exchanged during breathing is referred to as a **respirometer** or **spirometer** (Figure 23-10).

A respirometer consists of a weighted drum inverted over a chamber of water. The drum usually contains oxygen or air. A tube connects the air-filled chamber with the subject's mouth. During inspiration, air is removed from the chamber, the drum sinks, and an upward deflection is recorded by the stylus on the graph paper on the kymograph (rotating drum). During expiration, air is added, the drum rises, and a downward deflection is recorded. The record is called a *spirogram* (Figure 23-11). Spirometric studies measure lung capacities and rates and depths of ventilation for diagnostic purposes. Spirometry is usually indicated for individuals who exhibit labored breathing. It is also used in the diagnosis of respiratory disorders such as emphysema and bronchial asthma.

During normal quiet breathing, about 500 ml of air moves into the respiratory passageways with each inspiration. The same amount moves out with each expiration. This volume of air inspired (or expired) is called *tidal volume* (Figure 23-11). Actually, only about 350 ml of the tidal volume reaches the alveoli. The other 150 ml remains in the dead spaces of the nose, pharynx, larynx, trachea, and bronchi and is known as *dead air.*

By taking a very deep breath, we can inspire a good deal more than 500 ml. This excess inhaled air, called the *inspiratory reserve volume,* averages 3,100 ml above the 500 ml of tidal volume. Thus the respiratory system can pull in as much as 3,600 ml of air. If we inhale normally and then exhale as forcibly as possible, we should be able to push out 1,200 ml of air in addition to the 500 ml tidal volume. This extra 1,200 ml is called the *expiratory reserve volume.* Even after the expiratory reserve volume is expelled, a good deal of air still remains in the lungs because the lower intrathoracic pressure keeps the alveoli slightly inflated. This air, the *residual volume,* amounts to about 1,200 ml. Opening the thoracic cavity allows the intrathoracic pressure to equal the atmospheric pressure, forcing out the residual volume. The air remaining is called the *minimal volume.*

The presence of minimal volume can be demonstrated by placing a piece of lung in water and watching it float. Minimal volume provides a medical and legal tool for determining whether a baby was born dead or died after birth. Fetal lungs contain no air, and so the lung of a stillborn will not float in water.

Lung capacity can be calculated by combining various lung volumes. *Inspiratory capacity,* the total inspiratory ability of the lungs, is the sum of tidal volume plus inspiratory reserve volume (3,600 ml). *Functional residual capacity* is the sum of residual volume plus expiratory reserve volume (2,400 ml). *Vital capacity* is the sum of inspiratory reserve volume, tidal volume, and expiratory reserve volume (4,800 ml). Finally, *total lung capacity* is the sum of all volumes (6,000 ml).

Minute Volume of Respiration

The total air taken in during 1 minute is called the **minute volume of respiration.** It is calculated by

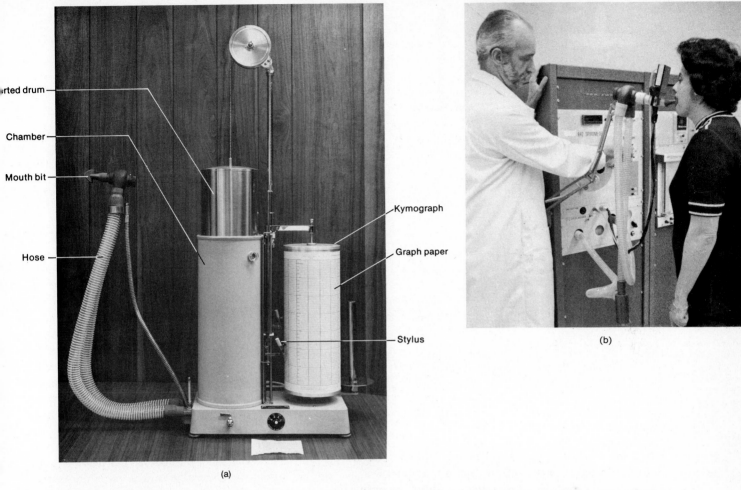

Figure 23–10 Respirometers. (a) Collins respirometer. This type of respirometer is the one commonly used in college biology laboratories. (Courtesy of Charles I. Foster, Vice President, Warren E. Collins, Inc., Braintree, Mass.) (b) Ohio 842 respirometer. This instrument is a highly sophisticated respirometer that utilizes a computerized mechanism for recording results. (Courtesy of Lenny Patti.)

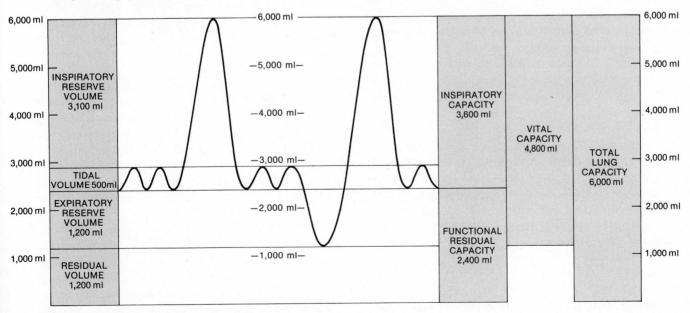

Figure 23–11 Spirogram of lung volumes and capacities.

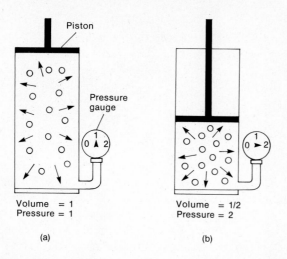

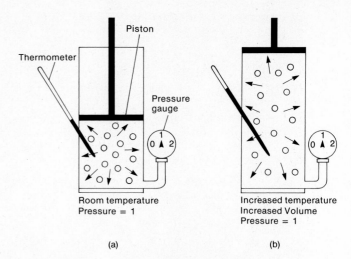

Figure 23–12 Boyle's law. The volume of a gas varies inversely with the pressure. If the volume is decreased to $1/4$, what happens to the pressure?

Figure 23–13 Charles' law. The volume of a gas is directly proportional to its absolute temperature, assuming that the pressure is constant.

multiplying the tidal volume by the normal breathing rate per minute:

$$\begin{array}{c} \text{Minute volume} \\ \text{of respiration} \end{array} = \underset{\substack{\text{(tidal} \\ \text{volume)}}}{500} \times \underset{\substack{\text{(average} \\ \text{rate/min)}}}{16}$$

$$= 8,000 \text{ ml/min}$$

The measurement of respiratory volumes and capacities is an essential tool for determining how well the lungs are functioning. For instance, during the early stages of emphysema, many of the alveoli lose their elasticity. During expiration they fail to snap inward, and consequently they fail to force out a normal amount of air. Thus the residual volume is increased at the expense of the expiratory reserve volume. Pulmonary infections can cause inflammation and an accumulation of fluid in the air spaces of the lungs. The fluid reduces the amount of space available for air and consequently decreases the vital capacity.

Exchange of Respiratory Gases

As soon as the lungs fill with air, oxygen moves from the alveoli to the blood, through the interstitial fluid, and finally to the cells. Carbon dioxide moves in just the opposite direction—from the cells, through interstitial fluid to the blood, and to the alveoli. To understand how respiratory gases are exchanged in the body, you need to know a few gas laws.

BOYLE'S LAW

Boyle's law states that the volume of a gas varies inversely with pressure, assuming that the temperature is

constant. Suppose we place a gas in a cylinder that has a movable piston and a pressure gauge (Figure 23-12). Initially the pressure is 1 atmosphere (atm). This pressure is created by the gas molecules striking the wall of the container. Now suppose we place weights on the piston to increase the external pressure. The gas molecules are compressed because the gas sample is now concentrated in a smaller volume. The gauge shows that the pressure doubles as the gas is compressed to half its volume. Now the same number of gas molecules are striking half as much wall area. The same number of molecules in half the space produces twice the pressure. Conversely, if the piston is raised to increase the volume, the pressure will decrease. Thus the volume of a gas varies inversely with pressure. Boyle's law applies to the operation of a bicycle pump and the blowing up of a balloon. Differences in pressure force air into our lungs when we inhale and force the air out when we exhale.

CHARLES' LAW

According to **Charles' law**, the volume of a gas is directly proportional to its absolute temperature, assuming that the pressure remains constant. Suppose the gas in our cylinder exerts an initial pressure of 1 atm (Figure 23-13) when the piston is halfway down. If the temperature of the gas in the cylinder is raised, the pressure on the cylinder wall will increase proportionately because the gas molecules are moving faster. As a result, the piston moves upward. As the space in the cylinder increases, the gas molecules have to travel a greater distance to strike the cylinder wall. The number of collisions will decrease and so will the pressure—to 1 atm. Notice that the pressures of the two cylinders are constant. However, the volume has increased in proportion to the temperature increase.

DALTON'S LAW

According to **Dalton's law,** each gas in a mixture of gases exerts its own pressure as if all the other gases were not present. This *partial pressure* is denoted as *p*. The total pressure of the mixture is simply calculated by adding all the partial pressures. Atmospheric air is a mixture of several gases – oxygen, carbon dioxide, nitrogen, water vapor, and a number of other gases which appear in such small quantities that we will ignore them. Atmospheric pressure, which is 760 mm Hg at standard conditions, is the sum of the pressures of all these gases:

$$\text{Atmospheric pressure (760 mm Hg)}$$
$$= p\,O_2 + p\,CO_2 + p\,N_2 + p\,H_2O$$

We can determine partial pressure by multiplying the percentage of the mixture the particular gas constitutes by the total pressure of the mixture. To find the partial pressure of oxygen in the atmosphere, multiply the percentage of atmospheric air composed of oxygen (21 percent) by the total atmospheric pressure (760 mm Hg):

$$\text{Atmospheric } p\,O_2 = 21\% \times 760 \text{ mm Hg}$$
$$= 159.60 \text{ or } 160 \text{ mm Hg}$$

Now suppose we want to calculate the atmospheric $p\,CO_2$. Since the percentage of CO_2 in the atmosphere is 0.04, we multiply this figure by 760 mm Hg. Thus:

$$\text{Atmospheric } p\,CO_2 = 0.04\% \times 760 \text{ mm Hg}$$
$$= 0.3040 \text{ or } 0.3 \text{ mm Hg}$$

The partial pressures of gases in your blood are very important in determining the movement of oxygen and carbon dioxide between the atmosphere and lungs, the lungs and blood, and the blood and body cells. When a mixture of gases diffuses across a semipermeable membrane, each gas diffuses from the area where its partial pressure is greater to the area where its partial pressure is less. Every gas is on its own and behaves as if the other gases in the mixture did not exist.

HENRY'S LAW

You have probably noticed that a bottle of soda makes a hissing sound when the cap is removed, and bubbles rise to the surface for some time afterward. The gas in carbonated beverages is carbon dioxide, and its ability to stay in solution (dissolve in the soda) depends on its partial pressure. The higher the partial pressure of a gas over a liquid, the more gas will stay in solution. Since the soda is bottled under pressure and capped, the CO_2 remains dissolved as long as the bottle is unopened.

Once you remove the cap, the pressure is released and the gas begins to bubble out. This phenomenon is explained by **Henry's law:** The quantity of a gas that will dissolve in a liquid is proportional to the partial pressure of the gas, when the temperature remains constant.

Henry's law also explains a condition called *caisson disease,* or the *bends.* Even though the air we breathe contains nitrogen, very little of it gets into the blood because it does not dissolve at body temperature. But when a diver breathes air under pressure, the quantity of dissolved nitrogen increases. If the diver surfaces too quickly, the pressure of the dissolved nitrogen increases above atmospheric pressure. As a result, nitrogen comes out of solution and tiny bubbles of it may find their way into body tissues. The bubbles may effect nerve impulses and cause severe pain. The only treatment is to reapply the pressure and then remove it slowly so that bubbles do not form. Divers avoid bends by breathing a mixture of helium and oxygen instead of air. Helium is only about 40 percent as soluble as nitrogen.

A major clinical application of Henry's law is the *hyperbaric chamber* (Figure 23-14). Using pressure to cause more of a gas to dissolve is an effective technique in treating patients who have been infected by anaerobic bacteria. (These bacteria, which cause diseases such as tetanus and gangrene, cannot live in the presence of oxygen – hence their name anaerobic.) A hyperbaric chamber contains oxygen at a pressure of 3 to 4 atm (2,280 to 3,040 mm Hg). The infected body tissues pick up the oxygen, and the bacteria are killed. Hyperbaric chambers may also be used for treating certain heart disorders and carbon monoxide poisoning.

External Respiration

External respiration is the exchange of oxygen and carbon dioxide between the alveoli and the blood (Figure 23-15a). During inspiration, atmospheric air containing oxygen is brought down into the alveoli. Meanwhile venous blood, which is low in oxygen and high in carbon dioxide, is pumped through the pulmonary artery into the capillaries overlying the alveoli. Exhibit 23-1 shows that the oxygen in the alveolar air has a partial pressure of 100 mm Hg and the oxygen in venous (deoxygenated) blood has a partial pressure of only 40 mm Hg. Oxygen moves down its partial pressure gradient from the alveoli to the blood until the blood's $p\,O_2$ reaches 95 mm Hg, the $p\,O_2$ of arterial blood. While the blood is being oxygenated, carbon dioxide is also moving down its partial pressure gradient. Thus carbon dioxide diffuses from the venous blood, where its partial pressure is 45 mm Hg, to the alveoli, where its partial pressure is 40 mm Hg.

Figure 23—14 Hyperbaric chamber. (Courtesy of Harvey Markinson, Sales Manager, Vacudyne Altair, Chicago Heights, Illinois.)

External respiration is aided by several anatomic adaptations. The total thickness of the alveolar-capillary (respiratory) membranes is only 0.004 mm. Thicker membranes would inhibit diffusion. The blood and air are also given maximum surface exposure to each other. The total surface area of the alveoli is about 540 sq ft, many more times the total surface area of the skin. Lying over the alveoli are countless capillaries—so many that 900 ml of blood is able to participate in gas exchange at any time. Finally, the capillaries are so narrow that the red blood cells must flow through them in single file. This feature gives each red blood cell maximum exposure to the available oxygen.

The efficiency of external respiration depends on several factors. One of the most important is altitude. As long as alveolar pO_2 is higher than venous blood pO_2, oxygen diffuses from the alveoli into the blood. As a person ascends in altitude, the atmospheric pO_2 decreases, the alveolar pO_2 correspondingly decreases, and less oxygen diffuses into the blood. The common symptoms of altitude sickness—shortness of breath, nausea, dizziness—are attributable to the low concentrations of oxygen in the blood. Another factor that affects external respiration is the total surface area available for O_2–CO_2 exchange. Any pulmonary disorder that decreases the functional surface area formed by the alveolar-capillary membranes decreases the efficiency of external respiration. A third factor that influences external respiration is the minute volume of respiration (tidal volume times rate of respiration per minute). Certain drugs, such as morphine, slow down the respiration rate, thereby decreasing the amounts of oxygen and carbon dioxide that can be exchanged between the alveoli and the blood.

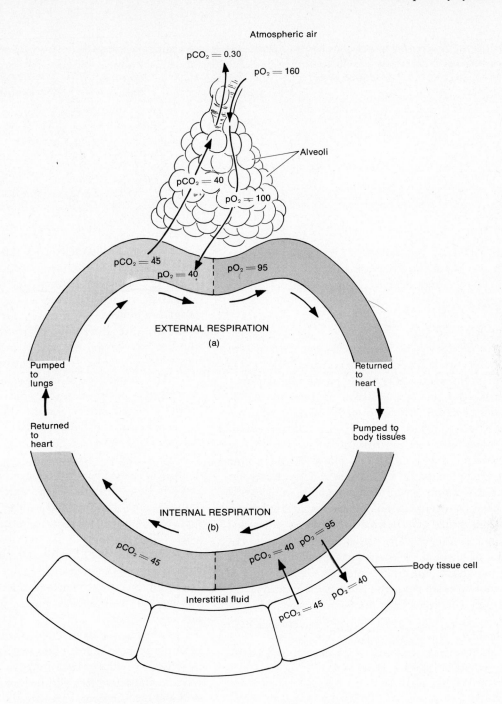

Atmospheric air

$pCO_2 = 0.30$

$pO_2 = 160$

Alveoli

$pCO_2 = 40$

$pO_2 = 100$

$pCO_2 = 45$

$pO_2 = 40$

$pO_2 = 95$

EXTERNAL RESPIRATION

(a)

Pumped to lungs

Returned to heart

Returned to heart

Pumped to body tissues

INTERNAL RESPIRATION

(b)

$pCO_2 = 45$

$pCO_2 = 40$ $pO_2 = 95$

Body tissue cell

$pCO_2 = 45$ $pO_2 = 40$

Interstitial fluid

Figure 23–15 Partial pressures involved in respiration. (a) External. (b) Internal. All pressures are in millimeters Hg.

Internal Respiration

As soon as the task of external respiration is completed, the blood moves through the pulmonary veins to the heart, where it is pumped out to the body tissues. In the capillaries of the body tissues a second exchange, called **internal respiration**, takes place. This is the exchange of O_2 and CO_2 between the blood and body tissues (see Figure 23-15b). Exhibit 23-1 shows that the pO_2 of body tissues (40 mm Hg) is much lower than the pO_2 of the arterial blood (95 mm Hg). On the other hand, the

pCO_2 of body tissues (45 mm Hg) is higher than the pCO_2 of the arterial blood (40 mm Hg). As you might expect, the two gases move down their concentration gradients. Oxygen diffuses from the blood through the interstitial fluid into the body tissues, and carbon dioxide diffuses from the body tissues through the interstitial fluid into the blood until the blood attains partial pressures that are typical of venous blood. The blood now returns to the lungs before it can exchange more gas with body tissues.

Exhibit 23–1 PARTIAL PRESSURES OF OXYGEN AND CARBON DIOXIDE*

	Atmospheric Air	Alveolar Air	Venous Blood	Arterial Blood	Body Tissues
pO_2	160	100	40	95	40
pCO_2	0.30	40	45	40	45

*All pressures are approximate under normal conditions.

Transport of Respiratory Gases

When oxygen and carbon dioxide enter the blood, certain physical and chemical changes occur that aid gas exchange.

OXYGEN

When oxygen enters the blood, it dissolves in the plasma (Figure 23-16). When 0.5 ml of oxygen has dissolved in 100 ml of blood, the pO_2 of the blood approximately equals the pO_2 inside the alveoli. Because the partial pressures are equal, oxygen normally could not continue to move into the blood. However, the cells of the body need much more oxygen than this to survive. In fact, they need 20 ml oxygen/100 ml blood. To get around this problem, most of the oxygen quickly leaves the plasma and combines with the hemoglobin of red blood cells. Oxygen that has become attached to hemoglobin is no longer a free gas. It cannot behave as a gas and consequently cannot affect partial pressure. In this way, the pO_2 of the blood is lowered again and more oxygen from the alveoli can diffuse into the plasma. Most of these molecules are captured by hemoglobin until all the hemoglobin molecules are bound to oxygen. At this point, the pO_2 of the alveoli and the plasma is equalized and the blood leaves the lungs.

The chemical union between oxygen and hemoglobin is symbolized as follows:

$$\text{Hb} + \text{O}_2 \underset{\substack{\text{low } pO_2 \\ \text{high H}^+ \\ \text{high temp}}}{\overset{\substack{\text{high } pO_2 \\ \text{low } pCO_2}}{\rightleftharpoons}} \text{HbO}_2$$

Reduced hemoglobin (uncombined hemoglobin) Oxygen Oxyhemoglobin (combined hemoglobin)

The reduced hemoglobin represents hemoglobin that has not combined yet with oxygen. Oxyhemoglobin is the compound formed by the union of oxygen and hemoglobin. As you can see, this is a reversible reaction. When the pO_2 is high, as it is in the lungs, there are more oxygen molecules available to make contact with the hemoglobin. Thus more molecules of oxyhemoglobin can be formed.

When the blood reaches the tissue cells of the body, three factors combine to split the oxyhemoglobin apart. The first factor is simply the low pO_2 in the cells. When an oxygen molecule separates from the oxyhemoglobin molecule, it moves into the tissue cells before it has a chance to recombine with the hemoglobin. All the free oxygen gas also moves into the tissue, so no other oxygen molecules are available to take its place.

The second factor is acidity. Oxyhemoglobin quickly splits apart in an acid solution. This acid comes from the large quantity of carbon dioxide moving from tissue cells through the interstitial fluid into the blood. As the carbon dioxide is taken up by the blood, much of it is temporarily converted into carbonic acid. Thus a high pCO_2 encourages the oxyhemoglobin to release its oxygen. Lactic acid, the reaction product of contracting muscles, also increases blood acidity. Thus active muscles are able to obtain oxygen more quickly than resting ones.

The third factor that encourages the blood to give up its oxygen is an increase in temperature. Heat energy is a by-product of the metabolic reactions of all cells, and contracting muscle cells release an especially large amount of heat.

Splitting the oxyhemoglobin molecule is another example of how homeostatic mechanisms adjust body activities to cellular needs. Active cells require more oxygen, and active cells liberate more acid and heat. The acid and heat, in turn, stimulate the oxyhemoglobin to release its oxygen.

CARBON DIOXIDE

Venous blood contains about 56 ml carbon dioxide/100 ml blood. Like oxygen, carbon dioxide is carried by the blood in several forms (see Figure 23-16). The smallest percentage, about 9 percent, is dissolved in plasma. A somewhat higher percentage, about 27 percent, combines with the protein part of the hemoglobin to form carbaminohemoglobin. This reaction may be represented as follows:

$$\text{Hb} + \text{CO}_2 \rightleftharpoons \text{HbNHCOOH}$$

Hemoglobin Carbon dioxide Carbaminohemoglobin

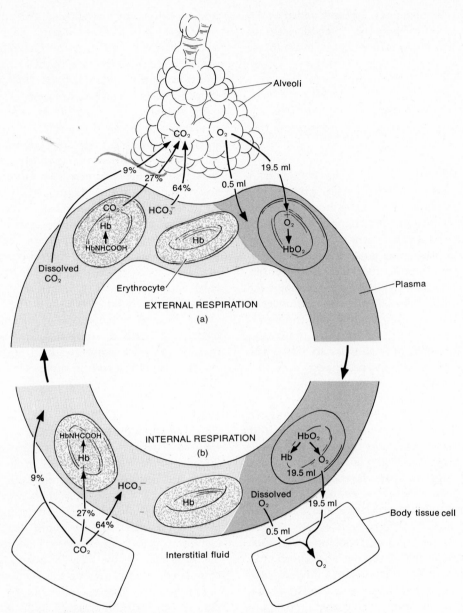

Figure 23—16   Carriage of respiratory gases in respiration. (a) External. (b) Internal.

The greatest percentage of carbon dioxide, about 64 percent, is converted into the bicarbonate ion (HCO_3^-) in the following way:

$$CO_2 + H_2O \rightleftharpoons H_2CO_3 \rightleftharpoons H^+ + HCO_3^-$$

Carbon dioxide	Water	Carbonic acid	Hydrogen ion	Bicarbonate ion

As the carbon dioxide diffuses from tissues into blood plasma and then into red blood cells, an enzyme called carbonic anhydrase stimulates the major portion of the gas to combine with water to form *carbonic acid.* Inside red blood cells, carbonic acid dissociates into H^+ and

HCO_3^- ions. The H^+ ions provide the acid stimulus for the release of oxygen from oxyhemoglobin. Some of the HCO_3^- ions remain in the cell and combine with potassium, the chief positive ion of intracellular fluid, to form potassium bicarbonate ($KHCO_3$). The majority of the bicarbonate ions, however, diffuse out into the plasma and combine with sodium, the principal positive ion of extracellular fluid, to form sodium bicarbonate ($NaHCO_3$).

When the blood reaches the lungs, events reverse. The high pO_2 causes oxygen to replace the carbon dioxide on the hemoglobin molecule. Bicarbonate breaks apart and releases CO_2. Finally, the CO_2 that has been traveling dissolved in the plasma and the CO_2 that has been released from the reversed reactions diffuse down the partial pressure gradient and into the alveolar sacs.

Control of Respiration

Respiration is controlled by several mechanisms that help the body maintain homeostasis.

NERVOUS CONTROL

The size of the thorax is affected by the action of the respiratory muscles. These muscles contract and relax in turn as a result of nerve impulses transmitted to them from centers in the brain. The area from which nerve impulses are sent to respiratory muscles is located in the medulla and is referred to as the **respiratory center.** This center is functionally divided into two regions: the *inspiratory center,* which causes the inspiratory muscles to contract and thus brings on inspiration, and the *expiratory center,* which inhibits the inspiratory center and thereby allows the inspiratory muscles to relax. The respiratory center, along with its "pacemaker" centers in the pons, regulates the rhythm of respiration in the following manner.

An area of the pons, called the *apneustic center,* continually sends stimulatory impulses to the inspiratory center. The stimulated inspiratory center, in turn, sends impulses over motor neurons in the phrenic and intercostal nerves that stimulate the respiratory muscles to contract, and inspiration occurs. As soon as the lungs are filled, the inspiratory impulses are shut off by two controls.

The first set of controls involves the stretch receptors in the lung tissue. When the lungs expand to a critical point, the stretch receptors are stimulated and impulses are sent along the vagus nerves to the expiratory center. The expiratory center then sends out inhibitory impulses to the inspiratory center. The inspiratory muscles relax and expiration follows. As air leaves the lungs during expiration, the lungs are deflated and the stretch receptors are no longer stimulated. Thus the inspiratory center is no longer inhibited, and a new respiration begins. These events are called the *Hering-Breuer reflex.* The Hering-Breuer reflex controls the depth and rhythm of respiration. It also prevents the lungs from inflating to the point of bursting.

Should the Hering-Breuer reflex fail to operate, breathing will still continue. If the vagus nerves are severed, inspirations are longer and deeper than normal, but expiration is eventually initiated by a backup control. This second control is provided by the *pneumo-taxic center* of the pons. When the inspiratory center is stimulated, it sends impulses to the pneumotaxic center as well as to the respiratory muscles. After a delay, the

pneumotaxic center sends impulses to the expiratory center, which inhibits the inspiratory center.

The term applied to normal quiet breathing is *eupnea.* Eupnea is a function of the Hering-Breuer reflex, and it involves shallow, deep, or combined shallow and deep breathing. Shallow, or chest, breathing is called *costal breathing.* It consists of an upward and outward movement of the chest as a result of contraction of the intercostal muscles. Deep, or abdominal, breathing is called *diaphragmatic breathing.* It consists of the outward movement of the abdomen as a result of the contraction and descent of the diaphragm.

The respiratory center has connections with the cerebral cortex, which means we can voluntarily alter our pattern of breathing. We can even refuse to breathe at all for a short time. Voluntary control is protective because it enables us to prevent water or irritating gases from entering the lungs. The ability to stop breathing is limited by the buildup of CO_2 in the blood, however. When the pCO_2 increases to a certain level, the inspiratory center is stimulated, impulses are sent to inspiratory muscles, and breathing resumes whether or not the person wishes. It is impossible for anyone to kill himself by holding his breath.

CHEMICAL AND PRESSURE STIMULI

Chemical stimuli, particularly CO_2, O_2, and H^+, determine how fast we breathe. All three stimuli may act directly on the respiratory center or on chemoreceptors located in the aortic and carotid bodies. Probably the main chemical stimulus that alters respirations is CO_2. Under normal circumstances, arterial blood pCO_2 is 40 mm Hg. If there is even a slight increase in pCO_2 – a condition called *hypercapnia* – chemoreceptors in the medulla and in the carotid and aortic bodies are stimulated. Stimulation of the chemoreceptors causes the inspiratory center to become highly active, and the rate of respiration increases. This increased rate, *hyperventilation,* allows the body to expel more CO_2 until the pCO_2 is lowered to normal. Now let us consider the opposite situation. If arterial pCO_2 is lower than 40 mm Hg, the receptors are not stimulated and stimulatory impulses are not sent to the inspiratory center. Consequently, the center sets its own moderate pace until CO_2 accumulates and the pCO_2 rises above 40 mm Hg. A slow rate of respiration is called *hypoventilation.*

The oxygen receptors are sensitive only to large drops in the pO_2. If arterial pO_2 falls from a normal of 100 mm Hg to 70 mm Hg, the oxygen receptors become stimulated and send impulses to the inspiratory center. But if the pO_2 falls much below 70 mm Hg, the cells of the respiratory center suffer oxygen starvation and do not respond well to any chemical receptors. They send fewer impulses to the respiratory muscles, and the respiration rate decreases or breathing ceases altogether.

Within limits, any decrease in arterial blood pH stimulates chemoreceptors in the carotid and aortic bodies in the medulla. This stimulation results in increased respiration. The carotid and aortic sinuses also contain pressoreceptors that are stimulated by a rise in blood pressure. Although these pressoreceptors are concerned mainly with the control of circulation, they help control respiration. For example, a sudden rise in blood pressure decreases the rate of respiration, and a drop in blood pressure brings about an increase in the respiratory rate. Other factors that control respiration are:

1 A sudden cold stimulus such as plunging into cold water causes a temporary cessation of breathing called *apnea.*

2 A sudden, severe pain brings about apnea, but a prolonged pain triggers the general adaptation syndrome and increases respiration rate.

3 Stretching the anal sphincter muscle increases the respiratory rate. This technique is sometimes employed to stimulate respiration during emergencies.

4 Irritation of the pharynx or larynx by touch or chemicals brings about an immediate cessation of breathing followed by coughing.

HEART AND LUNG RESUSCITATION

A serious decrease in respiration or heart rate presents an urgent crisis because the body's cells cannot survive long if they are starved of oxygenated blood. In fact, if oxygen is withheld from the cells of the brain for 4–6 minutes, brain damage or death will result. Heart-lung resuscitation is the artificial reestablishment of normal or near normal respiration and circulation. The two simplest techniques for heart-lung resuscitation are exhaled air ventilation and external cardiac compression. Both techniques can be administered at the site of emergency, and both are highly successful. They can be used for any heart or respiratory failure, whether the cause be drowning, strangulation, carbon monoxide or insecticide poisoning, overdose of a drug or anesthesia, electrocution, or myocardial infarction. However, the success of heart-lung resuscitation is directly related to speed and efficiency. Delay may be fatal.

Exhaled Air Ventilation

A technique for reestablishing respiration is **exhaled air ventilation.** The first step is immediate opening of the airway by tilting the victim's head backward as far as it will go without being forced. The tilted position opens the upper air passageways to their maximum size (Figure 23-17). If the patient does not resume spontaneous

breathing after the head has been tilted backward, artificial ventilation must be given mouth to mouth or mouth to nose. In the more usual mouth-to-mouth method the victim's nostrils are pinched together with the thumb and index finger. The rescuer then opens his mouth widely, takes a deep breath, makes a tight seal with his mouth around the patient's mouth, and blows in about twice the amount the patient normally breathes. He then removes his mouth and allows the patient to exhale passively. This cycle is repeated approximately 12 times per minute for adults. Atmospheric air contains about 21 percent O_2 and a trace of CO_2. Exhaled air still contains about 16 percent O_2 and 5 percent CO_2. This amount is more than adequate to maintain a victim's blood pO_2 and pCO_2 at normal levels if air is given at the prescribed rate and amount.

If the rescuer observes the following three signs, ventilation is adequate:

1 The chest rises and falls with every breath.

2 The lungs can be felt to resist as they expand.

3 Air can be heard escaping during exhalation.

External Cardiac Compression

External cardiac compression, or **closed-chest cardiac compression (CCCC),** consists of the application of rhythmic pressure over the sternum (see Figure 23-17). The rescuer places the heels of the hands on the lower half of the sternum and presses down firmly and smoothly at least 60 times a minute. This action compresses the heart and produces an artificial circulation because the heart lies almost in the middle of the chest between the lower portion of the sternum and the spine. When properly done, external cardiac compression can produce systolic blood pressure peaks of over 100 mm Hg. It can also bring carotid arterial blood flow up to 35 percent of normal.

Complications that can occur from the use of cardiac compression include fracture of the ribs and sternum, laceration of the liver, and the formation of fat emboli. They can be minimized by adhering to the following precautions:

1 Never compress over the xiphoid process at the tip of the sternum. This bony prominence extends down over the abdomen, and pressure on it may cause laceration of the liver, which can be fatal.

2 Never let your fingers touch the patient's ribs when you compress. Keep your fingers off the patient, and place the heel of your hand in the middle of the patient's chest over the lower half of the sternum.

3 Never compress the abdomen and chest simultaneously since this action traps the liver and may rupture it.

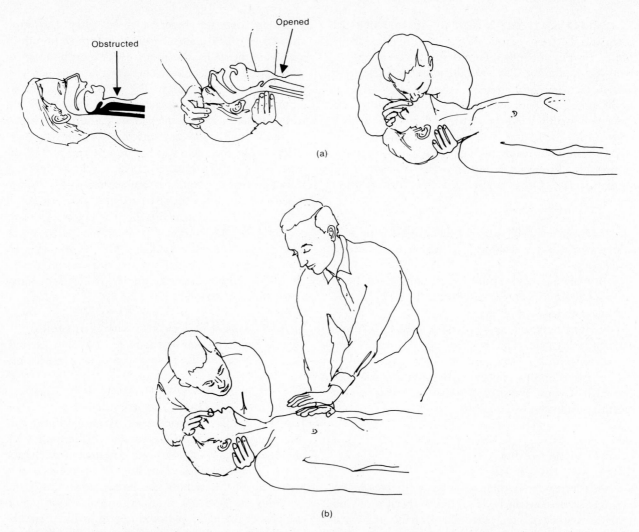

Figure 23–17 Heart-lung resuscitation. (a) Exhaled air ventilation. Shown on the left is the procedure for immediate opening of the airway. Shown on the right is the procedure for mouth-to-mouth respiration. (b) External cardiac compression technique in conjunction with exhaled air ventilation.

4 Never use sudden or jerking movements to compress the chest. Compression should be smooth, regular, and uninterrupted, with 50 percent of the cycle compression and 50 percent relaxation.

Compression of the sternum produces some artificial ventilation but not enough for adequate oxygenation of the blood. Therefore exhaled air ventilation must always be used with it. This combination constitutes **heart-lung resuscitation.** When there are two rescuers, the best technique is to have one rescuer apply at least 60 cardiac compressions per minute. The other rescuer should exhale into the patient's mouth between every fifth and sixth compression. The sequential steps in emergency heart-lung resuscitation must be continued uniformly and without interruption until the patient recovers or is pronounced dead.

Food Choking

Food choking (café coronary), the sixth leading cause of accidental death, is often mistaken for myocardial infarction. However, it can be recognized easily. The victim cannot speak or breathe; he may become panic stricken and run from the room. He becomes pale, then deeply cyanotic, and collapses. Without intervention, death occurs in 4–5 minutes.

The food or other obstructing object causing asphyxiation may lodge in the back of the throat or enter the trachea to occlude the airway. Tracheotomy, even

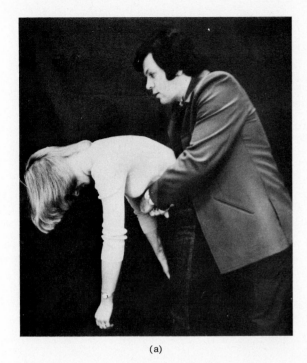

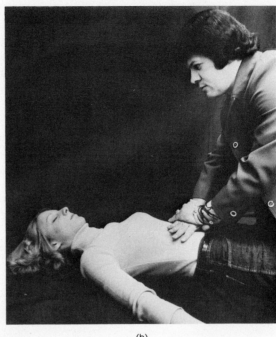

(a)   (b)

Figure 23–18   Heimlich maneuver. The principle employed in this anti-choke first aid method is the application of force to compress the air in the lungs to expel objects in the air passageways. (a) If the victim is upright, stand behind the person and place both arms around the waist just above the belt line. Grasp your right wrist with your left hand and allow the victim's head, arms, and upper torso to fall forward. Next, the victim's abdomen is rapidly and strongly compressed. If another person is present, he or she can assist by removing the ejected object from the victim's mouth. (b) If the victim is recumbent, the rescuer may rapidly and strongly force the victim's diaphragm upward by applying pressure as shown. (Courtesy of Donald Castellaro and Deborah Massimi.)

when a physician performs it, can be hazardous in a nonclinical setting. An instrument for removing food from the back of the throat has been developed but is seldom at hand in the presenting emergency. However, there is a first aid procedure that does not require special instruments and can be performed by any informed layman. This procedure is called the **Heimlich maneuver.**

In all probability, food choking occurs during inspiration, which causes the bolus of food to be sucked against the opening into the larynx. At the time of the accident, the lungs are therefore expanded. Even during normal expiration, however, some tidal air (500 ml) and the entire expiratory reserve volume (1,200 ml) are present in the lungs.

Pressing one's fist upward into the epigastrium elevates the diaphragm suddenly – compressing the lungs in the rib cage and increasing the air pressure in the tracheobronchial tree. This pressure is forced out through the trachea and will eject the food (or object) blocking the airway. The action can be simulated by

inserting a cork in a compressible plastic bottle and then squeezing it suddenly – the cork flies out because of the increased pressure. Figure 23-18 shows how to administer the Heimlich maneuver.

SMOKING

As part of ordinary breathing, many irritating substances are inhaled. Almost all pollutants, including inhaled smoke, have an irritating effect on the bronchial tubes and lungs and may be regarded as stresses or irritating stimuli.

Close examination of the epithelium of a bronchial tube reveals three kinds of cells (Figure 23-19). The surface cells are columnar cells that contain cilia. At intervals between the ciliated columnar cells are the mucus-secreting goblet cells. The bottom of the epithelium normally contains a row of basal cells above the basement membrane. The basal cells divide continuously, replacing the ciliated columnar epithelium as they wear down and are sloughed off. The bronchial

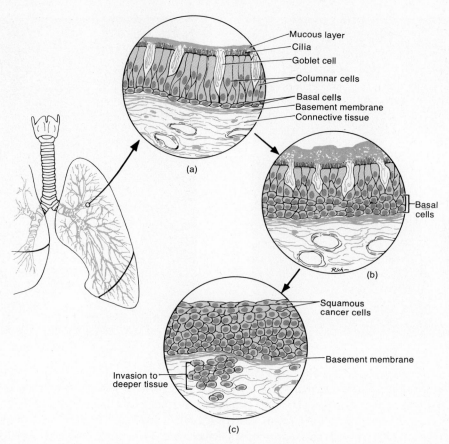

Mucous layer
Cilia
Goblet cell
Columnar cells
Basal cells
Basement membrane
Connective tissue

(a)

Basal cells

(b)

Squamous cancer cells

Basement membrane

Invasion to deeper tissue

(c)

Figure 23–19   Effects of smoking on the respiratory epithelium. (a) Microscopic view of the normal epithelium of a bronchial tube. (b) Initial response of the bronchial epithelium to irritation by pollutants. (c) Advanced response of the bronchial epithelium.

epithelium is important clinically because a common lung cancer, **bronchogenic carcinoma,** starts in the walls of the bronchi.

The constant irritation by inhaled smoke and pollutants causes an enlargement of the goblet cells of the bronchial epithelium. They respond by secreting excessive mucus. The basal cells also respond to the stress by undergoing cell division so fast that the basal cells push into the area occupied by the goblet and columnar cells. As many as 20 rows of basal cells may be produced. Many researchers believe that if the stress is removed at this point, the epithelium can return to normal.

If the stress persists, more and more mucus is secreted and the cilia become less effective. As a result, mucus is not carried toward the throat but remains trapped in the bronchial tubes. The individual then develops a "smoker's cough." Moreover, the constant irritation from the pollutant slowly destroys the alveoli, which are replaced with thick, inelastic connective tissue. Mucus that has accumulated becomes trapped in

the air sacs. Millions of sacs rupture – reducing the diffusion surface for the exchange of oxygen and carbon dioxide. The individual has now developed emphysema. If the stress is removed at this point, there is little chance for improvement. Alveolar tissue that has been destroyed cannot be repaired. But removal of the stress can stop further destruction of lung tissue.

If the stress continues, the emphysema gets progressively worse, and the basal cells of the bronchial tubes continue to divide and break through the basement membrane. At this point the stage is set for bronchogenic carcinoma. Columnar and goblet cells disappear and may be replaced with squamous cancer cells. If this happens, the malignant growth spreads throughout the lung and may block a bronchial tube. If the obstruction occurs in a large bronchial tube, very little oxygen enters the lung and disease-producing bacteria thrive on the mucoid secretions. In the end, the patient may develop emphysema, carcinoma, and a host of infectious diseases. Treatment involves surgical removal of the diseased lung. However, metastasis of the growth through the lymphatic or blood system may result in new growths in other parts of the body such as the brain and liver.

Other factors may be associated with lung cancer. For instance, breast, stomach, and prostate malignancies can metastasize to the lungs. People who apparently have not been exposed to pollutants do occasionally develop bronchogenic carcinoma. However, the occurrence of bronchogenic carcinoma is probably over 20 times higher in heavy cigarette smokers than it is in non-smokers.

NEBULIZATION

Many of the previously mentioned respiratory disorders are treated by means of a comparatively new method of treatment called *nebulization.* This procedure is the administering of medication, in the form of droplets that are suspended in air, to selected areas of the respiratory tract. The patient inhales the medication as a fine mist. Droplet size is directly related to the number of droplets suspended in the mist. Smaller droplets (approximately 2 μm in diameter) can be suspended in greater numbers than can large droplets and will reach the alveolar ducts and sacs. The larger droplets (approximately 7-16 μm in diameter) will be deposited mostly in the bronchi and bronchioles. Droplets of 40 μm and larger will be deposited in the upper respiratory tract—the mouth, pharynx, trachea, and main bronchi. Nebulization therapy can be used with many different types of drugs, such as chemicals that relax the smooth muscle of the respiratory passageways, chemicals which reduce the thickness of mucus, and antibiotics.

DISORDERS

HAY FEVER
An allergic reaction to the proteins contained in foreign substances such as plant pollens, dust, and certain foods is called **hay fever.** Allergic reactions are a special antigen-antibody response that initiate either a localized or a systemic inflammatory response. In hay fever the response is localized in the respiratory membranes. The membranes become inflamed, and a watery fluid drains from the eyes and nose.

BRONCHIAL ASTHMA
Bronchial asthma is a reaction, usually allergic, characterized by attacks of wheezing and difficult breathing. Attacks are brought on by spasms of the smooth muscles that lie in the walls of the smaller bronchi and bronchioles, causing the passageways to close partially. The patient has trouble exhaling, and the alveoli may remain inflated during expiration. Usually the mucous membranes that line the respiratory passageways become irritated and secrete excessive amounts of mucus that may clog the bronchi and bronchioles and worsen the attack. About three out of four asthma victims are allergic to edible or airborne substances as common as wheat or dust. Others are sensitive to the proteins of harmless bacteria that inhabit the sinuses, nose, and throat. Asthma might also have a psychosomatic origin.

EMPHYSEMA
In **emphysema,** the alveolar walls lose their elasticity and remain filled with air during expiration. The name means "blown up" or "full of air." Reduced forced expiratory volume is the first symptom. Later, alveoli in other areas of the lungs are damaged. The lungs become permanently inflated because they have lost elasticity. To adjust to the increased lung size, the size of the chest cage increases. The patient has to work to exhale. Oxygen diffusion does not occur as easily across the damaged alveolar membrane, blood pO_2 is somewhat lowered, and any mild exercise that raises the oxygen requirements of the cells leaves the patient breathless. As the disease progresses, the alveoli are replaced with thick fibrous connective tissue. Even carbon dioxide does not diffuse easily through this fibrous tissue. If the blood cannot buffer all the carbonic acid that accumulates, the blood pH drops or unusually high amounts of carbon dioxide may dissolve in the plasma. High carbon dioxide levels are toxic to brain cells. Consequently, the inspiratory center becomes less active and the respiration rate slows down, further aggravating the problem. The compressed and damaged capillaries around the deteriorating alveoli may no longer be able to receive blood. As a result, a backup pressure increases in the pulmonary trunk and the right ventricle overworks as it attempts to force blood through the remaining capillaries.

Emphysema is generally caused by a long-term irritation. Air pollution, occupational exposure to industrial dust, and cigarette smoke are the most common irritants. Chronic bronchial asthma also may produce alveolar damage. Cases of emphysema are becoming more and more frequent in the United States. The irony is that the disease can be prevented and the progressive deterioration can be stopped by eliminating the harmful stimuli.

PNEUMONIA

The term **pneumonia** means an acute infection or inflammation of the alveoli. In this disease the alveolar sacs fill up with fluid and dead white blood cells, reducing the amount of air space in the lungs. (Remember that one of the cardinal signs of inflammation is edema.) Oxygen has difficulty diffusing through the inflamed alveoli, and the blood pO_2 may be drastically reduced. Blood pCO_2 usually remains normal because carbon dioxide always diffuses through the alveoli more easily than oxygen does. If all the alveoli of a lobe are inflamed, the pneumonia is called *lobar pneumonia*. If only parts of the lobe are involved, it is called *lobular,* or *segmental, pneumonia.* If both the alveoli and the bronchial tubes are included, it is called *bronchopneumonia.*

The most common cause of pneumonia is the pneumococcus bacterium, but other bacteria or a fungus may be the source of trouble. Viral pneumonia is caused by several viruses, including the influenza virus.

HYALINE MEMBRANE DISEASE (HMD)

Sometimes called glassy-lung disease or infant respiratory distress syndrome (RDS), **hyaline membrane disease (HMD)** is responsible for approximately 20,000 newborn infant deaths per year. Before birth, the respiratory passages are filled with fluid. Part of this fluid is amniotic fluid inhaled during respiratory movements in utero. The remainder is produced by the submucosal glands and the goblet cells of the respiratory epithelium.

At birth, this fluid-filled airway must become an air-filled airway, and the collapsed primitive alveoli (terminal sacs) must expand and function in gas exchange. The success of this transition depends largely on the pulmonary surfactant—a mixture of lipoproteins that lowers surface tension in the fluid layer lining the primitive alveoli once air enters the lungs. Surfactant is present in the fetus's lungs as early as the twenty-third week. By 28 to 32 weeks, however, the amount of surfactant is great enough to prevent alveolar collapse during breathing. Surfactant is produced continuously by alveolar cells. The presence of surfactant can be detected by amniocentesis.

Although in a normal, full-term infant the second and subsequent breaths require less respiratory effort than the first, breathing is not completely normal until about 40 minutes after birth. The entire lung is not inflated fully with the first one or two breaths. In fact, for the first 7 to 10 days, small areas of the lungs may remain uninflated.

In the newborn whose lungs are deficient in surfactant, the effort required for the first breath is essentially the same as that required in normal newborns. However, the surface tension of the alveolar fluid is 7 to 14 times higher than the surface tension of alveolar fluid with a monomolecular layer of surfactant. Consequently, during expiration after the first inspiration, the surface tension of the alveoli increases as the alveoli deflate. The alveoli collapse almost to their original uninflated state.

Idiopathic RDS usually appears within a few hours after birth. Affected infants show difficult and labored breathing with withdrawal of the intercostal and subcostal spaces. Death may occur soon after onset of respiratory difficulty or may be delayed for a few days, although many infants survive. At autopsy, the lungs are underinflated and areas of atelectasis are prominent. (In fact, the lungs are so airless they sink in water.) If the infant survives for at least a few hours after developing respiratory distress, the alveoli are often filled with a fluid of high-protein content that resembles a hyaline (or glassy) membrane. RDS occurs frequently in premature infants and also in infants of diabetic mothers, particularly if the diabetes is untreated or poorly controlled.

A new treatment currently being developed called PEEP—positive end expiratory pressure—could reverse the mortality rate from 90 percent deaths to 90 percent survival. This treatment consists of passing a tube through the air passage to the top of the lungs to provide needed oxygen-rich air at continuous pressures of up to 14 mm Hg. Continuous pressure keeps the baby's alveoli open and available for gas exchange.

MEDICAL TERMINOLOGY

Apnea   Absence of respirations.

Asphyxia   Oxygen starvation due to low atmospheric oxygen or interference with ventilation, external respiration, or internal respiration.

Atelectasis (*ateles* = incomplete; *ektasis* = stretching)   A collapsed lung or portion of a lung.

Bronchitis (*bronch* = bronchus, trachea)   Inflammation of the bronchi and bronchioles.

Cheyne-Stokes respiration   Irregular breathing beginning with shallow breaths that increase in depth and rapidity, then decrease and cease altogether for 15–20 seconds. The cycle repeats itself again and again. Cheyne-Stokes is normal in infants. It is also often seen just before death from pulmonary, cerebral, cardiac, and kidney disease and is referred to as the "death rattle."

Diphtheria   An acute bacterial infection that causes the mucous membranes of the oropharynx, nasopharynx, and larynx to enlarge and become leathery. Enlarged membranes may obstruct airways and cause death from asphyxiation.

Dyspnea (*dys* = painful, difficult)   Labored breathing (short-winded).

Hypoxia   Reduction in oxygen supply to cells.

Influenza   Viral infection that causes inflammation of respiratory mucous membranes as well as fever.

Orthopnea   Inability to breathe in a horizontal position.

Pneumothorax (*pneumo* = lung)   Air in pleural space causing collapse of the lung. Most common cause is surgical opening of chest during heart surgery, making intrathoracic pressure equal atmospheric pressure.

Pulmonary edema   Excess amounts of interstitial fluid in the lungs producing cough and dyspnea. Common in failure of the left side of the heart.

Pulmonary embolism   Presence of a blood clot or other foreign substance in a pulmonary arterial vessel stopping circulation to a part of the lungs.

Rales   Sounds sometimes heard in the lungs that resemble bubbling or rattling. May be caused by air or an abnormal secretion in the lungs.

Respirator   A metal chamber that entombs the chest; also called an "iron lung." Used to produce inspiration and expiration in patient with paralyzed respiratory muscles. Pressure inside the chamber is rhythmically alternated to suck out and push in chest walls.

STUDY OUTLINE

ORGANS

1 Respiratory organs include the nose, pharynx, larynx, trachea, bronchi, and lungs.

2 They act with the cardiovascular system to supply oxygen and remove carbon dioxide from the blood.

NOSE

1 The external portion is made of cartilage and skin and lined with mucous membrane; openings to the exterior are the external nares.

2 The internal portion communicates with the nasopharynx (through the internal nares) and the paranasal sinuses.

3 The nose is divided into cavities by a septum. Anterior portions of the cavities are called the vestibules.

4 The nose is adapted for warming, moistening, and filtering air, for olfaction, and for speech.

PHARYNX

1 The pharynx, or throat, is a muscular tube lined by a mucous membrane.

2 The anatomic regions are nasopharynx, oropharynx, and laryngopharynx.

3 The nasopharynx functions in respiration. The oropharynx and laryngopharynx function both in digestion and in respiration.

LARYNX

1 The larynx is a passageway that connects the pharynx with the trachea.

2 Prominent cartilages are the thyroid, or Adam's apple, the epiglottis, which prevents food from entering the larynx, and the cricoid, which connects the larynx and trachea.

3 The larynx contains true vocal cords that produce sound. Taut cords produce high pitches, and relaxed cords produce low pitches.

TRACHEA

1 The trachea extends from the larynx to the primary bronchi.

2 It is composed of smooth muscle and C-shaped rings of cartilage and is lined with a ciliated mucous membrane.

BRONCHI

1 The bronchial tree consists of primary bronchi, secondary bronchi, bronchioles, and terminal bronchioles. Walls of bronchi contain rings of cartilage; walls of bronchioles do not.

2 A developed picture of the tree is called a bronchogram.

LUNGS

1 Lungs are paired organs in the thoracic cavity. They are enclosed by the pleural membrane. The parietal pleura is the outer layer; the visceral pleura is the inner layer.

2 The right lung has three lobes. The left lung has two lobes and a depression, the cardiac notch. Each lobe consists of lobules, which contain lymphatics, arterioles, venules, terminal bronchioles, respiratory bronchioles, alveolar ducts, alveolar sacs, and alveoli.

3 Gas exchanges occur across the alveolar-capillary membranes.

RESPIRATION
VENTILATION

1 Inspiration occurs when intrapulmonic pressure falls below atmospheric pressure. Contraction of the diaphragm and intercostal muscles increases the size of the thorax and decreases the intrathoracic pressure. Decreased intrathoracic pressure causes a decreased intrapulmonic pressure.

2 Expiration occurs when intrapulmonic pressure is higher than atmospheric pressure. Relaxation of diaphragm and intercostal muscles increases intrathoracic pressure, which causes an increased intrapulmonic pressure.

AIR VOLUMES EXCHANGED

1 Among the air volumes exchanged in ventilation are tidal volume, inspiratory reserve, expiratory reserve, residual volume, and minimal volumes.

MINUTE VOLUME OF RESPIRATION

1 The minute volume of respiration is the total air taken in during 1 minute (tidal volume times 16 respirations per minute).

EXCHANGE OF RESPIRATORY GASES

1 The partial pressure of a gas is the pressure exerted by that gas in a mixture of gases. It is symbolized by p.

2 Boyle's law states that the volume of a gas varies inversely with pressure, assuming that the temperature is constant.

3 Charles' law indicates that the volume of a gas is directly proportional to its absolute temperature, assuming that the pressure remains constant.

4 According to Dalton's law, each gas in a mixture of gases exerts its own pressure as if all the other gases were not present.

5 Henry's law states that the quantity of a gas that will dissolve in a liquid is proportional to the partial pressure of the gas, when the temperature remains constant.

EXTERNAL RESPIRATION

1 In internal and external expiration O_2 and CO_2 move from areas of their higher partial pressure to areas of their lower partial pressure.

2 External respiration is aided by a thin alveolar-capillary membrane, a large alveolar surface area (about 540 square feet), and a rich blood supply.

INTERNAL RESPIRATION

1 Internal respiration takes place in the capillaries of the body tissues, and is the exchange of O_2 and CO_2 between the blood and body tissues.

TRANSPORT OF RESPIRATORY GASES

1 In each 100 ml of oxygenated blood, there are 20 ml of $O_2 - 0.5$ ml is dissolved in plasma and 19.5 ml is carried with hemoglobin as oxyhemoglobin (HbO_2).

2 In each 100 ml of deoxygenated blood, there are 56 ml of CO_2. About 9 percent of CO_2 is dissolved in plasma, about 27 percent combines with hemoglobin as carbaminohemoglobin ($HbNHCOOH$), and about 64 percent is converted to the bicarbonate ion (HCO_3^-).

CONTROL OF RESPIRATION

1 Nervous control is regulated by the respiratory centers in the medulla and pons, which control the rhythm of respiration. The Hering-Breuer reflex controls the depth and rhythm of respiration.

2 Chemical control is regulated by chemical stimuli (CO_2, O_2, and H^+ ions) in the blood that stimulate chemoreceptors in the carotid sinuses and the aorta as well as the medulla itself.

3 Pressoreceptors in the carotid and aortic bodies also influence rate of respiration.

HEART AND LUNG RESUSCITATION

1 Exhaled air ventilation is used to reestablish respiration. External cardiac compression is used to reestablish circulation.

2 The Heimlich maneuver is a new and effective first aid procedure in case of food choking.

SMOKING

1 Pollutants, including smoke, act as stresses on the epithelium of the bronchi and lungs. Constant irritation results in excessive secretion of mucus and rapid division of bronchial basal cells.

2 Additional irritation may cause retention of mucus in bronchioles, loss of elasticity of alveoli, and less surface area for gaseous exchange.

3 In the final stages, bronchial epithelial cells may be replaced by cancer cells. The growth may block a bronchial tube and spread throughout the lung and other body tissues.

NEBULIZATION

1 Nebulization is a comparatively new method of introducing different types of drugs, chemicals, and antibiotics directly into the respiratory tracts.

REVIEW QUESTIONS

1 What organs make up the respiratory system? What function do the respiratory and cardiovascular systems have in common?

2 Describe the structures of the external and internal nose and describe their functions in filtering, warming, and moistening air.

3 What is the pharynx? Differentiate the three regions of the pharynx and indicate their roles in respiration.

4 Describe the structures of the larynx and explain how they function in respiration and voice production.

5 Describe the location and structure of the trachea. What is tracheotomy? Intubation?

6 What is the bronchial tree? Describe its structure. What is a bronchogram?

7 Where are the lungs located? Distinguish the parietal pleura from the visceral pleura. What is pleurisy?

8 Define each of the following parts of a lung: base, apex, costal surface, medial surface, hilum, root, cardiac notch, and lobe.

9 What is a lobule of the lung? Describe its composition and function in respiration.

10 Indicate several ways in which the respiratory organs are structurally adapted to carry on their respiratory functions.

11 What are the basic differences among ventilation, external respiration, and internal respiration?

12 Discuss the basic steps involved in inspiration and expiration. Be sure to include values for all pressures involved.

13 What is a respirometer? Define the various lung volumes and capacities. How is the minute volume of respiration calculated?

14 Define the partial pressure of a gas. How is it calculated?

15 Define Boyle's law, Charles' law, Dalton's law, and Henry's law.

16 What are the partial pressures of oxygen and carbon dioxide in the atmosphere, alveolar air, arterial blood, body tissues, and venous blood?

17 Construct a diagram to illustrate how and why the respiratory gases move during external and internal respiration.

18 How are oxygen and carbon dioxide carried by the blood?

19 Discuss how the inspiratory and expiratory centers are related to the Hering-Breuer reflex. What is the role of the apneustic center and the pneumotaxic center in controlling respiration?

20 Explain how the following chemical stimuli affect respiration: pCO_2, pO_2, and H^+ ions.

21 How do pressoreceptors affect the control of respiration?

22 How does the control of respiration demonstrate the principle of homeostasis?

23 What is the objective of heart and lung resuscitation? What cautions must be taken in exhaled air ventilation and external cardiac compression? Why?

24 Describe the steps involved in the Heimlich maneuver.

Chapter 24
The
Digestive
System

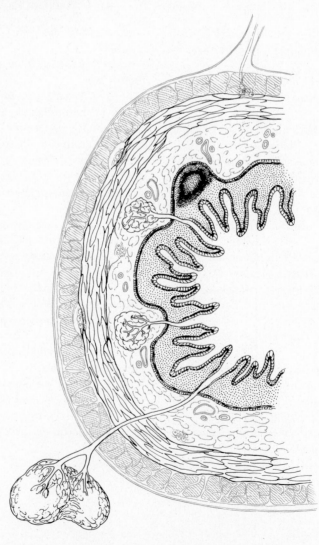

STUDENT OBJECTIVES

■ Define digestion as a chemical and mechanical process.

■ Identify the organs of the alimentary canal and the accessory organs of digestion.

■ Describe the structure of the wall of the alimentary canal.

■ Define the mesentery, lesser omentum, greater omentum, and falciform ligament as extensions of the peritoneum.

■ Describe the role of the mouth in mechanical digestion.

■ Identify the location of the salivary glands.

■ Define the function of saliva in digestion.

■ Define the action of salivary amylase.

■ Describe the mechanisms that regulate the secretion of saliva.

■ Identify the parts of a typical tooth.

■ Compare deciduous and permanent dentitions.

■ Discuss the sequence of events involved in swallowing.

■ Describe the structural features of the stomach and the relationship between these features and digestion.

■ Compare mechanical and chemical digestion in the stomach.

■ Describe the factors that control the secretion of gastric juice.

■ Describe the relationship of the pancreas to digestion.

■ Define the role of the liver and gallbladder in digestion.

■ Describe the structural features of the small intestine that adapt it for digestion and absorption.

■ Describe the mechanisms involved in the hormonal control of digestion in the stomach and small intestine.

■ Describe the digestive activities of the small intestine that reduce carbohydrates, proteins, and fats to their final products.

■ Describe the mechanical movements of the small intestine.

■ Define absorption.

■ Compare the fates of absorbed nutrients.

■ Describe the structural features of the large intestine that adapt it for absorption, formation of feces, and elimination.

■ Describe the mechanical movements of the large intestine.

■ Describe the processes involved in the formation of feces.

■ Discuss the mechanisms involved in defecation.

■ List the causes and symptoms of dental caries and periodontal disease.

■ Contrast the location and effects of gastric and duodenal ulcers.

■ Describe cirrhosis as a disorder of an accessory organ of digestion.

■ Describe the location of tumors of the gastrointestinal tract.

■ Define medical terminology associated with the digestive system.

We all know that food is vital to life. Food is required for the chemical reactions that occur in every cell – both those that synthesize new enzymes, cell structures, bone, and all the other components of the body and those that release the energy needed for the building processes. However, the vast majority of foods we eat are simply too large to pass through the plasma membranes of the cells. Therefore, chemical and mechanical **digestion** must occur first.

CHEMICAL AND MECHANICAL DIGESTION

Chemical digestion is a series of catabolic reactions that break down the large carbohydrate, lipid, and protein molecules which we eat into molecules that are usable by body cells. These products of digestion are small enough to pass through the walls of the digestive organs, into the blood and lymph capillaries, and eventually into the body's cells. **Mechanical digestion** consists of various movements that aid chemical digestion. Food must be pulverized by the teeth before it can be swallowed. Then the smooth muscles of the stomach and small intestine churn the food so it is thoroughly mixed with the enzymes that catalyze the reactions.

The digestive system therefore prepares food for consumption by the cells. It does this through five basic activities:

1 Ingestion, or eating, which is taking food into the body.

2 Peristalsis, the movement of food along the digestive tract.

3 Mechanical and chemical digestion.

4 Absorption, the passage of digested food from the digestive tract into the circulatory and lymphatic systems for distribution to cells.

5 Defecation, the elimination of indigestible substances from the body.

GENERAL ORGANIZATION

The organs of digestion are traditionally divided into two main groups. First is the **gastrointestinal (GI) tract** or **alimentary canal,** a continuous tube running through the ventral body cavity and extending from the mouth to the anus (Figure 24-1). The relationship of the digestive organs to the nine regions of the abdominopelvic cavity may be reviewed in Figure 1-5. The length of a tract taken from a cadaver is about 9 m (30 ft). In a living person it is somewhat shorter because the muscles in its walls are in a state of tone. Organs composing the gastrointestinal tract include the mouth, pharynx, esophagus, stomach, small intestine, and large intestine.

The GI tract contains the food from the time it is eaten until it is digested and prepared for elimination. Muscular contractions in the walls of the GI tract break down the food physically by churning it. Secretions produced by cells along the GI tract break down the food chemically.

The second group of organs composing the digestive system consists of the **accessory organs** – the teeth, tongue, salivary glands, liver, gallbladder, pancreas, and appendix. Teeth are cemented to bone, protrude into the GI tract, and aid in the physical breakdown of food. The other accessory organs except for the tongue lie totally outside the tract and produce or store secretions that aid in the chemical breakdown of food. These secretions are released into the tract through ducts.

General Histology

The walls of the GI tract, especially from the esophagus to the anal canal, have the same basic arrangement of tissues. The four tunics or coats of the tract from the inside out are the mucosa, submucosa, muscularis, and serosa or adventitia (Figure 24-2).

The **tunica mucosa,** or inner lining of the tract, is a mucous membrane attached to a thin layer of visceral muscle. Two layers compose the membrane: a *lining epithelium,* which is in direct contact with the food, and an underlying layer of connective tissue called the *lamina propria.* Under the lamina propria are two thin layers of visceral muscle called the *muscularis mucosa.*

The epithelial layer is composed of nonkeratinized cells that are stratified in the mouth and esophagus but are simple throughout the rest of the tract. The functions of the stratified epithelium are protection and secretion. The functions of the simple epithelium are secretion and absorption. However, the lack of keratin allows some absorption to occur in all parts of the tract.

The lamina propria is made of loose connective tissue containing many blood and lymph vessels and scattered lymph nodules. This layer supports the epithelium, binds it to the muscularis mucosa, and provides it with a blood and lymph supply. The blood and lymph vessels are the avenues by which nutrients in the tract reach the other tissues of the body. The lymph tissue also protects against disease. Remember that the GI tract is in contact with the outside environment and contains food which often carries harmful bacteria. Unlike the skin, the mucous membrane of the tract is not protected from bacterial entry by keratin. The lamina propria also contains glandular epithelium that secretes products necessary for chemical digestion.

The muscularis mucosa contains visceral muscle fibers that throw the mucous membrane of the intestine into small folds which increase the digestive and

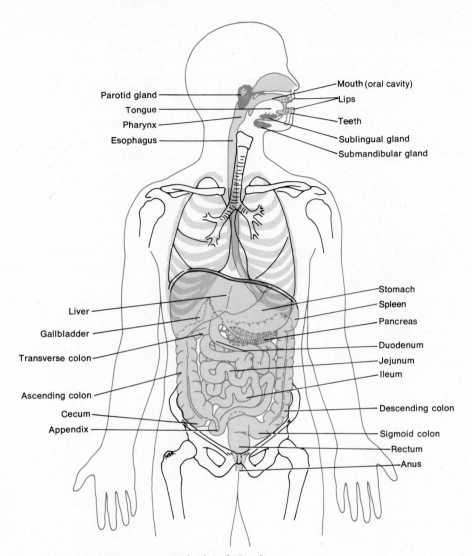

Parotid gland
Tongue
Pharynx
Esophagus

Mouth (oral cavity)
Lips
Teeth
Sublingual gland
Submandibular gland

Liver
Gallbladder
Transverse colon
Ascending colon
Cecum
Appendix

Stomach
Spleen
Pancreas
Duodenum
Jejunum
Ileum
Descending colon
Sigmoid colon
Rectum
Anus

Figure 24-1   Organs of the digestive system and related structures.

absorptive area. With one exception, which will be described later, the other three coats of the intestine contain no glandular epithelium.

The **tunica submucosa** consists of dense connective tissue binding the tunica mucosa to the tunica muscularis. It is highly vascular and contains a portion of the *plexus of Meissner,* which is part of the autonomic nerve supply to the muscularis mucosa.

The **tunica muscularis** of the mouth, pharynx, and esophagus consists in part of skeletal muscle that produces voluntary swallowing. Throughout the rest of the tract, the muscularis consists of smooth muscle that is generally found in two sheets: an inner ring of circular fibers and an outer sheet of longitudinal fibers. Contractions of the smooth muscles help to break down food physically, mix it with digestive secretions, and propel it through the tract. The muscularis also contains the

major nerve supply to the alimentary tract – the *plexus of Auerbach,* which consists of fibers from both autonomic divisions.

The **tunica serosa,** the outermost layer of the canal, is a serous membrane composed of connective tissue and epithelium. This covering, also called the visceral peritoneum, is worth discussing in detail.

Peritoneum

The **peritoneum** is the largest serous membrane of the body. Serous membranes are associated with the heart (pericardium), lungs (pleura), and other thoracic organs. Serous membranes consist of a layer of simple squamous epithelium (called mesothelium) and an underlying supporting layer of connective tissue. The *parietal peritoneum* lines the walls of the abdominal

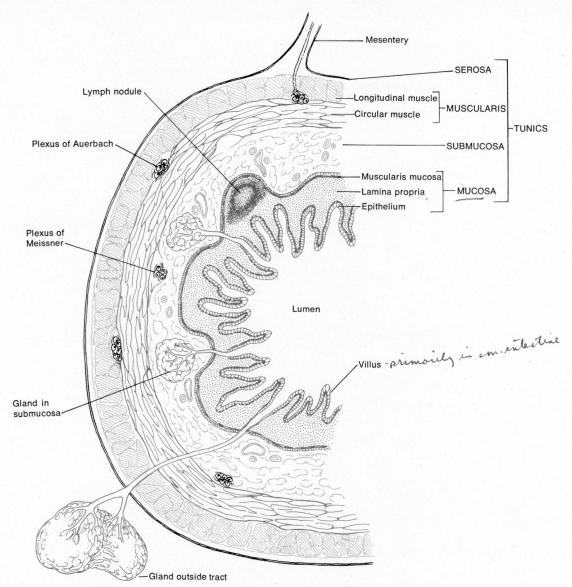

Figure 24-2 Gastrointestinal tract seen in cross section.

cavity. The *visceral peritoneum* covers some of the organs and constitutes their serosa. The space between the parietal and visceral portions of the peritoneum is called the *peritoneal cavity.*

Unlike the two other serous membranes of the body, the pericardium and pleura, the peritoneum contains large folds that weave in between the viscera. The folds bind the organs to each other and to the walls of the cavity and contain the blood and lymph vessels and the nerves that supply the abdominal organs. One extension of the peritoneum is called the **mesentery** and is an outward fold of the serous coat of the intestines (Figure 24-3). Attached to the posterior abdominal wall

is the tip of the fold. The mesentery binds the small intestine to the wall. A similar fold of parietal peritoneum, called the **mesocolon,** binds the large intestine to the posterior body wall. It also carries blood vessels and lymphatics to the intestines.

Other important peritoneal folds are the falciform ligament, the lesser omentum, and the greater omentum. The **falciform ligament** attaches the liver to the anterior abdominal wall and diaphragm. The **lesser omentum** arises as two folds in the serosa of the stomach and duodenum suspending the stomach and duodenum from the liver. The **greater omentum** is a large fold in the serosa of the stomach that hangs down like an apron

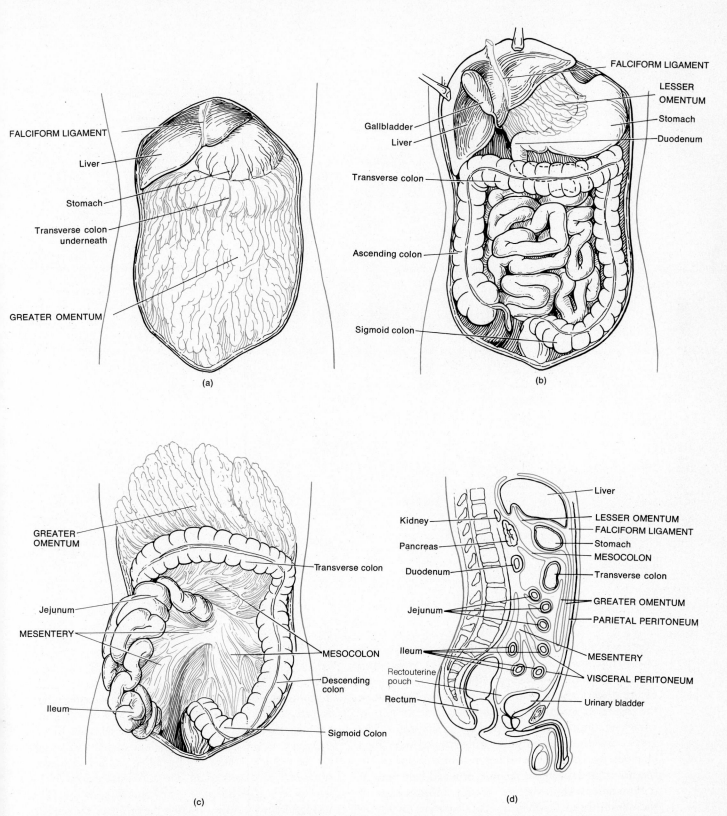

Figure 24–3 Extensions of the peritoneum. (a) Greater omentum. (b) Lesser omentum. The liver and gallbladder have been lifted. (c) Mesentery. The greater omentum has been lifted. (d) Sagittal section through the abdomen and pelvis indicating the relationship of the peritoneal extensions to each other.

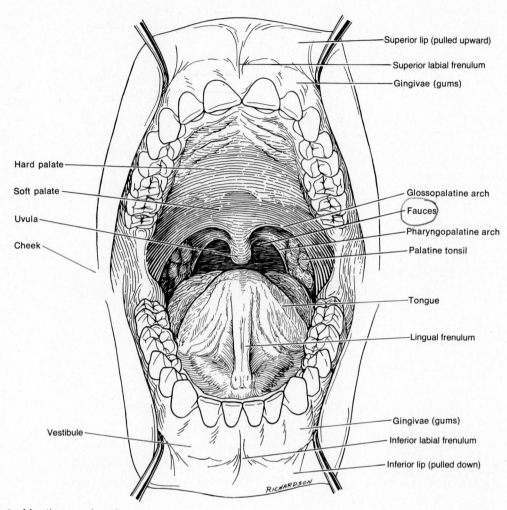

Superior lip (pulled upward)

Superior labial frenulum

Gingivae (gums)

Hard palate

Soft palate

Uvula

Cheek

Glossopalatine arch

Fauces

Pharyngopalatine arch

Palatine tonsil

Tongue

Lingual frenulum

Gingivae (gums)

Inferior labial frenulum

Inferior lip (pulled down)

RICHARDSON

Vestibule

Figure 24—4 Mouth or oral cavity.

over the front of the intestines. It then passes up to part of the large intestine (the transverse colon), wraps itself around it, and finally attaches to the parietal peritoneum of the posterior wall of the abdominal cavity. Because the greater omentum contains large quantities of adipose tissue, it commonly is called the "fatty apron." The greater omentum contains numerous lymph nodes. If an infection occurs in the intestine, plasma cells formed in the lymph nodes combat the infection and help prevent it from spreading to the peritoneum. Inflammation of the peritoneum (*peritonitis*) is a serious condition because the peritoneal membranes are continuous with each other, enabling the infection to spread to all the organs in the cavity.

ORGANS

Mouth or Oral Cavity

The **mouth**, also referred to as the **oral** or **buccal cavity,** is formed by the cheeks, hard and soft palates, and tongue (Figure 24-4). Forming the lateral walls of the oral cavity are the cheeks—muscular structures covered on the outside by skin and lined by stratified squamous nonkeratinized epithelium. The anterior portions of the cheeks terminate in the superior and inferior lips. The lips are fleshy folds surrounding the orifice of the mouth. They are covered on the outside by skin and on the inside by a mucous membrane. The transition zone where the two kinds of covering tissue meet is called the

papillae

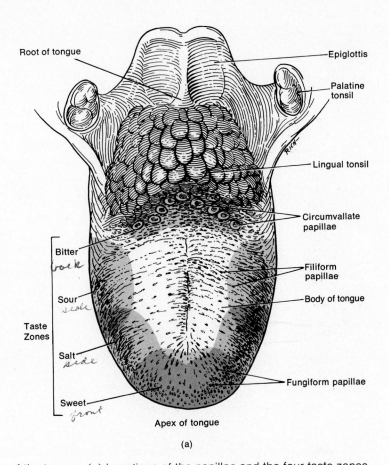

Root of tongue

Epiglottis

Palatine tonsil

Lingual tonsil

Circumvallate papillae

Bitter
back

Filiform papillae

Sour
side

Body of tongue

Taste Zones

Salt
side

Sweet
front

Fungiform papillae

Apex of tongue

(a)

Figure 24–5 Dorsum of the tongue. (a) Locations of the papillae and the four taste zones.

Taste buds are joined to the brain by cranial nerves 5th 7th 9th, 10th. They are modified epithelial cells surrounded by nerve endings

vermilion—this portion of the lips is not keratinized and the color of the blood in the underlying blood vessels is visible through the transparent surface layer of the vermilion. The inner surface of each lip is attached to its corresponding gum by a midline fold of mucous membrane called the *labial frenulum*. The orbicularis oris muscle and connective tissue lie between the external integumentary covering and the internal mucosal lining. During chewing the cheeks and lips help to keep food between the upper and lower teeth. They also assist in speech.

The *vestibule* of the oral cavity is bounded externally by the cheeks and lips and internally by the gums and teeth. The *oral cavity proper* extends from the vestibule to the *fauces,* the opening between the oral cavity and the pharynx or throat.

The **hard palate**—the anterior portion of the roof of the mouth—is formed by the maxillae and palatine bones and is lined by mucous membrane. The **soft palate** forms the posterior portion of the roof of the mouth. It is an arch-shaped muscular partition between the oropharynx and nasopharynx and is lined by mucous

membrane. Hanging from the middle of the lower border of the soft palate is a fingerlike muscular process called the *uvula.* On either side of the base of the uvula are two muscular folds that run down the lateral side of the soft palate. Anteriorly, the *glossopalatine arch* runs downward, laterally, and forward to the side of the base of the tongue. Posteriorly, the *pharyngopalatine arch* projects downward, laterally, and backward to the side of the pharynx. The palatine tonsils are situated between the arches, and the lingual tonsils are situated at the base of the tongue. At the posterior border of the soft palate, the mouth opens into the oropharynx through the fauces.

The **tongue,** together with its associated muscles, forms the floor of the oral cavity. It is composed of skeletal muscle covered with mucous membrane (Figure 24-5). The extrinsic muscles of the tongue originate outside the tongue and insert into it. They move the tongue from side to side and in and out and maneuver food for chewing and swallowing. They also form the floor of the mouth and hold the tongue in position. The intrinsic muscles originate and insert within the tongue,

attached to hyoid bone & mandible & styloid process of temporal bone

(b)

Figure 24–5 (cont.) Dorsum of the tongue. (b) Scanning electron micrograph of filiform papillae at a magnification of 100×. (Courtesy of Fisher Scientific Company and S.T.E.M. Laboratories, Inc., Copyright 1975.)

and they alter the shape of the tongue for speech and swallowing. The lingual frenulum, a fold of mucous membrane in the midline of the undersurface of the tongue, aids in limiting the movement of the tongue posteriorly. If the lingual frenulum is too short, tongue movements are restricted, speech is faulty, and the person is said to be "tongue-tied." These functional problems can be corrected by cutting the lingual frenulum.

The upper surface and sides of the tongue are covered with **papillae,** projections of the lamina propria covered with epithelium. Taste buds are located in some papillae. *Filiform papillae* are conical projections distributed in parallel rows over the anterior two-thirds of the tongue and contain no taste buds. *Fungiform papillae* are mushroomlike elevations distributed among the filiform papillae and are more numerous near the tip of the tongue. They appear as red dots on the surface of the tongue, and most of them contain taste buds. *Circumvallate papillae* are arranged in the form of an inverted V on the posterior surface of the tongue, and all of them

rough texture of tongue

contain taste buds. Note the taste zones of the tongue in Figure 24-5a.

Salivary Glands

Saliva is a fluid that is continuously secreted by glands in or near the mouth. Ordinarily, just enough saliva is secreted to keep the mucous membranes of the mouth moist. But when food enters the mouth, secretion increases so the saliva can lubricate, dissolve, and chemically break down the food. The mucous membrane lining the mouth contains many small glands, the *buccal glands,* that secrete small amounts of saliva. However, the major portion of saliva is secreted by the **salivary glands,** which lie outside the mouth and pour their contents into ducts that empty into the oral cavity. There are three pairs of salivary glands: the parotid, submandibular (submaxillary), and sublingual glands (Figure 24-6). The *parotid glands* are located under and in front of the ears between the skin and the masseter

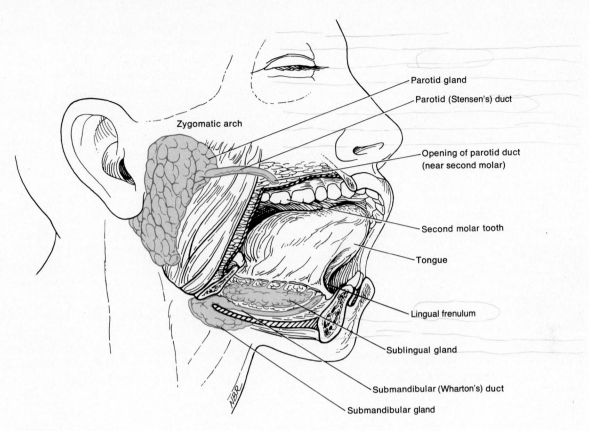

Figure 24–6 Salivary glands.

muscle. Each secretes into the oral cavity vestibule via a duct, called Stensen's duct, that pierces the buccinator muscle to open into the vestibule opposite the upper second molar tooth. The *submandibular glands* are found beneath the base of the tongue in the posterior part of the floor of the mouth. Their ducts (Wharton's ducts) run superficially under the mucosa on either side of the midline of the floor of the mouth and enter the oral cavity proper just behind the central incisors. The *sublingual glands* are anterior to the submandibular glands, and their ducts open into the floor of the mouth in the oral cavity proper. The parotid glands are compound tubuloacinar glands, whereas the submandibular and sublinguals are compound acinar glands.

SALIVA

The fluids secreted by the buccal glands and the three pairs of salivary glands constitute **saliva**. Amounts of saliva secreted daily vary considerably but range from 1,000 to 1,500 ml. Chemically, saliva is 99.5 percent water and 0.5 percent solutes. Among the solutes are salts – chlorides, bicarbonates, and phosphates of sodium and potassium. Some dissolved gases and various organic substances including urea and uric acid, serum albumin and globulin, mucin, the bacteriolytic enzyme lysozyme, and the digestive enzyme amylase are also present.

The water in saliva provides a medium for dissolving foods so they can be tasted and digestive reactions can take place. The chlorides in the saliva activate the amylase. The bicarbonates and phosphates buffer chemicals that enter the mouth and keep the saliva at a slightly acidic pH of 6.35 to 6.85. Urea and uric acid are found in saliva because the saliva-producing glands (like the sweat glands of the skin) help the body to get rid of wastes. Mucin is a protein that forms mucus when dissolved in water. Mucus lubricates the food so it can be easily turned in the mouth, formed into a ball or bolus, and swallowed. The enzyme lysozyme destroys bacteria, thereby protecting the mucous membrane from infection and the teeth from decay.

DIGESTION

Depending on the cells the gland contains, each saliva-producing gland supplies different ingredients to saliva. The parotids contain cells that secrete a thin watery serous liquid containing the enzyme salivary amylase. The submandibular glands contain cells similar to those found in the parotids plus some mucous cells. Therefore, they secrete a fluid that is thickened with mucus but still contains quite a bit of enzyme. The sublingual glands contain mostly mucous cells, so they secrete a much thicker fluid that contributes only a small amount of enzyme to the saliva.

The enzyme salivary amylase initiates the breakdown of polysaccharides (carbohydrates)—this is the only chemical digestion that occurs in the mouth. Carbohydrates are starches and sugars and are classified as either monosaccharides, disaccharides, or polysaccharides. Monosaccharides are small molecules containing several carbon, hydrogen, and oxygen atoms—an example is glucose. Disaccharides consist of two monosaccharides linked together; polysaccharides are chains of three or more monosaccharides. The vast majority of carbohydrates we eat are polysaccharides. Since only monosaccharides can be absorbed into the bloodstream, ingested disaccharides and polysaccharides must be broken down. The function of *salivary amylase* is to break the chemical bonds between some of the monosaccharides that make up the polysaccharides. In this way, the enzyme breaks the long-chain polysaccharides into shorter polysaccharides called *dextrins.* Given sufficient time, salivary amylase also can break down the dextrins into the disaccharide maltose. Food usually is swallowed too quickly for more than 3 to 5 percent of the carbohydrates to be reduced to disaccharides in the mouth. However, salivary amylase in the swallowed food continues to act on polysaccharides for another 15 to 30 minutes in the stomach before the stomach acids eventually inactivate it.

CONTROL OF SECRETION

Normally, moderate amounts of saliva are continuously secreted to keep the mucous membranes moist and to lubricate the movements of the tongue and lips during speech. The saliva is then swallowed and reabsorbed to prevent fluid loss. Dehydration, however, causes the salivary and buccal glands to cease secreting saliva to conserve water. The subsequent feeling of dryness in the mouth promotes sensations of thirst. This phenomenon is also noted during fear or anxiety when sympathetic stimulation dominates. Food stimulates the glands to secrete heavily. When food is taken into the mouth, chemicals in the food stimulate the taste receptors. Rolling a dry, indigestible object over the tongue produces friction, which also may stimulate the receptors. Impulses are conveyed from the receptors to two salivary centers in the brain stem. Returning autonomic impulses from one of the centers activate the secretion of saliva from the parotid glands, while returning autonomic impulses from the other center activate the submandibular and sublingual glands.

The smell, sight, touch, or sound of food preparation also stimulates increased saliva secretion. These stimuli constitute psychological activation and involve learned behavior. Memories stored in the cerebral cortex that associate the stimuli with food are stimulated. The cortex sends impulses to the brain stem via extrapyramidal pathways, and the salivary glands are activated. Psychological activation of the glands has some benefit to the body because it allows the mouth to start chemical digestion as soon as the food is ingested. Salivation is entirely under nerve control.

Saliva continues to be secreted heavily some time after food is swallowed. This flow of saliva washes out the mouth and dilutes and buffers the chemical remnants of irritating substances.

Teeth

The **teeth** or **dentes,** are located in sockets of the alveolar processes of the mandible and maxillae. The alveolar processes are covered by the *gingivae (gums),* which extend slightly into each socket (Figure 24-7) forming the *gingival sulcus.* The sockets are lined by the *periodontal ligament,* which is attached to the socket walls and the cemental surface of the roots. Thus it anchors the teeth in position and also acts as a shock absorber to dissipate the forces of chewing.

A typical tooth consists of three principal portions. The **crown** is the portion above the level of the gums. The **root** consists of one to three projections embedded in the socket. The **cervix** is the constricted junction line of the crown and the root.

Teeth are composed primarily of *dentin,* a bonelike substance that gives the tooth its basic shape and rigidity. The dentin encloses a cavity. The enlarged part of the cavity, the *pulp chamber,* lies in the crown and is filled with *pulp,* a connective tissue containing blood vessels, nerves, and lymphatics. From the pulp chamber, narrow extensions of the pulp cavity run through the root of the tooth and are called *root canals.* Each root canal has an opening at its base, the *apical foramen.* Through the foramen enter blood vessels bearing nourishment, lymphatics affording protection, and nerves providing sensation. The dentin of the crown is covered by *enamel* that consists primarily of calcium phosphate and calcium carbonate. Enamel is the hardest substance in the body and protects the tooth from the wear of chewing. It is also a barrier against acids that easily dissolve the dentin. The dentin of the root is covered by *cementum,* another bonelike substance. Pyorrhea is inflammation of the periodontal ligament and adjacent gums. A prolonged, severe case of pyorrhea can weaken the periodontal ligament, erode the alveolar bone, and thereby loosen the tooth.

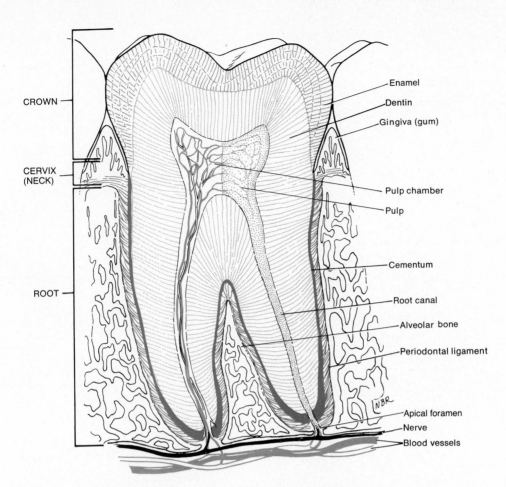

CROWN

CERVIX (NECK)

ROOT

Enamel
Dentin
Gingiva (gum)

Pulp chamber
Pulp

Cementum

Root canal

Alveolar bone

Periodontal ligament

Apical foramen
Nerve
Blood vessels

Figure 24–7 Parts of a typical tooth seen in sagittal section through a molar.

DENTITIONS

Everyone has two **dentitions,** or sets of teeth. The first of these – the deciduous teeth, milk teeth, or baby teeth – begin to erupt at about 6 months of age, and one pair appears at about each month thereafter until all 20 are present. Figure 24-8a illustrates the deciduous teeth. The incisors, which are closest to the midline, are chisel-shaped and adapted for cutting into food. Next to the incisors, moving posteriorly, are the canines or cuspids, which have a pointed surface called the cusp. Canines are used to tear and shred food. The incisors and canines have only one root apiece. Behind them lie the first and second primary molars, which have four cusps and two or three roots. Upper molars have three roots; lower molars have two roots. The molars crush and grind food.

All the deciduous teeth are lost – generally between 6 and 12 years of age – and are replaced by the permanent dentition (Figure 24-8b). The permanent dentition contains 32 teeth that appear between the age of 6 and adulthood. It resembles the deciduous dentition with the following exceptions: the deciduous molars are replaced with premolars or bicuspids that have two cusps

and one root (upper first bicuspids have two roots and are used for crushing and grinding). The permanent molars erupt into the mouth behind the bicuspids. They do not replace any primary teeth and erupt as the jaw grows to accommodate them – the first molars at age 6, the second molars at age 12, the third molars after age 18. The human jaw is becoming increasingly smaller and often does not afford enough room behind the second molars for the eruption of the third molars or wisdom teeth. In this case, the third molars remain embedded in the alveolar bone and are said to be "impacted." Most often they cause pressure and pain and must be surgically removed. In some individuals, however, evolution has caught up with this problem – third molars may be dwarfed in size or may not develop at all.

DIGESTION

Through chewing, or mastication, the teeth pulverize food and mix it with saliva. As a result, the food is reduced to a soft, flexible mass, called a *bolus,* that is easily swallowed. Exhibit 24-1 summarizes the digestion that occurs in the mouth.

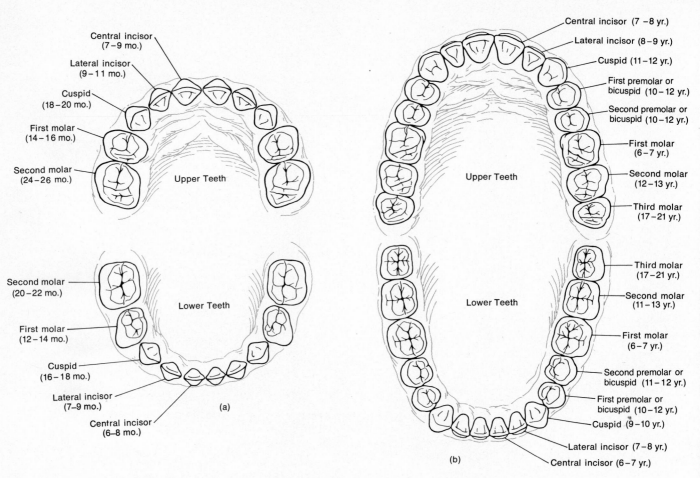

Figure 24–8 Dentitions and times of eruptions. The times of eruptions are indicated in parentheses. (a) Deciduous dentition. (b) Permanent dentition.

Deglutition

Swallowing, or **deglutition,** moves food from the mouth to the stomach. Swallowing starts with the bolus on the upper side of the tongue. Then the tip of the tongue rises and presses against the palate (Figure 24-9). The **bolus** slides to the back of the mouth and is pulled through the fauces by muscles that lie in the oropharynx. During this period the respiratory passageways close, and breathing is temporarily interrupted. The soft palate and uvula move upward to close off the nasopharynx, and the larynx is pulled forward and upward under the tongue. As the larynx rises, it meets the epiglottis, which seals off the glottis. The movement of the larynx also pulls the vocal cords together, further sealing off the respiratory tract, and widens the opening between the laryngopharynx and esophagus. The bolus passes through the laryngopharynx and enters the esophagus in 1 second. The respiratory passageways then reopen and breathing resumes.

The **esophagus,** the third organ involved in deglutition, is a muscular, collapsible tube that lies behind the trachea. It is about 23 to 25 cm (10 inches) long and begins at the end of the laryngopharynx, passes through the mediastinum in front of the vertebral column, pierces the diaphragm through an opening called the esophageal hiatus, and terminates in the upper portion of the stomach.

The esophagus does not produce digestive enzymes and does not carry on absorption. It does, however, secrete mucus and transport food to the stomach.

Food is pushed through the esophagus by muscular movements called **peristalsis** (Figure 24-10). Peristalsis is a function of the tunica muscularis. In the section of the esophagus lying just above and around the top of the bolus, the circular muscle fibers contract. The contraction constricts the esophageal wall and squeezes the bolus downward. Meanwhile, longitudinal fibers lying around the bottom of the bolus and just below it also contract. Contraction of the longitudinal fibers shortens this lower section, pushing its walls outward so it can receive the bolus. The contractions are repeated in a wave that moves down the esophagus, pushing the food

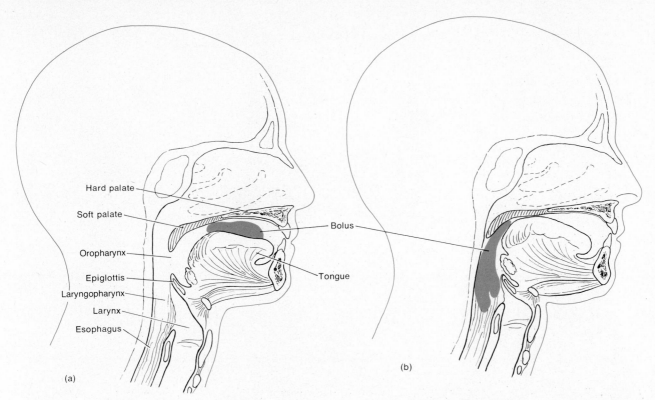

Hard palate

Soft palate

Oropharynx

Epiglottis

Laryngopharynx

Larynx

Esophagus

Bolus

Tongue

(a)

(b)

Figure 24—9 Deglutition. (a) Position of structures prior to swallowing. (b) During swallowing, the tongue rises against the palate, the nose is closed off, the larynx rises, the epiglottis seals off the larynx, and the bolus is passed into the esophagus.

Exhibit 24—1 DIGESTION IN MOUTH

Structure	Activity	Result	Structure	Activity	Result
Cheeks	Keep food between teeth during mastication.	Foods are uniformly chewed.	**Buccal glands**	Secrete saliva.	Lining of mouth and pharynx moistened and lubricated.
Lips	Keep food between teeth during mastication.	Foods are uniformly chewed.	**Salivary glands**	Secrete saliva.	Same as above. Saliva softens, moistens, and dissolves food, coats food with mucin, cleanses mouth and teeth. Salivary amylase reduces polysaccharides to dextrins and the disaccharide maltose.
Tongue Extrinsic muscles	Move tongue from side to side and in and out.	Maneuver food for mastication and deglutition (swallowing).			
Intrinsic muscles	Alter shape of tongue.	Deglutition.			
Taste buds	Serve as receptors for food stimulus.	Nerve impulses from taste buds to brain to salivary glands stimulate secretion of saliva.	**Teeth**	Cut, tear, and pulverize food.	Solid foods are reduced to smaller particles for swallowing.

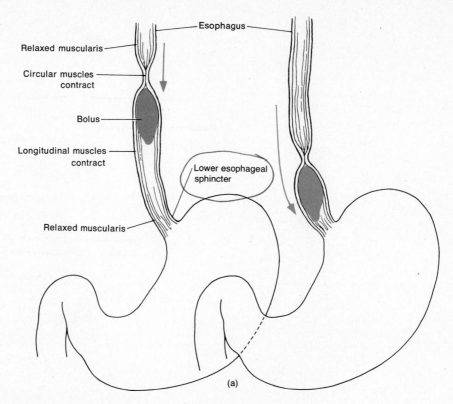

Esophagus

Relaxed muscularis

Circular muscles
contract

Bolus

Longitudinal muscles
contract

Lower esophageal
sphincter

Relaxed muscularis

(a)

Figure 24–10　Peristalsis. (a) Diagrammatic representation.

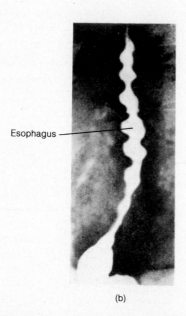

Esophagus

(b)

Figure 24–10 (cont.)　Peristalsis. (b) Anteroposterior projection of peristalsis made during fluoroscopic examination while a patient was swallowing a barium "meal." (Courtesy of Lester W. Paul and John H. Juhl, *The Essentials of Roentgen Interpretation*, 3d ed., Harper & Row, Publishers, Inc., New York, 1972.)

toward the stomach. Passage of the bolus is further facilitated by glands secreting mucus. The passage of solid or semisolid food from the mouth to the stomach takes 4 to 8 seconds. Very soft foods and liquids pass through in about 1 second.

Just above the level of the diaphragm, the esophagus is slightly narrowed. This narrowing has been attributed to a physiological sphincter in the inferior part of the esophagus known as the *lower esophageal* or *gastroesophageal sphincter.* A **sphincter** is an opening that has a thick circle of muscle around it. The lower esophageal sphincter relaxes during swallowing and thus aids the passage of the bolus from the esophagus into the stomach. The movement of the diaphragm against the stomach during breathing presses on the stomach and helps prevent the regurgitation of gastric contents from the stomach to the esophagus.

Exhibit 24-2 summarizes the digestion-related activities of the pharynx and esophagus.

Stomach

The **stomach** is a J-shaped enlargement of the GI tract directly under the diaphragm in the epigastric, umbilical, and left hypochondriac regions of the abdomen. The superior portion of the stomach is a continuation of the esophagus. The inferior portion empties into the duodenum, the first part of the small intestine. Within each individual, the position and size of the stomach vary continually. For instance, the diaphragm pushes the stomach downward with each inspiration and pulls it upward with each expiration. Empty, it is about the size of a large sausage, but the stomach can stretch itself to accommodate large amounts of food.

ANATOMY

The stomach is divided into four areas: cardia, fundus, body, and pylorus (Figure 24-11). The *cardia* surrounds the lower esophageal sphincter, and the rounded portion above and to the left of the cardia is the *fundus.* Below the fundus, the large central portion of the stomach is called the *body.* The narrow, inferior region is the *pylorus.* The concave medial border of the stomach is called the *lesser curvature,* and the convex lateral border is the *greater curvature.* The pylorus communicates with the duodenum of the small intestine via a sphincter called the *pyloric valve.*

Two abnormalities of the pyloric valve can occur in infants. *Pylorospasm* is characterized by failure of the muscle fibers encircling the opening to relax normally. Ingested food does not pass easily from the stomach to the small intestine, the stomach becomes overly full, and the infant vomits frequently to relieve the pressure. Pylorospasm is treated by adrenergic drugs that relax the muscle fibers of the valve. *Pyloric stenosis* is a narrowing of the pyloric valve caused by a tumorlike mass that apparently is formed by enlargement of the circular muscle fibers. It must be surgically corrected.

The stomach wall is composed of the same four basic layers as the rest of the alimentary canal, with certain modifications. When the stomach is empty, the mucosa lies in large folds that can be seen with the naked eye. These folds are called *rugae.* As the stomach fills and distends, the rugae gradually smooth out and disappear. Microscopic inspection of the mucosa reveals a layer of simple columnar epithelium containing many narrow openings that extend down into the lamina propria. These pits—*gastric glands*—are lined with three kinds of secreting cells: zymogenic, parietal, and mucous. The zymogenic, or *chief cells,* secrete the principal gastric enzyme called *pepsinogen.* Hydrochloric acid, which activates the pepsinogen, is produced by the *parietal cells.* The *mucous cells* secrete mucus and the intrinsic factor—a substance involved in the absorption of vitamin B_{12}. Secretions of the gastric glands are called *gastric juice.*

The submucosa of the stomach is composed of loose areolar connective tissue, and it connects the mucosa to the muscularis. The muscularis, unlike that in

gastric = stomach

Exhibit 24–2 DIGESTIVE ACTIVITIES OF PHARYNX AND ESOPHAGUS

Structure	Activity	Result
Pharynx	Deglutition	Food is passed from oropharynx to laryngopharynx and into esophagus. Air passageways are closed off
Esophagus	Peristalsis	Bolus is forced down esophagus into stomach.
	Secretion of mucus	Bolus passes smoothly down esophagus.

other areas of the alimentary canal, has three layers of smooth muscle: an outer longitudinal layer, a middle circular layer, and an inner oblique layer. This arrangement of fibers allows the stomach to contract in a variety of ways to churn food, break it into small particles, mix it with gastric juice, and pass it to the duodenum. The serosa covering the stomach is part of the visceral peritoneum. At the lesser curvature the two layers of the visceral peritoneum come together and extend upward to the liver as the lesser omentum. At the greater curvature, the visceral peritoneum continues downward as the greater omentum hanging over the intestines.

DIGESTION

Several minutes after food enters the stomach, gentle, rippling, peristaltic movements called *mixing waves* pass over the stomach every 15 to 25 seconds. These waves macerate food, mix it with the secretions of the digestive glands, and reduce it to a thin liquid called *chyme.* Few mixing waves are observed in the fundus, which is primarily a storage area. Foods may remain in the fundus for an hour or more without becoming mixed with gastric juice. During this time, salivary digestion continues.

The principal chemical activity of the stomach is to begin the digestion of proteins. In the adult, digestion is achieved primarily through the enzyme *pepsin.* Pepsin breaks certain peptide bonds between the amino acids making up proteins. Thus a protein chain of many amino acids is broken down into fragments of amino acids—long fragments called *proteoses;* somewhat shorter ones are *peptones.* Pepsin is most effective in the very acidic environment of the stomach (pH of 1). It

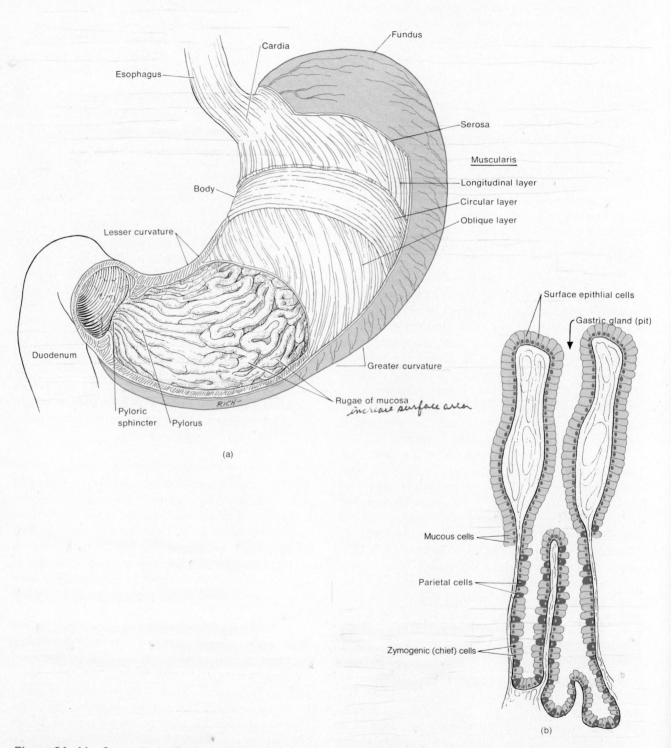

Cardia

Fundus

Esophagus

Serosa

Muscularis

Longitudinal layer

Circular layer

Oblique layer

Body

Lesser curvature

Duodenum

Greater curvature

Pyloric sphincter

Pylorus

Rugae of mucosa

increase surface area

RICH-

(a)

Surface epithlial cells

Gastric gland (pit)

Mucous cells

Parietal cells

Zymogenic (chief) cells

(b)

Figure 24–11   Stomach. (a) External and internal anatomy. (b) Diagram of gastric glands from the fundic wall.

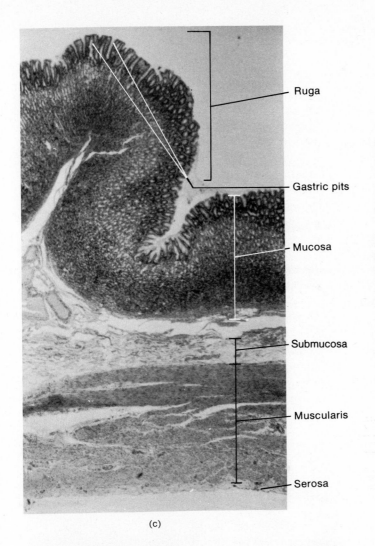

- Ruga
- Gastric pits
- Mucosa
- Submucosa
- Muscularis
- Serosa

(c)

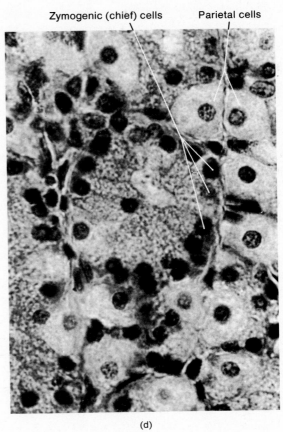

Zymogenic (chief) cells   Parietal cells

(d)

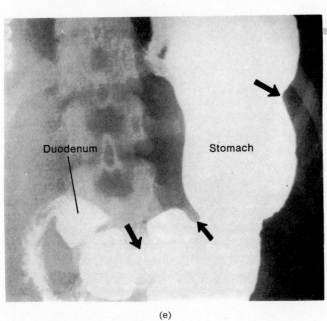

Duodenum   Stomach

(e)

constriction

Figure 24–11 (cont.) Stomach. (c) Histology of the fundic wall at a magnification of 6×. (d) Enlarged aspect of parietal and chief cells at a magnification of 160×. (Courtesy of Victor B. Eichler, Wichita State University.) (e) Anteroposterior projection of a normal stomach. Note the peristaltic waves indicated by the arrows. (Courtesy of Lester W. Paul and John H. Juhl, *The Essentials of Roentgen Interpretation,* 3d ed., Harper & Row, Publishers, Inc., New York, 1972.)

becomes inactive in an alkaline environment. What keeps pepsin from digesting the protein in stomach cells along with the food? First of all, pepsin is secreted in an inactive form called *pepsinogen,* so it cannot digest the proteins in the zymogenic cells that produce it. When pepsinogen comes in contact with the hydrochloric acid secreted by the parietal cells, it is converted to active pepsin. Once pepsin has been activated, the stomach cells are protected by mucus. The mucus coats the mucosa and forms a barrier between the gastric juice and the cells. Sometimes the mucus fails to do its job, and the pepsin and hydrochloric acid eat a hole in the stomach wall known as a *gastric ulcer.*

Another enzyme of the stomach is *gastric lipase.* Gastric lipase splits the butterfat molecules found in milk. This enzyme operates best at a pH of 5 to 6 and has a limited role in the adult stomach. Adults rely exclusively on an enzyme found in the small intestine to digest fats.

As digestion proceeds in the stomach, more vigorous peristaltic waves begin at about the middle of the stomach, pass downward, reach the pyloric valve, and sometimes go into the duodenum. The movement of chyme from the stomach into the duodenum depends on a pressure gradient between the two organs. When the pressure in the stomach (intragastric pressure) is greater than that in the duodenum (intraduodenal pressure), chyme is forced into the duodenum. Peristaltic waves are largely responsible for increased intragastric pressure. It is estimated that 2–5 ml of chyme are passed into the duodenum with each peristaltic wave. When intraduodenal pressure exceeds intragastric pressure, the pyloric valve closes and prevents the regurgitation of chyme from the duodenum to the stomach. The stomach empties all its contents into the duodenum 2 to 6 hours after ingestion. Food rich in carbohydrate leaves the stomach in a few hours. Protein foods are somewhat slower, and emptying is slowest after a meal containing large amounts of fat. The stomach wall is impermeable to the passage of most materials into the blood, so most substances are not absorbed until they reach the small intestine. However, the stomach does participate in the absorption of some water and salts, certain drugs, and alcohol.

CONTROL OF SECRETION

The secretion of gastric juice is regulated by both nervous and hormonal mechanisms. Seeing, smelling, tasting, touching, hearing the sounds of food preparation, or thinking of food stimulate the cerebral cortex to send impulses to the medulla. The medulla relays impulses over the parasympathetic fibers in the vagus nerve to stimulate the gastric glands to secrete. This psychic stimulation prepares the stomach for digestion.

Once the food reaches the stomach, both nervous and hormonal mechanisms ensure that gastric secretion continues. Food of any kind stimulates receptors in the walls of the stomach. These receptors send impulses over a reflex arc of vagal fibers to the medulla and back to the gastric glands, and they may send messages directly to the glands as well. Emotions such as anger, fear, and anxiety may slow down digestion in the stomach because they stimulate the sympathetic nervous system, which in turn inhibits the impulses of the parasympathetic fibers. Protein foods stimulate the pyloric mucosa to secrete a hormone called *gastrin.* Gastrin is absorbed into the bloodstream, circulated through the body and finally reaches its target cells, the gastric glands, where it stimulates secretion of large amounts of digestive enzymes and hydrochloric acid.

Some investigators believe that when proteoses and peptones leave the stomach and enter the duodenum, they stimulate the intestinal mucosa to release intestinal gastrin, a hormone that stimulates the gastric glands to continue their secretion. However, this mechanism produces relatively small amounts of gastric juice. Exhibit 24-3 summarizes the chief activities of the stomach.

The next step in the breakdown of food is digestion in the small intestine. Chemical digestion in the small intestine depends not only on its own secretions but on activities of three organs outside the alimentary canal: the pancreas, liver, and gallbladder.

Pancreas

The **pancreas** is a soft, oblong tubuloacinar gland about 12.5 cm (6 inches) long and 2.5 cm (1 inch) thick. It lies posterior to the greater curvature of the stomach and is connected by a duct (sometimes two) to the duodenum (Figure 24-12). The pancreas is divided into a head, body, and tail. The *head* is the expanded portion near the C-shaped curve of the duodenum. Moving superiorly and to the left of the head are the centrally located *body* and the terminal tapering *tail.* The pancreas is made up of small clusters of glandular epithelial cells. Some clusters, called *islets of Langerhans,* form the endocrine portions of the pancreas and consist of alpha and beta cells that secrete glucagon and insulin. The other masses of cells, called *acini,* are the exocrine portions of the organ. Secreting cells of the acini release a mixture of digestive enzymes called *pancreatic juice,* which is dumped into small ducts attached to the acini. Pancreatic juice eventually leaves the pancreas through a large main tube called the *pancreatic duct,* or *duct of Wirsung.* In most people the pancreatic duct unites with the common bile duct from the liver and gallbladder and

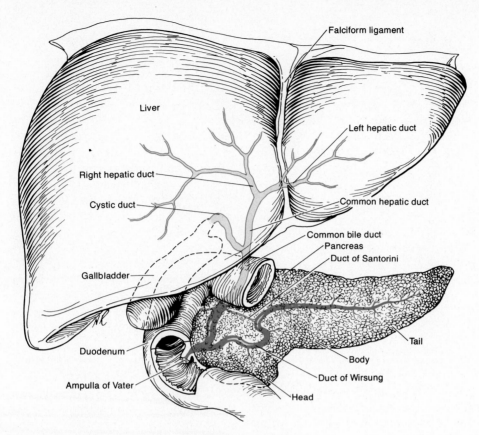

Figure 24–12   Pancreas in relation to liver, gallbladder, and duodenum.

Exhibit 24–3   GASTRIC DIGESTION

Structure	Activity	Result
Mucosa		
Rugae	Provide large surface area for stretching of stomach.	Allow for distension of stomach.
Mucous cells	Secrete mucus.	Prevents digestion of stomach wall.
	Secrete intrinsic factor.	Required for erythrocyte formation.
Zymogenic cells	Secrete pepsinogen.	Its active form (pepsin) digests proteins into proteoses and peptones.
Parietal cells	Secrete hydrochloric acid.	In its presence, pepsinogen is converted into pepsin.
Muscularis	Mixing waves.	Macerate food, mix it with gastric juice, and reduce food to chyme.
	Peristaltic waves.	Force chyme through pyloric valve into duodenum.
Lower esophageal sphincter	Regulates passage of bolus from esophagus into stomach.	Prevents backflow of food from stomach to esophagus.
Pyloric valve	Opens to permit passage of chyme into duodenum.	Prevents backflow of food from duodenum to stomach.

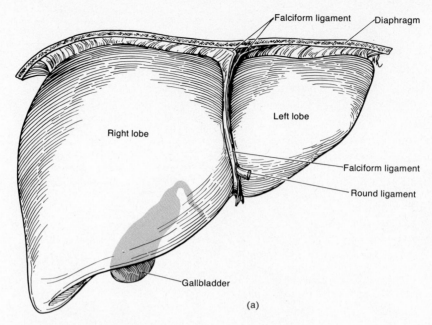

Figure 24-13 Liver. (a) External anatomy in anterior view.

enters the duodenum in a small, raised area called the *ampulla of Vater.* An accessory duct, the *duct of Santorini,* may also lead from the pancreas and empty into the duodenum about 2.5 cm (1 inch) above the ampulla of Vater.

The functions of the pancreas are twofold. The acini secrete enzymes that digest food in the small intestine, and the alpha and beta cells secrete the hormones glucagon and insulin, which regulate blood sugar level.

Liver

The **liver** performs so many vital functions that we cannot live without it:

1 The liver manufactures the anticoagulant heparin and most of the other plasma proteins.

2 The reticuloendothelial cells of the liver phagocytose worn-out red blood cells and some bacteria.

3 Liver cells contain enzymes that either break down poisons or transform them into less harmful compounds. When amino acids are burned for energy, for example, they leave behind toxic nitrogenous wastes that are converted to urea by the liver cells. Moderate amounts of urea are harmless to the body and are easily excreted by the kidneys and sweat glands.

4 Newly absorbed nutrients are collected in the liver. It can change any excess monosaccharides into glycogen or fat, both of which can be stored. In addition it can transform glycogen, fat, and protein into glucose and vice versa, depending on the body's needs.

5 The liver stores glycogen, copper, iron, and vitamins A, D, E, and K. It also stores some poisons that cannot be broken down and excreted. (High levels of DDT are found in the livers of animals, including humans, who eat sprayed fruits and vegetables.)

6 The liver manufactures bile, which is used in the small intestine for the emulsification and absorption of fats.

The liver is the largest single organ in the body, weighing about 1.4 kg (4 lb) in the average adult. It is located under the diaphragm and occupies most of the right hypochondrium and part of the epigastrium of the abdomen. The liver is covered largely by peritoneum and completely by a dense connective tissue layer that lies beneath the peritoneum. Anatomically, the liver is divided into two principal lobes—the **right lobe** and the **left lobe**—separated by the *falciform ligament* (Figure 24-13). The right lobe, besides the main lobe, also has associated with it an inferior *quadrate lobe* and a posterior *caudate lobe.* The falciform ligament is a reflection of the parietal peritoneum, which extends from the undersurface of the diaphragm to the superior surface of the liver as visceral peritoneum where it separates the two principal lobes of the liver. In the free border of the falciform ligament is the *ligamentum teres (round ligament).* It extends from the liver to the umbilicus. The ligamentum teres is a fibrous cord homologous to the umbilical vein of the fetus.

The lobes of the liver are made up of numerous functional units called *lobules,* which may be seen under a microscope. A lobule consists of cords of *hepatic* (liver)

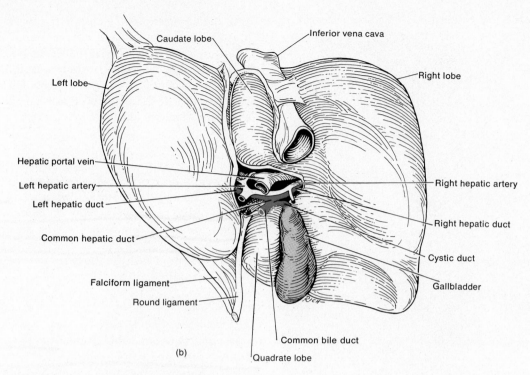

Figure 24–13 (cont.) Liver. (b) External anatomy in posteroinferior view.

cells arranged in a radial pattern around a *central vein.* Between the cords are endothelial-lined spaces called *sinusoids* through which blood passes. The sinusoids are also partly lined with phagocytic cells, termed *Kupffer cells,* that destroy worn-out white and red blood cells and bacteria.

The liver receives a double supply of blood. From the hepatic artery it obtains oxygenated blood, and from the hepatic portal vein it receives deoxygenated blood containing newly absorbed nutrients. Branches of both the hepatic artery and the hepatic portal vein carry the blood into the sinusoids of the lobules, where oxygen, most of the nutrients, and certain poisons are extracted by the hepatic cells. Nutrients are stored or used to make new materials. The poisons are stored or detoxified. Products manufactured by the hepatic cells and nutrients needed by other cells are secreted back into the blood. The blood then drains into the central vein and eventually passes into a hepatic vein. Unlike the other products of the liver, bile normally is not secreted into the bloodstream.

Bile is manufactured by the hepatic cells and secreted into *bile capillaries* or *canaliculi* that empty into small ducts. These small ducts eventually merge to form the larger *right* and *left hepatic ducts,* which unite to leave the liver as the *common hepatic duct.* Further on, the common hepatic duct joins the *cystic duct* from the gallbladder. The two tubes become the *common bile duct,* which empties into the duodenum at the ampulla of

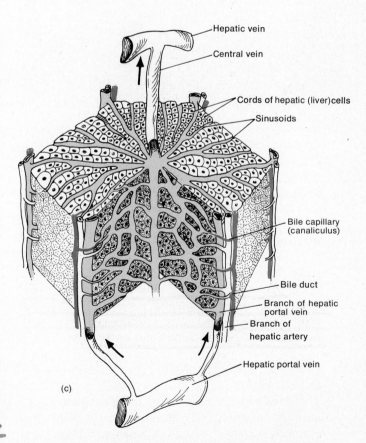

Figure 24–13 (cont.) Liver. (c) Diagrammatic representation of the microscopic appearance of a lobule.

Handwritten margin note (rotated, left side): LIVER - MANUFACTURES BILE FOR BREAKDOWN OF FATS IN SM. INT. GALL BLADDER - STORES BILE

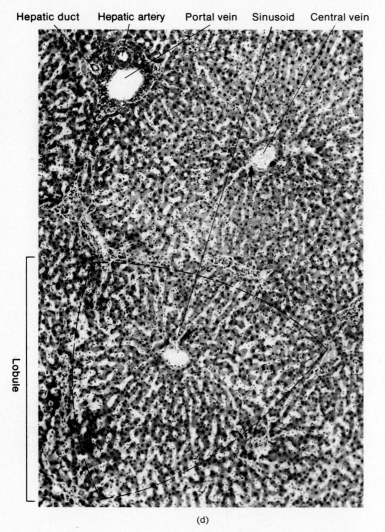

Hepatic duct   Hepatic artery   Portal vein   Sinusoid   Central vein

Lobule

(d)

Figure 24–13 (cont.) Liver. (d) Photomicrograph of a lobule at a magnification of 65×. The broken line indicates the boundary of a lobule and the arrows indicate sinusoids emptying into the central vein. (Courtesy of Edward J. Reith, from *Atlas of Descriptive Histology,* by Edward J. Reith and Michael H. Ross, Harper & Row, Publishers, Inc., New York, 1970.)

Vater. The *sphincter of Oddi* is a valve in the common bile duct. When the small intestine is empty the sphincter closes, and the backed-up bile overflows into the cystic duct to the gallbladder, where it is stored.

Gallbladder

The **gallbladder** is a sac located along the underside of the liver (Figure 24-12). Its inner walls consist of a mucous membrane arranged in rugae resembling those of the stomach. When the gallbladder fills with bile, the rugae allow it to expand to the size and shape of a pear. The middle, muscular coat of the wall consists of smooth muscle fibers. Contraction of these fibers ejects the bile into the cystic duct. The outer coat is the visceral peritoneum. Figure 24-14 presents a cross section of the abdomen at the level of the pancreas. Note the relationship of the digestive organs to one another.

Small Intestine

The major portions of digestion and absorption occur in a long tube called the **small intestine.** The small intestine begins at the pyloric valve of the stomach, coils through the central and lower part of the abdominal cavity, and eventually opens into the large intestine. It averages 2.5 cm (1 inch) in diameter and about 6.35 m (21 ft) in length.

ANATOMY

The small intestine is divided into three segments: duodenum, jejunum, and ileum. The *duodenum,* the broadest part, originates at the pyloric valve of the stomach and extends about 25 cm (10 inches) until it merges with the jejunum. The *jejunum* is about 2.5 m (8 ft) long and extends to the ileum. The final portion of the small intestine, the *ileum,* measures about 3.6 m (12 ft) and joins the large intestine at the *ileocecal valve.* A roentgenogram of the normal small intestine is shown in Figure 24-15a.

The wall of the small intestine is composed of the same four tunics or coats that make up most of the GI tract. However, both the mucosa and the submucosa are modified to allow the small intestine to complete the processes of digestion and absorption. The mucosa contains many pits lined with glandular epithelium. These pits – the *intestinal glands,* or *crypts of Lieberkühn* – secrete the intestinal digestive enzymes. The submucosa of the duodenum contains *Brunner's glands,* which secrete mucus to protect the walls of the small intestine from the action of the enzymes.

Since almost all the absorption of nutrients occurs in the small intestine, its walls need special equipment. The epithelium covering and lining the mucosa consists of simple columnar epithelium. Some of the epithelial cells have been transformed to goblet cells, which secrete additional mucus. The rest contain *microvilli* – fingerlike projections of the plasma membrane. Digested nutrients diffuse more quickly into the intestinal wall because the microvilli increase the surface area of the plasma membrane.

The mucosa lies in a series of *villi* – projections 0.5 to 1 mm high giving the intestinal mucosa its velvety appearance. The enormous number of villi (4 to 5

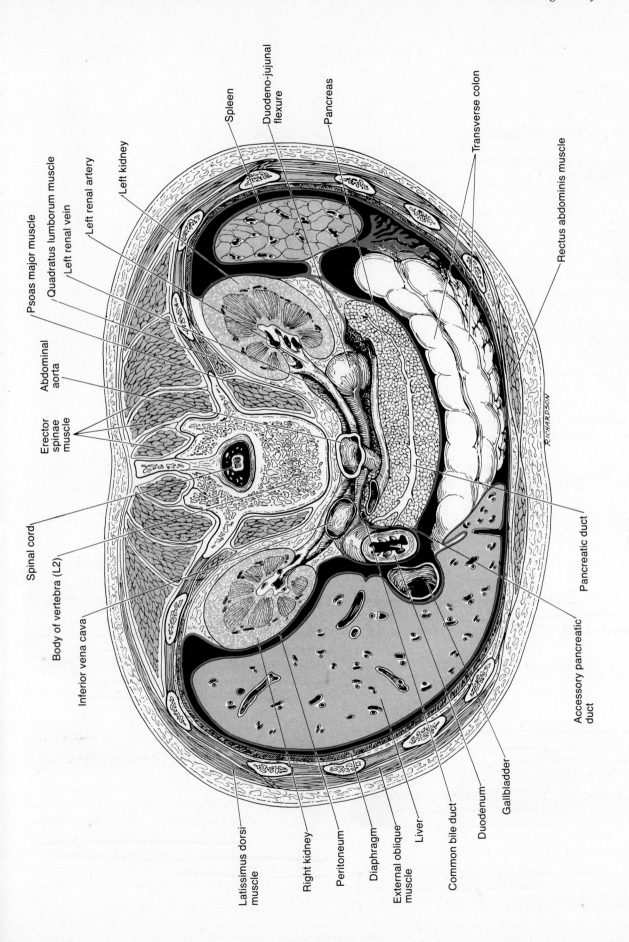

Figure 24-14 Abdomen seen in cross section at the level of the pancreas.

million) vastly increases the surface area of the epithelium available for the epithelial cells specializing in absorption. Each villus has a core of lamina propria, the connective tissue layer of the mucosa. Embedded in this connective tissue are an arteriole, a venule, a capillary network, and a *lacteal* (lymphatic vessel). Nutrients that diffuse through the epithelial cells which cover the villus are able to pass through the capillary walls and the lacteal and enter the blood.

In addition to the microvilli and villi, a third set of projections called *plicae circulares* further increases the surface area for absorption. The plicae are permanent deep folds in the mucosa and submucosa. Some of the folds extend all the way around the intestine, and others extend only part way around.

The muscularis of the small intestine consists of two layers of smooth muscle. The outer, thinner layer contains longitudinally arranged fibers. The inner, thicker layer contains circularly arranged fibers. Except for a major portion of the duodenum, the serosa (or visceral peritoneum) completely covers the small intestine. The histological aspects of the small intestine are shown in Figure 24-16 on pages 562–563.

There is an abundance of lymphatic tissue in the walls of the small intestine. Single lymph nodules, called *solitary lymph nodules,* are most numerous in the lower part of the ileum. Aggregated lymph nodules, referred to as *Peyer's patches,* are also most numerous in the ileum.

CHEMICAL DIGESTION

The digestion of carbohydrates, proteins, and lipids in the small intestine requires secretions from the pancreas, liver, and intestinal glands.

Secretions   Each day the liver secretes 800 to 1,000 ml (almost 1 qt) of the yellow, brownish, or olive-green liquid called *bile.* Bile consists of water, bile salts, bile acids, a number of lipids, and two pigments called biliverdin and bilirubin. Bile is partially an excretory product and partially a digestive secretion. When red blood cells are broken down, iron, globin, and bilirubin are released. The iron and globin are recycled, but the bilirubin is excreted into the bile ducts. Bilirubin eventually is broken down in the intestines, and its breakdown products give feces their color. If the liver is unable to remove bilirubin from the blood or if the bile ducts are obstructed, large amounts of bilirubin circulate through the bloodstream and collect in other tissues, giving the skin and eyes a yellow color called *jaundice.* Other substances found in bile aid in the digestion of fats and are required for their absorption.

Each day the pancreas produces 1,200 to 1,500 ml (about 1 to 1½ qt) of a clear, colorless liquid called *pancreatic juice.* Pancreatic juice consists mostly of water,

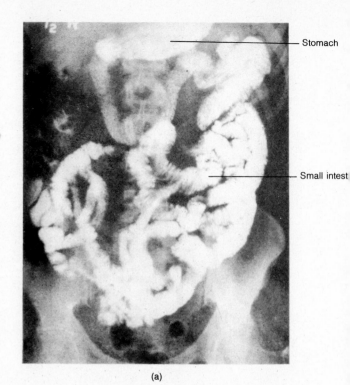

(a)

Figure 24–15 Small intestine. (a) Anteroposterior projection of the normal small intestine ½ hour after taking a barium "meal." (Courtesy of Lester W. Paul and John H. Juhl, *The Essentials of Roentgen Interpretation,* 3d ed., Harper & Row, Publishers, Inc., New York, 1972.)

some salts, sodium bicarbonate, and enzymes. The sodium bicarbonate gives pancreatic juice a slightly alkaline pH (7.1 to 8.2) that stops the action of pepsin from the stomach and creates the proper environment for the enzymes in the small intestine. The enzymes in pancreatic juice include a carbohydrate-digesting enzyme, several protein-digesting enzymes, and the only active fat-digesting enzyme in the adult body.

The intestinal juice, or *succus entericus,* is a clear yellow fluid secreted in amounts of about 2 to 3 liters (2 to 3 qt) a day. It has a pH of 7.6, which is slightly alkaline, and contains water, mucus, and enzymes that complete the digestion of carbohydrates and proteins.

■ *Control of Secretion*   The small intestine starts to prepare itself for digestion as soon as food enters the mouth. Taste buds send impulses to the brain when they are stimulated. One way the brain responds is by sending impulses over the vagus nerve that stimulate the pancreas to secrete its juice. Thus pancreatic enzymes enter the small intestine within 1 to 2 minutes after food is ingested.

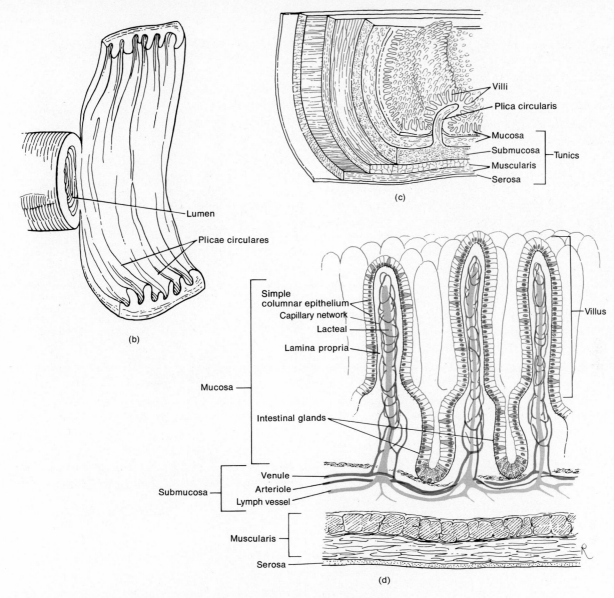

Villi
Plica circularis
Mucosa
Submucosa ⎫
Muscularis ⎬ Tunics
Serosa ⎭

(c)

Lumen

Plicae circulares

(b)

Simple
columnar epithelium
Capillary network
Lacteal
Lamina propria

Mucosa

Villus

Intestinal glands

Submucosa
Venule
Arteriole
Lymph vessel

Muscularis

Serosa

(d)

Figure 24–15 (cont.) Small intestine. (b) Section of small intestine cut open to expose plicae circulares. (c) Villi in relation to the tunics of the small intestine. (d) Enlarged aspect of several villi.

When the chyme leaves the stomach and enters the duodenum, chemicals in the chyme stimulate endocrine cells in the duodenal mucosa to secrete a number of hormones. Most of the hormones stimulate secretion of digestive juices. One of them, called enterogastrone, slows down stomach contractions. The hormonal control of digestion is summarized in Exhibit 24-4, page 565.

■ *Digestive Process* When chyme reaches the small intestine, the carbohydrates and proteins have been only partly digested and are not ready for absorption. Lipid digestion has not even begun. Digestion in the small intestine continues as follows:

1 Carbohydrates In the mouth, some polysaccharides are broken down into dextrins containing several monosaccharide units (Figure 24-17, p. 564). Even

though the action of salivary amylase may continue in the stomach, few polysaccharides are reduced to disaccharides by the time chyme leaves the stomach. *Pancreatic amylase,* an enzyme in pancreatic juice, breaks dextrins into the disaccharide maltose. Sucrose and lactose, two other disaccharides, are ingested. Next, three enzymes in the intestinal juice digest the disaccharides into monosaccharides. *Maltase* splits maltose into two molecules of glucose. *Sucrase* breaks sucrose into a molecule of glucose and a molecule of fructose. *Lactase* digests lactose into a molecule of glucose and a molecule of galactose. This process completes the digestion of carbohydrates.

2 Proteins Protein digestion starts in the stomach, where most of the proteins are fragmented by the action of pepsin into short chains of amino acids called

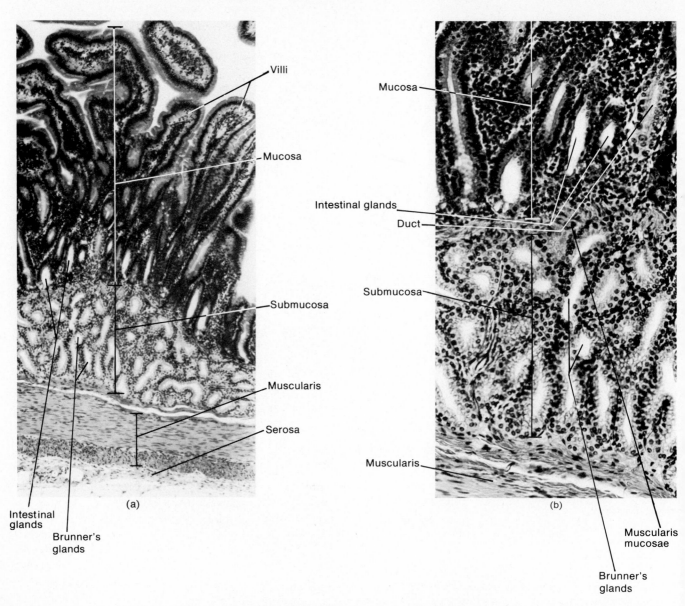

Villi

Mucosa

Submucosa

Muscularis

Serosa

Intestinal glands

Brunner's glands

(a)

Mucosa

Intestinal glands

Duct

Submucosa

Muscularis

(b)

Muscularis mucosae

Brunner's glands

Figure 24–16   Histology of the small intestine. (a) Section through the duodenum showing various tunics at a magnification of 15×. (b) Enlarged aspect of intestinal glands at a magnification of 40×.

peptones and proteoses (Figure 24-18, p. 565). Enzymes found in pancreatic juice continue the digestion. *Trypsin* digests any intact proteins into peptones and proteoses, breaks the peptones and proteoses into dipeptides (containing only two amino acids), and breaks some of the dipeptides into single amino acids. *Chymotrypsin* duplicates trypsin's activities. *Carboxypeptidase,* the third enzyme, reduces whole or partly digested proteins to amino acids. To prevent these enzymes from digesting the proteins in the cells of the pancreas, they are secreted in inactive forms – trypsin as *trypsinogen,* activated by an intestinal enzyme called *enterokinase;* chymotrypsin as

chymotrypsinogen, activated in the small intestine by trypsin; and carboxypeptidase as *procarboxypeptidase,* also activated in the small intestine by trypsin. Protein digestion is completed by several intestinal enzymes grouped together under the name *erepsin,* which converts all the remaining dipeptides into single amino acids. Single amino acids can be absorbed.

3 Lipids   In an adult, almost all lipid digestion occurs in the small intestine. The first step in the process is the *emulsification* of neutral fats (triglycerides), which is a function of bile. Neutral fats, or just simply fats, are the most abundant lipids in the diet. They are called

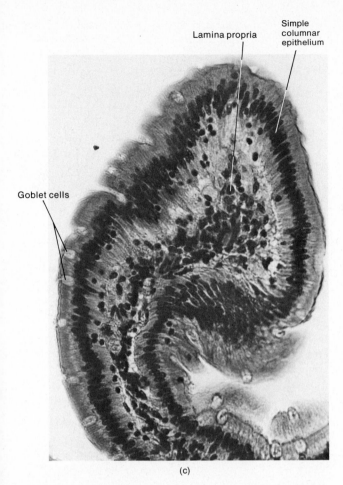

Lamina propria

Simple columnar epithelium

Goblet cells

(c)

Figure 24–16 (cont.) Histology of the small intestine. (c) Enlarged aspect of a villus at a magnification of 80×. (Courtesy of Victor B. Eichler, Wichita State University.)

triglyceride because they consist of a molecule of glycerol and three molecules of fatty acid (Figure 24-19, p. 566). Bile salts break the globules of fat into droplets (emulsification) so the fat-splitting enzyme can get at the lipid molecules. In the second step, *pancreatic lipase,* an enzyme found in pancreatic juice, hydrolyzes each fat molecule into fatty acids, glycerol, and glycerides, end products of fat digestion. Glycerides consist of glycerol with one or two fatty acids still attached and are known as monoglycerides and diglycerides, respectively.

MECHANICAL DIGESTION

In the small intestine, three distinct movements occur as a result of contractions of the longitudinal and circular muscles. These movements are rhythmic segmentation, pendular movements, and propulsive peristalsis. Rhythmic segmentation and pendular movements are strictly localized contractions in areas containing food. The two

movements mix the chyme with the digestive juices and bring every particle of food into contact with the mucosa for absorption. They do not push the intestinal contents along the tract. *Rhythmic segmentation* starts with the contractions of circular muscle fibers in a portion of the intestine, an action that constricts the intestine into segments. Next, muscle fibers that encircle the middle of each segment also contract, dividing each segment into two. Finally, the fibers that contracted first relax, and each small segment unites with an adjoining small segment so that large segments are reformed. This sequence of events is repeated 12 to 16 times a minute, sloshing the chyme back and forth. *Pendular movements* consist of alternating contractions and relaxations of the longitudinal muscles. The contractions cause a portion of the intestine to shorten and lengthen, spilling the chyme back and forth.

The third movement, *propulsive peristalsis,* propels the chyme onward through the intestinal tract. Peristaltic movement in the intestine is similar to that in the esophagus. In the intestine, these waves may be as slow as 5 cm (2 inches)/minute or as fast as 50 cm (20 inches)/second.

ABSORPTION

All the chemical and mechanical phases of digestion from the mouth down through the small intestine are directed toward changing foods into forms that can diffuse through the epithelial cells lining the mucosa into the underlying blood and lymph vessels. The diffusible forms are monosaccharides (glucose, fructose, and galactose), amino acids, fatty acids, glycerol, and glycerides. Passage of these digested nutrients from the alimentary canal into the blood or lymph is called **absorption.**

About 90 percent of all absorption takes place throughout the length of the small intestine. The other 10 percent occurs in the stomach and large intestine. Absorption of materials in the small intestine occurs specifically through the villi and depends on diffusion, facilitated diffusion, osmosis, and active transport. Monosaccharides and amino acids are absorbed into the blood capillaries of the villi and transported in the bloodstream to the liver via the hepatic portal system. Fatty acids, glycerol, and glycerides do not enter the bloodstream immediately. They cluster together and become surrounded by bile salts to form water-soluble particles called *micelles.* These are absorbed into the intestinal epithelial cell. Here they enter the smooth endoplasmic reticulum where triglycerides are resynthesized. The triglycerides, along with small quantities of phospholipids and cholesterol, are then organized into protein-coated lipid droplets called *chylomicrons.* The

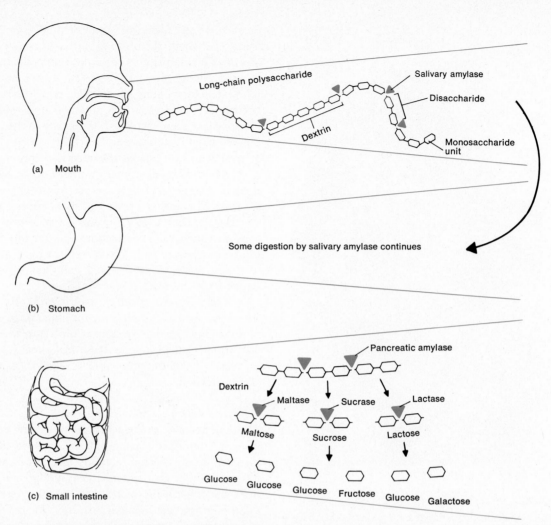

Figure 24–17 Digestion of carbohydrates. (a) In the mouth. (b) Some breakdown continues in the stomach. (c) In the small intestine.

protein coat keeps the chylomicrons suspended and prevents them from sticking to each other or to the walls of the lymphatics or blood vessels. Small chylomicrons leave the intestinal cells and enter the blood capillaries in the villus. Larger chylomicrons enter the lacteal in the villus and are transported by way of lymphatic vessels to the thoracic duct and enter the cardiovascular system at the left subclavian vein. Finally they arrive at the liver through the hepatic artery.

Most of the products of carbohydrate, protein, and lipid digestion are processed by the liver before they are delivered to the other cells of the body. Large amounts of water, electrolytes, mineral salts, and some vitamins also are absorbed in the small intestine.

In summary, then, the principal chemical activity of the small intestine is to digest all foods into forms that

are usable by body cells. (See Exhibit 24-5.) Any undigested materials that are left behind are processed in the large intestine.

Large Intestine

The overall functions of the large intestine are the completion of absorption, the manufacture of certain vitamins, the formation of feces, and the expulsion of feces from the body.

ANATOMY

The **large intestine** is about 1.5 m (5 ft) in length and averages 6.5 cm (2.5 inches) in diameter. It extends from the ileum to the anus and is attached to the posterior abdominal wall by its *mesocolon* of visceral peritoneum.

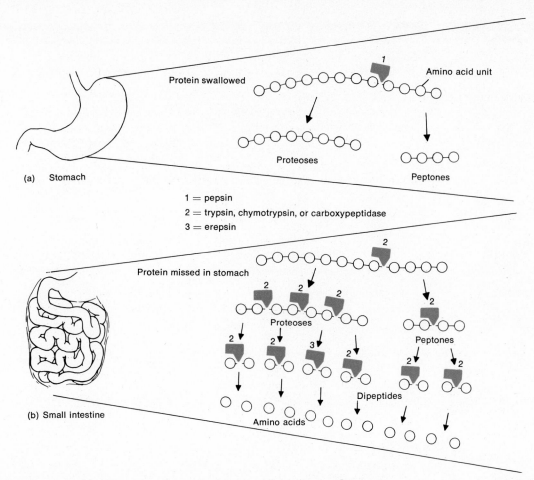

Figure 24–18 Digestion of proteins. (a) In the stomach. (b) In the small intestine.

Exhibit 24–4 HORMONAL CONTROL OF DIGESTION IN STOMACH AND SMALL INTESTINE

Hormone	Where Produced	Stimulant	Action
Stomach gastrin	Pyloric mucosa	Partially digested proteins	Causes gastric glands to secrete gastric juice.
Intestinal gastrin	Intestinal mucosa	Partially digested proteins	Same as above.
Secretin	Intestinal mucosa	Acidity of chyme	Stimulates secretion of pancreatic juice rich in bicarbonate and promotes production of bile by liver.
Cholecystokinin (pancreozymin)	Intestinal mucosa	Fats	Stimulates secretion of pancreatic juice rich in enzymes.
Enterocrinin	Intestinal mucosa	Acidity of chyme	Stimulates secretion of succus entericus.
Cholecystokinin (pancreozymin)	Intestinal mucosa	Combination of acid and fat	Causes ejection of bile from gallbladder and opening of sphincter of Oddi.
Enterogastrone	Intestinal mucosa	Fats	Inhibits secretion of gastric juice and decreases gastric motility.

Exhibit 24—5   DIGESTION AND ABSORPTION IN SMALL INTESTINE

Structure	Description	Function
Pancreas (pancreatic juice)		
Trypsin	Protein-digesting enzyme activated by the intestinal enzyme enterokinase	Digests intact proteins into proteoses and peptones. Digests partially digested proteins into dipeptides plus some amino acids.
Chymotrypsin	Protein-digesting enzyme activated by trypsin	Same as above.
Carboxypeptidase	Protein-digesting enzyme activated by trypsin	Reduces proteins to amino acids.
Pancreatic amylase	Enzyme that digests carbohydrates	Converts dextrins into disaccharide maltose.
Pancreatic lipase	Fat-splitting enzyme	Converts neutral fat (triglyceride) into fatty acids, glycerol, and glycerides.
Liver (bile)	Bile salts	Emulsifies neutral fats in preparation for digestion by pancreatic lipase. Bile salts also allow products of neutral fat digestion to be absorbed.

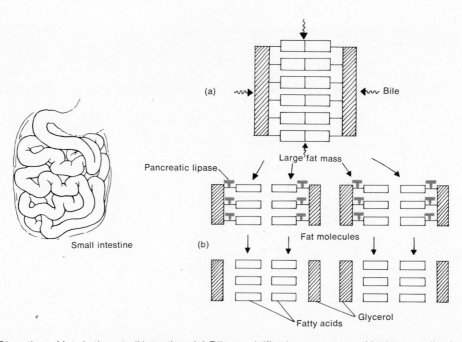

Figure 24—19   Digestion of fats in the small intestine. (a) Bile emulsifies large masses of fat into smaller fragments so that pancreatic lipase can break down the fat molecules. (b) The enzyme pancreatic lipase breaks fats into fatty acids, glycerol, and glycerides.

Exhibit 24-5 (cont.)

Structure	Description	Function
Small intestine		
Mucosa and submucosa		
Intestinal glands	Secrete succus entericus	
Maltase		Converts disaccharide maltose into mono-saccharide glucose.
Sucrase		Converts disaccharide sucrose into mono-saccharides glucose and fructose.
Lactase		Converts disaccharide lactose into mono-saccharides glucose and galactose.
Erepsin		Changes dipeptides into amino acids.
Microvilli	Projections of plasma membranes of intestinal epithelial cells	Increase surface area for absorption.
Villi	Fingerlike projections of mucous membrane	Serve as sites for absorption of digested foods and increase absorptive area.
Plicae circulares	Circular folds of mucosa and submucosa	Increase surface area for digestion and absorption.
Goblet cells	Secrete mucus	Lubricates foods and protects mucosa.
Intestinal glands	Secrete intestinal digestive juices	Digest carbohydrates and proteins.
Brunner's glands	Secrete mucus	Lubricates foods and protects mucosa.
Solitary lymph nodules	Lymphatic tissue associated with small intestine	Filter lymph.
Peyer's patches	Same as above	Same as above.
Muscularis		
Rhythmic segmentation	Alternating contractions of circular fibers produce segmentation and resegmentation of portions of small intestine	Mixes chyme with digestive juices and brings food into contact with mucosa for absorption.
Pendular movement	Contractions of longitudinal muscle pull portions of intestine forward and backward	Same as for rhythmic segmentation.
Propulsive peristalsis	Waves of contraction and relaxation of circular and longitudinal muscle passing length of small intestine.	Moves chyme forward.

Structurally, the large intestine is divided into four principal regions: cecum, colon, rectum, and anal canal (Figure 24-20).

The opening from the ileum into the large intestine is guarded by a fold of mucous membrane called the *ileocecal valve.* This structure allows materials from the small intestine to pass into the large intestine. Hanging below the ileocecal valve is the *cecum,* a blind pouch about 6 cm (2 or 3 inches) long. Attached to the cecum is a twisted, coiled tube, measuring about 8 cm (3

inches) in length, called the *vermiform appendix (vermis* = worm). The visceral peritoneum of the appendix, called the *mesoappendix,* attaches the appendix to the inferior part of the ileum and adjacent part of the posterior abdominal wall. Inflammation of the appendix is called *appendicitis.*

The open end of the cecum merges with a long tube called the *colon.* The colon is divided into ascending, transverse, descending, and sigmoid portions. The *ascending colon* ascends on the right side of the abdomen,

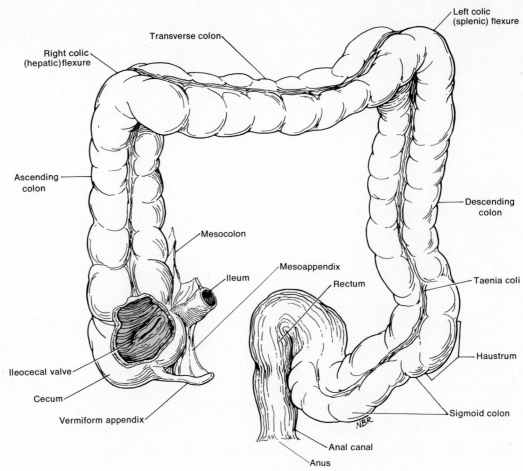

Figure 24–20   Large intestine.

reaches the undersurface of the liver, and turns abruptly to the left. Here it forms the *right colic (hepatic) flexure*. The colon continues across the abdomen to the left side as the *transverse colon*. It curves beneath the lower end of the spleen on the left side as the *left colic (splenic) flexure* and passes downward to the level of the iliac crest as the *descending colon*. The *sigmoid colon* begins at the left iliac crest, projects inward to the midline, and terminates as the rectum at about the level of the third sacral vertebra.

The *rectum,* the last 20 cm (7 to 8 inches) of GI tract, lies anterior to the sacrum and coccyx. The terminal 2 to 3 cm of the rectum is called the *anal canal* (Figure 24-21). Internally, the mucous membrane of the anal canal is arranged in longitudinal folds called *anal columns* that contain a network of arteries and veins. Inflammation and enlargement of the anal veins is known as *hemorrhoids* or *piles.* The opening of the anal canal to the exterior is called the *anus.* It is guarded by an internal sphincter of smooth muscle and an external sphincter of skeletal muscle. Normally the anus is closed except during the elimination of the wastes of digestion.

The wall of the large intestine differs from that of the small intestine in several respects. No villi or permanent circular folds are found in the mucosa, which does, however, contain simple columnar epithelium with numerous goblet cells. These cells secrete mucus that lubricates the colonic contents as they pass through the colon. Solitary lymph nodes also are found in the mucosa. The submucosa of the large intestine is similar to that found in the rest of the alimentary canal. The muscularis consists of an external layer of longitudinal muscles and an internal layer of circular muscles. Unlike other parts of the digestive tract, the longitudinal muscles do not form a continuous sheet around the wall but are broken up into three flat bands called *taeniae coli.* Each band runs the length of the large intestine. Tonic contractions of the bands gather the colon into a series of pouches called *haustra,* which give the colon its puckered appearance. The serosa of the large intestine is part of the visceral peritoneum.

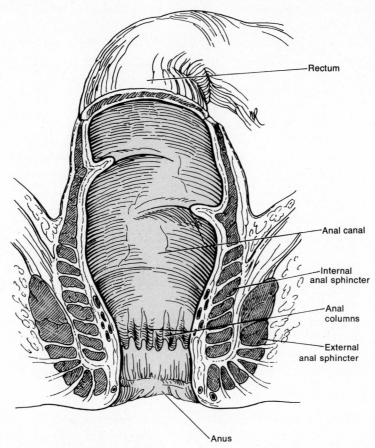

Rectum

Anal canal

Internal anal sphincter

Anal columns

External anal sphincter

Anus

Figure 24–21 Anal canal seen in longitudinal section.

ACTIVITIES

The principal activities of the large intestine are concerned with mechanical movements, absorption, and the formation and elimination of feces.

■ *Movements* Movements of the colon begin when substances enter through the ileocecal valve. Since chyme moves through the small intestine at a fairly constant rate, the time required for a meal to pass into the colon is determined by gastric evacuation time. As food passes through the ileocecal valve, it fills the cecum and accumulates in the ascending colon.

One movement characteristic of the large intestine is *haustral churning*. In this process the haustra remain relaxed and distended while they fill up. When the distension reaches a certain point, the walls contract and squeeze the contents into the next haustrum. Peristalsis also occurs, although at a slower rate than in other portions of the tract (3 to 12 contractions per minute). A final type of movement is *mass peristalsis,* a strong

peristaltic wave that begins in about the middle of the transverse colon and drives the colonic contents into the rectum. Food in the stomach initiates this reflex action in the colon. Thus mass peristalsis usually takes place three or four times a day, during a meal or immediately after.

■ *Absorption and Feces Formation* By the time the intestinal contents arrive at the large intestine, digestion and absorption are almost complete. By the time the chyme has remained in the large intestine 3 to 10 hours, it has become solid or semisolid as a result of absorption and is now known as *feces*. Chemically, feces consist of water, inorganic salts, epithelial cells from the mucosa of the alimentary canal, bacteria, products of bacterial decomposition, and undigested parts of food not attacked by bacteria.

Mucus is secreted by the glands of the large intestine, but no enzymes are secreted. The mucus serves as a lubricant to aid the movement of the colonic

Exhibit 24–6 DIGESTIVE ACTIVITIES OF LARGE INTESTINE

Structure	Action	Function
Mucosa	Secretes mucus	Lubricates colon and protects mucosa.
	Absorbs water and other soluble compounds	Water balance is maintained and feces become solidified. Vitamins and minerals are absorbed, and toxic substances are sent to liver to be detoxified.
	Bacterial action	Undigested carbohydrates, proteins, and amino acids are broken down so they can be expelled in feces or absorbed and detoxified by liver. Certain vitamins are synthesized.
Muscularis	Haustral churning	Haustra fill and contract, moving contents from haustra to haustra.
	Peristalsis	Contractions of circular and longitudinal muscles continually move contents along length of colon.
	Mass peristalsis	Strong peristaltic wave forces contents into sigmoid colon and rectum.
	Defecation	Contractions in sigmoid colon and rectum rid body of feces.

materials and acts as a protective covering for the mucosa.

Chyme is prepared for elimination in the large intestine by the action of bacteria. These bacteria ferment any remaining carbohydrates and release hydrogen, carbon dioxide, and methane gas. They also convert remaining proteins to amino acids and break down the amino acids into simpler substances: indole, skatole, hydrogen sulfide, and fatty acids. Some of the indole and skatole is carried off in the feces and contributes to its odor. The rest is absorbed. Bacteria also decompose bilirubin, the breakdown product of red blood cells that is excreted in bile, to simpler pigments which give feces their brown color. Intestinal bacteria also aid in the synthesis of several vitamins needed for normal metabolism, including some B vitamins and vitamin K.

Although most water absorption occurs in the small intestine, the large intestine absorbs enough to make it an important organ in maintaining the body's water balance. Intestinal water absorption is greatest in the cecum and ascending colon. The large intestine also absorbs inorganic solutes plus some products of bacterial action, including vitamins and large amounts of indole and skatole. The indole and skatole are transported to the liver, where they are converted to less toxic compounds and excreted in the urine.

■ *Defecation* Mass peristaltic movement pushes fecal material into the rectum. The resulting distension of the rectal walls stimulates pressure-sensitive receptors, initiating a reflex for **defecation**, which is emptying of the rectum. Contraction of the longitudinal rectal muscles shortens the rectum, thereby increasing the pressure inside it. The pressure forces the sphincters open, and the feces are expelled through the anus. Voluntary contractions of the diaphragm and abdominal muscles aid defecation by increasing the pressure inside the abdomen, which pushes the walls of the sigmoid colon and rectum inward. If defecation does not occur, the feces remain in the rectum until the next wave of mass peristalsis again stimulates the pressure-sensitive receptors, creating the desire to defecate.

Activities of the large intestine are summarized in Exhibit 24-6.

DISORDERS

DENTAL CARIES

Dental caries, or tooth decay, involve a gradual demineralization (softening) of the enamel and dentin. If this condition remains untreated, various microorganisms may invade the pulp, causing inflammation and infection with subsequent death (necrosis) of the dental pulp and abscess of the alveolar bone surrounding the root's apex. Such teeth are treated by root canal therapy.

The process of dental caries is initiated when bacteria act on carbohydrates deposited on the tooth, giving off acids which demineralize the enamel. Microbes that digest carbohydrates include two bacteria, *Lactobacillus acidophilus* and *Streptococcus mutans.* Research suggests that the streptococci break down carbohydrates into *dental plaque,* a polysaccharide that adheres to the tooth surface. When other bacteria digest the plaque, acid is produced. Saliva cannot reach the tooth surface to buffer the acid because the plaque covers the teeth.

PERIODONTAL DISEASES

Periodontal disease is a collective term for a variety of conditions characterized by inflammation and degeneration of the gingivae, alveolar bone, periodontal ligament, and cementum. The initial symptoms are enlargement and inflammation of the soft tissue and bleeding gums. Without treatment, the soft tissue may deteriorate and the alveolar bone may be resorbed, causing loosening of the teeth and receding of the gums.

Periodontal diseases are frequently caused by local irritants, such as bacteria, impacted food, and cigarette smoke, or by a poor "bite." The latter may put a strain on the tissues supporting the teeth. Periodontal diseases may also be caused by allergies, vitamin deficiences, and a number of systemic disorders, especially those that affect bone (blood dyscrasias), connective tissue, or circulation.

PEPTIC ULCERS

An **ulcer** is a craterlike lesion in a membrane. Ulcers that develop in areas of the alimentary canal exposed to acid gastric juice are called *peptic ulcers.* Peptic ulcers occasionally develop in the lower end of the esophagus. However, most of them occur on the lesser curvature of the stomach, in which case they are called *gastric ulcers,* or in the first part of the duodenum, where they are called *duodenal ulcers.*

Hypersecretion of acid gastric juice seems to be the immediate cause of duodenal ulcers. In gastric ulcer patients, because the stomach walls are highly adapted to resist gastric juice through their secretion of mucus, the cause may be hyposecretion of mucus. Hypersecretion of pepsin also may contribute to ulcer formation.

Among the factors believed to stimulate an increase in acid secretion are emotions, certain foods or medications (alcohol, coffee, aspirin), and overstimulation of the vagus nerve. Normally, the mucous membrane lining the stomach and duodenal walls resists the secretions of hydrochloric acid and pepsin. In some people, however, this resistance breaks down and an ulcer develops.

The danger inherent in ulcers is the erosion of the muscular portion of the wall of the stomach or duodenum. This erosion could damage blood vessels and produce fatal hemorrhage. If an ulcer erodes all the way through the wall, the condition is called *perforation.* Perforation allows bacteria and partially digested food to pass into the peritoneal cavity, producing peritonitis.

CIRRHOSIS

Cirrhosis is a chronic disease of the liver in which the parenchymal (functional) liver cells are replaced by fibrous connective tissue, a process called *stromal repair.* Often there is a lot of replacement with adipose connective tissue as well. The liver has a high ability for parenchymal regeneration, so stromal repair occurs whenever a parenchymal cell is killed or cells are damaged continuously for a long time. These conditions could be caused by *hepatitis* (inflammation of the liver), certain chemicals that destroy liver cells, parasites that infect the liver, and alcoholism.

TUMORS

Both benign and malignant **tumors** occur in all parts of the gastrointestinal tract. The benign growths are much more common, but malignant tumors are responsible for 30 percent of all deaths from cancer in the United States. For early diagnosis, complete routine examinations are necessary. Cancers of the mouth usually are detected through routine dental checkups.

A regular physical checkup should include rectal examination. Fifty percent of all rectal carcinomas are within reach of the finger, and 75 percent of all colonic carcinomas can be seen with the sigmoidoscope. Both the fiberoptic sigmoidoscope and the more recent fiberoptic endoscope are flexible tubular instruments composed of a light and many tiny glass fibers. They allow visualization, magnification, biopsy, electrosurgery, and even photography of the entire length of the gastrointestinal tract. The greatest contribution of colonoscopy may be identification and removal of malignant polyps of the colon (gastric polypectomy) before invasion of the bowel wall or lymphatic metastasis. It has proved to be a safe and effective treatment that avoids the significant expense, risk, and discomfort of major surgery. Colonoscopy may be the greatest advance yet toward lowering the death rate from cancer of the colon. Unfortunately, this type of cancer has shown a considerable increase in incidence over the last 20 years.

Another test in a routine examination for intestinal disorders is the filling of the gastrointestinal tract with barium, which is either swallowed or given in an enema. Barium, a mineral, shows up on x rays the same way that calcium appears in bones. Tumors as well as ulcers can be diagnosed this way. The only definitive treatment of gastrointestinal carcinomas, if they cannot be removed using the endoscope, is surgery.

MEDICAL TERMINOLOGY

Cholecystitis Inflammation of the gallbladder that often leads to infection. Some cases are caused by obstruction of the cystic duct with bile stones. Stagnating bile salts irritate the mucosa. Dead mucosal cells provide a medium for bacteria.

Colitis An inflammation of the colon and rectum. Inflammation of the mucosa reduces absorption of water and salts, producing watery, bloody feces, and—in severe cases—dehydration and salt depletion. Irritated muscularis spasms produce cramps.

Colonoscopy The endoscopic examination of the entire colon.

Colostomy An incision of the colon to create an artificial opening or "stoma" to the exterior. This opening serves as a substitute anus through which feces are eliminated. A temporary colostomy may be done to allow a badly inflamed colon to rest and heal. If the rectum is removed for malignancy, the colostomy provides a permanent outlet for feces.

Constipation Infrequent or difficult defecation.

Diarrhea Frequent defecation of liquid feces.

Flatus Excessive amounts of air (gas) in the stomach or intestine, usually expelled through the anus. If the gas is expelled through the mouth, it is called *belching* (burping). Flatus may result from gas released during the breakdown of foods in the stomach or from swallowing air or gas-containing substances such as carbonated drinks.

Heartburn A burning sensation in the region of the esophagus and stomach. It may result from regurgitation of gastric contents into the lower end of the esophagus or from distension stemming from retention of regurgitated food and gastric contents in the lower esophagus.

Hepatitis (*hepato* = liver) A liver inflammation. It may be caused by organisms such as viruses, bacteria, and protozoa or by the absorption of materials, such as carbon tetrachloride and certain anesthetics and drugs, that are toxic to liver cells.

Hernia Protrusion of an organ or part of an organ through a membrane or cavity wall, usually the abdominal cavity. *Diaphragmatic* or *hiatal hernia* is the protrusion of the lower esophagus, stomach, or intestine into the thoracic cavity through the hole in the diaphragm that allows passage of the esophagus. *Umbilical hernia* is the protrusion of abdominal organs through the navel area of the abdominal wall. *Inguinal hernia* is the protrusion of the hernial sac containing the intestine into the inguinal opening. It may extend into the scrotal compartment, causing strangulation of the herniated part.

Lesion Any pathological or traumatic discontinuity of tissue or loss of function of a part.

Mumps Viral disease causing painful inflammation and enlargement of the salivary glands, particularly the parotids. In adults, the sex glands and pancreas may be involved. Inflammation of testes may cause male sterility.

Nausea Discomfort preceding vomiting. Possibly, it is caused by distension or irritation of the gastrointestinal tract, most commonly the stomach.

Pancreatitis Inflammation of the pancreas. The pancreas secretes active trypsin instead of trypsinogen, and the trypsin digests the pancreatic cells and blood vessels.

Vomiting Expulsion of stomach (and sometimes duodenal) contents through the mouth by reverse peristalsis. The abdominal muscle walls forcibly empty the stomach.

CHEMICAL AND MECHANICAL DIGESTION

1 Chemical digestion is a series of catabolic reactions that break down the large carbohydrate, lipid, and protein molecules of food into molecules that are usable by body cells.

2 Mechanical digestion consists of movements that aid chemical digestion.

GENERAL ORGANIZATION

1 The organs of digestion are usually divided into two main groups.

2 First is the gastrointestinal (GI) tract, or alimentary canal, a continuous tube running through the ventral body cavity from the mouth to the anus.

3 The second group consists of the accessory organs: teeth, tongue, salivary glands, liver, gallbladder, pancreas, appendix.

GENERAL HISTOLOGY

1 The basic arrangement of tissues in the alimentary canal from the inside outward is the mucosa, submucosa, muscularis, and serosa.

PERITONEUM

1 Extensions of the peritoneum include the mesentery, lesser omentum, greater omentum, falciform ligament, and mesocolon.

ORGANS

MOUTH OR ORAL CAVITY

1 The mouth is formed by the cheeks, palates, lips, and tongue, which aid mechanical digestion.

2 The tongue, together with its associated muscles, forms the floor of the oral cavity. It is composed of skeletal muscle covered with mucous membrane.

3 The upper surface and sides of the tongue are covered with papillae. Some papillae contain taste buds.

SALIVARY GLANDS

1 The major portion of saliva is secreted by the salivary glands, which lie outside the mouth and pour their contents into ducts that empty into the oral cavity.

2 There are three pairs of salivary glands: the parotid, submandibular (submaxillary), and sublingual glands.

3 The salivary glands produce saliva that lubricates food and starts the chemical digestion of carbohydrates.

TEETH

1 The teeth, or dentes, project into the mouth and are adapted for mechanical digestion.

2 A typical tooth consists of three principal portions: crown, root, and cervix.

3 Teeth are composed primarily of dentin covered by enamel—the hardest substance in the body.

DEGLUTITION

1 Both pharynx and esophagus assume a role in deglutition, or swallowing.

2 When a bolus is swallowed, the respiratory tract is sealed off and the bolus moves into the esophagus.

3 Peristaltic movements of the esophagus pass the bolus into the stomach.

STOMACH

1 The stomach begins at the bottom of the esophagus and ends at the pyloric valve.

2 Adaptations of the stomach for digestion include rugae that permit distension; glands that produce mucus, hydrochloric acid, and enzymes; and a three-layered muscularis for efficient mechanical movement.

3 Nervous and hormonal mechanisms initiate the secretion of gastric juice.

4 Proteins are chemically digested into peptones and proteoses through the action of pepsin in the stomach.

5 The stomach also stores food, produces the intrinsic factor, and carries on some absorption.

PANCREAS, LIVER, GALLBLADDER

1 Pancreatic acini produce enzymes that enter the duodenum via the pancreatic duct. Pancreatic enzymes digest proteins, carbohydrates, and fats.

2 Cells of the liver produce bile, which is needed to emulsify fats.

3 Bile is stored in the gallbladder and passed into the duodenum via the common bile duct.

SMALL INTESTINE

1 This organ extends from the pyloric valve to the ileocecal valve.

2 It is very highly adapted for digestion and absorption. Its glands produce enzymes and mucus, and its wall contains microvilli, villi, and plicae circulares.

3 The enzymes of the small intestine digest carbohydrates, proteins, and fats into the end products of digestion: monosaccharides, amino acids, fatty acids, and glycerol.

4 The entrance of chyme into the small intestine stimulates the secretion of several hormones that coordinate the secretion and release of bile, pancreatic juice, and intestinal juice and inhibit gastric activity.

5 Mechanical digestion in the small intestine involves rhythmic segmentation, pendular movements, and propulsive peristalsis.

6 Absorption is the passage of the end products of digestion from the alimentary canal into the blood or lymph.

7 Absorption in the small intestine occurs through the villi. Monosaccharides and amino acids pass into the blood capillaries, small aggregations (chylomicrons) of fatty acids and glycerol pass into the blood capillaries, and large chylomicrons enter the lacteal.

LARGE INTESTINE

1 This organ extends from the ileocecal valve to the anus.

2 Mechanical movements of the large intestine include haustral churning, peristalsis, and mass peristalsis.

3 The large intestine functions in the synthesis of several vitamins and in water absorption, leading to feces formation.

4 The elimination of feces from the large intestine is called defecation. Defecation is a reflex action aided by voluntary contractions of the diaphragm and abdominal muscles.

REVIEW QUESTIONS

1 Define digestion. Distinguish between chemical and mechanical digestion.

2 In what respect is digestion an important component of your homeostatic mechanism?

3 Identify the organs of the alimentary canal in sequence. How does the alimentary canal differ from the accessory organs of digestion?

4 Describe the structure of each of the four tunics of the alimentary canal.

5 What is the peritoneum? Describe the location and function of the mesentery, lesser omentum, greater omentum, and falciform ligament.

6 What structures form the oral cavity? How does each structure contribute to digestion?

7 Make a simple diagram of the tongue. Indicate the location of the papillae and the four taste zones.

8 Describe the location of the salivary glands and their ducts. What are buccal glands?

9 Briefly explain the mechanisms that control saliva secretion.

10 Describe the composition of saliva and the role of each of its components in digestion. What is the pH of saliva?

11 By means of a labeled diagram, outline the action of salivary amylase in the mouth.

12 What are the principal portions of a typical tooth? What are the functions of each part?

13 Compare deciduous and permanent dentitions with regard to numbers of teeth and times of eruption.

14 Contrast the functions of incisors, cuspids, premolars, and molars. What is pyorrhea?

15 What is a bolus? How is it formed?

16 Define deglutition. List the sequence of events involved in passing a bolus from the mouth to the stomach.

17 Describe the location of the stomach. List and briefly explain the anatomical features of the stomach.

18 Distinguish between pyloric stenosis and pylorospasm.

19 What is the importance of rugae, zymogenic cells, parietal cells, and mucous cells in the stomach?

20 What is chyme? Why are protein-digesting enzymes secreted in an inactive form?

21 By means of a labeled diagram, outline the action of pepsin in the stomach.

22 Describe the action of gastric lipase in the infant stomach.

23 What is a gastric ulcer? How is it formed?

24 What forces move chyme through the pyloric valve into the duodenum?

25 What factors control the secretion of gastric juice?

26 Where is the pancreas located? Describe the duct system connecting the pancreas to the duodenum.

27 What are pancreatic acini? Contrast their functions with those of the islets of Langerhans.

28 Where is the liver located? What are its principal functions?

29 Draw a labeled diagram of a liver lobule.

30 How is blood carried to and from the liver?

31 Once bile has been formed by the liver, how is it collected and transported to the gallbladder for storage?

32 Where is the gallbladder located? How is it connected to the duodenum?

33 What are the subdivisions of the small intestine? How are the coats of the small intestine adapted for digestion and absorption?

34 Describe the movements in the small intestine.

35 By means of a labeled diagram, outline the chemical digestion that occurs in the small intestine.

36 List the hormones, and their actions, that control digestion in the stomach and small intestine.

37 Why is the small intestine considered the principal area of the digestive tract?

38 Define absorption. How are the end products of carbohydrate and protein digestion absorbed? How are the end products of fat digestion absorbed?

39 Suppose you have just eaten a roast beef sandwich with butter. Describe or diagram the chemical changes that occur in the sandwich as it passes through the mouth, stomach, and small intestine. Name the enzymes involved and the glands that secrete them. Include the role of bile. Remember that roast beef is a protein, bread is a carbohydrate, and butter is a fat.

40 What routes are taken by absorbed nutrients to reach the liver?

41 What are the principal subdivisions of the large intestine? How does the muscularis of the large intestine differ from that of the rest of the digestive tract?

42 Describe the mechanical movements that occur in the large intestine.

43 Explain the activities of the large intestine that change its contents into feces.

44 Define defecation. How does it occur?

Chapter 25
Metabolism

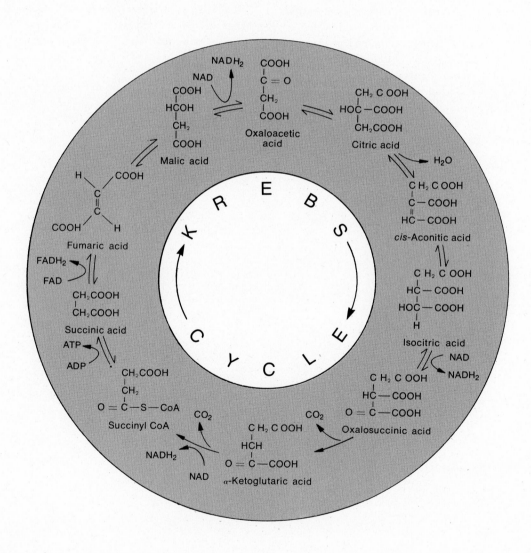

■ Define a nutrient and list the functions of the six classes of nutrients.

■ Define metabolism and contrast the physiological effects of catabolism and anabolism.

■ Describe the fate of glucose as it is catabolized via glycolysis, the Krebs cycle, and the electron transport system.

■ Define glycogenesis as an example of glucose anabolism into glycogen.

■ Define glycogenolysis as an example of glycogen catabolism into glucose.

■ Define gluconeogenesis as a conversion of lipids and proteins into glucose.

■ Describe fat storage in adipose tissue.

■ Explain how glycerol may be converted to carbohydrate.

■ Describe the catabolism of fatty acids via beta oxidation and ketogenesis.

■ Define ketosis and list its effects on the body.

■ Define lipogenesis as the synthesis of lipids from glucose and amino acids.

■ Describe the mechanism involved in protein synthesis.

■ Discuss the catabolism of amino acids, noting especially their conversion into acetyl coenzyme A, pyruvic acid, and acids of the Krebs cycle.

■ Describe the hormonal control of metabolism by contrasting the roles of insulin, glucagon, HGH, ACTH, TSH, epinephrine, and sex hormones.

■ Compare the sources, functions, and importance of minerals in metabolism.

■ Define a vitamin and differentiate between fat-soluble and water-soluble vitamins.

■ Compare the sources, functions, deficiency symptoms, and disorders of the principal vitamins.

■ Explain how heat is measured and produced by the body.

■ Describe how the caloric value of foods is determined.

■ Explain how the basal metabolic rate is measured.

■ Describe loss and conduction of body heat.

■ Explain how normal body temperature is maintained.

Nutrients are chemical substances in food that provide energy, form new body components, or assist body processes. There are six classes of nutrients: carbohydrates, lipids, proteins, minerals, vitamins, and water. Carbohydrates, proteins, and lipids are the raw materials for reactions in cells. The cells either break them down to release energy or use them to build new structures and new regulatory substances such as hormones and enzymes. Some minerals and many vitamins are used by enzyme systems that catalyze the reactions undergone by carbohydrates, proteins, and lipids. Water has four major functions. It acts as a reactant in hydrolysis reactions, as a solvent and suspending medium, as a lubricant, and as a coolant.

METABOLISM

In its broadest sense, **metabolism** refers to all the chemical activities of the body. Since chemical reactions either release or require energy, the body's metabolism may be thought of as an energy-balancing act. Accordingly, metabolism has two phases, catabolism and anabolism.

Catabolism

Catabolism is the term for processes that provide energy. Digestion is a catabolic process because the breaking of bonds releases energy. However, digestion occurs in a cavity outside the cells. Thus the energy is not directly available to the cells and is mostly dissipated as heat. This chapter concerns only the catabolic processes occurring in cells. Catabolism in cells consists of three steps, the first two of which are connected.

The first step, *oxidation or cellular respiration,* is the breakdown of absorbed nutrients resulting in a release of energy. Oxidation reactions are, therefore, decomposition reactions. Glucose is the body's favorite nutrient for oxidation, but fats and proteins are also oxidized. The second step in catabolism is the manufacture of ATP from ADP. This synthetic reaction utilizes the energy obtained from oxidation and provides a method for storing it. The final step in catabolism is the decomposition of ATP, releasing great quantities of energy.

Anabolism

Anabolism is just the opposite of catabolism – a series of synthetic reactions whereby small molecules are built up into larger ones that form the body's structural and functional components. Anabolic reactions are building processes in the cells. They require energy, and the energy is supplied by catabolic reactions of the body. One example of an anabolic process is the formation of

peptide bonds between amino acids – thereby building up the amino acids into the protein portions of cytoplasm, enzymes, and antibodies. Fats also participate in the body's anabolism. For instance, fats can be built up into the lipids that form the middle layer of the plasma membrane. They are also part of the steroid hormones.

CARBOHYDRATE METABOLISM

During the process of digestion, carbohydrates are hydrolyzed to become the simple sugars – glucose, fructose, and galactose – which are then absorbed into the capillaries of the villi of the small intestine and carried through the hepatic portal vein to the liver, where fructose and galactose are converted to glucose. Thus the story of carbohydrate metabolism is really the story of glucose metabolism.

Since glucose is the body's most direct source of energy, the fate of absorbed glucose depends on the body cells' energy needs. If the cells require immediate energy, the liver releases some of the glucose back into the bloodstream so it can be oxidized by the cells. The glucose not needed for immediate use is handled in several ways. First, the liver can convert excess glucose to glycogen and then store it. Skeletal muscle cells can also store glycogen. Second, if the glycogen storage areas are filled up, the liver cells can transform the glucose to fat that can be stored in adipose tissue. Later, when the cells need more energy, the glycogen and fat can be converted back to glucose and oxidized. Third, excess glucose can be excreted in the urine. Normally, this happens only when a meal containing mostly carbohydrates and no fats is eaten. Without the inhibiting effect of fats, the stomach empties its contents quickly and the carbohydrates are all digested at the same time. As a result, large numbers of monosaccharides suddenly flood into the bloodstream. Since the liver is unable to process all of them simultaneously, the blood glucose level rises and the conditions of hyperglycemia may result in glucose in the urine.

Glucose Catabolism

The oxidation of glucose is also known as **cellular respiration.** It occurs in every cell in the body and provides the cell's chief source of energy. The complete oxidation of glucose occurs in three successive phases: glycolysis, the Krebs cycle, and the electron transport system.

GLYCOLYSIS

The term **glycolysis** refers to a series of chemical reactions that convert glucose into *pyruvic acid (glyco =*

sugar; *lysis* = breakdown). The overall reaction for glycolysis can be written like this:

$$C-C-C-C-C-C \rightarrow 2 \ C-C-C \ + \ 2ATP$$

Glucose   Pyruvic
                                 acid

Many details of the intermediate reactions for glycolysis are shown in Figure 25-1a. Notice that the glucose molecule containing six carbon atoms is eventually broken down into two molecules of pyruvic acid, both of which contain three carbon atoms. Since the reactions leading to pyruvic acid formation are decomposition reactions, energy is released. Most of this energy goes into making two molecules of ATP that are used by the cell. The rest of the energy is expended as heat energy, some of which helps to maintain body temperature. Glycolysis occurs in the cytoplasm of the cell. It is an *anaerobic* process—it does not require oxygen.

The remaining steps in the oxidation of glucose are *aerobic*—they do require oxygen. The immediate fate of pyruvic acid, then, depends on the availability of oxygen. When the body is exercising strenuously, glycolysis occurs so rapidly that the lungs and blood cannot supply enough oxygen to break down all the pyruvic acid. The excess is converted to lactic acid. Moderate amounts of lactic acid are buffered by the body so that the pH is not significantly disturbed. In addition, several protective mechanisms prevent excessive amounts of the acid from building up. First, the liver changes lactic acid back to pyruvic acid and then to glucose. Second, rapid cellular respiration increases the blood $p\text{CO}_2$, thus increasing the respiratory rate. Eventually, the person becomes so short of breath that he is forced to stop exercising. Finally, lactic acid itself contributes to muscle fatigue and makes the person want to rest. After the exercise has stopped, the person breathes heavily until the blood $p\text{CO}_2$ returns to normal and the cells receive enough oxygen to break down the pyruvic acid. This is the oxygen debt. When the debt has been repaid, the lactic acid is changed back to pyruvic acid. The pyruvic acid then goes through a transitional process involving a special substance called coenzyme A before entering the Krebs cycle.

COENZYME A

An enzyme is a protein. However, many enzymes are attached to nonprotein compounds. If a nonprotein group detaches from the enzyme and acts as a carrier molecule, it is called a *coenzyme*. Enzymes work together with their coenzymes. The enzyme catalyzes the reaction, and the coenzyme attaches to the end product of the reaction and carries it to the next reaction. Each step in the oxidation of glucose requires a different enzyme and often a coenzyme as well. We are interested in only one

coenzyme at this point: a substance called *coenzyme A (CoA)*.

During the transitional step between glycolysis and the Krebs cycle, pyruvic acid is prepared for entrance into the cycle. Essentially, pyruvic acid is converted to a two-carbon compound by the loss of carbon dioxide. This two-carbon fragment, called an acetyl group, attaches itself to coenzyme A, and the whole complex is called *acetyl coenzyme A*. It is in this form that pyruvic acid enters the Krebs cycle (Figure 25-1b).

KREBS CYCLE

The *Krebs cycle*, or *citric acid cycle*, is a series of reactions in the matrix of the mitochondria of cells (Figure 25-1c). Coenzyme A carries an acetyl group to a mitochondrion, detaches itself, and goes back into the cytoplasm to pick up another fragment. The acetyl group combines with a substance called oxaloacetic acid to form citric acid.

It is at this point that we begin our study of the Krebs cycle. As the various acids move through the cycle, they undergo a number of changes, all controlled by specific enzymes. One of these changes is called *decarboxylation*, a process in which a chemical compound loses a molecule of carbon dioxide (CO_2). The α-ketoglutaric acid in the Krebs cycle has one less molecule of CO_2 than the oxalosuccinic acid. In other words, when oxalosuccinic acid is decarboxylated it becomes α-ketoglutaric acid. This is what the arrow followed by CO_2 means in the Krebs cycle. By losing a molecule of CO_2 and picking up a molecule of CoA, α-ketoglutaric acid is converted into succinyl CoA. Thus α-ketoglutaric acid has been decarboxylated.

Look at Figure 25-1 again. Another place where CO_2 is produced as a result of decarboxylation is between pyruvic acid and acetyl coenzyme A. The molecules of CO_2 thus formed by decarboxylation leave the mitochondria, diffuse through the cytoplasm of the cell to the plasma membrane, and then diffuse into the blood by internal respiration. Eventually the CO_2 is transported by the blood to the lungs by external respiration. Once it reaches the lungs, it is exhaled.

Let's examine the Krebs cycle again. As various acids move through the cycle, they are also involved in a series of *oxidation-reduction reactions*. When a molecule is *oxidized*, it loses hydrogen atoms. When a molecule is *reduced*, it gains hydrogen atoms. Every oxidation is coupled with a reduction. Isocitric acid in the Krebs cycle is converted to oxalosuccinic acid by losing two hydrogen atoms. In other words, it is oxidized. The hydrogen atoms are picked up by a coenzyme called *nicotinamide adenine dinucleotide (NAD)*. Since NAD picks up the hydrogen atoms, it is reduced and represented as NADH_2. This is what the arrow means between NAD and NADH_2. Notice the other points in

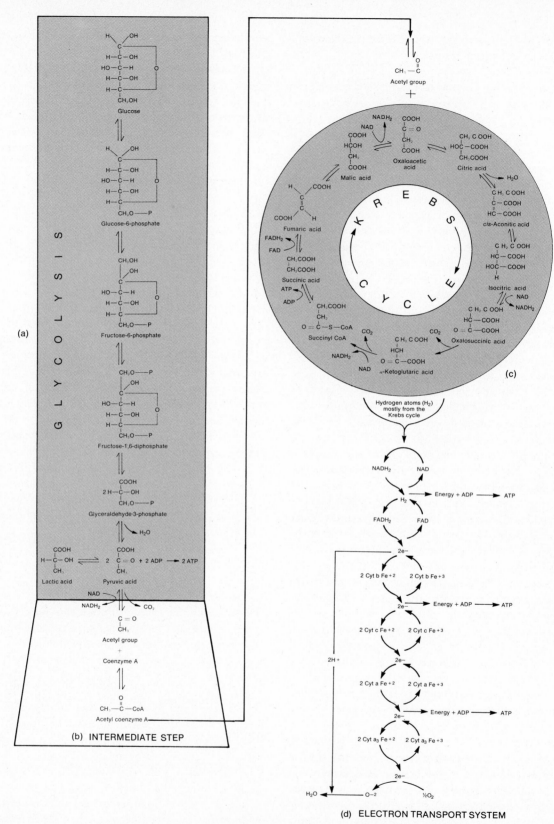

Figure 25–1   Complete oxidation of glucose. (a) Glycolysis. (b) Formation of acetyl CoA. (c) The Krebs cycle. (d) The electron transport system. See Figure 25–3 to see how glycolysis and the Krebs cycle are related to the metabolism of other nutrients.

the Krebs cycle and just before the cycle where coenzymes pick up hydrogen atoms:

1 Between pyruvic acid and acetyl coenzyme A, where pyruvic acid is oxidized and NAD is reduced to $NADH_2$.

2 Between isocitric acid and oxalosuccinic acid.

3 Between α-ketoglutaric acid and succinyl CoA, where α-ketoglutaric acid is oxidized and NAD is reduced to $NADH_2$.

4 Between succinic acid and fumaric acid, where succinic acid is oxidized and another coenzyme called *flavin adenine nucleotide* (FAD) is reduced to $FADH_2$.

5 Between malic acid and oxaloacetic acid, where malic acid is oxidized and NAD is reduced to $NADH_2$.

During these oxidation-reduction reactions, the acids lose hydrogen atoms. The hydrogen atoms are picked up by coenzymes and pass through the electron transport system, the final phase of cellular respiration.

ELECTRON TRANSPORT SYSTEM

The *electron transport system* is a series of reactions that occur on the cristae of mitochondria. Its purpose is to liberate the energy in the hydrogen atoms so it may be transferred to ATP and stored.

The first step in the electron transport system involves the transfer of hydrogen atoms mostly from the Krebs cycle from $NADH_2$ to FAD (Figure 25-1d). In the process, $NADH_2$ is oxidized to NAD and FAD is reduced to $FADH_2$. The importance of the hydrogen transfer from $NADH_2$ to FAD is that it releases energy. This energy is used to produce ATP from ADP. This is what the arrow between ADP and ATP means. The hydrogen atoms of $FADH_2$, however, do not stay intact. They ionize into hydrogen ions (H^+) and electrons (e^-) according to the following reaction:

$$H \rightarrow H^+ + e^-$$

Hydrogen atom   Hydrogen ion   Electron

In the next step of the electron transport system, the electrons (e^-) from the hydrogen atoms are passed from $FADH_2$ to a substance called cytochrome b. A *cytochrome* is a protein that contains an iron portion capable of alternating between a reduced form (Fe^{2+}) and an oxidized form (Fe^{3+}). At the same time, the H^+ ions are released by $FADH_2$. In the process of transfer, $FADH_2$ is oxidized to FAD. The electrons are successively passed from one cytochrome to another—from cytochrome b to cytochrome c to cytochrome a and finally to cytochrome a_3 (cytochrome oxidase). Each cytochrome in the electron transport system is alter-

nately reduced as it picks up electrons and oxidized as it gives up electrons. Because of the involvement of cytochromes in the electron transport system, it is also known as the *cytochrome system*.

In the process of electron transfer between cytochromes, more energy is liberated and then stored in ATP. The formation of ATP from the liberated energy occurs between cytochrome b and cytochrome c and between cytochrome a and cytochrome a_3. At the end of the electron transport system, the electrons are passed to oxygen, which becomes negatively charged. This oxygen, required for the complete breakdown of glucose, is supplied to your body cells through inspiration, external respiration, and internal respiration. The oxygen combines with the H^+ ions previously released from $NADH_2$ and forms water.

In the electron transport system, ATP is formed at three different places. Moreover, some ATP is formed directly in the Krebs cycle—between succinyl coenzyme A and succinic acid—without passing through the electron transport system. Some ATP is also formed during glycolysis. The bulk of ATP is formed in the electron transport system, however. In sum, the complete oxidation of a single molecule of glucose results in the production of 38 ATP molecules, 6 CO_2 molecules, and 12 H_2O molecules. The complete oxidation of glucose, then, can be summarized as follows:

$$C_6H_{12}O_6 + 6O_2 \rightarrow 6CO_2 + 12H_2O + 38ATP$$

Glucose   Oxygen   Carbon   Water   Adenosine
                     dioxide   triphosphate

Glycolysis, the Krebs cycle, and especially the electron transport system provide all the ATP for cellular activities. And because the Krebs cycle and electron transport system are aerobic processes, the cells cannot carry on their activities for long without sufficient oxygen.

Glucose Anabolism

Eventually, most of the glucose in the body is catabolized to supply energy. However, glucose participates in a number of anabolic reactions. One is the synthesis of one large molecule of glycogen from many glucose molecules. Another is the manufacture of glucose molecules from the breakdown products of lipids and proteins.

STORAGE

If glucose is not needed immediately for energy, it is combined with many other molecules of glucose to form a long-chain molecule called glycogen. This process is

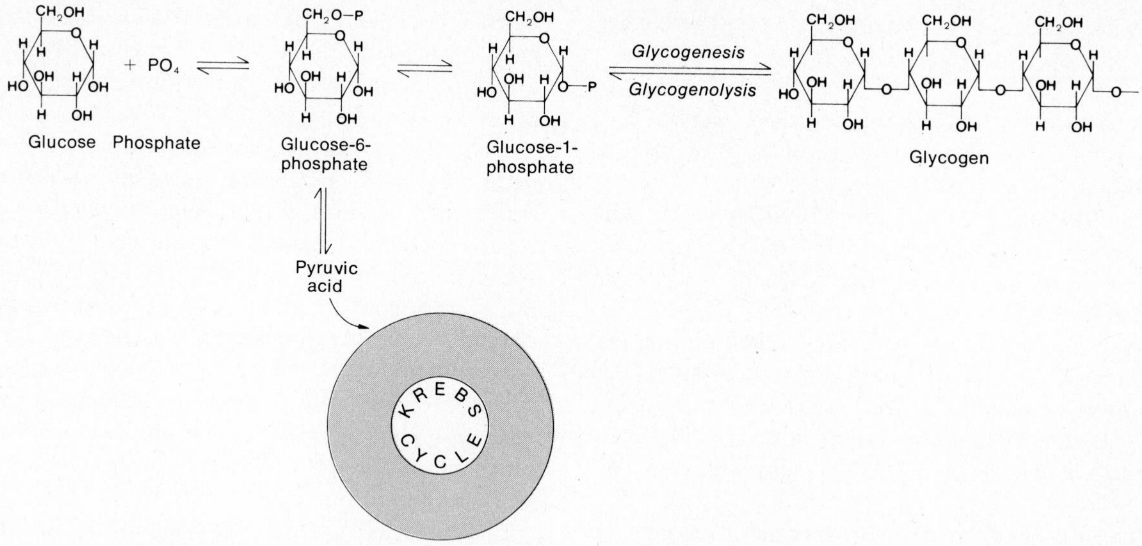

Figure 25–2 Glycogenesis and glycogenolysis. When glucose is needed, skeletal muscle cells can change glycogen into glucose-6-phosphate. This substance can be converted to pyruvic acid and enter the Krebs cycle in any body cell for oxidation to carbon dioxide and water. Liver cells also have the necessary enzymes to change glycogen all the way back to glucose, which can enter the blood for distribution to body cells.

called **glycogenesis** (Figure 25-2). *Glyco* means sugar; *genesis* means origin. The body can store about 400 g (1 lb) of glycogen in the liver and skeletal muscle cells. When the body needs energy, the glycogen stored in the liver is broken down into glucose and released into the bloodstream to be transported to the cells where it will be catabolized. The process of converting glycogen back to glucose is called **glycogenolysis** (*lysis* = breakdown). Glycogenolysis usually occurs between meals.

LIPID AND PROTEIN CONVERSION

When your liver runs out of glycogen, it is time to eat. If you do not eat, your body starts catabolizing fats and proteins. Actually, the body normally catabolizes some of its fats and a few of its proteins. But large-scale fat and protein catabolism does not happen unless you are starving, eating meals that contain very few carbohydrates, or suffering from an endocrine disorder.

For fats and proteins to be catabolized, they must be converted in the liver to glucose, a process called **gluconeogenesis.** Figure 25-3 shows how gluconeogenesis is related to other metabolic reactions. Notice that fats and proteins can be transformed into compounds that also are intermediate products in the catabolism of glucose. These compounds can be used by most cells to produce ATP. Glycerol, one of the products of fat digestion, may be converted into glyceraldehyde-3-phosphate. The amino acid alanine may be converted into pyruvic acid. Also notice that

most of the reactions in Figure 25-3 are reversible. Thus the intermediate products of glucose catabolism also can be resynthesized into glucose. This is exactly what happens during gluconeogenesis, which is therefore an anabolic process. Only the liver has the proper enzymes for initiating gluconeogenesis, although in some cases enzymes in other cells are needed to complete the process. Since many of the reactions in Figure 25-3 are reversible, many conversions of carbohydrates, fats, and proteins may also occur.

LIPID METABOLISM

Lipids are second to carbohydrates as a source of energy. More frequently they are used as building blocks to form essential structures.

Storage

When fats are eaten, they are digested into glycerol, fatty acids, and glycerides. As soon as the glycerol, fatty acids, and glycerides are absorbed by the intestinal epithelial cells, they recombine to form fat molecules and are transported as chylomicrons. Small chylomicrons enter the blood capillaries of the villi. Large chylomicrons enter the lacteals of the villi for transport to the left subclavian vein. If the body has no immediate need for the fat, it is stored in adipose tissue. About 50 percent of your stored fat is deposited in subcutaneous tissue—ap-

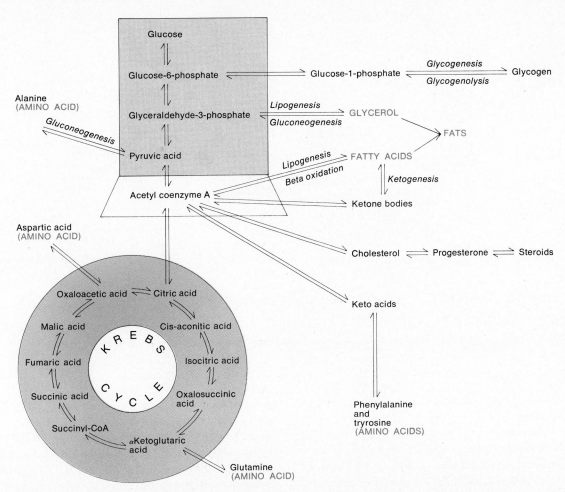

Figure 25—3 Metabolism of carbohydrates, lipids, and proteins. Note especially that glucose, fatty acids and glycerol, and amino acids must be converted into glucose breakdown products before they are catabolized. Mechanisms by which one type of nutrient is converted into another are indicated above and below the directional arrows. The catabolism of carbohydrates, lipids, and proteins is considered separately in Figure 25—1, 25—4, and 25—5, respectively.

proximately 12 percent around the kidneys, 10 to 15 percent in the omenta, 20 percent in genital areas, and 5 to 8 percent between muscles. Fat is also stored behind the eyes, in the furrows of the heart, and in the folds of the large intestine. Fat does not remain stationary until needed for energy. It is continually released from storage, transported through the blood, and redeposited in other adipose tissue cells. Some researchers estimate that as much as one-half of the total body-fat reserve changes position daily.

Catabolism

Fats stored in fat depots constitute the largest reserve of energy. The body can store much more fat than it can glycogen—hundreds of pounds of it in the obese.

Moreover, the energy yield of fats is more than twice that of carbohydrates. Nevertheless, fats are only the body's second-favorite source of energy because they are more difficult to catabolize than carbohydrates.

GLYCEROL

When fat molecules are metabolized, first they are separated into glycerol and fatty acids in the liver. The glycerol and fatty acids are then catabolized separately (see Figure 25-4). Glycerol is converted easily by the liver cells to a compound called glyceraldehyde-3-phosphate, one of the compounds also formed during the catabolism of glucose. The liver cells then transform glyceraldehyde-3-phosphate into glucose and release it to be catabolized by other cells. This is one example of gluconeogenesis. Glyceraldehyde-3-phosphate is an inter-

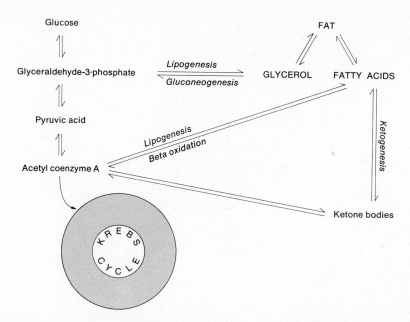

Figure 25–4 Metabolism of lipids. Glycerol may be converted to glyceraldehyde-3-phosphate and enter the Krebs cycle. Fatty acids undergo beta oxidation and ketogenesis and enter the Krebs cycle via acetyl coenzyme A.

mediate product in the conversion of glycerol to glucose and is also a link in the conversion of glucose to fats.

FATTY ACIDS

Fatty acids, however, are catabolized differently. The first step in fatty-acid catabolism involves a series of reactions called **beta oxidation.** Enzymes in the liver cells remove pairs of carbon atoms from the long chain of carbon atoms comprising a fatty-acid molecule. The liver cells convert some of the two-carbon fragments into acetyl CoA, which can be converted to glucose and catabolized by any cell. Most of the fragments, however, are converted into *acetone* or substances classified as *keto acids. Ketone bodies* is a collective term for keto acids and acetone; the formation of ketone bodies is called *ketogenesis.* The liver does not have the proper enzymes for converting the ketone bodies into an intermediate glucose product, so it releases them into the bloodstream. The ketone bodies are then catabolized, via the Krebs cycle, by the other body cells that have an enzyme which converts ketone bodies into acetyl coenzyme A. Whether the process is direct or indirect, acetyl CoA links fatty acid with glucose metabolism.

KETOSIS

Ketone bodies are normal intermediate products of fatty-acid metabolism. But because the body prefers glucose as a source of energy, they are generally produced in very small quantities (1.5 to 2.0 mg/100 ml blood). When the number of ketone bodies in the blood rises

above normal – a condition called **ketosis** – the keto acids must be buffered by the body. If too many accumulate, they use up the body's buffers and the blood pH falls. Thus extreme or prolonged ketosis can lead to acidosis, or abnormally low blood pH. The causes of ketosis include starvation, low-carbohydrate diets, and metabolic abnormalities. One of the most common abnormalities is diabetes mellitus, or insufficient insulin. Insulin is required for more than minimal glucose consumption by the cells, and it discourages fat catabolism. When a diabetic becomes seriously insulin-deficient, one of the telltale signs is a sweet smell of acetone on the breath.

Anabolism

Liver cells can synthesize lipids from glucose and amino acids through a process called **lipogenesis.** The steps in the conversion of glucose to lipids are just the reverse of the steps that transform fats into glucose. The links are glyceraldehyde-3-phosphate, which can be converted to glycerol, and acetyl CoA, which can be converted to fatty acids. Amino acids are transformed to lipids in the same way, but first they must be converted to glucose or to a glucose-breakdown product. The resulting glycerol and fatty acids can undergo anabolic reactions to become fat that can be stored or go through a series of anabolic reactions that produce other lipids. Such lipids may become the middle layer of the plasma membrane, part of the steroid hormones, or some other component of

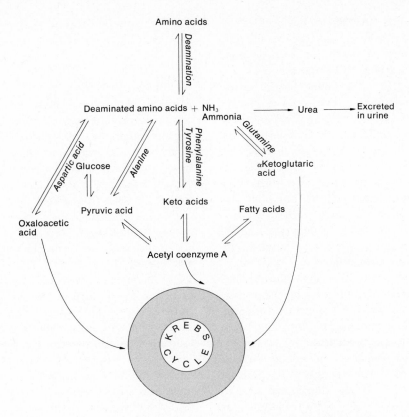

Figure 25—5   Metabolism of proteins. The deaminated amino acid may follow several pathways (see Figure 25—3).

the body. Exhibit 2-3 lists some of the many ways in which lipids are used.

PROTEIN METABOLISM

Proteins are primarily body builders. Generally, the body uses very little protein for energy—as long is it ingests sufficient carbohydrates and fats or has a supply of stored fat.

Catabolism

A certain amount of protein catabolism occurs in the body each day, although much of this is only partial catabolism. Proteins are extracted from worn-out cells and broken down into free amino acids. Some amino acids are converted into other amino acids, peptide bonds are reformed, and new proteins are made.

If other energy sources are used up, the body can catabolize large amounts of protein to carbon dioxide and water. The first step in the catabolism of amino acids consists of removing the amino group (NH_2) from the amino acid (Figure 25-5), a process called *deamination* that occurs in the liver. The liver cells then convert the NH_2 to ammonia (NH_3) and finally to urea that is excreted in the urine. The fate of the remaining part of

the amino acid depends on what kind of amino acid it is. Figure 25-3 shows that some amino acids are converted to keto acids and then to acetyl CoA, other amino acids are changed to pyruvic acid, and still others are converted to acids of the Krebs cycle.

Anabolism

Protein anabolism involves the formation of peptide bonds between amino acids to produce new proteins. Protein anabolism, or synthesis, is carried out on the ribosomes of almost every cell in the body, directed by the cells' DNA and RNA. The synthesized proteins are the primary constituents of cell structures, enzymes, antibodies, and many glandular secretions. Because proteins are a primary ingredient of most cell structures, high-protein diets are essential during the growth years, during pregnancy, and when tissue has been damaged by disease or injury.

CONTROL

The metabolism of carbohydrates, lipids, and proteins is controlled by a number of hormones. Two major hormones, insulin and glucagon, are secreted in response to blood glucose level. Under normal circumstances, the blood sugar level ranges from 80 to 120 mg/100 ml blood. After a meal, the level may rise to 120–130 mg/100 ml. The increase stimulates the pancreatic beta cells to secrete insulin, which pushes the blood glucose level down to normal in a few hours. Insulin lowers blood sugar by stimulating the liver cells to convert glucose to glycogen and fat, stimulating all other cells to catabolize glucose, and inhibiting the catabolism of fats and proteins. When the body fails to produce enough insulin, the blood sugar is somewhat lowered as the glucose is excreted in the urine. Thus sugar in the urine is a sign of untreated diabetes mellitus. The body also excretes glucose if a large amount of carbohydrates is absorbed all at once. In this case, the condition is called *alimentary glycosuria.* Alimentary glycosuria is temporary and is not considered pathological.

After a period of fasting, the blood glucose level falls below normal. Low blood glucose (hypoglycemia) stimulates the pancreatic alpha cells to release glucagon, which stimulates the liver to change glycogen to glucose (glycogenolysis) and to catabolize fats.

The anterior pituitary gland also assumes a role in the regulation of metabolism through the actions of HGH, ACTH, and TSH. HGH, the human growth hormone, encourages new tissue to be laid down by stimulating protein anabolism. HGH also stimulates the body cells to catabolize fats instead of carbohydrates. This action causes the blood sugar level to increase. Adrenocorticotropic hormone (ACTH) stimulates the adrenal cortex to increase its secretion of glucocorticoid hormones. The chief effect of the glucocorticoids is to increase the amount of energy available to cells during extreme stress. Glucocorticoids stimulate the liver to accelerate the breakdown of proteins into amino acids, which the liver converts to glucose by gluconeogenesis. Since the resulting glucose is released into the bloodstream, glucocorticoids are hyperglycemic. The third anterior pituitary hormone, thyroid-stimulating hormone (TSH), stimulates the thyroid to release thyroxin, a hormone that encourages glucose catabolism and is therefore hypoglycemic.

The adrenal medulla, the ovaries, and the testes also produce hormones that affect metabolism. Epinephrine and norepinephrine, the adrenal medulla hormones, accelerate the conversion of liver glycogen to glucose and are therefore hyperglycemic. Testosterone, the male hormone, and progesterone, the female hormone, encourage protein anabolism. Progesterone is secreted in copious amounts during pregnancy, when uterine and mammary tissues are growing.

Hormones are the primary regulators of metabolism. However, hormonal control is ineffective without the proper minerals and vitamins. Some minerals and many vitamins are components of the enzyme systems that catalyze the metabolic reactions.

MINERALS

Minerals are inorganic substances. They may appear in combination with each other or in combination with organic compounds. Minerals constitute about 4 percent of the total body weight, and they are concentrated most heavily in the skeleton. Minerals known to perform functions essential to life include calcium, phosphorus, sodium, chlorine, potassium, magnesium, iron, sulfur, iodine, manganese, cobalt, copper, and zinc. Other minerals – aluminum, silicon, arsenic, nickel – are present in the body, but their functions have not yet been determined.

Calcium and phosphorus form part of the structure of bone. But since minerals do not form long-chain compounds, they are otherwise poor building materials. Their chief role is to help regulate body processes. Calcium, iron, magnesium, and manganese are constituents of some coenzymes. Magnesium also serves as a catalyst for the conversion of ADP to ATP. Without these minerals, metabolism would stop and the body would die. Minerals such as sodium and phosphorus work in buffer systems. Sodium helps regulate the osmosis of water and, along with other ions, is involved in the generation of nerve impulses. Exhibit 25-1 describes the functions of some minerals vital to the body. Note that the body generally uses the ions of the minerals rather than the nonionized form. Some minerals, such as chlorine, are toxic or even fatal to the body if ingested in the nonionized form.

VITAMINS

Organic nutrients required in minute amounts to maintain growth and normal metabolism are called **vitamins.** Unlike carbohydrates, fats, or proteins, vitamins do not provide energy or serve as building materials. The essential function of vitamins is the regulation of physiological processes. Accordingly, some vitamins act as enzymes, others are precursors (forerunners) of enzymes, and still others serve as coenzymes.

Most vitamins cannot be synthesized by the body. One source of vitamins is ingested foods – for example, vitamin C in citrus fruits. Another source is vitamin pills. Other vitamins, such as vitamin K, are produced by bacteria in the gastrointestinal tract. The body can assemble some vitamins if the raw materials called *provitamins* are provided. Vitamin A is produced by the body from the provitamin carotene, a chemical present in spinach, carrots, liver, and milk. No single food

Exhibit 25—1 MINERALS VITAL TO THE BODY

Mineral	Comments	Importance
Calcium	Most abundant cation in body. Appears in combination with phosphorus in ratio of 2:1.5. About 99 percent is stored in bone and teeth. Remainder stored in muscle, other soft tissues, and blood plasma. Blood calcium level controlled by thyrocalcitonin and parathyroid hormone. Most is excreted in feces and small amount in urine. Recommended daily intake for children is 1.2–1.4 gm and 800 mg for adults. Good sources are milk, egg yolk, shellfish, green leafy vegetables.	Formation of bones and teeth, blood clotting, and normal muscle and nerve activity.
Phosphorus	About 80 percent found in bones and teeth. Remainder distributed in muscle, brain cells, blood. More functions than any other mineral. Most excreted in urine, small amount eliminated in feces. Recommended daily intake is 1.2–1.4 gm for children and 1,200 mg for adults. Good sources are dairy products, meat, fish, poultry, nuts.	Formation of bones and teeth. Constitutes a major buffer system of blood. Plays important role in muscle contraction and nerve activity. Component of many enzymes. Involved in transfer and storage of energy (ATP). Component of DNA and RNA.
Iron	About 66 percent found in hemoglobin of blood. Remainder distributed in skeletal muscles, liver, spleen, enzymes. Normal losses of iron occur by shedding of hair, epithelial cells, and mucosal cells and in sweat, urine, feces, and bile. Recommended daily intake for children is 7–12 mg and for adults 10–15 mg. Sources are meat, liver, shellfish, egg yolk, beans, legumes, dried fruits, nuts, cereals.	As component of hemoglobin, carries O_2 to body cells. Component of coenzymes involved in formation of ATP from catabolism.
Iodine	Essential component of thyroxin. Excreted in urine. Estimated daily requirement: 0.15–0.30 mg. Adequate sources are iodized salt, seafoods, cod-liver oil, and vegetables grown in iodine-rich soils.	Required by thyroid gland to synthesize thyroxin and triiodothyronine, hormones that regulate metabolic rate.
Copper	Some stored in liver and spleen. Most excreted in feces. Daily requirement about 2 mg. Good sources include eggs, whole-wheat flour, beans, beets, liver, fish, spinach, asparagus.	Required with iron for synthesis of hemoglobin. Component of enzyme necessary for melanin pigment formation.
Sodium	Most found in extracellular fluids, some in bones. Excreted in urine and perspiration. Recommended daily intake of NaCl (table salt) is 5 gm.	As most abundant cation in extracellular fluid, strongly affects distribution of water through osmosis. Part of bicarbonate buffer system.
Potassium	Principal cation in intracellular fluid. Most is excreted in urine. Recommended daily intake not known. Normal food intake supplies required amounts.	Functions in transmission of nerve impulses and muscle contraction.
Chlorine	Found in extracellular and intracellular fluids. Principal anion of extracellular fluid. Most excreted in urine. Normal intake of NaCl supplies required amounts.	Assumes role in acid–base balance of blood, water balance, and formation of HCl in stomach.
Magnesium	Component of soft tissues and bone. Excreted in urine and feces. Suggested minimum daily intake: 250–350 mg. Widespread in various foods.	Required for normal functioning of muscle and nervous tissue. Participates in bone formation. Constituent of many coenzymes.

Exhibit 25-1 (cont.)

Mineral	Comments	Importance
Sulfur	Constituent of many proteins (such as insulin) and some vitamins (thiamine and biotin). Excreted in urine. Good sources include beef, liver, lamb, fish, poultry, eggs, cheese, beans.	As component of hormones and vitamins, regulates various body activities.
Zinc	Important component of certain enzymes. Widespread in many foods.	Necessary for normal growth.
Fluorine	Component of bones, teeth, other tissues.	Appears to improve tooth structure.
Manganese	Distribution throughout body similar to that of copper. Human daily requirement about 2.5 mg.	Activates several enzymes. Needed for hemoglobin synthesis. Required for growth, reproduction, lactation.
Cobalt	Constituent of vitamin B_{12}.	As part of B_{12}, required for stimulation of erythropoeisis.

contains all the required vitamins – one of the best reasons for eating a balanced diet. The term *avitaminosis* refers to a deficiency of any vitamin in the diet.

On the basis of solubility, vitamins are divided into two principal groups: fat-soluble and water-soluble. *Fat-soluble* vitamins are absorbed along with digested dietary fats by the small intestine as micelles. In fact, they cannot be absorbed unless they are ingested with some fat. Fat-soluble vitamins are generally stored in cells, particularly liver cells, so reserves can be built up. Examples of fat-soluble vitamins are vitamins A, D, E, and K. *Water-soluble* vitamins, by contrast, are absorbed along with water in the gastrointestinal tract and dissolve in the body fluids. Excess quantities of these vitamins are excreted in the urine. Thus the body does not store water-soluble vitamins well. Examples of water-soluble vitamins are the B vitamins and vitamin C. Exhibit 25-2 lists the principal vitamins, their sources, functions, and related disorders.

METABOLISM AND HEAT

We will now consider the relationship of foods to body heat, mechanisms of heat gain and loss, and the regulation of body temperature.

Measuring Heat

Heat is a form of energy. That energy can be measured as **temperature** and expressed in units called calories. A *calorie (cal)* is the amount of heat energy required to raise the temperature of 1 g of water 1°C. Since the calorie is a small unit, the *Calorie* or *kilocalorie* is used instead. A

Calorie is equal to 1,000 cal and is defined as the amount of heat required to raise the temperature of 1,000 g of water 1°C. The Calorie is the unit we use to determine the heating value of foods and measure body metabolism rates.

Calories

The apparatus used to determine the caloric value of foods is called a *calorimeter* (Figure 25-6). A weighed sample of a dehydrated food is burned completely in an insulated metal container. The energy released by the burning food is absorbed by the container and then transferred to a known value of water that surrounds the container. The change in the water's temperature is directly related to the number of Calories released by the food. Knowing the caloric value of foods is important – if we know the amount of energy the body uses for various activities, we can adjust our food intake. In this way we can control body weight by taking in only enough Calories to sustain our activities.

Production of Body Heat

Most of the heat produced by the body comes from oxidation of the food we eat. The rate at which this heat is produced – the **metabolic rate** – is also measured in Calories. Among the factors that affect your metabolic rate are the following:

1 **Exercise** During strenuous exercise, your metabolic rate may increase as much as 40 times your normal rate.

2 Nervous system In a stress situation, your sympathetic nervous system is stimulated and the nerves release norepinephrine, which increases the metabolic rate of body cells. Strong sympathetic stimulation may increase the metabolic rate by 160 times, but only for a few minutes.

3 Hormones In addition to norepinephrine, two other hormones affect metabolic rate: epinephrine produced by the adrenal glands and thyroid hormones produced by the thyroid gland. Epinephrine is secreted in stress situations. Increased secretions of thyroid hormones increase the metabolic rate.

4 Body temperature The higher your body temperature, the higher your metabolic rate. Each 1°C rise in temperature increases the rate of biochemical reactions by about 10 percent. In fact, the metabolic rate may be doubled if you have a high fever.

Since many factors affect metabolic rate, it is difficult to measure and compare your metabolic rate with someone else's. Certain conditions must be controlled. The person should not exercise for 30 minutes to an hour before the measurement is taken. Also, the individual must be completely at rest, but awake, and air temperature should be comfortable. Finally, the person should fast for at least 12 hours and body temperature should be normal. All these conditions produce the *basal state* – a state in which factors that could affect metabolic rate have been greatly reduced or eliminated. **Basal metabolic rate** (BMR) is a measure of the rate at which your body breaks down foods (and therefore releases heat). It is also a measure of how much thyroxin your thyroid gland is producing, since thyroxin regulates the rate of food breakdown and is not a controllable factor under basal conditions.

Metabolic Rate

At one time the most convenient way to measure basal metabolic rate was to do it indirectly by measuring oxygen consumption. If a given amount of food releases a given amount of heat energy when it is oxidized, it must combine with a given amount of oxygen. Thus, by measuring the amount of oxygen needed for the metabolism of foods, we can determine how many Calories are produced. The amount of heat energy released when 1 liter of oxygen combines with carbohydrates is 5.05 Calories; with fats, the heat released is 4.70 Calories; with proteins, the heat released is 4.60 Calories. The average of the three values is 4.825 Calories. Therefore, every time a liter of oxygen is consumed, 4.825 Calories are produced.

The instrument used to determine metabolic rate indirectly is called a **respirometer** or **spirometer**

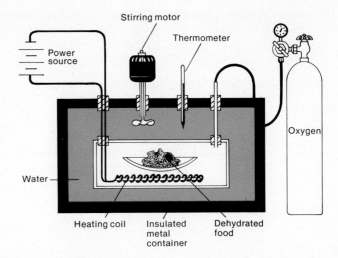

Figure 25–6 Parts of a typical calorimeter.

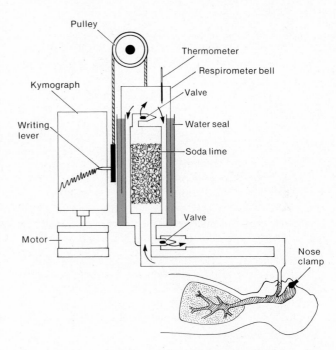

Figure 25–7 Respirometer for measuring basal metabolic rate.

(Figure 25-7). The subject places the mouthpiece in his mouth and breathes in and out of an inverted bell that rides up and down in a water bath. The respirometer bell contains O_2 that is breathed back and forth into the lungs. The O_2 is gradually taken into the blood and in its place CO_2 is exhaled. The exhaled air passes through a valve into a container of soda lime, which absorbs the CO_2, and then passes back into the respirometer bell. As oxygen is consumed, the level of the respirometer bell

Exhibit 25-2　THE PRINCIPAL VITAMINS

Vitamin	Comment and Source	Function	Related Symptoms and Deficiency Disorders
Fat-soluble			
A	Formed from provitamin carotene (and other provitamins) in intestinal tract. Requires bile salts and fat for absorption. Stored in liver. Sources of carotene and other provitamins include yellow and green vegetables; sources of vitamin A include fish, liver oils, milk, butter.	Maintains general health and vigor of epithelial cells.	Deficiency results in atrophy and keratinization of epithelium, leading to dry skin and hair, increased incidence of ear, sinus, respiratory, urinary, and digestive infections, inability to gain weight, drying of cornea with ulceration (xerophthalmia), nervous disorders, and skin sores.
		Essential for formation of rhodopsin, light-sensitive chemical in rods of retina.	Night blindness or decreased ability for dark adaptation.
		Growth of bones and teeth by apparently helping to regulate activity of osteoblasts and osteoclasts.	Slow and faulty development of bones and teeth.
D	In presence of sunlight, provitamin D_3 (derivative of cholesterol) converted to vitamin D in the skin. Dietary vitamin D requires moderate amounts of bile salts and fat for absorption. Stored in tissues to slight extent. Most excreted via bile. Sources include liver oils of bony fish, egg yolk, fortified milk.	Essential for absorption and utilization of calcium and phosphorus from gastrointestinal tract. May work with parathyroid hormone that controls calcium metabolism.	Defective utilization of calcium by bones leads to rickets in children and osteomalacia in adults. Possible loss of muscle tone.
E (tocopherols)	Stored in liver, adipose tissue, and muscles. Requires bile salts and fat for absorption. Sources include fresh nuts and wheat germ, seed oils, green leafy vegetables.	Believed to inhibit catabolism of certain fatty acids that help form cell structures, especially membranes. Involved in formation of DNA, RNA, and red blood cells. Believed to help protect liver from toxic chemicals like carbon tetrachloride.	Catabolism of certain fatty acids on exposure to oxygen (fatty acids in membranes of red blood cells may cause hemolytic anemia). Deficiency also causes muscular dystrophy in monkeys and sterility in rats.
K	Produced in considerable quantities by intestinal bacteria. Requires bile salts and fat for absorption. Stored in liver and spleen. Other sources include spinach, cauliflower, cabbage, liver.	Coenzyme believed essential for synthesis of prothrombin by liver and thus for normal blood clotting. Also known as antihemorrhagic vitamin.	Delayed clotting time results in excessive bleeding.

Exhibit 25-2 (cont.)

Vitamin	Comment and Source	Function	Related Symptoms and Deficiency Disorders
Water-soluble			
B₁ (thiamine)	Rapidly destroyed by heat. Not stored in body. Excessive intake eliminated in urine. Sources include whole-grain cereals, eggs, pork, nuts, liver, yeast.	Acts as coenzyme for 24 different enzymes involved in carbohydrate metabolism of pyruvic acid to CO_2 and H_2O. Essential for synthesis of acetylcholine.	Improper carbohydrate metabolism leads to buildup of pyruvic and lactic acids and insufficient energy for muscle and nerve cells. Deficiency leads to two syndromes: (1) *Beriberi*—partial paralysis of smooth muscle of GI tract causing digestive disturbances, skeletal muscle paralysis, atrophy of limbs. (2) *Polyneuritis*—reflexes related to kinesthesia are impaired, impairment of sense of touch, decreased intestinal motility, stunted growth in children, poor appetite.
B₂ (riboflavin)	Not stored in large amounts in tissues. Most is excreted in urine. Small amounts supplied by bacteria of GI tract. Other sources include yeast, liver, beef, veal, lamb, eggs, whole-wheat products, asparagus, peas, beets, peanuts.	Component of certain coenzymes concerned with carbohydrate and protein metabolism, especially in cells of eye, integument, mucosa of intestine, blood.	Deficiency may lead to improper utilization of oxygen resulting in blurred vision, cataracts, and corneal ulcerations. Also dermatitis and cracking of skin, lesions of intestinal mucosa, and development of one type of anemia.
Niacin (nicotinamide)	Derived from amino acid tryptophan. Sources include yeast, meats, liver, fish, whole-grain breads and cereals, peas, beans, nuts.	Essential component of coenzyme concerned with energy-releasing reactions. In lipid metabolism, inhibits production of cholesterol and assists in fat breakdown.	Principal deficiency is *pellagra*, characterized by dermatitis, diarrhea, and psychological disturbances.
B₆ (pyridoxine)	Formed by bacteria of GI tract. Stored in liver, muscle, brain. Other sources include salmon, yeast, tomatoes, yellow corn, spinach, whole-grain cereals, liver, yogurt.	May function as coenzyme in fat metabolism. Essential coenzyme for normal amino acid metabolism. Assists production of circulating antibodies.	Most common deficiency symptom is dermatitis of eyes, nose, and mouth. Other symptoms are retarded growth and nausea.
B₁₂ (cyanocobalamin)	Only B vitamin not found in vegetables; only vitamin containing cobalt. Absorption from GI tract dependent on HCl and intrinsic factor secreted by gastric mucosa. Sources include liver, kidney, milk, eggs, cheese, meat.	Coenzyme necessary for red blood cell formation, formation of amino acid methionine, entrance of some amino acids into Krebs cycle, and manufacture of choline (chemical similar in function to acetylcholine).	Pernicious anemia and malfunction of nervous system due to degeneration of axons of spinal cord.

Exhibit 25-2 (cont.)

Vitamin	Comment and Source	Function	Related Symptoms and Deficiency Disorders
Pantothenic acid	Stored primarily in liver and kidneys. Some produced by bacteria of GI tract. Other sources include kidney, liver, yeast, green vegetables, cereal.	Constituent of co-enzyme A essential for transfer of pyruvic acid into Krebs cycle, conversion of lipids and amino acids into glucose, and synthesis of cholesterol and steroid hormones.	Experimental deficiency tests indicate fatigue, muscle spasms, neuromuscular degeneration, insufficient production of adrenal steroid hormones.
Folic acid	Synthesized by bacteria of GI tract. Other sources include green leafy vegetables and liver.	Component of enzyme systems synthesizing purines and pyrimidines built into DNA and RNA. Essential for normal production of red and white blood cells.	Production of abnormally large red blood cells—macrocytic anemia.
Biotin	Synthesized by bacteria of GI tract. Other sources include yeast, liver, egg yolk, kidneys.	Essential coenzyme for conversion of pyruvic acid to oxaloacetic acid and synthesis of fatty acids and purines.	Mental depression, muscular pain, dermatitis, fatigue, nausea.
C (ascorbic acid)	Rapidly destroyed by heat. Some stored in glandular tissue and plasma. Sources include citrus fruits, tomatoes, green vegetables.	Exact role not understood. Promotes many metabolic reactions, particularly protein metabolism, including laying down of collagen in formation of connective tissue. As coenzyme may combine with poisons, rendering them harmless until excreted. Works with antibodies.	Scurvy: many symptoms related to poor connective tissue growth and repair including tender swollen gums, loosening of teeth (alveolar processes also deteriorate), poor wound healing, bleeding (vessel walls fragile because of connective tissue degeneration), and retardation of growth. Anemia and low resistance to infection of scurvy.

falls. The top of the bell is connected by a cord running from a pulley to a writing lever, which records the level of the gas container on a rotating kymograph. As the level of the respirometer bell falls, the writing lever falls. From the record on the kymograph, the amount of oxygen consumed in a given period of time can be measured.

Basal metabolic rate is usually expressed in Calories per square meter of body surface area per hour. Suppose you use 1.8 liters of oxygen in 6 minutes as recorded on a respirometer. This means that your oxygen consumption in an hour would be 18 liters (1.8 × 10). Your basal metabolic rate would be 18 × 4.825 Calories or 86.85 Calories per hour. To express the Calories per square meter of body surface, a standardized chart is used. Such a chart shows square meters of body surface relative to height in centimeters and weight in kilograms. If you weigh 75 kg and are 190 cm tall, your body surface area is 2 cm³. Your basal metabolic rate is equal to 86.85 Calories divided by 2, or 43.43 Calories per square meter per hour.

The normal basal metabolic rates for various age groups by sex are also listed in standardized charts. Suppose a 27-year-old male has a recorded basal metabolic rate of 45.7 Calories per square meter per hour. The chart would show that his basal metabolic rate should be about 40.3 – obviously above "normal." What does this mean? The recorded basal metabolic rate is 5.4 Calories or about 14 percent above normal. Since a basal metabolic rate between +15 or −15 is considered normal, this male is judged to be normal. Values above or below 15 may indicate an excess or deficiency of thyroid hormones. When the thyroid is secreting extreme quantities of thyroid hormones, the basal metabolic rate can go as high as +100. If, on the other hand, the thyroid is secreting very little of its thyroid hormones, the basal metabolic rate may be as low as −50.

The BMR is no longer widely used to measure thyroid function because of its unreliability. It is often elevated in nonthyroid diseases such as leukemia, emphysema, and congestive heart failure, for instance, and depressed in nonthyroid diseases such as vitamin B_1 deficiency and Addison's disease.

Another test for measuring thyroid function is the *protein-bound iodine* (PBI) *test*. Recall that thyroxin is transported in the blood in combination with a plasma protein called thyroxin-binding globulin (TBG). The PBI test measures the iodine bound to TBG, which is virtually all the iodine in the body that is found in thyroxin. Thus the PBI is a fairly good measure of the amount of circulating thyroxin.

The *serum thyroxin* (T_4) *test* for measuring thyroid function directly measures the level of thyroxin in the blood. It is becoming the most common test for thyroid function. Other tests of thyroid function are the T_3 *resin uptake test,* which indirectly measures the amount of triiodothyronine (T_3) in the blood, and the *radioactive iodine* (I^{131}) *uptake test,* which measures the amount of iodine used by the thyroid gland to synthesize the thyroid hormones.

Loss of Body Heat

Body heat is produced by the oxidation of foods we eat. This heat must be removed continually or body temperature would rise steadily. The principal routes of heat loss include radiation, conduction, convection, and evaporation.

RADIATION

Radiation is the transfer of heat from one object to another without physical contact. Your body loses heat by the radiation of heat waves to cooler objects nearby such as ceilings, floors, and walls. If these objects are at a higher temperature, you absorb heat by radiation. Incidentally, the air temperature has no relationship to the radiation of heat to and from objects. Skiers can remove their shirts in bright sunshine, even though the air temperature is very low, because the radiant heat from the sun is adequate to warm them. In a room at 70°F, about 60 percent of heat loss is by radiation.

CONDUCTION

Another method of heat transfer is *conduction.* In this process, body heat is transferred to a substance or object in contact with the body, such as chairs, clothing, and jewelry.

CONVECTION

When cool air makes contact with the body, it becomes warmed and is carried away by *convection* currents. Then more cool air makes contact with the body and is carried away. The faster the air moves, the faster the rate of conduction. About 15 percent of body heat is lost to the air by conduction and about 3 percent is conducted to nearby cooler objects such as clothing or jewelry.

EVAPORATION

Evaporation means the conversion of a liquid like water to a vapor. Water has a high heat of evaporation. The *heat of evaporation* is the amount of heat necessary to evaporate 1 g of water at 30°C. Because of water's high heat of evaporation, every gram of water evaporating from the skin takes with it a great deal of heat – about 0.58 Calorie per gram of water. Under normal conditions, about 22 percent of heat loss occurs through evaporation. The evaporation of only 150 ml of water per hour is enough to remove all the heat produced by the body under basal conditions. Under extreme conditions, about 4 liters (1 gal) of sweat is produced each hour and this volume can remove 2,000 Calories of heat from the body. This is approximately 32 times the basal level of heat production. The rate of evaporation is inversely related to relative humidity, the ratio of the actual amount of moisture in the air to the greatest amount it can hold at a given temperature. The higher the relative humidity, the lower the rate of evaporation.

Temperature Regulation

If the amount of heat production equals the amount of heat loss, you maintain a constant body temperature of 98.6°F. If your heat-producing mechanisms generate more heat than is lost by your heat-losing mechanisms, your body temperature rises. If your heat-losing mechanisms are giving off more heat than is generated by heat-producing mechanisms, your temperature falls.

Body temperature is regulated by systems that attempt to keep heat production and heat loss in balance. One such regulatory system is found in the hypothalamus in a group of neurons in the anterior portion referred to as the *preoptic area*. If blood temperature increases, the neurons of the preoptic area fire impulses more rapidly. If something causes the blood's temperature to decrease, these neurons fire impulses more slowly. The preoptic area is adjusted to maintain normal body temperature and thus serves as your thermostat.

Impulses from the preoptic area are sent to other portions of the hypothalamus to control either heat production or heat loss. These other portions are known as the heat-losing center and the heat-promoting center. The *heat-losing center,* when stimulated by the preoptic area, sets into operation a series of responses that lower body temperature. The *heat-promoting center,* when stimulated by the preoptic area, sets into operation a series of responses that raise body temperature. The heat-losing center is mainly parasympathetic in function; the heat-promoting center is primarily sympathetic.

Suppose the environmental temperature is low or blood temperature falls below normal. Both stresses stimulate the preoptic area. The preoptic area, in turn, activates the heat-promoting center. In response, the heat-promoting center discharges impulses that automatically set into operation a number of responses designed to increase body heat and bring body temperature back up to normal. One such response is vasoconstriction. Impulses from the heat-promoting center stimulate sympathetic nerves that cause blood vessels of the skin to constrict. The net effect of vasoconstriction is to decrease the flow of warm blood from the internal organs to the skin, thus decreasing the transfer of heat from the internal organs to the skin. This reduction in heat loss helps raise the internal body temperature.

Another response triggered by the heat-promoting center is the sympathetic stimulation of metabolism. The heat-promoting center stimulates sympathetic nerves leading to the adrenal medulla. This stimulation causes the medulla to secrete epinephrine and norepinephrine into the blood. The hormones, in turn, bring about an increase in cellular metabolism, a reaction that also increases heat production.

Heat production is also increased by responses of skeletal muscles. For example, stimulation of the heat-promoting center causes stimulation of parts of the brain that increase muscle tone and hence heat production. In fact, shivering is caused by a high degree of muscle tone. As the muscle tone increases, the stretching of the agonist muscle initiates the stretch reflex and the muscle contracts. This contraction causes the antagonist muscle to stretch and it too develops a stretch reflex. The repetitive cycle – called shivering – increases the rate of heat production by several hundred percent.

A final body response that increases heat production is increased production of thyroxin. A cold environmental temperature causes the secretion of the regulating factor TRF produced by the preoptic area of the hypothalamus. The TRF in turn stimulates the anterior pituitary to secrete TSH, which causes the thyroid to release thyroxin into the blood. Since increased levels of thyroxin increase the metabolic rate, body temperature is increased.

Now suppose some stress raises body temperature above normal. The stress stimulates the preoptic area, which in turn stimulates the heat-losing center and inhibits the heat-promoting center. Instead of blood vessels in the skin constricting, they dilate. The skin becomes warm and the excess heat is lost to the environment. At the same time, the metabolic rate is decreased, muscle tone decreases, and there is no shivering. All these responses reverse the heat-promoting effects and bring body temperature down to normal.

When the body is subjected to high environmental temperatures or strenuous exercise, the high temperature of the blood signals the hypothalamus, which activates the heat-losing center. In response, impulses are sent out to the sweat glands of the skin and they produce more perspiration. As the perspiration evaporates from the surface of the skin, the skin is cooled and body temperature drops to normal.

Temperature Abnormalities

Most of us have probably had a *fever* – an abnormally high body temperature. The most frequent cause of fever is infection from bacteria (and their toxins) and viruses. Other conditions that result in fever are heart attacks, tumors, tissue destruction by x rays or trauma, and reactions to vaccines. The mechanism of fever production is not completely understood. However, it is believed that foreign proteins (antigens) affect the hypothalamus by setting the thermostat at a higher temperature. It has been shown that 0.001 g of protein from the bacterium that causes typhoid fever can set the thermostat as high as 110°F and the body temperature will continue to be regulated at this temperature until the protein is eliminated.

Suppose that disease-producing microbes set the thermostat at 103°F. Now the heat-promoting mechanisms (vasoconstriction, increased metabolism, shivering) are operating at full force. Thus even though body temperature is climbing higher than normal, say 101°F, the skin remains cold and shivering occurs. This condition, called a *chill,* is a definite sign that body temperature is rising. After several hours, body temperature reaches the setting of the thermostat and the chills disappear. But the body will continue to regulate temperature at 103°F until the stress is removed. When the stress is removed, the thermostat is reset at normal (98.6°F). Since body temperature remains high in the beginning, the heat-losing mechanisms (vasodilation and sweating) go into operation to decrease body temperature. The skin becomes warm and the person begins to sweat. This phase of the fever is called the *crisis* and indicates that body temperature is falling.

Up to a point, fever is beneficial. The high body temperature is believed to inhibit the growth of some bacteria and viruses. Moreover, heat speeds up the rate of chemical reactions. This increase may help body cells to repair themselves more quickly during a disease. As a rule, death results if body temperature rises to 112–114°F. (On the other end of the scale, death usually results when body temperature falls to 70–75°F.)

STUDY OUTLINE

METABOLISM

1 Nutrients are chemical substances in food that provide energy, act as building blocks in forming new body components, or assist body processes.

2 There are six major classes of nutrients: carbohydrates, lipids, proteins, minerals, vitamins, and water.

3 Metabolism has two phases: catabolism and anabolism.

4 Catabolism is the term for processes that provide energy. Digestion is a catabolic process because the breaking of bonds releases energy.

5 Anabolism consists of a series of synthetic reactions whereby small molecules are built up into larger ones that form the body's structural and functional components.

CARBOHYDRATE METABOLISM

1 Carbohydrate metabolism is primarily concerned with glucose.

2 Glucose is broken down via glycolysis (anaerobic), the Krebs cycle, and the electron transport system (aerobic) to produce energy in the form of ATP. Lack of oxygen results in oxygen debt.

3 Glucose is anabolized and stored as glycogen by the liver and skeletal muscle cells. The transformation of glucose to glycogen is called glycogenesis. Glycogen breakdown is called glycogenolysis. Conversion of lipids and proteins to glucose is referred to as gluconeogenesis.

LIPID METABOLISM

1 Fats are stored in adipose tissue, mostly in subcutaneous tissue.

2 In fat catabolism, glycerol is converted in the liver into glucose and fatty acids undergo beta oxidation and transformation into ketone bodies. Ketone bodies are transformed by nonliver cells to acetyl CoA, which enters the Krebs cycle. The presence of excess ketone bodies is called ketosis.

3 Synthesis of lipids from glucose and amino acids is called lipogenesis.

PROTEIN METABOLISM

1 Amino acids are built into proteins that serve as cell structures, enzymes, antibodies, and glandular secretions.

2 Protein catabolism involves the deamination of amino acids in the liver. Amino acids may then be converted to keto acids, pyruvic acid, and acids of the Krebs cycle.

CONTROL

1 The metabolism of carbohydrates, lipids, and proteins is controlled by a number of hormones.

2 Two major hormones, insulin and glucagon, are secreted in response to blood glucose level.

3 The anterior pituitary gland helps regulate metabolism through the actions of HGH, ACTH, and TSH.

4 The adrenal medulla, the ovaries, and the testes produce hormones that affect metabolism.

MINERALS

1 Minerals are inorganic substances that help regulate body processes.

2 Calcium and phosphorus are necessary for growth of bones and teeth. Iron and copper are used in the synthesis of hemoglobin. Iodine is necessary for thyroxin and triiodothyronine synthesis. Sodium is used in water balance and buffers. Potassium is necessary for nerve impulse transmission. Chlorine is required for acid–base balance. Magnesium is used for proper muscle and nerve functioning. Sulfur is a component of hormones.

VITAMINS

1 Vitamins are organic nutrients that regulate metabolism. Many function in enzyme systems.

2 Fat-soluble vitamins are absorbed with fats and include A, D, E, and K.

3 Water-soluble vitamins are absorbed with water and include the B vitamins and vitamin C.

4 Representative physiological functions are: vitamin A – healthy epithelium and vision; D – proper utilization of calcium; K – blood clotting; B_1, B_2, niacin, B_6 – regulation of energy metabolism; B_{12} and folic acid – blood cell formation.

METABOLISM AND HEAT

1 The Calorie is the unit of heat used to determine the caloric value of foods and measure body metabolism rates.

2 The apparatus used to determine the caloric value of foods is called a calorimeter.

3 Most body heat is a result of oxidation of the food we eat. The rate at which this heat is produced is known as the metabolic rate.

4 Metabolic rate is affected by exercise, the nervous system, hormones, and body temperature.

5 Measurement of the metabolic rate under basal conditions is called the basal metabolic rate (BMR).

6 The instrument used to determine metabolic rate indirectly is called a respirometer.

7 The principal methods of heat loss include radiation, conduction, convection, and evaporation.

8 A normal body temperature is determined by a delicate balance between heat-production and heat-loss mechanisms.

REVIEW QUESTIONS

1 Define a nutrient. List the six classes of nutrients and indicate the function of each.

2 What is metabolism? Distinguish between catabolism and anabolism and give examples of each.

3 Explain what happens to glucose during glycolysis and the Krebs cycle and the electron transport system.

4 What is the importance of lactic acid in glucose catabolism? Relate your answer to oxygen debt.

5 Define glycogenesis and glycogenolysis. Under what circumstances does each occur?

6 Why is gluconeogenesis important? Give specific examples to substantiate your answer.

7 Indicate some areas where fat is stored in the body.

8 What is a kilocalorie? Relate your definition to fat and carbohydrate catabolism.

9 Explain how glycerol participates in gluconeogenesis.

10 Define beta oxidation. What are ketone bodies?

11 What is ketosis? What is its significance to the body?

12 Define lipogenesis. Why is the process important?

13 Briefly describe the mechanism involved in protein synthesis.

14 What is the value of proteins to the body?

15 Define deamination. Explain the conversions that occur between amino acids and keto acids, pyruvic acid, and acids of the Krebs cycle.

16 Indicate the role of the following hormones in the control of metabolism: insulin, glucagon, HGH, ACTH, TSH, epinephrine, sex hormones.

17 What is a mineral? Briefly describe the functions of the following minerals: calcium, phosphorus, iron, iodine, copper, sodium, potassium, chlorine, magnesium, sulfur, zinc, fluorine, manganese, cobalt.

18 Define a vitamin. Explain how we obtain vitamins. Distinguish between a fat-soluble and a water-soluble vitamin.

19 What are the functions of vitamin A? Relate its functions to health of the epithelium, night blindness, and growth of bones and teeth.

20 How is sunlight related to vitamin D? What are the functions of vitamin D?

21 What is believed to be the principal physiological activity of vitamin E?

22 How does vitamin K function in blood clotting?

23 Relate the roles of vitamin B_1 to beriberi and polyneuritis.

24 How does vitamin B_2 function in the body?

25 Why does niacin deficiency cause pellagra?

26 Relate the role of vitamin B_6 to dermatitis.

27 How does vitamin B_{12} function in red blood cell formation?

28 What are the principal physiological effects of pantothenic acid?

29 What relationship exists between folic acid and macrocytic anemia?

30 What are the functions of biotin in the body?

31 Relate vitamin C deficiency to the symptoms of scurvy.

32 Define a calorie.

33 What is metabolic rate and how is it measured?

34 Name the principal routes or methods of heat loss.

35 How is a normal body temperature determined?

Chapter 26
The Urinary System

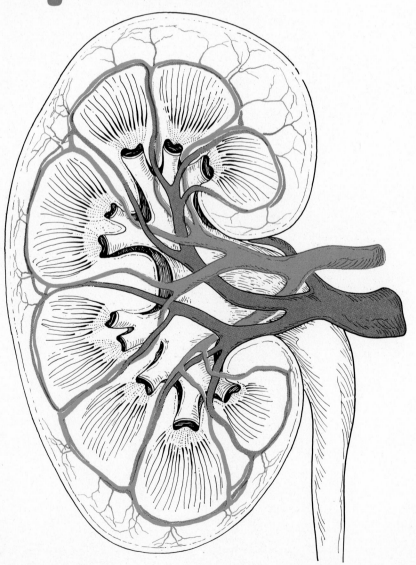

STUDENT OBJECTIVES

■ Identify the gross anatomical features of the kidneys.

■ Define the structural adaptations of a nephron for urine formation.

■ Describe the blood and nerve supply to the kidneys.

■ Describe the process of urine formation.

■ Define glomerular filtration, tubular reabsorption, and tubular secretion.

■ Compare the chemical composition of plasma, glomerular filtrate, and urine.

■ Define the forces that support and oppose the filtration of blood in the kidneys.

■ Discuss renal suppression as a disorder resulting from a decreased filtration pressure.

■ Describe the physiological role of tubular reabsorption.

■ Compare the obligatory and facultative reabsorption of water.

■ Describe tubular excretion as a mechanism of elimination and a control of blood pH.

■ Define kidney excretion of H^+ and NH_4^+ as a means of maintaining body pH while conserving bicarbonate.

■ Describe the countercurrent multiplier mechanism.

■ Compare the lungs, integument, and alimentary canal as organs of excretion that help maintain body pH.

■ Describe the effects of blood pressure, diet, temperature, and emotions on urine production.

■ List the physical characteristics of urine.

■ List the normal chemical constituents of urine.

■ Define albuminuria, glycosuria, hematuria, pyuria, ketosis, casts, and calculi.

■ Describe the structure and physiology of the ureters.

■ Describe the physiology of micturition.

■ Compare the causes of incontinence, retention, and suppression.

■ Describe the structure and physiology of the urethra.

■ Discuss the causes of ptosis, kidney stones, gout, glomerulonephritis, pyelitis, and cystitis.

■ Discuss the operational principle of hemodialysis.

■ Define medical terminology associated with the urinary system.

The metabolism of nutrients results in the production of wastes by body cells, including carbon dioxide and excess water and heat. Protein catabolism produces toxic nitrogenous wastes such as ammonia and urea. In addition, many of the essential ions such as sodium, chloride, sulfate, phosphate, and hydrogen tend to accumulate in the body. All the toxic materials and the excess essential materials must be eliminated.

The primary function of the **urinary system** is to keep the body in homeostasis by controlling the concentration and volume of blood. It does so by removing and restoring selected amounts of water and solutes. It also excretes selected amounts of various wastes. Two kidneys, two ureters, one urinary bladder, and a single urethra make up the system (Figure 26-1). The kidneys regulate the concentration and volume of the blood and remove wastes from the blood in the form of urine. Urine is excreted from each kidney through its ureter and is stored in the urinary bladder until it is expelled from the body through the urethra. Other systems that aid in waste elimination are the respiratory, integumentary, and digestive systems.

KIDNEYS

The paired **kidneys** are reddish organs that resemble kidney beans in shape. They are found just above the waist between the parietal peritoneum and the posterior wall of the abdomen. Since they are external to the peritoneal lining of the abdominal cavity, their placement is described as *retroperitoneal*. Relative to the vertebral column, the kidneys are located between the levels of the last thoracic and third lumbar vertebrae. The right kidney is slightly lower than the left because of the large area occupied by the liver.

External Anatomy

The average adult kidney measures about 11.25 cm (4 inches) long, 5.0 to 7.5 cm (2 to 3 inches) wide, and 2.5 cm (1 inch) thick. Its concave medial border faces the vertebral column. Near the center of the concave border is a notch called the *hilum* through which the ureter leaves the kidney. Blood and lymph vessels and nerves also enter and exit the kidney through the hilum. The hilum is the entrance to a cavity in the kidney called the *renal sinus*.

Three layers of tissue surround each kidney. The innermost layer, the *renal capsule*, is a smooth, transparent, fibrous membrane that can easily be stripped off the kidney and is continuous with the outer coat of the ureter at the hilum. It serves as a barrier against trauma and the spread of infection to the kidney. The second layer, the *adipose capsule*, is a mass of fatty tissue surrounding the renal capsule. It also protects the kidney from trauma and holds it firmly in place in the abdominal cavity. The outermost layer, the *renal fascia*, is a thin layer of fibrous connective tissue that anchors the kidneys to their surrounding structures and to the abdominal wall. Some individuals, especially thin people in whom either the adipose capsule or renal fascia is deficient, may develop *ptosis* (dropping) of one or both kidneys. Ptosis is dangerous because it may cause kinking of the ureter with reflux of urine and retrograde pressure. Ptosis of the kidneys below the rib cage also makes these organs susceptible to blows and penetrating injuries.

Internal Anatomy

A coronal (frontal) section through a kidney reveals an outer, reddish area called the *cortex* and an inner, reddish brown region called the *medulla* (Figure 26-2). The cortex is arbitrarily divided into an outer cortical zone and an inner juxtamedullary zone. Likewise, the medulla is divided into an outer zone (one-third) and an inner zone (two-thirds). Within the medulla are 8 to 18 striated, triangular structures termed *renal*, or *medullary*, *pyramids*. The striated appearance is due to the presence of straight tubules and blood vessels. The bases of the pyramids face the cortical area, and their apices, called *renal papillae*, are directed toward the center of the kidney. The cortex is the smooth-textured area extending from the renal capsule to the bases of the pyramids and into the spaces between them. The cortical substance between the renal pyramids forms the *renal columns*. Together the cortex and renal pyramids constitute the parenchyma of the kidney. Structurally, the parenchyma of each kidney consists of approximately 1 million microscopic units called nephrons, collecting ducts, and their associated vascular supply. Nephrons are the functional units of the kidney. They help form the urine and regulate blood composition.

In the renal sinus of the kidney is a large cavity called the *renal pelvis*. The edge of the pelvis contains cuplike extensions called the *major* and *minor calyces*. There are 2 or 3 major calyces and 7 to 13 minor calyces. Each minor calyx collects urine from collecting ducts of the pyramids. From the major calyces, the urine drains into the pelvis and out through the ureter.

Nephron

The physiological unit of the kidney is the **nephron** (Figure 26-3, p. 604). Essentially, a nephron is a *renal* *tubule* and its vascular component. A nephron begins as a double-walled globe, called *Bowman's (glomerular) capsule,* lying in the cortex of the kidney. The inner layer or wall of the capsule, known as the visceral layer, consists of epithelial cells called podocytes. The visceral layer surrounds a capillary network called the *glomerulus.* A space separates the inner wall from the outer one known as the parietal layer. The parietal layer is composed of simple squamous epithelium. Collectively, Bowman's capsule and the enclosed glomerulus constitute a *renal corpuscle.*

The visceral layer of Bowman's capsule and the endothelium of the glomerulus form an **endothelial-capsular membrane**. This membrane consists of the following parts in the order in which substances filtered by the kidney must pass through (Figure 26-4, p. 606).

1 *Endothelium of the glomerulus* This single layer of endothelial cells has completely open pores averaging 500–1,000 Å in diameter.

2 *Basement membrane* This membrane lies beneath the endothelium and contains no pores. It consists of fibrils in a mucopolysaccharide matrix. It serves as the dialyzing membrane.

3 *Epithelium of the visceral layer of Bowman's capsule* These epithelial cells, because of their peculiar shape, are called *podocytes.* The podocytes contain foot-like structures called *pedicels.* The pedicels are arranged parallel to the circumference of the glomerulus and cover the basement membrane except for spaces between them called *filtration slits* or *slit pores.*

The endothelial-capsular membrane filters water and solutes in the blood. Large molecules, such as proteins, and the formed elements in blood do not normally pass through it. The substances that are filtered pass into the space between the visceral and parietal layers of Bowman's capsule. The filtered fluid moves into the renal tubule.

Bowman's capsule opens into the *proximal convoluted tubule,* which also lies in the cortex. Convoluted means the tubule is coiled rather than straight; proximal signifies that this portion of the tubule originates at the Bowman's capsule. The wall of the proximal convoluted tubule consists of cuboidal epithelium with microvilli. These cytoplasmic extensions, like those of the small intestine, increase the surface area of reabsorption and secretion.

The next section of the tubule, the *descending limb of Henle,* dips into the medulla. It consists of squamous epithelium. The tubule then bends into a U-shaped structure called the *loop of Henle.* As the tubule straightens, it increases in diameter and ascends toward

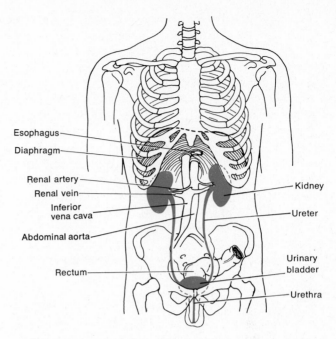

Figure 26–1   Organs of the male urinary system.

the cortex as the *ascending limb of Henle,* which consists of cuboidal and low columnar epithelium. In the cortex the tubule again becomes convoluted. Because of its distance from the point of origin at the Bowman's capsule, this section is referred to as the *distal convoluted tubule.* The cells of the distal tubule, like those of the proximal tubule, are cuboidal. Unlike the cells of the proximal tubule, however, the cells of the distal tubule have few microvilli. The distal tubule terminates by merging with a straight *collecting duct.* In the medulla, the collecting ducts receive the distal tubules of several nephrons, pass through the renal pyramids, and open into the calyces of the pelvis through a number of large *papillary ducts.* Cells of the collecting ducts are cuboidal; those of the papillary ducts are columnar.

Nephrons are frequently classified into two kinds. A *cortical nephron* usually has its glomerulus in the outer cortical zone, and the remainder of the nephron rarely penetrates the medulla. A *juxtamedullary nephron* usually has its glomerulus close to the corticomedullary junction, and other parts of the nephron penetrate deeply into the medulla (Figure 26-3).

Blood and Nerve Supply

The nephrons are partly responsible for removing wastes from the blood and regulating its fluid and electrolyte content. Thus they are abundantly supplied with blood vessels. The right and left *renal arteries* transport about

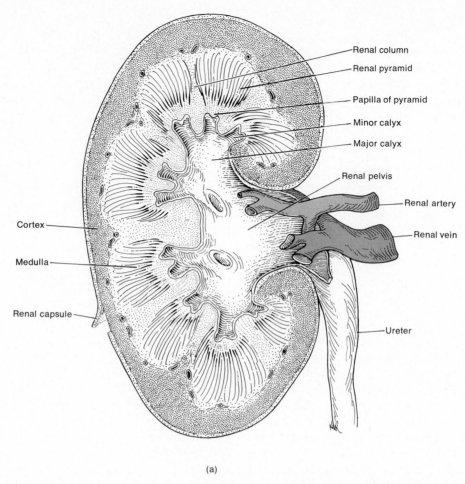

(a)

Figure 26–2   Kidney. (a) Coronal section through the right kidney illustrating gross internal anatomy.

one-fourth the total cardiac output to the kidneys (Figure 26-5, p. 607). Approximately 1,200 ml passes through the kidneys every minute. Before or immediately after entering through the hilum, the renal artery divides into several branches that enter the parenchyma and pass as the *interlobar arteries* between the renal pyramids in the renal columns. At the base of the pyramids, the interlobar arteries arch between the medulla and cortex and become known as the *arcuate arteries*. Divisions of the arcuate arteries produce a series of *interlobular arteries* which enter the cortex and divide into *afferent arterioles* (Figure 26-3). One afferent arteriole is distributed to each glomerular capsule, where the arteriole divides into the capillary network termed the *glomerulus*. The glomerular capillaries then reunite to form an *efferent arteriole,* leading away from the capsule, that is smaller in diameter than the afferent arteriole. This variation in diameter helps raise the glomerular pressure. The afferent-efferent arteriole situation is

unique because blood usually flows out of capillaries into venules and not into other arterioles. Each efferent arteriole of a cortical nephron divides to form a network of capillaries, called the *peritubular capillaries,* around the convoluted tubules. The efferent arteriole of a juxtamedullary nephron also forms peritubular capillaries. In addition, it forms long loops of thin-walled vessels called *vasa recti* that dip down alongside the loop of Henle into the medullary region of the papilla. The peritubular capillaries eventually reunite to form *interlobular veins*. The blood then drains through the *arcuate veins* to the *interlobar veins* running between the pyramids and leaves the kidney through a single *renal vein* that exists at the hilum. The vasa recta pass blood into the interlobular veins. From here, it goes to the arcuate veins, the interlobar veins, and then into the renal vein.

The nerve supply to the kidneys is derived from the *renal plexus* of the autonomic system. Nerves from the

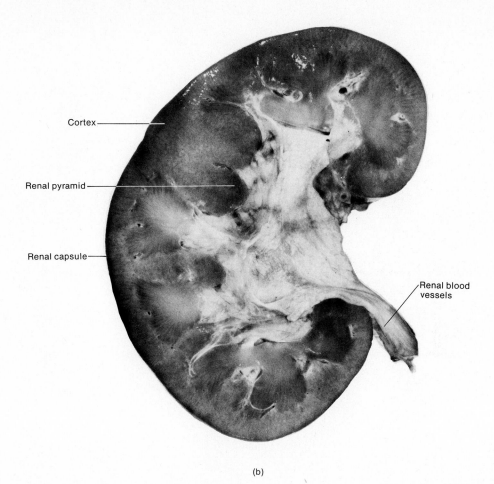

Cortex

Renal pyramid

Renal capsule

Renal blood vessels

(b)

Figure 26–2 (cont.) Kidney. (b) Photograph of a kidney in coronal section. (Courtesy of Carol Kerr, Letterman Army Medical Center, San Francisco.)

plexus accompany the renal arteries and their branches and are distributed to the vessels. Because the nerves are vasomotor, they regulate the circulation of blood in the kidney by regulating the diameters of the small blood vessels.

Juxtaglomerular Apparatus

As the afferent arteriole approaches the renal corpuscle, the smooth muscle cells of the tunica media become modified. Their nuclei become rounded (instead of elongated), and their cytoplasm contains granules (instead of myofibrils). Such modified cells are called *juxtaglomerular cells*. The cells of the distal convoluted tubule adjacent to the afferent arteriole become considerably narrower. These cells are known as the *macula densa*. Together with the modified cells of the afferent arteriole they constitute the **juxtaglomerular apparatus** (Figure 26-6, p. 608).

Physiology

The major work of the urinary system is done by the nephrons. The other parts of the system are primarily passageways and storage areas. Nephrons carry out three important functions. They control blood concentration and volume by removing selected amounts of water and solutes. They help regulate blood pH. And they remove toxic wastes from the blood. As the nephrons go about these activities, they remove many materials from the blood, return the ones that the body requires, and eliminate the remainder. The eliminated materials are collectively called urine. The entire volume of blood in the body is filtered by the kidneys 60 times a day. Urine formation requires three principal processes: glomerular filtration, tubular reabsorption, and tubular secretion.

GLOMERULAR FILTRATION

The first step in the production of urine is called **glomerular filtration.** Filtration – the forcing of fluids

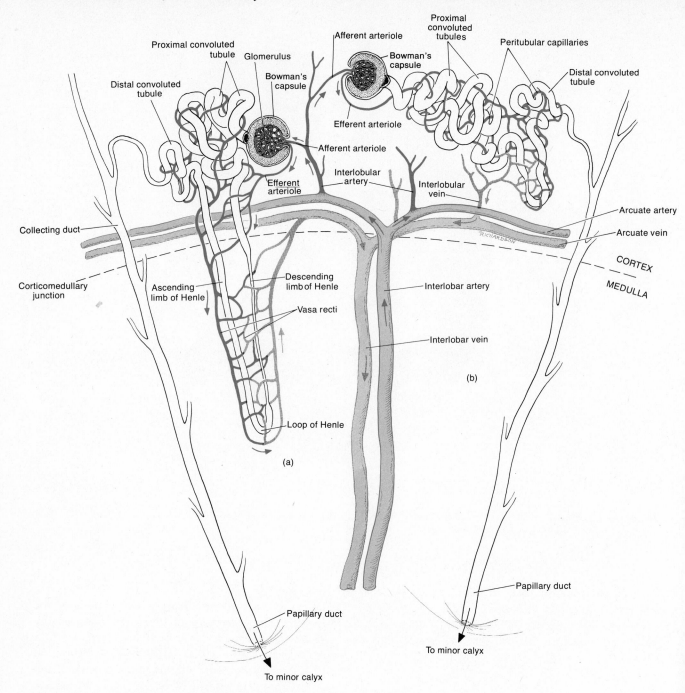

Figure 26–3   Nephrons. (a) Juxtamedullary nephron. (b) Cortical nephron.

and dissolved substances through a membrane by an outside pressure—occurs in the renal corpuscle of the kidneys across the endothelial-capsular membrane. When blood enters the glomerulus, the blood pressure forces water and dissolved blood components (plasma) through the walls of the capillaries, basement membrane, and on through the adjoining visceral wall of Bowman's capsule (Figure 26-7a, p. 609). The resulting fluid is the *filtrate*. In a healthy person, the filtrate consists of all the materials present in the blood except for the formed elements and most proteins, which are too large to pass through the endothelial-capsular-barrier. Exhibit 26-1 (p. 610) compares the constituents of plasma, glomerular filtrate, and urine during a 24-hour period. Although the values shown are typical, they vary considerably according to diet. The chemicals listed in plasma are those present in glomerular blood plasma before filtration. The chemicals listed in the filtrate immediately after Bowman's capsule are those that pass from the glomerular blood plasma through the endothelial-capsular membrane before reabsorption. The chemicals in the filtrate are the ones that have been filtered.

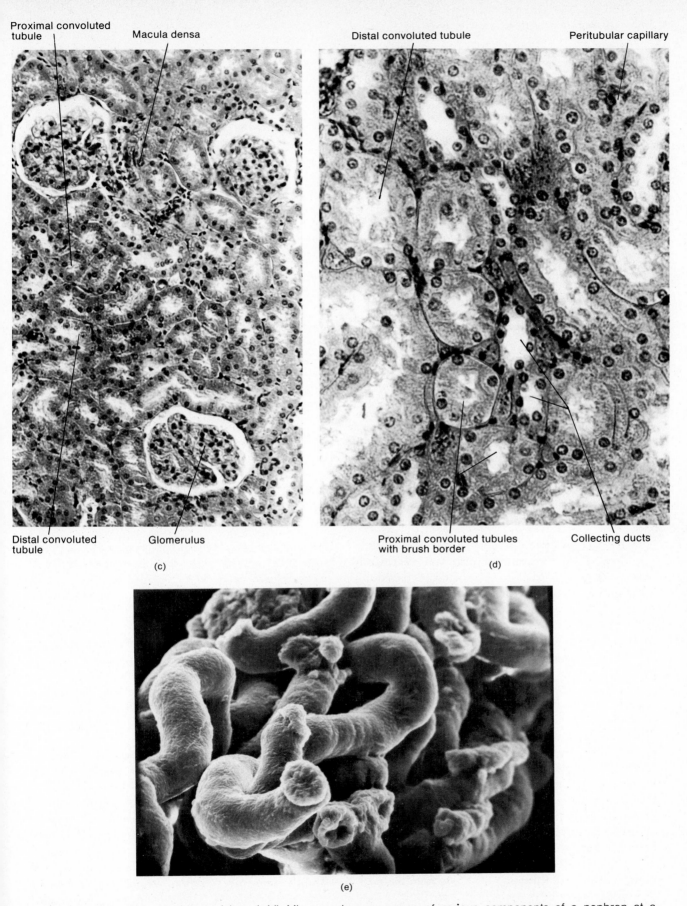

Proximal convoluted tubule

Macula densa

Distal convoluted tubule

Peritubular capillary

Distal convoluted tubule

Glomerulus

(c)

Proximal convoluted tubules with brush border

Collecting ducts

(d)

(e)

Figure 26–3 (cont.) Nephrons. (c) and (d) Microscopic appearance of various components of a nephron at a magnification of 100×. (Courtesy of Victor B. Eichler, Wichita State University.) (e) Scanning electron micrograph of renal tubules at a magnification of 500×. (Courtesy of Fisher Scientific Company and S.T.E.M. Laboratories, Inc., Copyright 1975.)

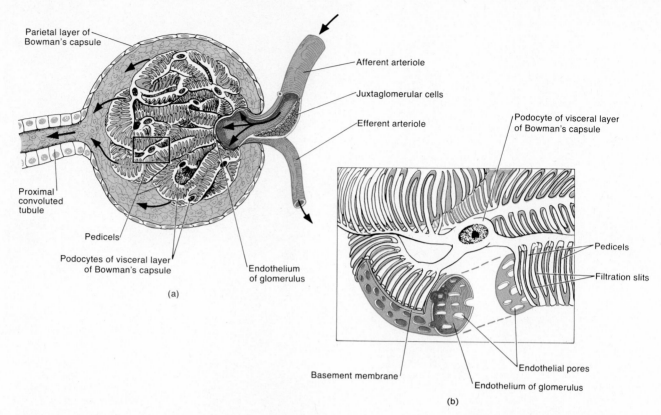

Parietal layer of Bowman's capsule

Afferent arteriole

Juxtaglomerular cells

Efferent arteriole

Podocyte of visceral layer of Bowman's capsule

Proximal convoluted tubule

Pedicels

Podocytes of visceral layer of Bowman's capsule

Endothelium of glomerulus

(a)

Pedicels

Filtration slits

Basement membrane

Endothelial pores

Endothelium of glomerulus

(b)

Figure 26–4 Endothelial-capsular membrane. (a) Overview of a renal corpuscle. (b) Enlarged aspect of a portion of the endothelial-capsular membrane.

Renal corpuscles are especially structured for filtering blood. First, each capsule contains a tremendous length of highly coiled glomerular capillaries presenting a vast surface area for filtration. Second, the endothelial-capsular membrane is structurally adapted for filtration. Although the endothelial pores generally do not restrict the passage of substances, the basement membrane permits the passage of smaller molecules. Thus water, glucose, vitamins, amino acids, small proteins, nitrogenous wastes, and ions pass into the Bowman's capsule. Large proteins and the formed elements in blood do not normally pass through the basement membrane. The filtration slits permit only the occasional passage of very small plasma proteins such as albumins. Third, the efferent arteriole is smaller in diameter than the afferent arteriole, so great resistance to the outflow of blood from the glomerulus is established. Consequently, blood pressure is higher in the glomerular capillaries than in other capillaries. Glomerular blood pressure averages 75 mm Hg, whereas the blood pressure of other capillaries averages only 30 mm Hg. Fourth, the endothelial-capsular membrane separating the blood from the space in Bowman's capsule is very thin (0.1 μm).

The filtering of the blood depends on a number of opposing pressures. The chief one is the *glomerular blood*

hydrostatic pressure. **Hydrostatic** (*hydro* = water) **pressure** is the force that a fluid under pressure exerts against the walls of its container. Glomerular blood hydrostatic pressure means the blood pressure in the glomerulus. This pressure tends to move fluid out of the glomeruli at a force averaging 75 mm Hg.

However, glomerular blood hydrostatic pressure is opposed by two other forces. The first of these, *capsular hydrostatic pressure,* develops in the following way. When the filtrate is forced into the space between the walls of Bowman's capsule, it meets with two forms of resistance: the walls of the capsule and the fluid that has partly filled the renal tubule. As a result, some filtrate is pushed back into the capillary. The amount of "push" is the capsular hydrostatic pressure. It usually measures about 20 mm Hg (Figure 26-7c).

The second force opposing filtration into the Bowman's capsule is the *blood osmotic pressure. Osmotic pressure* is the pressure that develops because of water movement into a contained solution. Suppose we place a cell in a hypotonic solution. As water moves from the solution into the cell, the volume inside the cell increases, forcing the cell membrane outward (Figure 26-7b). Osmotic pressure always develops in the solution with the higer concentration of solutes. Hydrostatic

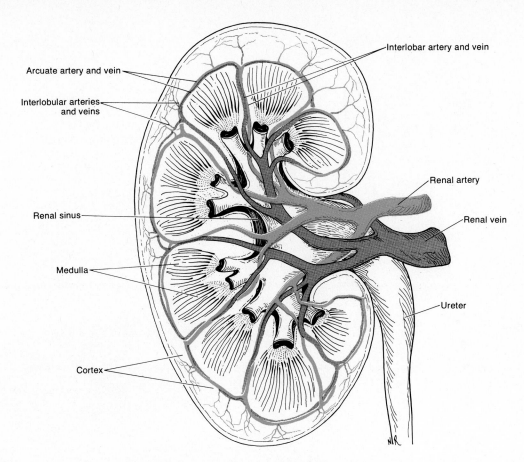

Arcuate artery and vein

Interlobar artery and vein

Interlobular arteries and veins

Renal artery

Renal sinus

Renal vein

Medulla

Ureter

Cortex

Figure 26–5   Blood supply of the right kidney seen in coronal section.

pressure develops because of a force outside a solution. Osmotic pressure develops because of the concentration of the solution itself. Since the blood contains a much higher concentration of proteins than the filtrate does, water moves out of the filtrate and back into the blood vessel. This blood osmotic pressure is normally about 30 mm Hg.

To determine how much filtration finally occurs, we have to subtract the forces that oppose filtration from the glomerular blood hydrostatic pressure. The net result is called the **effective filtration pressure**, which is abbreviated P_{eff}:

$$P_{\text{eff}} = \begin{pmatrix} \text{glomerular} \\ \text{blood} \\ \text{hydrostatic} \\ \text{pressure} \end{pmatrix} - \begin{pmatrix} \text{capsular} & \text{blood} \\ \text{hydrostatic} + \text{osmotic} \\ \text{pressure} & \text{pressure} \end{pmatrix}$$

By substituting the values just discussed, a normal P_{eff} may be calculated as follows:

$$P_{\text{eff}} = (75 \text{ mm Hg}) - (20 \text{ mm Hg} + 30 \text{ mm Hg})$$
$$= (75 \text{ mm Hg}) - (50 \text{ mm Hg})$$
$$= 25 \text{ mm Hg}$$

This means that a pressure of 25 mm Hg causes a normal amount of plasma to filter from the glomerulus into the Bowman's capsule. This is about 125 ml of filtrate per minute.

Certain conditions may alter these pressures and thus the P_{eff}. In some forms of kidney disease the glomerular capillaries become so permeable that the plasma proteins are able to pass from the blood into the filtrate. As a result, the capsular filtrate exerts an osmotic pressure that draws water out of the blood. Thus, if a capsular osmotic pressure develops, the P_{eff} will increase. At the same time, blood osmotic pressure decreases, further increasing the P_{eff}.

The P_{eff} also is affected by changes in the general arterial blood pressure. Severe hemorrhaging produces a drop in general blood pressure that also decreases the glomerular blood hydrostatic pressure. If the blood pressure falls to the point where the hydrostatic pressure in the glomeruli reaches 60 mm Hg, no filtration occurs because the glomerular blood hydrostatic pressure equals the opposing forces. Such a condition is called *renal suppression*.

A final factor that may affect the P_{eff} is the regulation of the size of the afferent and efferent arterioles. In this case, glomerular blood hydrostatic

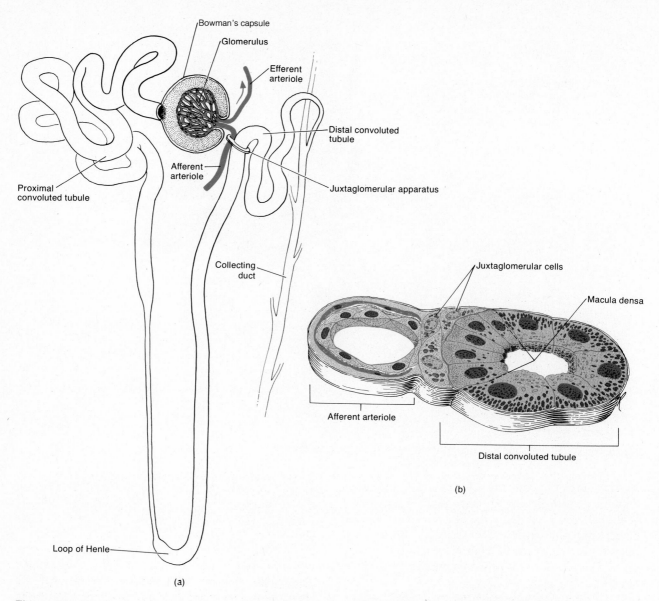

Figure 26—6 Juxtaglomerular apparatus. (a) External view. (b) The cells of the juxtaglomerular apparatus.

pressure is regulated separately from the general blood pressure. Sympathetic impulses and small doses of epinephrine cause constriction of both afferent and efferent arterioles. However, intense sympathetic impulses and large doses of epinephrine cause greater constriction of afferent than efferent arterioles. This intense stimulation results in a decrease in glomerular hydrostatic pressure even though blood pressure in other parts of the body may be normal or even higher than normal. Intense sympathetic stimulation is most likely to occur during the alarm reaction of the general adaptation syndrome. Blood may also be shunted away from the kidneys during hemorrhage.

TUBULAR REABSORPTION

The amount of filtrate that flows out of all the renal corpuscles of both kidneys every minute is called the glomerular filtration rate (GFR). In the normal adult, this rate is about 125 ml/minute—about 180 liters (45 gal) a day. But as the filtrate passes through the renal tubules, it is reabsorbed into the blood at a rate of about 123–124 ml/minute—about 178–179 liters a day. Thus only about 1 percent of the filtrate actually leaves the body—about 1–2 ml/minute or 1–2 liters/day. The movement of the filtrate back into the blood of the peritubular capillaries or vasa recta is called **tubular reabsorption.** Tubular reabsorption is carried out by

accomplishing dialysis is the kidney machine (Figure 26-14). A tube connects it with the patient's radial artery. The blood is pumped from the artery and through the tubes to one side of a semipermeable cellophane membrane. The other side of the membrane is continually washed with an artificial solution called the dialyzing solution. The blood that passes through the artificial kidney is treated with an anticoagulant. Only about 500 ml of the patient's blood is in the machine at a time. This volume is easily compensated for by vasoconstriction and increased cardiac output.

All substances (including wastes) in the blood except protein molecules and blood cells can diffuse back and forth across the semipermeable membrane. The electrolyte level of the plasma is controlled by keeping the dialyzing solution electrolytes at the same concentration found in normal plasma. Any excess plasma electrolytes move down the concentration gradient and into the dialyzing solution. If the plasma electrolyte level is normal, it is in equilibrium with the dialyzing solution and no electrolytes are gained or lost. Since the dialyzing solution contains no wastes, substances such as urea move down the concentration gradient and into the dialyzing solution. Thus wastes are removed and normal electrolyte balance is maintained.

A great advantage of the kidney machine is that nutrition can be bolstered by placing large quantities of glucose in the dialyzing solution. While the blood gives up its wastes, the glucose diffuses into the blood. Thus the kidney

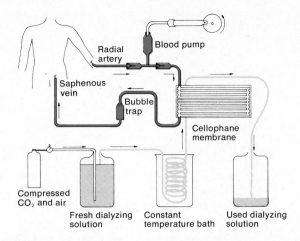

Figure 26–14 Operation of an artificial kidney. The blood route is indicated in color. The route of the dialyzing solution is indicated in gray.

machine beautifully accomplishes the principal function of the fundamental unit of the kidney—the nephron.

There are obvious drawbacks to the artificial kidney, however. The blood must be anticoagulated during dialysis, and a large amount of the patient's blood must flow through this apparatus to make it work. To date, no artificial kidney has been capable of becoming a permanent implant.

MEDICAL TERMINOLOGY

Cystoscope (*cyst* = bladder; *scope* = to view)   Instrument used for examination of the urinary bladder.

Dysuria (*dys* = painful)   Painful urination.

Nephrosis (*neph* = kidney)   Any disease of the kidney but usually one that is degenerative.

Oliguria (*olig* = scanty)   Scanty urine.

Polyuria (*poly* = much)   Excessive urine.

Stricture   Narrowing of the lumen of a canal or hollow organ, as the ureter or urethra.

Uremia (*emia* = condition of blood)   Toxic levels of urea in the blood resulting from severe malfunction of the kidneys.

STUDY OUTLINE

KIDNEYS
1 Three layers of tissue surround each kidney: the renal capsule, the adipose capsule, and the renal fascia.
2 Nephrons are the functional units of the kidneys. They help form urine and regulate blood composition.
3 A nephron is made up of a renal corpuscle and a long tubule composed of functionally different regions.

4 The kidneys form urine by glomerular filtration, tubular reabsorption, and tubular secretion.
5 The primary force behind filtration is hydrostatic pressure.
6 If the hydrostatic pressure falls to 50 mm Hg, renal suppression occurs because the glomerular blood hydrostatic

pressure exactly equals the opposing pressures (capsular hydrostatic pressure and blood osmotic pressure).

7 Most substances in plasma are filtered by the Bowman's capsule. Normally, blood cells and proteins are not filtered.

8 Tubular reabsorption retains substances needed by the body, including water, glucose, amino acids, and ions.

9 Around 80 percent of the reabsorbed water is returned by obligatory reabsorption, the rest by facultative reabsorption.

10 Chemicals not needed by the body are discharged into the urine by tubular secretion.

11 The kidneys help maintain blood pH by excreting H^+ and NH_4^+ ions. In exchange, the kidneys conserve sodium bicarbonate.

12 The ability of the kidneys to produce either hyperosmotic or hyposmotic urine is based on the countercurrent multiplier mechanism.

HOMEOSTASIS

1 The primary homeostatic function of the urinary system is to regulate the concentration and volume of blood by removing and restoring selected amounts of water and solutes. It also excretes wastes to maintain homeostasis.

2 Besides the kidneys, the lungs, integument, and alimentary canal assume excretory functions.

3 Urine volume is influenced by blood pressure, diet, diuretics, temperature, and emotions.

4 The physical characteristics of urine evaluated in a urinalysis are color, odor, turbidity, pH, and specific gravity.

5 Chemically, normal urine contains about 95 percent water and 5 percent solutes. The solutes include urea, creatinine, uric acid, indican, acetone bodies, salt, and ions.

6 Abnormal conditions diagnosed through urinalysis include albuminuria, glycosuria, hematuria, pyuria, ketosis, casts, and calculi.

URETERS, URINARY BLADDER, AND URETHRA

1 The paired ureters convey urine from the kidneys to the urinary bladder, mostly by peristaltic contraction.

2 The urinary bladder stores urine. The expulsion of urine from the bladder is called micturition.

3 Lack of control over micturition is called incontinence. Failure to void urine is referred to as retention. Inability of the kidneys to produce urine is called suppression.

4 The urethra extends from the floor of the bladder to the exterior and discharges urine from the body.

REVIEW QUESTIONS

1 What are the functions of the urinary system? What organs compose the system?

2 Describe the location of the kidneys. Why are they said to be retroperitoneal?

3 Prepare a labeled diagram that illustrates the principal external and internal features of the kidney.

4 What is a nephron? List and describe the parts of a nephron from the Bowman's capsule to the collecting duct.

5 How are nephrons supplied with blood?

6 What is glomerular filtration? Define the filtrate.

7 Set up an equation to indicate how effective filtration pressure is calculated. What is the cause of renal suppression?

8 What are the major chemical differences among plasma, filtrate, and urine?

9 Define tubular reabsorption. Why is the process physiologically important?

10 What chemical substances are normally reabsorbed by the kidneys?

11 Describe how glucose and sodium are reabsorbed by the kidneys. Where does the process occur?

12 How is chloride reabsorption related to sodium reabsorption?

13 Distinguish obligatory from facultative reabsorption of water. How is facultative reabsorption controlled?

14 Define tubular secretion. Why is it important?

15 Explain the mechanisms by which the kidneys help to control body pH.

16 Define the countercurrent multiplier mechanism.

17 Contrast the functions of the lungs, integument, and alimentary tract as excretory organs.

18 What is urine? Describe the effects of blood pressure, blood concentration, temperature, diuretics, and emotions on the volume of urine formed.

19 Describe the following physical characteristics of normal urine: color, turbidity, pH, and specific gravity.

20 Describe the chemical composition of normal urine.

21 Define each of the following: albuminuria, glycosuria, hematuria, pyuria, ketosis, casts, calculi.

22 Describe the structure and function of the ureters.

23 How is the urinary bladder adapted to its storage function? What is micturition?

24 Contrast the causes of incontinence, retention, and suppression.

25 Compare the urethra in the male and female.

Chapter 27
Fluid, Electrolyte, and Acid-Base Dynamics

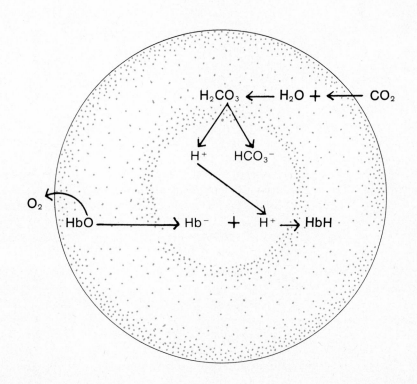

STUDENT OBJECTIVES

■ Define a body fluid.

■ Distinguish between intracellular fluid and extracellular fluid.

■ Define the processes available for fluid intake and fluid output.

■ Compare the mechanisms regulating fluid intake and fluid output.

■ Compare the effects of nonelectrolytes and electrolytes on body fluids.

■ Calculate the concentration of ions in a body fluid.

■ Contrast the electrolytic composition of the three major fluid compartments.

■ Describe electrolyte functions and regulation.

■ Define the factors involved in the movement of fluid between plasma and interstitial fluid and between interstitial fluid and intracellular fluid.

■ Define the relationship between electrolyte imbalance and fluid imbalance.

■ Describe the role of buffers, respirations, and kidney excretion in maintaining body pH.

■ Define acid–base imbalances and their effects on the body.

■ Describe appropriate treatments for acidosis and alkalosis.

The term **body fluid** refers to the body water and its dissolved substances. Fluid composes 45 to 75 percent of the body weight.

FLUID COMPARTMENTS

About two-thirds of the fluid is located in cells and is termed **intracellular fluid** or **ICF.** The other third, called **extracellular fluid**, or **ECF**, encompasses all the rest of the body fluids – interstitial fluid, plasma and lymph, cerebrospinal fluid, GI tract fluids, synovial fluid, the fluids of the eyes and ears, pleural, pericardial, and peritoneal fluids, and the glomerular filtrate. Body fluids are separated into distinct compartments whose walls are the semipermeable membranes provided by the plasma membranes of cells. A compartment may be as small as the interior of a single cell or as large as the combined interiors of the heart and vessels. Fluids exist in compartments, but keep in mind that they are in constant motion from one compartment to another. In the healthy individual, the volume of fluid in each compartment remains stable – another example of homeostasis.

Water comprises the bulk of all the body fluids. When we say that the body is in *fluid balance,* we mean it contains the required amount of water distributed to the various compartments according to their needs.

Osmosis is the primary way in which water moves in and out of body compartments. The concentration of solutes in the fluids is therefore a major determinant of fluid balance. Most solutes in body fluids are electrolytes – compounds that dissociate into ions. Fluid balance, then, means water balance, but it also implies electrolyte balance. The two are inseparable.

WATER

Water is by far the largest single constituent of the body, comprising from 45 to 75 percent of the total body weight. The percentage varies from person to person and depends primarily on the amount of fat present and age. Since fat is basically water-free, lean people have a greater proportion of water to total body weight than fat people. Water proportion also decreases with age. An infant has the highest amount of water per body weight. In an average adult male, water constitutes about 65 percent of the body weight, and in a female it constitutes about 55 percent. Females have more subcutaneous fat than males.

Fluid Intake and Output

The primary source of body fluid is water derived from ingested liquids (1,600 ml) and foods (700 ml) that has been absorbed from the alimentary canal. This water amounts to about 2,300 ml daily. Another source of fluid is metabolic water, the water produced through catabolism. This amounts to about 200 ml daily. Thus total fluid input averages about 2,500 ml/day.

There are several avenues of fluid output. The kidneys lose about 1,500 ml/day, the skin about 500 ml/day, the lungs about 300 ml/day, and the GI tract about 200 ml/day. Fluid output thus totals 2,500 ml/day. Under normal circumstances, fluid intake equals fluid output so the body maintains a constant volume of fluid.

Regulation of Intake

Fluid intake is regulated by thirst. When water loss is greater than water intake, the *dehydration* stimulates thirst through both local and general responses. Locally, it leads to a decrease in the flow of saliva that produces a dryness of the mucosa of the mouth and pharynx (Figure 27-1). Dryness is interpreted by the brain as a sensation of thirst. In addition, dehydration raises blood osmotic pressure. It is believed that receptors in the thirst center of the hypothalamus are stimulated by the increase in osmotic pressure and initiate impulses that also are interpreted as a sensation of thirst. The response, a desire to drink fluids, thus balances the fluid loss.

The initial quenching of thirst results from wetting the mucosa of the mouth and pharynx, but the major inhibition of thirst is believed to occur as a result of distension of the intestine. Apparently, stimulated stretch receptors in the walls of the intestine send impulses that inhibit the thirst center in the hypothalamus.

Regulation of Output

Under normal circumstances, fluid output is adjusted by ADH and aldosterone, both of which regulate urine production. Under abnormal conditions, other factors may influence output heavily. If the body is dehydrated, blood pressure falls, glomerular filtration rate decreases accordingly, and water is conserved. Conversely, excessive blood fluid results in an increase of blood pressure, glomerular filtration rate, and fluid output. Hypertension produces the same effect. Hyperventilation leads to increased fluid output through the loss of water vapor by the lungs. Vomiting and diarrhea result in fluid loss from the gastrointestinal tract. Finally, fever and destruction of extensive areas of the skin bring about excessive water loss through the skin.

ELECTROLYTES

The body fluids contain a variety of dissolved chemicals. Some of these chemicals are compounds with covalent bonds—that is, the elements that compose the molecule share electrons and do not form ions. Such compounds are called **nonelectrolytes.** Nonelectrolytes include most organic compounds, such as glucose, urea, and creatine. Other compounds, called **electrolytes,** have at least one ionic bond. When they dissolve in the body fluid, they dissociate into positive and negative ions. The positive ions are called cations; the negative ions are anions. Acids, bases, and salts are electrolytes. Most electrolytes are inorganic compounds, but a few are organic. For example, some proteins form ionic bonds. When the protein is put in solution, the ion detaches and the rest of the protein molecule carries the opposite charge.

Electrolytes serve three general functions in the body. First, many are essential minerals. Second, they control the osmosis of water between body compartments. Third, they help maintain the acid–base balance required for normal cellular activities.

Concentration

During osmosis, water moves to the area with the greater number of particles in solution. A particle may be a whole molecule or an ion. An electrolyte exerts a far greater effect on osmosis than a nonelectrolyte because an electrolyte molecule dissociates into at least two particles, both of them charged. Suppose the nonelectrolyte glucose and two electrolytes are placed in solution:

$$C_6H_{12}O_6 \xrightarrow{H_2O} C_6H_{12}O_6$$
Glucose

$$NaCl \xrightarrow{H_2O} Na^+ + Cl^-$$
Sodium chloride

$$CaCl_2 \xrightarrow{H_2O} Ca^{2+} + Cl^- + Cl^-$$
Calcium chloride

Notice that glucose does not break apart when dissolved in water. A molecule of glucose, therefore, contributes only one particle to the solution. Sodium chloride, on the other hand, contributes two ions, or particles, and calcium chloride contributes three. Thus calcium chloride has three times as great an effect on solute concentration as glucose. Just as important, once the electrolyte dissociates, its ions can attract other ions of the opposite charge. If equal amounts of Ca^{2+} and Na^+ are placed in solution, the calcium ion will attract twice as many chloride ions to its area as the sodium ion.

To determine how much effect an electrolyte has on concentration, we must look at the concentrations of its individual ions. The concentration of an ion is commonly expressed in **milliequivalents per liter** or **meq/liter**—the number of electrical charges in each liter of solution. The meq/liter equals the number of charges the ion carries times the number of ions in solution. Ion concentration can be calculated as shown in Exhibit 27-1.

Distribution

Figure 27-2 compares the principal chemical constituents of the three major fluid compartments: plasma, interstitial fluid, and intracellular fluid. The chief difference between plasma and interstitial fluid is that plasma contains quite a few protein anions, whereas interstitial fluid has hardly any. Since normal capillary membranes are practically impermeable to protein, the protein stays in the plasma and does not move out of the blood into the interstitial fluid. Plasma also contains more Na^+ ions but fewer Cl^- ions than the interstitial fluid. In most other respects the two fluids are similar.

Intracellular fluid varies considerably from extracellular fluid, however. In extracellular fluid, the most abundant cation is Na^+ and the most abundant anion is Cl^-. In intracellular fluid, the most abundant cation is K^+ and the most abundant anion is HPO_4^- (phosphate). Also, there are more protein anions in intracellular fluid than in extracellular fluid.

Functions and Regulation

SODIUM

Sodium (Na^+), the most abundant extracellular ion, represents about 90 percent of extracellular cations. Sodium is necessary for the transmission of impulses in nervous and muscle tissue. Its movement also plays a significant role in fluid and electrolyte balance.

Sodium loss from the body may occur through excessive perspiration, certain diuretics, and burns. Such a loss can result in *hyponatremia (natrium = sodium)*, a lower than normal blood sodium level. Hyponatremia is characterized by muscular weakness, headache, hypotension, tachycardia, and circulatory shock. Severe sodium loss can result in mental confusion, stupor, and coma.

The sodium level in the blood is controlled primarily by the hormone aldosterone from the adrenal cortex. Aldosterone acts on the distal convoluted kidney tubule cells and causes them to increase their reabsorption of sodium. The sodium thus moves from the filtrate back into the blood. Aldosterone is secreted in response to reduced blood volume or cardiac output, decreased extracellular sodium, increased extracellular potassium, and physical stress.

Exhibit 27—1   CALCULATING CONCENTRATION OF ION SOLUTION

The number of milliequivalents of an ion in each liter of solution is expressed by the following equation:

$$\text{meq/liter} = \frac{\text{milligrams of ion per liter of solution} \times \text{number of charges on one ion}}{\text{atomic weight of ion}}$$

The atomic weight of an element indicates how heavy it is compared with the element carbon. Dividing the total weight of the solute by its atomic weight tells us how many ions we have. For instance, the atomic weight of calcium is 40 whereas that of sodium is 23. Calcium is therefore a heavier element, and 100 g of calcium contains fewer atoms than does 100 g of sodium. The atomic weights of the elements can be found in a periodic table.

Using the preceding formula, we can calculate the meq/liter for calcium. In 1 liter of plasma there are normally 100 milligrams (mg) of calcium. Thus, by substituting this value in the formula, we arrive at:

$$\text{meq/liter} = \frac{100 \times \text{number of charges}}{\text{atomic weight}}$$

The atomic weight of calcium is 40, and its number of charges is 2. By substituting these values we arrive at:

$$\text{meq/liter} = \frac{100 \times 2}{40}$$

$$= 5 \text{ meq/liter for calcium}$$

Let us now find the meq/liter for sodium:

mg Na$^+$/liter of plasma = 3,300 mg/liter
Number of charges = 1
Atomic weight = 23

$$\text{meq/liter} = \frac{3,300 \times 1}{23}$$

$$= 143.0$$

Comparing the meq/liter of sodium with that of calcium, we can see that even though calcium has a greater number of charges than sodium, the body retains many more sodium ions than calcium ions. Therefore the milliequivalent for sodium in plasma is higher.

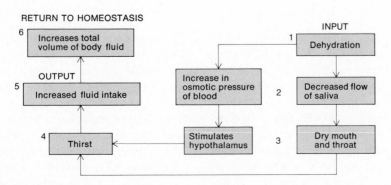

Figure 27—1   Regulation of fluid volume by the adjustment of intake to output.

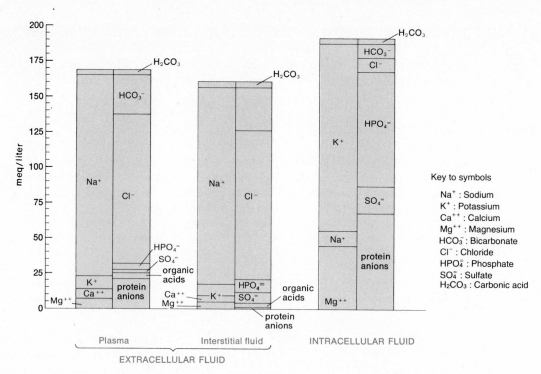

Figure 27–2 Comparison of electrolyte concentrations in plasma, interstitial fluid, and intracellular fluid. The height of each column represents the total electrolyte concentration.

CHLORIDE

Chloride (Cl^-) is mainly an extracellular anion. However, it can easily diffuse between the extracellular and intracellular compartments. This movement makes chloride important in regulating osmotic pressure differences between compartments. In the gastric mucosal glands, chloride combines with H^+ to form hydrochloric acid.

An abnormally low level of chloride in the blood, called *hypochloremia,* may be caused by excessive vomiting, dehydration, and certain diuretics. Symptoms include muscle spasms, alkalosis, depressed respirations, and even coma. The regulation of chloride is indirectly under control of aldosterone.

POTASSIUM

Potassium (K^+) is primarily an intracellular electrolyte. The most abundant cation in intracellular fluid, it helps maintain the fluid volume in cells and control pH. When potassium ions move out of the cell, they are replaced by sodium ions and hydrogen ions. This shift of hydrogen ions helps regulate pH. Potassium also assumes a key role in nervous and muscle tissue.

A lower than normal level of potassium, called *hypokalemia* (*kalium* = potassium), may result from vomiting, diarrhea, high sodium intake, and kidney disease. Symptoms include cramps and fatigue, flaccid

paralysis, mental confusion, increased urine output, shallow respirations, and changes in the electrocardiogram, including a lengthening of the Q-T interval and flattening of the T wave.

The blood level of potassium is under the control of mineralocorticoids, mainly aldosterone. The mechanism is exactly opposite that of sodium. When sodium concentration is low, aldosterone secretion increases and more sodium is reabsorbed. But when potassium concentration is high, more aldosterone is secreted and more potassium is excreted. This process occurs in the distal kidney tubular cells.

CALCIUM AND PHOSPHATE

Calcium (Ca^{2+}) and phosphate (HPO_4^-) are stored in bones and teeth and released when needed. Phosphate is principally an intracellular electrolyte; calcium is an extracellular electrolyte. Phosphate is an important structural component of bones and teeth. In addition, it is necessary for the formation of nucleic acids (DNA and RNA), the synthesis of high-energy compounds (ATP and creatine phosphate), and buffering reactions. Calcium is a structural component of bones and teeth. It is also required for blood clotting, neurotransmitter release, muscle contraction, and normal heartbeat.

Calcium and phosphate blood levels are regulated by several hormones. The parathyroid hormone (parathormone) from the parathyroid glands is released when the calcium blood level is low. Parathyroid hormone stimulates osteoclasts to release calcium and phosphate into the blood, increases the absorption of calcium and phosphate from the gastrointestinal tract, and causes renal tubular cells to excrete phosphate. Another hormone, thyrocalcitonin, from the thyroid gland, decreases the blood level of calcium by stimulating osteoblasts and inhibiting osteoclasts. In the presence of the hormone, osteoblasts remove calcium and phosphate from the blood and deposit them in bone. Estrogens and testosterone (female and male sex hormones, respectively) cause removal of calcium from bones when the blood calcium level is low.

MAGNESIUM

Magnesium (Mg^{2+}) is primarily an intracellular electrolyte. It activates enzyme systems needed to produce cellular energy by the breakdown of ATP into ADP and enzyme systems involved in essential reactions in the liver and bone tissue. Symptoms of magnesium deficiency include increased neuromuscular and central nervous system irritability leading to tremor, tetany, and possibly convulsions. Magnesium excess may cause central nervous system depression, coma, and hypotension. The control of magnesium is by aldosterone. Magnesium is excreted like sodium.

MOVEMENT OF BODY FLUIDS

Between Plasma and Interstitial Compartments

The movement of water between plasma and interstitial fluid occurs across capillary membranes (Figure 27-3) and is dependent on four principal pressures: (1) blood hydrostatic pressure, (2) interstitial fluid hydrostatic pressure, (3) blood osmotic pressure, and (4) interstitial fluid osmotic pressure.

Blood hydrostatic pressure is synonymous with the blood pressure in a capillary. In the glomeruli, it tends to force fluid out of the plasma compartment. In most capillaries, blood hydrostatic pressure is about 35 mm Hg at the arterial end of a capillary and about 15 mm Hg at the venous end. *Interstitial fluid hydrostatic pressure* is the pressure of the interstitial fluid against the cells of the tissue and especially the endothelial cells of the capillaries. It pushes the fluid out of the interstitial compartment and into the capillaries under a pressure of 2 mm Hg at the arterial end of a capillary and 1 mm Hg at the venous end.

Blood osmotic pressure pulls water into the plasma. It averages 25 mm Hg at both ends of the capillary.

Interstitial fluid osmotic pressure forces water into the interstitial compartment. It is zero at the arterial end of a capillary and 3 mm Hg at the venous end. The protein anions in plasma are chiefly responsible for the higher osmotic pressure of blood.

The difference between the two forces that move fluid out and the two forces that push it into the blood is called the **effective filtration pressure,** or P_{eff}. Effective filtration pressure determines the direction of fluid movement and is represented by the following equation:

$$P_{eff} = \begin{pmatrix} \text{blood} & \text{interstitial} \\ \text{hydrostatic} + \text{fluid osmotic} \\ \text{pressure} & \text{pressure} \end{pmatrix}$$
$$- \begin{pmatrix} \text{interstitial} & \text{blood} \\ \text{fluid hydrostatic} + \text{osmotic} \\ \text{pressure} & \text{pressure} \end{pmatrix}$$

By substituting the values already given, we can calculate the effective filtration pressure at the arterial end of a capillary as follows:

$$P_{eff} = (35 + 0) - (2 + 25)$$
$$= (35) - (27)$$
$$= 8 \text{ mm Hg}$$

This means that at the arterial end of a capillary, fluid moves from the plasma into the interstitial fluid because the sum of the blood hydrostatic pressure and interstitial fluid osmotic pressure forces is greater than the sum of the interstitial fluid hydrostatic pressure and blood osmotic pressure forces.

Now let us calculate the effective filtration pressure at the venous end of a capillary:

$$P_{eff} = (15 + 3) - (1 + 25)$$
$$= (18) - (26)$$
$$= -8 \text{ mm Hg}$$

Here the fluid moves from the interstitial fluid back into the plasma because the sum of the interstitial fluid hydrostatic pressure and blood osmotic pressure is greater than the sum of the blood hydrostatic pressure and interstitial fluid osmotic pressure.

Note that the amount of fluid leaving the plasma at the arterial end of a capillary equals the amount returned to plasma at the venous end. This fluid balance between plasma and interstitial fluid exists under normal conditions. The movement of water between plasma and interstitial fluid across capillary membranes due to the four pressures just described is referred to as *Starling's law of the capillaries.*

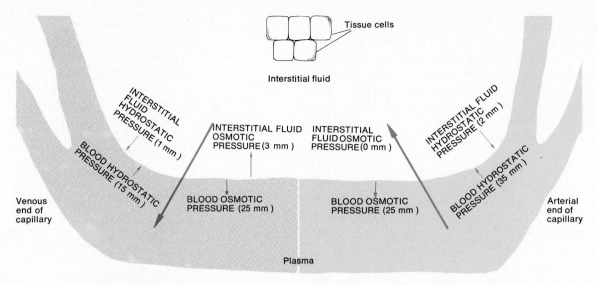

Figure 27–3 Movement of fluid between plasma and interstitial fluid. The heavy arrows indicate the directions of movement at the arterial and venous ends of the capillary.

Occasionally the balance between interstitial fluid and plasma is disrupted. *Edema,* the abnormal increase in interstitial fluid resulting in tissue swelling, is an example of fluid imbalance (Figure 27-4). One cause of edema, hypertension, raises the blood hydrostatic pressure. Another cause is inflammation. As part of the inflammatory response, capillaries become more permeable and allow proteins to leave the plasma and enter the interstitial fluid. Consequently, the blood osmotic pressure falls and the interstitial fluid osmotic pressure rises.

Between Interstitial and Intracellular Compartments

Water movement between the interstitial fluid and the intracellular fluid results from the same pressures that exist between the plasma and the interstitial fluid. Figure 27-3 shows that the intracellular fluid has a higher osmotic pressure than the interstitial fluid. In addition, the principal cation inside the cell is K^+ whereas the principal cation outside is Na^+. Normally, the higher intracellular osmotic pressure is balanced by forces that move water out of the cell, so the amount of water inside the cell does not change. When a fluid imbalance between these two compartments occurs, it is usually caused by a change in the Na^+ or K^+ concentration.

Sodium balance in the body normally is controlled by ADH and aldosterone. ADH regulates extracellular fluid electrolyte concentration by adjusting the amount of water reabsorbed into the blood by the kidney tubules. Aldosterone regulates extracellular fluid volume

by adjusting the amount of sodium reabsorbed by the blood from the kidney tubules. Certain conditions, however, may result in an eventual decrease in the sodium concentration in interstitial fluid. For instance, during sweating the skin excretes sodium as well as water. Sodium also may be lost through vomiting and diarrhea. Coupled with low sodium intake, these conditions can quickly produce a sodium deficit (Figure 27-5). The decrease in sodium concentration in the interstitial fluid lowers the interstitial fluid osmotic pressure and establishes an effective filtration pressure gradient between the interstitial fluid and the intracellular fluid. Water moves from the interstitial fluid into the cells, producing two results that can be quite serious.

The first result, an increase in intracellular water concentration called *overhydration,* is particularly disruptive to nerve cell function. In fact, severe overhydration, or *water intoxication,* produces neurological symptoms ranging from disoriented behavior to convulsions, coma, and even death. The second result of the fluid shift is a loss of interstitial fluid volume that leads to a decrease in the interstitial hydrostatic pressure. As the interstitial hydrostatic pressure drops, water moves out of the plasma, resulting in a loss of blood volume that may lead to shock.

ACID–BASE BALANCE

In addition to controlling water movement, electrolytes also help regulate the body's acid–base balance. The overall acid–base balance is maintained by controlling the H⁺ concentration of body fluids, particularly extra-cellular fluid. In a healthy person, the pH of the extracellular fluid is stabilized between 7.35 and 7.45. Homeostasis of this narrow range is essential to survival and depends on three major mechanisms: buffer systems, respirations, and kidney excretion.

Buffer Systems

Most **buffer systems** of the body consist of a weak acid and a weak base, and they function to prevent drastic changes in the pH of a body fluid by changing strong acids and bases into weak acids and bases. Buffers work within fractions of a second. Recall that a strong acid dissociates into H⁺ ions more easily than does a weak acid. Strong acids, therefore, lower pH more than weak ones because strong acids contribute more H⁺ ions. Similarly, strong bases raise pH more than weak ones because strong bases dissociate more easily into OH⁻ ions. The principal buffer systems of the body fluids are the carbonic acid–bicarbonate system, the phosphate system, the hemoglobin–oxyhemoglobin system, and the protein system.

CARBONIC ACID–BICARBONATE

The *carbonic acid-bicarbonate buffer system* is an important regulator of blood pH. This system is based on the weak acid carbonic acid (H_2CO_3) and a weak base, primarily sodium bicarbonate ($NaHCO_3$). The following equations illustrate the mechanism:

$$HCl + NaHCO_3 \rightleftharpoons NaCl + H_2CO_3$$

Hydrochloric acid (strong acid); Sodium bicarbonate (weak base of bicarbonate buffer system); Sodium chloride (salt); Carbonic acid (weak acid)

$$NaOH + H_2CO_3 \rightleftharpoons H_2O + NaHCO_3$$

Sodium hydroxide (strong base); Carbonic acid (weak acid of bicarbonate buffer system); Water; Sodium bicarbonate (weak base)

Because normal body processes tend to acidify the blood rather than make it more alkaline, the body needs more bicarbonate salt than it needs carbonic acid. In fact, when extracellular pH is normal (7.4), bicarbonate molecules outnumber carbonic acid by 20 to 1.

PHOSPHATE

The *phosphate buffer system* acts in essentially the same manner as the bicarbonate buffer system. Its two components are sodium dihydrogen phosphate

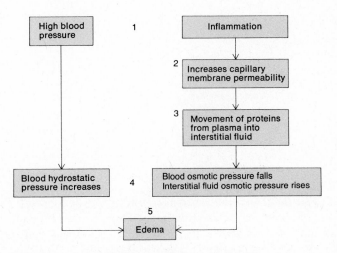

Figure 27–4   Conditions that produce edema.

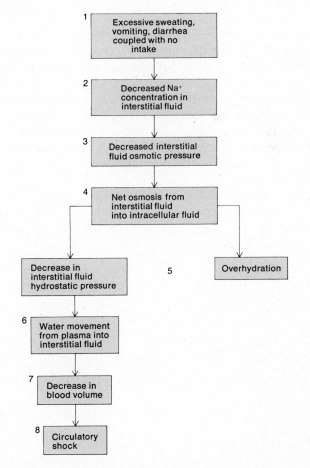

Figure 27–5   Interrelations between fluid imbalance and electrolyte imbalance.

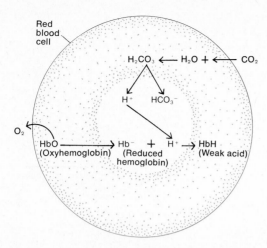

Figure 27–6 Hemoglobin–oxyhemoglobin buffer system. Oxyhemoglobin gives up its oxygen in an acid medium and buffers the acid. The HCO_3^- may remain in the cell and combine with K^+, or it may move out of the cell and combine with Na^+. In this way, much of the CO_2 is carried back to the lungs in the form of potassium bicarbonate or sodium bicarbonate.

(NaH_2PO_4) and sodium monohydrogen phosphate (Na_2HPO_4). The dihydrogen phosphate ion acts as the weak acid and is capable of buffering strong bases:

$$NaOH + NaH_2PO_4 \rightleftharpoons H_2O + Na_2HPO_4$$

Strong base   Weak acid   Water   Weak base

The monohydrogen phosphate ion acts as the weak base and is capable of buffering strong acids:

$$HCl + Na_2HPO_4 \rightleftharpoons NaCl + NaH_2PO_4$$

Strong acid   Weak base   Sodium chloride (salt)   Weak acid

The phosphate buffer system is an important regulator of pH both in red blood cells and in the kidney tubular fluids. NaH_2PO_4 is formed when excess H^+ ions in the kidney tubules combine with Na_2HPO_4. In this reaction, the sodium released from Na_2HPO_4 forms sodium bicarbonate $(NaHCO_3)$ and is passed into the blood. The H^+ ion that replaces sodium becomes part of the NaH_2PO_4 that is passed into the urine. This reaction is one of the mechanisms by which the kidneys help maintain pH by the acidification of urine.

HEMOGLOBIN–OXYHEMOGLOBIN

The *hemoglobin–oxyhemoglobin buffer system* is an effective method for buffering carbonic acid in the blood. When blood moves from the arterial end of a capillary to the venous end, the CO_2 given up by body cells enters the erythrocytes and combines with water to form carbonic acid (Figure 27-6). Simultaneously, oxyhemoglobin gives up its oxygen to the body cells, becomes reduced hemoglobin, and carries a negative charge. The hemoglobin anion attracts the H^+ from the carbonic acid and becomes an acid that is even weaker than carbonic acid. When the hemoglobin–oxyhemoglobin system is active, the exchange reaction that occurs shows why the erythrocyte tends to give up its oxygen when pCO_2 is high.

PROTEIN

The *protein buffer system* is the most abundant buffer in body cells and plasma. Proteins are composed of amino acids. An amino acid is an organic compound that contains at least one carboxyl group (COOH) and at least one amine group (NH_2). The carboxyl group acts like an acid and can dissociate in this way:

$$NH_2-\underset{\underset{H}{|}}{\overset{\overset{R}{|}}{C}}-COO^-H^+$$

The hydrogen ion is then able to react with any excess OH^- in the solution to form water. On the other hand, the amine group has a tendency to act as a base:

$$COOH-\underset{\underset{H}{|}}{\overset{\overset{R}{|}}{C}}-NH_3^+OH^-$$

The OH^- can dissociate, react with excess H^+ ions, and also form water. Thus proteins act as both acidic and basic buffers.

Respirations

Respirations also assume a role in maintaining the pH of the body. An increase in the CO_2 concentration in body fluids as a result of cellular respiration lowers the pH. This is illustrated by the following equation:

$$CO_2 + H_2O \rightleftharpoons H_2CO_3 \rightleftharpoons H^+ + HCO_3^-$$

Conversely, a decrease in the CO_2 concentration of body fluids raises the pH.

The pH of body fluids may be adjusted by a change in the rate of breathing, an adjustment that usually takes from 1 to 3 minutes. If the rate of breathing is increased, more CO_2 is exhaled and the blood pH rises. Slowing down the respiration rate means less CO_2 is exhaled, and the blood pH falls. Doubling the breathing rate increases the pH by about 0.23. Thus it can be increased from 7.4

Exhibit 27-2   MECHANISMS THAT MAINTAIN BODY pH

Mechanism	Comments
Buffer systems	Consist of weak acid and weak base. Prevent drastic changes in body fluid pH.
Carbonic acid–bicarbonate	Important regulator of blood pH.
Phosphate	Important buffer in red blood cells and kidney tubule cells.
Hemoglobin–oxyhemoglobin	Buffers carbonic acid in blood.
Protein	Most abundant buffer in body cells and plasma.
Respirations	Regulate CO_2 level of body fluids.
Increased	Raises pH.
Decreased	Lowers pH.
Kidneys	Excrete H^+ and NH_4^+ and conserve bicarbonate.

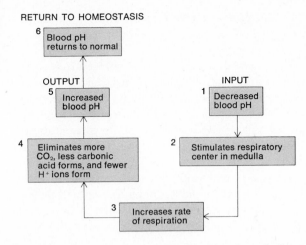

Figure 27–7   Relationship between pH and respirations.

to 7.63. Reducing the breathing rate to one-quarter its normal rate lowers the pH by 0.4. Thus it can be decreased from 7.4 to 7.0. If you consider that breathing rate can be altered up to eight times the normal rate, it should become obvious that alterations in the pH of body fluids may be greatly influenced by respiration.

The pH of body fluids, in turn, affects the rate of breathing (Figure 27-7). If, for example, the blood becomes more acidic, the increase in H^+ ions stimulates the respiratory center in the medulla and respirations increase. The same effect is achieved if the blood concentration of CO_2 increases. On the other hand, if the pH of the blood increases, the respiratory center is inhibited and respirations decrease, and a decrease in the CO_2 concentration of blood has the same effect. The respiratory mechanism normally can eliminate more acid or base than can all the buffers combined. A summary of the mechanisms that maintain body pH is presented in Exhibit 27-2.

Acid–Base Imbalances

The normal blood pH range is 7.35–7.45. Any considerable deviation from this value falls under the category of acidosis or alkalosis. A person has **acidosis** when the blood pH ranges from 7.35 down to 6.80. **Alkalosis** is a pH that ranges between 7.45 and 8.00. There are a number of causes of acidosis and alkalosis. *Respiratory acidosis* is caused by hypoventilation. It occurs as a result of any condition that decreases the movement of CO_2 from the blood to the alveoli of the lungs and therefore causes a buildup of CO_2, H_2CO_3, and H^+ ions. Possible causes include emphysema, pulmonary edema, injury to the respiratory center of the medulla, or disorders of the muscles involved in breathing. In uncompensated respiratory acidosis, the normal bicarbonate–carbonic acid ratio shifts from 20:1 to 10:1 or 8:1 and the blood pH decreases. *Respiratory alkalosis* is caused by hyperventilation–an increased minute volume of respiration in excess of that required to handle increased metabolic demands of the body. It therefore results in a decrease of CO_2. Conditions leading to respiratory alkalosis include oxygen deficiency due to high altitude, severe anxiety, and aspirin overdose–factors that stimulate the respiratory center. In uncompensated respiratory alkalosis, the normal bicarbonate–carbonic acid ratio shifts from 20:1 to 40:1 and the blood pH increases.

Metabolic acidosis results from an abnormal increase in acidic metabolic products other than CO_2 and from the loss of bicarbonate ion from the body. Ketosis is a good example of metabolic acidosis brought on by an increase in the production of acidic metabolic products. Acidosis due to loss of bicarbonate may occur with diarrhea and renal tubular dysfunction. In uncompensated metabolic acidosis, the ratio of bicarbonate to carbonic acid is 12.5:1. *Metabolic alkalosis* is caused by a nonrespiratory loss of acid by the body or excessive intake of alkaline drugs. Excessive vomiting of gastric contents results in a substantial loss of HCl and is probably the most frequent cause of metabolic alkalosis. In uncompensated metabolic alkalosis, the ratio of bicarbonate to carbonic acid is 31.6:1.

The principal physiological effect of acidosis is depression of the central nervous system through depression of synaptic transmission. If the blood pH falls below 7, depression of the nervous system is so acute that the individual becomes disoriented and comatose. In fact, patients with severe acidosis usually die in a state of coma. On the other hand, the major physiological effect of alkalosis is overexcitability of the nervous system through facilitation of synaptic transmission. The overexcitability occurs both in the central nervous system and in peripheral nerves. Because of the overexcitability, nerves conduct impulses repetitively even when not stimulated by normal stimuli, resulting in nervousness, muscle spasms, and even convulsions.

The primary treatments for respiratory acidosis are aimed at increasing the exhalation of CO_2. Excessive secretions may be suctioned out of the respiratory tract, and artificial respiration may be given. Treatment of metabolic acidosis consists of intravenous solutions of sodium bicarbonate and correcting the cause of acidosis.

The treatment of respiratory alkalosis is aimed at increasing the level of CO_2 in the body. One corrective measure is to have the patient rebreathe a mixture of his own CO_2 and O_2 from a paper bag. Treatment for metabolic alkalosis consists of giving a medication containing the chloride ion and correcting the cause of alkalosis.

STUDY OUTLINE

FLUID COMPARTMENTS

1 About two-thirds of the body's fluid is located in cells and is called intracellular fluid.

2 The other third is called extracellular fluid. It includes interstitial fluid, plasma and lymph, cerebrospinal fluid, GI tract fluids, synovial fluid, the fluids of the eyes and ears, pleural, pericardial, and peritoneal fluids, and the glomerular filtrate.

WATER

1 Water, together with substances dissolved in it, constitutes body fluid.

2 Primary sources of fluid intake are ingested liquids and foods and water produced by catabolism.

3 Avenues of fluid output are the kidneys, skin, lungs, and GI tract.

4 The stimulus for fluid intake is dehydration resulting in thirst sensations. The stimulus for fluid output is secretion of aldosterone and ADH.

ELECTROLYTES

1 Electrolytes are compounds that dissolve in body fluids and produce either cations (positive ions) or anions (negative ions).

2 Electrolyte concentration is expressed in milliequivalents per liter (meq/liter).

3 Plasma, interstitial fluid, and intracellular fluid contain varying kinds and amounts of electrolytes.

4 Electrolytes are needed for normal metabolism, proper fluid movement between compartments, and regulation of pH.

5 Sodium, calcium, and chloride are important extracellular electrolytes.

6 Potassium, phosphate, and magnesium are important intracellular electrolytes.

MOVEMENT OF BODY FLUIDS

1 At the arterial end of a capillary, fluid moves from plasma into interstitial fluid. At the venous end, fluid moves in the opposite direction. The movement of water between plasma and interstitial fluid across capillary membranes is referred to as Starling's law of the capillaries.

2 Fluid movement between interstitial and intracellular compartments depends on the movement of sodium and the secretion of aldosterone and ADH.

3 Fluid imbalance may lead to edema and overhydration.

ACID–BASE BALANCE

1 The overall acid–base balance of the body is maintained by controlling the H^+ concentration of body fluids, especially extracellular fluid.

2 The normal pH of extracellular fluid is 7.35–7.45.

3 Homeostasis of pH is maintained by buffers, respirations, and kidney excretion.

4 Acid–base imbalances are acidosis and alkalosis.

REVIEW QUESTIONS

1 Define body fluid. What are the body's principal fluid compartments? How are they separated?

2 What is meant by fluid balance? How are fluid balance and electrolyte balance related?

3 Describe the avenues of fluid intake and fluid output. Be sure to indicate volumes in each case.

4 Discuss the role of thirst in regulating fluid intake.

5 Explain how aldosterone and ADH adjust fluid output.

6 Define a nonelectrolyte and an electrolyte. Give specific examples of each.

7 Distinguish between a cation and an anion. Give several examples of each.

8 How is the ionic concentration of a fluid expressed? Calculate the ionic concentration of sodium in a body fluid.

9 Describe the functions of electrolytes in the body.

10 Describe some of the major differences in the electrolytic concentrations of the three fluid compartments in the body.

11 Name three important extracellular electrolytes.

12 List three important intracellular electrolytes.

13 Explain the forces involved in moving fluid between plasma and interstitial fluid. Summarize these forces by setting up an equation to express effective filtration pressure.

14 What is Starling's law of the capillaries?

15 Explain the factors involved in fluid movement between the interstitial fluid and the intracellular fluid.

16 Define edema. What are some of its causes?

17 Give an example of how electrolyte balance is related to fluid balance.

18 Explain how the following buffer systems help to maintain the pH of body fluids: carbonic acid–bicarbonate, phosphate, hemoglobin–oxyhemoglobin, and protein.

19 Describe how respirations are related to the maintenance of pH.

20 Briefly discuss the role of the kidneys in maintaining pH.

21 Define acidosis and alkalosis. Distinguish between respiratory and metabolic acidosis and alkalosis.

22 What are the principal physiological effects of acidosis and alkalosis?

23 How are acidosis and alkalosis treated?

Unit V
Continuity

This unit is designed to show you how the human organism is adapted for reproduction. It also traces the developmental sequence involved in pregnancy. Principles of inheritance and birth control conclude the unit.

Chapter 28
Reproduction

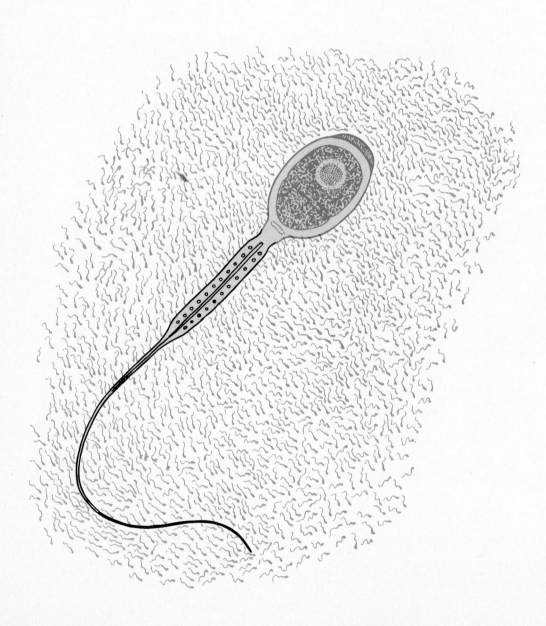

STUDENT OBJECTIVES

■ Define reproduction.

■ List the organs that comprise the male and female systems of reproduction.

■ Describe the role of the scrotum in protecting the testes.

■ Describe the testes as glands that produce sperm and the male hormone testosterone.

■ Describe the physiological effects of testosterone.

■ Trace the course of sperm cells through the system of ducts that lead from the testes to the exterior.

■ Contrast the functions of the seminal vesicles, prostate gland, and Cowper's glands in secreting constituents of seminal fluid.

■ Describe the chemical composition of seminal fluid.

■ Describe the penis as the organ of copulation.

■ Describe the ovaries as glands that produce ova and female sex hormones.

■ Describe the function of the uterine tubes in transporting a fertilized ovum to the uterus.

■ Identify the anatomical portions of the uterus and the ligaments that maintain its normal position.

■ Describe the principal events of the menstrual and ovarian cycles.

■ Discuss the physiological effects of estrogens and progesterone.

■ Correlate the activities of the menstrual and ovarian cycles.

■ Discuss the hormonal interactions that control the menstrual and ovarian cycles and menopause.

■ Describe the role of the vagina in the menstrual flow and copulation.

■ Identify the components of the vulva.

■ Define the anatomical boundaries of the perineum.

■ Describe the structure and development of the mammary glands.

■ Describe the symptoms and causes of disorders of the reproductive systems, including venereal diseases, prostate disorders, impotence, infertility, and menstrual abnormalities.

■ Describe ovarian cysts, endometriosis, and leukorrhea.

■ Discuss breast and cervical cancer. Describe special detection procedures such as mammography and thermography.

■ Define medical terminology associated with the reproductive systems.

Reproduction is the mechanism by which the thread of life is sustained. It is the process by which a single cell duplicates its genetic material, allowing an organism to grow and repair itself. In this sense, reproduction enables the individual organism to maintain its own life. But reproduction is also the process by which genetic material is passed from generation to generation. In this regard, reproduction maintains the life of the species.

The male and female reproductive systems are organized into organs that may be grouped by function. The testes and ovaries, also called gonads, function in the production of gametes—sperm cells and ova, respectively. The gonads also secrete hormones. The ducts transport, receive, and store gametes. Still other reproductive organs, called accessory glands, produce materials that support gametes.

MALE REPRODUCTIVE SYSTEM

The organs of the male reproductive system (Figure 28-1) are the testes, or male gonads, which produce sperm, a number of ducts that either store or transport sperm to the exterior, accessory glands that add secretions comprising the semen, and several supporting structures, including the penis.

Scrotum

The **scrotum** is a pouching of the abdominal wall consisting of loose skin and superficial fascia. It is the supporting structure for the testes. Externally, it looks like a single pouch of skin separated into lateral portions by a medial ridge called the *raphe.* Internally, it is divided by a septum into two sacs, each containing a single testis. The septum consists of superficial fascia and bundles of smooth muscle fibers: the *dartos.* The dartos is also found in the subcutaneous tissue of the scrotum. The testes are the organs that produce sperm. Sperm production and survival require a temperature that is lower than body temperature. Because the scrotum is isolated from the body cavities, it supplies an environment about 3°F below body temperature. Exposure to cold causes contraction of the smooth muscle fibers, moving the testes closer to the pelvic cavity where they can absorb body heat. This contraction causes the skin of the scrotum to wrinkle. Exposure to warmth reverses the process. A muscle in the spermatic cord, the cremaster muscle, also elevates the testes upon exposure to cold.

Testes

The **testes** are paired oval glands measuring about 5 cm (2 inches) in length and 2.5 cm (1 inch) in diameter. They weigh between 10 and 15 g (Figure 28-2a). During most of fetal life they lie in the pelvic cavity, but about 2 months prior to birth they descend into the scrotum. When the testes do not descend, the condition is referred to as *cryptorchidism.* Cryptorchidism results in sterility because the sperm cells are destroyed by the higher body temperature of the pelvic cavity. Undescended testes can be placed in the scrotum by administering hormones or by surgical means prior to puberty without ill effects.

The testes are covered by a dense layer of white fibrous tissue, the *tunica albuginea,* that extends inward and divides each testis into a series of internal compartments called *lobules.* Each lobule contains one to three tightly coiled tubules, the convoluted *seminiferous tubules,* that produce sperm by a process called *spermatogenesis.* A cross section through a seminiferous tubule reveals that it is packed with sperm cells in various stages of development (Figure 28-2b-d). The most immature cells, the *spermatogonia,* are located against the basement membrane. Toward the lumen in the center of the tube, one can see layers of progressively more mature cells. In order of advancing maturity, these are primary spermatocytes, secondary spermatocytes, and spermatids. By the time a **sperm cell,** or **spermatozoan,** has reached full maturity, it is in the lumen of the tubule and begins to be moved through a series of ducts. Embedded between the developing sperm cells in the tubules are *Sertoli cells* that produce secretions for supplying nutrients to the spermatozoa. Between the seminiferous tubules are clusters of *interstitial cells of Leydig.* These cells secrete the male hormone testosterone. Since the testes produce sperm and testosterone, they are both exocrine and endocrine glands.

SPERMATOZOA

Spermatozoa, once ejaculated, have a life expectancy of about 48 hours. A spermatozoan is highly adapted for reaching and penetrating a female ovum. It is composed of a head, a middle piece, and a tail (Figure 28-3). The head contains the nuclear material and the acrosome, which contains chemicals that effect penetration of the sperm cell into the ovum. Numerous mitochondria in the middle piece carry on the metabolism that provides energy for locomotion. The tail, a typical flagellum, propels the sperm along its way.

TESTOSTERONE

Secretions of the anterior pituitary gland assume a major role in the developmental changes associated with puberty. At the onset of puberty the anterior pituitary starts to secrete gonadotropic hormones called follicle-stimulating hormone (FSH) and interstitial cell stimulating hormone (ICSH). Their release is controlled from the hypothalamus by follicle-stimulating hormone re-

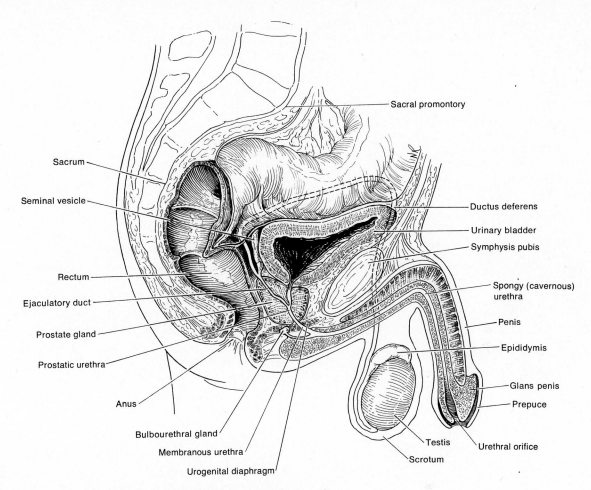

Figure 28–1   Male organs of reproduction seen in sagittal section.

leasing factor (FSHRF) and interstitial cell stimulating hormone releasing factor (ICSHRF). Once secreted, the gonadotropic hormones have profound effects on male reproductive organs. FSH acts on the seminiferous tubules to initiate spermatogenesis (Figure 28-4, p. 645). ICSH acts on the seminiferous tubules and further assists the tubules to develop mature sperm, but its chief function is to stimulate the interstitial cells of Leydig to secrete testosterone.

Testosterone has a number of effects on the body. It controls the development, growth, and maintenance of the male sex organs. It also regulates bone growth, protein anabolism, sexual behavior, sperm production, and the development of male secondary sex characteristics. These characteristics, which appear at puberty, include muscular development resulting in wide shoulders and narrow hips; body hair patterns that include pubic hair, axillary and chest hair (within hereditary limits), facial hair, and temporal hairline recession; and enlargement of the thyroid cartilage producing deepening of the voice.

The interaction of ICSH with testosterone illustrates the operation of another negative feedback system. ICSH stimulates the production of testosterone. But once the testosterone concentration in the blood reaches a certain level, it inhibits the release of ICSHRF by the hypothalamus. This inhibition, in turn, inhibits the release of ICSH by the anterior pituitary. Thus testosterone production is decreased. However, once the testosterone concentration in the blood decreases to a certain level, ICSHRF is released by the hypothalamus. This release stimulates the release of ICSH by the anterior pituitary. Finally, ICSH stimulates testosterone production. Thus the testosterone-ICSH cycle is complete. It is not yet clear how testosterone exerts a negative feedback inhibition of FSH secretion.

Ducts

When the sperm mature, they are moved through the convoluted seminiferous tubules to the **straight tubules.** The straight tubules lead to a network of ducts in

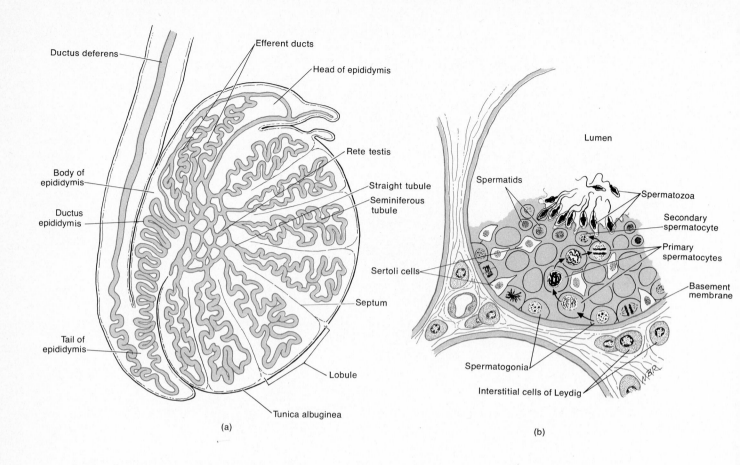

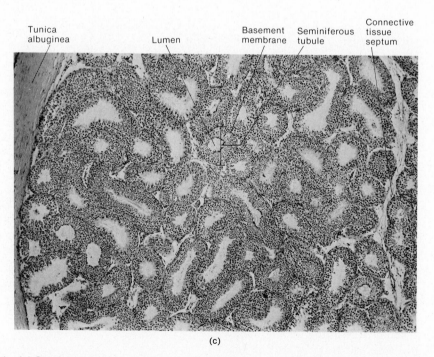

Figure 28–2   Testis. (a) Sectional view of a testis showing its system of tubes. (b) Cross section of a seminiferous tubule showing the stages of spermatogenesis. (c) Histology of several seminiferous tubules at a magnification of 65×. (Courtesy of Edward J. Reith, from *Atlas of Descriptive Histology,* by Edward J. Reith and Michael H. Ross, Harper & Row, Publishers, Inc., New York, 1970.)

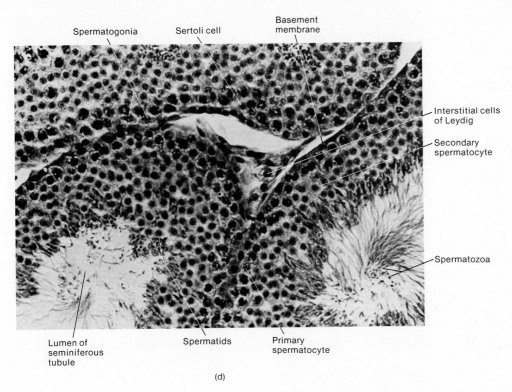

(d)

Figure 28–2 (cont.)   Testis. (d) Details of a seminiferous tubule at a magnification of 275×. (Courtesy of Victor B. Eichler, Wichita State University.)

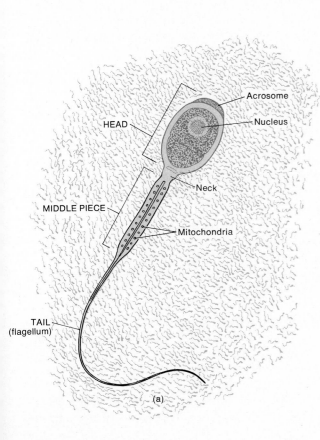

(a)

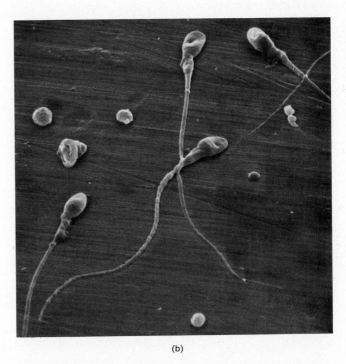

(b)

Figure 28–3   Spermatozoa. (a) Parts of a spermatozoan. (b) Scanning electron micrograph of several spermatozoa at a magnification of 2,000×. (Courtesy of Fisher Scientific Company and S.T.E.M. Laboratories, Inc., Copyright 1975.)

the testis called the **rete testis**. Some of the cells lining the rete testis possess cilia that probably push the sperm along. The sperm are next transported out of the testis through a series of coiled **efferent ducts** that empty into a single tube called the **ductus epididymis**. At this point, the sperm are morphologically mature (see Figure 28-2a).

EPIDIDYMIS

The two **epididymides** are comma-shaped organs. Each lies along the posterior border of the testis (see Figures 28-1 and 28-2) and consists mostly of a tightly coiled tube: the *ductus epididymis*. The larger, superior portion of the epididymis is known as the *head*. It consists of the efferent ducts that empty into the ductus epididymis. The *body* of the epididymis contains the ductus epididymis. The *tail* is the smaller, inferior portion. Within the tail, the ductus epididymis becomes the ductus deferens.

The ductus epididymis measures about 6 m (20 ft) in length and 1 mm in diameter. It is tightly packed within the epididymis, which measures only about 3.8 cm (1½ inches). The ductus epididymis is lined with pseudostratified columnar epithelium, and its wall contains smooth muscle. The free surfaces of the columnar cells contain long, branching microvilli called *stereocilia*. Functionally, the epididymis is the site of sperm maturation. It stores spermatozoa and propels them toward the urethra during ejaculation.

DUCTUS DEFERENS

Within the tail of the epididymis, the ductus epididymis becomes less convoluted, its diameter increases, and at this point it is referred to as the **ductus (vas) deferens** or **seminal duct** (see Figure 28-2a). The ductus deferens, about 45 cm long (18 inches), ascends along the posterior border of the testis, penetrates the inguinal canal, and enters the pelvic cavity, where it loops over the side and down the posterior surface of the urinary bladder. The dilated terminal portion of the ductus deferens is known as the *ampulla*. Histologically, the ductus deferens is lined with pseudostratified epithelium and contains a heavy coat of three layers of muscle. Peristaltic contractions of the muscular coat propel the spermatozoa toward the urethra during ejaculation.

Traveling with the ductus deferens as it ascends in the scrotum are the testicular artery, autonomic nerves, veins that drain the testes, lymphatics, and a small circular band of skeletal muscle called the *cremaster muscle*. These structures constitute the **spermatic cord**, a supporting structure of the male reproductive system. The cremaster muscle elevates the testes during sexual stimulation and exposure to cold. The spermatic cord passes through the *inguinal canal*, a slitlike passageway in the anterior abdominal wall just superior to the medial half of the inguinal ligament. The area of the inguinal canal and spermatic cord represents a weak spot in the abdominal wall, and it is frequently the site of a *hernia* – a rupture or separation of a portion of the abdominal wall resulting in the protrusion of a part of a viscus.

EJACULATORY DUCT

Posterior to the urinary bladder, each ductus deferens joins its **ejaculatory duct** (Figure 28-5). Each duct is about 2 cm (1 inch) long. Both ejaculatory ducts eject spermatozoa into the prostatic urethra. The urethra is the terminal duct of the system, serving as a common passageway for both spermatozoa and urine.

URETHRA

In the male, the **urethra** passes through the prostate gland, the urogenital diaphragm, and the penis. It measures about 20 cm (8 inches) in length and is subdivided into three parts (see Figures 28-1 and 28-7). The *prostatic portion* is 2 to 3 cm (1 inch) long and passes through the prostate gland. It continues inferiorly as the membranous portion as it passes through the urogenital diaphragm, a muscular partition between the two ischiopubic rami. The *membranous portion* is about 1 cm (½ inch) in length. After passing through the urogenital diaphragm, it is known as the *spongy (cavernous) portion* of the urethra. This portion is about 15 cm (6 inches) long. The spongy urethra enters the bulb of the penis and terminates at the external urethral orifice (see Figure 28-1).

Accessory Glands

Whereas the ducts of the male reproductive system store and transport sperm cells, the **accessory glands** secrete the liquid portion of semen. The paired **seminal vesicles** (Figure 28-5) are convoluted pouchlike structures, about 5 cm (2 inches) in length, lying posterior to and at the base of the urinary bladder in front of the rectum. They secrete the alkaline viscous component of semen rich in the sugar fructose and pass it into the ejaculatory duct. The seminal vesicles contribute about 60 percent of the volume of semen.

The **prostate gland** is a single, doughnut-shaped gland about the size of a chestnut (see Figure 28-5). It is inferior to the urinary bladder and surrounds the upper portion of the urethra. The prostate secretes an alkaline fluid that constitutes 13 to 33 percent of the semen into the prostatic urethra. The prostate may become enlarged or develop tumors in older men, obstructing urine flow and requiring surgical intervention.

The paired **bulbourethral**, or **Cowper's glands** are about the size of peas. They are located beneath the

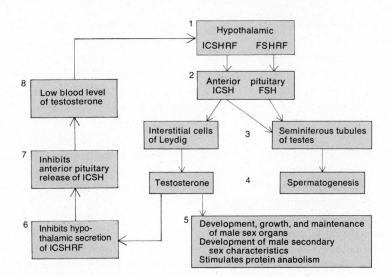

Figure 28-4   Secretion, physiological effects, and inhibition of testosterone.

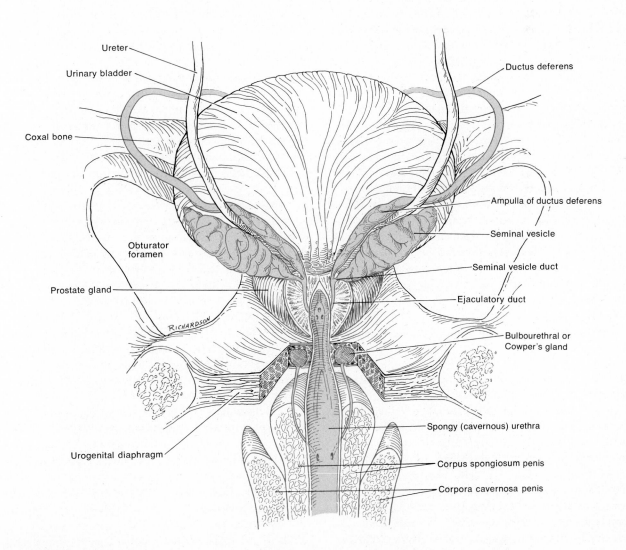

Figure 28-5   Relationships of some male reproductive organs: posterior view of urinary bladder.

prostate on either side of the membranous urethra (Figure 28-5). Like the prostate, the Cowper's glands secrete an alkaline fluid; their ducts open into the spongy urethra.

Semen

Semen, or **seminal fluid,** is a mixture of sperm and the secretions of the seminal vesicles, the prostate gland, and the bulbourethral glands. The average volume of semen for each ejaculation is 2.5–6 ml, and the average range of spermatozoa ejaculated is 50–100 million per milliliter. When the number of spermatozoa falls below approximately 20 million per milliliter, the male is likely to be sterile. Though only a single spermatozoan fertilizes an ovum, fertilization seems to require the combined action of a large number of spermatozoa. The ovum is enclosed by cells that present a barrier to the sperm. An enzyme called hyaluronidase is secreted by the acrosomes of sperm. Hyaluronidase is believed to dissolve intercellular materials of the cells covering the ovum, giving the sperm a passageway into the ovum.

Semen has a pH range of 7.35 to 7.50 – slightly alkaline. The prostatic secretion gives semen a milky appearance, and fluids from the seminal vesicles and bulbourethral glands give it a mucoid consistency. Semen provides spermatozoa with a transportation medium and nutrients. It also neutralizes the acid environment of the male urethra and the female vagina. Semen also activates sperm with enzymes after ejaculation.

Once ejaculated into the vagina, liquid semen coagulates rapidly due to a clotting enzyme produced by the prostate that acts on a substance produced by the seminal vesicle. This clot liquefies in a few minutes because of another enzyme produced by the prostate gland. It is not clear why semen coagulates and liquefies in this manner.

Penis

The **penis** is used to introduce spermatozoa into the vagina (Figure 28-6). The distal end of the penis is a slightly enlarged region called the *glans,* which means shaped like an acorn. Covering the glans is the loosely fitting *prepuce* or *foreskin.* Internally, the penis is composed of three cylindrical masses of tissue bound together by fibrous tissue. The two dorsally located masses are called the *corpora cavernosa penis.* The smaller ventral mass, the *corpus spongiosum penis,* contains the spongy urethra. All three masses of tissue are spongelike and contain blood sinuses. Under the influence of sexual stimulation, the arteries supplying the penis dilate, and

large quantities of blood enter the blood sinuses. Expansion of these spaces compresses the veins draining the penis so most entering blood is retained. These vascular changes result in an *erection.* The penis returns to its flaccid state when the arteries constrict and pressure on the veins is relieved. During ejaculation, the smooth muscle sphincter at the base of the urinary bladder is closed due to the higher pressure in the urethra caused by expansion of the corpus spongiosum penis. Thus urine is not expelled during ejaculation and semen does not enter the urinary bladder.

FEMALE REPRODUCTIVE SYSTEM

The female organs of reproduction (Figure 28-7) include the ovaries, which produce ova; the uterine or Fallopian tubes, which transport the ova to the uterus (or womb); the vagina; and external organs that constitute the vulva or pudendum. The mammary glands, or breasts, also are considered part of the female reproductive system.

Ovaries

The **ovaries,** or female gonads, are paired glands resembling unshelled almonds in size and shape. They are positioned in the upper pelvic cavity, one on each side of the uterus. The ovaries are maintained in position by a series of ligaments (Figure 28-8). They are attached to the broad ligament of the uterus, which is itself part of the parietal peritoneum, by a fold of peritoneum called the *mesovarium;* anchored to the uterus by the *ovarian ligament;* and attached to the pelvic wall by the *suspensory ligament.* These ligaments can be seen in Figure 28-9 (p. 650), which shows the pelvic organs viewed from above. Each ovary also contains a *hilum,* the point of entrance for blood vessels and nerves.

The microscope reveals that each ovary consists of the following parts (Figure 28-10, p. 650):

1 Germinal epithelium This layer of simple cuboidal epithelium covers the free surface of the ovary.

2 Tunica albuginea This is a capsule of collagenous connective tissue immediately deep to the germinal epithelium.

3 Stroma This is a region of connective tissue deep to the tunica albuginea. The stroma is composed of an outer, dense layer called the *cortex* and an inner, loose layer known as the *medulla.* The cortex contains ovarian follicles.

4 Ovarian follicles These are ova and their surrounding tissues in various stages of development.

5 Graafian follicle This endocrine gland is made up of a mature ovum and its surrounding tissues. The Graafian follicle secretes hormones called estrogens.

6 Corpus luteum This glandular body develops from a Graafian follicle after extrusion of an ovum (ovulation). The corpus luteum produces the hormones progesterone and estrogens.

The ovaries produce mature ova, discharge mature ova (ovulation), and secrete the female sexual hormones. The ovaries are analogous to the testes of the male reproductive system.

Uterine Tubes

The female body contains two **uterine,** or **fallopian, tubes,** which transport the ova from the ovaries to the uterus (see Figure 28-8). Measuring about 10 cm (4 inches) long, the tubes are positioned between the folds of the broad ligaments of the uterus. The funnel-shaped open end of each tube, called the *infundibulum,* lies close to the ovary but is not attached to it and is surrounded by a fringe of fingerlike projections called *fimbriae.* From the infundibulum the uterine tube extends inward and downward and attaches to the upper side of the uterus.

Histologically, the uterine tubes are composed of three layers. The internal *mucosa* contains ciliated columnar cells and secretory cells, which are believed to aid the nutrition of the ovum. The middle layer, the *muscularis,* is composed of a thick circular region of smooth muscle and an outer, thin, longitudinal region of smooth muscle. Wavelike contractions of the muscularis help move the ovum down into the uterus. The outer layer of the uterine tubes is a *serous membrane.*

About once a month a mature ovum ruptures from the surface of the ovary near the infundibulum of the uterine tube, a process called **ovulation.** The ovum is swept by the ciliary action of the epithelium of the infundibulum, which moves the ovum into the uterine tube. The ovum is then moved along the uterine tube by ciliary action supplemented by the wavelike contractions of the muscularis of the uterine tube. If the ovum is fertilized by a sperm cell, it usually occurs in the upper third of the uterine tubes. Fertilization may occur at any time up to 24 hours following ovulation. The ovum, fertilized or unfertilized, descends into the uterus within 7 days. Sometimes an ovum is fertilized while it is free in the pelvic cavity, and implantation may take place on one of the pelvic viscera. Pelvic implantations usually fail because the developing fertilized ovum does not make vascular connection with the maternal blood supply. On occasion, fertilized ova fail to descend to the uterus and implant in the uterine tubes. In this instance, the pregnancy must be terminated surgically before the tube ruptures. Both pelvic and tubular implantations are referred to as *ectopic pregnancies.*

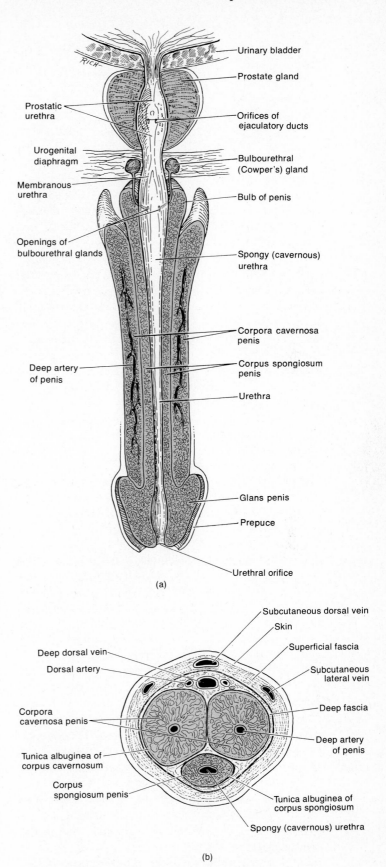

(a)

(b)

Figure 28–6 Internal structure of the penis. (a) View from the floor of the penis. (b) Cross section.

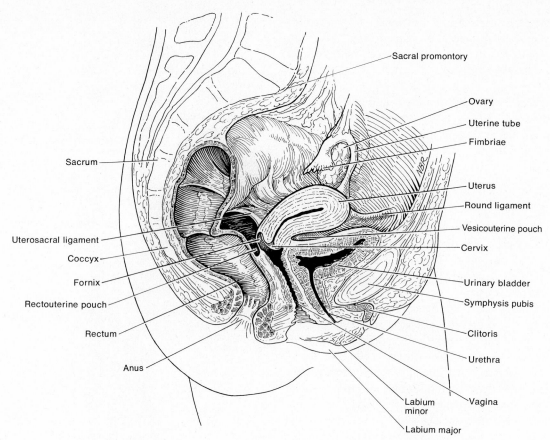

Figure 28–7   Female organs of reproduction seen in sagittal section.

Uterus

The site of menstruation, implantation of a fertilized ovum, development of the fetus during pregnancy, and labor is the **uterus**. Situated between the urinary bladder and the rectum, the uterus is an inverted, pear-shaped organ (see Figures 28-7 and 28-8). Before the first pregnancy, the adult uterus measures approximately 7.5 cm (3 inches) long, 5 cm (2 inches) wide, and 1.75 cm (1 inch) thick. Anatomical subdivisions of the uterus include the dome-shaped portion above the uterine tubes called the *fundus,* the major tapering central portion called the *body,* and the inferior narrow portion opening into the vagina called the *cervix.* Between the body and the cervix is a constricted region about 1 cm (½ inch) long: the *isthmus.* The interior of the body of the uterus is called the *uterine cavity,* and the interior of the narrow cervix is called the *cervical canal.* The junction of the uterine cavity with the cervical canal is the *internal os.* The *external os* is the place where the cervix opens into the vagina.

Normally the uterus is flexed between the uterine body and the cervix. In this position, the body of the uterus projects forward and slightly upward over the urinary bladder, and the cervix projects downward and backward, joining the vagina at nearly a right angle. Several structures that are either extensions of the parietal peritoneum or fibromuscular cords, referred to as ligaments, maintain the position of the uterus. The paired *broad ligaments* are double folds of parietal peritoneum attaching the uterus to either side of the pelvic cavity. Uterine blood vessels and nerves pass through the broad ligaments. The paired *uterosacral ligaments,* also peritoneal extensions, lie on either side of the rectum and connect the uterus to the sacrum. The *cardinal (lateral cervical) ligaments* extend below the bases of the broad ligaments between the pelvic wall and the cervix and vagina. These ligaments contain smooth muscle, uterine blood vessels, and nerves and are the chief ligaments supporting the uterus and keeping it from dropping down into the vagina. The *round ligaments* are bands of fibrous connective tissue between the layers of the broad ligament. They extend from a point on the uterus just below the uterine tubes to a portion of the external genitalia. Although the liga-

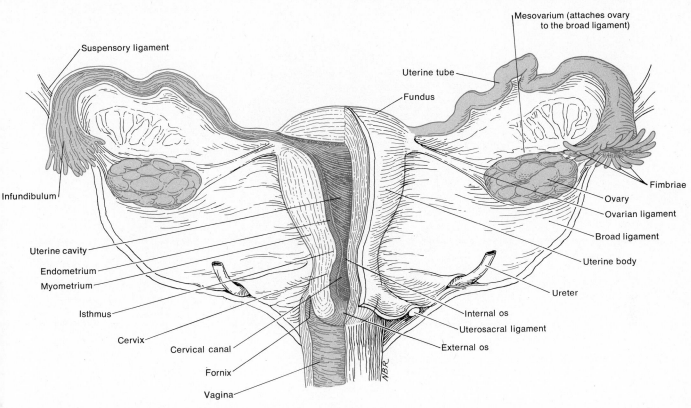

Figure 28–8 Uterus and associated female reproductive structures. The left side of the figure has been sectioned to show internal structures.

ments normally maintain the position of the uterus, they also afford the uterine body with some movement. As a result, the uterus may become malpositioned. A backward tilting of the uterus called *retroflexion* or a forward tilting called *anteflexion* may occur.

Histologically, the uterus consists of three layers of tissue. The outer layer, part of the parietal peritoneum, is referred to as the *serous layer* or *serosa.* Laterally, the serosa becomes the broad ligament. Anteriorly, it is reflected over the urinary bladder and forms a shallow pouch, the *vesicouterine pouch.* Posteriorly, it is reflected onto the rectum and forms a deep pouch, the *rectouterine pouch,* or *pouch of Douglas* – the lowest point in the pelvic cavity.

The middle layer of the uterus, the *myometrium,* forms the bulk of the uterine wall. This layer consists of smooth muscle fibers and is thickest in the fundus and thinnest in the cervix. During childbirth, coordinated contractions of the muscles help to expel the fetus from the body of the uterus.

The inner layer of the uterus, the *endometrium,* is a mucous membrane composed of two principal layers. The *stratum functionalis,* the layer closer to the uterine cavity, is shed during menstruation. The other layer, the *stratum basalis,* is permanent and produces a new functionalis following menstruation. The endometrium contains numerous glands.

Blood is supplied to the uterus by branches of the internal iliac artery called *uterine arteries.* Branches called *arcuate arteries* are arranged in a circular fashion underneath the serosa and give off *radial arteries* that penetrate the myometrium (Figure 28-11, p. 652). Just before the branches enter the endometrium, they divide into two kinds of arterioles. The *straight arteriole* terminates in the basalis and supplies it with the materials necessary to regenerate the functionalis. The *spiral arteriole* penetrates the functionalis and changes markedly during the menstrual cycle. The uterus is drained by the *uterine veins.*

Endocrine Relations

The principal events of the menstrual cycle can be correlated with those of the ovarian cycle and changes in the endometrium. All are hormonally controlled events.

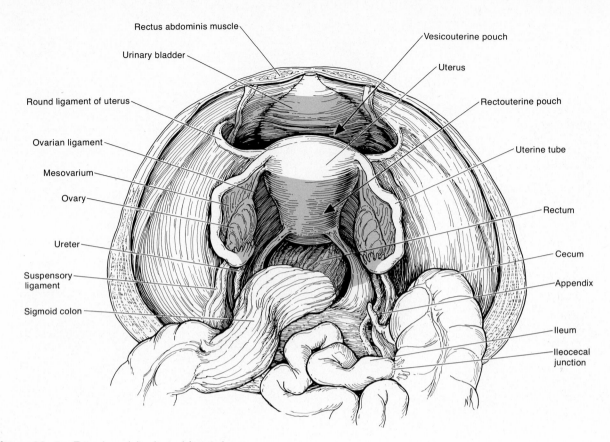

Figure 28–9   Female pelvis viewed from above.

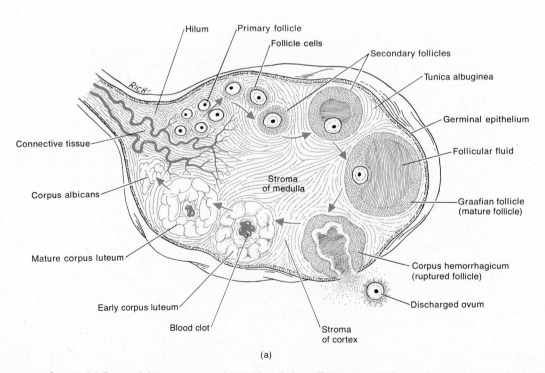

(a)

Figure 28–10   Ovary. (a) Parts of the ovary seen in sectional view. The arrows indicate the sequence of developmental stages that occur as part of the ovarian cycle.

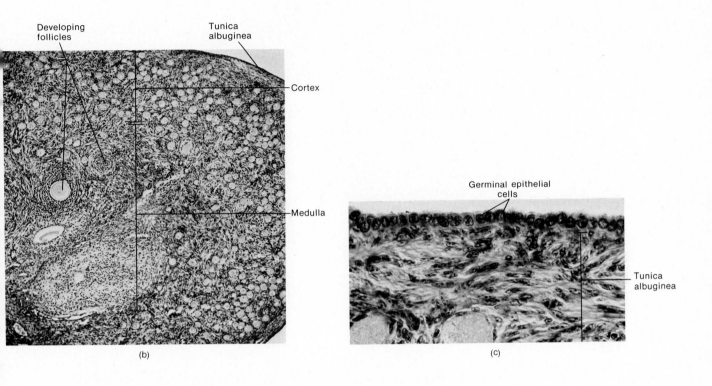

(b)

(c)

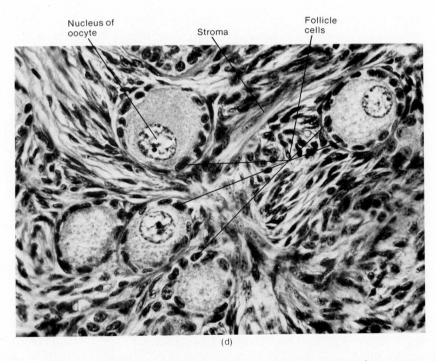

(d)

Figure 28–10 (cont.) Ovary. (b) Sectional view of a portion of the ovary at a magnification of 65×. (c) Enlarged aspect of the surface of the ovary at a magnification of 640×. (d) Primary follicles at a magnification of 640×. (Courtesy of Edward J. Reith, from *Atlas of Descriptive Histology,* by Edward J. Reith and Michael H. Ross, Harper & Row, Publishers, Inc., New York, 1970.)

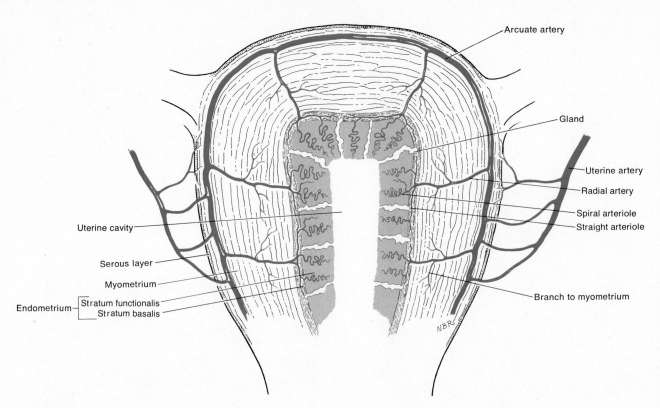

Figure 28—11   Blood supply to the uterus.

The **menstrual cycle** is a series of changes in the endometrium of a nonpregnant female. Each month the endometrium is prepared to receive a fertilized ovum. An implanted ovum eventually develops into an embryo and then a fetus, which normally remains in the uterus until delivery. If no fertilization occurs, a portion of the endometrium is shed. The **ovarian cycle** is a monthly series of events associated with the maturation of an ovum.

The menstrual cycle, ovarian cycle, and other changes associated with puberty in the female are controlled by regulating factors from the hypothalamus called the follicle-stimulating hormone releasing factor (FSHRF) and the luteinizing-hormone releasing factor (LHRF). FSHRF stimulates the release of follicle-stimulating hormone (FSH) from the anterior pituitary (Figure 28-12). FSH stimulates the initial development of the ovarian follicles and the secretion of estrogens by the follicles. LHRF stimulates the release of another anterior pituitary hormone – the luteinizing hormone (LH), which stimulates the development of ovarian follicles, brings about ovulation, and stimulates the production of both estrogens and progesterone by ovarian cells. The female sex hormones – estrogens and progesterone – affect the body in different ways. **Es-**

trogens, the hormones of growth, have four main functions. First is the development and maintenance of female reproductive structures, especially the endometrial lining of the uterus, secondary sex characteristics, and the breasts. The secondary sex characteristics include fat distribution to the breasts, abdomen, mons pubis, and hips; voice pitch; broad pelvis; and hair pattern. Second, they control fluid and electrolyte balance. Third, they increase protein anabolism. In this regard, estrogens are synergistic with human growth hormone (HGH). Fourth, they may increase the female sex drive. High levels of estrogens in the blood inhibit the release of FSHRF by the hypothalamus, which in turn inhibits the secretion of FSH by the anterior pituitary gland. This inhibition provides the basis for the action of one kind of contraceptive pill. **Progesterone,** the hormone of maturation, works with estrogens to prepare the endometrium for implantation and the breasts for milk secretion.

The duration of the menstrual cycle normally ranges from 24 to 35 days. For this discussion, we will assume an average duration of 28 days. Events occurring during the menstrual cycle may be divided into three phases: the menstrual phase, the preovulatory phase, and the postovulatory phase (Figure 28-13).

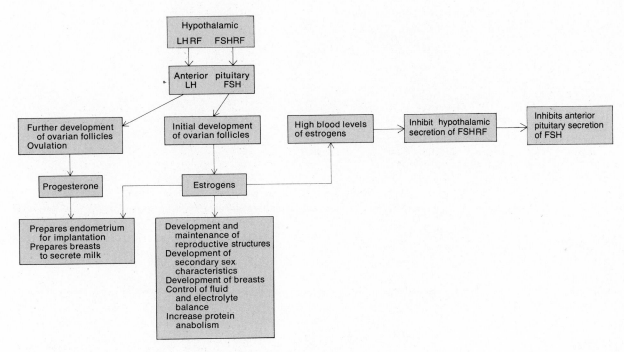

Figure 28–12 Physiological effects of estrogens and progesterone.

The *menstrual phase,* also called *menstruation* or the *menses,* is the periodic discharge of 25 to 65 ml of blood, tissue fluid, mucus, and epithelial cells. It lasts for approximately the first 5 days of the cycle. The discharge is associated with endometrial changes in which the functionalis layer degenerates and patchy areas of bleeding develop. Small areas of the functionalis detach one at a time (total detachment would result in hemorrhage), the uterine glands discharge their contents and collapse, and tissue fluid is discharged. The menstrual flow passes from the uterine cavity to the cervix, through the vagina, and ultimately to the exterior. Generally the flow terminates by the fifth day of the cycle. At this time the entire functionalis is shed and the endometrium is very thin because only the basalis remains.

During the menstrual phase, the ovarian cycle is also in operation. Ovarian follicles, called *primary follicles,* begin their development. At birth each ovary contains about 200,000 such follicles, each consisting of an ovum surrounded by a layer of cells. During the early part of each menstrual phase four or five primary follicles start to produce very low levels of estrogens. A clear membrane, the zona pellucida, also develops around the ova. Later in the menstrual phase (4–5 days) one primary follicle develops into a *secondary follicle* as the cells of the surrounding layer increase in number and differentiate, secreting follicular fluid. This fluid forces the ovum to the edge of the follicle. The production of estrogens by the secondary follicle elevates the estrogen level of the blood slightly. Ovarian follicle development is the result of FSHRF secretion by the hypothalamus, which in turn stimulates FSH production by the anterior pituitary. During this part of the ovarian cycle, FSH secretion is maximal. Although a number of follicles begin development each cycle, only one will attain maturity. The others undergo atresia (death).

The *preovulatory phase,* the second phase of the menstrual cycle, is the time between menstruation and ovulation. This phase of the menstrual cycle is more variable in length than the other phases. It lasts from days 6 to 13 in a 28-day cycle. During the preovulatory phase, a secondary follicle in the ovary matures into a *Graafian follicle,* a follicle ready for ovulation. During the maturation process, the follicle increases its estrogen production. Early in the preovulatory phase, FSH is the dominant hormone of the anterior pituitary, but close to the time of ovulation, LH is secreted in increasing quantities. Moreover, small amounts of progesterone may be produced by the Graafian follicle a day or two before ovulation.

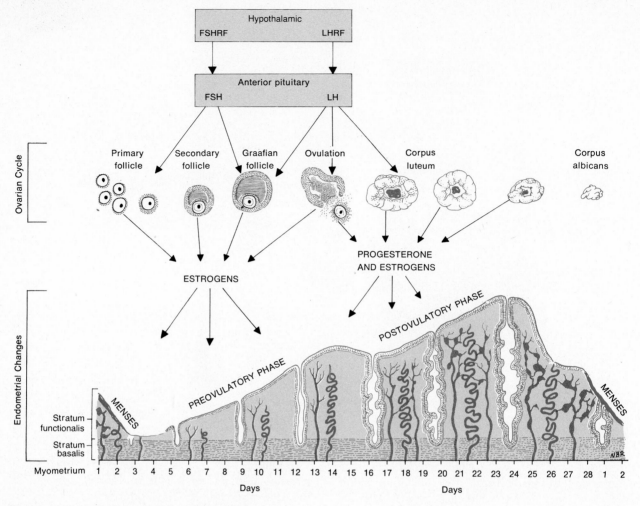

Figure 28—13 Menstrual and ovarian cycles.

FSH and LH stimulate the ovarian follicles to produce more estrogens, and this increase in estrogens stimulates the repair of the endometrium. Cells of the stratum basalis undergo mitosis and produce a new stratum functionalis. As the endometrium thickens, the short, straight endometrial glands develop and the arterioles coil and lengthen as they penetrate the functionalis. Because the proliferation of endometrial cells occurs during the preovulatory phase, the phase is also termed the *proliferative phase*. Still another name for this phase is the *follicular phase* because of increasing estrogen secretion by the developing follicle. Functionally, estrogen is the dominant ovarian hormone during this phase of the menstrual cycle.

Ovulation, the rupture of the Graafian follicle with release of the ovum into the pelvic cavity, occurs on day 14 in a 28-day cycle. Just prior to ovulation, the high estrogen level that developed during the preovulatory phase inhibits FSHRF by the hypothalamus. This, in turn, inhibits FSH secretion by the anterior pituitary.

Concurrently, LHRF is released by the hypothalamus. Thus LH secretion by the anterior pituitary is greatly increased. As FSH secretion is inhibited and LH and estrogen secretion increases, ovulation occurs. Following ovulation, the Graafian follicle collapses and blood within it forms a clot called the corpus hemorrhagicum. The clot is eventually absorbed by the remaining follicular cells. In time, the follicular cells enlarge, change character, and form the *corpus luteum,* or yellow body.

The *postovulatory phase* of the menstrual cycle is fairly constant in duration and lasts from days 15 to 28 in a 28-day cycle. It represents the time between ovulation and the onset of the next menses. Following ovulation, the level of estrogen in the blood drops slightly and LH secretion stimulates the development of the corpus luteum. The corpus luteum then secretes increasing quantities of estrogens and progesterone – the latter is responsible for preparing the endometrium to receive a fertilized ovum. Preparatory activities include secretory

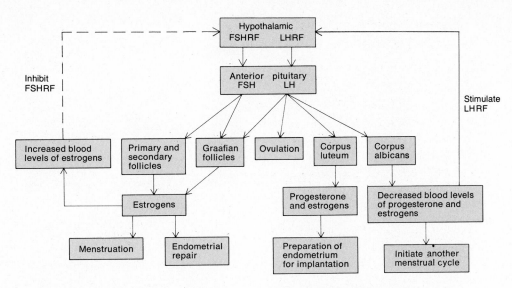

Figure 28-14   Summary of hormonal interactions of the menstrual and ovarian cycles.

activity of the endometrial glands that causes them to appear tortuously coiled, vascularization of the superficial endometrium, thickening of the endometrium, and an increase in the amount of tissue fluid. These preparatory changes are maximal about 1 week after ovulation, and they correspond to the anticipated arrival of the fertilized ovum. During the postovulatory phase, FSH secretion again gradually increases in response to FSHRF and LH secretion decreases as LHRF production decreases. The functionally dominant ovarian hormone during this phase is progesterone.

If fertilization and implantation do not occur, the rising levels of progesterone and estrogens from the corpus luteum inhibit LHRF and LH secretion. As a result the corpus luteum degenerates and becomes the *corpus albicans.* The decreased secretion of progesterone and estrogens by the degenerating corpus luteum then initiates another menstrual period. In addition, the decreased progesterone and estrogen levels in the blood bring about a new output of the anterior pituitary hormones—especially FSH in response to an increased output of FSHRF by the hypothalamus. Thus a new ovarian cycle is initiated. A summary of these hormonal interactions is presented in Figure 28-14.

If, however, fertilization and implantation do occur, the corpus luteum is maintained for about 4 months. For most of this time it continues to secrete estrogens and progesterone. The corpus luteum is maintained by *human chorionic gonadotropin,* a hormone produced by the developing placenta, until the placenta itself can secrete estrogen to support pregnancy and progesterone to support pregnancy and breast development for lactation.

The menstrual cycle normally occurs once each month from *menarche,* the first menses, to *menopause,* the last menses. The advent of menopause is signaled by the

climacteric—menstrual cycles become less frequent. The climacteric, which typically begins between ages 40 and 50, results from the failure of the ovaries to respond to the stimulation of gonadotropic hormones from the anterior pituitary. Some women experience hot flashes, copious sweating, headache, muscular pains, and emotional instability. In the postmenopausal woman there will be some atrophy of the ovaries, uterine tubes, uterus, vagina, external genitalia, and breasts.

The cause of menopause apparently is "burning out" of the ovaries. Throughout a woman's sexual life many of the primary ovarian follicles grow into Graafian follicles with each sexual cycle, and eventually most of them degenerate or are ovulated. At the age of approximately 45, therefore, few primary follicles remain to be stimulated by FSH and LH. As the number of primary follicles diminishes, the production of estrogens by the ovary decreases.

Vagina

The **vagina** serves as a passageway for the menstrual flow. It is also the receptacle for the penis during coitus, or sexual intercourse, and the lower portion of the birth canal. It is a muscular, tubular organ lined with mucous membrane and measures about 10 cm (4 inches) in length (Figures 28-7 and 28-8). Situated between the bladder and the rectum, it is directed upward and backward where it attaches to the uterus. A recess, called the *fornix,* surrounds the vaginal attachment to the cervix. The dorsal recess, called the posterior fornix, is larger than the ventral and two lateral fornices. The fornices make possible the use of contraceptive diaphragms. The mucosa of the vagina consists of stratified squamous epithelium and connective tissue that lies in a

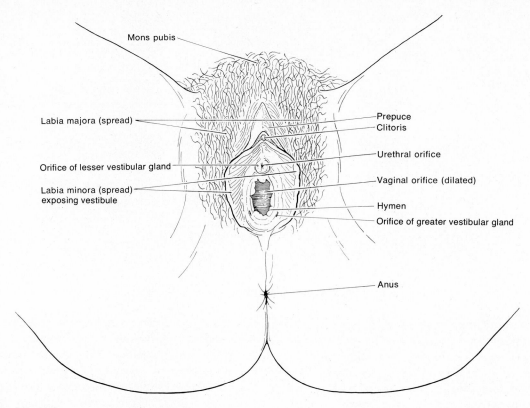

Mons pubis

Labia majora (spread)

Orifice of lesser vestibular gland

Labia minora (spread)
exposing vestibule

Prepuce
Clitoris

Urethral orifice

Vaginal orifice (dilated)

Hymen

Orifice of greater vestibular gland

Anus

Figure 28–15   Vulva.

series of transverse folds, the *rugae,* and is capable of a good deal of distension. The muscularis is composed of smooth muscle that can stretch considerably. This distension is important because the vagina receives the penis during sexual intercourse and serves as the lower portion of the birth canal. At the lower end of the vaginal opening *(vaginal orifice)* is a thin fold of vascularized mucous membrane called the *hymen,* which forms a border around the orifice, partially closing it (Figure 28-15). Sometimes the hymen completely covers the orifice, a condition called *imperforate hymen,* and surgery is required to open the orifice to permit the discharge of the menstrual flow. The mucosa of the vagina contains large amounts of glycogen that, upon decomposition, produce organic acids. These acids create a low-pH environment that retards microbial growth. However, the acidity is also injurious to sperm cells. Semen neutralizes the acidity of the vagina to ensure survival of the sperm.

Vulva

The term **vulva** or **pudendum** is a collective designation for the external genitalia of the female (see Figure 28-15).

The *mons pubis (veneris),* an elevation of adipose tissue covered by coarse pubic hair, is situated over the symphysis pubis. It lies in front of the vaginal and urethral openings. From the mons pubis, two longitudinal folds of skin, the *labia majora,* extend downward and backward. The labia majora, the female homologue of the scrotum, contain an abundance of adipose tissue and sebaceous and sweat glands; they are covered by hair on their upper outer surfaces. Medial to the labia majora are two folds of skin called the *labia minora.* Unlike the labia majora, the labia minora are devoid of hair and have few sweat glands. They do, however, contain numerous sebaceous glands.

The *clitoris* is a small, cylindrical mass of erectile tissue and nerves. It is located at the anterior junction of the labia minora. A layer of skin called the *prepuce,* or *foreskin,* is formed at the point where the labia minora unite and covers the body of the clitoris. The exposed portion of the clitoris is the *glans.* The clitoris is homologous to the penis of the male in that it is capable of enlargement upon tactile stimulation and assumes a role in sexual excitement of the female.

The cleft between the labia minora is called the *vestibule.* Within the vestibule are the hymen, vaginal orifice, urethral orifice, and the openings of several ducts.

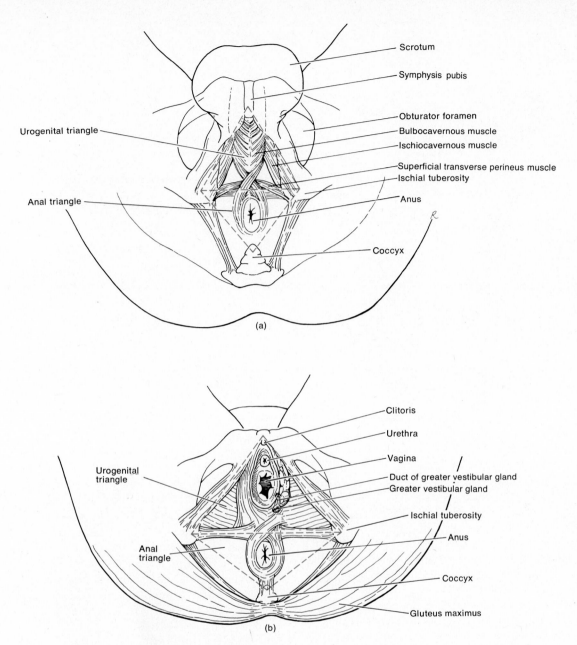

Scrotum

Symphysis pubis

Obturator foramen

Bulbocavernous muscle

Ischiocavernous muscle

Superficial transverse perineus muscle

Ischial tuberosity

Anus

Coccyx

Urogenital triangle

Anal triangle

(a)

Clitoris

Urethra

Vagina

Duct of greater vestibular gland

Greater vestibular gland

Ischial tuberosity

Anus

Coccyx

Gluteus maximus

Urogenital triangle

Anal triangle

(b)

Figure 28–16   Perineum. (a) Borders seen in the male. (b) Dissected view of some regions in the female.

The vaginal orifice occupies the greater portion of the vestibule and is bordered by the hymen. In front of the vaginal orifice and behind the clitoris is the *urethral orifice*. Behind and to either side of the urethral orifice are the openings of the ducts of the *lesser vestibular* or *Skene's glands*. These glands secrete mucus. On either side of the vaginal orifice itself are two small glands: the *greater vestibular* or *Bartholin's glands*. These glands open by a duct into the space between the hymen and labia minora and produce a mucoid secretion that supplements lubrication during sexual intercourse. The lesser vestibular glands are homologous to the male prostate. The

greater vestibular glands are homologous to the male bulbourethral or Cowper's glands.

Perineum

The **perineum** is the diamond-shaped area at the lower end of the trunk between the thighs and buttocks of both males and females (Figure 28-16). It is surrounded anteriorly by the symphysis pubis, laterally by the ischial tuberosities, and posteriorly by the coccyx. A transverse line drawn between the ischial tuberosities divides the perineum into an anterior *urogenital triangle* that con-

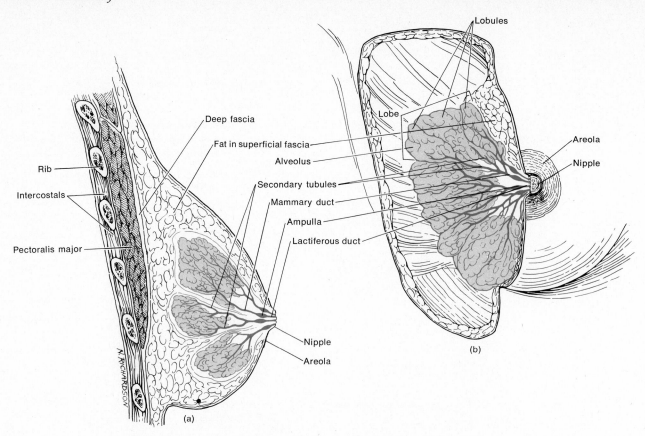

Figure 28–17   Mammary glands. (a) Sagittal section. (b) Front view partially sectioned.

tains the external genitalia and a posterior *anal triangle* that contains the anus. In the female, the region between the vagina and anus is known as the *clinical perineum*. If the vagina is too small to accommodate the head of an emerging fetus, the skin and underlying tissue of the clinical perineum tear. To avoid this, a small incision called an *episiotomy* is made in the perineal skin just prior to delivery.

Mammary Glands

The **mammary glands** are branched tubuloalveolar glands that lie over the pectoralis major muscles and are attached to them by a layer of connective tissue (Figure 28-17). Internally, each mammary gland consists of 15 to 20 *lobes,* or compartments, separated by adipose tissue. The amount of adipose tissue determines the size of the breasts. However, breast size has nothing to do with the amount of milk produced. In each lobe are several smaller compartments, called *lobules,* composed of connective tissue in which milk-secreting cells referred to as *alveoli* are embedded. Between the lobules are strands of connective tissue called the *suspensory ligaments of Cooper.* These ligaments run between the skin and deep fascia and support the breast. Alveoli are arranged in grapelike clusters. They convey the milk into a series of *secondary*

tubules. From here the milk passes into the *mammary ducts.* As the mammary ducts approach the nipple, expanded sinuses called *ampullae,* where milk may be stored, are present. The ampullae continue as *lactiferous ducts* that terminate in the *nipple.* Each lactiferous duct conveys milk from one of the lobes to the exterior, although some may join before reaching the surface. The circular pigmented area of skin surrounding the nipple is called the *areola.* It appears rough because it contains modified sebaceous glands.

At birth, both male and female mammary glands are undeveloped and appear as slight elevations on the chest. With the onset of puberty, the female breasts begin to develop – the mammary ducts elongate, extensive fat deposition occurs, and the areola and nipple grow and become pigmented. These changes are correlated with an increased output of estrogen by the ovary. Further mammary development occurs at sexual maturity with the onset of ovulation and the formation of the corpus luteum. During adolescence, lobules are formed and fat deposition continues, increasing the size of the glands. Although these changes are associated with estrogen and progesterone secretion by the ovaries, ovarian secretion is ultimately controlled by FSH, which is secreted in response to FSHRF by the hypothalamus.

The essential function of the mammary glands is milk secretion or lactation.

DISORDERS

VENEREAL DISEASES

The term *venereal* comes from Venus, goddess of love. **Venereal diseases** represent a group of infectious diseases that are spread primarily through sexual intercourse. After the common cold, venereal diseases are ranked as the most common communicable diseases in the United States.

Gonorrhea, more commonly known as "clap," is an infectious disease that primarily affects the mucous membrane of the urogenital tract, the rectum, and occasionally the eyes. The disease is caused by the bacterium *Neisseria gonorrhoeae.* Males usually suffer inflammation of the urethra with pus and painful urination. Fibrosis sometimes occurs in an advanced stage of gonorrhea, causing stricture of the urethra. There also may be involvement of the epididymis and prostate gland. In females, infection may occur in the urethra, vagina, and cervix, and there may be a discharge of pus. If the uterine tubes become involved, sterility and pelvic inflammation may result. Females can harbor the disease without any symptoms.

Discharges from the involved mucous membranes are the source of infection, and the bacteria are transmitted by direct contact, usually sexual. If the bacteria are transmitted to the eyes of a newborn in the birth canal, blindness may result. Administration of a 1 percent silver nitrate solution or penicillin in the infant's eyes prevents infection. Penicillin is the drug of choice for the treatment of gonorrhea in adults.

Syphilis is an infectious disease caused by the bacterium *Treponema pallidum.* It also is acquired through sexual contact. The early stages of the disease primarily affect the organs most likely to have made sexual contact—the genital organs, the mouth, and the rectum. The point where the bacteria enter the body is marked by a lesion called a *chancre.* In males, it usually occurs on the penis; in females, in the vagina or cervix. The chancre heals without scarring. Following the initial infection, the bacteria enter the bloodstream and spread throughout the body. In some individuals, an active secondary stage of the disease occurs, characterized by lesions of the skin and mucous membranes and often fever. The signs of the secondary stage also go away without medical treatment. During the next several years, the disease progresses without symptoms and is said to be in a latent phase. When symptoms again appear, anywhere from 5 to 40 years after the initial infection, the person is said to be in the tertiary stage of the disease. Tertiary syphilis may involve the circulatory system, skin, bones, viscera, and the nervous system.

MALE DISORDERS

PROSTATE

The prostate gland is susceptible to infection, enlargement, and benign and malignant tumors. Because the prostate surrounds the urethra, any of these disorders can obstruct the flow of urine. Prolonged obstruction may result in serious changes in the bladder, ureters, and kidneys.

Acute and chronic infections of the prostate gland are common in postpubescent males, many times in association with inflammation of the urethra. In **acute prostatitis** the prostate gland becomes swollen and tender. Appropriate antibiotic therapy, bed rest, and above-normal fluid intake are effective treatment.

Chronic prostatitis is one of the most common chronic infections in men of the middle and later years. On examination, the prostate gland feels enlarged, soft, and extremely tender. The surface outline is irregular and may be hard. This disease frequently produces no symptoms, but the prostate is believed to harbor infectious microorganisms responsible for some allergic conditions, arthritis, and inflammation of nerves (neuritis), muscles (myositis), and the iris (iritis).

An **enlarged prostate** gland occurs in approximately one-third of all males over age 60. The affected gland is two to four times larger than normal. The cause is unknown, and the enlarged condition usually can be detected by rectal examination.

Tumors of the male reproductive system primarily involve the prostate gland. Carcinoma of the prostate is the second leading cause of death from cancer in men in the United States. It is responsible for approximately 19,000 deaths annually. Its incidence is related to age, race, occupation, geography, and ethnic origin. Both benign and malignant growths are common in elderly men. Both types of tumors put pressure on the urethra, making urination painful and difficult. At times, the excessive back pressure destroys kidney tissue and gives rise to an increased susceptibility to infection. Therefore, even when the tumor is benign, surgery is indicated to remove the prostate or parts of it if the growth is obstructive and perpetuates urinary tract infections.

SEXUAL FUNCTIONAL ABNORMALITIES

Impotence is the inability of an adult male to attain or hold an erection long enough for normal intercourse. Impotence could be the result of physical abnormalities of the penis; systemic disorders such as syphilis; vascular disturbances; neurological disorders; or psychic factors such as fear of causing pregnancy, fear of venereal disease, religious inhibitions, and emotional immaturity.

Infertility, or **sterility,** is an inability to fertilize the ovum. It does not imply impotence. Male fertility requires viable spermatozoa, adequate production of spermatozoa by the testes, unobstructed transportation of sperm through the seminal tract, and satisfactory deposition in the vagina. The tubules of the testes are sensitive to many factors—x rays, infections, toxins, malnutrition—that may cause degenerative changes and produce male sterility.

If inadequate spermatozoa production is suspected, a sperm analysis should be performed. Analysis includes measuring the volume of semen, counting the number of sperm per milliliter, evaluating sperm motility 4 hours after ejaculation, and determining the percentage of abnormal sperm forms (not to exceed 20 percent).

FEMALE DISORDERS

MENSTRUAL ABNORMALITIES

Disorders of the female reproductive system frequently include menstrual disorders. This is hardly surprising because proper menstruation reflects not only the health of the uterus but the health of the glands that control it: the ovaries and the pituitary gland.

Amenorrhea is the absence of menstruation. If a woman has never menstruated, the condition is called *primary amenorrhea*. Primary amenorrhea can be caused by endocrine disorders, most often in the pituitary gland and hypothalamus, or by genetically caused abnormal development of the ovaries or uterus. *Secondary amenorrhea* is cessation of uterine bleeding in women who have previously menstruated. If pregnancy is ruled out, various endocrine disturbances are considered.

Dysmenorrhea is painful menstruation caused by contraction of the uterine muscles. A primary cause is believed to be inadequate progesterone, the hormone that prevents uterine contraction. It can also be caused by pelvic inflammatory disease, uterine tumors, cystic ovaries, or congenital defects.

Abnormal uterine bleeding includes menstruation of excessive duration and/or excessive amount, too frequent menstruation, intermenstrual bleeding, and postmenopausal bleeding. These abnormalities may be caused by disordered hormonal regulation, emotional factors, and systemic diseases.

OVARIAN CYSTS

Ovarian cysts are tumors of the ovary that contain fluid. Follicular cysts may occur in the ovaries of elderly women, in ovaries that have inflammatory diseases, and in menstruating females. They have thin walls and contain a serous albuminous material. Cysts may also arise from the corpus luteum or the endometrium. The endometrium is the inner lining of the uterus that is sloughed off in menstruation. *Endometriosis* occurs when the endometrial tissue grows outside the uterus. The tissue enters the pelvic cavity via the open Fallopian tubes and may be found in any of a dozen sites—on the ovaries, cervix, abdominal wall, and bladder. Causes are unknown. Endometriosis is common in women 30 to 40 years of age. Symptoms include premenstrual pain or unusual menstrual pain. The unusual pain is caused by the displaced tissue sloughing off at the same time the normal uterine endometrium is being shed during menstruation. Infertility can be a consequence. Treatment usually consists of hormone therapy or surgery. Endometriosis disappears at menopause or when the ovaries are removed.

INFERTILITY

Female infertility, or the inability to conceive, occurs in about 10 percent of married females in the United States. Once it is established that ovulation occurs regularly, the reproductive tract is examined for functional and anatomical disorders to determine the possibility of union of the sperm and the ovum in the oviduct.

BREASTS

The breasts of females are highly susceptible to cysts and tumors. Men are also susceptible to breast tumor, but certain breast cancers are 100 times more common in women.

In the female, the benign *fibroadenoma* is a common tumor of the breast. It occurs most frequently in young women. Fibroadenomas have a firm rubbery consistency and are easily moved about within the mammary tissue. The usual treatment is excision of the growth. The breast itself is not removed.

Breast cancer has the highest fatality rate of all cancers affecting women, but it is rare in men. In the female, breast cancer is rarely seen before age 30, and its occurrence rises rapidly after menopause. Breast cancer is generally not painful until it becomes quite advanced, so often it is not discovered early or, if noted, ignored. Any lump, no matter how small, should be reported to a doctor at once. If there is no evidence of *metastasis* (the spread of cancer cells to another part of the body), the treatment of choice is a *modified* or *radical mastectomy*. A radical mastectomy involves removal of the affected breast along with the underlying pectoral muscles and the axillary lymph nodes. Metastasis of cancerous cells is usually through the lymphatics or blood. Radiation treatment may follow the surgery to ensure the destruction of any stray cancer cells.

The mortality from breast cancer has not improved significantly in the last 50 years. Early detection—especially by breast self-examination and mammography—is still the most promising method to increase the survival rate.

It is estimated that 95 percent of breast cancer is first detected by the women themselves. Each month after the menstrual period the breasts should be thoroughly examined for lumps, puckering of the skin, or discharge.

Mammography is a sophisticated breast x-ray technique used to detect breast masses and determine whether they are malignant. The examination consists of two x-ray, right-angle views of each breast. Mammographic diagnosis of breast masses is 80 percent reliable. Mammographic x-ray prints are also used to guide surgeons performing mastectomies. As an aid in analyzing mammographic findings, a new x-ray technique, *xeroradiography,* is being used. In this photoelectric (rather than photochemical) method the x-ray image is reproduced on paper instead of film. Xeroradiography provides excellent soft-tissue detail and requires less radiation than film mammography.

Modern x-ray films, xeroradiography, and special x-ray techniques have reduced the problem of mammographic radiation. Ovaries are not exposed to radiation during mammography, and the technique can be used safely on pregnant women. Most cancer experts agree that mammography should be used only after a careful clinical examination and under limited conditions. *Thermography,* a method of measuring and graphically recording heat radiation emitted by the breast, is also frequently used in. conjunction with mammography. Tumors, both benign and malignant, emit more heat than nonaffected areas.

CERVICAL CANCER

Another common disorder of the female reproductive system is cancer of the uterine cervix. It ranks third in frequency after breast and skin cancers. **Cervical cancer** starts with a change in the shape of the cervical cells called *cervical dysplasia.* Cervical dysplasia is not a cancer in itself, but the abnormal cells tend to become malignant.

Cervical cancer, for the most part, is a venereal disease with a long incubation period. Inciting factors are not known, but herpesvirus type II has recently become suspect. Smegma and the DNA of spermatozoa have also been implicated. Cancer of the cervix (except for adenocarcinoma) does not occur in celibate women, and for unknown reasons it is rare in Jewish women.

Early diagnosis of cancer of the uterus is accomplished by the *Papanicolaou test,* or "Pap" smear. In this generally painless procedure, a few cells from the vaginal fornix (that part of the vagina surrounding the cervix) and the cervix are removed with a swab and examined microscopically. Malignant cells have a characteristic appearance and indicate an early stage of cancer, even before symptoms occur. Estimates indicate that the Pap smear is more than 90 percent reliable in detecting cancer of the cervix. Treatment of cervical cancer may involve complete or partial removal of the uterus, called a *hysterectomy,* or radiation treatment.

MEDICAL TERMINOLOGY

Castration   Excision of the testes or ovaries.

Colposcopy   Use of a low-power binocular microscope to examine vaginal, exocervical, and a portion of the endocervical mucosa under magnification inside the body.

Copulation   Sexual intercourse. Coitus refers to sexual intercourse between human beings.

Leukorrhea   A nonbloody vaginal discharge that may occur at any age and affects most women at some time.

Neoplasia   A condition characterized by the presence of new growths (tumors).

Oophorectomy   Excision of an ovary. Bilateral oophorectomy refers to the removal of both ovaries.

Pruritis   Itching.

Purulent   Containing or consisting of pus.

Salpingectomy   Excision of a uterine tube.

Salpingitis   Inflammation or infection of the uterine tube.

Smegma   The secretion, consisting principally of desquamated epithelial cells, found chiefly about the external genitalia and especially under the foreskin of the male.

Vaginitis   Inflammation of the vagina.

STUDY OUTLINE

MALE REPRODUCTIVE SYSTEM

1 The scrotum is a pouching of the abdominal wall that provides an appropriate temperature for the testes.

2 The major functions of the testes are sperm production and the secretion of testosterone.

3 FSH initiates spermatogenesis and ICSH stimulates the secretion of testosterone, which maintains the growth and development of the male reproductive organs.

4 Sperm cells are conveyed from the testes to the exterior through the convoluted seminiferous tubules, straight tubules, rete testis, efferent ducts, ductus epididymis, ductus deferens, ejaculatory duct, and urethra.

5 The seminal vesicles, prostate, and Cowper's glands secrete the liquid portion of semen.

6 Semen is a mixture of sperm and secreted liquids.

7 The penis serves as the organ of copulation.

FEMALE REPRODUCTIVE SYSTEM

1 The ovaries produce ova and secrete estrogens and progesterone.

2 The uterine tubes convey ova from the ovaries to the uterus and are the sites of fertilization.

3 The normal position of the uterus is maintained by a series of ligaments.

4 The uterus is associated with menstruation, implantation of a fertilized ovum, development of the fetus, and labor.

5 The function of the menstrual cycle is to prepare the endometrium each month for the reception of a fertilized egg.

6 The ovarian cycle produces a mature ovum each month.

7 FSH and FSHRF and LH and LHRF control the ovarian cycle. Estrogens and progesterone control the menstrual cycle.

8 The female climacteric is the time immediately before menopause.

9 Menopause is the cessation of the sexual cycles.

10 The vagina serves as a passageway for the menstrual flow, as the lower portion of the birth canal, and as the receptacle for the penis.

11 The vulva is a collective designation for the external genitalia of the female.

12 The perineum is a diamond-shaped area at the lower end of the trunk between the thighs and buttocks. The clinical perineum is the region from the vagina to the anus.

13 The mammary glands function in the secretion of milk.

REVIEW QUESTIONS

1 Define reproduction. List the male and female organs of reproduction.

2 Describe the function of the scrotum in protecting the testes from temperature fluctuations.

3 Describe the internal structure of a testis. Where are the sperm cells made?

4 Identify the principal parts of a spermatozoan.

5 Explain the effects of FSH and ICSH on the male reproductive system.

6 Describe the physiological effects of testosterone. How is the testosterone level in the blood controlled?

7 Trace the course of a sperm cell through the male system of ducts from the seminiferous tubules through the urethra.

8 What is the spermatic cord?

9 Briefly explain the functions of the seminal vesicles, prostate gland, and Cowper's glands.

10 What is seminal fluid? What is its function?

11 How is the penis structurally adapted as an organ of copulation?

12 How are the ovaries held in position in the pelvic cavity? What is ovulation?

13 What is the function of the uterine tubes? Define an ectopic pregnancy.

14 Diagram the principal parts of the uterus.

15 Describe the arrangement of ligaments that hold the uterus in its normal position. Explain the two major malpositions of the uterus.

16 Discuss the blood supply to the uterus. Why is an abundant blood supply important?

17 Define menstrual cycle and ovarian cycle. What is the function of each?

18 Briefly outline the major events of the menstrual cycle and correlate them with the events of the ovarian cycle.

19 Prepare a labeled diagram of the principal hormonal interactions involved in the menstrual and ovarian cycles.

20 Define female climacteric and menopause.

21 Explain some of the physiological effects of the climacteric and menopause.

22 What are the functions of the vagina?

23 List the parts of the vulva.

24 What is the perineum? Define episiotomy.

25 Describe the passage of milk from the alveoli cells of the mammary gland to the nipple.

Chapter 29
Development
and
Inheritance

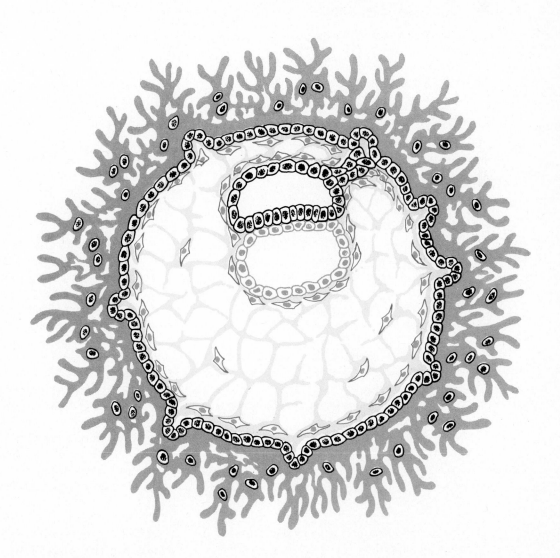

STUDENT OBJECTIVES

- Define meiosis.

- Contrast the events of spermatogenesis and oogenesis.

- Describe and discuss the role of the male and female in sexual intercourse.

- Describe the activities associated with fertilization and implantation.

- Discuss the formation of the primary germ layers, embryonic membranes, and placenta.

- List the body structures produced by the primary germ layers.

- Describe the function of the embryonic membranes.

- Describe the roles of the placenta and umbilical cord during embryonic and fetal growth.

- Discuss the principal body changes associated with fetal growth.

- Compare the sources and functions of the hormones secreted during pregnancy.

- Describe the three stages of labor.

- Describe the physiology of lactation.

- Define inheritance and describe the inheritance of PKU, sex, and color blindness.

- Contrast the various kinds of birth control and their effectiveness.

- Describe the procedure of amniocentesis and its value in preventing disease in the newborn.

- Describe the cause and symptoms of Down's syndrome.

- Define medical terminology associated with development and inheritance.

The *developmental processes* are a sequence of events starting with the fertilization of an egg and ending with the formation of a complete organism. As we look at this sequence, we will consider how reproductive cells are produced, the role of the male and female in sexual intercourse, and events associated with pregnancy, birth, and lactation. Finally, we will say a few words about inheritance.

GAMETE FORMATION

Chromosome Number

Each human being develops from the union of an ovum and a sperm. Ova and sperm are collectively called **gametes,** and they differ radically from all the other body cells in that they have only half the normal number of chromosomes in their nuclei. **Chromosome number** is the number of chromosomes in each nucleated cell that is not a gamete. Chromosome numbers vary from species to species. The human chromosome number is 46 – each brain cell, stomach cell, heart cell, and practically every other cell contains 46 chromosomes in its nucleus. In other words, there are 23 pairs of chromosomes in each cell other than a gamete. Two chromosomes that belong to a pair are called homologous chromosomes. The ovum or sperm has only one member of each pair. Of these 46 chromosomes, 23 contain the genes that are necessary for programming all the activities of the body. In a sense, the other 23 are a duplicate set. Another word for chromosome number is **diploid number** (*di-* = two), symbolized as *2n.*

A sperm containing 46 chromosomes that fertilizes an egg containing 46 chromosomes might be thought to create offspring with 92 chromosomes. In reality, the chromosome number does not double with each generation because of a special nuclear division called **meiosis.** Meiosis occurs in the process of producing sex cells. It causes a developing sperm or ovum to relinquish its duplicate set of chromosomes so that the mature gamete has only 23 – this is the **haploid number,** meaning "one-half" and symbolized *n.*

In the testes, the formation of haploid spermatozoa by meiosis is called **spermatogenesis.** In the ovary, the formation of a haploid ovum by meiosis is referred to as **oogenesis.**

Spermatogenesis

In humans, spermatogenesis takes about 2 ½ months. The seminiferous tubules are lined with immature cells called *spermatogonia* or sperm mother cells (Figure 29-1). Spermatogonia contain the diploid chromosome number and are the precursor cells for all the sperma-

tozoa the male will produce. At puberty, when the anterior pituitary secretes FSH in response to FSHRF from the hypothalamus, the spermatogonia embark on a lifetime of active division. As a result of their active mitosis, daughter cells are pushed inward toward the lumen of the seminiferous tubule. These cells lose contact with the basement membrane of the seminiferous tubule. And, as a result of certain developmental changes, these cells become known as *primary spermatocytes.* Primary spermatocytes, like spermatogonia, are diploid. The other daughter cells formed by mitosis of the spermatogonia remain near the basement membrane and form a reservoir of precursor cells.

Each primary spermatocyte enlarges before dividing. This nuclear division is the first of two that take place as part of meiosis. In the first, DNA is replicated and 46 chromosomes form and move toward the equatorial plane of the nucleus. There they line up by homologous pairs so that there are 23 pairs of chromosomes (each of two chromatids) in the center of the nucleus. The four chromatids of each homologous pair then twist around each other to form a *tetrad.* In a tetrad, portions of one chromatid are exchanged with portions of another. This process, called *crossing-over,* permits an exchange of genes among chromatids (Figure 29-2) that results in the recombination of genes. Thus the spermatozoa eventually produced may be genetically unlike each other and unlike the cell that produced them – hence the great variation among humans. Following crossing-over the spindle forms and the threads attach to the centromeres of the paired chromosomes. As the pairs separate, one member of each pair migrates to opposite poles of the dividing nucleus. The cells thus formed by the first nuclear division (reduction division) are called *secondary spermatocytes.* Each cell has 23 chromosomes – the haploid number. Each chromosome, however, is made up of two chromatids. Moreover, the genes of the chromosomes of secondary spermatocytes are rearranged as a result of crossing-over.

The second nuclear division of meiosis is equatorial division. There is no replication of DNA. The chromosomes (each of two chromatids) line up in single file around the equatorial plane, and the chromatids of each chromosome separate from each other. The cells thus formed from the equatorial division are called *spermatids.* Each contains half the original chromosome number, or 23 chromosomes, and is haploid. Each primary spermatocyte therefore produces four spermatids by meiosis (reduction division and equatorial division). Spermatids lie close to the lumen of the seminiferous tubule.

The final stage of spermatogenesis – *spermiogenesis* – involves the maturation of spermatids into spermatozoa. Each spermatid embeds in a Sertoli cell and develops a head with an acrosome and a flagellum (tail).

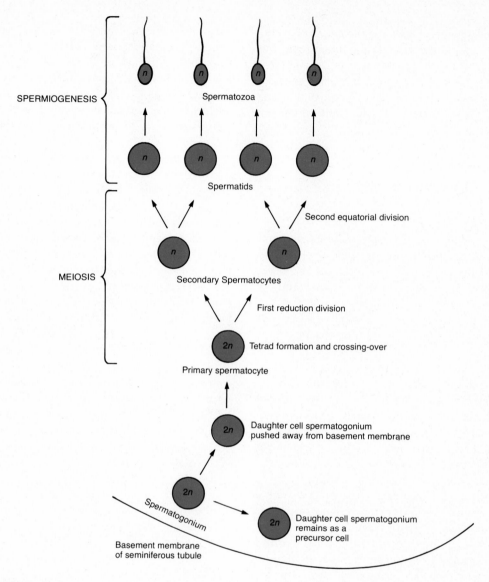

SPERMIOGENESIS

Spermatozoa

Spermatids

MEIOSIS

Second equatorial division

n

n

Secondary Spermatocytes

First reduction division

2*n* Tetrad formation and crossing-over

Primary spermatocyte

2*n* Daughter cell spermatogonium pushed away from basement membrane

2*n* Spermatogonium

2*n* Daughter cell spermatogonium remains as a precursor cell

Basement membrane of seminiferous tubule

Figure 29–1   Spermatogenesis.

Sertoli cells nourish the developing spermatids. Since there is no cell division in spermiogenesis, each spermatid develops into a single spermatozoan.

In summary, then, a single primary spermatocyte develops into four spermatozoa by meiosis and spermiogenesis. Spermatozoa enter the lumen of the seminiferous tubule and migrate to the ductus epididymis. After about 10 days, they complete their maturation and become capable of fertilizing an ovum. Many spermatozoa are probably stored in the ductus deferens.

Oogenesis

The formation of a haploid ovum by meiosis in the ovary is referred to as oogenesis. With some exceptions,

oogenesis occurs in essentially the same manner as spermatogenesis: It involves meiosis and maturation.

The precursor cell in oogenesis is a diploid cell called the *oogonium,* or egg mother cell (Figure 29-3). By the time the female fetus is ready for birth, the oogonia in the primary follicles have lost their ability to carry on mitosis. Such immature cells containing the diploid number of chromosomes are called *primary oocytes.* They remain in this stage until their follicular cells respond to FSH from the anterior pituitary, which in turn has responded to FSHRF from the hypothalamus. Starting with puberty, several follicles respond each month to the rising level of FSH. As the cycle proceeds and LH is secreted from the anterior pituitary, one of the follicles reaches a stage in which the diploid ovum, now a mature primary oocyte, undergoes its reduction division. Tetrad formation and crossing-over occur, and two cells of unequal size, both with 23 chromosomes of two chromatids each, are produced. The smaller cell is called the *first polar body* and is essentially a packet of discarded nuclear material. The larger cell, which receives most of the cytoplasm, is known as the *secondary oocyte.*

At this stage in the ovarian cycle, ovulation takes place. Since the secondary oocyte with its polar body and surrounding supporting cells is discharged at the time of ovulation, the "ovum" discharged is not yet mature. The discharged secondary oocyte enters the uterine tube and, if spermatozoa are present and fertilization occurs, the second division takes place: the equatorial division. The secondary oocyte produces two cells of unequal size, both of them haploid. The larger cell is called the *ootid;* the smaller is the *second polar body.* In time, the ootid develops into an *ovum,* or mature egg.

The first polar body may undergo another division. If it does, meiosis of the primary oocyte results in a single haploid ovum and three polar bodies. In any event, the polar bodies disintegrate. Thus in the female one oogonium produces a single ovum whereas each male spermatocyte produces four spermatozoa.

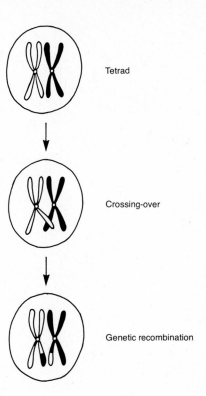

Figure 29–2 Crossing-over within a tetrad resulting in genetic recombination.

SEXUAL INTERCOURSE

Fertilization of an ovum is accomplished through *sexual intercourse* or *copulation,* or *coitus,* in which spermatozoa are deposited in the vagina. The male role in the sexual act starts with erection, the enlargement and stiffening of the penis. An erection may be initiated in the cerebrum by stimuli such as anticipation, memory, and visual sensation, or it may be a reflex brought on by stimulation of the touch receptors in the penis, especially in the glans. In any case, parasympathetic impulses that pass from the sacral portion of the spinal cord to the penis cause dilation of the arteries of the penis, allowing blood to fill the cavernous spaces of the spongy bodies. These impulses also cause the Cowper's glands to secrete mucus that affords some lubrication for intercourse. The major portion of lubricating fluid is produced by the female.

Tactile stimulation of the penis brings about emission and ejaculation. When sexual stimulation becomes intense, rhythmic sympathetic impulses leave the spinal cord at the levels of the first and second lumbar vertebrae and pass to the genital organs. These impulses cause peristaltic contractions of the ducts in the testes, the epididymides, and the seminal ducts that propel spermatozoa into the urethra—a process called *emission.* Simultaneously, peristaltic contractions of the seminal vesicles and prostate expel semen and prostatic fluid along with the spermatozoa. All these mix with the mucus of the Cowper's glands, resulting in the fluid called semen. Other rhythmic impulses sent from the spinal cord at the levels of the first and second sacral vertebrae reach the skeletal muscles at the base of the penis, and the penis expels the semen from the urethra to the exterior. The propulsion of semen from the urethra to the exterior constitutes an *ejaculation.* A number of sensory and motor activities accompany ejaculation, including a rapid heart rate, an increase in blood pressure, an increase in respiration, and pleasurable sensations. These activities, together with the muscular events involved in ejaculation, are referred to as an *orgasm.*

The female role in the sex act also involves erection, lubrication, and orgasm. Stimulation of the female, as in the male, depends on both psychic and tactile responses. Under appropriate conditions, stimulation of the female genitalia, especially the clitoris, results in *erection* and widespread sexual arousal. This response is controlled by parasympathetic impulses sent from the spinal cord to the external genitalia. These impulses also pass to the Bartholin's (greater vestibular) glands and vaginal mucosa, which secrete some of the *lubrication* during sexual intercourse. When tactile stimulation of the genitalia reaches maximum intensity, reflexes are initiated that cause the female *orgasm* or *climax.* Female

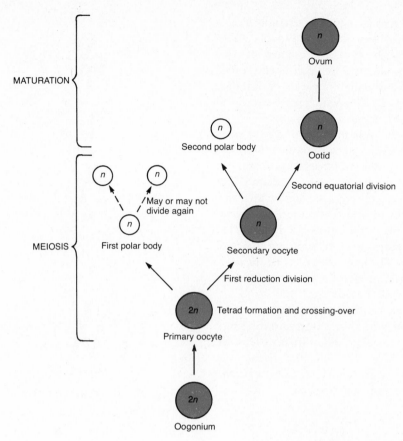

Figure 29-3 Oogenesis.

orgasm is somewhat like male ejaculation, except that there is no expulsion of semen, although there may be an increased secretion of cervical mucus. The perineal muscles contract rhythmically from spinal reflexes similar to those that occur in the male ejaculation. These same impulses also cause peristaltic movements of the uterus and uterine tubes, thus helping to transport the spermatozoa toward the ovum.

PREGNANCY

Once spermatozoa and ova are developed through meiosis and the spermatozoa are deposited in the vagina, pregnancy can occur. **Pregnancy** is a sequence of events including fertilization, implantation, embryonic growth, and, normally, fetal growth that terminates in birth.

Fertilization

The term **fertilization** is applied to the union of the sperm nucleus and the nucleus of the ovum. It normally occurs in the uterine tube when the ovum is about one-third of the way down the tube, usually within 24

hours after ovulation (Figure 29-4a). Peristaltic contractions and the action of cilia transport the ovum through the uterine tube. The mechanism by which sperm reach the uterine tube is still unclear. Some believe that sperm swim up the female tract by means of whiplike movements of their flagella; others believe sperm are transported by muscular contractions of the uterus. Their motility is probably a combination of both.

Sperm must remain in the female genital tract 4 to 6 hours before they are capable of fertilizing an ovum. During this time, the enzyme hyaluronidase is activated and secreted by the acrosomes of the spermatozoa. Hyaluronidase apparently dissolves parts of the membrane covering the ovum. Normally, only one spermatozoan fertilizes an ovum because once union is achieved, the ovum develops a fertilization membrane that is impermeable to the entrance of other spermatozoa. When the spermatozoan has entered the ovum, the tail is shed and the nucleus in the head develops into a structure called the *male pronucleus*. The nucleus of the ovum develops into a *female pronucleus*. After the pronuclei are formed, they fuse to produce a *segmentation nucleus* – a process termed fertilization. The segmentation nucleus contains 23 chromosomes from the male

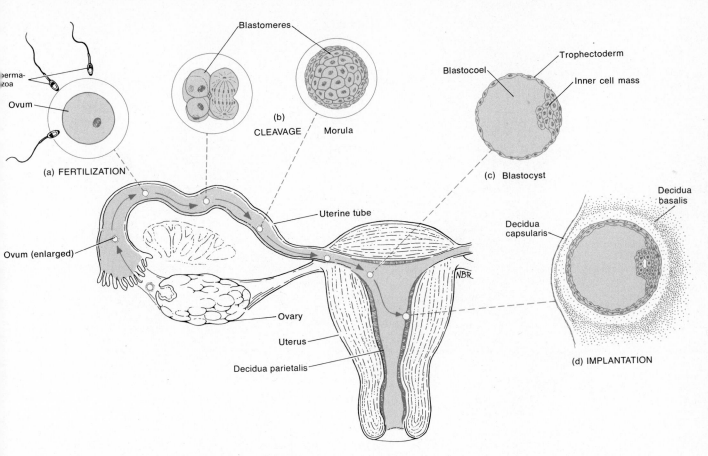

Figure 29—4 Fertilization, cleavage, and implantation of an ovum.

pronucleus and 23 chromosomes from the female pronucleus. Thus the fusion of the haploid pronuclei restores the diploid number. The fertilized ovum, consisting of a segmentation nucleus, cytoplasm, and enveloping membrane, is called a *zygote.*

Immediately after fertilization, rapid cell division of the zygote takes place (see Figure 29-4b). This early division of the zygote is called *cleavage.* The progressively smaller cells produced are called *blastomeres.* Successive cleavages produce a solid mass of cells, the *morula,* which is only slightly larger than the original zygote.

Implantation

As the morula descends through the uterine tube, it continues to divide and eventually forms a hollow ball of cells. At this stage of development the mass is referred to as a *blastocyst* (see Figure 29-4c). The blastocyst is differentiated into an outer covering of cells called the *trophectoderm* and an *inner cell mass,* and the internal cavity is referred to as the *blastocoel.* Whereas the trophectoderm ultimately will form the membranes composing the fetal portion of the placenta, the inner

cell mass will develop eventually into the embryo. About the fifth day after fertilization, the blastocyst enters the uterine cavity.

The attachment of the blastocyst to the endometrium occurs 7 to 8 days following fertilization and is called **implantation** (see Figure 29-4d). At this time, the endometrium is in its postovulatory phase. During implantation, the cells of the trophectoderm secrete an enzyme that enables the blastocyst to penetrate the uterine lining and become buried in the endometrium, usually on the posterior wall of the fundus or body of the uterus. The blastocyst becomes oriented so that the inner cell mass is toward the endometrium. Implantation enables the blastocyst to absorb nutrients from the glands and blood vessels of the endometrium for its subsequent growth and development.

Embryonic Period

The first 2 months of development are considered the **embryonic period.** During this period the developing human is called an *embryo.* After the second month it will be called a *fetus.* By the end of the embryonic period the rudiments of all the principal adult organs are present,

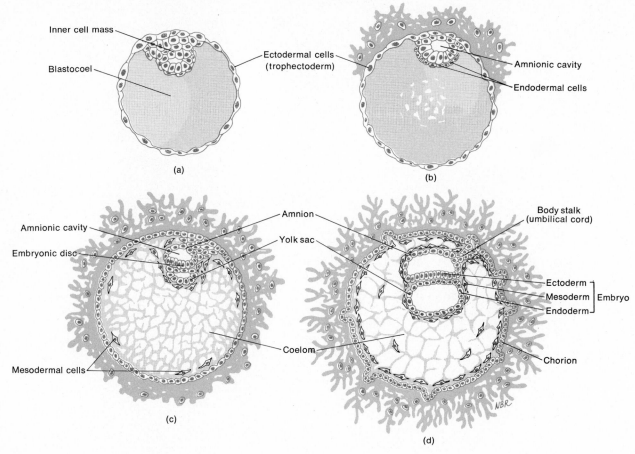

Inner cell mass

Blastocoel

Ectodermal cells
(trophectoderm)

Amnionic cavity

Endodermal cells

(a)

(b)

Amnionic cavity

Embryonic disc

Mesodermal cells

Amnion

Yolk sac

Coelom

Body stalk
(umbilical cord)

Ectoderm
Mesoderm } Embryo
Endoderm

Chorion

(c)

(d)

Figure 29–5   Formation of the three primary germ layers.

the embryonic membranes are developed, and the placenta is functioning. Let us now examine these events in detail.

BEGINNINGS OF ORGAN SYSTEMS

Following implantation, the inner cell mass of the blastocyst begins to differentiate into the three primary germ layers: ectoderm, endoderm, and mesoderm. The **primary germ layers** are the embryonic tissues from which all tissues and organs of the body will develop.

In the human being, the germ layers form so quickly that it is difficult to determine the exact sequence of events. Before implantation, a layer of *ectoderm* (the trophectoderm) already has formed around the blastocoel (Figure 29-5). The trophectoderm will become part of the chorion—one of the fetal membranes. Within 8 days after implantation, the inner cell mass moves downward so a space called the amnionic cavity lies between the inner cell mass and the trophectoderm. The bottom layer of the inner cell mass develops into an *endodermal* germ layer.

About the twelfth day after fertilization striking changes appear. A layer of cells from the inner cell mass has grown around the top of the amnionic cavity. These cells will become the amnion, another fetal membrane. The cells below the cavity are called the *embryonic disc*. They will form the embryo. The embryonic disc contains scattered ectodermal, mesodermal, and endodermal cells in addition to the endodermal layer. The cells of the endodermal layer have been dividing rapidly, so groups of them now extend downward in a circle. This circle is the yolk sac, another fetal membrane. The *mesodermal* cells also have been dividing, and many have left the area of the embryonic disc and can be seen around the structures that are becoming fetal membranes.

About the fourteenth day, the scattered cells in the embryonic disc separate into three distinct layers: the upper ectoderm, the middle mesoderm, and the lower endoderm. At this time the two ends of the embryonic disc draw together, squeezing off the yolk sac. The resulting cavity inside the disc is the endoderm-lined *primitive gut*. The mesoderm in the disc soon splits into

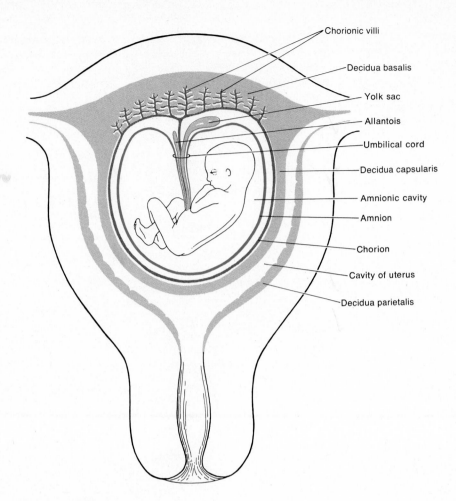

Chorionic villi

Decidua basalis

Yolk sac

Allantois

Umbilical cord

Decidua capsularis

Amnionic cavity

Amnion

Chorion

Cavity of uterus

Decidua parietalis

Figure 29—6 Embryonic membranes.

two layers, and the space between the layers becomes the coelom, or body cavity.

As the embryo develops, the endoderm becomes the epithelium lining the digestive tract and a number of other organs. The mesoderm forms the peritoneum, muscle, bone, and other connective tissue. The ectoderm develops into the skin and nervous system. Exhibit 29-1 provides more details about the fates of these primary germ layers.

EMBRYONIC MEMBRANES

During the embryonic period, the *embryonic membranes* form. These membranes lie outside the embryo and will protect and nourish the embryo and later on the fetus. The membranes are the yolk sac, the amnion, the chorion, and the allantois (Figure 29-6).

The *yolk sac* is an endoderm-lined membrane that encloses the yolk. In many species the yolk provides the primary or exclusive nutrient for the embryo, so the ova of these animals contain a great deal of yolk. However, the human embryo receives its nourishment from the endometrium. The human yolk sac is small, and during an early stage of development it becomes a nonfunctional part of the umbilical cord.

The *amnion* is a thin, protective membrane that initially overlies the embryonic disc. As the embryo grows, the amnion entirely surrounds the embryo and becomes filled with *amniotic fluid*. Amniotic fluid serves as a shock absorber for the fetus. The amnion usually ruptures just before birth and with its fluid constitutes the "bag of waters."

The *chorion* derives from the trophectoderm of the blastocyst and its associated mesoderm. It surrounds the embryo and, later, the fetus. Eventually the chorion becomes the principal part of the placenta, the structure through which materials are exchanged between mother and fetus. The amnion also surrounds the fetus and eventually fuses to the inner layer of the chorion.

The *allantois* is a small vascularized membrane. Later its blood vessels serve as connections in the placenta between mother and fetus. This connection is the umbilical cord.

Exhibit 29–1   STRUCTURES PRODUCED BY THE THREE PRIMARY GERM LAYERS

Endoderm	Mesoderm	Ectoderm
Epithelium of digestive tract and its glands	Skeletal, smooth, cardiac muscle	Epidermis of skin
Epithelium of urinary bladder and gallbladder	Cartilage, bone, other connective tissues	Hair, nails, skin glands
Epithelium of pharynx, auditory tube, tonsils, larynx, trachea, bronchi, lungs	Blood, bone marrow, lymphoid tissue	Lens of eye
	Endothelium of blood vessels and lymphatics	Receptor cells of sense organs
Epithelium of thyroid, parathyroid, thymus glands	Mesothelium of coelomic and joint cavities	Epithelium of mouth, nostrils, sinuses, oral glands, anal canal
Epithelium of vagina, vestibule, urethra, associated glands	Epithelium of kidneys and ureters	Enamel of teeth
Adenohypophysis	Epithelium of gonads and associated ducts	Entire nervous tissue, except adeno-hypophysis
	Epithelium of adrenal cortex	
	Stroma of most soft organs, except those of central nervous system	

PLACENTA AND UMBILICUS

Development of the placenta, the third major event of the embryonic period, is accomplished by the third month of pregnancy. The **placenta** has the shape of a flat cake when fully developed and is formed by the chorion of the embryo and a portion of the endometrium of the mother (Figure 29-7). It provides an exchange of nutrients and wastes between fetus and mother and secretes the hormones necessary to maintain pregnancy.

If implantation occurs, a portion of the endometrium becomes modified and is known as the **decidua**. The decidua includes all but the deepest layer of the endometrium and is shed when the fetus is delivered. Different regions of the decidua are named on the basis of their positions relative to the site of the implanted ovum. The *decidua parietalis* is the portion of the modified endometrium that lines the entire pregnant uterus, except for the area where the placenta is forming (Figure 29-6). The *decidua capsularis* is the portion of the endometrium that overlies the developing embryo (Figure 29-4d and Figure 29-6). The *decidua basalis* is the portion of the endometrium between the chorion and the muscularis of the uterus (Figure 29-4d and Figure 29-6). The decidua basalis becomes the maternal part of the placenta.

During embryonic life, fingerlike projections of the chorion, called *chorionic villi,* grow into the decidua basalis of the endometrium. These will contain fetal blood vessels of the allantois. They continue growing until they are bathed in maternal blood sinuses called *intervillous spaces.* Thus, maternal and fetal blood vessels are brought into close proximity. It should be noted, however, that maternal and fetal blood do not mix. Oxygen and nutrients from the mother's blood diffuse across the walls and into the capillaries of the villi. From the capillaries the nutrients circulate into the umbilical vein. Wastes leave the fetus through the umbilical arteries, pass into the capillaries of the villi, and diffuse into the maternal blood. The **umbilical cord** consists of an outer layer of amnion containing the umbilical arteries and umbilical vein, supported internally by mucous connective tissue from the allantois called Wharton's jelly. At delivery, the placenta detaches from the uterus and is referred to as the "afterbirth." At this time, the umbilical cord is severed, leaving the baby on its own. The scar that marks the site of the entry of the fetal umbilical cord into the abdomen is called the *umbilicus.*

Fetal Growth

During the **fetal period**, organs established by the primary germ layers grow rapidly. The organism takes on a human appearance. A summary of changes associated with the fetal period is presented in Exhibit 29-2.

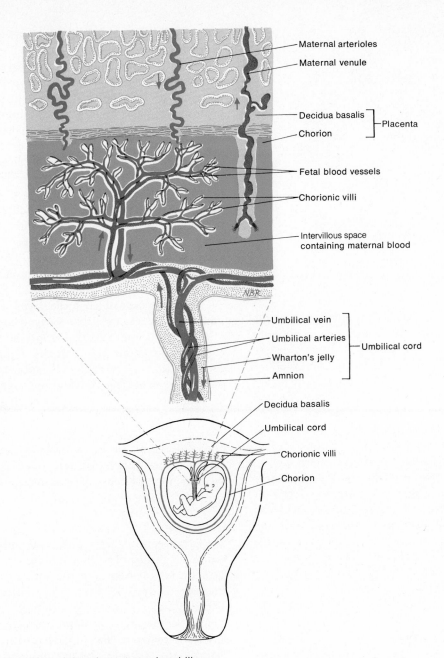

Figure 29-7   Structure of the placenta and umbilicus.

Hormones of Pregnancy

Following fertilization, the corpus luteum is maintained until about the fourth month of pregnancy. For most of this time it continues to secrete estrogens and progesterone. Both these hormones maintain the lining of the uterus during pregnancy and prepare the mammary glands to secrete milk. The amounts of estrogens and progesterone secreted by the corpus luteum, however, are only slightly higher than those produced after ovulation in a normal menstrual cycle. The high levels of estrogens and progesterone needed to maintain preg-

nancy and develop the breasts for lactation are provided by the placenta.

During pregnancy, the chorion of the placenta secretes a hormone called *human chorionic gonadotropin, or HCG.* This hormone is excreted in the urine of pregnant women from about the middle of the first month of pregnancy, reaching its peak of excretion during the third month. The HCG level decreases sharply during the fourth and fifth months and then levels off until childbirth. Excretion of HCG in the urine serves as the basis for some pregnancy tests. The primary

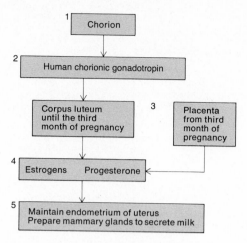

Figure 29-8   Hormones of pregnancy.

role of HCG seems to be to maintain the activity of the corpus luteum, especially with regard to continuous progesterone secretion—an activity necessary for the continued attachment of the fetus to the lining of the uterus (Figure 29-8).

The placenta begins to secrete estrogens and progesterone no later than the sixtieth day of pregnancy. They are secreted in increasing quantities until the time of birth. Once the placenta is established, the secretion of HCG is cut back drastically at about the fourth month. The corpus luteum then disintegrates—it is no longer needed because the placenta supplies the levels of estrogens and progesterone needed to maintain the pregnancy. The fetal hormones thus take over the management of the mother's body in preparation for parturition (birth) and lactation. Following delivery, estrogens and progesterone in the blood decrease to normal levels.

Parturition and Labor

The time the embryo or fetus is carried in the uterus is called *gestation.* The total human gestation period is 280 days from the beginning of the last menstrual period. The term **parturition** refers to birth. Parturition is preceded by a sequence of events commonly called **labor.** The onset of labor stems from a complex interaction of many factors, especially hormones. Just prior to birth, the muscles of the uterus contract rhythmically and forcefully. Both placental and ovarian hormones play a dominant role in these contractions. Since progesterone inhibits uterine contractions, labor cannot take place until its effects are diminished. At the end of gestation, however, there is just enough estrogen in the mother's blood to overcome the inhibiting effects of progesterone and labor commences. Oxytocin from

the posterior pituitary gland stimulates uterine contractions also.

Uterine contractions occur in waves, quite similar to peristaltic waves, that start at the top of the uterus and move downward. These waves expel the fetus. *True labor* begins when pains occur at regular intervals. The pains correspond to uterine contractions. As the interval between contractions shortens, the contractions intensify. Another sign of true labor is localization of pain in the back, which is intensified by walking. The final indication of true labor is the "show" and dilatation of the cervix. The "show" is a discharge of a blood-containing mucus that accumulates in the cervical canal during pregnancy. In *false labor,* pain is felt in the abdomen at long, irregular intervals. The pain does not intensify and is not altered significantly by walking. There is no "show" and no cervical dilatation.

The first stage of labor, the *stage of dilatation,* is the time from the onset of labor to the complete dilatation of the cervix (Figure 29-9). During this stage there are regular contractions of the uterus, a rupturing of the amniotic sac, and complete dilatation (10 cm) of the cervix. The next stage of labor, the *stage of expulsion,* is the time from complete cervical dilatation to delivery. In the final stage, the *placental stage,* the placenta or "afterbirth" is expelled a few minutes after delivery by powerful uterine contractions. These contractions also constrict blood vessels that were torn during delivery. In this way, the possibility of hemorrhage is reduced.

Lactation

The term **lactation** refers to the secretion of milk by the mammary glands. The major hormone in promoting lactation is prolactin from the anterior pituitary. It is released in response to the hypothalamic prolactin releasing factor (PRF). Even though prolactin levels increase as the pregnancy progresses, there is no milk secretion—estrogens and progesterone cause the hypothalamus to release prolactin inhibiting factor (PIF), which inhibits prolactin. Following delivery, the levels of estrogens and progesterone in the mother's blood decrease and the inhibition of prolactin is removed.

The principal stimulus in maintaining prolactin secretion during lactation is the sucking action of the infant. Sucking initiates impulses from receptors in the nipples to the hypothalamus. The impulses inhibit PIF production and prolactin is released by the anterior pituitary. The sucking action also initiates impulses to the posterior pituitary via the hypothalamus. These impulses stimulate the release of the hormone oxytocin by the posterior pituitary. This hormone moves milk from the alveoli of the mammary gland into the ducts, where it can be sucked. This process is referred to as milk

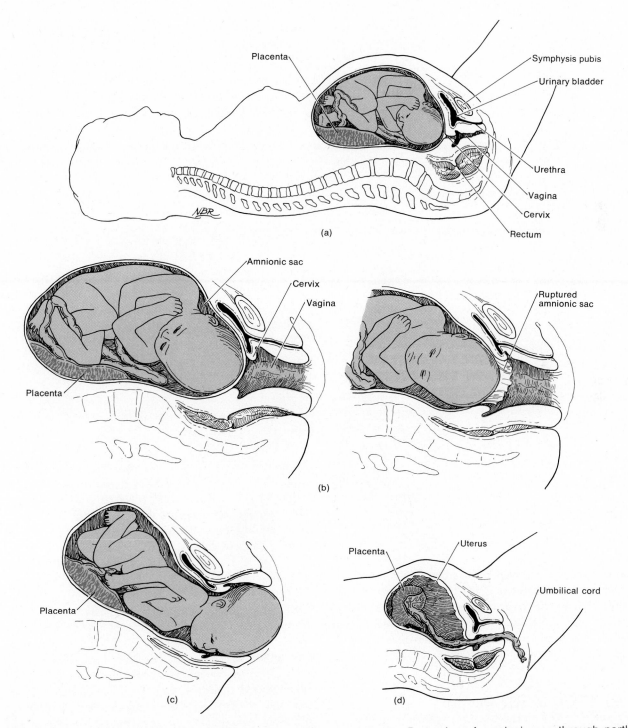

Figure 29–9 Parturition. (a) Fetal position prior to birth. (b) Dilatation. Protrusion of amnionic sac through partly dilatated cervix. Amnionic sac ruptured and complete dilatation of cervix. (c) Stage of expulsion. (d) Placental stage.

Exhibit 29–2   CHANGES ASSOCIATED WITH FETAL GROWTH

End of Month	Approximate Size and Weight	Representative Changes
1	0.6 cm ($^3/_{16}$ inch)	Eyes, nose, and ears not yet visible. Backbone and vertebral canal form. Small buds that will develop into arms and legs form. Heart forms and starts beating. Body systems begin to form.
2	3 cm (1 $^1/_4$ inches) 1 g ($^1/_{30}$ oz)	Eyes far apart, eyelids fused, nose flat. Ossification begins. Limbs become distinct as arms and legs. Digits are well formed. Major blood vessels form. Many internal organs continue to develop.
3	7.5 cm (3 inches) 28 g (1 oz)	Eyes almost fully developed but eyelids still fused, nose develops bridge, and external ears are present. Ossification continues. Appendages are fully formed and nails develop. Heartbeat can be detected. Body systems continue to develop.
4	18 cm (6$^1/_2$–7 inches) 113 g (4 oz)	Head large in proportion to rest of body. Face takes on human features and hair appears on head. Skin bright pink. Many bones ossified, and joints begin to form. Continued development of body systems.
5	25–30 cm (10–12 inches) 227–454 g ($^1/_2$–1 lb)	Head less disproportionate to rest of body. Fine hair (laguno hair) covers body. Skin still bright pink. Rapid development of body systems.
6	27–35 cm (11–14 inches) 567–681 g (1$^1/_4$–1$^1/_2$ lb)	Head becomes less disproportionate to rest of body. Eyelids separate and eyelashes form. Skin wrinkled and pink.
7	325–425 cm (13–17 inches) 1,135–1,362 g (2$^1/_2$–3 lb)	Head and body become more proportionate. Skin wrinkled and pink. Seven-month fetus (premature baby) is capable of survival.
8	41–45 cm (16$^1/_2$–18 inches) 2,043–2,270 g (4$^1/_2$–5 lb)	Subcutaneous fat deposited. Skin less wrinkled. Testes descend into scrotum. Bones of head are soft. Chances of survival much greater at end of eighth month.
9	50 cm (20 inches) 3,178–3,405 g (7–7$^1/_2$ lb)	Additional subcutaneous fat accumulates. Laguno hair shed. Nails extend to tips of fingers and maybe even beyond.

letdown. Lactation often prevents the occurrence of female sexual cycles for the first few months following delivery by inhibiting FSH and LH release by the anterior pituitary.

INHERITANCE

Inheritance is the passage of hereditary traits from one generation to another. It is the process by which you acquired your characteristics from your parents and will transmit your characteristics to your children. The branch of biology that deals with inheritance is called *genetics.*

Genotype and Phenotype

The nuclei of all human cells except gametes contain 23 pairs of chromosomes—the diploid number. One chromosome from each pair comes from the mother, and the other comes from the father. The two chromosomes that belong to a pair are called *homologous chromosomes.* Homologues contain genes that control the same traits. If a chromosome contains a gene for height, its homologue will contain a gene for height.

The relationship of genes to heredity is illustrated admirably by the disorder called phenylketonuria or PKU (see Figure 29-10). People with PKU are unable to manufacture the enzyme phenylalanine hydroxylase. It is believed that PKU is brought about by an abnormal gene, which can be symbolized as p. The normal gene will be symbolized as P. The chromosome concerned with directions for phenylalanine hydroxylase production will have either p or P on it. Its homologue will also have p or P. Thus every individual will have one of the following genetic makeups, or *genotypes:* PP, Pp, or pp. Although people with genotypes of Pp have the abnormal gene, only those with genotype pp suffer from the disorder because the normal gene dominates over and inhibits the abnormal one. A gene that dominates is called the *dominant gene,* and the trait expressed is said to be a dominant trait. The gene that is inhibited is called the *recessive gene,* and the trait expressed is called the recessive trait.

By tradition, we symbolize the dominant gene with a capital letter and the recessive one with a lowercase letter. If you have the same genes on homologous chromosomes (for example, PP or pp), you are said to be *homozygous* for the trait. If, however, the genes on homologous chromosomes are different (for example, Pp), you are said to be *heterozygous* for the trait. *Phenotype* refers to how the genetic makeup is expressed in the body. A person with Pp has a different genotype than one with PP, but both have the same phenotype – which in this case is normal production of phenylalanine hydroxylase.

To determine how gametes containing haploid chromosomes unite to form diploid fertilized eggs, special charts called *Punnett squares* are used. Usually, the male gametes (sperm cells) are placed at the side of the chart and the female gametes (ova) at the top (Figure 29-10). The four spaces on the chart represent the possible combinations of male and female gametes that could form fertilized eggs. Possible combinations are determined simply by dropping the female gamete on the left into the two boxes below it and dropping the female gamete on the right into the two spaces under it. The upper male gamete is then moved across to the two spaces in line with it, and the lower male gamete is moved across to the two spaces in line with it.

Several dominant and recessive traits inherited in human beings are listed in Exhibit 29-3.

Normal traits do not always dominate over abnormal ones, but genes for severe disorders are more frequently recessive than dominant. People who have severe disorders often do not live long enough to pass the abnormal gene on to the next generation. In this way, expression of the gene tends to be weeded out of the population. An exception is Huntington's chorea – a major disorder caused by a dominant gene and characterized by degeneration of nervous tissue, usually leading to mental disturbance and death. The first signs of Huntington's chorea do not occur until adulthood, very often after the person has already produced offspring.

Inheritance of Sex

Microscopic examination of the chromosomes in cells reveals that one pair differs in males and in females (Figure 29-11a). In females, the pair consists of two

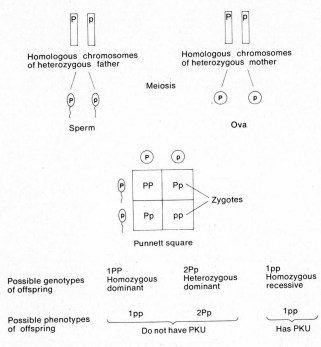

Figure 29–10   Inheritance of PKU.

rod-shaped chromosomes designated as X chromosomes. One X chromosome is present in males, but its mate is a hook-shaped structure called a Y chromosome. The XX pair in the female and the XY pair in the male are called the *sex chromosomes.* All other chromosomes are called *autosomes.*

The sex chromosomes are responsible for the sex of the individual (Figure 29-11b). When a spermatocyte undergoes meiosis to reduce its chromosome number, one daughter cell will contain the X chromosome and the other will contain the Y chromosome. Oocytes have no Y chromosomes and produce only X-containing ova. If the ovum is subsequently fertilized by an X-bearing sperm, the offspring normally will be female (XX). Fertilization by a Y sperm normally produces a male (XY).

Color Blindness and X-Linked Inheritance

The sex chromosomes also are responsible for the transmission of a number of nonsexual traits. Genes for these traits appear on X chromosomes, but many of these genes are absent from Y chromosomes. This feature produces a pattern of heredity that is different from the pattern described earlier. Let us consider color blindness. The gene for *color blindness* is a recessive one designated c. Normal vision, designated C, dominates. The C/c genes

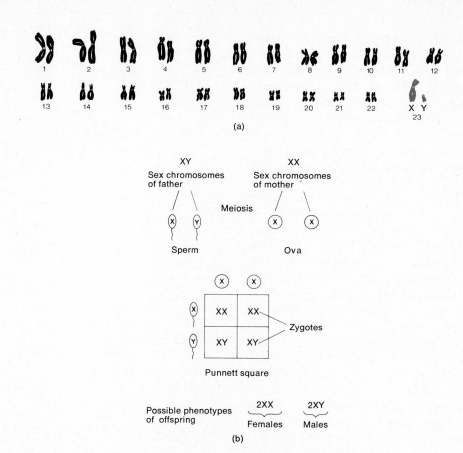

(a)

(b)

Figure 29–11 Inheritance of sex. (a) Human male chromosomes. Sex chromosomes are indicated in color. (b) Sex determination.

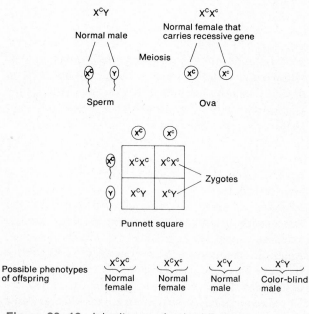

Figure 29–12 Inheritance of color blindness.

are located on the X chromosome. The Y chromosome, however, does not contain the segment of DNA that programs this aspect of vision. Thus the ability to see colors depends entirely on the X chromosomes. The genetic possibilities are:

$X^C X^C$	Normal female
$X^C X^c$	Normal female carrying the recessive gene
$X^c X^c$	Color-blind female
$X^C Y$	Normal male
$X^c Y$	Color-blind male

Only females who have two X^c chromosomes are color blind. In $X^C X^c$ females the trait is inhibited by the normal, dominant gene. Males, on the other hand, do not have a second X chromosome that would inhibit the trait. Therefore all males with an X^c chromosome will be color blind. The inheritance of color blindness is illustrated in Figure 29-12.

Traits inherited in the manner just described are called *X-linked traits*. Another X-linked trait is *hemophilia* – a condition in which the blood fails to clot or clots very slowly after an injury. It is a much more serious defect than color blindness because people with severe hemophilia can bleed to death from even a small cut. Like the trait for color blindness, hemophilia is caused by a recessive gene. If H represents normal clotting and h represents abnormal clotting, then X^hX^h females will have the disorder. Males with X^HY will be normal; males with X^hY will be hemophiliac. Actually, clotting time varies somewhat among hemophiliacs, so the condition may be affected by other genes as well.

A few other X-linked traits in human beings are nonfunctional sweat glands, certain forms of diabetes, some types of deafness, uncontrollable rolling of the eyeballs, absence of central incisors, night blindness, one form of cataract, white forelocks, juvenile glaucoma, and juvenile muscular dystrophy.

BIRTH CONTROL

Methods of **birth control** include removal of the gonads and uterus, sterilization, contraception, and abstinence. *Castration* (removal of the testes), *hysterectomy* (removal of the uterus), and *oophorectomy* (removal of the ovaries), are all absolute preventive methods. Once performed, these operations cannot be reversed and it is impossible to produce offspring. However, removal of the testes or ovaries has adverse effects because of the importance of these organs in the endocrine system. Generally these operations are performed only if the organs are diseased. Castration before puberty prevents the development of secondary sex characteristics.

One means of *sterilization* of males is *vasectomy* – a simple operation in which a portion of each ductus deferens is removed (Figure 29-13). An incision is made in the scrotum, the tubes are located, and each is tied in two places. Then the portion between the ties is cut out. Sperm production can continue in the testes, but the sperm cannot reach the exterior.

Sterilization in females generally is achieved by performing a *tubal ligation* – a similar operation on the uterine tubes. An incision is made into the abdominal cavity, the tubes are squeezed, and a small loop called a knuckle is made. A suture is tied tightly at the base of the knuckle and the knuckle is then cut. After 4 or 5 days the suture is digested by body fluids and the two severed ends of the tubes separate. The ovum thus is prevented from passing to the uterus, and the sperm cannot reach the ovum. Sterilization normally does not affect sexual performance or enjoyment.

Exhibit 29—3 HEREDITARY TRAITS IN HUMAN BEINGS

Dominant	Recessive
Curly hair	Straight hair
Dark hair	Light hair
Nonred hair	Red hair
Coarse body hair	Normal body hair
Normal skin pigmentation	Albinism
Brown eyes	Blue or gray eyes
Near or farsightedness	Normal vision
Normal hearing	Deafness
Normal color vision	Color blindness
Normal blood clotting	Hemophilia
Broad lips	Thin lips
Large eyes	Small eyes
Short stature	Tall stature
Polydactylism (extra digits)	Normal digits
Brachydactylism (short digits)	Normal digits
Syndactylism (webbed digits)	Normal digits
Normal muscle tone	Muscular dystrophy
Hypertension	Normal blood pressure
Diabetes insipidus	Normal excretion
Huntington's chorea	Normal nervous system
Normal mentality	Schizophrenia
Nervous temperament	Calm temperament
Average intellect	Genius or idiocy
Migraine headaches	Normal
Normal resistance to disease	Susceptibility to disease
Enlarged spleen	Normal spleen
Enlarged colon	Normal colon
A or B blood factor	O blood factor
Rh blood factor	No Rh blood factor

Another method of sterilizing women is the *laparoscopic* technique. After a woman receives local or general anesthesia, a harmless gas is introduced into her abdomen to create a gas bubble. The bubble expands the abdominal cavity and pushes the intestines away from the pelvic organs, permitting safe, easy access to the uterine tubes. Next the doctor makes a small incision at the lower rim of the navel and inserts a laparoscope to view the inside of the abdominal cavity and the uterine tubes. The tubes can be closed with this instrument or a second incision can be made at the pubic hairline to insert a cautery forceps. Once the uterine tubes are sealed, the instrument is removed, the gas is released, and the incision is covered with a bandage. After a few hours the patient can usually go home.

Contraceptives include all methods of preventing fertilization without destroying fertility – by natural,

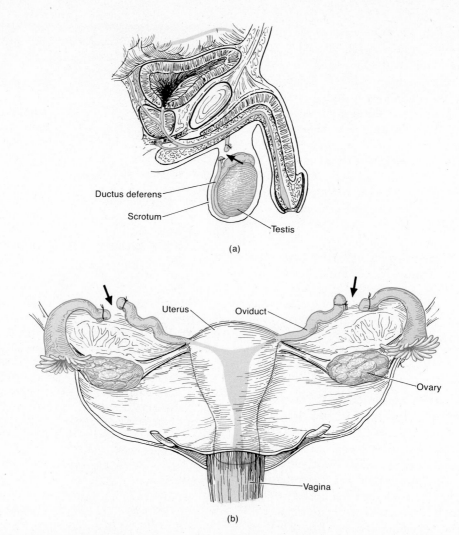

Figure 29–13   Sterilization. (a) Vasectomy. The ductus deferens of each testis is cut and tied after an incision is made into the scrotum. (b) Tubal ligation. Each uterine tube is cut and tied after an incision is made into the abdomen.

mechanical, and chemical means. The natural methods are complete or periodic abstinence. An example of periodic abstinence is the rhythm method, which takes advantage of the fact that a fertilizable ovum is available only during a period of 3–5 days in each menstrual cycle. During this time the couple refrains from intercourse. Few women have absolutely regular cycles, however. Moreover, some women occasionally ovulate during the "safe" times of the month, such as during menstruation.

Mechanical means of contraception include the condom used by the male and the diaphragm used by the female. The *condom* is a nonporous, elastic covering placed over the penis that prevents deposition of sperm in the female reproductive tract. The *diaphragm* (Figure 29-14) is a dome-shaped structure that fits over the cervix and is generally used in conjunction with a sperm-killing chemical. The diaphragm stops the sperm from passing into the cervix. The chemical kills the sperm cells.

Another mechanical method of contraception is an *intrauterine device (IUD)*. These small objects such as loops, spirals, and rings made of plastic, copper, or stainless steel are inserted into the cavity of the uterus (Figure 29-15). It is not clear how IUDs operate. Some investigators believe they cause changes in the uterus lining that, in turn, produce a substance which destroys either the sperm or the fertilized ovum. Some IUDs release contraceptive agents in minute amounts—one new device secretes progesterone.

Chemical means of contraception include the use of various foams, creams, jellies, suppositories, and douches that make the vagina and cervix unfavorable for sperm survival. *Oral contraception* (the pill) has found

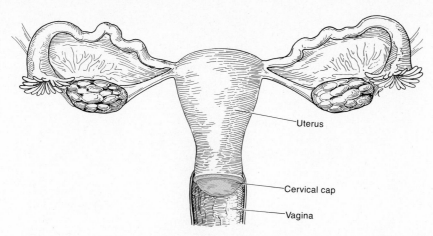

Uterus

Cervical cap

Vagina

Figure 29—14   Cervical cap diaphragm. This particular type of diaphragm fits directly over the cervix of the uterus. The thin spring around the margin of the diaphragm opens out, presses against the wall of the vagina, and stretches across the cervix.

rapid and widespread use. Although several pills are available, the most commonly used one is the combination pill. This pill contains a high concentration of progesterone and a low concentration of estrogens. These two hormones act on the anterior pituitary to decrease the secretion of FSH and LH by inhibiting FSHRF and LHRF production, respectively, by the hypothalamus. The low levels of FSH and LH are not adequate to initiate follicle maturation or ovulation. Consequently, pregnancy cannot occur in the absence of a mature ovum. A summary of contraceptive methods is presented in Exhibit 29-4.

DISORDERS

CHROMOSOMAL

Amniocentesis is a technique of withdrawing some of the amniotic fluid that bathes the developing fetus to diagnose genetic disorders. This fluid, which has many living cells floating in it, is removed by hypodermic needle puncture of the uterus, usually 16 to 20 weeks after conception. Cells in the fluid are examined for biochemical defects and abnormalities in chromosome number or structure. There are over 50 biochemical inheritable disorders and close to 300 chromosomal disorders that can be detected through amniocentesis—hemophilia, certain muscular dystrophies, Tay-Sachs disease, myelocytic leukemia, Klinefelter's and Turner's syndrome, sickle cell anemia, thalassemia, and cystic fibrosis. When both parents are known or suspected to be genetic carriers of any one of these disorders, amniocentesis is advised.

One chromosome disorder that may be diagnosed through amniocentesis is **Down's syndrome,** or mongolism (Figure 29-16). This disorder is characterized by mental retardation; retarded physical development (short stature and stubby fingers); distinctive facial structures (large tongue, broad skull, slanting eyes, and round head); and malformation of the heart, ears, and feet. Sexual maturity is rarely attained. Individuals with the disorder usually have 47 chromosomes instead of the normal 46. The extra chromosome is responsible for the syndrome. All the chromosomes of a person with Down's syndrome are in pairs except for the twenty-first pair. Here the chromosomes are present in triplicate.

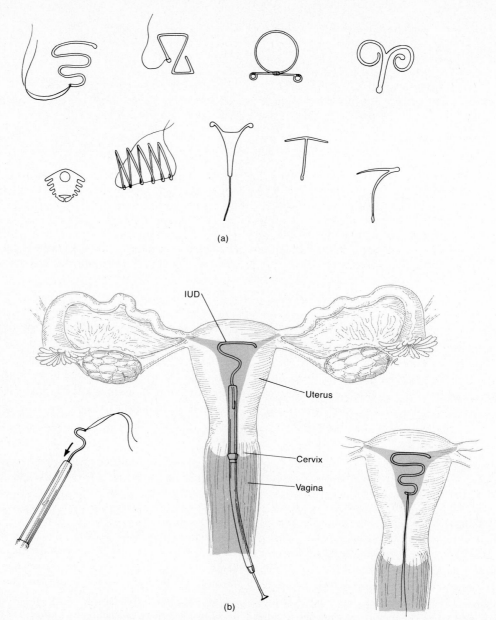

Figure 29–15 Intrauterine devices. (a) Representative designs of intrauterine devices. (b) Procedure for insertion. IUD's are inserted into a slightly open cervix. The device is threaded into a long, narrow-bore tube that is passed through the cervix. Once in position in the uterus, it spreads out to its former shape. Most IUD's have a thread or chain projecting into the vagina that may be detected by a finger (indicating the device is still in place) and that may be used for removal.

Figure 29–16 Down's syndrome. Appearance of chromosomes in an individual with Down's syndrome.

Exhibit 29–4 CONTRACEPTIVE METHODS

Method	Comments
Removal of gonads and uterus	Results in irreversible sterility. Generally performed if organs are diseased rather than as contraceptive methods because of importance of hormones produced by gonads.
Sterilization	Procedure involving severing ductus deferens in males and uterine tubes in females.
Natural contraception	Abstinence from intercourse during time of month woman is fertile. Under ideal circumstances, effectiveness in women with regular menstrual cycles may approach that of mechanical and chemical contraceptives. Extremely difficult to determine fertile period. Effectiveness can be greatly increased by measuring and recording body temperature each morning before getting up. Small rise in temperature will show that ovulation has occurred 1 or 2 days earlier and ovum can no longer be fertilized.
Mechanical contraception	
Condom	Thin, strong sheath of rubber or similar material worn by male to prevent sperm from entering vagina. Failures caused by sheath tearing or slipping off after climax. Rates in effectiveness with diaphragm if used correctly and consistently.
Diaphragm	Flexible rubber dome used with spermicidal cream or jelly. Inserted into vagina to cover cervix, providing barrier to sperm. Must be left in place at least 6 hours after intercourse and may be left in place as long as 24 hours. Must be fitted by physician and refitted every 2 years and after each pregnancy. Offers high level of protection; rate of two to three pregnancies per 100 women per year is estimated for consistent users. If motivation is weak, much higher pregnancy rates must be expected. Occasional failures caused by improper insertion or displacement during sexual intercourse.
Intrauterine devices	Small objects (loops, spirals, rings) made of plastic or stainless steel and inserted into uterus by physician. May be left in place indefinitely. Do not require continued attention by user. Some women cannot use them because of expulsion, bleeding, or discomfort. Not recommended for women who have not had child because uterus is too small and cervical canal too narrow. Infrequently, insertion may lead to inflammation of pelvic organs.
Chemical contraception	
Foams, creams, jellies, suppositories, vaginal douches	Sperm-killing chemicals inserted into vagina to coat vaginal surfaces and cervical opening. Provide protection for about 1 hour. Significantly effective when used alone, but believed to be more effective when used with diaphragm or condom.
Oral contraceptive	Except for total abstinence or surgical sterilization, most effective contraceptive known. Side effects include nausea, occasional light bleeding between periods, breast tenderness or enlargement, fluid retention, and weight gain. Should not be used by women who might be thrombosis-prone.

MEDICAL TERMINOLOGY

Abortion   Premature expulsion from the uterus of the products of conception – embryo or nonviable fetus.

Autosome   Any chromosome that is not a sex chromosome (not an X or Y chromosome). Humans have 22 pairs of autosomes.

Cautery   Application of a caustic (burning) substance or instrument for the destruction of tissue.

Cesarean section   Removal of the baby and placenta through an abdominal incision in the uterine wall.

Colpotomy   Incision of the vagina.

Culdoscopy   A female sterilization procedure where a culdoscope is used to view and seal the uterine tubes. The approach is through the vagina.

Endometrectomy   Removal of the uterine mucous lining; also called curettage. A variation of this procedure is dilation and curettage (D and C).

Hermaphroditism   Presence of both male and female sex organs in one individual.

Karyotype   The chromosomal elements typical of a cell, drawn in their true proportions, based on the average of measurements determined in a number of cells. Useful in judging whether or not chromosomes are normal in number and structure.

Laparoscope   A long thin instrument containing a light-reflecting material as well as a lens.

Lethal gene   A gene that, when expressed, results in death either in the embryonic state or shortly after birth.

Lochia   The discharge from the birth canal occurring after childbirth.

Mutation   A permanent heritable change in a gene that causes it to have a different effect than it had previously.

Puerperal Fever   This infectious disease of childbirth is also called puerperal sepsis and childbed fever. This disease results from an infection originating in the birth canal and affects the endometrium.

STUDY OUTLINE

GAMETE FORMATION
1 Ova and sperm are collectively called gametes.

2 Chromosome number is the number of chromosomes contained in each nucleated cell that is not a gamete.

3 The human chromosome number is 23 pairs – a total of 46 chromosomes (diploid).

4 Meiosis occurs in the process of producing sex cells. It causes a developing sperm or ovum to cut in half its duplicate set of chromosomes so that the mature gamete has only 23 chromosomes (haploid).

5 Spermatogenesis occurs in the testes. It is the formation of haploid spermatozoa by meiosis.

6 Oogenesis occurs in the ovaries. It is the formation of haploid ova by meiosis.

SEXUAL INTERCOURSE
1 The role of the male in the sex act involves erection, emission, and ejaculation.

2 The female role involves erection, lubrication, and climax.

PREGNANCY
1 Pregnancy is a sequence of events that includes fertilization, implantation, embryonic growth, fetal growth, and birth.

2 Fertilization, the union of the ovum and sperm, normally occurs in the uterine tubes.

3 The embedding of a fertilized ovum in the endometrium of the uterus is called implantation.

4 During embryonic growth, the primary germ layers and embryonic membranes are formed and the placenta is functioning.

5 The primary germ layers – ectoderm, mesoderm, and endoderm – form all tissues of the developing organism.

6 Embryonic membranes include the yolk sac, amnion, chorion, and allantois.

7 Fetal and maternal materials are exchanged through the placenta.

8 Pregnancy is maintained by the human chorionic gonadotropic hormone, estrogens, and progesterone secreted by the placenta.

9 The birth of a baby involves dilatation of the cervix, expulsion of the fetus, and delivery of the placenta.

10 Lactation, or the secretion of milk by the mammary glands, is influenced by estrogens and progesterone, prolactin, and oxytocin.

INHERITANCE

1 Inheritance is the passage of hereditary traits from one generation to another.

2 The genetic makeup of an organism is called its genotype. The traits expressed are called its phenotype.

3 Dominant genes control a particular trait; expression of recessive genes is inhibited by dominant genes.

4 Sex is determined by the Y chromosome of the male.

5 Color blindness and hemophilia primarily affect males because there are no counterbalancing dominant genes on the Y chromosomes.

BIRTH CONTROL

1 Methods include removal of gonads; sterilization; rhythm; use of condom, diaphragm, and intrauterine devices; chemicals that kill sperm; and the contraceptive pill.

2 Contraceptive pills of the combination type contain estrogens and progesterone in concentrations that decrease the secretion of FSH and LH and thereby inhibit ovulation.

REVIEW QUESTIONS

1 Define development. What is the importance of meiosis to development?

2 Compare the events associated with spermatogenesis and oogenesis.

3 Explain the role of the male's erection, emission, and ejaculation in the sex act. How do the female's erection, lubrication, and orgasm contribute to the sex act?

4 Define fertilization. Where does it normally occur? How is a morula formed?

5 What is implantation? How does the fertilized ovum implant itself?

6 Define the embryonic period. Describe some body structures formed by the ectoderm, mesoderm, and endoderm.

7 What is an embryonic membrane? Describe the functions of the four embryonic membranes.

8 Explain the importance of the placenta and umbilical cord to fetal growth.

9 Outline some of the major developmental changes during fetal growth.

10 Compare the sources and functions of estrogens, progesterone, and the human chorionic gonadotropic hormone. How does the placenta serve as an endocrine organ during pregnancy?

11 Define gestation. Distinguish between false and true labor.

12 Describe what happens during the stage of dilatation, the stage of expulsion, and the placental stage of delivery.

13 What is lactation? How is the female prepared for it?

14 Define inheritance. What is genetics?

15 Define the following terms: genotype, phenotype, dominant, recessive, homozygous, heterozygous.

16 What is a Punnett square?

17 List several dominant and recessive traits inherited in human beings.

18 Set up Punnett squares to show the inheritance of the following traits: sex, color blindness, hemophilia.

19 Briefly describe the following methods of birth control: removal of the gonads, sterilization, rhythm.

20 Distinguish between a condom, diaphragm, and IUD as methods of mechanical contraception.

21 List several examples and functions of chemical contraceptives.

22 Explain the operation of the combination contraceptive pill.

Appendix: Measuring the Human Body

MEASUREMENT DEFINED

When you **measure** something, you are comparing it with some standard scale to determine its *magnitude*. How long is it? How much does it weigh? How fast is it going? Some measurements are made directly by comparing the unknown quantity with the known unit of the same kind—for example, weighing a patient on a scale and taking the reading directly in pounds. Other measurements are indirect and thus must be done by calculation—for example, counting a person's blood cells in a certain number of squares on a microscope slide and then calculating the total blood count.

Regardless of how a measurement is taken, it always requires two things: a *number* and a *unit*. When recording the weight of a patient, you would not just say 145. You would have to say 145 pounds. That is, you would have to give both the number (145) and the unit (pounds). When you count blood cells, you report the measurement as 10,000 (number) white blood cells per cubic millimeter of blood (unit).

UNITS OF MEASUREMENT

All the units in use can be expressed in terms of one of three special units called *fundamental units*. These fundamental units have been established arbitrarily as length, mass, and time. All other units are *derived units*—they can always be written as some combination of the three fundamental units. Units are grouped into systems of measurement. The two principal systems of measurement commonly used in this country are the English and the metric systems. The apothecary system is used by physicians and pharmacists.

ENGLISH SYSTEM

The **English system** of measurement is essentially the same one used here. It is used in everyday household work, industry, and some fields of engineering. The fundamental units in the English system are the foot (length), the pound (mass), and the second (time). The standard of length is the *yard,* defined as the distance between two lines on a bronze bar kept at 62°F in the Office of the Exchequer in London. The yard is believed to have originated as the length of the stride of an average man. The standard of mass is the *pound*. It has been arbitrarily set as the weight of the bronze bar kept at the Office of the Exchequer. The standard of time in the English system is the *second.*

The basic problem with the English system is that there is no *uniform* progression from one unit to another. If the length of an object is 2½ yd and you want to convert it to feet, you have to multiply 2½ yd by 3 ft because there are 3 ft in a yard. If you want to convert the same length in yards to inches, you have to multiply 2½ yd by 3 by 12 (or 36) because there are 12 inches in a foot. In other words, to convert one unit of length to another, it is necessary to use *different* numbers each time. Conversions in the metric system are much easier since they are based on progressions of the number 10.

Exhibit A-1 lists English units of measurement.

METRIC SYSTEM

The **metric system**, introduced in France in 1790, is now used by all major countries, except the United States. Scientific observations are almost universally expressed in metric units.

Length

The standard of length in the metric system is the *meter* (m). It was originally defined in 1790 as one ten-millionth of the distance from the North Pole to the Equator. In 1889 it was redefined as the distance measured at 0°C between two lines on a bar of platinum iridium kept at the Bureau of Standards in France. Three facsimiles of this bar are also kept at the U.S. Bureau of Standards in Washington. The meter is equal to 39.37 inches.

A major advantage of using the metric system is that each unit is related to others by some factor of 10. Thus 1 m is the same as 10 decimeters (dm) or 100 centimeters (cm) or 1,000 millimeters (mm). If you were working with the English system of length, you would have to say that 1 yd is the same as 3 ft or 36 in. Figure A-1 illustrates the differences between metric and English conversions by comparing the metric meter stick and the English yardstick. Exhibit A-2 lists the metric units of length with English equivalents.

Notice in Exhibit A-2 that decimals like 1 Å = 0.0000000001 m are cumbersome to work with. In a case like this, we convert the decimal to an *exponential form,* which is a convenient way of expressing very small decimals or very large numbers in abbreviated form. The exponential form is used to conserve time and space. To avoid writing many zeros, we can express such numbers as powers of 10. The exponential notation has the form $M \times 10^n$. M is the one-digit number to the left of the decimal point, and n is a positive or negative value. You must remember two things when changing numbers into exponential form. First, determine M by moving the decimal point so you leave only one nonzero digit to the left of it. Second, determine n by counting the number of places you moved the decimal point. If you moved to the left, n is a positive number. If you moved to the right, n is negative. Thus another way of stating the 1 Å equals 0.0000000001 m is to write 1 Å = 1 ×

Exhibit A—1 ENGLISH UNITS OF MEASUREMENT

Fundamental or Derived Unit	Units and Equivalents
LENGTH	12 inches = 1 foot (ft) = 0.333 yard (yd) 3 ft = 1 yd 1,760 yd = 1 mile (mi) 5,280 ft = 1 mi
MASS	1 ounce (oz) = 28.35 grams (g); 1 g = 0.0353 oz 1 pound (lb) = 453 g = 16 oz; 1 kilogram (kg) = 2.205 lb 1 ton = 2,000 lb = 907 kg
TIME	1 second (sec) = 1/86,400 of a mean solar day 1 minute (min) = 60 sec 1 hour (hr) = 60 min = 3,600 sec 1 day = 24 hr = 1,440 min = 86,400 sec
VOLUME	1 fluid dram (fl dr) = 0.125 fluidounce (fl oz) 1 fl oz = 8 fl dr = 0.0625 quart (qt) = 0.008 gallon (gal) 1 qt = 256 fl dr = 32 fl oz = 2 pints (pt) = 0.25 gal 1 gal = 4 qt = 128 fl oz = 1,024 fl dr

Exhibit A—2 METRIC UNITS OF LENGTH AND SOME ENGLISH EQUIVALENTS

Metric Unit	Meaning of Prefix	Metric Equivalent	English Equivalent
1 kilometer (km)	kilo = 1,000	1,000 m	3,280.84 ft or 0.62 mi; 1 mi = 1.61 km
1 hectometer (hm)	hecto = 100	100 m	328 ft
1 dekameter (dam)	deka = 10	10 m	32.8 ft
1 meter (m)	Standard unit of length		39.37 inches or 3.28 ft or 1.09 yd
1 decimeter (dm)	deci = $\frac{1}{10}$	0.1 m	3.94 inches
1 centimeter (cm)	centi = $\frac{1}{100}$	0.01 m	0.394 inch; 1 inch = 2.54 cm
1 millimeter (mm)	milli = $\frac{1}{1,000}$	0.001 m = $\frac{1}{10}$ cm	0.0394 inches
1 micrometer (μm) [formerly micron (μ)]	micro = $\frac{1}{1,000,000}$	0.000,001 m = $\frac{1}{10,000}$ cm	3.94×10^{-5} inches
1 nanometer (nm) [formerly millimicron (mμ)]	nano = $\frac{1}{1,000,000,000}$	0.000,000,001 m = $\frac{1}{10,000,000}$ cm	3.94×10^{-8} inches
1 angstrom (Å)		0.000,000,000,1 m = $\frac{1}{100,000,000}$ cm	3.94×10^{-9} inches

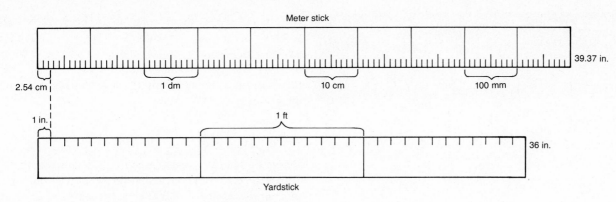

Figure A–1 Metric and English units of length.

10^{-10} m. Why? If you determine M by moving the decimal point so that only one nonzero digit is to the left of it, you have

$$0.0000000001.$$

Therefore $M = 1$. Now, since you have moved the decimal point 10 places to the right, $n = {}^{-10}$. Thus

$$1\,\text{Å} = 1 \times 10^{-10}\ \text{m}$$

Now do a problem on your own. The wavelength of yellow light is about 0.000059 cm per second. (Hereafter we will denote "per" in abbreviated form— for example, 0.000059 cm/second.) Convert the centimeters to exponential form. If your answer is 5.9×10^{-5}, you are ready to continue. If you got the wrong answer, reread the discussion.

Now suppose we are working with a very large number instead of a decimal. The same rules apply, but our exponential value will be positive rather than negative. Refer to Exhibit A-2. Note that 1 km equals 1,000 m. Even though 1,000 is not a cumbersome number, we can still easily convert it into exponential form. First move the decimal point so there is only one nonzero digit to the left of it to determine M:

$$1.000.$$

Now, because the decimal has been moved three places to the left, n equals ${}^{+3}$ or simply 3. Thus

$$1\,\text{km} = 1 \times 10^{3}\text{m}$$

Do another problem on your own. The speed of light is about 30,000,000,000 cm/second. Convert the centimeters to exponential form. Your answer should be 3×10^{10} cm.

Review Exhibit A-2 and note some common metric and English equivalents. Note also the exponential forms. Do you know what they mean? If you want to find out, convert them into decimals or large numbers.

Mass

Now let us look at the second fundamental unit of the metric system: mass. The standard unit of mass is the *kilogram* (kg). A kilogram is defined as the mass of an alloy of platinum (90 percent) and iridium (10 percent) kept at the Bureau of Standards in France. Mass is perhaps an unfamiliar term to you. *Mass* is simply the amount of matter an object contains. The mass of this textbook is the same whether it is measured in a laboratory, under the sea, on top of a mountain, or even on the moon. No matter where you take it, it still has the same quantity of matter. *Weight,* on the other hand, is a force determined by the pull of gravity on an object. This textbook will not have the same weight on earth as on the moon because of the differences in gravitation. In fact, the gravitational force on the moon is one-sixth that of earth. To simplify matters, however, weight and mass may be considered synonymous terms because the force of gravity on the surface of the earth is nearly constant. Thus weight remains nearly the same regardless of where the measurements are taken. Exhibit A-3 lists metric units of mass and some English equivalents.

The standard pound is defined in terms of the standard kilogram: The mass of a pound equals 0.4536 kg. Thus 1 kg = 2.205 lb, 453.6 g = 1 lb, and 28.35 g = 1 oz.

Time

The third fundamental unit of both the metric and the English systems is time. The standard of time is the *second.* The average length of all days throughout the

Exhibit A–3 METRIC UNITS OF MASS AND SOME ENGLISH EQUIVALENTS

Metric Unit	Metric Equivalent	English Equivalent
1 kilogram (kg)	1,000 g	2.205 lb
1 hectogram (hg)	100 g	
1 dekagram (dag)	10 g	
1 gram (g)	1 g	1 lb = 453.6 g 1 oz = 28.35 g
1 decigram (dg)	0.1 g	
1 centigram (cg)	0.01 g	
1 milligram (mg)	0.001 g	
1 microgram (μg)	0.000,001 g	

Exhibit A–4 METRIC UNITS OF VOLUME AND SOME ENGLISH EQUIVALENTS

Metric Unit	Metric Equivalent	English Equivalent
1 liter	1,000 ml	33.81 fl oz or 1.057 qt 946 ml = 1 qt
1 milliliter (ml)	0.001 liter	0.0338 fl oz; 30 ml = 1 fl oz 5 ml = 1 teaspoon
1 cubic centimeter (cm³)	0.999972 ml	0.0338 fl oz

year is called the mean solar day. There are 86,400 seconds in a mean solar day (24 × 60 × 60). Thus a second is defined as 1/86,400 of a mean solar day. Units of time are used in measuring pulse and heat rate, metabolic rate, x-ray exposure, and intervals between medications.

DERIVED UNITS

Derived units can always be written as some combination of the three fundamental units. Let us use volume as an example. Units of volume are derived from units of length. *Volume,* or capacity, in the English system may be expressed as cubic feet (ft³), cubic inches (inch³), and cubic yards (yd³). Volume in the metric system may be expressed in cubic units of length such as cubic centimeters (cm³) or in terms of the basic unit of volume, the *liter.* A liter is defined as the volume occupied by 100 g of pure water at 4°C. Since 1 cm³ of water at this temperature weighs 1 g, then 1,000 g of water occupies a volume of 1,000 cm³. This means that a liter is equal to 1,000 cm³ and 1 milliliter (ml) is equal to 1 cm³. Because of this relationship, many volume-measuring devices, such as hypodermic needles, may be graduated in milliliters or cubic centimeters. Exhibit A-4 lists metric units of volume and some English equivalents.

APOTHECARY SYSTEM

In addition to the English and metric systems, there is the **apothecary system.** This system is commonly used by physicians prescribing medications and by pharmacists preparing them. Exhibit A-5 lists the important units and equivalents of the apothecary system.

Exhibit A–5 APOTHECARY SYSTEM OF MASS AND VOLUME WITH METRIC EQUIVALENTS

Fundamental Unit	Apothecary Unit and Conversion	Metric Equivalent
MASS	60 grains = 1 dram (dr)	1 g = 15 grains
	8 dr = 1 oz	4 g = 1 dr
	12 oz = 1 lb	30 g = 1 oz
		1 kg = 32 oz
VOLUME	60 minims = 1 fluidram	1 ml (or cm³) = 15 minims
	8 fluidrams = 1 fluidounce	4 ml (or cm³) = 1 fluidram
	16 fluidounces = 1 pt	30 ml (or cm³) = 1 fluidounce
		500 ml (or cm³) = 1 pt
		1,000 ml (or cm³) = 1 qt

Frequently Used Conversions Based on Milligram

1,000 mg (1 g)	=	15 grains
600 mg (0.6 g)	=	10 grains
300 mg (0.3 g)	=	5 grains
60 mg (0.06 g)	=	1 grain
30 mg (0.03 g)	=	0.50 (1/2) grain
20 mg (0.02 g)	=	0.33 (1/3) grain
10 mg (0.01 g)	=	0.166 (1/6) grain
5 mg (0.005 g)	=	0.083 (1/12) grain
4 mg (0.004 g)	=	0.66 (1/15) grain
1 mg (0.001 g)	=	0.016 (1/60) grain
0.5 mg (0.005 g)	=	0.0083 (1/120) grain
0.1 mg (0.001 g)	=	0.0016 (1/600) grain

Glossary of Prefixes, Suffixes, and Combining Forms

It is possible to build many hundreds of medical words if you learn a few basic parts that can be combined in a variety of ways. A long complicated medical word will seem less difficult to analyze after you learn the meaning of these fundamental parts. Look at the word carefully, and say it out loud several times, so that each new word will become familiar to you.

A "prefix" is a part of a word that precedes the word root and changes its meaning. A "suffix" follows the word root and changes its meaning. A "combining form" is the basic word root such as *abdomin-*, referring to the belly region; and *aden-*, pertaining to a gland. Word roots are often followed by a vowel to facilitate pronunciation, as in *abdomino-* and *adeno-*, in which case they are then called combining forms.

We have also included here one of the methods used to learn the pronunciation of the medical terms we consider essential. These pronunciations are not the only ones. There is

and will always be some conflict between doctors and dictionaries regarding pronunciation of medical terms.

PRONUNCIATION KEY

Pronunciation appears in the parentheses immediately following the words. The strongest accented syllable appears in capital letters—for example, bī-LAT-er-al, TIB-ē-al. If the words are long, and if there is a secondary accent, it is noted by a double quote mark—for example, re″trō-per-i-ton-Ē-al.

Any additional secondary accents are noted by a double quote mark (″)—for example, stur″nō-klī″dō-MAS-toyd.

If the vowels are pronounced with a long sound, this is indicated by a *line above the vowel*. These vowels would be pronounced as follows:

ā as in *māke*, *ē* as in *bē*, *ī* as in *īvy*,
ō as in *pōle*, *ū* as in *pūre*

All other vowels not so marked are pronounced with the short sound as follows:

e as in *met* and *bet*, *i* as in *bit* and *sip*,
o as in *not*, *u* as *bud*

Other phonetics are as follows: *a* as in *father* is written *ah*, and the diphthong *oi* is written *oy*.

WORD ROOTS AND COMBINING FORMS

Many medical terms are "compound" words. That is, they are made up of more than one root or combining form. Examples of such compound words are leucocyte (white blood cell) and chemotherapy (treatment of disease by administering chemicals).

The following list includes some of the most commonly used word roots, prefixes, suffixes and combining forms used in making medical terms.

Acou-, Acu- hearing Acoustics (ah-KOOS-tiks), the science of sounds or hearing.

Acr-, Acro-, extremity Acromegaly (ak-rō-MEG-ah-lē), hyperplasia of the nose, jaws, fingers, and toes.

Aden-, Adeno- gland Adenoma (ad-i-NŌ-mah), a tumor with a glandlike structure.

Alg-, Algia- pain Neuralgia (nū-RAL-jē-ah), pain along the course of a nerve.

Angi- vessel Angiocardiography (an-jē-o-kar-dē-OG-rah-fē), roentgenography of the great blood vessels and heart after intravenous injection of radiopaque fluid.

Arthr-, Arthro- joint Arthropathy (ar-THRO-pah-thē), disease of a joint.

Aut-, Auto- self Autolysis (aw-TOL-li-sis), destruction of cells of the body by their own enzymes, even after death.

Bio- life, living Biopsy (BĪ-op-se), examination of tissue removed from a living body.

Blast- germ, bud Blastocyte (BLAS-tō-sīt), an embryonic or undifferentiated cell.

Blephar- eyelid Blepharitis (blef-ah-RĪ-tis), inflammation of the eyelids.

Bronch- trachea, windpipe Bronchoscopy (brong-KOS-kō-pē), direct visual examination of the bronchi.

Bucc- cheek Buccocervical (bū-kō-SER-vē-kal), pertaining to the cheek and neck.

Capit- head Decapitate (dē-KAP-i-tāte), to remove the head.

Carcin- cancer Carcinogenic (kar-sin-ō-JEN-ic), causing cancer.

Cardi-, Cardia-, Cardio- heart Cardiogram (KAR-dē-ō-gram), a recording of the force and form of the heart's movements.

Cephal- head Hydrocephalus (hī-drō-SEF-ah-lus), enlargement of the head due to an abnormal accumulation of fluid.

Cerebro- brain Cerebrospinal fluid (ser"i-brō-SPĪ-nal), fluid contained within the cranium and spinal canal.

Cheil- lip Cheilosis (kī-LŌ-sis), dry scaling of the lips.

Chole- bile, gall Cholecystogram, (kō-le-SIS-tō-gram), roentgenogram of the gallbladder.

Chondr-, Chondri-, Chondrio- cartilage Chondrocyte (KON-drō-sit), a cartilage cell.

Chrom-, Chromat-, Chromato- color Hyperchromic (hī-per-KRŌ-mik), highly colored.

Crani- skull Craniotomy (krā-nē-OT-ō-mē), surgical opening of the skull.

Cry-, Cryo- cold Cryosurgery (kr-IŌ-ser-jerē), surgical procedure using a very cold liquid nitrogen probe.

Cut- skin Subcutaneous (sub-kyoo-TĀ-nē-us), under the skin.

Cysti-, Cysto- sac or bladder Cystoscope (SIS-tō-skōp), instrument for interior examination of the bladder.

Cyt-, Cyto-, Cyte- cell Cytology (sī-TOL-ō-jē), the study of cells.

Dactyl-, Dactylo- digits (usually fingers, but sometimes toes) Polydactylia (pol-ē-DAK-til-ē-ah), above normal number of fingers or toes.

Derma-, Dermato-, skin Dermatosis (der-mah-TŌ-sis), any skin disease.

Entero- intestine Enteritis (en-ter-Ī-tis), inflammation of the intestine.

Gastr- stomach Gastrointestinal (gas"tro-in-TES-ti-nal), pertaining to the stomach and intestine.

Gloss-, Glosso- tongue Hypoglossal (hī-pō-GLOS-al), located under the tongue.

Hem-, Hemat- blood Hematoma (he-mah-TŌ-mah), blood beneath the skin.

Hepar-, Hepato- liver Hepatitis (hep-ah-TĪ-tis), inflammation of the liver.

Hist-, Histio- tissue Histology (his-TOL-ō-jē), the study of tissues.

Homeo-, Homo- unchanging, the same, steady Homeostasis (hō-mē-ō-STĀ-sis), achievement of a steady state.

Hyster- uterus Hysterectomy (his-ter-EK-tō-mē), surgical removal of the uterus.

Ileo-, ileum (part of intestine) Ileocecal valve (il″e-ō-SĒ-kal), folds at the opening between ileum and cecum.

Ilio- ilium (flaring portion of hipbone) Iliosacral (il″e-ō-SĀK-ral), pertaining to ilium and sacrum.

Lachry-, Lacri- tears Nasolacrimal (nā-zō-LAK-rē-mal), pertaining to the nose and lacrimal apparatus.

Leuco-, Leuko- white Leucocyte (LŪ-kō-sīt), white blood cell.

Lip-, Lipo- fat Lipoma (lī-PŌ-mah), a fatty tumor.

Mamm- breast Mammography (mam-OG-rah-fē), roentgenography of the mammary gland.

Mast- breast Mastitis (mas-TĪ-tis), inflammation of the mammary gland.

Meningo- membrane Meningitis (men-in-JĪ-tis), inflammation of the membranes of spinal cord and brain.

Metro- uterus Endometrium (en-dō-MĒ-trē-um), lining of the uterus.

Morpho- form, shape Morphology (mōr-FOL-ō-jē), study of form and structure of things.

Myelo- marrow, spinal cord Poliomyelitis (pō″lē-ō-mī-el-Ī-tis), inflammation of the gray matter of the spinal cord.

Myo- muscle Myocardium (mī-ō-KAR-dē-um), heart muscle.

Nephro- kidney Nephrosis (ne-FRŌ-sis), degeneration of kidney tissue.

Neuro- nerve Neuroblastoma (nū″-rō-BLAS-tō-mah), a malignant tumor of the nervous system composed of embryonic nerve cells.

Oculo- eye Binocular (bī-NOK-ū-lar), pertaining to the two eyes.

Odont- tooth Orthodontic (ōr-thō-DON-tik), pertaining to the proper positioning and relationship of the teeth.

Ophthalm- eye Ophthalmology (of-thal-MO-lō-jē), science of the eye and its diseases.

Oss-, Osseo-, Osteo-, bone Osteoma (os-tē-Ō-mah), a bone tumor.

Oto- ear Otosclerosis (ō″tō-skle-RŌ-sis), formation of bone in the labyrinth of the ear.

Patho- disease Pathogenic (path-ō-JEN-ik), giving origin to disease.

Phag-, Phago- to eat Phagocytosis (fa″gō-sī-TŌ-sis), a cell that eats microorganisms, other cells, and foreign particles.

Philic-, Philo- to like, have an affinity for Hemophilic (hē-mō-FI-lik), have an affinity for bleeding.

Phleb- vein Phlebitis (fleb-Ī-tis), inflammation of the veins.

Pneumo- lung Pneumothorax (nū-mō-THŌ-raks), air in the thoracic cavity.

Pod- foot Podiatry (pō-DĪ-a-trē), the diagnosis and treatment of foot disorders.

Procto- anus, rectum Proctoscopy (prok-TO-skō-pē), instrumental examination of the rectum.

Pseud-, Pseudo- false Pseudoangina (SŪ-dō-an-ji″nah), false angina.

Psycho- soul or mind Psychiatry (sī-KĪ-a trē), treatment of mental disorders.

Pyo- pus Pyuria (pī-ū-RĒ-ah), pus in the urine.

Scler-, Sclero- hard Arteriosclerosis (ar-tē″rē-ō-skle-RŌ-sis), hardening of the arteries.

Sep-, Septic- poison or toxic substance Septicemia (sep-ti-SĒ-mē-ah), presence of bacterial toxins in the blood (blood poisoning).

Soma-, Somato- body Somatotropic (sō-ma-tō-TRŌ-pik), having a stimulating effect on body growth.

Stasis-, Stat- stand still Homeostasis (hō-mē-ō-STĀ-sis), achievement of a steady state.

Therm- heat Thermometer (ther-MO-me-ter), instrument used to measure and record heat.

Viscer- organ Visceral (VIS-er-al), pertaining to the abdominal organs.

Prefixes

A-, An- without, lack of, deficient Aseptic (ā-SEP-tik), without infection; anesthesia (an-es-THĒ-zhē-ah), without sensation.

Ab- away from, from Abnormal (ab-NOR-mal), away from normal.

Ad- to, near, toward Adduction (Ad-DUK-shun), movement of an extremity toward the axis of the body.

Ambi- both sides Ambidextrous (am-bē-DEKS-trus), able to use either hand.

Ante- before Antepartum (an-tē-PAR-tum), before delivery of a baby.

Anti- against Anticoagulant (an″tē-kō-AG-ū-lant), a substance that prevents coagulation of blood.

Bi- two, double, both Biceps (BĪ-seps), a muscle with two heads of origin.

Brachi- arm Brachialis (brā-kē-AL-is), muscle for flexing forearm.

Brachy- short Brachyesophagus (brā-kē-e-SOF-ah-gus), short esophagus.

Brady- slow Bradycardia (brā-dē-KAR-dē-ah), abnormal slowness of the heartbeat.

Cata- down, lower, under, against Catabolism (ka-TAB-ō-lizm), metabolic breakdown into simpler substances

Circum- around Circumrenal (ser-kum-RĒ-nal), around the kidney.

Con- with, together Congenital (kon-JEN-i-tal), born with a defect.

Contra- against, opposite Contraception (kon-tra-SEP-shun), the prevention of conception.

Crypt- hidden, concealed Cryptorchidism (krip-TOR-ki-dizm), undescended or hidden testes.

Di-, diplo- two Diploid (DI-ployd), having double the haploid number of chromosomes.

Dis- separation, apart, away from Disarticulate (dis-ar-TI-kū-lāt), to separate at joint.

Dys- painful, difficult Dyspnea (disp-NĒ-ah), difficult breathing.

E-, Ec-, Ex- out from, out of Eccentric (eks-EN-trik), not located at the center.

Ecto-, Exo- outside Ectopic pregnancy (ek-TOP-ik), gestation outside the uterine cavity.

Em-, En- in or on Empyema (em-pī-E-mah), pus in a body cavity.

End-, Endo- inside Endocardium (en-dō-KAR-dē-um), membrane lining the inner surface of the heart.

Epi- upon, on above Epidermis (ep-i-DER-mis), outermost layer of skin.

Erythro- red Erythrocyte (e-RITH-rō-sīt), red blood cell.

Eu- well Eupnea (Ū-pnē-ah), normal breathing.

Ex-, Exo- out, away from Exocrine (EKS-ō-krin), excreting outwardly or away from.

Extra- outside, beyond, in addition to Extracellular (eks″tra-SEL-ū-lar), outside of the cell.

Galacto- milk Galactose (gal-AK-tōs), a milk sugar.

Glyco- sugar Glycosuria (glī-kō-SUR-ē-ah), sugar in the urine.

Gyn-, Gyne-, Gynec- female sex (women) Gynecology (gī″ne-KOL-o-jē), the study of the diseases of the female.

Hemi- half Hemiplegia (hem-ē-PLĒ-jē-ah), paralysis of only half of the body.

Heter-, Hetero- other, different Heterogeneous (he-ter-ō-JĒN-ē-ūs), composed of different substances.

Hydr- water Hydrocele (HĪ-drō-sēl), accumulation of fluid in a saclike cavity.

Hyper- beyond, excessive Hyperglycemia (hī″per-glī-SĒ-mē-ah), excessive amount of sugar in the blood.

Hypo- under, below, deficient Hypodermic (hī-pō-DER-mik), below the skin or dermis.

Idio- self, one's own, separate Idiopathic (id-ē-ō-PATH-ik), a disease without recognizable cause.

Inter- among, between Intercostal (in-ter-KOS-tal), between the ribs.

Intra- within, inside Intracellular (in″trah-SEL-ū-lar), inside the cell.

Iso- equal, like Isogenic (īso-JEN-ik), alike in morphological development.

Macro- large, great Macrophage (MAK-rō-faj), large phagocytic cell.

Mal- bad, abnormal Malnutrition (mal-nū-TRISH-un), lack of necessary food substances.

Mega-, Megalo- great, large Megakaryocyte (meg″ah-KAR-ē-ō-sīt), giant cell of bone marrow.

Meta- after, beyond Metacarpus (met-ah-KAR-pus), the part of the hand between the wrist and fingers.

Micro- small Microtome (MĪ-krō-tōm), instrument for preparing very thin slices of tissue for microscopic examination.

Necro- corpse, dead Necrosis (nē-KRŌ-sis), death of areas of tissue or bone surrounded by healthy parts.

Neo- new Neonatal (nē-ō-NĀT-al), pertaining to the first 4 weeks after birth.

Oo- **egg** Oocyte (Ō-ō-sīt), original egg cell.

Oligo- **small, deficient** Oliguria (ō-lig-UR-ē-ah), abnormally small amount of urine.

Ortho- **straight, normal** Orthopnea (or-THOP-nē-ah), inability to breathe in any position except when straight or erect.

Para- **near, beyond, apart from, beside** Paranasal (par-ah-NĀ-zal), near the nose.

Ped- **children** Pediatrician (pē-dē-at-TRISH-ē-an), a medical specialist in treatment of children's diseases.

Per- **through** Percutaneous (per-kyōō-TĀ-nē-us), through the skin.

Peri- **around** Pericardium (per-ē-KAR-dē-um), membrane or sac around the heart.

Poly- **much, many** Polycythemia (pol″ē-sī-THE-mē-ah), an excess of red blood cells.

Post- **after, behind** Postnatal (pōst--NĀ-tal), after birth.

Pre-, Pro- **before, in front of** Prenatal (prē-NĀ-tal), before birth.

Retro- **backward, located behind** Retroperitoneal (re″tro-per-i-ton-Ē-al), located behind the peritoneum.

Rhin- **nose** Rhinitis (rīNĪ-tis), inflammation of the nasal mucosa.

Semi- **half** Semicircular canals (semi-SER-kyōō- lar), canals in the shape of a half circle.

Sten- **narrow** Stenosis (ste-NŌ-sis), narrowing of a duct or a canal.

Sub- **under, beneath, below** Submucosa (sub-myōō-KŌ-sah), tissue layer under a mucous membrane.

Super- **above, beyond** Superficial (su-per-FISH-ē-al), confined to the surface, not through.

Supra- **above, over** Suprarenal (su-prah-RĒ-nal), adrenal gland above the kidney.

Sym-, Syn- **with, together, joined** Syndrome (SIN-drom), all the symptoms of a disease considered as a whole.

Tachy- **quick, rapid** Tachycardia (tak-ē-KAR-dē-ah), rapid heart action.

Tox-, Toxic- **poison** Toxemia (toks-Ē-mē-ah), poisonous condition of the blood.

Trans- **across, through, beyond** Transudation (trans-ū-DĀ-shun), oozing of a fluid through pores.

Tri- **three** Trigone (TRĪ-gōn), a triangular space, as at the base of the bladder.

Trich- **hair** Trichosis (tri-KO-sis), disease of the hair.

Zoo- **animal** Zoology (zō-OL-ō-jē), the study of animals.

Suffixes

-able **capable of, having ability to** Viable (VĪ-ah-bel), capable of living.

-ac, -al **pertaining to** Cardiac (KAR-dē-ak) (cardial), pertaining to the heart.

-ary **connected with** Ciliary (SIL-ē-ār-ē), resembling any hairlike structure.

-asis, -asia, -esis, -osis **condition or state of** Hemostasis (hē-mō-STĀ-sis), stopping of bleeding or circulation.

-cel, -cele **swelling, an enlarged space or cavity** Meningocele (men-in-JŌ-sēl), enlargement of the meninges.

-cid, -cide, -cis **cut or kill, destroy** Germicide (GERM-i-cīd), a substance that kills germs.

-ectasia, -ectasis **stretching, dilatation** Bronchiectasis (brong-kē-EK-tah-sis), dilatation of a bronchus or bronchi.

-ectomize, ectomy **cutting out, excision of, removal of** Thyroidectomy (thī-royd-EK-tō-mē), surgical removal of thyroid gland.

-emia **condition of blood** Lipemia (lī-PĒ-mē-ah), abnormally high concentration of fat in the blood.

-ferent **bear or carry** Efferent (Ef-er-ent), carrying away from a center.

-form **shape** Fusiform (FŪZ-i-form), spindle-shaped.

-gen **an agent that produces or originates** Pathogen (PATH-ō-jen), a microorganism or substance capable of producing a disease.

-genic **produced from, producing** Pyogenic (pī-ō-JEN-ik), produced from pus.

-gram **a record, that which is recorded** Electrocardiogram (ē-lek″trō-CAR-dē-ō-gram), a record of heart action.

-graph **an instrument for recording** Electroencephalograph (ē-lek″trō-en-SEF-ah-lō-graf), an instrument for recording electrical activity of the brain.

-ia **a state or condition** Hypermetropia (hī″per-me-TRŌ-pē-ah), condition of farsightedness.

-iatrics, iatry **medical practice specialties** Pediatrics (pē-dē-AT-riks), medical science relating to care of children and treatment of their diseases.

-ism **condition of state** Rheumatism (RŪ-mah-tizm), inflammation, especially of muscle and joints.

-itis inflammation Neuritis (nū-RĪ-tis), inflammation of a nerve or nerves.

-logy, -ology the study or science of Physiology (fiz-ē-OL-ō-jē), the study of function of body parts.

-lyso, -lysis solution, dissolve, loosening Hemolysis (hē-MOL-is-is), dissolution of the red blood cells.

-malacia softening Osteomalacia (os″tē-o-mal-Ā-she-ah), softening of bone.

-oma tumor Fibroma (fī-BRŌ-mah), a tumor composed mostly of fibrous tissue.

-ory pertaining to Sensory (SEN-sō-rē), pertaining to sensation.

-ose full of Adipose (AD-i-pōz), characterized by excessive amounts of fat.

-pathy disease Neuropathy (nū-RŌ-pah-thē), disease of the peripheral nervous system.

-penia deficiency Thrombocytopenia (throm″bo-sī-tō-PĒ-nē-ah), deficiency of thrombocytes in the blood.

-phobe, -phobia fear of, aversion for Hydrophobia (hī-drō-FŌ-bē-ah), fear of water.

-plasia, -plasty reconstruction of Rhinoplasty (RĪ-nō-plas-tē), surgical reconstruction of the nose.

-plegia, -plexy stroke, paralysis Apoplexy (AP-ō-plek-sē), sudden loss of consciousness, and paralysis.

-pnea to breathe Apnea (AP-nē-ah), temporary absence of respiration, following a period of overbreathing.

-poiesis formation of Hematopoiesis (hē-mat″ō-PĪ-es-is), formation and development of red blood cells.

-rrhea flow, discharge Diarrhea (dī-ah-RĒ-ah), abnormal frequency of bowel evacuation, the stools with a more or less fluid consistency.

-scope an instrument used to look into or examine a part Bronchoscope (BRONG-kō-skōp), instrument used to examine the interior of a bronchus.

-stomy creation of a mouth or artificial opening Tracheostomy (trā-kē-OST-ō-mē), creation of an opening in the trachea.

-tomy cutting into, incision into Appendectomy (apen-DEC-tō-mē), surgical removal of the appendix.

-trophy a state relating to nutrition Hypertrophy (hī-PER-trō-fē), a condition resulting from increase in size of cells.

-tropic turning toward, influencing, changing Gonadotropic (gō-na-dō-TRŌ-pik), influencing the gonads.

-uria urine Polyuria (pol-ē-Ū-rē-ah), excessive secretion of urine.

Glossary of Terms

Abatement (ah-BĀT-ment) A decrease in the seriousness of a disorder; a decreasing severity of pain or other symptoms.

Abdomen (ab-DŌ-men) The area between the diaphragm and pelvis.

Abduct (ab-DUKT) To draw away from the axis or midline of the body or one of its part.

Abortion (ab-ŌR-shun) The premature loss or removal of the embryo or nonviable fetus; any failure in the normal process of developing or maturing.

Abscess (AB-ses) A localized collection of pus and liquefied tissue in a cavity.

Absorption (ab-SŌRP-shun) The taking up of liquids by solids or of gases by solids or liquids.

Acapnia (a-KAP-nē-ah) Decrease in the normal amount of carbon dioxide in the blood.

Accommodation (ak-kom-mō-DĀ-shun) A change in the shape of the eye lens so that vision is more acute; an adjustment of the eye lens for various distances; focusing.

Accretion (ah-KRĒ-shun) A mass of material that has accumulated in a space or cavity; the adhesion of parts.

Acetabulum (as"i-TAB-yū-lum) The rounded cavity on the external surface of the innominate bone that receives the head of the femur.

Acetone (AS-e-tōne) **bodies** Organic compounds that may be found in excessive amounts in the urine and blood of diabetics. Found whenever too much fat in proportion to carbohydrate is being oxidized. Also called ketone bodies (KĒ-tōne).

Acetylcholine (a sēt-il-KŌ-lēn) A chemical transmitter substance that is liberated at synapses in the central nervous system. It stimulates skeletal muscle contraction.

Achilles (ah-KIL-ez) **reflex** Extension of foot and contraction of calf muscles following a tap upon tendon of Achilles.

Achilles tendon The tendon of the soleus and gastrocnemius muscles at the back of the heel.

Achlorhydria (ā-klōr-HĪ-drē-ah) Absence of hydrochloric acid in the gastric juice.

Acid (AS-id) A proton donor; excess hydrogen ions producing a pH less than 7.

Acidosis (as-i-DŌ-sis) A serious disorder in which the normal alkaline substances of the blood are reduced in amount.

Acoustic (ah-KOOS-tik) Pertaining to sound or the sense of hearing.

Acromion (ah-KRŌ-mē-on) The lateral triangular projection of the spine of the scapula, forming the point of the shoulder and articulating with the clavicle.

Actin (AK-tin) One of the contractile proteins in muscle fiber. The other protein is myosin.

Action potential An impulse or wave of negativity along a conducting neuron.

Actomyosin (ak-tō-MĪ-ō-sin) The combination of actin and myosin in a muscle cell.

Acuity (ak-Ū-i-tē) Clearness, or sharpness, usually of vision.

Acute (ak-ŪT) Having rapid onset, severe symptoms, and a short course, not chronic.

Adam's apple   The laryngeal prominence of the thyroid cartilage.

Adaptation (ad-ap-TĀ-shun)   The adjustment of the pupil of the eye to light variations.

Addison's (AD-i-sonz) **disease**   Disorder due to deficiency in the secretion of adrenocortical hormones.

Adenohypophysis (ad″i-nō-hī-POF-i-sis)   The anterior portion of the pituitary gland.

Adenoids (Ad-i-noyds)   The pharyngeal tonsils.

Adenosine triphosphate (ATP) (ad-EN-ō-sēn trī-FOS-fāte)   A compound containing sugar, adenine, nitrogen, and three phosphoric acids; the breakdown of ATP provides the energy for cellular work.

Adhesion (ad-HĒ-zhun)   Abnormal joining of parts to each other.

Ad libitum (ad-LIB-it-um)   At pleasure; the amount desired.

Adolescence (ad-ō-LES-ens)   The period from the beginning of puberty until adult life.

Adrenal (ad-RĒ-nal) **glands**   Two glands located superior to each kidney; also called the suprarenal glands.

Adrenalin (ad-REN-ah-lin)   Proprietary name for epinephrine; one of the active secretions of the medulla of the adrenal gland.

Adrenergic (ad-ren-ER-jik)   Applied to nerve fibers that, when stimulated, release norepinephrine (noradrenalin) at their terminations.

Adrenocorticotropic (ad-rēn-ō-kōr-ti-kō-TRŌP-ik) **hormone (ACTH)**   Hormone produced by the anterior pituitary gland, which influences the cortex of the adrenal glands.

Adsorption (ad-SORP-shun)   Process whereby a gas or a dissolved substance becomes concentrated at the surface of a solid or at the interfaces of a colloid system.

Adventitia (ad-ven-TISH-yah)   The outermost covering of a structure or organ.

Aerobic (ayer-Ō-bik)   Living only in the presence of oxygen.

Afferent (AF-er-ent)   Carrying impulses toward a center, nerves that transmit impulses toward the central nervous system.

Agglutination (ag-glū-tin-Ā-shun)   Clumping of microorganisms, blood corpuscles or particles; an immunity response; an antigen-antibody reaction.

Agglutinin (ag-GLŪ-tin-in)   A specific principle or antibody in blood serum of an animal affected with a microbic disease that is capable of causing the clumping of bacteria, blood corpuscles, or particles.

Agglutinogen (ag-GLŪ-tin-o-jen)   A genetically determined antigen located on the surface of erythrocytes. These proteins are responsible for the two major blood group classifications: the ABO group and the Rh system.

Agnosia (ag-NŌ-zē-ah)   A loss of the ability to recognize the meaning of stimuli from the various senses (visual, auditory, touch).

Agonist (AG-ō-nist)   The prime mover – the muscle directly engaged in contraction as distinguished from muscles that are relaxing at the same time.

Albinism (AL-bin-ism)   Abnormal, nonpathological, partial or total absence of pigment in skin, hair, and eyes.

Albumin (al-BŪ-min)   A protein substance found in nearly every animal or plant tissue and fluid.

Albuminuria (al-bū″min-UR-ēa)   Presence of albumin in the urine.

Aldosterone (al-DOS-tē-rōne)   Powerful salt-retaining hormone of the adrenal cortex.

Alimentary (al-i-MENT-ar-ē)   Pertaining to nutrition.

Alkaline (AL-ka-lin)   Containing more hydroxyl than hydrogen ions and producing a pH of more than 7.

Alkalosis (al-kah-LŌ-sis)   Increased bicarbonate content of the blood due to excess of alkalies or withdrawal of acid or chlorides from the blood.

Allantois (al-AN-tō-is)   A kind of elongated bladder between the chorion and amnion of the fetus.

Allergic (ah-LER-jik)   Pertaining to or sensitive to an allergen.

Alveolar (al-VĒ-ō-lar)   A small depression.

Alveolus (al-VĒ-ō-lus)   A small hollow or cavity; an air cell in the lungs.

Ameboid (am-Ē-boyd)   Having the appearance and characteristics of an ameba.

Amenorrhea (ā-men-ō-RĒ-ah)   Absence or suppression of menstruation.

Amino (am-Ē-no) **acids**   Any one of a class of organic compounds occurring naturally in plant and animal tissues and forming the chief constituents of protein. About 22 different ones are known.

Amniocentesis (am″nē-ō-sen-TĒ-sis)   Removal of amniotic fluid by inserting a needle transabdominally into the amniotic cavity.

Amnion (AM-nē-on)   The inner of the fetal membranes, a thin transparent sac that holds the fetus suspended in amniotic fluid. Also called the "bag of waters."

Amorphous (ā-MOR-fus)   Without definite shape or differentiation in structure; pertains to solids without crystalline structure.

Amphiarthrosis (am″fē-ar-THRO-sis)   A form of articulation midway between diarthrosis and synarthrosis, in which the articulating bony surfaces are separated by an elastic substance to which both are attached, so that the mobility is slight, but may be exerted in all directions.

Amphoteric (am-fō-TER-ik)   Affecting both red and blue litmus paper.

Ampulla (am-PŪ-lah)   A saclike dilatation of a canal.

Amyl nitrate (Ā-mil NĪ-trāte)   An organic compound that produces dilatation of the blood vessels when inhaled. Used in attacks of angina pectoris.

Anabolism (ah-NA-bō-lizm)   The building up of the body substance.

Anaerobe (an-Ā-er-ōb)   A microorganism that thrives best or lives only without oxygen.

Analgesia (an-al-JĒ-zē-ah)   Absence of normal sense of pain.

Anaphylactic (an″a-fil-AC-tik)   Pertaining to increasing susceptibility to any foreign protein introduced into the body; decreasing immunity.

Anastomosis (ah-nas-tō-MO-sis)   A communication between either blood vessels, lymphatics or nerves. An end-to-end union or joining together.

Anatomy (an-NAT-ō-mē)   The structure or study of structure of the body and the relationship of its parts to each other.

Androgen (AN-drō-jen)   Substance producing or stimulating male characteristics, such as the male hormone.

Anemia (ah-NĒM-ē-ah)   A decrease in certain elements of the blood, especially red cells and hemoglobin.

Aneurysm (AN-ū-rizm)   A saclike enlargement of a blood vessel caused by a weakening of the wall.

Angina pectoris (AN-ji-nah PEK-tō-ris)   An agonizing pain in the chest that may or may not involve heart or artery disease.

Angiography (an-jē-OG-rah-fē)   The injection of a contrast medium into the common carotid or vertebral artery that demonstrates the cerebral blood vessels in the x-rays, and may detect brain tumors with specific vascular patterns.

Angiotensin (an-jē-ō-TEN-sin)   A blood substance that produces vasoconstriction.

Angstrom (Å) (ANG-strum)   One tenth of a millimicron or about one two hundred fifty millionth of an inch.

Anion (AN-ī-un)   An ion, carrying a negative charge.

Anisocytosis (an″eso-sī-TO-sis)   An abnormal condition in which there is a lack of uniformity in the size of red blood cells.

Ankylose (ANG-ke-lōs)   Immobilization of a joint by pathological or surgical process.

Anomaly (ah-NOM-ah-lē)   An abnormality that may be a developmental (congenital) defect; a variant from the usual standard.

Anorexia (an-ō-REK-sē-ah)   Loss of appetite; anorexia nervosa is a nervous condition in which a person may become seriously weakened from lack of food.

Anoxemia (an-oks-ĒM-ē-ah)   Lack of oxygen in the blood.

Anoxia (an-OK-sē-ah)   Deficiency of oxygen.

Antepartum (an-tē-PAR-tum)   Before delivery of the child; occurring (to the mother) before childbirth.

Anterior (an-TĒ-rē-ōr)   In front of or the ventral surface.

Antibiotic (an-tī-bī-OT-ic)   Destructive of living things, especially chemicals produced by living organisms that act against disease-producing organisms.

Antibody (AN-ti-bodē)   A specific substance produced by and in an animal or person that produces immunity to the presence of an antigen.

Antidiuretic (an-tī-dīur-ET-ik)   Substance that inhibits urine formation.

Antigen (AN-ti-jen)   Any substance that when introduced into the tissues or blood induces the formation of antibodies or reacts with them.

Antimetabolite (an-tī-met-AB-ō-līt)   Any substance resembling a normal metabolite, but foreign to the body, which competes with, replaces, or antagonizes the regular metabolite.

Antioxidants (an-tī-OKS-i-dants)   Substances that inhibit oxidation.

Antrum (AN-trum)   Any nearly closed cavity or chamber, especially one within a bone, such as a sinus.

Anulus fibrosus (AN-ū-lus fī-BRO-sus)   A ring of fibrous tissue and fibrocartilage that encircles the pulpy substance (nucleus pulposus) of an intervertebral disc.

Anuria (an-U-rē-ah)   Absence of urine formation.

Anus (Ā-nus)   The distal end and outlet of the rectum.

Aorta (ā-OR-tah) The main systemic trunk of the arterial system of the body; emerges from the left ventricle.

Aperture (AP-er-chur) An opening or orifice.

Apex (A-peks) The pointed end of a conical structure.

Aphasia (a-FĀ-szē-ah) Loss of ability to express oneself properly through speech or loss of verbal comprehension.

Apocrine (AP-o-krin) **gland** A type of gland in which the secretory products gather at the free end of the secreting cell and are pinched off, along with some of the cytoplasm, to become the secretion, as in mammary glands.

Apocrine (AP-ō-krin) Pertaining to cells that lose part of their cytoplasm while secreting.

Aponeurosis (ap″ō-nū-RŌ-sis) A sheetlike layer of connective tissue joining a muscle to the part that it moves or functioning as a sheath enclosing a muscle.

Appendage (ah-PEN-dij) A part or thing attached to the body.

Aqueduct (AK-we-duct) A canal or passage, especially for the conduction of a liquid.

Aqueous (ĀK-wē-us) **humor** The water fluid that fills the anterior cavity of the eye.

Arachnoid (ar-AK-noyd) The middle of the three coverings (meninges) of the brain.

Arachnoid villi (VIL-ē) Berrylike tufts of arachnoid that protrude into the superior sagittal sinus and through which the cerebrospinal fluid enters the bloodstream; arachnoid granulations.

Arbor vitae (AR-bōr VĒ-tay) The treelike appearance of the white matter tracts of the cerebellum when seen in midsagittal section. Also a series of branching ridges within the cervix of the uterus.

Arch of aorta (ā-OR-tah) The most superior portion of the aorta, lying between the ascending and descending segments of the aorta; the brachiocephalic, left common carotid, and left subclavian arteries are the three branches of the aortic arch.

Areola (ah-RĒ-ō-lah) Any tiny space in a tissue; the pigmented ring around the nipple of the breast.

Areolar (ah-RĒ-ō-lar) A type of connective tissue.

Arrhythmia (ah-RITH-mē-ah) Irregular heart action causing absence of rhythm.

Arteriogram (ar-TĒR-ē-ō-gram) A roentgenogram of an artery obtained by injecting radiopaque substances into the blood.

Arthrology (ar-THROL-o-jē) The scientific study or description of joints.

Arthrosis (ar-THRŌ-sis) A joint of articulation.

Articulate (ar-TIK-ū-lāt) To join together as a joint to permit motion between parts.

Arytenoid (ar-IT-en-oyd) Resembling a ladle.

Ascites (as-SĪ-tēz) Serous fluid in the peritoneal cavity.

Aseptic (ā-SEP-tik) Free from any infectious or septic material.

Asphyxia (as-FIX-ē-ah) Unconsciousness due to interference with the oxygen supply of the blood.

Aspirate (AS-pir-āte) To remove by suction.

Astereognosis (as-ter″-ē-ōg-NŌ-sis) Inability to recognize objects or forms by touch.

Asthenia (as-THĒ-nē-ah) Lack or loss of strength; debility.

Astigmatism (ah-STIG-mah-tizm) An irregularity of the lens or cornea of the eye causing the image to be out of focus and producing faulty vision.

Astrocyte (AS-trō-sīt) A neuroglial cell having a star shape.

Ataxia (a-TAKS-ē-ah) A lack of muscular coordination, lack of precision.

Atherosclerosis (ath″erō-skle-RŌ-sis) A disease involving mainly the lining of large arteries, in which yellow patches of fat are deposited, forming plaques that decrease the size of the lumen. Also called arteriosclerosis.

Atresia (ah-TRĒ-zē-ah) The abnormal closure of a passage, or the absence of a normal body opening.

Atrium (Ā-trē-um) A cavity or sinus.

Atrophy (AT-rō-fē) A wasting away or decrease in size of a part, due to a failure, abnormality of nutrition, or lack of use.

Audiovisual (aw-dē-ō-VIZH-u-al) Stimulating the senses of both hearing and sight (used in describing aids such as slides and motion pictures in teaching).

Aura (AW-rah) A feeling or sensation that precedes an epileptic seizure or any paroxysmal attack (like those of bronchial asthma).

Auricle (AW-ri-kul) The flap or pinna of the ear; an appendage of an atrium of the heart.

Auscultation (aws-kul-TĀ-shun) Examination by listening to sounds in the body.

Autopsy (AW-top-sē) The examination of the body after death; a postmortem study of the corpse. Also necropsy.

Axilla (ak-SIL-ah) The small hollow beneath the arm where it joins the body at the shoulder; the armpit.

Azygos (AZ-ĭ-gos) An anatomical structure that is not paired; occurring singly.

Bactericide (bak-TĒ-ri-cīd) An agent that kills bacteria.

Bacteriophage (bac-TĒ-rē-ō-faj) A nonspecific agent capable of destroying bacteria.

Baroreceptor (bar-ō-rē-SEP-tōr) Receptor stimulated by pressure change.

Bartholin's (BAR-tō-linz) **glands** Two small mucous glands located one on each side of the vaginal opening.

Basal (BĀ-zel) **metabolic** (met-ah-BOL-ik) **rate** **(BMR)** The rate of metabolism measured under standard or basal conditions.

Base A nonacid, a proton acceptor, characterized by excess of OH ion and a pH greater than 7.

Basement membrane A thin layer of amorphous substance underlying the epithelium of mucous surfaces.

Basophil (BĀ-sō-fil) A white blood cell characterized by a pale nucleus and large, densely basophilic granules.

Bel A measure of sound intensity.

Benign (bē-NĪN) Not malignant.

Bicipital (bī-CIP-i-tal) Having two heads, as in muscle.

Bifurcate (bī-FUR-kāt) Having two branches or divisions; forked.

Bilateral (bī-LAT-er-al) Pertaining to two sides of the body.

Bile (bīl) A secretion of the liver.

Biliary (BIL-ē-a-rē) Relating to bile, the gallbladder, or the bile ducts.

Bilirubin (bil-i-ROO-bin) The orange or yellowish pigment in bile.

Biliverdin (bil-i-VER-din) A greenish pigment in bile formed in the oxidation of bilirubin.

Biopsy (BĪ-op-sē) Removal of tissue or other material from the living body for examination, usually microscopic.

Blastocyst (BLAS-tō-sist) A stage in the development of a mammalian embryo.

Blastomere (BLAS-tō-mēr) One of the cells resulting from the cleavage of segmentation of a fertilized ovum.

Blastula (BLAS-tū-lah) An early stage in the development of a zygote.

Blepharism (BLEF-ah-rizm) Spasm of the eyelids; continuous blinking.

Blind spot Special area in the retina at the end of the optic nerve in which there are no light receptor cells.

Blood–brain barrier A special mechanism that prevents the passage of materials from the blood to the cerebrospinal fluid and brain.

Bolus (BŌ-lus) A soft, rounded mass, usually food, that is swallowed.

Bradycardia (brā-dē-KAR-dē-ah) Slow heart rate.

Bronchiectasis (brong″-kē-EK-tah-sis) A chronic disorder in which there is a loss of the normal elastic tissue and expansion of lung air passages. Characterized by difficult breathing, coughing, expectoration of pus, and foul breath.

Bronchiole (BRONG-kē-ōl) The smaller divisions of the bronchi.

Bronchogram (BRONG-kō-gram) A roentgenogram of the lungs and bronchi.

Bronchoscope (BRONG-kō-skōp) An instrument used to examine the interior of the large tubes (bronchi) of the lungs.

Bronchus (BRONG-kus) One of the two large branches of the trachea.

Brownian movement Random movement of particles distinguished from motility of living microorganisms.

Buccal (BŪK-al) Pertaining to the cheek or mouth.

Buffer (BU-fer) A substance that tends to preserve a certain hydrogen-ion concentration by adding acids or bases.

Bunion (BUN-yun) Inflammation and thickening of the bursa of a toe joint.

Bursa (BUR-sah) A sac or pouch of synovial fluid located at friction points, especially about joints.

Cachexia (kah-KEK-sē-ah) A state of ill health, malnutrition, and wasting.

Calcify (KAL-si-fī) To harden by deposits of calcium salts.

Calcitonin (kal-si-TŌN-in) Hormone secreted by the parafollicular cells of the thyroid gland.

Calculus (KAL-kū-lus) A stone formed within the body, as in the gallbladder, kidney, or urinary bladder.

Calorie (KAL-ō-rē) A unit of heat. The small calorie is the standard unit and is the amount of heat necessary to raise 1 g of water 1°C. The large Calorie (kilocalorie) is used in metabolic and nutrition studies; it is the amount of heat necessary to raise 1 kg of water 1°C.

Calyx (KĀL-iks)   Any cuplike division of the kidney pelvis.

Canal (kan-AL)   A narrow tube, channel, or passageway.

Canaliculus (kan-al-IK-ū-lus)   A small channel or canal, as in bones, where they connect the lacunae.

Cancellous (KAN-sel-us)   Having a reticular or latticework structure, as in spongy tissue of bone.

Cancer (KAN-ser)   A malignant tumor of epithelial origin tending to infiltrate and give rise to new growths or metastases; also called carcinoma (kar-si-NŌ-mah).

Capillary (KAP-i-lar-ē)   A microscopic blood vessel located between an arteriole and venule.

Carbohydrate (car-bō-HĪD-rāt)   An organic compound containing carbon, hydrogen, and oxygen in a particular amount and arrangement and comprised of sugar subunits.

Carbuncle (KAR-bun-kal)   A hard, pus-filled, and painful inflammation involving the skin and underlying connective tissues. Similar to boils but larger and more deeply rooted with several openings.

Caries (KĀ-rēz)   Decay of a tooth or bone.

Carotid (kah-ROT-id)   The main artery in the neck extending to the head.

Carpal (KAR-pul)   Pertaining to the carpus or wrist.

Casein (KĀS-ēin)   The main protein in milk, seen in milk curds.

Cast   A solid mold of a part; can originate from different areas of the body and be composed of various materials.

Castration (kas-TRĀ-shun)   The removal of the testes or ovaries.

Catalysis (kat-AL-is-is)   Decomposition produced chemically by a substance not affected by the reaction.

Cataract (KAT-ah-rakt)   Loss of transparency of the crystalline lens of the eye, or its capsule, or both.

Catheter (KATH-i-ter)   A tube that can be inserted into a body cavity through a canal to remove fluids such as urine or blood; also a tube inserted into a blood vessel through which opaque materials are injected into blood vessels or the heart for x-ray visualization.

Cation (KAT-ī-on)   An ion with a positive charge.

Cauda equina (KAW-dah ē-KWĪ-nah)   A taillike collection of roots of spinal nerves at the inferior end of the spinal canal.

Caudal (KAW-dal)   Pertaining to any taillike structure; inferior in position.

Cecum (SĒ-kum)   A blind pouch at the proximal end of the large intestine to which the ileum is attached.

Celiac (SĒ-li-ak)   Pertaining to the abdomen.

Cellulitis (sel-ū-LĪ-tis)   Inflammation of cellular or connective tissue, especially the subcutaneous tissue.

Cellulose (SEL-ū-lōs)   A fibrous form of carbohydrate constituting the main structural support of plant tissues.

Centigrade (SEN-ti-grād)   A unit of measurement consisting of 100 gradations between the boiling and freezing points.

Centimeter (SEN-ti-mēt-er)   The hundredth part of a meter.

Centrifugation (sen-tri-fū-GĀ-shun)   The process of separating heavier materials from lighter ones such as blood cells from plasma.

Cephalic (se-FA-lik)   Pertaining to the head; superior in position.

Cerumen (se-RŪ-men)   Ear wax.

Cervix (SER-viks)   Neck; any constricted portion of an organ, especially the lower cylindrical part of the uterus.

Chalazion (kah-LĀ-zē-on)   A small tumor of the eyelid.

Chemotherapy (kēm″ō-THER-a-pē)   The treatment of illness or disease by chemicals.

Chiasm (KĪ-azm)   A crossing; especially the crossing of the optic nerve fibers.

Chiropractic (kī″-rō-PRAK-tik)   A system of treating disease by using one's hands to manipulate body parts, mostly the vertebral column.

Choana (kō-AN-a)   Funnel-shaped, like the posterior openings of the nasal fossa.

Cholecystectomy (kō″-le-sis-TEK-tō-mē)   Surgical removal of the gallbladder.

Cholesterol (kō-LES-ter-ol)   An organic sterol found in many parts of the body.

Cholinergic (kōl-in-ER-jik)   Applied to nerve endings of the nervous system that liberate acetylcholine at a synapse.

Cholinesterase (kōl″in-ES-ter-ās)   An enzyme that hydrolyzes acetylcholine.

Chorion (KŌR-ē-on)   The outermost fetal membrane; serves a protective and nutritive function.

Chromatography (krō″mat-OG-rah-fē)   Separating chemical substances and particles by differential movement through a two-phase system.

Chronic (KRO-nik)   Long-term; applied to a disease that is not acute.

Chyle (kīl)   The milky fluid found in the lacteals of the small intestine after digestion.

Chyme (kīm)   The semifluid mixture of partly digested food and digestive secretions found in the stomach and small intestine during digestion of a meal.

Cicatrix (SIK-ah-triks)   A scar left by a healed wound.

Ciliary (SIL-ē-ār-ē)   Pertaining to any hairlike processes; an eyelid, an eyelash.

Circadian (ser-KĀD-ē-an)   Daily; occurring on a 24-hour cycle.

Circle of Willis   A ring of arteries forming an anastomosis at the base of the brain between the internal carotid and vertebral arteries and arteries supplying the brain.

Circumcision (ser-kum-SIZH-un)   Removal of the foreskin, the fold over the glans penis.

Cirrhosis (si-RŌ-sis)   A liver disorder in which the parenchymal cells are destroyed and replaced by connective tissue.

Clitoris (KLIT-or-is)   An erectile organ of the female that is homologous to the male penis.

Cochlea (KŌK-lē-ah)   A winding cone-shaped tube forming a portion of the inner ear and containing the organ of hearing.

Coitus (KŌ-i-tus)   Sexual intercourse; also coition or copulation.

Colostomy (kō-LOS-tō-mē)   The surgical creation of a new opening from the colon to the body surface.

Colostrum (kō-LOS-trum)   The first milk secreted at the end of pregnancy.

Colposcope (KOL-pō-skōp)   An instrument used to examine the vagina; a vaginal speculum.

Coma (KŌ-mah)   Profound unconsciousness from which one cannot be roused.

Conjunctivitis (kon-junk″-ti-VĪ-tis)   Inflammation of the delicate membrane covering the eyeball and lining the eyelids.

Contralateral (kon″trah-LAT-er-al)   Situated on, or pertaining to, the opposite side.

Convoluted (CON-vō-lū-ted)   Rolled together or coiled.

Corpus (KOR-pus)   The principal part of any organ; any mass or body.

Costal (KOS-tal)   Pertaining to a rib.

Cramp   A spasmodic, especially a tonic, contraction of one or many muscles, usually painful.

Craniotomy (krā-nē-OT-ō-mē)   Any operation on the skull, as for surgery on the brain or decompression of the fetal head in difficult labor.

Crenation (krē-NĀ-shun)   The conversion of normally shaped red blood corpuscles into shrunken, knobbed, starry forms, as when blood is mixed with a salt solution of 5 percent strength.

Cretinism (KRĒ-tin-izm)   Severe congenital thyroid deficiency during childhood leading to physical and mental retardation.

Cryosurgery (CRĪ-ō-ser-jer-ē)   The destruction of tissue by application of extreme cold, as in the destruction of lesions in the thalamus for the treatment of Parkinsonism.

Curare (kū-RAH-rē)   A toxic extract that paralyzes muscle by acting on the motor end plates.

Cutaneous (kyōō-TĀ-nē-us)   Pertaining to the skin.

Cyanosis (sī-an-O-sis)   Slightly bluish or dark purple discoloration of the skin and the mucous membrane due to an oxygen deficiency.

Cystoscope (SIS-tō-skōp)   An instrument used to examine the inside of the urinary bladder.

Cystitis (sis-TĪ-tis)   Inflammation of the urinary bladder.

Cytopenia (sī-tō-PĒ-nē-ah)   Reduction or lack of cellular elements in the circulating blood.

Debility (dē-BIL-i-tē)   Weakness of tonicity in functions or organs of the body.

Deciduous (dē-SID-ū-us)   Anything that is cast off at maturity, especially the first set of teeth.

Decubitus (dē-KYŌŌ-be-tus)   The lying-down position; a decubitus ulcer is caused by pressure when a patient is confined to bed for a long period of time.

Defecation (def-e-KĀ-shun)   The discharge of excreta (feces) from the rectum.

Deglutition (dē-glū-TISH-un)   The act of swallowing.

Dehydration (dē-hīd-RĀ-shun)   A condition due to excessive water loss from the body or its parts.

Deleterious (del-e-TĒR-ē-us)   Harmful; noxious.

Dens (denz)   Tooth.

Dentine (DEN-tēn)   The osseous tissues of a tooth enclosing the pulp cavity.

Dentition (den-TI-shun)   The eruption of teeth; the number, shape, and arrangement of teeth. Also called teething.

Dermatology (der-mah-TOL-ō-jē)   The medical specialty dealing with diseases of the skin.

Dermatome (DER-mah-tōm)   An instrument for incising the skin or cutting thin transplants of skin; the cutaneous area developed from one embryonic spinal cord segment and receiving most of its innervation from one spinal nerve.

Detritus (de-TRĪ-tus)   Any broken-down or degenerative tissue or carious matter.

Diagnosis (dī-ag-NŌ-sis)   Recognition of disease states from symptoms, inspection, palpation, posture, reflexes, general appearance, abnormalities, and other means.

Dialysis (dī-AL-i-sis)   The process of separating crystalloids (smaller particles) from colloids (larger particles) by the difference in their rates of diffusion through a semipermeable membrane.

Diapedesis (dī″ah-pe-DĒ-sis)   The passage of white blood cells through intact blood vessel walls.

Diaphragm (DĪ-a-fram)   Any partition that separates one area from another, especially the dome-shaped skeletal muscle between the thoracic and abdominal cavities.

Diarthrosis (dī-ar-THRŌ-sis)   An articulation in which opposing bones move freely, as in a hinge joint.

Diastole (dī-AS-tō-lē)   The relaxing dilatation period of the heart muscle, especially of the ventricles.

Diathermy (DĪ-ah-ther-mē)   The generation of heat in body tissues using high-frequency electric currents.

Differentiation (dif″er-en-shē-Ā-shun)   Acquirement of specific functions different from those of the original general type.

Digitalis (dij-i-TAL-is)   The dried leaf of foxglove used in the treatment of heart disease.

Dilate (DĪ-lāte)   To expand or swell.

Diplopia (dip-LŌ-pē-ah)   Double vision.

Dissect (DĪ-sekt)   To separate tissues and parts of a cadaver (corpse) or an organ for anatomical study.

Distal (DIS-tal)   Farthest from the center, from the medial line, or from the trunk.

Diuretic (dī-ūr-ET-ik)   Any agent that increases the secretion or lack of absorption of urine.

Diurnal (dī-UR-nal)   Daily.

Diverticulum (dī-ver-TIK-ū-lum)   A sac or pouch in the walls of a canal or organ, especially in the colon.

Dorsal (DOR-sal)   Pertaining to the back.

Dropsy (DROP-sē)   A condition rather than a disease: abnormal accumulation of water in the tissues and cavities.

Dysentery (DIS-en-ter-ē)   A painful disorder due to intestinal inflammation and accompanied by frequent, loose, bloody stools.

Dysfunction (dis-FUNK-shun)   Absence of complete normal function.

Dysmenorrhea (dis″men-o-RĒ-ah)   Painful or difficult menstruation.

Dystrophia (dis-TRŌ-fē-ah)   Progressive weakening of a muscle.

Ectopic (ek-TOP-ik)   Out of the normal location.

Eczema (EK-ze-mah)   A skin rash characterized by itching, swelling, blistering, oozing, and scaling of the skin.

Edema (ē-DĒ-mah)   An abnormal accumulation of fluid in the body tissues.

Effusion (ef-Ū-zhun)   The escape of fluid from the lymphatics or blood vessels into a cavity or into tissues.

Electrolyte (ē-LEK-trō-līt)   Any compound that separates into ions when dissolved in water.

Embolism (EM-bō-lizm)   Obstruction or closure of a vessel by a transported blood clot, a mass of bacteria, or other foreign material.

Embryo (EM-brē-ō)   The young of any organism in an early stage of development; in humans between the second and eighth weeks inclusively.

Emesis (EM-e-sis)   Vomiting.

Emphysema (em-fi-SĒ-mah)   A swelling or inflation of air passages with resulting stagnation of air in parts of the lungs; loss of elasticity in the alveoli.

Emulsification (ē-mul″si-fi-CĀ-shun)   The dispersion of large fat globules in the intestine to smaller uniformly distributed particles.

Endogenous (en-DŌJ-en-us)   Growing from or beginning within the organism.

Endoscope (EN-dō-skōp)   An instrument used to look inside hollow organs such as the stomach (gastroscope) or urinary bladder (cystoscope).

Endosteum (en-DOS-tē-um)   The membrane that lines the medullary cavity of bones.

Enuresis (en-ūr-Ē-sis)   Involuntary discharge of urine, complete or partial, after age 3.

Enzyme (EN-zīm)   A substance that effects the speed of chemical changes; an organic catalyst, usually a protein.

Epidemiology (ep″-i-DĒ-mē-ol″-ō-jē)   Medical science concerned with the occurrence and distribution of disease in human populations.

Epiphysis (ē-PIF-i-sis)   The end of a long bone usually larger in diameter than the shaft (the diaphysis); epiphyses is the plural form.

Episiotomy (ē-piz-ē-OT-ō-mē)   Incision of the clinical perineum at the end of second stage of labor to avoid tearing.

Epistaxis (ep-ē-STAKS-is)   Hemorrhage from the nose; nosebleed.

Epithelium (ep-i-THĒ-lē-um)   The tissue that forms glands or the outer part of the skin and lines blood vessels, hollow organs, and passages that lead externally from the body; epithelial (adjective).

Eructation (e-ruk-TĀ-shun)   The forceful expulsion of gas from the stomach; belching.

Erythema (er-e-THĒ-mah)   Skin redness usually caused by engorgement of the capillaries in the lower layers of the skin.

Erythropoiesis (ē-rith″rō-poi-Ē-sis)   The production of red blood cells.

Estrogen (ES-trō-jen)   Any substance that induces estrogenic activity or stimulates the development of secondary female characteristics; female hormones.

Etiology (ē-tē-OL-ō-jē)   The study of the causes of disease, including theories of origin and the organisms, if any, involved.

Euphoria (ū-FŌR-ē-ah)   A subjectively pleasant feeling of well-being marked by confidence and assurance.

Euthanasia (ū-than-Ā-zē-ah)   The proposed practice of ending a life in case of incurable disease.

Eversion (ē-VER-zhun)   Turning outward.

Exacerbation (eks-as-er-BĀ-shun)   An increase in the severity of symptoms or of disease.

Excrement (EKS-kre-ment)   Material cast out from the body as waste, especially fecal matter.

Excretion (eks-KRĒ-shun)   The elimination of waste products from the body. It can refer to the expulsion of any matter—whether from a single cell or from the entire body—or to the matter excreted.

Exhumation (eks-hū-MĀ-shun)   The procedure of taking the body out of the earth after burial.

Exogenous (ex-OJ-en-us)   Originating outside an organ or part.

Exophthalmos (ek-sof-THAL-mus)   An abnormal protrusion or bulging of the eyeball.

Exteroceptor (eks-ter-ō-SEP-tor)   A sense organ adapted for the reception of stimuli from outside the body.

Extravasation (eks-trah-va-SĀ-shun)   The process of escaping from a vessel into the tissues, especially blood, lymph, or serum.

Exudate (EKS-ū-dat)   Escaping fluid or semifluid material that oozes from a space that may contain serum, pus, and cellular debris.

Falciform (FAL-si-form)   Sickle-shaped; as in falciform ligament.

Fascia (FASH-ē-ah)   A fibrous membrane covering, supporting, and separating muscles.

Febrile (FEB-ril)   Feverish; pertaining to a fever.

Feces (FĒ-sēz)   Material discharged from the bowel and made up of bacteria, secretions, and food residue; stool.

Fenestration (fen-es-TRĀ-shun)   Surgical opening made into the labyrinth of the ear for some conditions of deafness.

Fetus (FĒ-tus)   The latter stages of the developing young of an animal. In human beings, the child in utero from the third month to birth.

Fibrillation (fī-bre-LĀ-shun)   Irregular twitching of individual muscle cells (fibers) or small groups of muscle fibers preventing effective action by an organ or muscle.

Fibrinolysis (fī-brin-OL-i-sis)   Action of a proteolytic enzyme that converts insoluble fibrin into a soluble substance.

Fibroblast (FĪ-bro-blast)   A flat, long connective tissue cell that forms the fibrous tissues of the body.

Fibrosis (fī-BRŌ-sis)   Abnormal formation of fibrous tissue.

Filtration (fil-TRĀ-shun)   The passage of a liquid through a filter or membrane that acts like a filter.

Fimbriae (FIM-brē-ē)   Fringelike structures, especially the lateral ends of the uterine tubes (oviducts).

Fissure (FISH-ūr)   A groove, fold, or slit that may be normal or abnormal.

Fistula (FIS-tū-lah)   An abnormal passage between two organs or between an organ cavity and the outside.

Flaccid (FLAK-sid)   Relaxed, flabby, or soft; lacking muscle tone.

Flagellum (fla-JEL-um)   A hairlike, motile process on the extremity of a bacterium or protozoon. Plural form is flagella.

Flatus (FLĀ-tus)   Gas or air in the digestive tract; commonly used to denote passage of gas rectally.

Fluoroscope (floo-Ō-ro-skōp)   An instrument for visual observation of the body (heart, bowels) by means of x ray.

Follicle (FOL-i-kul)   A small secretory sac or cavity.

Fontanel (fon-tah-NEL)   A soft area in a baby's skull; a membrane-covered spot where bone formation has not yet occurred.

Foramen (fōr-Ā-men)   A passage or opening; a communication between two cavities of an organ or a hole in a bone for passage of vessels or nerves.

Fossa (FOS-ah)   A furrow or shallow depression.

Fovea (FŌV-ē-ah)   A pit or cuplike depression.

Frenulum (FREN-ū-lum)   A small fold of mucous membrane that connects two parts and limits movement.

Fulminate (FUL-min-āte)   To occur suddenly with great intensity.

Fundus (FUN-dus)   The part of a hollow organ farthest from the opening.

Furuncle (FU-rung-kal)   A boil; painful nodule caused by bacteria. It is usually due to infection and inflammation of a hair follicle or oil gland.

Gamete (GAM-ēt)   A male or female reproductive cell; the spermatozoan or ovum.

Gangrene (GANG-grēn)   Death of tissue accompanied by bacterial invasion and putrefaction; usually due to blood vessel obstruction.

Gavage (gah-VAHZH)   Feeding through a tube passed through the esophagus and into the stomach.

Gene (jēn)   One of the biological units of heredity; an ultramicroscopic, self-reproducing DNA particle located in a definite position on a particular chromosome.

Genitalia (jen-i-TĀ-lē-ah)   Reproductive organs.

Genotype (JĒN-ō-tīp)   The basic hereditary combination of genes of an organism.

Germicide (JER-mi-cīde)   Any agent that destroys disease-producing organisms or pathogens.

Gerontology (je-ron-TOL-ō-gē)   The study of old age.

Gestation (jes-TĀ-shun)   The period of intrauterine fetal development.

Gingivitis (jin-je-VĪ-tis)   Inflammation of the gums.

Glaucoma (glaw-KŌ-mah)   An eye disorder in which there is increased pressure due to an excess of fluid within the eye.

Glomerulus (glō-MER-ū-lus)   A rounded mass of nerves or blood vessels, especially the microscopic tuft of capillaries that is surrounded by the expanded part of each kidney tubule.

Glucosuria (gloo-kō-SŪ-rē-ah)   Abnormal amount of sugar in the urine.

Goiter (GOY-ter)   An enlargement of the thyroid gland.

Gonad (GŌ-nad)   A term referring to the female sex glands (ovaries) and the male sex glands (testes).

Gradient (GRĀ-dē-ent)   A slope or gradation in the body; difference in concentration or electrical charges across a semipermeable membrane.

Gravida (GRA-vi-da)   A pregnant woman.

Groin (groyn)   The depression between the thigh and the trunk; the inguinal region.

Gynecology (gīn″-e-KOL-o-jē)   The branch of medicine dealing with the study and treatment of disorders of the female reproductive system.

Gyrus (JĪ-rus)   One of the tortuous elevations (convolutions) of the cerebral cortex region of the brain. Plural form is gyri.

Half-life   The time required for a radioactive substance to lose one-half its energy.

Haustra (HAWS-tra)   The sacculated elevations of the colon.

Hematoma (hēm″-ah-TŌ-mah)   A tumor or swelling filled with blood.

Hematuria (hēm-at-Ū-re-ah)   Blood in the urine.

Hemocytometer (hēm-o-sī-TOM-e-ter)   An instrument used in counting blood cells.

Hemolysis (hē-MOL-i-sis)   The rupture of red blood cells with release of hemoglobin into the plasma.

Hemophilia (hē-mō-FĒL-ē-ah)   A hereditary blood disorder where there is a deficient production of certain factors involved in blood clotting, resulting in excessive bleeding into joints, deep tissues, and elsewhere.

Hemorrhoids (HEM-ō-royds)   Dilated or varicosed blood vessels (usually veins) in the anal region; also called piles.

Hemostat (HĒ-mō-stat)   An agent or instrument used to prevent the flow or escape of blood.

Hernia (HER-nē-ah)   The protrusion or projection of an organ or part of an organ through the wall of the cavity containing it.

Heterosexual (het″-er-o-SEK-shu-al)   Relating to the opposite sex.

Hilum (HĪ-lum)   An area, depression, or pit where blood vessels and nerves enter or leave the organ.

Homogeneous (hō-mō-JĒN-ē-us)   Having similar or the same consistency and composition throughout.

Homologous (hō-MOL-ō-gus)   Similar in structure and origin, but not necessarily in function.

Homosexual (hō″-mō-SEK-shu-al)   Pertaining to the same sex. Someone who is sexually attracted to another individual of the same sex.

Hordeolum (hor-DĒ-ō-lum)   Inflammation of a sebaceous gland of the eyelid; a sty.

Hyaluronidase (hī″al-ū-RON-i-dāse)   An enzyme that breaks down hyaluronic acid, increasing the permeability of connective tissues by dissolving the substances that hold body cells together.

Hydrocele (HĪ-drō-seal)   A fluid-containing sac or tumor. Specifically, a collection of fluid formed in the space along the spermatic cord and in the scrotum.

Hydrophobia (hī″-drō-FŌ-bē-ah)   Rabies. Characterized by severe muscle spasms when attempting to drink water. An abnormal fear of water.

Hydrostatic (hī-drō-STAT-ik)   Pertaining to the pressure of liquids in equilibrium and that exerted on liquids.

Hypercapnia (hī-per-KAP-nē-ah)   An abnormal amount of carbon dioxide in the blood.

Hyperemia (hī-per-Ē-mē-ah)   An excess of blood in an area or part of the body.

Hyperplasia (hī-per-PLĀ-zē-ah)   An abnormal increase in the number of normal cells in a tissue or organ, increasing its size.

Hypertrophy (hī-PER-trō-fē)   An excessive enlargement or overgrowth of an organ or part.

Hypoxia (hī-POKS-ē-ah)   Lack of adequate oxygen; also anoxia.

Immunity (im-Ū-ni-tē)   The state of being resistant to injury, particularly by poisons, foreign proteins, and invading parasites.

Impetigo (im-pe-TĪ-gō)   A contagious skin disorder characterized by pustular eruptions.

Implantation (im-plan-TĀ-shun)   The insertion of a tissue or a part into a new part of the body. The attachment of the preembryonic ball of cells (blastula) into the lining of the uterus.

Impotence (IM-pō-tens)   Weakness; inability to copulate; failure to maintain an erection.

Inanimate (in-AN-i-mat)   Lacking life.

Incontinence (in-KON-tin-ens)   Inability to retain urine, semen, or feces, through loss of spincter control.

Insertion (in-SER-shun)   The manner or place of attachment of a muscle to the bone that it moves.

In situ (in SĪ-tū)   In position.

Integument (in-TEG-ū-ment)   A covering, especially the skin.

Intercellular (in-ter-SEL-ū-lar)   Between the cells of a structure.

Intracellular (in-tra-SEL-ū-lar)   Within cells.

Intubation (in-tu-BĀ-shun)   Insertion of a tube into the larynx through the glottis for entrance of air or to dilate a stricture.

Intussusception (in″tus-sus-SEP-shun)   The infolding (invagination) of one part of the intestine within another segment.

In utero (in Ū-ter-ō)   Within the uterus.

Invagination (in-vaj-in-Ā-shun)   The pushing of the wall of a cavity into the cavity itself.

In vitro (in VIT-rō)   In a glass, as in a test tube.

In vivo (in VIV-ō)   In the living body.

Ipsilateral (ip-sē-LAT-er-al)   On the same side, affecting the same side of the body.

Ischemia (is-KĒM-ē-ah)   A lack of sufficient blood to a part due to obstruction of circulation.

Isotonic (ī-sō-TON-ik)   Having equal tension or tone; having equal osmotic pressure between two different solutions or between two elements in a solution.

Isotope (Ī-sō-tōpe)   A chemical element that has the same atomic number as another but a different atomic weight. Radioactive isotopes change into other elements with the emission of certain radiations.

Jaundice (JAWN-dis)   A condition characterized by yellowness of skin, white of eyes, mucous membranes, and body fluids.

Karyotype (KAR-ē-ō-tīp)   Chromosome characteristics of an individual or a group of cells.

Keratin (KER-a-tin)   A special insoluble protein found in the hair, nails, and other horny tissues of the epidermis.

Ketosis (kē-TŌ-sis)   Abnormal condition marked by excessive production of ketone bodies.

Kilogram (KIL-ō-gram)   Equivalent to 1,000 g; about 2.2 lb avoirdupois.

Kinesthesia (kin-es-THĒ-szē-ah)   Ability to perceive extent, direction, or weight of movement; muscle sense.

Kinesiology (kin-e″-sē-OL-ō-jē)   The study of the movement of body parts.

Krebs cycle   The citric acid cycle; a series of energy-yielding steps in the catabolism of carbohydrates.

Kyphosis (kī-FŌ-sis)   An increased curvature of the chest giving a hunchback appearance.

Labium (LĀ-be-um)   A lip; liplike. Plural form is labia.

Labyrinth (LAB-i-rinth)   Intricate communicating passageways, especially the internal ear.

Lacrimal (LAK-rim-al)   Pertaining to tears.

Lactation (lak-TĀ-shun)   The period of milk release; suckling in mammals.

Lacteal (LAK-te-al)   Related to milk; one of many intestinal lymph vessels that absorb fat from digested food.

Lacuna (la-KŪ-nah)   A small, hollow space, such as that found in bones, in which the osteoblasts lie. Plural form is lacunae.

Lamina (LAM-in-ah)   A thin, flat layer or membrane, as the flattened part of either side of the arch of a vertebra. Plural form is laminae.

Laryngoscope (lar-INJ-ō-skōp)   An instrument for examining the larynx.

Latent (LAY-tent) **period**   The period elapsing between the application of a stimulus and the response.

Lateral (LAT-er-al)   To the side.

Lesion (LĒ-zhun)   Any local diseased change in tissue formation.

Leukemia (lū-KĒ-mē-ah)   A cancerlike disease of the blood-forming organs characterized by a rapid and abnormal increase in the number of white blood cells plus many immature cells in the circulating blood.

Leukocyte, leucocyte (LŪ-kō-sīt)   A white blood cell.

Leukocytosis (lū-kō-sī-TŌ-sis)   An increase in the number of white blood cells, characteristic of many infections and other disorders.

Leukopenia (lū-kō-PĒ-nē-ah)   A decrease of the number of white blood cells below 5,000 per cubic millimeter.

Leukoplakia (lū-kō-PLĀ-kē-ah)   A disorder in which there are white patches in the mucous membranes of the tongue, gums, and cheeks.

Libido (li-BĒ-dō)   The sexual drive, conscious or unconscious.

Lipoma (li-PŌ-mah)   A fatty tissue tumor, usually benign.

Liter (LĒ-ter)   Volume occupied by 1 kg of water at standard atmospheric pressure; equivalent to 1.057 qt.

Lobe (lōbe)   A curve or rounded projection.

Lordosis (lor-DŌ-sis)   Abnormal anterior convexity of the spine.

Lumbar (LUM-bar)   Region of the back and side between the ribs and pelvis; loins.

Lumen (LŪ-men)   The space within an artery, vein, intestine, or tube.

Lymphocyte (LIM-fō-sīt)   A type of white blood cell, found in lymph nodes, associated with the immune system.

Macula (MAK-ū-lah)   A discolored spot or a colored area.

Malaise (ma-LĀYZ)   Discomfort, uneasiness, indisposition, often indicative of infection.

Malignant (mah-LIG-nant)   Referring to diseases that tend to become worse and cause death; especially the invasion and spreading of cancer.

Manometer (man-OM-e-ter)   A device used for determining liquid or gaseous pressure.

Mastication (mas″ti-KĀ-shun)   Chewing.

Meatus (mē-Ā-tus)   A passage or opening, especially the external portion of a canal.

Medial (MĒ-dē-al)   Relating to the midline.

Median (MĒ-ē-an)   A vertical plane dividing the body into right and left halves; situated in the middle.

Melanin (MEL-an-in)   The dark pigment found in some parts of the body such as the skin.

Melanoma (mel-an-Ō-mah)   A tumor containing melanin, usually dark.

Menopause (MEN-o-pawz)   The termination of the menstrual cycles.

Metabolism (me-TAB-ō-lizm)   The physical and chemical changes or processes by which living substance is maintained, producing energy for the use of the organism.

Metastasis (me-TA-sta-sis)   The transfer of disease from one organ or part of the body to another.

Microcephalus (mī-krō-SEF-ah-lus)   A fetus or individual with an abnormally small head.

Microgram (MĪ-krō-gram)   One one-millionth of a gram or $\frac{1}{1,000}$ of a milligram.

Micron (MĪ-kron) One one-millionth of a meter or ⅟₁,₀₀₀ of a millimeter, ⅟₂₅,₀₀₀ of an inch.

Micturition (mik-tu-RI-shun) The act of expelling urine from the bladder; urination.

Millimeter (MIL-i-mē-ter) One one-thousandth of a meter, about ⅟₂₅ inch.

Mittelschmerz (MIT-el-shmerz) Abdominopelvic pain that supposedly indicates the release of an egg from the ovary.

Milliliter (MIL-ē-lē-ter) One-thousandth of a liter; equivalent to 1 cm³.

Morbid (MOR-bid) Diseased; pertaining to disease.

Mucin (MŪ-sin) A protein found in mucus.

Mucus (MŪ-kus) The thick fluid secretion of the mucous glands and mucous membranes.

Multiparous (mul-TIP-ar-us) Having borne more than one child.

Myology (mī-OL-ō-jē) The science or study of the muscles and their parts.

Myopia (mī-Ō-pē-ah) Defect in vision so that objects can only be seen distinctly when very close to the eyes; nearsightedness.

Narcosis (nar-KŌ-sis) Unconscious state due to narcotics.

Nebulization (ne″bŭl-i-ZĀ-shun) Treatment with spray method.

Necrosis (ne-KRŌ-sis) Death of a cell or group of cells as a result of disease or injury.

Neonatal (nē-ō-NĀ-tal) Pertaining to the first 4 weeks after birth.

Neoplasm (NĒ-ō-plazm) A mass of new, abnormal tissue; a tumor.

Nephritis (ne-FRĪ-tis) Kidney inflammation.

Nulliparous (nul-LIP-ar-us) Never having borne a child.

Nystagmus (nis-TAG-mus) Constant, involuntary, rhythmic movement of the eyeballs; horizontal, rotary, or vertical.

Occlusion (o-KLOO-zhun) The act of closure or state of being closed.

Odontoid (ō-DON-toyd) Toothlike.

Olfactory (ōl-FAK-tō-rē) Pertaining to smell.

Oligospermia (ol-i-gō-SPER-me-ah) A deficiency of spermatozoa in the semen.

Oogenesis (ō-ō-JEN-e-sis) Formation and development of the ovum.

Ophthalmic (of-THAL-mik) Pertaining to the eye.

Orbit (OR-bit) The bony pyramid-shaped cavity of the skull that holds the eyeball.

Organelle (or-gan-EL) A tiny specific particle of living material present in most cells and serving a specific function.

Orgasm (OR-gazm) A state of highly emotional excitement; especially that which occurs at the climax of sexual intercourse.

Orifice (OR-i-fis) Any aperture or opening.

Ossicle (OS-i-kul) Any small bone; as the three tiny bones of the ear.

Ossification (os″i-fi-KĀ-shun) Formation of bone substance.

Osteomyelitis (os″tē-ō-mī-i-LĪ-tis) Inflammation of bone marrow or of the bone and marrow.

Osteoporosis (os″tē-ō-pō-RŌ-sis) Increased porosity of bone.

Ostium (OS-tē-um) Any small opening; especially entrance into a hollow organ or canal.

Otic (Ō-tik) Pertaining to the ear.

Ovulation (ō-vū-LĀ-shun) The discharge of a mature egg cell (ovum) from the follicle of the ovary.

Ovum (Ō-vum) The female reproductive or germ cell; an egg cell.

Oxidation (ok-si-DĀ-shun) The combining of oxygen with substances such as organic molecules in tissues; the loss of electrons.

Pacchionian (pak-kē-Ō-nē-an) **bodies** Small growths of the arachnoid tissue of the cerebrum.

Palate (PAL-at) The horizontal structure separating the mouth and the nasal cavity; the roof of the mouth.

Palliative (PAL-ē-ah-tiv) Serving to relieve or alleviate without curing.

Palpate (PAL-pāt) To examine by touch; to feel.

Papilla (pah-PIL-ah) Any small projection or elevations; plural form is papillae.

Parenchyma (par-EN-ki-mah) The essential parts of any organ concerned with its function.

Parenteral (par-EN-ter-al) Situated or occurring outside the intestines; as by a subcutaneous method.

Paries (PĀ-rēs) The enveloping wall of any structure, especially hollow organs.

Parous (PA-rus) Having borne at least one child.

Paroxysm (PAR-oks-sizm) A sudden periodic attack or recurrence of symptoms of a disease.

Parturition (par-tū-RISH-un) Act of giving birth to young; childbirth, delivery.

Pathogenesis (path″-ō-JEN-e-sis) The development of disease or a morbid or pathological state.

Pectoral (PEK-tō-ral) Pertaining to the chest or breast.

Pediatrician (pē″dē-a-TRI-shun) A physician who specializes in the care and treatment of children and their illnesses.

Perineum (per-i-NĒ-um) The pelvic floor; the space between the anus and the scrotum in the male and between the anus and the vulva in the female.

Periphery (pe-RIF-er-ē) Outer part or a surface of the body; part away from the center.

Peritonitis (per-i-tōn-Ī-tis) Inflammation of the peritoneum, the membranous coat lining the abdominal cavity and covering the viscera.

pH The symbol commonly used in expressing hydrogen-ion concentration.

Phalanges (fah-LAN-jēz) Bones of a finger or toe.

Phlebotomy (fleb-OT-ō-mē) The cutting of a vein to allow the escape of blood.

Pilonidal (pī-lo-NĪ-dal) Containing hairs resembling a tuft inside a cyst or sinus.

Pinna (PIN-nah) The projecting part of the external ear; the auricle.

Pinocytosis (pi″nō-sī-TŌ-sis) The absorption of liquids by cells.

Pleura (PLOŌR-ah) The serous membrane that enfolds the lungs and lines the walls of the chest and diaphragm.

Plexus (PLEK-sus) A network of nerves, veins, or lymphatic vessels.

Podiatry (pō-DĪ-ah-trē) The diagnosis and treatment of foot disorders.

Pons (ponz) A process of tissue connecting two or more parts; the portion of the brain stem between the medulla and midbrain.

Postpartum (pōst-PAR-tum) After parturition; occurring after the delivery of a baby.

Premonitory (prē-MON-i-tō-rē) Giving previous warning; as premonitory symptoms.

Presbyopia (prez-bē-Ō-pē-ah) Defect of vision due to advancing age; loss of elasticity of the lens of the eye.

Pressor (PRES-or) Stimulating the activity of a function, especially vasomotor, usually accompanied by an increase in blood pressure.

Primigravida (prī-mi-GRAV-ida) A woman pregnant for the first time.

Primipara (prī-MIP-a-ra) A woman who has given birth to her first child. Usage is not uniform.

Primordial (prī-MŌR-dē-al) Existing first; especially primordial egg cells in the ovary.

Proctology (prok-TOL-ō-jē) The branch of medicine that treats the rectum and its disorders.

Progeny (PROJ-e-nē) Refers to offspring or descendants.

Prognosis (prog-NŌ-sis) A forecast of the probable results of a disorder; the outlook for recovery.

Prolapse (PRŌ-laps) A dropping or falling down of an organ, especially the uterus or rectum.

Proliferation (pro-lif″er-Ā-shun) Rapid and repeated reproduction of new parts, especially cells.

Prosthesis (PROS-the-sis) Replacement of a missing part by an artificial substitute.

Protuberance (pro-TŪ-ber-ans) A part that is prominent beyond a surface, like a knob.

Proximal (PROK-si-mal) Near the point of origin; referring to the nearest part.

Psychosomatic (sī-kō-sō-MA-tik) Pertaining to the relationship between mind and body.

Puberty (PŪ-ber-tē) Period of life at which the reproductive organs become functional.

Puerperium (pu″-er-PER-ē-um) The period or state of confinement after childbirth. Usually 4 to 6 weeks.

Pulmonary (PUL-mōn-ary) Concerning or affected by the lungs.

Pulse (puls) Throbbing caused by the regular recoil and alternate expansion of an artery; the periodic thrust felt over arteries in time with the heartbeat.

Pus The liquid product of inflammation containing leucocytes, or their remains, and debris of dead cells.

Pyorrhea (pī-ō-RĒ-ah) A discharge or flow of pus, especially the tooth sockets and the tissues of the gums.

Pyrexia (pī-REK-sē-ah) A condition in which the temperature is above normal.

Radiography (rā-dē-OG-ra-fē) The making of x-ray pictures.

Ramus (RĀ-mus) A branch, especially a nerve or blood vessel. Plural form is rami.

Receptor (rē-SEP-tor) A nerve ending that receives a stimulus.

Recumbent (rē-KUM-bent) Lying down.

Regimen (REJ-i-men) A strictly regulated scheme of diet, exercise, or activity designed to achieve certain ends.

Regurgitation (rē-gur-ji-TĀ-shun) Return of solids or fluids to the mouth from the stomach; flowing backward of blood through incompletely closed heart valves.

Relapse (rē-LAPS) The return of a disease weeks or months after its apparent cessation.

Renal (RĒ-nal) Pertaining to the kidney.

Resuscitation (rē-sus-i-TĀ-shun) Act of bringing a person back to full consciousness.

Reticulum (rē-TIK-ū-lum) A network.

Retraction (rē-TRAK-shun) A shortening; the act of drawing backward or state of being drawn back.

Retroversion (re-trō-VER-zhun) A turning backward of an entire organ, especially the uterus.

Rhinology (rī-NOL-ō-jē) The study of the nose and its disorders.

Rickets (RIK-ets) Condition affecting children and often causing deformities; results from vitamin D deficiency.

Roentgen (RENT-gen) The international unit of radiation; a standard quantity of x or gamma radiation.

Salpingitis (sal-pin-JĪ-tis) Inflammation of the uterine (Fallopian) tube or of the auditory (Eustachian) tube.

Sarcoma (sar-KŌ-mah) A connective tissue tumor, often highly malignant.

Sciatica (sī-AT-ik-ah) Inflammation and pain along the sciatic nerve; felt at the back of the thigh running down the inside of the leg.

Sclerosis (skle-RŌ-sis) A hardening with loss of elasticity of the tissues.

Scoliosis (skō-lē-Ō-sis) An abnormal lateral curvature from the normal vertical line of the spine.

Sebum (SĒ-bum) A fatty secretion of the sebaceous glands of the skin.

Sella turcica (SEL-ah TUR-si-kah) A saddlelike depression on the middle upper surface of the sphenoid bone enclosing the pituitary gland.

Senescence (sen-ES-ens) The process of growing old; the period of old age.

Senility (se-NIL-i-tē) A loss of mental ability in old people. The state of being old.

Sepsis (SEP-sis) A morbid condition resulting from the presence of pathogenic bacteria and their products.

Septum (SEP-tum) A wall dividing two cavities.

Sigmoid (SIG-moyd) Shaped like the Greek letter sigma (S).

Simmond's (SIM-mundz) **disease** A condition in which atrophy of the pituitary gland causes extreme weight loss.

Sinus (SĪ-nus) A hollow in a bone or other tissue; a channel for blood; any cavity having a narrow opening.

Sinusoid (SĪ-nus-oyd) A blood space in certain organs as the liver or spleen.

Spasm (spazm) An involuntary, convulsive, muscular contraction.

Specific gravity The weight of a substance compared with an equal volume of water (represented by 1.000).

Spermatogenesis (sper″mah-tō-JEN-e-sis) The formation and development of the spermatozoa.

Spermicidal (sper-mi-SĪ-dal) An agent that kills spermatozoa.

Sphincter (SFINK-ter) A circular muscle constricting an orifice.

Sphygmomanometer (sfig″mō-man-OM-e-ter) An instrument for measuring arterial blood pressure.

Spirometer (spī-ROM-et-er) An apparatus used to measure air capacity of the lungs.

Sputum (SPŪ-tum) Substance ejected from the mouth containing saliva and mucus.

Squamous (SKWĀ-mus) Scalelike.

Stasis (STĀ-sis) Stagnation or halt of normal flow of fluids, as blood, urine, or of the intestinal mechanism.

Sterility (ster-IL-it-ē) Infertility; absence of reproductive power.

Stratum (STRĀ-tum) A layer.

Stricture (STRIK-tur) A local contraction of a tubular structure.

Stroma (STRŌ-mah)   The tissue that forms the ground substance, foundation, framework of an organ, as opposed to its functional parts.

Sulcus (SUL-kus)   A groove or depression between parts, especially a fissure between the convolutions of the brain. Plural form is sulci.

Suppuration (sup-ū-RĀ-shun)   The process of pus formation.

Susceptibility (su-sep″-ti-BIL-i-tē)   Lack of resistance of a body to the deleterious or other effects of an agent such as pathogenic microorganisms.

Suture (SU-chur)   A type of joint, especially in the skull, where bone surfaces are closely united.

Symphysis (SIM-fi-sis)   A line of union; a cartilagenous joint such as that between the bodies of the pubic bones.

Symptom (SIMP-tum)   A specific recognizable abnormality.

Syncytium (sin-SISH-i-um)   A multinucleated mass of protoplasm produced by the merging of cells.

Syndrome (SIN-drōm)   A group of abnormalities that occur together in a characteristic pattern; the complete picture of a disease.

Systemic (sis-TEM-ik)   Affecting the whole body; generalized.

Systole (SIS-tō-lē)   Heart muscle contraction, especially that of the ventricles.

Tactile (TAK-til)   Pertaining to the sense of touch.

Teratogen (ter-A-tō-jen)   A substance capable of producing an anomaly or teratoma.

Teratology (ter-at-OL-ō-jē)   The branch of science dealing with the study of monsters.

Teratoma (ter-a-TŌ-ma)   A congenital tumor containing embryonic elements of all three primary germ layers, as hair or teeth.

Tetany (TET-an-ē)   A nervous condition characterized by intermittent or continuous tonic muscular contractions of the extremities.

Therapy (THER-a-pē)   The treatment of disease or disorder.

Thoracic (thor-A-sik)   Pertaining to the chest.

Thrombocyte (THROM-bō-sīt)   A tiny particle found in the circulating blood, a blood platelet thought to be part of the mechanism of blood clotting.

Thrombophlebitis (throm″bō-fle-BĪ-tis)   A disorder in which inflammation of a vein wall is followed by the formation of a blood clot (thrombus).

Thymectomy (thy-MEK-tō-mē)   Surgical removal of the thymus.

Tinnitus (tin-Ī-tus)   A ringing or tinkling sound in the ears.

Toxic (TOK-sik)   Pertaining to poison; poisonous.

Trabecula (tra-BEK-ū-lah)   A fibrous cord of connective tissue serving as supporting fiber by forming a septum extending into an organ from its wall or capsule.

Trace elements   Organic elements normally found in minute quantities in blood and tissues; examples are fluorine, copper, and manganese.

Transplantation (trans-plan-TĀ-shun)   The transfer or implantation of body tissue from one part of the body to another or from one person to another.

Transverse (trans-VERS)   Lying across; crosswise.

Trauma (TRAW-mah)   An injury or wound that may be produced by external force or by shock, as in psychic trauma.

Tumor (TŪ-mor)   A swelling or enlargement.

Ulcer (UL-ser)   An open lesion of the skin or a mucous membrane of the body with loss of substance and necrosis of the tissue.

Umbilical (um-BIL-i-kal)   Pertaining to the umbilicus or navel.

Umbilicus (um-BIL-i-kus)   A small scar on the abdomen that marks the former attachment of the umbilical cord to the fetus; the navel.

Uremia (ū-RĒ-mē-ah)   A toxic condition from urea and other waste products in the blood.

Urticaria (ur-ti-KĀ-rē-ah)   A skin reaction to certain foods, drugs, or other substances to which a person may be allergic; hives.

Uvula (Ū-vū-lah)   A soft, fleshy mass, especially the V-shaped pendant part, descending from the soft palate.

Varicocele (VAR-i-kō-sēl)   A twisted vein; especially the accumulation of blood in the veins of the spermatic cord.

Varicose (VAR-i-kōs)   Pertaining to an unnatural swelling, as in the case of a varicose vein.

Vas   A vessel or duct.

Vascular (VAS-kū-lar)   Pertaining to or containing many vessels.

Ventral (VEN-tral)   Pertaining to the anterior or front side of the body; opposite of dorsal.

Vertigo (VUR-ti-go)   Sensation of dizziness.

Vesicle (VES-i-kal)   A small bladder or sac containing liquid.

Vestibule (VES-tib-ūl)   A small space or cavity at the beginning of a canal, especially the inner ear, larynx, mouth, nose, vagina.

Villus (VIL-lus)   One of the short vascular hairlike processes found on certain membranous surfaces. Plural form is villi.

Viscosity (vis-KOS-i-tē)   The state of being sticky or thick.

Wheal (hwēl)   Elevated lesion of the skin.

Bibliography

BIBLIOGRAPHY FOR UNIT I

Abston, S. "Burns in Children," *Clinical Symposia* 28, Ciba Pharmaceutical Company, 1976.

Allison, A. "Lysosomes and Disease," *Scientific American,* November 1967.

Artz, C. P. "Severe Burns: Current Concepts of Specialized Care," *Modern Medicine,* 30 April 1973.

Atlas of Human Anatomy. New York: Barnes & Noble (Harper & Row), 1970.

Baer, R. L., and W. B. Shelley. "Psoriasis – Disfiguring but Treatable," *Medical World News,* 7 April 1975.

Basmajian, J. V. *Grant's Method of Anatomy.* 9th ed. Baltimore: Williams & Wilkins, 1975.

Berlin, N. I. "Research Strategy in Cancer: Screening, Diagnosis, Prognosis," *Hospital Practice,* January 1975.

Berns, M. W., and D. E. Rounds. "Cell Surgery by Laser," *Scientific American,* February 1970.

"Beyond Survival: Toward A Healthy View of Human Aging," *Modern Medicine,* 1 November 1976.

Bickers, D. R., and A. Kappas. "Metabolic and Pharmacologic Properties of the Skin," *Hospital Practice,* May 1974.

Bowen, W. H. "Prospects for the Prevention of Dental Caries," *Hospital Practice,* May 1974.

Burger, M. M. "Surface Properties of Neoplastic Cells," *Hospital Practice,* July 1973.

Campbell, A. M. "How Viruses Insert Their DNA into the DNA of the Host Cell," *Scientific American,* December 1976.

Caplan, R. M. "Three Bothersome Skin Infections," *Modern Medicine,* 4 February 1974.

Christensen, J. B., and I. R. Telford. *Synopsis of Gross Anatomy,* 2d ed. New York: Harper & Row, 1972.

Clemente, C. D. *Anatomy: A Regional Atlas of the Human Body.* Philadelphia: Lea & Febiger, 1975.

Cohen, S. N. "The Manipulation of Genes," *Scientific American,* July 1975.

Daniels, F., Jr. "Saving Sun Worshippers from Their God," *Medical Opinion,* July 1976.

Drake, T., and H. Maibach. "Taking the Heartbreak out of Psoriasis," *Modern Medicine,* 1 August 1975.

Dubois, E. L., and N. Talal. "Lupus: The New Great Imitator," *Medical World News,* 14 June 1976.

Engelman, D. M., and P. B. Moore. "Neutron-Scattering Studies of the Ribosome," *Scientific American,* October 1976.

Folkman, J. "The Vascularization of Tumors," *Scientific American,* May 1976.

Frank, S. B. "Uncommon Aspects of Common Acne," *Cutis,* 1974.

Frenay, A. C., Sr. *Understanding Medical Terminology,* 4th ed. St. Louis: Catholic Hospital Association, 1970.

Gardner, E., D. J. Gray, and R. O'Rahilly. *Anatomy: A Regional Study of Human Structure.* 4th ed. Philadelphia: Saunders, 1975.

Goldberg, N. D. "Cyclic Nucleotides and Cell Function," *Hospital Practice,* May 1974.

Goldman, R. *Principles of Medical Science.* New York: McGraw-Hill, 1973.

Gordon, R., et al. "Image Reconstruction from Projections," *Scientific American,* October 1975.

Gray, H. *Anatomy of the Human Body.* Edited by C. M. Goss. 29th ed. Philadelphia: Lea & Febiger, 1973.

Green, M. "Viral Cell Transformation in Human Oncogenesis," *Hospital Practice,* September 1975.

Hamilton, W. J., G. Simon, and S. G. I. Hamilton. *Surface and Radiological Anatomy.* 5th ed. London: Macmillan, 1975.

Herzenberg, L. H., et al. "Fluorescence-Activated Cell Sorting," *Scientific American,* March 1976.

Hoffman, J. F. "Ionic Transport across the Plasma Membrane," *Hospital Practice,* October 1974.

Hollingshead, W. H. *Textbook of Anatomy.* 3d ed. New York: Harper & Row, 1974.

Holtzman, E. "The Biogenesis of Organelles," *Hospital Practice,* March 1974.

Holvey, D. N., and J. H. Talbott (eds.). *The Merck Manual.* 12th ed. Rahway, N.J.: Merck Sharp & Dohme, 1972.

Human Physiology and the Environment in Health and Disease. p. 6: "Aging." Readings from *Scientific American.* San Francisco: Freeman, 1975.

Kaminester, L. H. "Sunlight, Skin Cancer, and Sunscreens," *J. Amer. Med. Assoc.,* 30 June 1975.

——. "Warts: Another Look at a Nuisance," *Consultant,* September 1976.

Kligman, A. M. "An Overview of Acne," *J. Invest. Dermatol.,* 1974.

Lane, C. "Rabbit Hemoglobin from Frog Eggs," *Scientific American,* August 1976.

Leaf, A. "Unusual Longevity: The Common Denominators," *Hospital Practice,* October 1973.

Lederberg, J. "DNA Splicing: Will Fear Rob Us of Its Benefits?" *Prism,* November 1975.

Luria, S. E. "Colicins and the Energetics of Cell Membranes," *Scientific American,* December 1975.

Maniatis, T., and Mark Ptashne. "A DNA Operator-Repressor System," *Scientific American,* January 1976.

Markert, C. L., and H. Ursprung. *Developmental Genetics.* Englewood Cliffs, N.J.: Prentice-Hall, 1971.

Montagna, W. *Advances in Biology of Skin.* Vols. 1–12. New York: Appleton-Century-Crofts, 1972.

——. "The Skin," *Scientific American,* February 1965.

Neutra, M., and C. P. LeBlond. "The Golgi Apparatus," *Scientific American,* February 1969.

"New Dressing Aids Burn Healing," *J. Amer. Med. Assoc,* 26 July 1976.

Nicoll, P. A., et al. "The Physiology of the Skin," *Ann. Rev. Physiol.,* 1972.

Nomura, M. "Ribosomes," *Scientific American,* October 1969.

Pansky, B., and E. L. House. *Review of Gross Anatomy.* 3d ed. New York: Macmillan, 1975.

Peterson, H. D. "Burn Sepsis–Management of the Severe Burn," *Forum on Infection,* September 1976.

Raff, M. C. "Cell-Surface Immunology," *Scientific American,* May 1976.

Rao, D. B.: "Management of Dermal and Decubitus Ulcers," *Drug Therapy,* October 1976.

Romanes, G. J. *Cunningham's Textbook of Anatomy.* 11th ed. London: Oxford University Press, 1972.

Roseman, S. "Sugars of the Cell Membrane," *Hospital Practice,* January 1975.

Rosenberg, E. *Cell and Molecular Biology.* New York: Holt, 1971.

Ross, R., and P. Bornstein. "Elastic Fibers in the Body," *Scientific American,* June 1971.

Sanders, B. B., Jr., and G. S. Stretcher. "Warts–Diagnosis and Treatment," *J. Amer. Med. Assoc.,* 28 June 1976.

Satir, B. "The Final Steps in Secretion," *Scientific American,* October 1975.

Satir, P. "How Cilia Move," *Scientific American,* October 1974.

——. "Scanning Electron Microscopy," *Medical World News,* 7 December 1973.

Shalita, A. R. "Acne Vulgaris: Not Curable but Treatable," *Modern Medicine,* 1 August 1975.

Shimkin, M. B. *Science and Cancer.* U.S. Dept. of Health, Education, and Welfare, Publication No. (NIH) 74-568, 1973.

Sisson, T. R. C. "Photopharmacology: Light as Therapy," *Drug Therapy,* August 1976.

"Skin Cancer: 3,000,000 New Cases in 1976," *J. Amer. Med. Assoc.,* 26 April 1976.

Solomon, A. K. "The State of Water in Red Cells," *Scientific American,* February 1971.

Taber, C. W. *Taber's Cyclopedic Medical Dictionary.* 8th ed. Philadelphia: Davis, 1959.

Temin, H. M. "RNA-Directed Synthesis," *Scientific American,* January 1972.

Tillery, B. C. "The Acne Victim: Time to Pay Off a Past-Due Debt," *Modern Medicine,* 1 August 1975.

Toner, P. G., and K. E. Carr. *Cell Structure: An Introduction to Biological Electron Microscopy.* 2d ed. Baltimore: Williams & Wilkins, 1971.

Tortora, G. J. *Principles of Human Anatomy.* San Francisco: Canfield, 1977.

"Virus Theory of Cancer: A Lumbering Turtle, The," *Drug Therapy,* October 1976.

Weaver, R. F. "The Cancer Puzzle," *National Geographic,* September 1976.

Wessells, N. K., and W. J. Rutter. "Phases in Cell Differentiation," *Scientific American,* March 1969.

Wurtman, R. J. "The Effects of Light on the Human Body," *Scientific American,* July 1975.

BIBLIOGRAPHY FOR UNIT II

Aufranc, O. E., and R. H. Turner. "Total Replacement of the Arthritic Hip," *Hospital Practice,* October 1971.

Bethlem, J. *Muscle Pathology.* New York: American Elsevier, 1970.

Bingham, R. "Rheumatoid Disease: Has One Investigator Found Its Cause and Its Cure?" *Modern Medicine,* 15 February 1976.

Blau, S. P. "How to Select Therapy in Rheumatoid Arthritis," *Consultant,* September 1976.

Bones, Joints and Muscles of the Human Body: A Programmed Text for Physical Therapy Aides. New York: Macmillan, 1970.

Bourne, G. W. *The Biochemistry and Physiology of Bone.* 2d ed. New York: Academic Press, 1972.

Close, R. I. "Dynamic Properties of Mammalian Skeletal Muscles," *Physiol. Rev.,* 1972.

Cohen, C. "The Protein Switch of Muscle Contraction," *Scientific American,* November 1975.

Cutler, P., and J. T. Harrington, Jr. "Therapeutic Seminar 3: A Practical Guide to the Treatment of Arthritis," *Current Prescribing,* May 1975.

Ehrlich, G. E. "Learning How to Fight Rheumatoid Arthritis," *Medical Tribune,* 12 May 1976.

—————. "Physical Clues to the Rheumatic Diseases," *Consultant,* August 1976.

Garn, S. "Bone-Loss and Aging," in *The Physiology and Pathology of Human Aging.* New York: Academic Press, 1975.

Harris, W. H., and R. P. Heaney. *Skeletal Renewal and Metabolic Bone Disease.* Boston: Little, Brown, 1970.

Hollander, J. L., and D. J. McCarty, Jr. *Arthritis and Allied Conditions.* 8th ed. Philadelphia: Lea & Febiger, 1972.

Hoyle, G.: "How Is Muscle Turned On and Off ?" *Scientific American,* April 1970.

Intramuscular Injections. New York: Wyeth Laboratories, 1972.

Margaria, R. "The Sources of Muscular Energy," *Scientific American,* March 1972.

Melnick, A. "Helping the Child Stricken with Juvenile Rheumatoid Arthritis," *Medical Opinion,* August 1974.

Mouratoff, G. J. "Pharmacotherapy of Rheumatoid Arthritis," *Modern Medicine,* 7 August 1972.

Murray, J. M., and A. Weber. "The Cooperative Action of Muscle Proteins," *Scientific American,* February 1974.

"Muscular Dystrophy Meeting," *MD,* September 1976.

"Prosthetic Knuckle Enables Deformed Hands to Grip and Pinch," *J. Amer. Med. Assoc.,* 22 September 1975.

Riiegg, J. C. "Smooth Muscle Tone," *Physiol. Rev.,* 1971.

Rodnan, G. P. (ed.). "Primer on the Rheumatic Diseases," *J. Amer. Med. Assoc.,* April 1973.

Rogoff, B., and J. Sergent. "Rheumatoid Arthritis," *Hospital Medicine,* October 1975.

Sampson, P. "Replacement of Some Arthritic Joints Is Possible – But Still a Major Job," *J. Amer. Med. Assoc.,* 29 March 1976.

Skosey, J. L. "Rheumatoid Arthritis: Overlooked, Underestimated, and Confusing," *Consultant,* July 1974.

Smythe, H. "Nonsteroidal Therapy in Inflammatory Joint Disease," *Hospital Practice,* September 1975.

Steinbach, H. L., and R. H. Gold. "Pyogenic Infections of Bone: A Roentgenologic Guide," *Hospital Medicine,* July 1972.

Thompson, G. R. "Diagnosis and Treatment of Septic Arthritis," *Hospital Medicine,* June 1976.

Tonomuna, Y. *Muscle Protein, Muscle Contraction and Cation Transport.* Baltimore: University Park Press, 1972.

Tronzo, R. G. "Bone: Self-repairing, Self-renewing," *Consultant,* April 1972.

Weissmann, G. "The Molecular Basis of Acute Gout," *Hospital Practice,* July 1971.

Wiltse, L. L., et al. "Chymopapain Chemonucleolysis in Lumbar Disk Disease." *J. Amer. Med. Assoc.,* 3 February 1975.

BIBLIOGRAPHY FOR UNIT III

"Associated Agent Lymphotoxin among Clues to Multiple Sclerosis," *J. Amer. Med. Assoc.,* 13 September 1976.

Basmajian, J. V. "Biofeedback: The Clinical Tool behind the Catchword," *Modern Medicine,* 1 October 1976.

Baum, J. L. "Therapeutic Uses of Soft Contact Lenses," *Scientific American,* August 1973.

Best, C. H., and N. B. Taylor. *Best and Taylor's Physiological Basis of Medical Practice.* 9th ed. Baltimore: William & Wilkins, 1973.

"Biofeedback Training for Childbirth," *Medical World News,* 30 June 1975.

Brown, G. M. "Psychiatric and Neurologic Aspects of Endocrine Disease," *Hospital Practice,* August 1975.

Catt, K. J. *An ABC of Endocrinology.* Boston: Little, Brown, 1971.

Conomy, J. P. "A Most Useful Test in Evaluating Coma," *Consultant,* September 1976.

Cook, A. W. "Electrical Stimulation in Multiple Sclerosis," *Hospital Practice,* April 1976.

Daftary, A. V., and R. I. H. Wang. "Dopamine – The 'Neglected' Neurotransmitter," *Drug Therapy,* June 1976.

Dean, G. "The Multiple Sclerosis Problem," *Scientific American,* July 1970.

DeWied, D. "Hormonal Influences on Motivation, Learning, and Memory Processes," *Hospital Practice,* January 1976.

Dey, F. L. "Auditory Fatigue and Predicted Permanent Hearing Defects from Rock-and-Roll Music," *New England J. Med.,* 26 February 1970.

Diassi, P. A., et al. "Endocrine Hormones," *Ann. Rev. Pharmacol.,* 1970.

Donald, P. J. "Guide to the Diagnosis and Management of Eustachian Otitis," *Hospital Medicine,* March 1976.

"Exploring the Frontiers of the Mind," *Time,* 14 January 1974.

Field, J. (ed.). "Endocrinology," in *Handbook of Physiology.* Baltimore: Williams & Wilkins, 1972.

Grollman, A. "What to tell Patients about Prostaglandins," *Consultant,* October 1976.

Guillemin, R., and R. Burgus. "The Hormones of the Hypothalamus," *Scientific American,* November 1972.

Guyton, A. C. *Textbook of Medical Physiology.* 5th ed. Philadelphia: Saunders, 1976.

Hamwi, G. J. "Nutrition and Diseases of the Endocrine Glands," *Amer. J. Clin. Nutr.,* 1970.

Harpen, R. *Human Senses in Action.* Baltimore: Williams & Wilkins, 1972.

Heimer, L. "Pathways in the Brain," *Scientific American,* July 1971.

Henahan, J. F. "What's New on Heads and Their Aches?" *Modern Medicine,* 1 April 1976.

Herring, M. "Timely Action Urged in Tay-Sachs Pregnancy," *Medical Tribune,* 26 February 1975.

"Hydrocephalus is All Wrapped Up," *Medical World News,* 10 August 1973.

"Hydrocephalus: Cranial Wrap Provides Hope," *Medical Tribune,* 18 July 1973.

Ignelzi, R. J., and W. M. Kirsch. "Follow-up Analysis of Ventriculoperitoneal and Ventriculoatrial Shunts for Hydrocephalus," *J. Neurosurg.,* 1975.

Kandel, E. R. "Nerve Cells and Behavior," *Scientific American,* July 1970.

Kaufman, H. E. "Corneal Transplantation: A Progress Report," *Hospital Practice,* July 1973.

Kolansky, H., and W. T. Moore. "Toxic Effects of Chronic Marihuana Use," *J. Amer. Med. Assoc.,* 2 October 1972.

Kopin, I. J. "Catecholamines, Adrenal Hormones, and Stress," *Hospital Practice,* March 1976.

Krieger, D. T. "The Hypothalamus and Neuroendocrinology," *Hospital Practice,* September 1971.

Krieger, D. T. "The Neuroendocrinology Series," *Hospital Practice,* April 1975.

Kushnick, T., and N. Diamond. "Mass Screening in New Jersey for Tay-Sachs Gene," *Med. Soc. New Jersey,* April 1976.

Laros, R. K., Jr., et al. "Prostaglandins," *Amer. J. Nurs.,* June 1973.

Lester, H. A. "The Response to Acetylcholine," *Scientific American,* February 1977.

Lewis, J. A. "Violence and Epilepsy," *J. Amer. Med. Assoc.,* 16 June 1975.

Llinás, R. R. "The Cortex of the Cerebellum," *Scientific American,* January 1975.

Lipscomb, D. M. "Ear Damage from Exposure to Rock and Roll Music," *Arch. Otalarying.,* November 1969.

Loken, M. K., and M. Frick. "Scanning the Brain: Radionuclide Scintigraphy," *Modern Medicine,* 1 May 1976.

Luria, A. R. "The Functional Organization of the Brain," *Scientific American,* March 1970.

Masland, R. L. "The Diagnosis and Treatment of Little Seizures," *Hospital Medicine,* January 1976.

McEwen, B. S. "Interactions between Hormones and Nerve Tissue," *Scientific American,* July 1976.

———. "The Brain As a Target Organ of Endocrine Hormones," *Hospital Practice,* May 1975.

Mines, S. "Background on Biofeedback: An Overlong Look," *Modern Medicine,* 15 January 1975.

Miyamoto, H., et al. "Antibodies to Vaccinia and Measles Viruses in Multiple Sclerosis Patients," *Arch. Neurol.,* 1976.

"Multiple Sclerosis: Explorations of Cause," *J. Amer. Med. Assoc.,* 14 June 1976.

Norenberg, D. D. "Tracking Down the Cause of Dizziness," *Medical Opinion,* May 1976.

O'Malley, B. W. "Hormones, Genes, and Cancer," *Hospital Practice,* July 1975.

O'Malley, B. W., and W. T. Schrader. "The Receptors of Steroid Hormones," *Scientific American,* February 1976.

Pappenheimer, J. R. "The Sleep Factor," *Scientific American,* August 1976.

Patten, B. M. "The Ancient Art of Memory," *Arch. Neurol.,* 1972.

Pettigrew, J. D. "The Neurophysiology of Binocular Vision," *Scientific American,* August 1972.

Pines, M. "Memory: How It Can be Created and Destroyed," *Modern Medicine,* 22 July 1974.

———. "The Memory Code: Is It Beyond Comprehension?" *Modern Medicine,* 5 August 1974.

Pribam, K. H. "The Neurophysiology of Remembering," *Scientific American,* January 1969.

Rassmussen, H., and M. M. Pechet. "Calcitonin," *Scientific American,* October 1970.

Robertson, D. M., and H. B. Dinsdale. *The Nervous System: Structure and Function in Disease.* Baltimore: Williams & Wilkins, 1972.

Ross, J. "The Resources of Binocular Perception," *Scientific American,* March 1976.

Russman, B. S. "Convulsive Seizures in Infancy and Childhood," *Pediatric Neurology,* May 1976.

Selye, H. *Stress in Health and Disease.* London: Butterworths, 1976.

———. *Stress without Distress.* Philadelphia: Lippincott, 1974.

———. *The Stress of Life.* New York: McGraw-Hill, 1976.

Stellar, S., and Y. Bhandari. "Results of Present-Day Therapy of Parkinson's Disease," *J. Med. Soc. New Jersey,* April 1976.

Stent, G. S. "Cellular Communication," *Scientific American,* September 1972.

"Stimulator over Cerebellum Controls Intractable Epilepsy," *Medical Tribune,* 5 September 1973.

Thomas, R. C. "Electrogenic Sodium Pump in Nerve and Muscle Cells," *Physiol. Rev.,* 1972.

"Treating Common Sleep Disorders," *Medical World News,* 5 April 1976.

"Trigeminal Neuralgia Surgery Refined," *Medical World News,* 17 May 1974.

Vander, A. J. *Renal Physiology.* New York: McGraw-Hill, 1975.

VanHeyningen, R. "What Happens to the Human Lens in Cataract," *Scientific American,* December 1975.

Villee, D. *Human Endocrinology.* Philadelphia: Saunders, 1975.

Werblin, F. S. "The Control of Sensitivity in the Retina," *Scientific American,* October 1970.

Whittaker, V. P. "Membranes in Synaptic Function," *Hospital Practice,* April 1974.

Wilentz, J. S. *Senses of Man.* New York: Apollo Editions, 1971.

Williams, R. *Textbook of Endocrinology.* 5th ed. Philadelphia: Saunders, 1974.

Yahr, M. D. "Brain Tumors," *Hospital Medicine,* September 1973.

Young, R. W. "Visual Cells," *Scientific American,* October 1970.

Young, W. R. "The Enduring Mystery of Dyslexia," *Reader's Digest,* 1976.

BIBLIOGRAPHY FOR UNIT IV

Affonso, D. "Continuous Positive Airway Pressure." *Amer. J. Nurs.,* April 1976.

Audrey, Burgess. *The Nurse's Guide to Fluid and Electrolyte Balance.* New York: McGraw-Hill, 1970.

Avery, M. E., et al. "The Lung of the Newborn Infant," *Scientific American,* April 1973.

Bains, M. S., and E. J. Beattie. "Cardiac Tamponade," *Hospital Medicine,* October 1976.

Baron, H. C. "Valvular Incompetence and Varicose Veins," *Hospital Medicine,* April 1976.

Benditt, E. P. "The Origin of Atherosclerosis," *Scientific American,* February 1977.

Bergman, A. B. "Sudden Infant Death Syndrome," *Pediatric Annals,* November 1974.

Bergman, S. G. "The Atherosclerosis Problem," *Clinical Pediatrics,* January 1975.

Bergsagel, D. E. "Plasma-Cell Myeloma," *Drug Therapy,* September 1976.

Bierenbaum, M. L. "A Review of Some Epidemiological Factors in Coronary Heart Disease," *J. Med. Soc. New Jersey,* December 1975.

Borland, J. L. "Rational Management of Peptic Ulcer Disease," *Hospital Practice,* July 1976.

Bowen, W. H. "Prospects for the Prevention of Dental Caries," *Hospital Practice,* May 1974.

Brown, M. R., and C. B. Lillibridge. "When to Think of Celiac Disease," *Clinical Pediatrics,* January 1975.

Burkitt, D. P., et al. "Dietary Fiber and Disease," *J. Amer. Med. Assoc.,* 19 August 1974.

Capra, J. D., and A. B. Edmundson. "The Antibody Combining Site," *Scientific American,* January 1977.

Carey, L. C. "Shock: Differential Diagnosis and Immediate Treatment," *Hospital Medicine,* May 1975.

Cerami, A., and C. M. Peterson. "Cyanate and Sickle-Cell Disease," *Scientific American,* April 1975.

Chandra, P., and W. R. S. North. "Randomised Controlled Trial of Yoga and Bio-Feedback in Management of Hypertension," *Lancet,* 1975.

Child, J., et al. "Blood Transfusions," *Amer. J. Nurs.,* September 1972.

"Childhood Obesity," *Pediatrics Digest* (Special Issue), June 1976.

"Chronic Kidney Disease: Why You Need A Special Diet," *Drug Therapy,* June 1976.

Chung, Y. C., and T. Matsumoto. "Surgical Staging in Hodgkin's Disease – For a Better Prognosis," *Consultant,* July 1976.

Connell, A. M. "Dietary Fiber and Diverticular Disease," *Hospital Practice,* March 1976.

Cooper, M. D., and A. R. Lawton, III. "The Development of the Immune System," *Scientific American,* November 1974.

Crelin, E. S. "Development of the Lower Respiratory System," *Clinical Symposia* 27, CIBA Pharmaceutical Company 1975.

Cunningham, D. J. *Textbook of Anatomy.* Edited by G. J. Romanes. London: Oxford University Press, 1971.

"Cystic Fibrosis Said to Be Masked in Most Patients," *Pediatric News,* February 1975.

Danowski, T. S. "The Management of Obesity," *Hospital Practice,* April 1976.

Daughaday, C. C., and S. D. Douglas. "Phagocytes," *Pediatric Annals,* Immunology in Infancy and Childhood, June 1976.

Davenport, H. W. "Why the Stomach Does Not Digest Itself," *Scientific American,* January 1972.

"Developer of Heimlich Hug Elucidates Physiologic Basis," *Medical Tribune,* 7 May 1975.

"Dramatic Advances against Kidney Diseases," *Medical Tribune,* 23 June 1976.

Dwyer, J. T. "How to Counsel Today's Vegetarians," *Medical Opinion,* March 1975.

Effler, D. B. "Myocardial Revascularization," *J. Amer. Med. Assoc.,* 23 February 1976.

Friedberg, D. Z., and S. B. Litwin. "The Medical and Surgical Management of Patients with Congenital Heart Disease," *Clinical Pediatrics,* April 1976.

Garvey, J. "Infant Respiratory Distress Syndrome," *American Journal of Nursing,* April 1975.

"Giving Up Smoking: How the Various Programs Work," *Medical World News,* 1 November 1976.

Grace, W. J. "Guide to the Management of the Complications of Acute Myocardial Infarction," *Hospital Medicine,* November 1975.

Graham, D. Y. "Update on Peptic Ulcer Therapy," *Consultant,* February 1976.

Griffith, L. S. C. "The Proper Use of Coronary Arteriography," *Modern Medicine,* 15 February 1975.

Grimes, O. F. "Detecting, Evaluating, and Removing Gastric Polyps," *Modern Medicine,* March 1976.

Grollman, A. "Body Fluids and Electrolytes," *Consultant,* May 1976.

————. "How Drugs Work: The Immunosuppressive Agents," *Consultant,* August 1976.

Hanson, R. F., and W. C. Duane. "Cholesterol Gallstones – The Search for a Cause Yields a New Treatment," *Modern Medicine,* 15 September 1975.

Haubrich, W. S. "Gastric Ulcer," *Hospital Medicine,* January 1976.

Heimlich, H. J. "A Life-Saving Maneuver to Prevent Food Choking," *J. Amer. Med. Assoc.,* 27 October 1975.

Hurst, J. W. (ed.). *The Heart: Arteries and Veins.* 3d ed. New York: McGraw-Hill, 1974.

"Hypertension: Getting-And-Keeping-It under Control," *Medical World News,* 22 March 1976.

Janick, J., et al. "The Cycles of Plant and Animal Nutrition," *Scientific American,* September 1976.

Kadlub, J. J., and K. G. Kadlub. "Toward Breaching the SIDS Impasse," *Pediatrics Digest,* March 1976.

Kaplan, A. A., and W. M. Ludwig. "A Guide to the Diagnosis and Management of Cholelithiasis," *Hospital Medicine,* April 1976.

Kelley, W. N. "Current Therapy of Gout and Hyperuricemia," *Hospital Practice,* May 1976.

Keren, D. F., and A. J. Grindon. "Blood Components Instead of Blood," *Drug Therapy,* April 1976.

Kettel, L. J. "Acute Respiratory Acidosis," *Hospital Medicine,* February 1976.

Kretchmer, N. "Lactose and Lactase," *Scientific American,* October 1972.

Kyriakos, M. "Malignant Tumors of the Small Intestine," *J. Amer. Med. Assoc.,* 5 August 1974.

Laird, W. P. "Childhood and Diet as Related to Atherosclerosis," *Clinical Pediatrics,* May 1975.

Laragh, J. H. "An Approach to the Classification of Hypertensive States," *Hospital Practice,* January 1974.

Levin, B. "Early Detection of Colon Cancer," *Consultant,* July 1976.

Lieber, C. S. "Alcohol and Malnutrition in the Pathogenesis of Liver Disease," *J. Amer. Med. Assoc.,* 8 September 1975.

————. "The Metabolism of Alcohol," *Scientific American,* March 1976.

Luy, M. L. M. "The Roughage Rage: Would More Fiber in Our Diet Improve Health?" *Modern Medicine,* 1 April 1976.

Lynn, H. B. "A Re-evaluation of Splenectomy," *Pediatric Annals,* Traumatic Injuries in Children, October 1976.

Manfredi, F. "Acid-Base Problems Originating in the Lungs," *Medical Opinion,* March 1976.

Margen, S., and L. H. Allen. "What to Look for in the Nutritional Examination," *Consultant,* March 1976.

McKechnie, J. C. "Diverticular Disease and Diet," *Consultant,* July 1976.

Meares, E. M. "New Concepts in Treating Prostatic Disease," *Modern Medicine,* 18 March 1974.

Merrill, J. P. "Curable Hypertension," *Hospital Medicine,* April 1976.

Moser, R. H. (ed.). "Standards for Cardiopulmonary Resuscitation (CPR) and Emergency Cardiac Care (ECC)," *J. Amer. Med. Assoc.,* 18 February 1974.

"National Study Focuses on Gallstone-Dissolving Drug," *Medical World News,* 26 July 1976.

O'Connor, J. J. "Cryosurgery for Hemorrhoids: Quick Recovery, Few Complications," *Consultant,* February 1976.

Page, H. "Our Cells Learn Sensitivities – and Transfer Them," *Modern Medicine,* 15 March 1976.

Paradise, J. L., and C. D. Bluestone. "Toward Rational Indications for Tonsil and Adenoid Surgery," *Hospital Practice,* February 1976.

Perloff, J. K. "When to Think of Bypass Surgery for Angina," *Consultant,* October 1976.

Polish, E. "Jaundice: Guide to Diagnosis," *Hospital Medicine,* October 1976.

Reuben, D. *The Save Your Life Diet.* New York: Random House, 1975.

Roberts, J. A. "A Guide to the Urologic Examination," *Hospital Medicine,* February 1976.

"Roughage in the Diet," *Medical World News,* 6 September 1974.

Schoenfield, L. J. "Clinical Aspects of the Chenodeoxycholic Acid Trial," *Hospital Practice,* August 1974.

Simone, J. V. "Childhood Leukemia: The Changing Prognosis," *Hospital Practice,* July 1974.

Scrimshaw, N. S., and V. R. Young. "The Requirements of Human Nutrition," *Scientific American,* September 1976.

Shinya, H., and W. I. Wolff. "Colonoscopic Polypectomy: Technique and Safety," *Hospital Practice,* September 1975.

Shneour, E. A. "Good Nutrition Should Begin at Conception," *Modern Medicine,* 15 April 1974.

Schwachman, H. "Changing Concepts of Cystic Fibrosis," *Hospital Practice,* January 1974.

Sun, D. C. H. "Guide to the Diagnosis and Management of Hiatal Hernia," *Hospital Medicine,* November 1975.

Tarazi, R. C. "The Heart in Hypertension: Its Load and Its Role," *Hospital Practice,* December 1975.

"The Origins of Obesity," *Am. J. Dis. Child.,* May 1976.

"The Sugar Conspiracy," *Drug Therapy,* October 1976.

Tumen, H. J. "Alcoholic Liver Disease," *Hospital Medicine,* September 1974.

Vander, A. J. *Renal Physiology.* New York: McGraw-Hill, 1975.

Walker, W. J. "Curbing Risk Factors Has Helped Reduce U.S. Coronary Deaths," *Modern Medicine,* 1 June 1976.

Wallerstein, R. O. "Role of the Laboratory in the Diagnosis of Anemia," *J. Amer. Med. Assoc.,* 2 August 1976.

Wanner, A., and K. R. Ratzan. "Chronic Bronchitis and Emphysema," in *Forum on Infection.* New York: Biomedical Information Corp., 1975.

Westfall, U. A. "Electrical and Mechanical Events in the Cardiac Cycle," *Am. J. Nurs.,* February 1976.

Williams, M. H., Jr. "Answers to Questions on Respiratory Emergencies," *Hospital Medicine,* October 1976.

Wilson, S. M., and O. H. Beahrs. "Colorectal Malignancies: A Diagnostic Guide," *Hospital Medicine,* April 1976.

Winick, M. "Childhood Obesity," *Nutrition Today,* May/June 1974.

Wolff, W. I. "Colonoscopy and Polypectomy," *Consultant,* June 1976.

Young, J. R. "Axioms on Thrombophlebitis and Phlebothrombosis," *Hospital Medicine,* October 1976.

BIBLIOGRAPHY FOR UNIT V

Bundey, S. "Importance of Genetic Counseling," *Modern Medicine,* 1 October 1973.

——. "Typical Problems Encountered in Genetic Counseling," *Modern Medicine,* 15 October 1973.

Burnett, L. S., et al. "An Evaluation of Abortion: Techniques and Protocols," *Hospital Practice,* August 1975.

Capraro, V. J., and M. B. Gallego. "Breast Disorders," *Pediatrics Annals,* Adolescent Gynecology, January 1975.

Centerwall, W. R., and J. L. Murdoch. "Human Chromosome Analysis," *American Family Physician,* April 1975.

Crile, G., Jr. "Let's Stop Scaring Women Away from Mammography," *Modern Medicine,* 15 October 1976.

Davidson, J. M. "Hormones and Sexual Behavior in the Male," *Hospital Practice,* September 1975.

Di Benedetto, R. J., et al. "The Physiology of the Uterine Tube," *J. Med. Soc. New Jersey,* July 1975.

"Early Detection of Breast Cancer with Mammography," *RN,* December 1974.

Eckert, C. "How to Evaluate and Manage Breast Lumps," *J. Amer. Med. Assoc.,* 24 November 1975.

Fehr, P. E. "Guidelines for Prescribing in Pregnancy," *Modern Medicine,* 15 June 1976.

Friedmann, T. "Prenatal Diagnosis of Genetic Disease," *Scientific American,* November 1971.

Gardner, E. J. *Principles of Genetics.* 4th ed. New York: Wiley, 1972.

Hill, E. C. "Carcinoma of the Cervix: Diagnostic Guide," *Hospital Medicine,* May 1976.

Jones, O. W. "Where We Stand Today in Prenatal Diagnosis," *Medical Opinion,* May 1976.

Kaback, M. M., and J. S. O'Brien. "Tay-Sachs Prototype for Prevention of Genetic Disease," *Hospital Practice,* March 1973.

Kandall, S. R. "Control of Bacterial Infection in the Nursery," *Pediatric Annals,* Neonatal Morbidity, February 1976.

Langer, A., et al. "Amniotic Fluid Analysis in Prenatal Diagnosis," *J. Med. Soc. New Jersey,* July 1975.

Leis, H. P., Jr., et al. "Diagnosis of Breast Cancer," *Hospital Medicine,* November 1974.

Mays, E. T., et al. "Hepatic Changes in Young Women Ingesting Contraceptive Steroids," *J. Amer. Med. Assoc.,* 16 February 1975.

"Menopause—The Normal Disorder," *Drug Therapy,* September 1976.

Michael, R. P. "Hormones and Sexual Behavior in the Female," *Hospital Practice,* December 1975.

"Midtrimester Amniocentesis for Prenatal Diagnosis," *J. Amer. Med. Assoc.,* 27 September 1976.

Nadler, H. L. "Prenatal Detection of Genetic Disorders," *Modern Medicine,* 24 June 1974.

Page, E. W., et al. *Human Reproduction: The Core Content of Obstetrics, Gynecology and Prenatal Medicine.* Philadelphia: Saunders, 1972.

Peck, D. R., and R. M. Lowman. "Mammography," *J. Amer. Med. Assoc.,* 18 October 1976.

Piver, M. S. "Chemotherapy for Gynecologic Malignancies," *Drug Therapy,* October 1976.

"Research on Immunity to Syphilis," *Infectious Diseases,* August 1974.

Resnick, M. I. "How to Find and Stage Prostatic Carcinoma," *Modern Medicine,* 15 October 1976.

Rowley, P. T. "Genetic Screening: Whose Responsibility?" *J. Amer. Med. Assoc.,* 26 July 1976.

Rugh, R., and L. B. Shettles. *From Conception to Birth.* New York: Harper & Row, 1971.

Schwartz, G. F. "A Plea for Sensible Breast Biopsy," *Medical Opinion,* March 1975.

Segal, S. J. "The Physiology of Human Reproduction," *Scientific American,* September 1974.

Seligmann, J. "New Science of Birth," *Newsweek,* 15 November 1976.

Shearman, R. P. (ed.). *Human Reproductive Physiology.* Oxford: Blackwell, 1972.

Shocket, B. R. "Medical Aspects of Sexual Dysfunction," *Drug Therapy,* June 1976.

Strax, P. "Control of Breast Cancer through Mass Screening," *J. Amer. Med. Assoc.,* April 1976.

"The Second-Generation IUD's—Progestasert," *Current Prescribing,* January 1976.

"Three Methods of Tubal Occlusion—And All Look Good," *Modern Medicine,* 15 September 1976.

Tietze, C., and S. Lewit, "Abortion," *Scientific American,* January 1969.

"Treating Menopausal Women and Climacteric Men," *Medical World News,* 28 June 1974.

Wils, C. "Genetic Load," *Scientific American,* March 1970.

Index